FATIGUE '96

Volume I

Editors
G. Lütjering and H. Nowack

Proceedings of the
Sixth International Fatigue Congress

6-10 May 1996
Berlin, Germany

PERGAMON

U.K.   Elsevier Science Ltd, The Boulevard, Langford Lane, Kidlington, Oxford,
                        OX5 1GB, U.K.

U.S.A.   Elsevier Science Inc., 660 White Plains Road, Tarrytown, New York
                        10591-5153, U.S.A.

JAPAN   Elsevier Science Japan, Higashi Azabu 1-chome Building 4F, 1-9-15
                        Higashi Azabu, Minato-ku, Tokyo 106, Japan

First Edition 1996

Library of Congress Cataloging in Publication Data
A catalog record for this book is available from the
Library of Congress

British Library Cataloguing in Publication Data
A catalogue record for this book is available from the
British Library

ISBN 0 08 0422683

*In order to make this volume available as economically and as rapidly   as
possible the authors' typescripts have been reproduced in their original forms.
This method unfortunately has its typographical limitations but it is hoped that
they in no way distract the reader.*

Printed and bound in Great Britain by BPC Wheatons Ltd, Exeter

Conference Co-Chairmen

G. Lütjering   Germany
H. Nowack   Germany

International Steering Committee

A.   Blom   Sweden (Chairman)
J.-P. Baïlon   Canada
J.   Dickson   Canada
M.   Jono   Japan
P.   Lukas   Czech Republic
R.   Ritchie   USA
T.   Tanaka   Japan

Fatigue '96
Local Organizing Committee

D. Aurich
E. Bauerfeind
W. Bergmann
R. Hettwer
B. Jaenicke
I. Maslinksi
J. Völker

International Advisory Board

M.   Anglada   Spain
J.-P. Baïlon   Canada
V.   Bicego   Italy
A.   Blom   Sweden
D.   Davidson   USA
J.   Dickson   Canada
P.   Haagensen Norway
M.   Jono   Japan
J.   Knott   United Kingdom
I.   Le May   Canada
T.   Lindley   United Kingdom
J.   Llorca   Spain
P.   Lukas   Czech Republic
G.   Lütjering Germany
A.   McEvily   USA
P.   Neumann   Germany
J.   Newman   USA
H.   Nowack   Germany
P.   Petit   France
L.   Remy   France
R.   Ritchie   USA
O.   Romaniv   Ukraine
E.   Starke   USA
R.   Stickler   Austria
T.   Tanaka   Japan
Z.   Wang   China
R.   Wanhill   The Netherlands
X.   Wu   China
M.   Yan   China

Honorary Fellows and Life Members of IFC

L. Forsyth
P. Coffin
P. Paris
H. Kitagawa †
C. Beevers †
J. Schijve

INTERNATIONAL FATIGUE SERIES

1981   STOCKHOLM, SWEDEN

1984   BIRMINGHAM, U.K.

1987   CHARLOTTESVILLE, U.S.A

1990   HONOLULU, HAWAII

1993   MONTREAL, CANADA

INTERNATIONAL FATIGUE CONGRESS

1996   BERLIN, GERMANY

PREFACE

The first five conferences of this series on fatigue were held under the name International Conferences on Fatigue and Fatigue Threshold. The series began in 1981 in Stockholm, Sweden, followed up by conferences in Birmingham, U.K. (1984) and Charlottesville, USA, in 1987. The Society of Materials Research of Japan organized the conference in 1990 and moved it to Honolulu, Hawaii. At the 1993 conference in Montreal, Canada, the decision was made to change the name for future conferences to International Fatigue Congress. Therefore, now between May 6 and May 10, 1996, it is the Sixth International Fatigue Congress which is being held in Berlin.

The conference site is in East Berlin, close to the heart of the capital of the former German Democratic Republic. It was chosen deliberately to draw the attention of the participants to the times before the reunification of Germany, to a living situation which can still be sensed in the area around the Berlin Congress Center (BCC).

The aim of the Sixth International Fatigue Congress, besides covering the entire field of fatigue, was to promote the intimate connection between basic science and engineering application by the selection of appropriate session topics.

Fatigue is the main cause of failure of engineering structures and components. Making reliable fatigue predictions is highly difficult because knowledge about fatigue mechanisms in all stages of the fatigue process must be developed much further. In addition, the decreasing availability of raw materials and energy resources forces engineers to continually reduce the weight of constructions. This congress presents research results also particularly for new materials, including composites. Researchers, on the other hand, are confronted with the engineering demands. Furthermore, the overwhelming development which is presently taking place in the field of computer software and hardware dealing with fatigue problems is outlined along with the directions of future developments in all areas of fatigue.

Close to 300 papers are published in the proceedings, including nearly 30 overview and keynote papers covering the various session topics. The proceedings should therefore serve as a comprehensive review of the fatigue field at the present state-of-the-art, suitable for scientists, engineers, and students.

The organization of the Sixth International Fatigue Congress and the editorial work for the proceedings was performed by the DVM, which is celebrating its 100th anniversary in 1996. The local arrangements in conjunction with the conference were conducted by the Local Organizing Committee, which is gratefully acknowledged. Members of the International Steering Committee and the International Advisory Board made valuable contributions to the preparations for the conference, which is highly appreciated. Finally, the Co-chairmen would like to express their special thanks to Mrs. I. Maslinski, Managing Director of the DVM, and to Mrs. C. Hardey for their devoted help in making the conference and publication of the proceedings possible.

Gerd Lütjering and Horst Nowack
FATIGUE '96 Co-Chairmen

CONTENTS
VOLUME I

OVERVIEW LECTURES

CYCLIC STRESS STRAIN BEHAVIOR AND MODELS

MICROMECHANISMS IN FATIGUE

INITIATION AND PROPAGATION OF SHORT CRACKS

INTRINSIC AND EXTRINSIC INFLUENCES ON CRACK PROPAGATION

THRESHOLD VALUE FOR ENGINEERING APPLICATIONS

MACROCRACKS AND DAMAGE TOLERANCE DESIGN

VARIABLE AMPLITUDES AND PREDICTION

CORROSION FATIGUE, DATA AND DESIGN

CONTENTS
VOLUME II

INFLUENCE OF HIGH TEMPERATURES

MICROSTRUCTURAL DESIGN FOR FATIGUE RESISTANCE

MULTIAXIAL BEHAVIOR

SIGNIFICANCE OF FEM IN FATIGUE ANALYSIS

FATIGUE AND FAILURE OF COMPONENTS

PROBABILISTIC LIFE PREDICTION METHODS

COMPUTERIZED FATIGUE ANALYSIS

FATIGUE DATA BANKS

SURFACE TECHNOLOGIES FOR FATIGUE IMPROVEMENT

CONTENTS
VOLUME III

ADVANCED FATIGUE PROPERTIES OF METAL MATRIX COMPOSITES

PERFORMANCE OF POLYMERS AND POLYMER MATRIX COMPOSITES

CYCLIC FATIGUE OF CERAMICS AND CERAMIC MATRIX COMPOSITES

FATIGUE MECHANISMS OF INTERMETALLICS

FATIGUE OF BIOMATERIALS

CASE STUDIES, ESPECIALLY AEROSPACE, AUTOMOTIVE, POWER GENERATION

ULTRASONIC FATIGUE

EXPERIMENTAL TECHNIQUES AND EQUIPMENT

AUTHOR INDEX

SUBJECT INDEX

OVERVIEW LECTURES

FATIGUE RELATED STRUCTURAL INTEGRITY ISSUES

A. F. BLOM

The Aeronautical Research Institute of Sweden (FFA)
P.O. Box 11021, S-16111 Bromma, Sweden and
The Royal Institute of Technology; S-100 44 Stockholm, Sweden

ABSTRACT

Design for structural integrity requires knowledge of loads, stresses, material properties, fatigue damage modelling, manufacturing, structural testing, NDI and statistical aspects. Here, methodology for verification and management of damage tolerance of critical components in new fighter aircraft is briefly described. Also, some case studies of fatigue problems in existing structures are summarised. Examples are taken from the aircraft industry, nuclear sector, wind energy industry and welded structures in general.

KEYWORDS

Structural integrity; fatigue; damage tolerance; loads; stresses; fracture mechanics; structural testing; aircraft structures; composite materials; nuclear industry; wind energy; welded structures.

INTRODUCTION

The methodology currently used in Sweden for fatigue management and verification of airframes is described. Applications from the new fighter aircraft JAS39 Gripen are included in order to illustrate the various concepts being considered. The present paper discusses the handling of load sequences and load spectra development, stress analyses and fracture mechanics analyses, fatigue crack growth modelling, component and full scale testing, service load monitoring regarding both the dedicated test aircraft, which is used to verify basic load assumptions, and also the individual load tracking programme developed for the new fighter.

After this description of structural integrity work on a new product, some comments on fatigue problems of old structures will be given. These comments are based on case studies of the older fighter aircraft 37 Viggen as well as from the nuclear industry, wind energy conversion systems and the general transportation and mechanical engineering industry. The aim is to show that certain aspects always emerge irrespective of the industry or component studied. Safety issues and cost, however, lead to very different design and certification procedures. Because of space limitations, no pictures or figures will be included in this paper, but many examples will be shown during the lecture.

AIRCRAFT

Whilst fatigue is recognised as the primary failure mechanism in most structural components, the consequences of such failures (in terms of lives and money) are probably more obvious and severe for aircraft than in any other area. Fatigue and damage tolerance of aircraft are dominant factors throughout the design process, the manufacturing, the structural testing (component level and full-scale), the initial test flights and during the subsequent usage of the aircraft. Without adequate knowledge of relevant loading conditions, resulting stress distributions and material properties no

3

fatigue assessment can be achieved. Further on, such work has to be attached to a specific design philosophy - can the actual component be inspected during service, in the case of damage - can it be repaired or replaced, and in the case of failure, will that cause loss of the entire aircraft? Questions like these, and many more specific detail problems, are dealt with in existing regulations and specifications for military aircraft. It is far beyond the scope of the present paper to discuss these specifications in any detail (there are also other ones treating, for example, flight and operations tests, vibrations, sonic fatigue and non destructive inspections). Instead, the purpose here is to briefly discuss fatigue management and verification of airframes, with emphasis on the new Swedish fighter JAS39 Gripen. A more detailed description of these issues is given in Ref. (1).

THE SAAB JAS39 GRIPEN AIRCRAFT

The basic military concept behind the JAS39 aircraft is that each individual aircraft and each individual pilot should provide multi-mission capability, i.e. instantly be able to perform air-defence as well as air-to-surface attacks or reconnaissance, without any changes to the aircraft itself. This multi-mission capability will result in an operational flexibility allowing concentration of units in a manner previously not technically possible. The Gripen is designed to fit into the Swedish Air Force's base system. This means that the JAS 39 is capable of taking off from and landing on road bases. Flight preparation at wartime bases should be possible by conscripted personnel under field conditions. The aircraft is aerodynamically unstable in certain parts of its operational envelope. Stability is obtained by means of a triplex digital fly by wire control system with a triplex analogue back up system.

Approximate Data of the Aircraft

Engine: One Volvo Flygmotor RM12 (= General Electric F404) producing 8 tons thrust
Weight: 8 tons   Speed: Supersonic at all altitudes
Length: 14 m   Wingspan: 8 m

Wing: Multi-spar design with skins and spars made of CFRP-laminates.
             Root ribs made of Al-alloy. The structure is bolted together.

Canard, Fin and Control surfaces:
             Full core aluminium honeycomb with CFRP laminate skins.
             Some rudders of Al-alloy.

Fuselage: With exception for a few doors it is all metallic; 7475 skins, 7075 extrusions and 7010 forgings.

Most of the attachment fittings joining the various parts of the aircraft are made of 7010 forgings.

The CFRP laminates were initially made of Ciba-Geigy toughened epoxy with HTA 7 fibres. Presently also Hercules systems are qualified to be used.

The Design Requirements

Design limit load factor = 9.0
Ultimate safety factor   = 1.5
CFRP laminates not allowed to buckle below 150%LL

The design and testing requirements essentially agree with U.S.A.F. military specifications.

Damage tolerance requirements for the JAS 39 Gripen

Only so-called critical parts are required to comply with damage tolerance requirements. A part is classified as critical if its failure alone may cause the loss of an aircraft. The requirements are essentially identical with USAF Mil-A-83444, Ref. (2). The main difference is that the residual strength requirement is always 120%LL.

Analysis Goal: 3000 hrs. inspection-free service life
Verification Goal: 4000 hrs. inspection-free service life
Minimum Requirement: No detail to have a shorter inspection interval than 400 hrs.

Damage tolerance verification policy

Various fatigue and damage tolerance issues for this fighter aircraft have already been presented, e.g. Ref. (3). Previous experience from a damage tolerance study on the Saab AJ37 Viggen aircraft led to the following policy:

> The damage tolerance verification is to be done on full-scale assemblies, where the critical part is correctly built into its surrounding structure.

> The critical parts are to be manufactured according to series production standard. (Example: Die forgings are not to be substituted by hand forgings or plate material).

> Fatigue testing of critical parts is not allowed to be eliminated just because a damage tolerance verification testing is carried out (Reason: The damage tolerance test is biased, because it interrogates the structure only at the points where artificial flaws have been made. A correct fatigue test does not suffer from that kind of bias.). The fatigue testing shall cover at least four inspection free lives. i.e., 4 x 4000 hrs.

> The damage tolerance verification testing shall be done using realistic flight by flight simulation testing, and shall cover at least two inspection free service lives, i.e., 2 x 4000 hrs.

> A critical part having initial artificial flaws in accordance with Mil-A-83444 and which survives 2 x 4000 hrs. of realistic service life simulation followed by a residual strength testing to 120%LL is considered to have fulfilled the verification goal: 4000 hrs. of inspection-free service life.

> If significant crack growth occurs during the damage tolerance testing, a safe inspection interval is to be established based on crack growth observations.

LOADS

A conceptual model is used which covers design and sizing, structural test verification and service monitoring and involves engineering activities such as mission analysis, external loads analysis, structural analysis and stress analysis.

Design parameters in the mission analysis originate from estimated threats and expected usage and are expressed as a sequence of flights and ground conditions. The conditions are defined by flight mechanics parameters such as load factor, roll rate, speed, control surface deflections, thrust, fuel burn, weapons etc.

Each set of flight parameters defines a certain flight condition. Determining the external loads for those conditions (manoeuvres) at different aircraft configurations requires analysis of e.g. structural dynamic response, aerodynamic pressure distributions at different speeds, angles of attack etc. Ground loads analysis include such events as landing impact, taxiing, braking, turning etc.

The load analysis makes use of techniques like the finite element method to predict dynamic transient response, e.g. landing, computational fluid dynamics for prediction of aerodynamic pressure fields and six degree of freedom flight mechanics model with control system logics to predict in-flight manoeuvres. Numerical predictions are supported by wind-tunnel tests of models.

The global spectrum approach

For fatigue and damage tolerance sizing and test verification, a methodology referred to as the global spectrum approach, in which the design loading of the complete aircraft is defined, has been developed. In this approach the description of each unique manoeuvre is defined as a sequence of instantaneous and balanced load cases. Sequential manoeuvres build up unique missions and sequences of missions build up the expected usage. The load cases consist of linear combinations of unit load cases (including aerodynamic loading, inertia loads, ground loads etc.) for the finite element model of the complete airframe in a hierarchic way. All information about load case structure and their occurrences in the sequences are stored in a global sequence database. Through this methodology results local load sequences, in terms of loads, stresses, displacements etc., which can be directly used for crack growth calculations or counted, by means of the rain-flow algorithm, in order to be used for classical fatigue analysis or planning of structural test programmes.

This database is updated continuously as a project develops and as flight test data becomes available, The current global sequence for the JAS39A Gripen aircraft consists of 35 different types of unique missions built up by more than 13 000 unique load cases. These unique missions are combined into 60 deterministic sub sequences (defined on basis of pilot training programs and consisting of 12 different taxi load sequences, 35 different flight load sequences and 13 different landing load sequences) which are then combined to a sequence covering 313 complete missions (taxi out, flying, landing, taxi in) containing a total of more than 500 000 load cases. A unique repetition sequence representing an actual service load history of these 313 flights is equivalent to 200 flh and, thus, consists of more than 500 000 states.

The internal loads distribution are obtained by solving a finite element model of the complete aircraft using the external loads representative for the unit load cases in the global sequence description. The finite element models of the single- and twin-seated version of the JAS39 Gripen aircraft are solved for 460 unit load cases and stored in a structural analysis database.

Having access to the global sequence database and the structural analysis database, the engineer is able to obtain local load or stress sequences and rain-flow counted spectra for any member element of the finite element model of the complete aircraft. This is done by an interactive software system which is connected to the two databases. The global spectrum program delivers load sequences for all the loading actuators of the various structural tests. It also delivers the basis for comparison with the results from the loads monitoring registrations on each individual aircraft.

Load calibration

One test aircraft (Test A/C 39.2) is dedicated to the loads survey programme. About 500 strain gauges have been installed. The aircraft has been subjected to about 150 calibration load cases with the purpose to accurately be able to deduce interface loads between wing and fuselage, wing and control surfaces and fin and fuselage. Also shear, torque and bending at three wing sections and at four fuselage sections, as well as bending and torque in the canard pivot, landing gear loads and loads from external stores are measured.

Service loads monitoring

For the Swedish fighter JAS39 Gripen, every aircraft will be equipped with a service loads monitoring system. The following six loads will be recorded:

1	Vertical acceleration at C.G. = Load factor, n_z	
2	Canard pivot:	Bending, M_X
3	" -	Torque, M_y
4	Wing:	Fwd attachment bending, M_X
5	"	Aft attachment force, S_z
6	Fin:	Aft attachment bending, M_X

With the exception for the first one, the forces will be measured by means of calibrated strain gauge bridges. Each individual aircraft will be calibrated while flying in detail specified manoeuvres. Nominally identical strain gauge installations on the loads survey aircraft (Test A/C 39.2), on the major static test structure, on the major fatigue test structure, on two canard pivot test articles and on two fin plus rear fuselage test articles will add confidence to the calibration procedure.

Recorded data will be analysed on board each aircraft using the range pair range count algorithm. After transferring to a ground based computer, comparisons will be made with the corresponding local spectra as delivered from the global spectrum software system.

STRESS SPECTRA AND FRACTURE MECHANICS

Models used for assessment of fatigue life and damage tolerance of fighter aircraft vary from rather simple to highly complex depending on factors such as; if the structural component is primary or secondary, local stress levels, inspectability etc. Typically, simple models are used initially and more sophisticated analysis is resorted to in complex geometries and when safety margins are lower. This is described in more detail in Refs. (1, 4 - 5).

Elimination of insignificant stress cycles

Due to the basics of the global spectrum approach, described above, very many small load cycles are generated. It is therefore advisable to omit certain small cycles which from a fatigue viewpoint have no practical meaning. The system contains interactive facilities to remove insignificant states. By defining omission conditions connected to the earlier defined sequences the reduction of the sequences is immediately displayed. For example, for a lower flange region the original sequence consists of 7813920 states for 3000 flight hours. By omitting cycles with ranges smaller than 20 kN (approximately 10% of the maximum range) the total sequence is reduced by 97.1% to 224250 states. The omission technique adopted also works when more than one load sequence is considered through a lowest common denominator scheme.

The software has features to facilitate identification of load cases. This can be done by retaining load case identification numbers all the way through the rain flow count operations, omission procedures etc. This technique has proven to be most important when planning for structural testing. For JAS39 every unique load case of the global sequence has been assigned a 9 digits code number according to a systematic scheme.

Stress intensities and fcg

The global finite element model predicts the internal loads distribution and the different types of entities that can be obtained depend on mesh and type of elements. The most simple case is to extract a scalar sequence, e.g. a normal force sequence from a flange element. If more complex elements are used, e.g. beams, solids etc., more stress components are available. Since each state in the global sequence represents a certain load case, i.e. a certain combination of the components, each change from one state to the next state also means a change of the stress intensity factor to another. This complicates the situation if the stress components are not correlated. However, frequently it is possible to obtain a stress intensity factor solution which is applicable for the complete sequence from a linear combination of normal force and bending moments.

Sometimes when predicted crack growth lives initially do not meet the target lives, more sophisticated local modelling can be made and connected to the global model and the global sequence. For example, for the lower flange region a local stress model using solid elements has been solved using the p-version of the finite element method., Ref. (6). The stress intensity factor for a crack in this section is derived using three dimensional weight function technique.

Damage tolerance analyses are made for large number of assumed cracks in primary structure. For the JAS39A aircraft more than 1000 crack sites have been analysed. The type of analysis ranges from simplified ones using basic stress intensity factor solutions to more or less advanced ones using p-version FEM either with the crack incorporated in the mesh or by applying 3D weight functions to stress distributions for the uncracked structure. Each analysis is summarised on a damage tolerance analysis sheet.

STRUCTURAL TESTING

Here, we will discuss the testing carried out to substantiate the analytical methodologies used. Such tests include simple coupon tests to study influence of load spectrum truncation etc., component tests to verify numerical predictions and to reveal any unexpected problems, and full scale fatigue and damage tolerance testing of an entire aircraft. The overall test programme for the new fighter aircraft goes on for more than 10 years.

The damage tolerance analysis creates the necessary conditions for structural integrity. This integrity also needs to be demonstrated in structural testing. Besides from testing for obtaining data for predictions, three main levels of testing are detail testing, major component testing and final full scale verification testing. These three levels are described below.

Detail testing is mainly performed early in the sizing work. It is used to verify detail design of vital structural members and to qualify the application of prediction methods to typical structural configurations. The lower flange region of the main wing attachment frames were studied in that way. Six different crack/geometry configurations were tested.

The slow crack growth damage tolerance requirement is applied to all flight safety critical components - even small size units such as tension bolts, hinge bolts for the control surface attachments, control actuators and other small details of the control system. Primary bolts are verified to tolerate semicircular surface flaws with 1.3 (+0.2/-0) mm radius. The verification testing of the tension bolts for the wing and fin attachments have proven very satisfactory damage tolerance.

Major component testing is done for early fatigue and damage tolerance verification. The key point in these tests is that a critical part is tested while properly installed in its nearest boundary structure. This test will spot fatigue critical areas and demonstrate the stable growth of those natural cracks that may initiate and of any artificially made cracks.

Damage tolerance testing of large components involve very many initial flaws, for the rear fuselage tested together with fin and rudder more than 100 flaws were introduced.

The final verification of the fatigue and damage tolerance performance is made with a complete airframe tested for several service lives. The test arrangement of the static test article consisted of 85 independently controlled load channels. A second airframe for fatigue and damage tolerance testing is now installed in the same rig and testing will be subjected to at least four service lives of fatigue testing.

The test program for structural verification of the JAS39 aircraft is basically carried out in the following way. Component tests for verification of both fatigue and damage tolerance are first subjected to two service lives of fatigue testing, then artificial flaws are introduced, whereafter the test is subjected to a further 2 service lives of fatigue testing. Finally, a residual strength test is carried out with the purpose to verify a load capacity in excess of 120% LL. The pure damage tolerance verification tests are performed with artificial flaws introduced from the very beginning and are subjected to two lifetimes of fatigue testing followed by the residual strength test.

Before deliveries of aircraft to the Swedish Air Force, a full scale static test will be performed (this is already completed) and a full scale fatigue test (currently running) must have simulated one design life.

COMPOSITE MATERIALS

An overview of work done in the U.S.A. and recommended certification procedures are summarised for composite structures in Ref. (7). This reference states that draft USAF damage tolerance design requirements for composites are conceptually equivalent to MIL-A-87221. Recognition to the unique property characteristics of composites leads to significantly different defect/damage assumptions for composites as compared to metals. Of the assumed flaw/damage types, scratches, delamination and impact damage, the last one dominates design as it is the most severe. The recommended compliance to the draft requirements is summarised in Ref. (10) to allow no significant damage growth in two design lives. This is due to rapid unstable growth after growth initiation. This requirement eliminates in-service inspections. Finally, no full-scale test validation is required as subcomponent test are considered to accurately represent damage tolerance behaviour of full-scale structures.

For the new Swedish fighter JAS 39 Gripen a certification procedure similar to the one mentioned above is being used for the composite structures (wing skins and spars, elevons, leading edge flaps, canards, air inlets, fin and rudder and air brakes). All CRFP-laminate structures were verified by static testing before the first flight. The structures were tested dry, as received from manufacturing. Smaller components such as elevons and rudder were tested at high (+85°C) as well as low (-40°C) temperature besides at room temperature. The wing, the fin and the canard were tested at room temperature besides at room temperature. The wing, the fin and the canard were tested at room temperature only. As all verification testing was done with dry laminates, the requirement was increased to 1.2 times 150%=180% limit load. Subsequently, all those structures have also passed four design lives of fatigue testing without any fatigue damage.

CURRENT TRENDS

Current development trends concerning loads activities is to incorporate data from more flight systems, influenced by flight conditions, e.g. the gearbox etc., into the global sequence description. Other improvements of the global spectrum approach will be to early verify the mission analysis by a closer connection to the flight simulator. By letting different pilots solve a certain flight task, the variability in manoeuvres can be evaluated early in the project.

Trends regarding structural analysis is to create more detailed models of vital structures and to incorporate those into the global finite element model as substructures. Special facilities have been implemented into the loads sequence handling system for identification of the most significant load cases. Much more detailed models are being assessed by a parametric approach early on in the analysis phase, to account for the fact that certain dimensions are determined very early and are difficult to modify later on.

Regarding crack growth modelling, retardation has yet not been used for certification purposes. However, recent advances will be utilised for the C-version of the JAS39 Gripen aircraft, to decide which old tests can still be used for the new design.

Other areas which will be studied in more detail within the next few years include: crack growth in mechanical joints, multiaxial loading, constraint effects in three dimensional crack growth, and probabilistic aspects of damage tolerance and the entire structural design methodology.

THE 37 VIGGEN AIRCRAFT

The Swedish fighter aircraft 37 Viggen was designed some 30 years ago on a safe life and/or fail safe basis. Some major parts of the aircraft have been re-assessed in terms of a damage tolerance evaluation, Refs. (1, 4-5). The aim of this assessment has been to ensure structural safety, and to investigate the possibilities for extending the original life of the aircraft. This can be carried out because the total life of the aircraft will depend on its durability while the damage tolerance ensures, at all stages, the structural safety of the aircraft.

Here, there is no room to discuss the work performed in any detail. However, a number of rather interesting results emerge from the study.

Because of the original safe life design, resulting in significantly higher nominal stresses than in the JAS39 Gripen aircraft, very extensive finite element analyses have been necessary in order to obtain accurate stress distributions in critical sections. These stress distributions have subsequently been used for the evaluation of 3D stress intensity factors. Also, high demands have been placed on the accuracy of the crack growth predictions. Hence, extensive validation of the crack growth prediction technique has been required. This was done both on coupon level and on full scale built up structures, e.g. the main wing attachment frame assemblage, the fin etc.

Load measurements over the past many years have shown that the utilised design spectrum is more severe than the average usage spectrum of the aircraft. This aspect as well as the difference in stresses between a free structural component tested and the same component built into the aircraft are accounted for. The outcome of all analytical and experimental work is that safe periods of crack growth have been established. Inspection intervals have been determined, assuming that the critical locations are depot or base level inspectable, to half of the safe crack growth periods.

NUCLEAR INDUSTRY

Rotor Cracking

Cracks were found in a number of rotors in turbo generators used at several of the Swedish nuclear reactor plants. These rotors are some seven or eight m long, weighing around 70 or 80 metric tons and costing some 10 MUSD in manufacturing. The main problem though, is that manufacturing requires close to two years which would lead to drastic loss in energy production and, hence, economy. Thus, there is a strong motivation for continued operation of the rotors, with or without modifications to the rotors.

The first result from this case study was a common one. It was found that the cause for the observed cracking was a manufacturing mistake leading to a too small notch radius. Rotor safety and rate of continued crack growth, however, was not so easy to determine. The loading of the actual crack location consists of many millions of small stress cycles due to rotating bending (at 50 Hz), but also very large compressive loads due to start/stop cycles (a shrink fit joint with small tensile stress during operation). Superposed on these loads are certain temperature fluctuations. The total loading of the crack site is biaxial.

A large programme was initiated to verify certain assumptions of mechanisms for fatigue cracking, accurate load measurements and detailed finite element modelling of stresses and stress intensity factors. The work also included tear down of one rotor to check the performed NDI, specimen testing and fractography.

The entire work will be published in the open literature within the next year or so. Here, it is enough to say that original NDI (mainly ultrasonics) was found to have significantly overestimated the actual crack sizes. The mechanism for crack initiation was initiation due to the residual tensile stresses formed at the notch due to compressive loading. Subsequent crack growth occurred at decelerating rates until arrest. Many combinations of blocks of small cycles were tested between large compressive overloads as well as studying the effects of variation in the magnitude of the different loads.

Cracks in Ignalina TK-channels

The Ignalina nuclear power plant in Lithuania is the largest reactor in the world. The design is of the same type as that of the failed Chernobyl reactor. The Swedish Nuclear Inspectorate is working together with Lithuanian authorities and the Ignalina staff to maintain safe operation of the plant. During NDI work a number of cracks were found in TK-channels of both reactor blocks. These channels (more than 1000) contain the fuel cells (some seven m long) which once subjected to radiation must be kept cooled to stay stable. The load operation of fuel cells involves a complicated procedure whereby a special machine is attached to the TK-channel, water is applied for cooling, the channel is opened and the old fuel cell removed and the new one inserted. Then the channel is closed, water turned off and the machine removed. During this operation there is both thermal and mechanical loading of uncertain magnitude applied to the cracked TK-channel. Question - are the observed flaws fatigue cracks or other defects and in any case are they potentially detrimental?

Field measurements were performed by strain gauging different TK-channels during the entire load operation. These measurements revealed unexpectedly high temperatures (some 280 centigrades) and very high cooling and heating rates. Mechanical bending/torsion loads were lower and it was deemed that existing flaws were introduced by poor welding quality. Mechanical loads in themselves would not be dangerous to the integrity of the channels, but the effect of the very high temperature rates is still being considered by modelling of the entire loading process.

WIND ENERGY CONVERSION SYSTEMS

<u>Operational Experience from the Swedish Prototypes</u>

Two large wind energy conversion systems (WECS) were built in Sweden in the early eighties. One plant, Näsudden 2 MW WECS, was designed with a concrete tower and steel blades. The other one, Maglarp 3 MW WECS, was designed with a metal tower and glass-fibre/polyester composite blades. Both plants are large designs with blade diameters of 75 m (21 tons per blade) and 78.2 m (13.5 tons per blade), respectively. The design, fatigue analyses, fatigue testing and operational experience of these two prototypes were described in some detail in Ref. (8).

Based on the experience from Näsudden WECS it was decided to change from steel to fibre reinforced plastics for future designs. This was due to both extensive inspection requirements and large weight (causing drawbacks also for the machinery and the nacelle). Similar experiences have been made also in Germany and elsewhere. Now, it is interesting to note that the original design of the blades was based on cumulative damage type of calculations together with appropriate design rules for welded structures. It appeared early that due to the special type of loading on large blades (different from small blades where bending dominate) early cracking would be very slow whereas once the gravitational loads start to contribute to fcg the cracks would grow rapidly to failure. An inspection programme was initiated based on fracture mechanics predictions. Essentially cracking occurred as anticipated and TIG repair of cracked structure made it possible to continue operation.

The Maglarp composite blades showed only cosmetical damage mainly in terms of some dirt accumulation on the blade surface and erosion of the blades. Secondary structural parts required some maintenance and repair effort, but this was largely due to an unfavourable lay-up angle of the surface layer. This can easily be rectified in future design concepts.
Both projects indicate the essential need for knowledge of actual loading on the WECS. Also, it became very clear that although the composite blades behaved rather well, there is a need for more accurate understanding of failure mechanisms and cumulative damage accumulation in such materials.

<u>Failure of Small Wind Energy Blades</u>

Several failures of blades in operating small WECS around the world (in one case the blade failure caused total failure of the entire WECS) led to replacement or rejection of some one hundred manufactured glass/polyester blades. A court case followed where a number of questions were raised by the involved parties (mainly Danish companies). Several legal and economical aspects should not be disclosed but the technical investigation followed standard procedures, starting with the assumed load spectra and looking in detail into the design procedure including stress analyses, utilised material data and even the design requirements of the country of blade origin. Extensive testing of material properties were also undertaken on specimens from the broken blades. These tests involved the measurement of glass content, Barcol hardness, residual styrene content, delamination strength, compressive strength and tensile strength. It was demonstrated that the design followed good engineering requirements and that no deficiency in the utilised materials existed, but rather that lack of quality control caused large variations in hardening of the polyester. The failure mechanism of the blades was then attributed to local delaminations followed by compressive failure of the blades. It is worth pointing out that material deficiencies are only a scarce reason to service failures of structural components. In most of the author's experience the reason for failures was found to be too high local stresses, frequently because of poor knowledge of operating loads and sometimes because of poor local design (small notch radii).

TRANSPORTATION AND OTHER INDUSTRIES

Although several differences exist between various types of transportation systems, there are also many similarities in terms of how structural integrity related issues are being handled. Whether rail bound systems, earth moving equipment and other heavy entrepreneurial transportation devices, lorries, ships or personal cars are considered, the same type of activity as outlined above for aircraft must be considered. Knowledge of field usage, loads, stresses etc. are necessary for any fatigue design or analysis irrespective of which design concept is being used.

Although additional problems in terms of corrosion and high temperature behaviour may appear in process industry, there are still many similarities in terms of fatigue problems with both the transportation and the general mechanical engineering industries. What has perhaps sometimes been overlooked is the strong effect of many manufacturing processes on fatigue behaviour under service conditions. Both welding, casting, gas and other types of cutting, and sometimes also forging control the defect distributions of the final products. Mostly, these defects are severe enough to completely remove fatigue initiation period for the actual component. This leads to several aspects of which one particularly important point is that the correlation between fatigue strength and static material strength disappears. In fact, frequently high strength materials show worse fatigue life results than lower strength materials once defects are introduced by any means. It is beyond the scope of this paper to discuss these aspects in any detail, but some general comments on welded structures are given below, since weld related fatigue problems are basically generic and common in very many industrial products.

Another general rather interesting observation the author has made is that several products used rather successfully appear to be badly designed and ought to fail. In the observed cases, the explanation frequently appeared to be due to the industrial cleaning process used prior to painting. This cleaning, or blasting, typically has similar intensities to those of shot-peening used to introduce beneficial compressive stresses. Failures have been observed once that cleaning process was removed or significantly altered, supporting the assumption of beneficial effects of such process.

Welded Structures

Fatigue design of welded joints has developed significantly over the past two decades and there now exist several codes regulating such design. Still, there is a need for systematic experimental studies of welded joint behaviour under spectrum loading. This is to improve the accuracy in life predictions but also to assess certain areas which are not completely understood. These include the behaviour at very long fatigue life, the thickness effect, the relaxation and redistribution of residual stresses under spectrum loading, life improvement techniques, correlation of weld procedures with defect distributions and resulting fatigue life, and further development of fracture mechanics design rules for welded joints.

Based on results from an internordic programme, Refs. (9, 10), a number of interesting points emerge. Firstly, it is obvious that fatigue behaviour of welded joints subjected to spectrum loading is a highly complex area. Then, we note that long life behaviour is poorly understood and it is certainly worth observing that several existing codes appear to be unconservative in the long life regime. Clearly, long life behaviour is sensitive to initial conditions, i.e. the actual welding process and resulting local geometries and defect distributions. Hence, performing spectrum testing at long lives (which is obviously tedious and therefore expensive) may primarily be a test of the actual welding procedure of that test specimen only. An improved statistical data base is desperately needed, but this may only be achieved by further laborious testing unless an analytical means is devised for reducing the applied load spectra and thereby speeding up the actual testing. This appears as fairly straight forward as long as it is recognised that there must be no assumption of any fatigue limit under spectrum conditions. This discussion is directly relevant also to spectrum testing of life improved welded joints, where otherwise impossibly long testing times may be needed. Such improvement techniques are manifold and include shot peening, hammer peening, TIG-dressing etc., and combinations of these. In Sweden the same participants, joined with some new companies, as in the study discussed here, are now starting spectrum testing of welded joints treated with various improvement techniques. The study also focuses on how improvement techniques could be used to make utilisation of high strength steels possible in fatigue loaded components where there is a need for high static strength.
The study presented in Refs. (9, 10) also revealed a new type of defect, cold laps, which is believed to be prevalent under MAG-welding but may be of less importance under manual metal arc welding. Careful examination of local weld geometries and initial defect distributions revealed much interesting information but it was still not possible to correlate the initial defect distribution to the resulting fatigue lives. Fracture mechanics analyses of welded joints are particularly interesting as initial defects are always present, there is no fatigue initiation phase. It is also very interesting to try to use fracture mechanics to classify welds in complicated structures, thus making it possible to use codes for other geometries than certain standard ones. However, it appears that there is yet a long way

to go before fracture mechanics characterisation of weldments becomes mature, at least for engineering purposes.

The performed spectrum loading revealed several interesting results just in terms of actual fatigue lives. However, what may be equally interesting was the significant stress relaxation (50% or more) which occurred early in the fatigue life (within 8% of total specimen life) at stress levels with lives around 10 million load cycles. This was found by repetitive X-ray diffraction studies and indicates that post welding treatment (annealing) is probably not cost effective since the tensile residual stresses are likely to relax very quickly anyhow. This is restricted, however, to loading conditions similar to those verified here.

In a new internordic activity which has just started there are several new goals. The aim is to introduce high strength steels and high speed weld procedures (Rapid Arc and Rapid Melt) into conventionally manufactured components. Optimisation of weld parameters as well as structural geometry is included throughout the project. Many different sub tasks are involved and specifically, regarding fatigue behaviour, the influence of various life improvement techniques, e.g. shot peening and TIG-dressing, on fatigue life is to be studied on both specimens and full-scale components. Applications are made to various box beams for transportation vehicles and cranes, load actuator attachments etc., and are to include life cycle cost analyses for these designs before and after optimisation. This new study is expected to end in the late 1997 when results should be made available in the open literature. Anybody interested in this study could contact the present author, who is co-ordinating the effort, for further information. One specific example from this study is presented at the present conference, Ref. (11), and shows the beneficial effect of TIG-dressing on the fatigue strength both under constant amplitude and spectrum loading. Also, in this latter work, is described an analytical model capable of predicting the general influence of stress level, crack length, stress concentrations and residual stresses on the fatigue life of both untreated and TIG-dressed specimens.

Cracking of Hydraulic Press

The largest press ever manufactured by one particular Swedish company, showed fatigue cracks very early during service life. The manufacturer was surprised, since the same design had been successfully used for decades, in smaller presses. The cracking was very costly, since the plant using the press for manufacturing of stainless tubes was entirely dependent on function of the press. Thus, two questions emerged; why did the press crack and how could it be altered without immediately obtaining new cracking? As to the answer of the first question, there may be several factors to consider. The scale up in size automatically gives rise to larger inertial forces and also a larger volume for cracks to initiate in, i.e. a statistical aspect. However, also the residual stresses increase since such stresses need certain thicknesses to develop fully. As for the remedy, repair welding was performed but would not be adequate in itself. A detailed stress analysis was also made, and as a result the local design was changed by introducing two new side supports, changing the load path around the critical location and thus reducing the stresses, in this case in the order of some 30 - 40 %.

Fatigue of Defibrating Equipment for Paper Industry

A new design of defibrating process equipment was invented with a huge potential for increase in productivity. However, this design introduced pulsating forces, causing cracking of the component. Various designs were tried and a welded approach was decided. Despite modifications, cracks continued to appear early in life and it was decided to investigate how best to improve the component which is several m large and rather costly. In this case, the practical solution was to introduce all possible improvements simultaneously as time rather than cost was more important. Hence, the boundary conditions were modified to reduce local stresses at critical location, local geometry was similarly changed to obtain lower stresses, the material was changed to another stainless steel, the weld process (originally MAG) was changed to TIG-welding and shot-peening was applied to the entire product. It would be of academic interest to sometimes try to separate the positive effect of all these steps, but in the end this is not deemed important for the manufacturer.

CONCLUSIONS

Structural design of military airframes is driven by damage tolerance requirements. Durability, mostly assessed in terms of conventional fatigue, is achieved by proper inspection procedures, repairs and/or replacements. Structural integrity can only be achieved by detailed knowledge of flight profiles and resulting load spectra, stress distributions and material properties.

The methodology currently used in Sweden for fatigue management and verification of airframes has been described. Models used for assessment of fatigue life and damage tolerance of our fighter aircraft vary from rather simple to highly complex depending on various factors. Typically, simple models are used initially and more sophisticated analysis is resorted to in complex geometries and when safety margins are lower.

Analyses must always be complemented by testing, from coupon specimen level up to full-scale testing. Fatigue management and verification also involve service load monitoring regarding both dedicated test aircraft, used to verify basic load assumptions, and also individual load tracking programmes developed for new fighters.

Some case studies from other fields than aerospace were included to illustrate that the same aspects as discussed for aircraft always emerge, when considering structural integrity, irrespective of the industry or component studied. Safety issues and cost, however, obviously lead to very different design and certification procedures.

Acknowledgements: The author is grateful to all those colleagues who contributed to the various projects cited in the text.

REFERENCES

1. Blom, A.F. and Ansell, H., "Fatigue Management and Verification of Airframes", in An Assessment of Fatigue Damage and Crack Growth Prediction Techniques, pp. 12:1 - 12:25, AGARD Report 797, March 1994.
2. Military Specification MIL-A-83444 (USAF), Airplane Damage Tolerance Requirements, July 1974.
3. Ansell, H. and Johansson, T., "Design Parameters and Variability in the Aircraft Sizing Process - Aspects of Damage Tolerance", in Durability and Structural Reliability of Airframes, Vol. 1, Ed. A.F. Blom, EMAS Ltd, Warley, U.K., 1993, pp. 619 - 640.
4. Palmberg, B., Olsson, M.-O., Boman, P.-O. and Blom, A.F., "Damage Tolerance Analysis and Testing of the Fighter Aircraft 37 Viggen", in Proc. 17th ICAS Congress, International Council of the Aeronautical Sciences, AIAA, Washington, D.C., U.S.A., 1990, pp. 909-917.
5. Palmberg, B., Olsson, M.-O., Boman, P.-O. and Blom, A.F., "Damage Tolerance Assessment of the Fighter Aircraft 37 Viggen Main Wing Attachment", J. of Aircraft, pp. 377-381, 1993.
6. Andersson, B., Falk, U., Babuska , I., "Accurate and Reliable Determination of Edge and Vertex Stress Intensity Factors in Three-Dimensional Elastomechanics", in Proc. 17th ICAS Congress, International Council of the Aeronautical Sciences, AIAA, Washington, D.C., U.S.A., 1990, pp. 1730-1746.
7. Whitehead, R.S., "Certification of Primary Composite Aircraft Structures", in Proc. 14th ICAS Symposium, Ed. D.L. Simpson, EMAS Ltd., Warley, U.K., 1987, pp. 585-617.
8. Blom, A.F., Svenkvist, P. and Thor, S.-E., "Fatigue Design of Large Wind Energy Conversion Systems and Operational Experience from the Swedish prototypes", J. of Wind Engng and Industrial Aerodynamics, 34, pp. 45-76, 1990.
9. Blom, A.F., ed., (1993). Fatigue under Spectrum Loading and in Corrosive Environments, EMAS Ltd., Warley, U.K., 472 pages.
10. Blom, A.F., "Spectrum Fatigue Behaviour of Welded Joints", Int. J. Fatigue, in press.
11. Lopez Martinez, L., Blom, A.F. and Wang, G.S., "Fatigue Behaviour of TIG Improved Welds", in this conference.

SOME ASPECTS OF THE INFLUENCE OF MICROSTRUCTURE ON FATIGUE.

J. PETIT and J. MENDEZ

Laboratoire de Mécanique et de Physique des Matériaux, URA CNRS 863
ENSMA, Site du Futuroscope, Chasseneuil du Poitou, BP. 109
86960 FUTUROSCOPE Cedex, France.

ABSTRACT

An overview on some aspects of the role of microstructure on fatigue life and fatigue crack propagation is given. The respective influence of microstructural features, environment and temperature for metallic alloys is discussed.

KEYWORDS

Microstructure ; fatigue life ; fatigue propagation ; environment ; temperature.

INTRODUCTION

The fracture fatigue behaviour of metals and metallic alloys have been shown widely influenced by material composition and by various microstructural parameters. Numerous articles can be found in the literature on these topics. Some excellent reviews can be mentioned (Cazaud et al., 1969 ; Fine, 1980, François, 1989) but it would be a task beyond the scope of this paper to refer to all of them.
An overview will be just given on some aspects of the role of microstructure on cyclic behaviour, crack initiation and crack propagation with a special attention to the respective influence of some intrinsic and extrinsic factors and of coupled effects between them.

CYCLIC DEFORMATION MECHANISMS AND FATIGUE LIFE

Low cycle fatigue

The homogeneity of cyclic plastic distribution is one of the predominant factors that have to be achieved for improving Low Cycle Fatigue (LCF) resistance of metallic materials.
It is generally admitted that planar slip FCC alloys are more resistant to fatigue than wavy slip alloys (Laird and Feltner, 1967 ; Grosskreutz, 1972). This is however controversed by recent works on Cu-Al and Cu-Zn monocrystalline or polycrystalline materials (Lukas et al., 1993 ; Mendez and Legendre, 1996). Figure 1 illustrates the effect of planar slip by comparing the LCF resistance of Cu and Cu-Zn 70/30 materials. For a given grain size the fatigue life in vacuum is clearly higher for the α brass. However, in air, at low plastic amplitudes both the materials have the same life ; the effect of environment on fatigue cracking is therefore enhanced by planar slip. Nevertheless, a beneficial effect of planar slip has been noted for other fcc materials. For example, an addition of nitrogen to austenitic stainless steels inhibits cross slip of dislocations and favours planar slip. This is accompanied by an improvement of the material to high temperature low cycle fatigue (Nilsson (1984).
Generally the differences in the slip mode are directly associated with variations in the Stacking Fault Energy. However, other factors favouring planar slip as dynamic strain ageing or short range order have also been evoked (Vogt et al., 1990). In this way, planar arrangements have been observed in

different carbon steels or in austenitic stainless steels in the temperature range 200 - 450°C (Mughrabi, 1993 ; Mendez et al., 1992). As illustrated in Fig. 2 the fatigue life in vacuum of a 316L type austenitic stainless steel, under cyclic plastic strain control, is observed to increase in the temperature range 200-400°C. Concurrently, planar arrangements are observed and their formation is accompanied by a typical cyclic hardening characterized by a strong and continuous increase of the flow stress and no evidence of saturation. This enhancement of the fatigue resistance between 200°C and 500°C due to planar slip occurs in spite of a considerable strong increase of the stress amplitude. The maximum of fatigue resistance results of the competition between the degree of planarity and the stress level. As a consequence, the highest fatigue life is reached at 300°C whilst the dynamic strain ageing phenomenon is maximum at 400°C (Gerland et al., 1993). Now, the intrinsic effect of the slip mode is masked in air by the enhancement of environmental effects at high temperatures. The beneficial role of planar deformation only appears in this environment through a negligible drop in the fatigue resistance of the 316L material between 20°C and 300°C.

In the LCF range, crack initiation in ductile materials occurs early in the life (N/N$_F$ < 0.1 ~ 0.2) and fatigue lifetimes are therefore mainly determined by the crack growth period. This has been established for various materials having fatigue lives between 10^3 and more than 10^5 cycles, in air as in vacuum (Mendez and Violan, 1991). Figure 3 shows the evolution of the surface length (2c) of the fatal crack with the fraction of life N/N$_F$ for 316L specimens cycled in vacuum at different temperatures under déformation conditions very close to those presented in figure 2. These results are related to different deformation modes according to the temperature level, wavy slip at 20°C or 600°C, planar slip at 200°C or 400°C. It is interesting to note that all the experimental data are regrouped in a single 2c - N/N$_F$ curve. Such results which has been confirmed for other ductile materials in the LCF regime (Magnin et al., 1989 ; Mendez and Violan, 1991) shows an homologous behaviour of the main crack irrespective of the mode of slip or microcracking. This means that the beneficial effect of planar slip concerns in the same way crack initiation and crack growth.

It seems that planar arrangements do not allow simply the degree of plastic reversibility to be enhanced but favours particularly the homogeneity of cyclic plastic deformation. This was observed in the 316L type alloy evoked here at 300° and 400°C and was pointed out also by Laird and coworkers on Cu-Al single crystals (Mendez et al., 1992 ; Laird, 1993).
An example of the favourable role of homogeneous deformation is given in a paper presented in this conference concerning dual-phase steels (Mateo et al., 1996). The α' phase, formed at 400°C by spinoïdal decomposition, hardens the ferritic phase and, below a critical strain amplitude, avoids the deformation to be localized in the α phase. In this case, in air as in vacuum a significant increase of fatigue life is obtained comparatively to the unaged material which can be explained by the difficulty for cracks to grow through the ferrite. In this type of materials the difference of behaviour between the ferrite and the austenite can induce significant morphological texture effects in LCF resistance. Differences in fatigue lifetime associated with the relative orientation of both the α and γ phases with regard to the stress axis, can be correlated to the number of α/γ phase boundaries encountered by the main crack during its growth (Perdriset, 1994).Another example concerns metastable austenitic stainless steels type AISI 304L in which martensitic transformations induced by cyclic hardening at low temperature can be exploited to increase the fatigue resistance of this material at room temperature (Maier et al., 1993).

The detrimental effect of grain size in LCF has frequently been reported for copper (Christ et al.,1988 ; Lukas and Kunz, 1986). It is noticeable that such a strong effect of the grain size is associated with intergranular cracking mechanisms. It has been shown that the decrease of fatigue life with grain size in copper is mainly due to an acceleration of crack growth in air as in vacuum (Gotte et al., 1996). In contrast, a recent work on an aluminium magnesium Al-2.63 Mg alloy, exhibiting transgranular damage, shows limited effect of grain size in the LCF range (Turnbull and de los Rios, 1995).

High cycle fatigue

Fig. 4 shows that fatigue strength in copper also increases in the high cycle fatigue (HCF) range when the grain size decreases. Moreover fatigue limit is lower when the grain is coarser or the environment is more aggressive (Mendez, 1984). On the Al-2.63 Mg alloy evoked before (Turnbull and de los Rios, 1995), the endurance stress is a function of the inverse square root of the polycrystal grain size as described by a Hall-Petch type relation.The grain size contribution to the endurance strength is shown to be similar to its contribution to the cyclic yield stress.
In two-phase alloys, cracking mechanisms are related to the inhomogeneity of stresses and strains due to the misfit between the phases or to strain localization into the softer phase. Recent studies on

different $\alpha + \beta$ or β forged Titanium alloys have emphasized the role of microstructural features on the successive stages of damage determining fatigue life (Wagner and Lütjering, 1988).

In the pseudo elastic regime, S-N curves are mainly determined by the yield and tensile strength. However materials with similar yield strengths can exhibit significant differences in fatigue life related to crack nucleation characteristics (Awadé, 1995).

The effect of different microstructural parameters concerning alpha phase amount or grain morphology have been evaluated on $(\alpha + \beta)$ TA6V alloys by investigating the initiation and growth of the naturally initiated microcraks (Demulsant and Mendez, 1996). Isolated coarse lamellar structures favours a rapide initiation of large cracks 200-300 µm long. For fine structures initiation also occurs early in the life, but in this case propagating upto 200-300 µm the very small initiated cracks must consume an important number of cycles. The difference in fatigue resistance is therefore related here to the initial crack length since no significant influence of microstructure on crack growth is observed and that the number of cycles for crack initiation is very limited (Fig. 5). The slip length appears here as one of the predominant factors which should be minimized in order to achieve maximum fatigue strength. A few zones with coarse structures are sufficient to produce dramatic drop in the fatigue resistance to crack initiation. In contrast a fine microstructure may induce the initiation of a considerable number of microcracks which accelerates crack growth by coalescence. It is interesting to point out that in such high strength materials, crack initiation can represent a large fraction of the fatigue life even in the "low cycle fatigue range" with fatigue lives around 10^4 cycles : the larger the microstructural element, the shorter the number of cycles to initiate a crack and the larger its initial length (Demulsant, 1994 ; Awadé and Mendez, 1995) (Fig. 6) This behaviour explains the differences in fatigue life and endurance limit which in titanium alloys is clearly associated with the non initiation of cracks. Another interesting result is that crack initiation mechanisms may be highly assisted by the environment, in particular the α lamellae interface cracking. Consequently an increase of temperature may have a detrimental effect in air whilst it has a beneficial effect in vacuum (Awadé, 1995).

FATIGUE CRACK PROPAGATION

The fatigue propagation of long cracks in metallic alloys has been shown dependent on various microstructural parameters. A selection of demonstrative examples provided by the literature and authors results will give illustrations of the respective role of grain size, alloy phases, texture and age-hardening. An evaluation of some parameters is made when tests are carried out under inert environment and when crack closure corrections are performed, because the environment may affect the fatigue behaviour in different ways and the contribution of crack closure may depend on microstructure.

Grain size.

Grains boundaries can constitute high obstacle to the movement of dislocations, thus increasing the flow stress according to the Hall-Petch relation. Fatigue process being basically the result of cumulated plastic deformation, can be affected by hardening processes, and can also interact with grain boundaries. Numerous studies have shown that the decrease of grain size improves the fatigue limit but conversely reduces the resistance against crack propagation. For example, Haberz et al., 1993 have clearly demonstrated in ARMCO iron with different grain sizes ranging between 3 and 3000 µm, that the nominal stress intensity threshold is strongly influenced by the grain size (Fig. 7). But after correction for closure, the effective threshold is not very significantly affected. Consequently, the change of the threshold is mainly caused by the increase of the roughness induced closure effect (Suresh. et al., 1994) with increasing grain size. Such observations are in accordance with many others, but the range of grain size here explored is exceptional. Finally, it comes out from this work that there is no substantial influence of grain size on the effective behaviour of a stage II crack.

Texture

The texture can modify the nominal crack propagation (R = 0.1 in air) as illustrated in figure 8 for a 2090 - T8X Aluminium Lithium alloy (Vankateswara et al., 1991). Faster crack growth and lower threshold are observed in the 1.6 mm T83 thin sheet compared to the 12.7 mm T81 thick plate for tests performed in the LT orientation. But after closure correction both materials present the same effective behaviour. The more zig-zaging crack path in the thick plate only increases the contribution of the roughness induced closure (or non closure) as illustrated in figure 2.

Alloy phases

Two examples have been selected to illustrate the influence of alloy phases. Crack propagation data in duplex (iced water quenched after 1 h à 760°C) and normalized AISI-1018 are plotted in figure 9 (Minakawa et al. 1982). Obviously the resistance of the duplex structure against crack propagation for a fatigue test performed at a R ratio of 0.05, is very much higher than that of the normalized material. It could be attractive to relate such performance and high threshold level to the increase in the yield stress (427 MPa and 255 MPa respectively for duplex and normalized conditions). But after closure correction, the effective propagation of both microstructures is identical. Hence the differences in the nominal curves must be attributed to the difference existing in the contribution of crack closure. Here again the microstructure does not affect the effective stage II propagation.

The next example (figure 10) shows the effect of retained metastable β phase on fatigue crack propagation characteristics of forged bars of a Ti 6246 Titanium alloy (Niinomi et al., 1993). The crack propagation rates in the aged material (6 h at 863 K of solutionizing) are substantially lower than in the as-solutionized microstructure, with a threshold range decreased of more than 50 %. But in this case, even after closure correction, there is still a large difference between the two microstructures. This example shows that when the propagation mechanism in itself is changed from one microstructure to the other, the effective behaviour is also modified.

Aged conditions

An illustration of the coupled influence of microstructure and environment still existing after closure correction is given in figure 11 on a 7075 alloy in two aged conditions tested in ambient air and high vacuum (Petit et al., 1994). The peakaged matrix contains shearable Guinier-Preston zones and shearable precipitates which promote a localization of the plastic deformation within a single slip system in each individual grain along the crack front. The overaged matrix contains larger and less coherent precipitates which favour a wavy slip mechanism (Lindigkeit et al., 197 ; Kirby et al., 1979 ; Petit, 1984). In vacuum, the peak-aged T651 condition leads to a highly retarded crystallographic propagation (so called stage I - like regime) while the overaged T7351 condition gives a conventional stage II propagation. In ambient air, the single slip mechanism which is still operative in the peak-aged alloy, is assumed to offer a preferential path for hydrogen embrittlement which leads to a strongly accelerated propagation. Conversely ambient air has little influence on the stage II regime as observed in the overaged alloy. It can be noticed that the influence of ambient air has inverted the ranking of the propagation curves with respect to the microstructures. It can be underlined that a crystallographic propagation is faster than stage II in air, but is highly retarded in vacuum. These results indicate a high sensitivity to environment of the slip mechanisms, specially near the grain boundaries.

Intrinsic fatigue crack growth

The above examples have shown that if one intends to analyse the specific role of microstructure it would be useful to analyse the crack propagation behaviour of the material in conditions where the influence of crack closure and environment are eliminated, that is to say to examine the intrinsic fatigue crack growth behaviour.

On the basis of numerous experimental data obtained in high vacuum on technical Aluminium alloys with various aging conditions, on Aluminium based single crystals and on steels and Titanium alloys it has been shown (Petit et al., 1994) that the intrinsic fatigue crack growth can be described according to three characteristic regimes (see example of Al alloys in figure 12).

The faster intrinsic stage I, has been identified on single crystals of Al-Zn-Mg alloys with a peak-aged microstructure which favors crystallographic propagation along a PSB (persistent slip bands) which develop in a {111} planes pre-oriented for single slip. This regime is also observed on various materials in the early growth of microstructural short cracks.

The intermediate intrinsic stage II is commonly observed on polycrystals and single crystals when crack propagation proceeds at macroscopic scale along planes normal to the loading direction. Such propagation is induced by microstructure wich promote homogenous deformation and wavy slip as large or non coherent precipitates or small grains size. The figure 13 illustrates a typical change from a near threshold stage I to a mid ΔK stage II propagation in an Al-Zn-Mg single crystal.

The slowest regime, or intrinsic stage I-like propagation corresponds to a crystallographic crack growth observed near the threshold in polycrystals or in the early stage of growth of naturally initiated microcracks when ageing conditions or low stacking fault energy generate heterogeneous deformation along single slip systems, within individual grains (see example in figure 14) . Crack branching and crack deviation mechanisms (Suresh, 1985) and barrier effect of grain boundaries (de los Rios, 1985), are assumed to lower the stress intensity factor at the crack tip of the main crack.

The stage II regime is in accordance with a propagation law derived from the models initially proposed by Mc Clintock 1963, Rice 1965 or Weertman 1966:

$$da/dN= A/D^* \ (\Delta K_{eff}/\mu)^4 \qquad\qquad (1)$$

where A is a dimensionless parameter, μ the shear modulus and D* the critical cumulated displacement leading to rupture over a crack increment ahead of the crack tip.

Intrinsic data for well identified stage II propagation are plotted in figure 15 in a da/dN vs $\Delta K_{eff}/E$ (E : Young Modulus) diagram for a wide selection of Al alloys, in figure 16 for a selection of steels and some data for a TA6V Ti alloy with the mean curve for Al alloys.
This diagram constitutes an excellent validation of the above relation and confirm that the LEFM concept is very well adapted to describe the intrinsic growth of a stage II crack which clearly appears to be nearly independent on the alloy composition, the microstructure (when it does not introduce a change in the deformation mechanism), the grain size, and hence the yield stress. The predominant factor is the Young modulus of the matrix, and the slight differences existing between the three base metals can be interpreted as some limited change in D* according to the alloy ductility (Petit, 1984).
The stage I-like regime cannot be rationalized using the above relation (Fig. 17).The retardation is well marked when the number of available slip systems is limited (Ti alloys) or can be nearly absent when some secondary slip systems can be activated near the boundaries (Xu et al., 1991 in Al-Li alloys).

Environmentally assisted propagation

Following the rationalization of intrinsic stage II propagation as presented above, some similar rationalization of FCG in air could be expected after correction for crack closure and temperature effects ($\Delta K_{eff}/E$). Figure 18 presents a compilation of stage II propagation data obtained in ambient air for almost the same alloys as in vacuum (see, Fig. 15). Obviously no rationalization does exist in air. The sensitivity to air environment is shown strongly dependent as well on base metals, addition elements, and microstructures (see 7075 alloy in three different conditions) as on R ratio and growth rate. However a typical common critical rate range at about 10^{-8} m/cycle can be pointed out for all materials. This critical step is associated to stress intensity factor ranges at which the plastic zone size at the crack tip is of the same order as grain or sub-grain diameters. In addition there is a general agreement to consider that, for growth rates lower than this critical range, crack propagation results from a step-by-step advance mechanism instead of a cycle-by-cycle progression as generally observed in the Paris regime in air.
A comprehensive model has been established for environmentally assisted crack growth (Petit et al., 1984, 1993, 1994) as schematically illustrated in figure 19 :
– at growth rates higher than a critical rate (da/dN)$_{cr}$ which depends upon several factors as surrounding partial pressure of water vapour, load ratio, test frequency, chemical composition and microstructure, the crack growth mechanism is assisted by water vapour adsorption but it is still controlled by plasticity as in vacuum.
– at growth rates lower than (da/dN)$_{cr}$, an Hydrogen assisted crack growth mechanims becomes operative, Hydrogen being provided by adsorbed water vapour when some critical conditions are fulfilled.
At room temperature and for conventional test frequencies of 20 to 50 Hz, (da/dN)$_{cr}$ is about 10^{-8} m/cycle as pointed out in figure 10. As recently described by Henaff et al., 1995, the adsorption assisted stage II propagation verify relation (1), adsorption being just assumed to reduce the cumulated displacement D* in accordance with Lynch approach (Lynch, 1988). The modelling of the hydrogen-assisted propagation has to be developed by the introduction of the coupled effect of two concurrent mechanisms, i.e. Hydrogen action and plastic cyclic accumulation, which can be strongly affected by environment and temperature.

High temperature

Fatigue crack propagation can be strongly affected by temperature and coupled environmental effects. For example, the crack propagation behaviour of Ti alloys explored at 300°C (ambient air, R = 0.1, 35

Hz) have shown a sensitivity to alloy composition and microstructure. But after closure correction, most of the differences are shown to vanish (Sarrazin-Baudoux, 1996). As pointed out at room temperature, when there is no important change in the crack growth mechanism, the stage II regime is poorly sensible to microstructure. However, at 500°C, even after closure correction, a substantial difference can be observed between Ti 6246 and Ti 6242 with a better effective resistance of Ti 6242 suggesting some change in the response of Ti 6246 due to the temperature (Fig. 20). Experiments performed in different environments and at low frequency support a dominant embrittling effect of water vapour with the development of a corrosion-fatigue propagation which can be described using the relation (Lesterlin, 1995).

$$da/dN = 0.5 \text{ x } \Delta CTOD \qquad\qquad (2)$$

On going experiments on Ti 6242 and further literature results on studies presently carried out by several researchers on Ti 1100 or MI814, would provide more information to enlarge this analysis.

At higher temperatures creep-fatigue can induce transgranular or intergranular propagation with respect to microstructure and environment. For example, at 750°C serrated grain boundaries on a PM Astroloy (Figure 21) strongly improve the resistance against crack propagation by changing the fracture process from an intergranular to a mixed mode (inter-transgranular) propagation. It can be also noticed the absence of significant influence of environment in the present experimental conditions (Loyer-Danflou et al., 1993).

CONCLUSION

In this overview of the influence of microstructure on the fatigue behaviour of metallic alloys some aspects have been emphasized :
– the homogeneity of the plastic deformation is a critical process for crack initiation and crack propagation. Intrinsically it depends on alloy composition, size and morphology of grains, phase distribution, size and coherence of precipitates, and texture.
– environment (air compared to high vacuum) and temperature can strongly modify the deformation processes and the propagation mechanisms. An important role of water vapour on crack propagation has been underlined and three intrinsic crack propagation regimes have been clearly identified.
– Microstructure has little influence on the effective stage II propagation for a given based matrix metal, but it can strongly modify the closure contribution.
– high temperature, hold time and air environment favour fast intercrystalline creep-fatigue propagation.

REFERENCES

ASM "Materials Science Seminar" (1979) St Louis, ASM pub.
Awadé, E. (1995). Rôle des facteurs microstructuraux sur l'endommagement par fatigue de l'alliage de Titane bétacez. Effet de la température et de l'environnement. Doctorat Thesis, Poitiers, France.
Awadé, E. and J. Mendez (1996). Microstructure and fatigue processes relationships in the betacez alloy at room and elevated temperatures. Proc. 8th World Conf. on Titanium, To be published.
Cazaud, R., G. Pomey, P. Rabbe, Ch. Janssen (1969). "La fatigue des métaux" Dunod, Paris.
Christ, H.J., H. Mughrabi and C. Witting-Link (1998). Cyclic deformation behaviour, microstructure and fatigue crack initiation of copper polycrystals fatigued in air and in vacuum. "Basic mechanisms in fatigue of metals". P. Lukas and J. Polak Eds. Elsevier, 83-92.
De los Rios, E.R., J. Hussain, J. Mohamed and K.J. Miller (1985). A micro-mechanics analysis for short fatigue crack growth, Fat. fract. Engng. Mater. Struct.., Vol.8, n° 1, 49-63.
Demulsant, X. and J. Mendez (1996). Microstructure effects on small fatigue crack initiation and growth in Ti6Al4V alloys. Fatigue Engng. Mat. Struct. in press.
Démulsant, F. (1994). Facteurs microstructuraux gouvernant l'amorçage et la croissance des fissures de fatigue dans les alliages de titane. Doctorate Thesis, Université de Poitiers, France.
François, D. (1989). The influence of the microstructure on fatigue, Advances in Fatigue Science and Technology, C. Maura Branco et al. eds., Kluwer Acad. Pub., 23-76.
Gerland, M., J. Mendez, J. Lepinoux and P. Violan (1993). Dislocation structure and corduroy contrast in a 316L alloy fatigued at (0.3-0.5) Tm. Mater. Sci. Engng., A164, 226-229.
Gotte, L., J. Mendez, P. Villechaise and P. Violan. Effect of grain size on low cycle fatigue of copper polycrystals. Proc. "Fatigue 96".
Grosskrentz, J.C. (1972). Strengthening and fracture in fatigue, Metal Trans. 3, 1255-1262.

Haberz, K., R. Pippan and H.P. Stüwe (1993). The threshold of stress intensity range in iron. Proceeding of Fatigue 93, Vol. 1, J.P. Baïlon and J.I. Dickson eds.525-530.

Hénaff, G., K. Marchal and J. Petit (1995). On fatigue crack propagation enhancement by a ga zeous atmosphere : experimental and theoretical aspects, Acta Metall. Mater., Vol. 43, n° 8, 2931-2942.

Kirby, B.R. and C.J. Beevers (1979). Slow fatigue crack growth and threshold behaviour in air and vacuum of commercial Aluminium alloys, Fatigue Engng. Mater. Struct. Vol.1, 203-215.

Laird, C. (1993). Cyclic plasticity in face centered cubic materials. "Fatigue 93". J.P. Baïlon and J.I. Dickson Ed., EMAS, Vol.I, 57-70.

Land, C. and C.E. Feltner (1967). The Coffin-Manson law in relation to slip character. Trans. Metal Soc. AIME 239, 1074-1084.

Lesterlin, S.(1995). Influence de l'environnement et de la température sur la fissuration par fatigue des alliages de titane. Doctorate Thesis, Université de Poitiers, France.

Lindigkeit, J., G. Terlinde, A. Crysler and G. Lütjering (1979). The effect of grain size on the fatigue crack propagation behaviour of aged-hardened alloys in inert and corrosive environment, Acta Metal., vol. 27, N° 11, 1717-1726.

Loyer-Danflou, H. , M. Marty, A. Walder, J. Mendez and P. Violan (1993). The effect of environment and structure on creep fatigue crack propagation in a P/M astroloy, Journal de Physique IV, Colloque C7, Vol. 3, 359-370.

Lukas, P. and L. Kunz (1986). Mechanisms of near-threshold fatigue crack propagation and high cycle fatigue in copper. Acta technica CSAV n°4, 460-488.

Lukas, P. and L. Kunz and J. Krejci (1993). Fatigue behaviour of planar slip single crystals. "Fatigue 93". J.P. Baïlon and J.I. Dickson Ed., EMAS, Vol.I, 71-76.

Lynch, S.P., (1988). Acta metall. 36, 2639.

Magnin, T., C. Ramade, J. Lepinoux and L.P. Kubin (1989). Low cycle fatigue mechanisms of f.c.c. and b.c.c. polycrystals Homologous behaviour ?. Mater. Sci. Engng. Vol. A118, 41-51.

Maier, H.J., B. Donth, M. Bayerlein and H. Mughrabi (1993). Strength enchancement of 304L stainless steel by fatigue-induced low-temperature martensitic transformation. "Fatigue 93" J.P. Baïlon and J.I. Dickson Ed., EMAS, Vol.I,85-90

Mateo, A. P. Violan, L.Llanes, J. Mendez and M. Anglada. Fatigue life of duplex stainless steels : influence of ageing and environment. in "Fatigue'96".

Mc Clintocks, F.A. (1963). Fracture of Solids, Inter science pub., 65-102.

Mendez, J. (1984). Doctorat ès Sciences Université de Poitiers.

Mendez, J. and P. Violan (1991). Processus d'endommagement et cumul des dommages en fatigue plastique. Journées de Printemps de la SFM. Ed. Revue de Métallurgie, Paris, n° 5, 58-70.

Mendez, J., L. Legendre, To be published.

Mendez, J., P. Violan, R. Alain and M. Gerland (1992). Crack initiation and growth in a 316L stainless steel cycled between 20°C and 600°C in vacuum. ECF9. Reliability and Structural Integrity of Advanced Materials. S. Sedmak, A. Sedmak, D. Ruzic, Ed., EMAS VOL.1, 439-444.

Minakawa, K., Y. Matsuo and A.J. Mc Evily (1982). The influence of a Duplex Microstructure in Steels on Fatigue Crack Growth in the near-threshold region, Metall. Trans. A.,Vol.13A, 439-445.

Morris E. Fine (1980). "Fatigue resistance of metals". Metall. Trans. A., 11A, 365-379.

Mughrabi, H. (1993). Cyclic plasticity and fatigue of metals. Proc. Euromet 93, Journal de Physique IV, Vol.3, 659-668.

Niinomi, M., T. Kobayashi and A. Shimokawa (1993). Effect of retained metastable β phase on fatigue crack propagation characteristics of α+β type Titanium alloys, Proceeding of Fatigue 93, Vol. II, J.P. Baïlon and J.I. Dickson eds.663-668.

Nilsson, J.O. (1984). The influence of nitrogen on high temperature low cycle fatigue behaviour of austenitic stainless steels. Fatigue Engng. Mater. Struct. Vol. 7, n° 1, 55-64.

Perdriset, F. (1994). Influence de l'azote sur les mécanismes de fissuration en fatigue-corrosion d'un acier inoxydable austéno-ferritique de type Z3 CND 2205. Doctorate Thesis, Université de Lille, France.

Petit, J. (1984). Some aspects of near-threshold crack growth : microstructural and environmental effects, Fatigue crack growth threshold concepts, D. Davidson et al. eds., TMS AIME pub., 3-24.

Petit, J. and G. Hénaff (1993). A survey of near-threshold fatigue crack propagation : mechanisms and modelling, Fatigue 93, J.P. Baïlon and J.I. Dickson Ed., EMAS pub. 503-512.

Petit, J., J. de Fouquet and G. Henaff (1994). "Influence of ambient atmosphere on fatigue crack growth behaviour of metals", Handbook of Fatigue Crack Propagation in Metallic Structures. A. Carpinterie ed. Elsevier pub., vol. 2, 1159-1204.

Rice, J.P. (1965). ASTM STP 416.

Sarrazin-Baudoux, C., S. Lesterlin and J. Petit (1996). Fatigue behaviour of Titanium alloys at room temperature and 300°C in ambient air and in high vacuum, Proc. Fatigue 96.

Suresh, S. and R.O. Ritchie (1984). Near-threshold fatigue crack propagation : a perspective on the role of crack closure, in Fatigue Crack Growth Threshold Concepts, D. Davidson et al. eds., TMS AIME pub. 227-262.

Suresh, S.(1985). Fatigue crack deflection and fracture surface contact : micromechanical models,Metall. Trans. A., vol. 16A, 249-260.

Turnbull, A. and E.R. de los Rios (1995). The effect of grain size on fatigue crack growth in an aluminium magnesium alloy. Fatigue Fract. Engng. Mat. Struct. Vol. 18, n°11, 1355-1366.

Vankateswara, K.T. and R.O. Ritchie (1991). Fatigue of Aluminium-Lithium, Center for Advanced Materials, L.B.L. 30176, Univ. of California.

Vogt, J-B., T. Magnin and J. Foct. (1990). Factors influencing planar slip during fatigue in f.c.c. stainless steels. "Fatigue 90". H. Kitagawa and T. Tanaka, Ed., MCEP, Vol.1, 87-92.

Wagner, L. and G. Lütjering (1988). Propagation of small fatigue cracks in Ti alloys. 6th World Conference on Titanium. P. Lacombe, R. Tricot, G. Beranger. Eds. SFM. Les Editions de Physique, Vol. 1, 345-350.

Weertman, J. (1966). Int. J. Fract. Mech.2, 460-467.

Xu, Y.B., L. Wang, Y. Zhang, Z.G. Wang and Q.Z. Hu (1991). Metal Trans., 22A, 723-729.

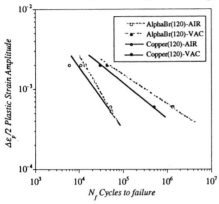

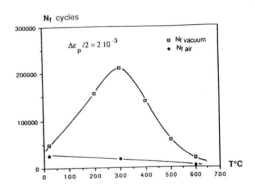

Fig. 1. Low cycle fatigue of copper and 70/30 α brass. Grain size 120 µm. Room temperature. Air and vacuum (Mendez and Legendre, 1996).

Fig. 2. Number of cycles to failure as a function of temperature of a 316L type stainless steel fatigued in plastic train control. (Gerland et al., 1993).

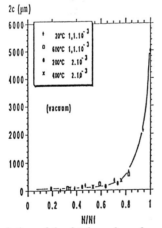

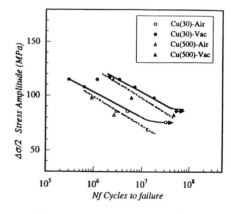

Fig. 3. Evolution of the fatal crack surface length versus the fraction of life of a 316L type stainless steel cycled in plastic strain control in the temperature range 20 - 600°C (Mendez et al. 1992)

Fig. 4. Effect of grain size on the high cycle fatigue of copper in air and vacuum. (J. Mendez, 1984).

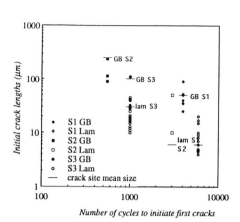

Fig.5. Effect of the nature (GBα or α platelet interfaces) and initiation sites dimensions on the number of cycles to initiation and initial crack lengths. Betacez titanium alloy with three different microstructures. $\Delta\sigma/2 = \pm 850$ MPa. Air, 0.25 Hz (Awadé, 1995)

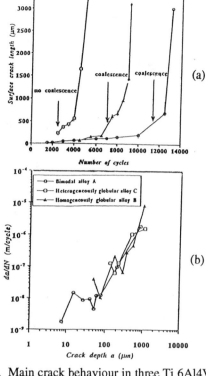

Fig. 6. Main crack behaviour in three Ti 6Al4V alloys. $\Delta\sigma/2 = \pm 750$ MPa. Room temperature, air, 0.25 Hz. (a) Coarse microstructures favours the initiation of large cracks. A high crack density accelerates crack growth by coalescence.
(b) No significant effect of microstructure is observed on small crack growth (Demulsant and Mendez, 1996).

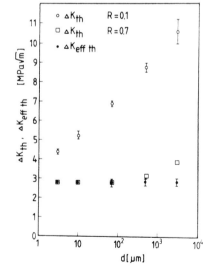

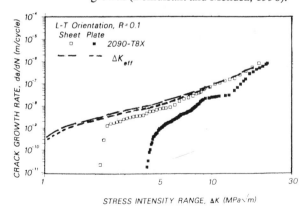

Fig. 7. Threshold and effective threshold as functions of the grain size in ARMCO steel (Haberz et al., 1993).

Fig. 8. Fatigue crack propagation in 1.6 mm-thin sheet and 12.7 mm-thick plate of 2090 T8X Al-Li alloy. (Vankateswara et al. 1991).

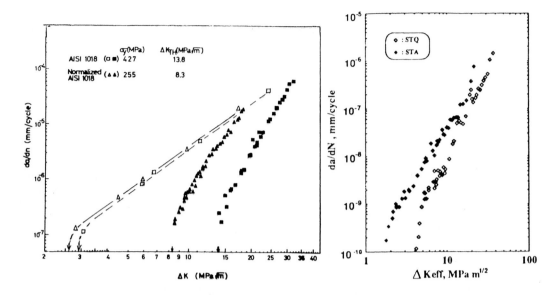

Fig. 9. Rate of growth vs ΔK for AISI 1018 in Duplex and normalized conditions. Effective data (open symbols) are plotted using crack opening measurements from (Minakawa et al., 1982)

Fig. 10. Relationships between da/dN and ΔK$_{eff}$ in solution treated specimen and aged specimen of Ti-6Al-2Sn-4Zr-6Mo forged bars (Niinomi et al., 1993).

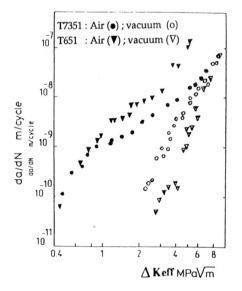

Fig. 11. Illustration of interaction between microstructure and environment on a 7075 alloy in two ageing conditions tested in ambient air and high vacuum.

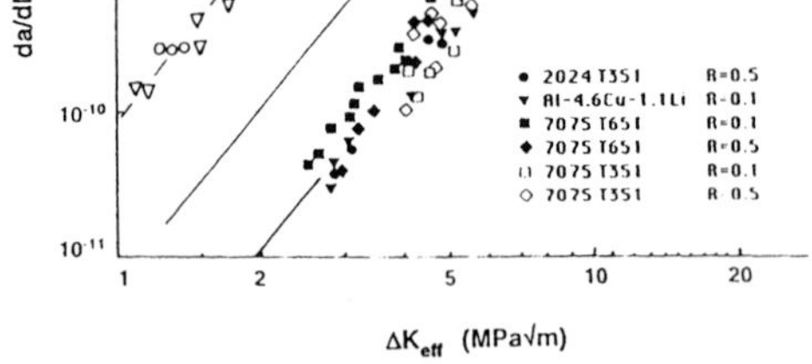

Fig.12. Illustration of the three intrinsinc propagation regimes for Al alloys (mean curve from fig. 15 for stage II).

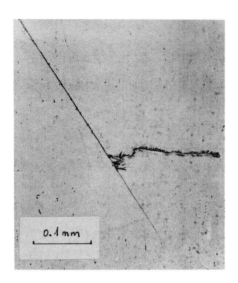

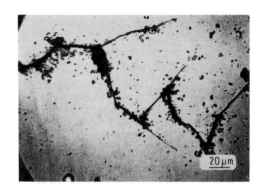

Fig. 13. Stage I to stage II transition in a peak aged single crystal of Al-4.5 % Zn - 12.5 % Mg preoriented for single slip (high vacuum, R = 0.1, 35 Hz).

Fig. 14. Stage I-like propagation in 2024T351 tested in high vacuum (R = 0.5, da/dN = 2.10^{-11} m/cycle).

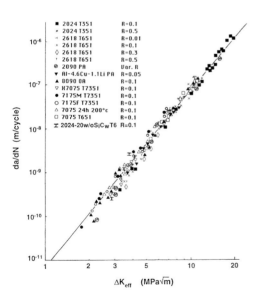

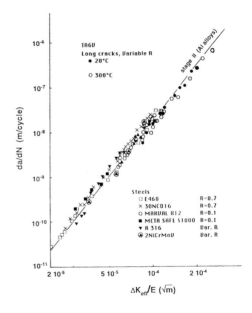

Fig. 15. Intrinsic state II propagation. Al based alloys.

Fig. 16. Intrinsic stage II propagation. Steels and TA6V alloy compared to mean curves for Al alloys after rationalization in term of $\Delta K_{eff}/E$.

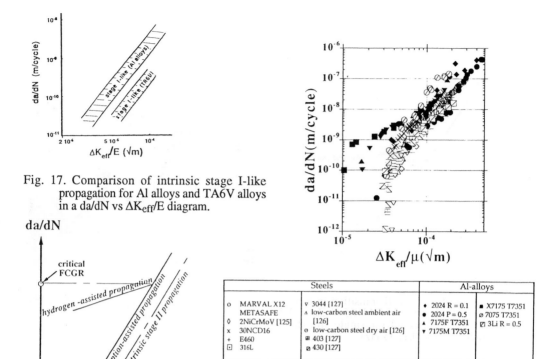

Fig. 17. Comparison of intrinsic stage I-like propagation for Al alloys and TA6V alloys in a da/dN vs $\Delta K_{eff}/E$ diagram.

Fig. 19. Schematic illustration of environmentally assisted stage II fatigue crack growth mechanisms.

Fig. 18. Effective data in terms of $\Delta K_{eff}/\mu$ (μ = shear modulus) for steels and Al alloys.

	Steels		Al-alloys	
o MARVAL X12	v 3044 [127]		♦ 2024 R = 0.1	■ X7175 T7351
METASAFE	▵ low-carbon steel ambient air		● 2024 P = 0.5	⌀ 7075 T7351
◊ 2NiCrMoV [125]	[126]		▲ 7175F T7351	▨ 3Li R = 0.5
x 30NCD16	⊖ low-carbon steel dry air [126]		▼ 7175M T7351	
+ E460	⊞ 403 [127]			
⊡ 316L	⊘ 430 [127]			

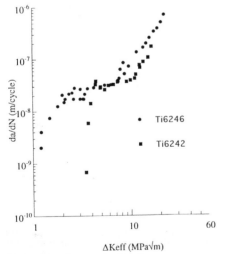

Fig. 20. Effective crack propagation at 500°C of two Ti alloys (R = 0.1, 35 Hz).

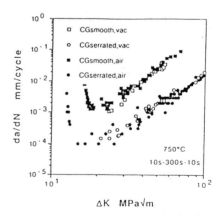

Fig. 21. Effect of grain boundary morphology (smooth or serrated boundaries) on creep-fatigue crack propagation in air and high vacuum at 750°C in a PM Astroloy.

Small Fatigue Cracks in Advanced Materials

Keisuke TANAKA and Yoshiaki AKINIWA

Department of Mechanical Engineering, Nagoya University,
Furo-cho, Chikusa-ku, Nagoya 464-01, Japan

ABSTRACT

The anomalous propagation behavior of small fatigue cracks has been ascribed to three reasons: (1) premature crack closure or crack-tip shielding, (2) large-scale yielding, and (3) microstructural inhomogeneities. The effective stress intensity range ΔK_{eff} is a crack driving force parameter for small cracks whose anomalous propagation behavior comes from premature crack closure. For cracks with large-scale yielding, the J integral range, ΔJ, is an appropriate parameter. The microstructurally small crack propagates fast just after nucleation, and then decelerates due to barriers such as grain boundaries and second phases. This type of the behavior is commonly observed in various advanced metallic materials, metal-matrix composites and ceramics. The fatigue limit of smooth specimens is determined by the propagation condition of small cracks across microstructural barriers.

KEYWORDS

Fatigue; Small Fatigue Cracks; Crack Closure; Microstructure; Plasticity; Advanced Materials; Composites; Ceramics

INTRODUCTION

Linear elastic fracture mechanics (LEFM) is a well-established discipline for assessing the strength and durability of structural components containing cracks and subjected to cyclic loading. The use of ΔK-based conventional methodology, however, is questionable when the crack length is physically small or less than about 2mm. Physically small fatigue cracks often show fast, irregular growth rates, and the life assessment of a structure based on the large-crack fracture mechanics often results in a non-conservative estimate.

During the last two decades, numerous works have been conducted to understand the anomalous propagation behavior of small fatigue cracks. Main reasons proposed for the anomalous propagation behavior are the following three: (1) crack-tip shielding or crack closure, (2) large-scale yielding or macroscopic plasticity, and (3) microstructural inhomogeneities.

In the present paper, the mechanics of small fatigue cracks is first described, and then applied to the propagation of small fatigue cracks in advanced metallic materials, metal-matrix composites, and ceramics. The state-of-art of small fatigue crack studies will be reviewed and several possible directions of future studies will be proposed. The understanding of the interaction of small fatigue cracks with microstructures will be essential for the development of these advanced materials.

MECHANICS OF SMALL FATIGUE CRACKS

Large Cracks and Small Cracks

A unique correlation between the crack propagation rate and the range of the applied value of the stress intensity factor, ΔK, is based on the similitude concept. Figure 1 illustrates a fatigue crack propagating under cyclic loading. The K-singularity field spreads over the region with the dimension of about one-tenth of the crack length. When the crack is large and when the physical (chemical) process of fatigue crack growth occurs within the K-singularity field, the rate of crack propagation is unique. On the other hand, when the crack length is small, the K-similitude breaks down.

The breakdown of the K-similitude due to the smallness of the crack can be induced through several mechanisms. The following two are primarily mechanisms in the case of fatigue crack propagation in non-corrosive environment (Tanaka, 1987, Ritchie *et al.*, 1986).

(1) Breakdown of the mechanical similitude. The ΔK value loses its meaning as a crack driving force parameter for the following tow cases. The one is large-scale yielding. The plastic zone size is large compared with the crack length. The other is premature crack closure. The crack closure (or crack-tip shielding) is not fully developed because of the short crack wake. These cracks are called *mechanically small cracks*.

(2) Breakdown of the microstructural similitude. When the crack length is on the order of the material microstructure, e.g., the grain size, plastic deformation near the crack tip and the crack growth direction are very much influenced by the material microstructure. The assumption of macroscopic continuum is violated. These small cracks are named *microstructurally small cracks*.

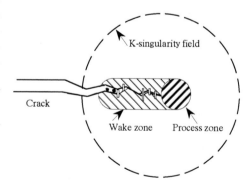

Crack-Tip Shielding or Crack Closure

The applied value of SIF, K, often does not

Fig. 1 Real crack model.

characterize the near-tip stress-strain field of real cracks because of crack-tip shielding. Crack-tip shieldings by fiber bridging, particle wedging, and fracture surface contact in the wake region of a crack are shown in Fig.1. The true characterizing parameter for the near-tip field of a real crack with crack-tip shielding is

$$K_{tip} = K - K_s \quad (1)$$

where K_s is the amount of reduction in SIF due to shielding (Ritchie, 1988). Under cyclic loading, the range of K_{tip} will be a true crack driving force.

$$\Delta K_{tip} = \Delta K_{app} - \Delta K_s \quad (2)$$

For large cracks, K_s is constant or a function of only K, so that K can be used instead of K_{tip} as an apparent crack driving force.

In metal fatigue, the crack closure is the main shielding mechanism. Since crack closure is only operating at the minimum stress, the K_{tip} value at the minimum stress is

$$K_{tipmin} = K_{cl} \quad (3)$$

At the maximum stress, it is

$$K_{tipmax} = K_{max} \quad (4)$$

The range of K_{tip} is

$$\Delta K_{tip} = \Delta K_{eff} = K_{max} - K_{cl} \quad (5)$$

and

$$\Delta K_s = \Delta K_{cl} = K_{cl} - K_{min} \quad (6)$$

The principal mechanisms for crack closure are plasticity-induced, roughness-induced, and oxide-induced ones.

The effective stress intensity factor, ΔK_{eff}, defined by Eq.(5) has been successfully used for predicting the propagation behavior of mechanically small cracks emanating from notches and defects. Tanaka *et al.* (1994) studied a small crack near the tip of a sharp notch in aluminum alloy 2024-T6 reinforced with 20 volume percent of SiC particles. The crack propagation rate is correlated to the applied range ΔK in Fig. 2(a), where the data for long cracks and of unreinforced materials are shown for comparison. The propagation rate of short cracks is higher than that predicted from the da/dN-ΔK relation for long cracks. At low stress amplitudes, short cracks decelerate and

(a)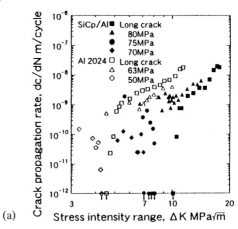

(b)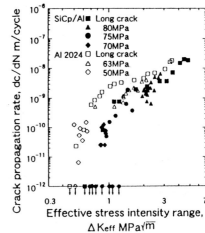

(c)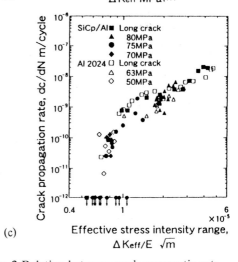

Fig. 2 Relation between crack propagatio rate and stress intensity range for notched specimens of aluminum alloys unreinforced and reinforced with SiC particles.

finally stop.   In Fig. 2(b), the crack propagation rate is correlated to the effective stress intensity range ΔK_{eff}. The relation between da/dN and ΔK_{eff} is identical for short and long cracks.   The difference in the da/dN-ΔK_{eff} relation between unreinforced and reinforced materials may come from the difference in the stiffness as shown in Fig. 2(c).

Large-Scale Plasticity Effect

Small cracks can be accelerated by large-scale crack-tip yielding or macroscopic plasticity such as small cracks in smooth specimens under low-cycle strain-controlled fatigue.   This acceleration is caused by a reduced amount of crack closure and by large-scale cyclic yielding at the crack tip.   For these cracks, ΔJ integral can be regarded as a controlling parameter. Figure 3(a) shows the relation between da/dN and ΔK_{eff} for small semi-elliptical surface cracks in aluminum alloys unreinforced and reinforced with 20 volume percent SiC particles (Tanaka et al., 1995).   The crack propagation rate in reinforced material under strain-controlled tests is very high because of macroscopic plasticity.   When the rate da/dN (m/cycle) is correlated to $\Delta J/E$ (N/m), all the data including the results of small cracks emanating from notches follow the following relation:

$$da/dN = 1.90 \times 10^6 (\Delta J/E)^{1.60} \qquad (7)$$

Microstructural Effect

Inhomogeneous microstructures induce fast and irregular propagation of small fatigue cracks nucleated in smooth specimens.   The propagation behavior of these small cracks is not unique even under the same value of ΔK_{eff}. The resistance to the propagation of small cracks or slip bands will be different depending on the material inhomogeneities or blocked by microstructural barriers.

Figure 4 illustrates microstructurally small cracks, whose slip bands are blocked by three kinds of material barriers: (a) the grain boundary, (b) the second phase particles, and (c) reinforced fibers.

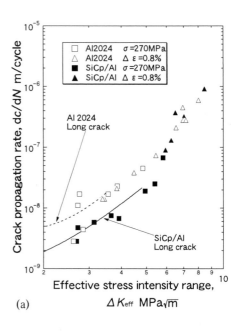

(a)

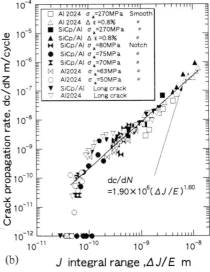

(b)

Fig. 3 Crack propagation rate correlated to the effective stress intensity range and the J integral range for aluminum alloys unreinforced and reinforced with SiC particles.

The interaction of a propagating fatigue crack with grain boundaries has been modeled by several investigators. A fatigue crack formed in a most favorably oriented grain (i.e. in a large grain with a high orientation factor situated at the specimen surface) will first show crystallographic Stage I growth. The crack-tip slip band will be blocked when it hits the grain boundaries of less-favorably oriented grains.

Tanaka *et al.* (1986) used the continuously distributed dislocation theory to analyze the interaction between the crack-tip slip band and the grain boundary by assuming the isotropic material and the slip band collinear to the crack. Once the frictional stress in each grain is given, the size of the slip band zone and the crack-tip opening displacement (CTOD) can be determined for the cases of the equilibrium slip band and the blocked slip band. By assuming the crack propagation rate is a power function of the range of CTOD, a Monte Carlo simulation was conducted by Tanaka *et al.* (1992) for fatigue crack propagation in the grain structure whose grain size and frictional stress were random variables. A small crack growth fast because it is nucleated in favorably oriented grains, and then decelerated as it hits the grain boundary. The propagation rate of small cracks approaches to that of large cracks as it grows. The yield stress in the material at the crack tip becomes equal to the macroscopic yield stress.

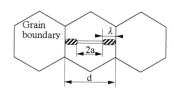

(a) Grain boundary blocking.

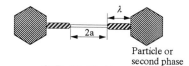

(b) Particle blocking.

Edwards and Zhang (1994) assumed the propagation rate was proportional to the plastic zone size in a similar model described above. They proposed that the plane stress state near the specimen surface was responsible for fast growth rates of small cracks.

In composite materials, reinforced particles or fibers act as the barrier to small crack propagation as illustrated in Fig. 4(b) and (c).

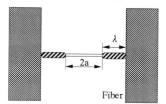

(c) Fiber blocking.

Fig. 4 Materials barriers to small fatiguwe crack propagation.

Threshold for Small Crack Propagation

The fatigue limit of smooth specimens of most metals and metallic composites is determined by the condition of the propagation of nucleated small cracks. Figure 5 illustrates the tip of a fatigue crack at the threshold. The dislocations emitted from the crack tip is blocked by material barriers. The back stress of dislocations reduces the true value of the stress intensity factor at the crack tip as

$$K_{tip} = \Delta K_{eff} - K_{disl} \qquad (8)$$

where K_{disl} is the crack-tip shielding by dislocations in the frontal process zone. The condition of the threshold will be that K_{tip} is equal to the critical SIF for dislocation emission.

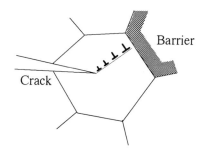

Fig. 5 Threshold condotion for fatigue crack propagation.

Tanaka *et al.* (1986) proposed an alternate criterion for the threshold, which is the condition that the crack-tip slip band cross over the barrier, such as the grain boundary. From this condition, the fatigue limit, σ_{wo}, is given by

$$\sigma_{wo} = (2\sigma_f/\pi)\cos^{-1}(a/c) + \Delta K_c^m/\sqrt{\pi c} \qquad (9)$$

where a is the crack length, c is the crack length plus the slip band length, and ΔK_c^m is the strength of blocking. If a and c are proportional to the grain size, d, we obtain the following Petch-type relation:

$$\sigma_{wo} = \sigma_0 + k_f/\sqrt{d} \qquad (10)$$

where σ_0 corresponds to the frictional stress and k_f to the strength of blocking. Figure 6 shows the relation between the fatigue limit and the inverse square root of the grain size of steels by Kinefuchi *et al.* (1992). Solution hardening and precipitation hardening by Si, P and Cu increase the constant term in Eq. (10) corresponding to the frictional stress, while the second phase such as martensite increases the k_f term corresponding to the strength of blocking.

To predict the fatigue threshold of notched components, the estimation of the amount of crack closure is significant. Short cracks formed at the notch root stop at a constant value of ΔK_{eff}. Tanaka and Akiniwa (1988) proposed the resistance curve (R-curve) method for predicting the fatigue threshold of notched components.

SMALL FATIGUE CRACKS
IN METALLIC MATERIALS

Small fatigue cracks nucleated in smooth specimens often show Stage I crystallographic growth. The Stage I growth of small cracks often blocked by the grain boundary. Okazaki *et al.* (1990) examined the propagation behavior of small fatigue cracks in directionally-solidified Al-Li alloy at 453K (The 0.2% proof stress of the material at 453K is 151 MPa.). Figure 7(a) shows the behavior of a small crack propagating on the slip plane (111) to the slip direction [101] under the stress range $\Delta\sigma = 190$ MPa with a

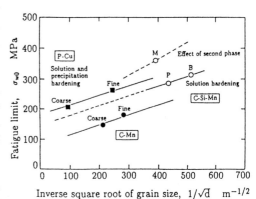

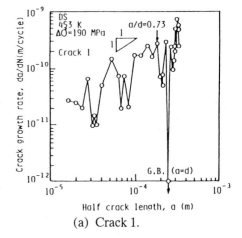

Fig. 6 Relation between fatigue limit and inverse square root of grain size for steels.

(a) Crack 1.

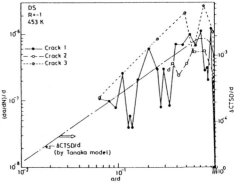

(b) Cracks 1, 2, 3.

Fig. 7 Fatigue crack propagation rate and ΔCTSD as a function of crack length in directionally soloidified Al-Li alloy.

stress ratio of R=-1. The crack propagation rate increases in proportion to the crack length until the crack-tip slip band is blocked by the grain boundary. The crack deceleration starts at the crack length of about 70 percent of the grain size.

For the propagation of Stage I fatigue cracks on the slip plane, the range of the crack-tip sliding displacement, ΔCTSD, will control the crack growth rate. By using the blocked slip band model proposed by Tanaka *et al.* and assuming the crack propagation rate proportional to ΔCTSD, the crack propagation rate is compared with ΔCTSD in Fig. 7(b). The crack propagation rate of cracks 1, 2, and 3 is nicely predicted by the model.

After crossing the first grain boundary, fatigue cracks show irregular growth behavior in polycrystalline Al-Li alloys. The crack path is very much crystallographic. The growth rate of those small cracks is higher than that predicted from the da/dN-ΔK relation of long cracks, mainly because of the lack of crack closure. The relation between da/dN and ΔK for small fatigue cracks is close to the da/dN-ΔK_{eff} relation for large cracks. As the crack length increases, it approaches to the da/dN-ΔK relation of large cracks.

Similar growth behavior of small cracks has been reported for a variety of metallic materials, including nickel based superalloys, titanium alloys, aluminum alloys, and steels. The phase boundary impedes the propagation of small cracks in two phase materials. In intermetallic compounds, such as titanium aluminide alloys, small cracks are blocked by the lamellar structure perpendicular to the crack propagation direction.

Figure 8 illustrates a summary of the propagation behavior of small fatigue cracks compared with the relation of da/dN with ΔK or ΔK_{eff} for large cracks. For Stage I propagation, the crack growth rate can be faster than the da/dN-ΔK_{eff} relation, because the crack is nucleated in favorably oriented grains (soft spots) and the crack propagation path is on the slip plane. After fatigue cracks begin to propagate in Stage II manner, the crack propagation rate initially follows the da/dN-ΔK_{eff} relation, because of the lack of crack closure. As cracks get larger, the crack propagation rate approaches to the da/dN-ΔK relation. The rate dip or the retardation of the small crack propagation rate is caused by material barriers such as grain boundaries and phase boundaries. Microstructurally small cracks tend to grow faster in the coarser-grained materials, because of the larger amount of micro-plasticity or the less degree of grain boundary blocking (Brown *et al.*, 1986).

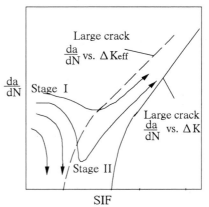

Fig. 8 Small fatigue crack propagation in smooth specimens.

SMALL FATIGUE CRACKS IN METAL-MATRIX COMPOSITES

Most popular discontinuous metal matrix composites (MMC) are aluminum alloys reinforced with SiC particles or whiskers, or with Al_2O_3 particles or short fibers. Because of higher modulus and lower density of reinforced materials, significant improvement in specific strength and stiffness can be obtained. The site of fatigue crack initiation changes depending on the matrix-particle system and their processing method. Figure 9 presents the classification of the crack nucleation sites reported for metallic materials containing inclusions (Tanaka *et al.*, 1982). The same classification can be applied to discontinuous MMCs. Some large-sized SiC particles are cracked before fatigue tests. The fatigue crack initiation takes place ahead of the cracked particle (Type A). For weakly bonded particles, the interface between the particle and the matrix will be cracked either by the applied stress or by the stress concentration due to pile-up dislocations (Type B). For strongly bonded particles, fatigue cracks are nucleated in the slip band formed by the stress concentration of the particles (Type C). For fine SiC particulate-reinforced aluminum alloys made by powder metallurgy, fatigue cracks are initiated in the matrix near particles or in the vicinity of the cluster of particles.

The variation of propagation rate of microstructurally small cracks is more pronounced in composites than in unreinforced materials. Figure 10 shows the relation between the crack propagation rate and the crack length (or ΔK) for molten-metal processed SiC particle-reinforced aluminum alloy (Kumai *et al.*, 1992). The scatter bands of the small-crack data relative to the long-crack data are similar to those of unreinforced materials reported previously. A small crack was often arrested when the crack tip reached SiC particles. Zero-growth data points were no longer obtained for crack depths greater than 200μ m, and the results begin to converge.

The crack growth rate will decrease when the crack tip approaches to the uncracked SiC particles, because of the blocking of the slip band or the decrease in the SIF range. The crack will be attracted to cracked or decohered particles.

(a) Type A
Slip band crack from cracked particle

(b) Type B
Interface debonding

(c) Type C
Slip band crack from non-cracked particle

Fig. 9 Clasification of fatigue crack initiation in particulate-reinforced metal-matrix composites.

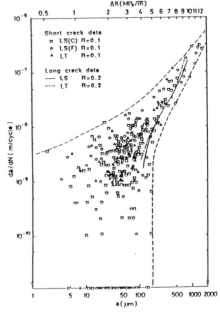

Fig. 10 Relation between crack propagation rate and crack length for smooth specimens of aluminum alloy reinforced with SiC particles.

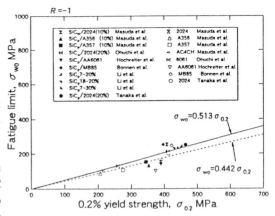

Fig. 11 Relation between fatigue limit and yield strength for aluminum alloys unreinforced and reinforced with SiC particles or whiskers.

The fatigue strength of aluminum alloys can be improved by the reinforcement. In Fig.11, the fatigue limit σ_{w0} of smooth specimens of aluminum alloys unreinforced and reinforced with SiC particles or whiskers is plotted against 0.2% yield strength $\sigma_{0.2}$. The relation is

$$\sigma_{w0} = 0.442\,\sigma_{0.2} \tag{11}$$

for unreinforced material (the dashed line), and

$$\sigma_{w0} = 0.513\,\sigma_{0.2} \tag{12}$$

for reinforced material (the solid line).

If the fatigue limit of composites σ_{wc} is determined by the fatigue limit of the matrix σ_{wM} and if σ_{wM} is proportional to the matrix yield strength σ_{YM} with a proportional constant β, the fatigue limit σ_{wc} is given by

$$\sigma_{wc} = \sigma_{wM}/P = \beta\,\sigma_{YM}/P = \beta\,\sigma_{YC} \tag{13}$$

where P is the dimensional constant (less than one) correlating the average stress in the matrix to that of the composite.

Under the above assumption, the proportional constant for the relation between σ_{wc} and σ_{YC} for composites is the same as that for unreinforced materials. The increase in σ_{w0} for composite materials is caused by the increase in yield strength. Since the proportional constant for composites is larger than that for unreinforced materials in the experimental data shown in Fig.11, an additional improvement is achieved by the reinforcement. In composites, non-propagating cracks have been observed just below the fatigue limit, and the reinforced particles yield stronger barriers for small fatigue crack propagation (Li *et al.*, 1995).

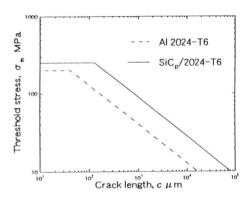

Fig. 12 Fatigue failure diagrams for the reinforced and unreinforced materials.

Since the threshold SIF range for the propagation of large cracks as well as the fatigue limit of smooth specimens can be improved by particulate reinforcement, the composite becomes less sensitive to notches or defect. Figure 12 shows the failure diagram of pre-cracked materials. The safe zone expands for every crack length by reinforcement.

The initiation and early propagation of small fatigue cracks in long-fiber reinforced Ti-alloys (SCS-6/Ti-15-3) have been studied by Gou *et al.* (1995). Figure 13 illustrates the early fatigue damage in the composite. The first damage for the composite started from a cracking of the reaction layer followed by fiber

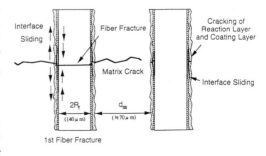

Fig. 13 Schematic representations of the early damage in the long-fiber reinforced composite.

fracture.   The matrix cracking initiated near the broken fiber.   The fiber ahead of the matrix crack tip fractured as the matrix crack approached to the fiber.

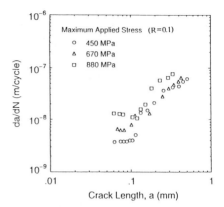

Fig. 14 Relation between crack propagation rate and crack length.

The change of the propagation rate of small cracks with the half crack length on the surface is shown in Fig.14, where the crack length is the sum of the broken fiber width (70 μ m) and the matrix crack length.   The crack propagation first decreases and then increases with crack length.   The transition to the acceleration region takes place at a crack length of about 150 μ m, i.e. when the second fiber fracture occurs at the matrix crack tip.

The crack propagation rate in the initial deceleration region is dependent on the applied stress.   Several non-propagating cracks were observed and their half length was less than 150 μ m.

In the case of long cracks, the matrix fatigue cracking proceeds without fracture of fiber, resulting in the crack-tip shielding by bridged fibers.   On the other hand, the propagation of small cracks is accompanied by fiber fracture.

SMALL FATIGUE CRACKS IN CERAMICS

The critical crack length of ceramics for unstable fracture is usually less than 1mm because of a low value of fracture toughness.   The propagation behavior of small cracks in ceramics is extremely significant.

Steffen *et al.* (1991) observed the propagation behavior of naturally occurring small (<100 μ m) surface crack in magnesia-partially-stabilized zirconia ceramics (Mg-PZT) under cyclic loading.   As in metal fatigue, small cracks propagate at stress intensity levels significantly smaller than ΔK_{th}, and the crack propagation rate decreases with crack extension.   They ascribed the anomalously high propagation rate to the lack of the crack-tip shielding due to phase transformation.

Ueno and Kishimoto (1995) examined the propagation behavior of naturally occurring small cracks in sintered silicone nitride (Si_3N_4).   The average grain size was about 0.48 μ m and the average aspect ratio of the elongated grain was 6.8.   Figure 15 shows the relation between the surface crack length 2a and the number of stress cycles normalized by

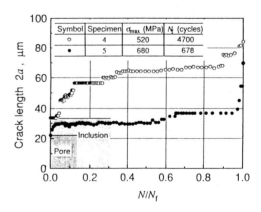

Fig. 15 Relation between crack length and stress cycles.

that for fracture, N/N_r. A crack is nucleated from an inclusion in specimen No.4, while from a pore in specimen No.5. Both cracks propagate fast in the very early period, and then decelerate, followed by an acceleration just before fracture. As seen in the figure, the crack propagates by repeating growth and arrest in the middle deceleration period.

In Fig. 16, the crack propagation rate is plotted against the maximum stress intensity factor calculated by assuming semi-circular crack shape. The figure also presents the data for large cracks in CT specimen and small artificial semi-elliptical surface crack. The average propagation rate is denoted by the circles, and the scatter band of the crack propagation rate is also shown. The small cracks grow much faster than large cracks and artificial cracks.

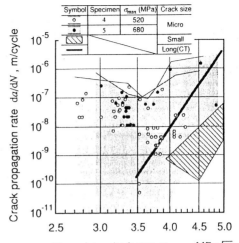

Fig. 16 Relation between crack propagation rate and maximum stress intensity factor.

Similar crack propagation behavior was observed in a polycrystalline Al_2O_3. Ueno and Kishimoto (1995) ascribed the fast propagation of small cracks to the inappropriate use of the stress intensity factor, not to the lack of crack-tip shielding. They claimed the stress at a distance of twice the grain size as a critical parameter for fatigue crack propagation of microstructurally small cracks in ceramics. The non-singular stress term of the crack-tip stress field is significant for very short cracks because the fracture process at the crack tip takes place in the zone with the finite dimension. As engineering approaches to these problems, the point stress model, the mean stress model or the fictitious crack length model has been used.

CONCLUDING REMARKS

The anomalous behavior of small fatigue crack propagation is caused by the breakdown of the similitude concept in the ΔK-based conventional fracture mechanics. This dissimilitude occurs mainly through three reasons: (1) premature crack closure or crack-tip shielding, (2) large-scale yielding, and (3) microstructural inhomogeneities. When the crack length is large compared with microstructure and when the crack closure is not fully developed, the effective stress intensity range ΔK_{eff} can be used as a crack driving force. For cracks with large-scale yielding, the J integral range, ΔJ, is an appropriate parameter. Those small cracks are called mechanically small cracks.

The propagation of microstructurally small cracks is highly irregular because of microstructural inhomogeneities. The nucleated crack propagates fast and then decelerates due to microstructural barriers such as grain boundaries and second phases. The propagation behavior of microstructurally small cracks is common in various advanced metallic materials, metal-matrix composites and ceramics. The fatigue limit of smooth specimens is determined by the propagation condition of small cracks across microstructural barriers. Further developments of micro or meso-mechanical analysis of small crack propagation are necessary.

REFERENCES

Brown, C. W. and J. E. King (1986). The relevance of microstructural influences in the short crack region to overall fatigue resistance, *Small Fatigue Cracks*, Edited by R. O. Ritchie and J. Lankford, 73-95.

Edwards, L. and Y. H. Zhang (1994). Investigation of small fatigue cracks - II. A plasticity based model of small fatigue crack growth. *Acta Metall. Mater.*, *43*, 1423-1431.

Gou, S. Q., Y. Kagawa and K. Honda (1995). Observation of short fatigue crack growth process in SiC-fiber reinforced Ti-15-3 alloy composite. To be published in Metall. Mater. Trans.

Kinefuchi, M., T. Yokomaku and K. Tanaka (1992). *Proc. 21st Symp. Fatigue*, Soci. Mater. Sci. Japan., 167-170.

Kumai, S. J. E. King and J. F. Knott (1992). Fatigue crack growth behavior in molten-metal processed SiC particle-reinforced aluminum alloys. *Fatigue Fract. Engng. Mater. Struct.*, *15*, 1-11.

Li, X. and H. Misawa (1995). Silicone carbide particles in small fatigue crack initiation and propagation behavior of SiC particulate reinforced Al-alloys. *Trans. JSME*, *61*, 1940-1945.

Okazaki, M., S. Yamahira and Y. Kojima (1990). Observation of first stage high-cycle fatigue crack growth behavior using directionally-solidified Al-Li alloy at high temperature. *J. Mater. Sci. Japan*, *39*, 549-555.

Ritchie, R. O. and J. Lankford (1986). Overview of the small crack problem. *Small Fatigue Cracks*, Edited by R. O. Ritchie and J. Lankford, 1-5.

Ritchie, R. O. (1988). Mechanisms of fatigue crack propagation in metals, ceramics and composite: Role of crack-tip shielding. *Mater. Sci. Eng.*, *A103*, 15-28.

Steffen, A. A. R. H. Danskardt and R. O. Ritchie (1991). Cyclic fatigue life and crack-growth behavior of microstructurally small cracks in magnesia-partially-stabilized zirconia ceramics, *J. Am. Ceram. Soci.*, *74*, 1259-1268.

Tanaka, K. and T. Mura (1982). A theory of fatigue crack initiation at inclusions. *Met. Trans. A 13A*, 117-123.

Tanaka, K. Y. Akiniwa, Y. Nakai and R. P. Wei (1986). Modeling of small fatigue crack growth interacting with grain boundary. *Eng. Fract. Mech.*, *24*, 803-819.

Tanaka, K. (1987). Mechanisms and mechanics of short fatigue crack propagation. *JSME Inter. J.*, *35*, 1-13.

Tanaka, K. and Y. Akiniwa (1988). Resistance-curve method for predicting propagation threshold of short fatigue cracks at notches. *Eng. Fract. Mech.*, *30*, 863-876.

Tanaka, K., M. Kinefuchi and T. Yokomaku (1992). Modeling of statistical characteristics of the propagation of small fatigue cracks. *Small Fatigue Cracks*, Edited by K. J. Miller and E. R. de los Rios, Mech. Eng. Publ., 351-368.

Tanaka, K., Y. Akiniwa, K. Shimizu and G. Matsubara (1994). Fatigue crack propagation from notch in SiC particulate-reinforced aluminum alloy. *Trans. JSME*, *A60*, 1143-1149.

Tanaka, K., Y. Akiniwa and K. Shimizu (1995). Small fatigue crack propagation in SiC particulate-reinforced aluminum alloy. *Trans. JSME*, *A61*, 1190-1196.

Ueno, A. and H. Kishimoto (1995). Fatigue crack initiation and propagation in ceramics, To be published in Fracture Mechanics of Ceramics, 11 and 12.

THERMAL AND MECHANICAL FATIGUE OF
MULTI-LAYERED AND GRADED MATERIALS

S. Suresh

Department of Materials Science and Engineering
Massachusetts Institute of Technology
Cambridge, MA 02139, USA

Extended Abstract

New theoretical formulations, computational results, and experimentally validated micromechanisms are developed in this study on the evolution of stresses, the accumulation of plastic strains, and the initiation of damage and fracture in thin-film and thick multi-layers subjected to cyclic variations in temperature and/or mechanical loads. Particular attention is devoted to the design of layer geometry in one or more layers with a view to improving the thermal and mechanical fatigue characteristics. Detailed experiments and computations are also performed on the thermal and mechanical fatigue response of multi-layered structures wherein one or more layers comprises a compositionally graded composite microstructure.

The applications for which the results of the present work on multi-layers are pertinent, with appropriate modifications, include thermal-barrier and wear-resistant coatings for structural components, thin-films of metallization and passivation layers on Si substrates that are used in microelectronics, layered structures resulting from such joining processes as welding, diffusion bonding and explosion-cladding, and magnetic multi-layers used in storage devices. The compositionally graded multi-layered structures are of considerable interest in view of their current or potential use in such applications as surface coatings for engineering components, solid oxide fuel cells, and thin-film layers (grown by molecular beam epitaxy or CVD) in microelectronics and optoelectronics for controlling the density, spacing and mobility of misfit/threading dislocations.

The first task in the modeling of thermal fatigue in a general metal-ceramic bi-layered structure of arbitrary layer thickness involves the identification of critical temperatures at which distinct transitions occur in the evolution and spread of cyclic plasticity (Suresh et al., 1994). It is shown that there exist four critical temperatures which signify four demarcation lines in fatigue response: (1) the temperature change ΔT (from a certain reference temperature or the processing temperature) at which plastic deformation commences in one of the layers, (2) the temperature change beyond which a reversal of temperature causes reversed yielding, (3) the value of ΔT at which the entire metallic layer becomes fully plastic, and (4) the value of ΔT at which reversed yielding occurs in the entire metallic layer. It is shown that these temperature limits help design the limits of thermal cycling for layered structures. The changes in curvature of the layered solid, arising as a result of thermal mismatch strains during temperature excursions, are used as a gage of thermo-elastoplastic deformation. Accurate laser-scanning techniques which are capable of monitoring in-situ the changes in curvature of the layered solid during temperature changes of up to 900°C are used to check the validity of the theory (Shen and Suresh, 1995). Experiments on layered systems of practical interest and model materials, such as Si-Al, Si-Al-SiO$_2$, Ni-Al$_2$O$_3$, Al-Al$_2$O$_3$, and Cu-Al$_2$O$_3$, are used to guide the rate-independent plasticity models and the rate-dependent creep models.

The theory is also extended to include critical temperatures for the onset of plastic flow or failure in multi-layered structures comprising a compositionally graded interlayer between homogeneous layers (Giannakopoulos et al., 1995; Finot et al., 1996). The theory is compared with experiments on model system of Ni-Ni$_x$(Al$_2$O$_3$)$_{1-x}$-Al$_2$O$_3$. Thermal fatigue experiments involving measurements of cyclic curvature evolution and crack initiation are also made for this compositionally graded tri-layered system (Finot et al., 1996).

The formulations for the estimation of thermal stresses, accumulated plastic strains, and the evolution of curvature seen for a bilayer or a trilayer are also extended to a general multi-layered solid (Finot and Suresh, 1996) for which the inelastic deformation arises from coupled electrical-mechanical effects (as in piezoelectric multi-layers), differential shrinkage (as in tape-cast metals and ceramics), epitaxial strains (as in thin-films used in microelectronics), thermal mismatch, or mechanical loading. For plate or beam geometries of multi-layers subject to plane stress, plane strain, generalized plane strain or equal biaxial stress, these formulations can be cast into a form where cyclic damage and crack initiation during thermal and mechanical fatigue can be simulated using software used in conjunction with personal computers. Such analyses using personal computer software (Finot and Suresh, 1994) and finite element codes (Giannakopoulos et al., 1995; Finot and Suresh, 1996) simulate stresses, accumulated plastic strains, curvature, steady-state creep deformation, and failure initiation for a variety of fatigue loading conditions, and provide guidelines for the geometric design of multi-layers with sharp interfaces.

A major objective in the use of multi-layered materials is to tailor the properties of a component for specific functions, such as minimization of thermal residual stresses at critical locations, the suppression of plastic flow for a given temperature excursion, or the improvement of damage tolerance. Design maps and engineering diagrams for a variety of compositionally graded structures intended for structural and microelectronics applications are developed in this study, where the thermomechanical response, including the thermal fatigue behavior, is optimized for different geometrical combinations and compositional gradients of the layered structure (Giannakopoulos et al., 1995; Finot and Suresh, 1996). Both small deformation and large deformation, which can give rise to buckling and instability, are addressed.

Finally, the mechanics and micromechanics of fatigue fracture at arbitrary angles to interfaces in multi-layered structures subject to externally imposed cyclic loads are investigated. Systematic fatigue experiments and detailed finite element simulations on model metal-metal and metal-ceramic layered systems have been conducted in this work (Suresh et al., 1992, 1993; Sugimura et al., 1995a, 1995b; Kim et al., 1995). These studies reveal that (a) the ability of the fatigue crack to penetrate the interface is strongly influenced by the plastic mismatch of the materials (with no elastic mismatch) surrounding the interface, (b) different trends pertaining to crack-tip shielding and amplification result for different combinations of elasticity and plasticity mismatch, and (c) compositional gradation of the interlayer has a significant effect on the fatigue crack growth resistance of the multi-layer. The practical implications of these results are demonstrated with industrial examples.

References

Giannakopoulos, A.E., Suresh, S., Finot, M. and Olsson, M. (1995). Elastoplastic analysis of thermal cycling: Layered materials with compositional gradients. Acta Metall. Mater. 43, 1335-1354.

Finot, M. and Suresh, S. (1994). MultiTherm^{TM}, Software for thermal and mechanical analysis for use with a personal computer. Copyright MIT, Cambridge, MA.

Finot, M. and Suresh, S. (1996). Small and large deformation of thick and thin-film multi-layers: Effects of layer geometry, plasticity and compositional gradients. J. Mech. Phys. Solids, in press.

Finot, M., Suresh, S., Bull, C. and Sampath, S. (1996). Curvature changes during thermal cycling of Ni-Alumina compositionally graded multi-layered materials. Mater. Sci. Eng. A. 205, 59-71.

Kim, A.S., Suresh, S. and Shih, C.F. (1995). Fracture normal to interfaces with homogeneous and graded composition. Int. J. Solids Struct., submitted.

Shen, Y.-L. and Suresh, S. (1995). Elastoplastic deformation of multilayered materials during thermal cycling. J. Mater. Res. 10, 1200-1215.

Sugimura, Y., Lim, P.G., Shih, C.F. and Suresh, S. (1995a). Fracture normal to a bimaterial interface: effects of plasticity on crack-tip shielding and amplification. Acta Metall. Mater. 43, 1157-1169.

Sugimura, Y., Grondin, L. and Suresh, S. (1995b). Fatigue crack growth at arbitrary angles to bimaterial interfaces. Scripta Metall. Mater. 33, 2007-2012.

S. Suresh, A.E. Giannakopoulos, and M. Olsson (1994). Elastoplastic analysis of thermal cycling: layered materials with sharp interfaces. J. Mech. Phys. Solids, 42, 979-1018.

Suresh, S., Sugimura, Y. and Ogawa, T. (1993). Fatigue cracking in materials with brittle surface coatings. Scripta Metall. Mater. 29, 237-242.

Suresh, S., Sugimura, Y. and Tschegg, E.K. (1992). Growth of a fatigue crack approaching a perpendicularly-oriented bimaterial interface. Scripta Metall. Mater. 27, 1189-1194.

INTERFACE FATIGUE-CRACK GROWTH
IN LAYERED MATERIALS

JIAN-KU SHANG

Department of Materials Science and Engineering
University of Illinois, Urbana, IL 61801, USA

ABSTRACT

Performance of many layered materials depends critically on the properties of bi-materials interfaces. In this paper, current work on interface fatigue crack growth is reviewed. A flexural peel technique is introduced and complete sets of fracture-mechanics solutions are given for select bi-materials combinations. Interface fatigue-crack growth studies based on this technique are summarized. Micromechanisms and micromechanics pertinent to the effects of load-mix, interface microstructure, and interface morphology on interface fatigue-crack growth are discussed.

INTRODUCTION

Layered materials differ from conventional monolithic materials in that a substantial volume of the material resides on the interface. Because of the high interface/volume ratio, the mechanical behavior of layered materials depends strongly on the mechanical response of the interface. Under fatigue loading, the crack resistance of the interface can be degraded so that subcritical crack growth occurs at crack-driving forces far below the fracture toughness of the interface.

In this paper, recent studies on fatigue crack growth along bi-material interfaces in layered sandwich specimens are reviewed. First, key concepts of interface fracture mechanics are introduced. A critical assessment of interface fracture mechanics specimens then follows. Current understanding of interface resistance to fatigue crack growth is summarized, based mostly on recent studies on solder interfaces and polymer-metal interfaces. The paper is concluded by an example of fatigue life-prediction on adhesive joints.

ELEMENTS OF INTERFACE FRACTURE MECHANICS

For an interface crack between two materials with different elastic constants (Fig. 1), the crack tip field is given as [1,2]:

$$\sigma_{yy} + i\,\sigma_{xy} \mid \theta = 0 = \frac{K r^{i\varepsilon}}{\sqrt{2\pi r}} \qquad r \to 0 \qquad (1)$$

where
$$\varepsilon = \frac{1}{2\pi} \ln\left(\frac{\frac{\kappa_i}{\mu_i} + \frac{1}{\mu_o}}{\frac{\kappa_o}{\mu_o} + \frac{1}{\mu_i}}\right) \qquad \kappa = \begin{cases} 3 - 4\nu & \text{plane stress} \\ 3 - \nu/1 + \nu & \text{plane strain} \end{cases}$$

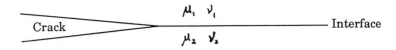

Fig. 1. Geometry of an interface crack problem.

μ is the shear modulus and υ is the Poisson's ratio. The scale parameter, K, depends on the external stress, T, the geometry, elastic properties of the two solids, and a characteristic length, L (e.g., crack length) as following

$$K = Y\,T\,\sqrt{L}\,L^{-i\varepsilon}\,e^{i\varphi} \qquad (2)$$

where Y is a dimensionless real positive quantity and φ is the phase angle of the quantity, $KL^{-i\varepsilon}$. For homogeneous materials, $\varepsilon = 0$, K becomes the stress intensity factor, and φ, the phase angle of the external loading. When $\varepsilon \neq 0$, as is the general case for a bi-material interface, Mode-I and Mode-II stress intensity factors, K_I, K_{II}, can be defined as

$$K\,r^{i\varepsilon} = K_I + i\,K_{II} \qquad (3)$$

and the local phase angle at the crack tip as

$$\phi = \tan^{-1}\!\left(\frac{\sigma_{xy}}{\sigma_{yy}}\right) = \tan^{-1}\!\left(\frac{K_{II}}{K_I}\right) = \varphi - \varepsilon \ln\!\left(\frac{L}{r}\right) \qquad (4)$$

Unlike in the homogenous material, where the Mode-I and Mode-II components of the crack tip stress field can be readily separated, Eqn (4) shows the coupling of material constants with external loading condition in determining the local crack tip field. Because of this coupling, it becomes necessary to describe the crack tip field in terms of two parameters such as the strain energy release rate, G, and the local phase angle ϕ or the components of the complex stress intensity factor, K_I and K_{II}, with the strain energy release rate related to the magnitude of the scale parameter, $|K|$, by

$$G = \left(\frac{1-\upsilon 1}{\mu 1} + \frac{1-\upsilon 2}{\mu 2}\right)\frac{|K|^2}{4\,\cosh^2(\pi\varepsilon)} \qquad (5).$$

In applications of interface-fracture mechanics concepts to interface fracture problems, two simplifications are often made. One is to use a sandwich specimen where the interlayer is made to be much smaller than the other dimensions of the specimen [3]. In this case, the phase angle φ can be written as the sum of the phase angle of the external loading, ψ, and a bi-material correction ω, which depends on material constants and the characteristic length, L:

$$\varphi = \psi + \omega.$$

The local phase angle then becomes:

$$\phi = \psi + \omega + \varepsilon\ln\!\left(\frac{r}{L}\right) \qquad (6)$$

Another simplification is to take ε to be zero because the numerical value of the bimaterial constant, ε, is very small (for most materials, $|\varepsilon| < 0.08$). Under this assumption, the strain energy release rate, G, may be broken into Mode-I and Mode-II components, G_I and G_{II}. G_I and G_{II} can then be calculated using fracture mechanics methods established for the homogenous materials, such as the finite element analysis based on crack closure method [4], and G_{II}/G_I ratio be used in place of the local phase angle to represent the mixed mode loading condition ahead of the interface crack. The use of G_I and G_{II} has been widely accepted in the adhesion community.

INTERFACE-FRACTURE MECHANICS TECHNIQUES

Experimental studies on fracture behavior of interfaces within the framework of interface fracture mechanics have been rather limited both in quantity and in scope. While cracks are assumed to be lying on the interface in most theoretical studies, actually placing a well-defined crack on the interface and maintaining the crack growth on the interface turned out to be most difficult. Over the last three decades, a number of interface-fracture specimens have been suggested and these specimens are summarized in Fig. 2. Among these specimens, the four-point bend specimen was very useful for fast fracture studies, but for crack growth studies, the layered double cantilever beam specimen (LDCB) or its slight variation has almost always been chosen because of its simple fracture mechanics condition [5,6].

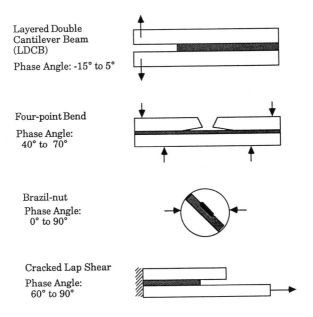

Fig. 2 Specimens proposed for interface crack studies.

Layered Double Cantilever Beam Specimen

Recent analysis indicated that this specimen provides a loading condition close to Mode-I, with the phase angle in the range of -15° to 10° for most materials [3]. If the interlayer is much thinner than the outer beams, the solution for the homogenous material can be used to calculate the crack driving force, the strain energy release rate, for the interface crack. As with other sandwich specimens, the interlayer remains bonded to one substrate after the fracture so that the complication from residual stress in the interlayer is avoided. To facilitate the introduction of an initial crack, a chevron notch is often used as the crack-starter.

Despite its popularity, LDCB specimen suffers several major drawbacks. The first of them is the difficulty in starting an interface crack. Cracks started from the notches in LDCB specimen do not necessarily lay on the interface. The second problem is even if a precrack is initiated on one of the interfaces, the subsequent growth of the interface crack is not necessarily confined onto the same interface. Instead, as Fig. 3 illustrates, the crack tends to alternate its path between two similar interfaces. The alternating crack path can lead to larger scattering in crack growth rate measurements because of the uncertainty in locating the crack tip, and more importantly, to overestimating the fatigue crack growth resistance such as the fatigue threshold by as much as more than a factor of two as shown in Fig. 4.

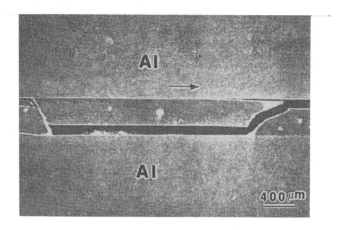

Fig. 3 Alternating crack path in LDCB specimen.

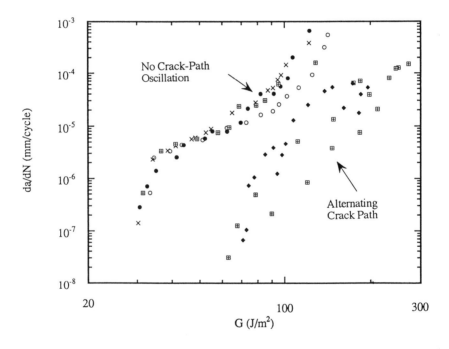

Fig. 4 Fatigue crack growth in LDCB specimens.

Two sets of solutions have been devised to overcome the drawbacks of the standard LDCB specimen. For the crack-starting problem, placing a pre-existing flaw on one interface has found wide acceptance. Another solution to the problem is to use the flexural peel specimen described below to grow a long fatigue crack on the interface and then load the cracked specimen in the double cantilever geometry. The alternating crack path problem has been solved either by strengthening one of the interfaces relative to the other (precracked) interface or by using the flexural peel specimen.

Flexural Peel Specimen

The flexural peel specimen is given in Fig. 5. It consists of three layers made of two different materials. The two outer layers are made of the same material with elastic properties, E_o and ν_o, and bonded by a second material with elastic properties, E_i and ν_i. The thickness of three layers are t_1, h, and t_2 from the top. The width of the specimen is B. An initial crack of length, a, is introduced onto the lower interface by inserting a weakly bonded region at the end of the interlayer. The specimen is fixed at one end and loaded at the other end by an external load, P. The distance from the loading line to the end of the upper beam is L.

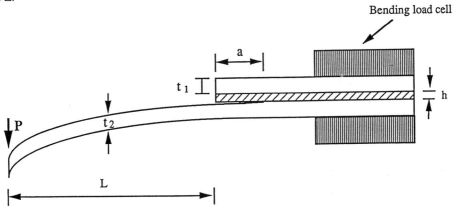

Fig. 5. Flexural peel specimen.

The fracture mechanics solutions for the specimen have been obtained by the analytical J-integral formulation and finite element calculations, and verified by the experimental calibrations. The strain energy release rate is given as [7]

$$G = \frac{P^2(L+a)^2}{2E_o B}(\frac{1}{I} - \frac{1}{I^*})$$

(7)

where I and I* are the moments of inertia of the lower beam and the composite beam respectively. The loading phase angle depends on the elastic constants of two materials and the thicknesses of three layers. For epoxy interlayer and Al outerlayers, the phase angle ranges from 20° to 70° at a characteristic distance of one fiftieth of the interlayer thickness, as shown in Fig. 6.

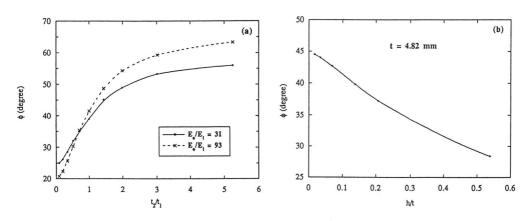

Fig. 6. Phase angles of flexural peel specimen as functions of a) beam thickness and b) layer thickness.

Because the upper beam at the cracked end is let free, the flexural peel specimen avoids crack development on the upper interface and thus solves the alternating crack path problem seen in the LDCB specimen. The loading arm (the lower beam) of the specimen may be extended so that the crack stability can be maintained by performing the subcritical crack growth experiment under displacement control. As Fig. 6 indicates, a wide range of crack tip loading condition may be achieved by selecting the thickness of the two outer beams without modifications to either the loading arrangement or the specimen preparation procedure. The specimen has been used successfully to study crack growth behavior of many polymer-metal, metal-metal, and metal-ceramic interfaces [8-14]. The results related to fatigue crack growth are summarized below.

INTERFACE RESISTANCE TO FATIGUE CRACK GROWTH

Although it has been widely assumed that the fracture energy of the interface be the same as the interfacial energy, experimental studies have shown that real fracture energies are orders of magnitude higher for most interfaces [15]. The large discrepancy comes from various non-linear irreversible processes at the crack tip. Under repeated cyclic loading, these irreversible processes can degrade significantly. The degradation process may differ so much for different microstructural and mechanical variables that the magnitude of the fatigue resistance of the interface is only a small fraction of the fracture toughness of the interface, and becomes highly dependent on the loading condition, interface morphology, interfacial microstructure, and the layer thickness.

Dependence on Loading Condition

While it is to be expected that the fatigue behavior of an interface should vary with external loading condition as in the bulk solid, for interface cracks, the coupling of material constants and external loading may cause the crack tip loading condition to differ for different material interfaces even the external loading is kept the same. Therefore, it is necessary to understand how the fatigue crack growth behavior may depend on local crack tip condition rather than simply the external loading condition.

It is possible to achieve a wide range of loading conditions using the brazil-nut specimen, four-point bend specimen and flexural peel specimen. However, for both brazil nut and four-point bend specimens, it is hard to make two crack fronts grow evenly and the crack surfaces are compressed by the external loading so that it is difficult to locate the crack tip. The results summarized below were obtained using the flexural peel specimen where the loading condition was systematically varied by selecting the thicknesses of the two outer beams following the calculations in Fig. 6.

The dependence of interface fatigue-crack growth on crack-tip loading condition is shown in Fig. 7 for a polymer-metal interface. Increasing the shear component of the crack tip loading or the phase angle reduced sharply the fatigue crack growth rate along the interfaces, resulting in higher fatigue crack growth resistance for the interface. The effect of the loading condition is seen throughout the entire range of the crack-growth rates examined. The effect of the loading condition on the interface fatigue crack growth in the near-threshold regime is given in Fig. 8 and can be expressed by the following empirical relationship between the fatigue threshold, ΔG_{th}, and the phase angle, ϕ:

$$\Delta G_{th} = 23 \, e^{\tan\phi} \quad (J/m^2) \qquad (8).$$

The relationship in Eqn. (8) may be understood in terms of the plastic work dissipated in the polymer interlayer. At low phase angles ($\phi < 30°$), the polymer failed by extensive cavitation at the crack tip. The plastic work dissipation can be estimated from [9]:

$$G_s = \sigma_f \, \varepsilon_f \, R_p \qquad (9),$$

where σ_f is the tensile strength, ε_f, the tensile failure strain, and R_p, the plastic zone size at the crack tip. In contrast, at high phase angles, the polymer failed by laminar shear and the plastic work became:

$$G_s = \tau_f \, S \qquad (10)$$

where τ_f is the shear strength and S is the shear step length measured from the failure surface [9].

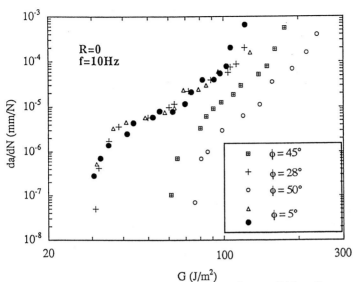

Fig. 7. Fatigue crack growth behavior of epoxy/Al interface.

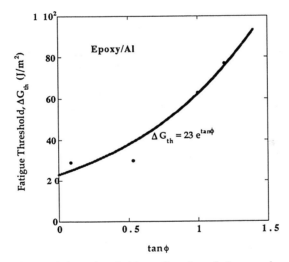

Fig. 8. Fatigue threshold as a function of phase angle.

Role of Interface Morphology

Another distinct feature of the interface crack growth behavior is that the crack growth depends on the interface morphology for if the fatigue crack follows the interface, the crack path is then prescribed by the morphology of the interface. This point is well demonstrated in a copper-solder interface. The interface was formed by the reaction of liquid solder alloy with copper, which produced intermetallic cells on the copper surface [11,12]. The cells initially assumed the shape of a spherical cap and grew into a pancake shape after reaching a hemisphere. The cell growth can be interrupted by controlling the cooling rate after solder reflowing to produce the effect of the cooling rate on the fatigue crack growth along the solder-copper interface shown in Fig. 9. In the near-threshold regime, faster cooling resulted in higher fatigue threshold whereas just the opposite was observed in the fast crack growth regime (near the instability).

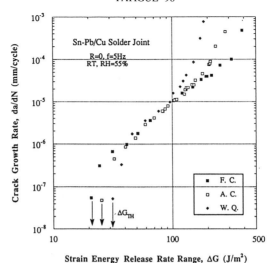

Fig. 9. Fatigue crack growth behavior of solder-copper interfaces under furnace-
cooling (F.C.), air-cooling (A.C.), and water-quenching (W.Q.) conditions.

The opposite effects of the cooling rate at the two extremes of Fig. 9 further emphasized the significance of cyclic degradation. For the solder-copper interfaces, the non-linear toughening process in the fast fracture regime was degraded so much that it became overshadowed in the near-threshold regime by crack-surface sliding process. The crack sliding was promoted by the rough nature of the interface and by the small crack opening at the low strain energy release rates. A crack-sliding model based on the cellular interface morphology indicated that the fatigue threshold was comparable to the sliding resistance of the interface, G_S [14]:

$$G_S = \frac{4\pi\alpha\tau^2L}{1+4\varepsilon^2}\left(1 - \frac{\pi}{3\sqrt{3}(1+R)^2}\right) \tag{11}$$

where α and ε are bimaterial constants, τ is the shear strength of the solder alloy, L is the sliding zone length, and R is the roughness of the interface.

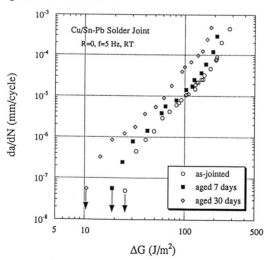

Fig. 10. Fatigue crack growth behavior of solder-copper interfaces under different aging conditions.

Dependence on Interfacial Microstructure

Just as important as the interface morphology, the interfacial microstructure also plays a significant role in determining the crack path and influencing the fatigue crack growth resistance. At the interface between Sn-Pb solder alloy and copper, the growth of Sn-Cu intermetallic compounds consumes Sn preferentially, resulting in the accumulation of a minor Pb-rich phase at the interface. The amount of the minor phase can be controlled by aging the interface to different stages [11]. Consequently, the interface fatigue crack growth behavior becomes strongly dependent on the aging condition as shown in Fig. 10.

The effect of the minor phase can be explained in terms of the changes in the flow property of the interfacial microstructure. In the as-reflowed condition (no aging), the flow property of the interfacial microstructure is the same as that of the eutectic microstructure. When slightly aged, the flow property of the interfacial microstructure is reduced to that of a coarsened solder microstructure. Upon severe aging, a continuous Pb-rich network is formed and the property of the interfacial microstructure is best represented by the flow property of the Pb-rich phase. Since the plastic work dissipation at the crack tip is directly related to the flow property of the interlayer as shown in Eqns. (9) and (10), the fatigue crack growth resistance as measured by the fatigue threshold is proportional to the microhardness of the interfacial microstructure (Fig. 11).

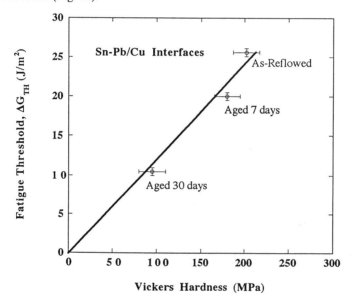

Fig. 11. Fatigue threshold as a function of the microhardness of interfacial microstructure.

Effect of Layer Thickness

In layered materials or structures, the layer thickness may affect the interface crack growth in two different ways, i.e., by altering the crack tip loading condition, and by limiting the volume of plastically deforming material. The effect of the former is shown in Fig. 6 for the layered flexural peel specimen. Reducing the thickness of the interlayer increases the phase angle, which should lead to an apparently higher interface resistance to fatigue crack growth according to Fig. 7. On the other hand, decreasing the interlayer thickness reduces the volume of plastic dissipation, and should result in lower interface resistance to fatigue crack growth. As a result of the competition between these two opposing forces, the fatigue crack growth followed a rather complex dependence on the layered thickness as shown in Fig. 12 for a polymer-metal interface [16]. The fatigue threshold increased with the layer thickness from 40 μm to 330 μm but decreased with further increase in the layer thickness. A more extensive study using lap shear specimens has shown that the fatigue life of the specimens seems to peak at a layer thickness of ~300 μm [17].

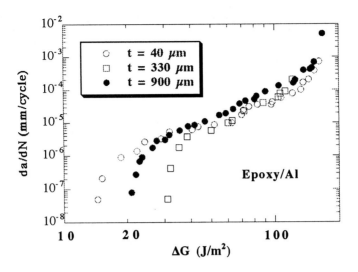

Fig. 12. Dependence of fatigue crack growth on layer thickness,t.

LIFE PREDICTION OF ADHESIVE JOINTS

In the final section of this paper, an example is given to demonstrate the application of the interface fracture mechanics to predicting the fatigue life of adhesive joints where the interface crack growth precedes the final failure. The model joint used to illustrate and test the interface-fracture mechanics methodology was a single lap shear joint, which has been a preferred test specimen for many adhesive studies. The dimensions of the specimen are given in Fig. 13. Two 1-in wide Al5754 sheets were bonded by a toughened epoxy adhesive, XD4600. The Al-sheets were given a proprietary chromate treatment before the adhesive was applied. The surface treatment assured consistent interface bonding for all specimens (as indicated by a consistent lap shear strength of 14 MPa ±5%). The specimens using a sinusoidal waveform at a frequency of 10 Hz and a load-ratio of 0.1 in the room air.

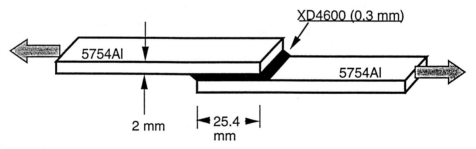

Fig. 13. Dimensions of single lap shear specimens.

The life prediction procedure followed the general outline of the damage-tolerance approach developed for metals. When applied to adhesive joints, this approach requires integrating the fatigue crack growth curve, da/dN = f(G) in Fig. 7, for cracked lap shear specimens as following:

$$N_f = \int_{a_i}^{a_f} \frac{da}{f(G)} \qquad (12)$$

where a_i and a_f are the initial and final crack sizes. Taking the initial crack size as the thickness of the adhesive and the final crack length as half of the overlap length, the fatigue lives of adhesive joints were calculated for different applied loads and are compared with experimental data in Fig. 14 [18]. With the exception of one point, which was from a specimen containing a large pore on the interface, the model prediction serves as the lower bound of the experimental data.

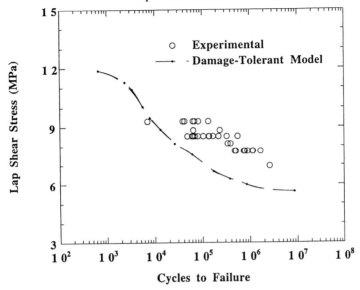

Fig. 14. Comparison of the predicted lives with experimental data for lap shear joints.

CONCLUDING REMARKS

Current research on fatigue crack growth along interfaces is reviewed. Our understanding of fatigue crack growth behavior along interfaces is just at its beginning. While the general fracture mechanics framework has been in place for some time, the major difficulty appears to be experimental. The problem with crack path was examined and its solutions were suggested. Recent studies based on newly developed flexural peel specimen have uncovered several key elements in determining fatigue crack behavior of interfaces, namely, crack tip loading condition, interface morphology, interfacial microstructure, and layer thickness. The application of interface-fatigue analysis to bonded structures has shown that the interface-fracture mechanics approach has great promise in solving the complex problem of interfacial fatigue.

ACKNOWLEDGMENT

The author would like to acknowledge the support of the Ford Motor Company through its University Research Program and the Federation of Advanced Materials Industries program at the University of Illinois.

REFERENCES

1.   J. R. Rice, *J. Appl. Mech.*, **55**, 98 (1988).
2.   J. W. Hutchinson and Z. Suo, in *Advances in Applied Mechanics*, Eds., Hutchinson and Wu, **29**, 64 (1991).
3.   Z. Suo and J. W. Hutchinson, *Mater. Sci. Engng.*, **A107**, 135 (1989).
4.   I. S. Raju, *Engng. Fract. Mech.*, **28**, 251 (1987).

5.   S. Mostovoy and E. J. Ripling, *Adhesion Science and Technology*, **9 B**, Ed., L. H. Lee, Plenum Press, New York, 513 (1975).

6.   R. M. Cannon, B. J. Dalgleish, R. H. Dauskardt, T. S. Oh and R. O. Ritchie, *Acta metall.*, **39**, 2145 (1991).

7.   Z. Zhang and J. K. Shang, *Metall. Mater. Trans. A*, **27 A**, 205 (1996).

8.   G. Liu and J. K. Shang, *Metall. Mater. Trans. A*, **27 A**, 213 (1996).

9.   Z. Zhang and J. K. Shang, *Metall. Mater. Trans. A*, **27 A**, 221 (1996).

10. D. Yao, Z. Zhang, and J. K. Shang, ASME Sym. Mech. Surface Mount Assembly, ASME paper 95-WA/EEP-12, 1995.

11. Z. Zhang and J. K. Shang, *Metall. Mater. Trans. A*, **26 A**, 2677 (1995).

12. D. Yao and J. K. Shang, *IEEE Transactions: CPMT*, February 1996, in press.

13. D. Yao and J. K. Shang, ASME Sym. Mech. Surface Mount Assembly, ASME paper 95-WA/EEP-10, 1995.

14. D. Yao and J. K. Shang, ASME Sym. Mech. Surface Mount Assembly, ASME paper 95-WA/EEP-13, 1995.

15. A. G. Evans, in *Metal/Ceramic Interfaces*, Eds., M. F. Ashby, M. Rühle, and A. G. Evans (1991).

16. Z. Zhang and J. K. Shang, unpublished research, University of Illinois, 1995.

17. S. Koppikar and J. K. Shang, unpublished research, University of Illinois, 1995.

18. J. K. Shang and R. A. Chernenkoff, Soc. Auto. Eng., SAE Cong., Detroit, 1996.

CYCLIC STRESS STRAIN BEHAVIOR AND MODELS

CYCLIC STRESS-STRAIN BEHAVIOUR, MICROSTRUCTURE AND FATIGUE LIFE

H. MUGHRABI

Institut für Werkstoffwissenschaften, Lehrstuhl I, Universität Erlangen-Nürnberg, Martensstr. 5, 91058 Erlangen, FRG

ABSTRACT

Selected cases of cyclic stress-strain behaviour are reviewed and discussed. First, examples are given for the shapes of hysteresis loops with respect to athermal and thermal stress components and to Masing and non-Masing behaviour in dependence on the cyclic prehistory. It is pointed out that the slim hysteresis loops of several high-strength materials exhibit non-negligible effects of non-linear elasticity which can complicate the loop analysis. The cyclic deformation and fatigue behaviour, including the dependence on temperature, of copper and α-iron as typical model representatives of f.c.c and b.c.c crystal structure, respectively, are compared. Distinct differences are pointed out. In the final section, specific aspects of the fatigue behaviour of some engineering materials are discussed: 1) An example is given for the increasing fatigue-induced martensitic transformation in a metastable austenitic stainless steel at low temperatures. 2) The problem of the applicability of the Manson-Coffin and the Basquin laws to the description of fatigue lives in the case of high-strength materials is illustrated with some examples. 3) Some effects of mean stress on cyclic creep and fatigue life of steels are discussed.

KEYWORDS

Cyclic stress-strain behaviour, hysteresis loops, Masing behaviour, non-linear elasticity effects, temperature dependence, fatigue life, Manson-Coffin law, Basquin law, fatigue-induced martensitic transformation, mean stresses, cyclic creep.

INTRODUCTION

When a material is subjected to repeated cyclic loading, its failure by fatigue represents the event of most concern. While failure can occur unexpectedly, in many cases – especially so in ductile materials – a characteristic cyclic stress-strain response precedes the final event of failure. Hence, the study of the cyclic stress-strain behaviour of materials is an important aspect of fatigue analysis. The aim is to link features of cyclic deformation such as cyclic hardening or softening and saturation to fatigue life. The microstructural processes occurring during cyclic deformation represent another important aspect of metal fatigue that deserves special attention when aiming for a deeper understanding of the cylic stress-strain behaviour and its role in fatigue life.

In this spirit, the present survey focusses on the importance of the cyclic stress-strain behaviour of metallic materials and its relation to microstructure with respect to fatigue life. Thereby, attention will be paid to affects of crystal stucture and test conditions, and distinct differences between ductile and high-strength materials will be considered. Both face-centred cubic (f.c.c.) and body-centred cubic (b.c.c.) model materials such as copper and α-iron and different types of engineering materials will be dealt with. Briefly, the following topics will be addressed:

a) The hysteresis loop
b) The cyclic stress-strain curve
c) Effects of fatigue at low and at high temperatures
d) Fatigue life
e) The role of microstructure and its stability during cyclic deformation

THE HYSTERESIS LOOP-SHAPE AND MASING BEHAVIOUR

The hysteresis loop of a cyclically stressed material contains valuable information on the details of cyclic (micro)yielding. Its width at zero stress is equal to the plastic strain range $\Delta\varepsilon_{pl}$ (twice the plastic strain amplitude) which controls fatigue life in many cases of low-cycle fatigue through the Manson-Coffin law. In terms of modelling, it is important to note that the second derivative of the stress-strain behaviour (in relative co-ordinates) permits to extract the distribution of obstacle strengths from the shape of the hysteresis loop, compare, for example, Polák (1991) and Christ (1991). In more refined treatments, it is also possible to separate the effects of athermal (internal) stress σ_G and effective thermal stress σ^*, stemming from the thermally activated glide process. An example from the work of Polák et al. (1982) is shown in Fig. 1.

Another interesting aspect of the shape of hysteresis loops is whether so-called Masing behaviour is obeyed. Masing behaviour (cf. Asaro, 1975) implies that a unique cyclic stress-strain relationship describing forward and reverse straining exists. According to Wüthrich (1982), such a unique stress-strain relationship is a prerequisite for the applicability of the cyclic J-integral concept to fatigue crack propagation. In practice, Masing behaviour is found to be obeyed only in special cases. An example from the work of Bayerlein et al. (1987) on AISI 304 L stainless steel is shown in Fig. 2. In this case, the saturated hysteresis loops obtained in single-step strain-controlled tests do not exhibit Masing behaviour (Fig. 2a), whereas the hysteresis loops observed in an incremental step test do (Fig. 2b). As a consequence, the cyclic stress-strain curves differ for both cases. The explanation for this behaviour is that, in single-step tests, materials exhibiting marked cyclic hardening develop different dislocation microstructures characteristic of the applied strain amplitude, whereas, in an incremental step test, a more or less "stabilized" and "constant" dislocation microstructure, corresponding to an amplitude somewhat below the upper amplitude of the strain block, prevails during each cycle, irrespective of the current strain amplitude. The implication of this behaviour is that, under variable amplitude conditions, Masing behaviour can be assumed to hold in good approximation, even for ductile materials with non-negligible cyclic hardening. In the particular case of the metastable austenitic stainless steel AISI 304 L, the formation of fatigue-induced martensite and its effect on cyclic hardening must also be considered (Bayerlein et al., 1987, 1989). In addition, attention must be paid to the glide mode (Christ, 1991, Christ and Mughrabi, 1992). High-strength materials in which only negligible microstructural changes occur during cyclic stressing can in some cases show Masing behaviour even in single-step tests and thus exibit a more or less unique cyclic stress-strain curve which can be suitably used as a mechanical equation of state in assessments of fatigue life. In general, it is advisable to check from case to case whether Masing behaviour can be assumed and whether a unique cyclic stress-strain behaviour exists (Christ, 1991; Christ and Mughrabi, 1992).

Another interesting feature of the shape of the hysteresis loop is frequently found in the case of high-strength materials exhibiting narrow loops with high stress levels and very little plasticity. Figure 3 shows an example, taken from work on the bainitically hardened roller bearing steel SAE 52100 (Sommer et al., 1991). Figure 3a shows the sickle-shaped hysteresis loop observed in a plot of stress σ versus the plastic strain $\varepsilon_t - \sigma/E_o$ in the linear Hooke approximation (ε_t: total strain; E_o: Young's modulus in the limit of vanishing stress). It could be shown that the unusual shape of the hysteresis loop is caused by the non-negligible non-linearity of the elastic strain at high stresses. With an extended Hooke's law ($\sigma = E_o \cdot \varepsilon_{el} + k \cdot \varepsilon_{el}^2$, ε_{el}: elastic strain, k: constant), implying a stress-dependent differential Young's modulus $E_D(\sigma) = \sqrt{(E_o^2 + 4k\sigma)}$, the hysteresis loop can be replotted in the form of σ versus the correct plastic strain ε_{pl}. With this procedure a loop of customary shape is obtained, as shown in Fig. 3b (with $E_o = 206$ GPa and $k = -730$ GPa). Similar observations have been made on other high-strength materials such as the high-temperature titanium alloy IMI 834 (Renner, 1995) and a particulate-reinfor-

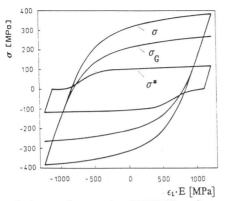

Fig. 1. Hysteresis loop of a low-carbon steel at 213 K. Plot of stress σ vs. $\varepsilon_t \cdot E$ (ε_t: total strain, E: Young's modulus). The course of the athermal stress σ_G and the effective stress σ^* are also shown. After Polák *et al.* (1982).

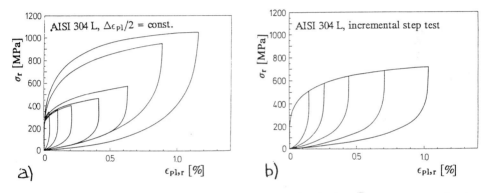

Fig. 2. Hysteresis loops of the austenitic stainless steel AISI 304 L at room temperature in plots of relative stress σ_r vs. relative plastic strain $\varepsilon_{pl,r}$, referred to the lower tip of the loops. After Bayerlein *et al.* (1987). a) Single-step test. b) Incremental step test.

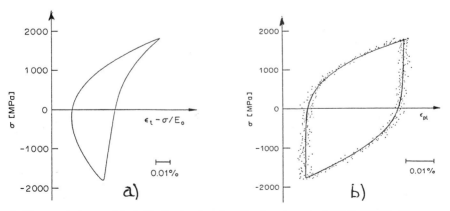

Fig. 3. Hysteresis loop of bainitically hardened roller bearing steel SAE 52100. Stress range $\Delta\sigma = 3600$ MPa, 293 K. a) Sickle-shaped hysteresis loop in plot σ vs. $\varepsilon_t - \sigma/E_o$. b) Reconstructed hysteresis loop in plot σ vs ε_{pl}. After Sommer *et al.* (1991).

ced aluminium-alloy-matrix composite (Biermann *et al.*, 1995). A consequence of such behaviour is that special care must be taken in the analysis of such slim hysteresis loops with the aim of separating correctly the elastic and plastic contributions to the total strain.

COMPARISON OF THE CYCLIC DEFORMATION AND FATIGUE BEHAVIOUR OF THE "MODEL MATERIALS" COPPER AND α-IRON

Copper

The cyclic deformation behaviour of f.c.c. metals like copper is characterized by a remarkable cyclic hardening capability due to easy dislocation multiplication and accumulation and by a rather well-defined state of cyclic saturation. These features are well represented in the results of an extensive study on copper polycrystals (Bayerlein, 1991). At low temperatures, the cyclic hardening increment increases strongly, due to reduced dynamic recovery. This is reflected in cyclic hardening curves measured at different temperatures (Fig. 4a) and in a plot of saturation stress amplitude $\Delta\sigma_s/2$ versus temperature T for copper polycrystals fatigued at $\Delta\varepsilon_{pl} = 4.4 \cdot 10^{-4}$ in Fig. 4b. The enhanced cyclic hardening at low temperatures is the dominant cause of the strong increase of stress below room temperature. By comparison, the thermal effective stress contribution is small. Figure 5 shows cyclic stress-strain curves of copper polycrystals at three different temperatures around, above and below room temperature. While the shapes of the curves are similar, the shift to higher stress values at lower temperatures is evident. These curves can be described in good approximation by a power law $\Delta\sigma/2 = \text{const.} \cdot (\Delta\varepsilon_{pl}/2)^n$ with $n \approx 0.27 - 0.28$ (except at 107 K for $\Delta\varepsilon_{pl}/2 < 10^{-3}$, where $n \approx 0.11$).

The dependence of fatigue life (number of cycles to failure: N_f) of copper on plastic strain amplitude obeys a Manson-Coffin law, as reported earlier (cf., e.g., Lukáš *et al.*, 1974). For the same temperatures as in Fig. 5 the Manson-Coffin plots shown in Fig. 6 are found for tests in vacuum, i. e. in the absence of environmental effects. The fatigue ductility exponent of the Manson-Coffin relationship is found to be approximately -0.4. In the following, it will be interesting to contrast these results against those of comparable studies on pure α-iron.

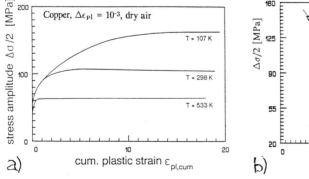
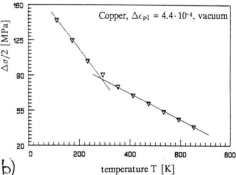

Fig. 4. Cyclic hardening and saturation behaviour of copper polycrystals. a) Plot of stress amplitude $\Delta\sigma_s/2$ vs. cummulative plastic strain for $\Delta\varepsilon_{pl} = 10^{-3}$ at different temperatures. b) Dependence of saturation stress amplitude $\Delta\sigma/2$ on temperature T for copper polycrystal fatigued at $\Delta\varepsilon_{pl} = 4.4 \cdot 10^{-4}$. After Bayerlein (1991).

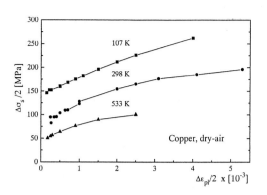

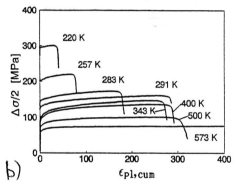

Fig. 5. Cyclic stress-strain curves $\Delta\sigma_s/2$ vs $\Delta\varepsilon_{pl}/2$
for copper polycrystals fatigued at
different temperatures.
After Bayerlein (1991).

Fig. 6. Manson-Coffin plots of copper poly
crystals fatigued at different temperatures
in vacuum. After Bayerlein (1991)
and Bayerlein and Mughrabi (1992).

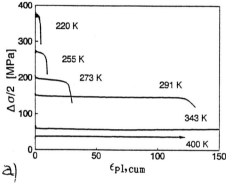

Fig 7. Cyclic hardening curves of α-iron polycrystals fatigued at different temperatures at
$\dot{\varepsilon}_{pl} = 10^{-3}\,s^{-1}$. After Sommer (1992) and Mughrabi et al. (1994).
a) $\Delta\varepsilon_{pl} = 2 \cdot 10^{-4}$. b) $\Delta\varepsilon_{pl} = 2 \cdot 10^{-3}$.

α-Iron

The deformation of b.c.c. metals like α-iron differs characteristically from that of f.c.c. metals.
Dislocation glide is thermally activated and controlled by the formation of double kinks on screw
dislocations (Seeger, 1981). As a consequence, the increase of the thermal effective stress component
σ^* at low temperatures is appreciable and depends also markedly on plastic strain rate $\dot{\varepsilon}_{pl}$. In the low-
temperature range of thermally activated dislocation glide, small strain amplitudes can be
accommodated by quasi-reversible to-and-fro motion of the mobile non-screw dislocation segments,
while the mobility of the screw dislocations and hence dislocation multiplication are strongly impeded,
cf. Mughrabi et al. (1981). The following results on α-iron were obtained in the framework of a
comprehensive study on pure (and on carburized) α-iron polycrystals (Sommer, 1992; Mughrabi et al.
1994; Sommer et al., 1995). Here, only some results obtained on pure, decarburized α-iron (interstitial
content smaller than 1 wt. ppm) will be reported.

Cyclic hardening at low $\Delta\varepsilon_{pl}$ $(2 \cdot 10^{-4})$ is negligible at all temperatures investigated (220 K < T < 400 K).
Still, the stress level increases drastically with decreasing temperature, as shown in Fig. 7a. At higher

$\Delta\varepsilon_{pl}$ $(2\cdot10^{-3})$, cyclic hardening is more pronounced (Fig. 7b). Still it is clear that the major part of the increased stress level at lower temperatures is due to the enhanced thermal effective stress level σ^*. The athermal cyclic hardening increment $\sigma_{G,h}$ is largest at an intermediate temperature of about 350 K (Fig. 8), due to the fact that, while dislocation multiplications is impeded at lower temperatures, the steady-state dislocation density at higher temperatures (at which dislocation multiplication is easier) is decreased due to increasing dynamic recovery. The shapes of the cyclic stress-strain curves and their dependence on temperature, shown in Fig. 9, are somewhat unusual and can be explained as follows. At low temperatures and (very) low $\Delta\varepsilon_{pl}$, it is remarkable that the stress level increases with decreasing $\Delta\varepsilon_{pl}$. This could be related to the fact that, in this range of $\Delta\varepsilon_{pl}$, quasi-reversible cyclic deformation is strongly inhomogeneous and confined to only a small fraction of the grains in which, however, the effectively acting local strain rate is hence considerably in excess of the imposed strain rate. As a consequence, the effective thermal stress level σ^* is enhanced markedly. Next, the "plateau" observed around $\Delta\varepsilon_{pl} \approx 10^{-3}$ at 220 K, 256 K and 291 K is assigned to the quasi-reversible to-and fro motion of mobile non-screw dislocation segments as in earlier work, cf. Mughrabi et al. (1981). The increase of the stress above $\Delta\varepsilon_{pl} = 10^{-3}$ reflects the athermal cyclic hardening under conditions of increasing dislocation multiplication. Only at 400 K is a cyclic stress-strain curve observed which is qualitatively similar to that of copper, exhibiting a continuous increase of athermal stress level (σ_G) with increasing plastic strain amplitude.

Finally, we consider the fatigue lives. Figure 10 shows Manson-Coffin plots for the four temperatures investigated. It will be noted that a Manson-Coffin-type relationship is obeyed only in the range of higher $\Delta\varepsilon_{pl}$. At lower $\Delta\varepsilon_{pl}$ ($\leq 5\cdot10^{-4}$), an unusual behaviour is found: N_f decreases with decreasing $\Delta\varepsilon_{pl}$. In this range of $\Delta\varepsilon_{pl}$, rather high σ^*-values and hence rather large elastic strain ranges $\Delta\varepsilon_{el}$ occur. As a consequence, one finds that a Basquin-like relationship ($\Delta\varepsilon_{el}/2$ vs. N_f) describes the fatigue lives in the range of $\Delta\varepsilon_{pl} = 10^{-4}$ to $5\cdot10^{-4}$ and for <u>different</u> low temperatures quite satisfactorily (Fig. 11). While this result may be surprising, it emphasizes the importance of high effective (thermal) stresses σ^* with respect to fatigue life.

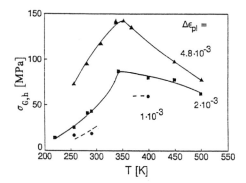

Fig. 8. Temperature dependence of athermal cyclic hardening increment $\sigma_{G,h}$ of fatigued α-iron polycrystals ($\dot{\varepsilon}_{pl} = 10^{-3}s^{-1}$). After Sommer (1992) and Mughrabi et al. (1994).

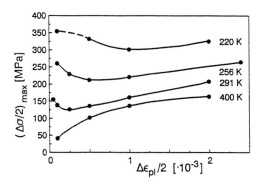

Fig. 9. Cyclic stress-strain curves of α-iron polycrystals fatigued at different temperatures ($\dot{\varepsilon}_{pl} = 10^{-3}s^{-1}$). After Sommer (1992) and Mughrabi et al. (1994).

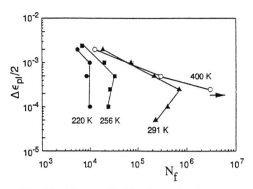

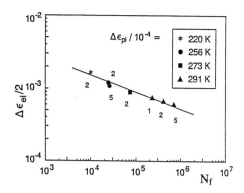

Fig. 10. Manson-Coffin plots of α-iron polycrystals fatigued at different temperatures ($\dot{\varepsilon}_{pl} = 10^{-3}\text{s}^{-1}$). After Sommer (1992) and Mughrabi et al. (1994).

Fig. 11. Basquin-plot of α-iron polycrystals fatigued at low plastic strain amplitudes ($10^{-4} < \Delta\varepsilon_{pl} < 5\cdot10^{-4}$, $\dot{\varepsilon}_{pl} = 10^{-3}\text{s}^{-1}$) at different temperatures. After Sommer (1992) and Sommer et al. (1995).

SPECIFIC ASPECTS OF FATIGUE OF SOME ENGINEERING MATERIALS

Deformation-Induced Martensitic Transformation in a Metastable Austenitic Stainless Steel

The driving force of deformation-induced martensitic transformation during the fatigue of metastable austenitic stainless steels increases both with increasing plastic strain amplitude (Bayerlein et al., 1987, 1989) and, in particular, with decreasing temperature. Figure 12 shows some cyclic hardening curves of the austenitic stainless steel AISI 304 L at $\Delta\varepsilon_{pl} = 2.52\cdot10^{-2}$ at different temperatures (Maier et al., 1993). With decreasing temperature, a marked increase of the rate of cyclic hardening and of the (saturation) stress attained is apparent. The major reason for this behaviour lies in the combined enhancement of work hardening both by dislocations and by the fatigue-induced martensite at the lower temperatures. The fatigue lives at the lower temperatures are reduced as a consequence of the increased stress amplitude at given plastic strain amplitude. In a formal sense, the fatigue lives could be described crudely in terms of a Basquin-type relationship.

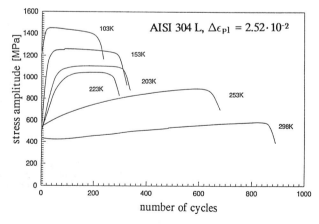

Fig. 12. Cyclic hardening curves of the austenitic stainless steel AISI L 304 for $\Delta\varepsilon_{pl} = 2.52\cdot10^{-2}$ at different temperatures. After Maier et al. (1993).

Manson-Coffin and Basquin Plots of Fatigue Lives

It is common practice to describe fatigue lives N_f in terms of the Manson-Coffin and the Basquin laws in the low-cycle and high-cycle fatigue regimes, respectively:

$$\Delta\varepsilon_{pl} / 2 = \varepsilon_f' \cdot (2 \cdot N_f)^c \qquad \text{(low-cycle fatigue)}$$

$$\Delta\sigma / 2 = \sigma_f' \cdot (2 \cdot N_f)^b \qquad \text{(high-cycle fatigue)}$$

Here, the fatigue strength coefficient σ_f' and the fatigue ductility coefficient ε_f' are, in good approximation, equal to the true fracture stress and strain, respectively, cf. Morrow (1964). The fatigue ductility exponent c is typically -0.6 and the fatigue strength exponent b lies around - 0.1.

Several authors have noted that, in some high-strength materials, deviations from the Manson-Coffin law commonly occur at low $\Delta\varepsilon_{pl}$, cf. Polák (1991) and Sommer (1992), with the tendency towards shorter fatigue lives. An example for a deviation from the Manson-Coffin-law in the low-cycle fatigue regime was given earlier for α-iron fatigued at low temperatures (Figs. 10 and 11), and it was shown that, in a formal sense, the fatigue lives could be described by a Basquin-type relationship. However, inspection of Fig. 11 shows that, in this case, unusually large values of the fatigue strength coefficient and the fatigue strength exponent (b = - 0.22) are found.

As another example, we quote the results obtained on the fatigue lives of the bainitically hardened roller bearing steel SAE 52100 by Christ et al. (1992). In this study, it was found that the fatigue lives which were all in the low-cycle range (up to a few thousand cycles) could be described formally in terms of a Manson-Coffin law

$$\Delta\varepsilon_{pl} / 2 = 18 \cdot (2 \cdot N_f)^{-1.63}$$

or, alternatively, in terms of a Basquin law

$$\Delta\sigma / 2 = 7700 \cdot (2 \cdot N_f)^{-0.20}.$$

The shortcomings of such representations of fatigue life are that the "constants" ε_f', c, σ_f', and b are found to assume unusual (too high) values. Similar examples referring to other materials such as the titanium alloy IMI 834 (Renner, 1995) could be cited. It is clear that, at present, there exists no unique satisfactory description of fatigue lives of the kind of materials discussed here.

Effects of Mean Stress and Cyclic Creep on Fatigue Life

In practice, static and symmetric cyclic loads are frequently superposed. Hence, it is of considerable interest to study the effects of mean stresses on the cyclic deformation and fatigue behaviour of materials. Systematic studies of fatigue with positive and negative mean stress σ_m, including investigations of the effects of the cyclic hardening or softening and cyclic creep on fatigue life, have been performed only in recent years. In these works, in particular the ferritic normalized steel SAE 1045 (Glaser, 1988; Glaser et al., 1991; Wamukwamba, 1992) and the austenitic stainless steel AISI 304 L (Wamukwamba, 1992) have been studied in detail. The work of Wamukwamba was performed not only at room temperature but also at higher temperatures (up to 650°C for AISI 304 L and up to 375°C for SAE 1045) in order to study also the effects of dynamic strain ageing. Here, we confine ourselves for the sake of simplicity to some results obtained by Wamukwamba (1992) on the austenitic stainless steel AISI 304 L at room temperature.

First, it is to be noted that in the stress-controlled tests with zero mean stress the austenitic stainless steel AISI 304 L exhibits cyclic primary hardening at all temperatures. Figure 13 shows cyclic hardening curves obtained at room temperature without and with positive/negative mean stresses σ_m. This behaviour differs from that of the ferritic steel SAE 1045 which exhibits an incubation period followed by cyclic softening for small values of $|\sigma_m|$ (Glaser et al., 1991). The stainless steel specimens experience cyclic creep strains which are significantly larger for positive than for negative mean stresses σ_m, as shown in Fig. 14. In Fig. 15a the dependence of the plastic strain amplitude in "saturation" on mean stress σ_m is shown for two stress amplitudes. The corresponding dependence of fatigue life on mean stress is plotted in Fig. 15b. It is interesting to note that, while negative mean stresses enhance fatigue life in all cases (as long as the yield stress in compression is not exceeded), positive mean stresses increase the fatigue life only for the smaller stress amplitude ($\Delta\sigma/2 = 250$ MPa) but decrease it at higher stress amplitude ($\Delta\sigma/2 = 300$ MPa). The latter behaviour is attributed to more dominant cyclic creep at the larger stress amplitude, causing earlier failure.

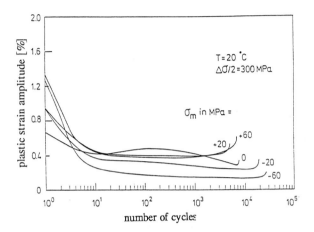

Fig. 13. Cyclic hardening curves, $\Delta\varepsilon_{pl}/2$ vs. number of cycles N, of austenitic stainless steel AISI 304 L at room temperature, $\Delta\sigma/2 = 300$ MPa, mean stresses -60 MPa < σ_m < +60 MPa. After Wamukwamba (1992).

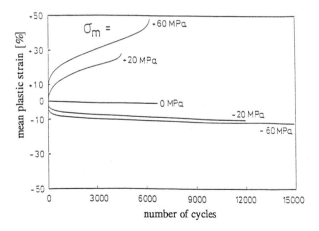

Fig. 14. Cyclic creep curves of austenitic stainless steel AISI 304 L in stress-controlled tests ($\Delta\sigma/2 = 300$ MPa) at room temperature with mean stresses -60 MPa < σ_m < +60 MPa. After Wamukwamba (1992).

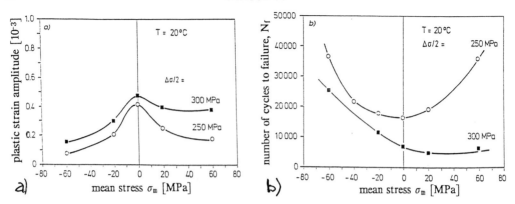

Fig. 15. Influence of mean stresses σ_m on a) plastic strain amplitude, and b) fatigue life of austenitic stainless steel AISI 304 L in stress-controlled tests at room temperature. After Wamukwamba (1992).

The behaviour of the ferritic steel SAE 1045 differs characteristically from that of the austenitic steel AISI 304 L in serveral respects (Wamukwamba, 1992). Here, we note in particular that compressive mean stresses always led to an increase of fatigue life in the case of SAE 1045, cf. Glaser (1988) and Glaser *et al.* (1991), and that fatigue life, even with positive/negative mean stresses, could be represented quite satisfactorily in a Manson-Coffin-type diagram, as shown in Fig. 16, albeit with different slopes of the linearly fitted lines.

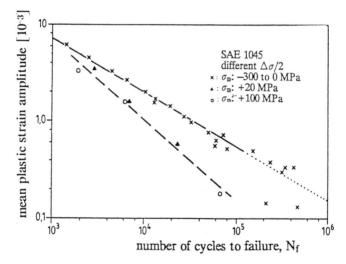

Fig. 16. Manson-Coffin representation of fatigue lives of the ferritic normalized steel SAE 1045 in stress-controlled tests at different stress amplitudes and mean stresses at room temperature. After Glaser *et al.* (1991).

CONCLUDING REMARKS

The few examples discussed in this review reflect impressively the complexity of fatigue behaviour, when considering different materials and different testing conditions. While microstructural interpretations can be offered in most cases, one is still far from a comprehensive consistent picture of metal fatigue and generally valid descriptions of cyclic stress-strain response and fatigue life.

ACKNOWLEDGMENTS

The financial support of Deutsche Forschungsgemeinschaft and Volkswagen-Stiftung over the years is acknowledged gratefully. The author expresses his gratitude to Mr. H. W. Höppel and Mr. M. Ott for their help in the preparation of the manuscript and to the (former) members of his research group who contributed to this work and in particular to Dr. M. Bayerlein, Dr. C. Sommer, Dr. C. K. Wamukwamba and to Prof. H. J. Christ for the use of some unpublished data.

REFERENCES

Asaro, R. J. (1975). Elastic-plastic memory and kinematic-type hardening. *Acta Metall.*, 23, 1255-1265.

Bayerlein, M. (1991). *Wechselverformungsverhalten und Ermüdungsrißbildung von vielkristallinem Kupfer.* Doctorate Thesis, Universität Erlangen-Nürnberg. VDI Verlag GmbH, Düsseldorf.

Bayerlein, M., H. J. Christ, and H. Mughrabi (1987). A critical evaluation of the incremental step test. In: *Proc. 2nd. Conf. on Low Cycle Fatigue and Elasto-Plastic Behaviour of Materials* (K. T. Rie, ed.), pp. 149-154. Elsevier, Amsterdam.

Bayerlein, M. and H. Mughrabi (1992). Fatigue crack initiation and early crack growth in copper polycrystals – effects of temperature and environment. In: *Short Fatigue Cracks*, ESIS 13 (K. J. Miller and E. R. de los Rios, eds.), pp. 55-82. Mechanical Engineering Publications, London.

Biermann, H., G. Beyer and H. Mughrabi (1995). Spannungs-Dehnungs-Hysteresekurven wechsel-verformter Proben eines Metallmatrix-Verbundwerkstoffes. To appear in: *Proc. of DGM-Conf. Verbundwerkstoffe und Werkstoffverbunde,* Bayreuth.

Christ, H.-J. (1991). *Wechselverformung von Metallen.* Springer Verlag, Berlin, Heidelberg, New York.

Christ, H.-J. and H. Mughrabi (1992). Microstructure and fatigue. In: *Proc. Third Int. Conf. on Low Cycle Fatigue and Elasto-Plastic Behaviour of Materials* (K. T. Rie, ed.), pp. 56-69. Elsevier Applied Science, London and New York.

Christ, H. J., C. Sommer, H. Mughrabi, A. P. Voskamp, J. M. Beswick and F. Hengerer (1992). Fatigue behaviour of three variants of the roller bearing steel SAE 52100. *Fatigue Fract. Engng. Mater. Struct.*, 15, 855-870.

Glaser, A. (1988). *Mittelspannungseinfluß auf das Verformungsverhalten von Ck 45 und 42 CrMo 4 bei spannungs- und dehnungskontrollierter homogen-einachsiger Schwingungsbeanspruchung.* Doctorate Thesis, Universität Karlsruhe (TH).

Glaser, A., D. Eifler and E. Macherauch (1991). Einfluß positiver und negativer Mittelspannungen auf das zyklische Spannungs-Dehnungsverhalten von normalisiertem Ck45. *Mat.-wiss. u. Werkstofftech.*, 22, 266-274.

Lukáš, P., M. Klesnil and J. Polák (1974). High cycle fatigue life of metals. *Mater Sci. Eng.*, 15, 239-245.

Maier, H. J., B. Donth, M. Bayerlein, H. Mughrabi, B. Meier and M. Kesten (1993). Optimierte Festigkeitssteigerung eines metastabilen austenitischen Stahles durch wechselverformungsinduzierte Martensitumwandlung bei tiefen Temperaturen. *Z. Metallk.*, 84, 820-826.

Morrow, J. D. (1964). Cyclic plastic strain energy and fatigue of metals. In: *Internal Friction, Damping and Cyclic Plasticity.* ASTM STP 378, pp. 45-84. American Society for Testing and Materials, Philadelphia.

Mughrabi, H., K. Herz and X. Stark (1981). Cyclic deformation and fatigue behaviour of α-iron mono- and polycrystals. *Int. J. Fract.*, 17, 193-220.

Mughrabi, H., C. Sommer and D. Lochner (1994). On the fatigue behaviour of α-iron polycystals and its dependence on temperature and carbon content. In: *Proc. 10th Int. Conf. on the Strength of Materials* (ICSMA 10, H. Oikawa, K. Maruyama, S. Takeuchi and M. Yamaguchi, eds.), pp. 481-484. The Japanese Institute of Metals.

Polák, J. (1991). Cyclic Plasticity and Low Cycle Fatigue Life of Metals. *Materials Science Monographs*, Vol. 63. Elsevier, Amsterdam.

Polák, J., M. Klesnil and J. Helešik (1982). The hysteresis loop. 2. Analysis of the loop shape. *Fatigue Fract. Engng. Mater. Struct.*, 5, 33-44.

Renner, H. (1995). Unpublished work.

Renner, H., H. Kestler and H. Mughrabi (1995). Influence of heat treatment and microstructure on the low cycle fatigue properties of the hot forged near-α titanium alloy IMI 834. To appear in this volume. (*Fatigue 96*).

Seeger, A. (1981). The temperature and strain-rate dependence of the flow stress of body-centred cubic metals: a theory based on kink-kink interactions. *Z. Metallk.*, 72, 369-380.

Sommer, C., H. J. Christ and H. Mughrabi (1991). Non-linear elastic behaviour of the roller bearing steel SAE 51200 during cyclic loading. *Acta Metall. Mater.*, 39, 1177-1187.

Sommer, C. (1992). *Wechselverformungsverhalten und Ermüdungsrißbildung von α-Eisenviel-kristallen.* Doctorate Thesis, Universität Erlangen-Nürnberg.

Sommer, C., D. Lochner and H. Mughrabi (1995). Publication in preparation.

Wüthrich, C. (1982). The extension of the J-integral concept to fatigue cracks. *Int. J. Fract.*, 20, R35-R37.

Wamukwamba, C. K. (1992). *Einfluß der Temperatur und der Mittelspannung auf das Wechsel-verformungsverhalten der Stähle Ck45 und X3 CrNi 18 9.* Doctorate Thesis, Universität Erlangen-Nürnberg.

CYCLIC DEFORMATION BEHAVIOR OF COPPER SINGLE CRYSTALS ORIENTED FOR DOUBLE AND MULTIPLE SLIP

Bo Gong[1], Zhirui Wang[1], Zhongguang Wang[2] and Yiwei Zhang[2]

1. Department of Metallurgy and Materials Science, University of Toronto
Toronto, Ontario, Canada
2. State Key Laboratory for Fatigue and Fracture of Materials, Institute of Metal Research
Chinese Academy of Science, Shenyang, P. R. China

ABSTRACT

The cyclic deformation of copper single crystals oriented for single slip ($[\bar{1}25]$), double slip ($[034]$ and $[\bar{1}17]$) and multiple slip ($[001]$) have been investigated. Cyclic hardening and saturation behaviors of the double and multiple slip oriented crystals significantly differ from those of the single slip oriented crystals and strongly depend on their orientations. The $[034]$ crystal shows a shorter plateau in the cyclic stress-strain curve than that of the $[\bar{1}25]$ crystal, while the $[\bar{1}17]$ and $[001]$ crystals exhibit no plateau. Surface observation revealed that the formation of deformation bands is a general phenomenon in the double and multiple slip oriented crystals. The mechanical behaviors have been rationalized in accord with dislocation structures and dislocation reactions between slip systems.

KEYWORDS

Cu, single crystal; double and multiple slip; cyclic deformation; dislocation reaction, persistent slip band.

INTRODUCTION

The cyclic deformation behavior of f. c. c. single crystals have been extensively studied in the past two or three decades in order to understand the fundamental fatigue mechanisms (Basinski et al., 1992, Laird, 1986). The most systematic work has been focused on copper single crystals oriented for single slip. It is now well-known that the cyclic stress-strain (CSS) curve of Cu crystals with single slip orientations consists of three stages of saturation with a characteristic plateau in the center (Mughrabi, 1978). In the plateau region, persistent slip bands (PSBs) develop and PSBs frequently act as preferential sites for the development of fatigue cracks.

In polycrystalline materials, due to the deformation incompatibility between adjacent grains caused by misorientation, every grain is in general subjected to a complicated and multiaxial loading condition in reality and consequently the fundamental research with polycrystals has not been as successful. Therefore, the investigation of cyclic deformation behavior of double and multiple slip oriented crystals has become more necessary so as to correlate single crystal behavior to polycrystal behavior. Unfortunately, work in this area has been very limited (Jin et al ., 1984, Lepisto et al., 1986). In this paper, we will report our recent work on the cyclic deformation behavior of Cu single crystals oriented for double and multiple slip, including cyclic hardening behavior, feature of CSS curve, slipping characteristics, and dislocation structures. The results will be interpreted in terms of dislocation reaction between slip systems.

EXPERIMENTAL DETAILS

Copper single crystals were grown from OFHC copper of 99.999% purity by the Czochralski method. The orientations of the crystal bars are shown in the crystallographic triangle in Fig. 1. The [$\bar{1}$25] and [001] crystals are of single and multiple slip orientations, respectively. The [034] and [$\bar{1}$17] crystals belong to double slip orientation. The fatigue specimens were carefully spark machined from the crystals to have a gauge section of 6x4x16 mm^3 for the [001] orientation and 7x5x16 mm^3 for the rest of orientations. Before the fatigue test, specimens were annealed at 800°C for 1 hr, and then electropolished for surface observation.

Cyclic deformation tests were performed on either a MTS fatigue testing machine (for the [001] crystal) or a Schenck servohydraulic testing machine (for the rest orientations) at room temperature under a constant plastic strain control mode. The plastic shear strain amplitude (γ_{pl}) ranged from 8x10^{-5} to 1.4x10^{-2}. A triangle waveform signal was used with a frequency of 0.1-0.4 Hz. All specimens were cycled at a specifically selected constant plastic strain amplitude till the end of the test except that some specimens of the [$\bar{1}$25] crystal were tested with a step-ascending mode. The surfaces of the fatigued specimens were observed by a light microscope and SEM after testing. For the [001] crystal, development of slip bands was also examined *in situ* by a microscope attached on the testing machine. Thin slices were cut from the fatigued [034] and [001] specimens along certain crystallographic planes for TEM observation.

RESULTS

Cyclic hardening

The cyclic hardening curves for the four orientations tested at an approximately same γ_{pl}, obtained by plotting shear stress (τ) against cumulative shear strain ($\gamma_{pl, cum}=4N\gamma_{pl}$, N is the cyclic number), are compared in Fig. 2. The mean initial cyclic hardening coefficient, $\theta_{0.2}$, evaluated from $\theta_{0.2}=\Delta\tau/\Delta\gamma_{pl,cum}$ by taking $\Delta\gamma_{pl, cum} =0.2$ against γ_{pl} (Mughrabi, 1978), is plotted in Fig. 3. As is generally observed, all crystals display an initial hardening stage and then come into a stable state. The initial hardening of the [034] double slip crystal is apparently similar to that of the [$\bar{1}$25] crystal, especially at low strain amplitude levels ($\gamma_{pl}<10^{-3}$). The [$\bar{1}$17] and [001] crystals, however, show a much more rapid initial hardening stage, but their behavior seems similar to each other.

CSS curve

The CSS curves for single, double and multiple slip crystals, obtained by plotting the saturation stress against the stain amplitude, are shown in Fig. 4. The curve for the [$\bar{1}$25] crystal shows a plateau in the resolved shear strain range of 8x10^{-5}<γ_{pl}<6.8x10^{-3} with an average saturation stress of 27.3 MPa, very close to the results of previous studies (Mughrabi, 1978). The CSS curve for the [034] crystal also shows a plateau with an average value of 27.7 MPa. But the upper limit of the plateau has moved to a lower γ_{pl} value of approximately 4.3x10^{-3}. For the [$\bar{1}$17] and [001] crystals, there is no noticeable plateau in their CSS curves. The saturation stresses are much higher and monotonously increase with increasing γ_{pl}.

Surface feature

The slip band patterns on the surfaces were found to be strongly dependent on both applied strain amplitude and crystal orientation. In general, the patterns consist of two sets of PSBs, one is primary and the other is secondary. Only at sufficiently high amplitudes, three or four sets of PSBs could be observed in localized regions under microscope. For the double slip crystals, i.e. the [034] and [$\bar{1}$17] crystals, secondary slip became pronounced when $\gamma_{pl}>1x10^{-3}$ (Gong *et al.*, 1995). For the [001] crystal, however, this could happen as long as $\gamma_{pl}>5\times10^{-4}$. Figure 5 shows a multiple slip pattern observed in the [001] crystal fatigued at $\gamma_{pl}=1.8\times10^{-3}$.

Another important feature on the surfaces is the appearance of deformation bands (DBs). Usually two type of DBs (DBI and DBII) simultaneously form, and their habit planes are approximately perpendicular to each other. The formation of DBs is also strain amplitude dependent. For the double slip crystals, the critical value was found to be 1×10^{-3}. For the [001] crystal, it is about 6×10^{-4}. The typical DBs formed in the [034] crystal are shown in Fig. 6.

In situ observation of the [001] crystal surface during cycling further revealed that slip bands appeared at very early stage of cyclic deformation and some of them may become inactive after a certain amount of hardening. The formation of secondary slip bands is roughly corresponding to the maximum in cyclic hardening curve.

Dislocation structure

For the [034] crystal, the dislocation structure in different specimens but all tested at $\gamma_{pl}<1\times10^{-3}$ are basically same, consisting of vein and ladder structures. As the imposed strain amplitude increased to 3.4×10^{-3}, dislocation structure also changed and contains a large fraction of uncondensed labyrinth structure except vein and ladder structures. When γ_{pl} increased to 5.2×10^{-3}, extended $(\bar{1}01)$ walls and cell structures were observed. For the [001] crystal, however, the dislocation structure consists of homogenous labyrinth structure over all the strain amplitudes investigated. And the volume fraction of the wall structure even if not well condensed increases with decreasing γ_{pl}. Fig. 7 shows the microstructures of the [001] crystal fatigued at $\gamma_{pl}=4.8\times10^{-4}$.

DISCUSSION

The fact that the cyclic hardening behavior strongly depends on the crystal orientation is actually in a good agreement with the theory of dislocation reaction between different slip systems. For the [034] double slip crystal, the primary and critical slip systems are equally stressed. Dislocation interaction between these two systems, giving a product of sessile jog, is not expect to be strong (Hirth *et al.*, 1982). For the [$\bar{1}17$] double slip crystal, the primary and conjugate dislocations would strongly react, producing a Lomer-Cottrell lock (Hirth *et al.*, 1982). Therefore, an orientation on the 001/$\bar{1}11$ boundary of the standard crystallographic triangle would show much higher hardening rates than one on the 001/011 boundary. Results from the present study has confirmed this theory again. Since the cyclic hardening behavior of the [001] crystal is similar to that of the [$\bar{1}17$] crystal, we believe that such hardening is mainly caused by the formation of Lomer-Cottrell lock.

The plateau phenomenon in the CSS curve observed in single crystals oriented for single slip has been well interpreted in term of "two phase" model (Winter, 1974). For the [034] crystal, if $\gamma_{pl}<10^{-3}$, the dislocation structure still belongs to "two phase" structure. Thus, it should be expected to observe a plateau in the CSS curve with this orientation. For the [001] crystal, dislocation structure is exclusively of labyrinth structure, even at low strain amplitudes. The "two phase" model apparently must break down. Therefore, we are not surprised to see no plateau in the CSS curve. On the other hand, it has been found that the CSS curve of the [001] crystal perfectly fits the power law function of copper polycrystals (Lukas, 1985), but corrected by the Taylor factor (M=3.06), i. e. we have $\tau=k'(\gamma_{pl})^n$, where k' and n are 146 MPa and 0.205 respectively. The curve derived from the equation is plotted by the dotted line in Fig. 4. This implies that many fatigue phenomena of polycrystals may and should be explained with those of the single crystals oriented for multiple slip.

The present study has also demonstrated that the formation of DBs is a general phenomenon in double and multiple slip oriented crystals. Such a behavior seems to be a mechanism to accommodate the severe deformation produced on different planes by more than one slip systems as detailed investigation revealed that these DBs have characteristics of kink band, i.e. the rotation of one region to the other occurred with a habit plane being ($\bar{1}01$). Kinking was usually observed in unidirectional deformation and related to geometrical softening of the active slip systems (Crocker et al, 1975). Under cyclic deformation, if strain amplitude is lower and glide is highly reversible, kinking would not form. This is the case for low amplitude fatigue of single crystal oriented for single slip. For double and multiple slip crystals, however, because of the dislocation reaction between slip systems, reversibility of glide is expected to be poor. The irreversible strain would accumulate during cyclic deformation and result eventually in the formation of DBs. It is our belief that strong DB formation would have a pronounced influence on the development of fatigue crack in these crystals.

CONCLUSION

Based on the present study, the following conclusions have been reached:
1. The cyclic hardening and softening behavior of single crystals oriented for double and multiple slip greatly differ from those of single slip oriented ones, and strongly depend on their orientations and dislocation reactions.
2. The crystal with an orientation on the $001/\bar{1}11$ boundary of the crystallographic triangle shows much higher cyclic hardening rate than that on the $001/011$ boundary.
3. The crystal with an orientation on the $001/011$ boundary shows a plateau in the CSS curve shorter than that of the single slip orientation, while one on the $00\bar{1}/\bar{1}11$ boundary displays no plateau.
4. The [001] crystal exhibit a CSS curve similar to that of polycrystals.
5. The formation of deformation bands is a general phenomenon in fatigued single crystals oriented for double and multiple slip.

ACKNOWLEDGMENT

This work is a part of the joint research project between the University of Toronto, Canada, and the Institute of Metal Research, Chinese Academy of Science, China. This project is financially supported by NSERC of Canada and NSFC of China under grant numbers of OGP 0046427 and 19392300-4, respectively.

REFERENCES

Basinski, Z. S. and S. J. Basinski (1992). Fundamental aspects of low amplitude cyclic deformation in face-centered cubic crystals. *Progress in Materials Science* ., **36**, 89-148.
Crocker, A. G. and J. S. Abell. The crystallography of deformation kinking. *Philos. Mag.*, **33**, 305-310.
Gong, B., Z. G. Wang and Y. W. Zhang (1995). The cyclic deformation behavior of Cu single crystal oriented for double slip. *Mater. Sci. Eng.*, **A194**, 171-178.
Hirth, H. P. and J. Lothe (1982).*Theory of dislocations*, second edition, Wiley, New York.
Jin, N. Y. and A. T. Winter (1984). Cyclic deformation of copper single crystals oriented for double slip. *Acta Metall.*, **32**, 989-995.
Jin, N. Y. and A. T. Winter (1984). Dislocation structure in cyclically deformed [001] copper crystals. *Acta Metall.*, **32**, 1173
Laird, C., P. Charsley and H. Mughrabi (1986). Low energy dislocation structures produced by cyclic deformation. *Mater. Sci. Eng.*, **81**, 433-450.
Lepisto, T. K. and P. O. Kettunen (1986). Comparison of the cyclic stress-strain behavior of single- and [111] multiple-slip-oriented copper single crystals. *Mater. Sci. Eng.*, **83**, 1-15.
Lukas, P. and L. Kunz (1985). Is there a plateau in the cyclic stress-strain curves of polycrystalline copper? *Mater. Sci. Eng.*, **74**, L1-L5
Mughrabi, H (1978). The cyclic hardening and saturation behavior of copper single crystals. *Mater. Sci. Eng.*, **33**, 207-223.
Winter, A. T. (1974). A model for the fatigue of copper at low plastic strain amplitude. *Philos. Mag.*, **30**, 719-38 .

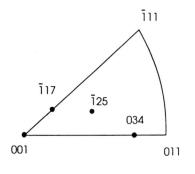

Fig. 1. Orientations of single crystals investigated.

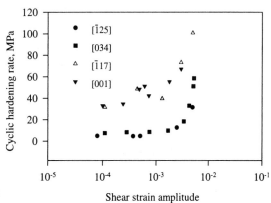

Fig. 3. Initial cyclic hardening rate ($\theta_{0.2}$) versus plastic shear strain amplitude (γ_{pl}).

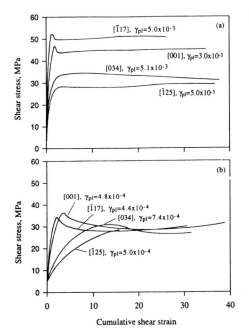

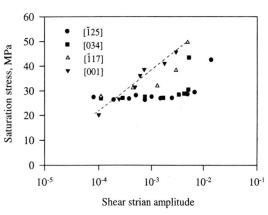

Fig. 4. Cyclic stress-strain curves of Cu single crystals with various orientations.

Fig. 2. Cyclic hardening curves of Cu single crystals with various orientations. (a) $\gamma_{pl}=3.0\times10^{-3}$-$5.1\times10^{-3}$; (b) $\gamma_{pl}=4.4\times10^{-4}$-$7.4\times10^{-4}$.

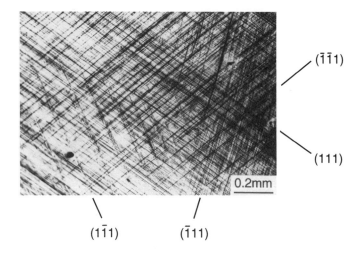

$(\bar{1}\bar{1}1)$

(111)

$(1\bar{1}1)$ $(\bar{1}11)$

Fig. 5. Multiple slip pattern formed in the [001] crystal, $\gamma_{pl}=1.8\times10^{-3}$. The corresponding slip planes are denoted in the figure.

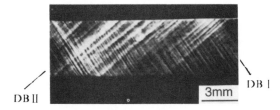

DB II   DB I

Fig. 6. Deformation bands formed in the [034] crystal, $\gamma_{pl}=3.4\times10^{-3}$.

Fig. 7. Dislocation structures in the fatigued [001] crystal, $\gamma_{pl}=4.8\times10^{-4}$. The foil plane is $(\bar{1}11)$.

A MICROSTRUCTURE - BASED APPROACH FOR PREDICTING STABILISED HYSTERESIS LOOPS

A. Schwab, C. Holste and J. Bretschneider

Technische Universität Dresden, Fakultät Mathematik und Naturwissenschaften, Institut für Physikalische Metallkunde, D-01062 Dresden, Germany

ABSTRACT

For the quantitative description of the stress-strain dependence along stabilised hysteresis loops a model is proposed which considers the heterogeneous dislocation distribution in the individual grains of cyclically deformed face- centred cubic (f.c.c.) metals. On the basis of the grain-dependent dislocation structure observed in polycrystalline nickel hysteresis loops are predicted and compared with those determined experimentally.

KEYWORDS

Fatigue, polycrystals, hysteresis loop, microstructure, dislocations.

INTRODUCTION

The study of stabilised hysteresis loops is a prerequisite for understanding of plastic behaviour of fatigued f.c.c. polycrystals during cyclic deformation, because the hysteresis loop reflects the statistical processes of the plastic deformation on all structure scale levels.

Up to now there is no satisfactory approach to calculate the stabilised hysteresis loop on the basis of microscopic glide mechanisms, although there are some concepts:

1. One group of authors (cf. for example Estrin,1991) starts from rate equations for the change of dislocation densities during cyclic deformation. The connection between structure parameters (for example dislocation densities) and mechanical values are given by simple constitutive equations. The stress- strain relation (hysteresis branch) is obtained by solving a system of differential equations. This procedure does not taken into account the heterogeneous arrangement of dislocations and the kinetic reversible dislocation movement.

2. Another group of authors (Holste et al., 1980, Polak et al., 1982, Mughrabi, 1983) considers the specimen as a composite of mesoscopic material elements. Each material element is characterized by a typical mechanical response with a critical flow stress. The macroscopic hysteresis loop is obtained by averaging the mechanical stress-strain behaviour of the individual mesoscopic elements. In these considerations the distribution function of the critical flow stresses is not expressed by microscopic parameters.

A first attempt to correlate the critical flow stress of the mesoscopic elements with geometrical parameters of the dislocation structure was done for fatigued nickel single crystals (Holste et al., 1990). The results of this investigation were in a fairly good agreement with experimental data.

Therefore, in the present work we describe tentatively stabilised hysteresis loops of f.c.c. polycrystals in terms of a composite model based on the dislocation microstructure. The heterogeneity of plastic deformation is taken into account at a mesoscopic scale (extension of dislocation dense regions and dislocation poor regions in a grain) and at a grain size scale.

THE ELASTOPLASTIC BEHAVIOUR OF AN INDIVIDUAL GRAIN

Phenomenological Relations for Calculating the Stress- Strain Relation

According to the relatively small plastic strains that occur in cyclic deformation only one slip system is assumed to be active in the saturation region.
In this case the mechanical response of an individual grain can be characterized by the relation between the resolved shear stress τ and the resolved shear strain γ in the active slip system. In order to calculate this relation for an individual grain with a heterogeneous dislocation microstructure the following assumptions were made:
1. An individual grain consists of mesoscopic volume elements with critical flow stresses τ_c which in tension and compression differ only by the sign.
Two groups of mesoscopic elements are distinguished:
a) dislocation poor regions (PRs), for example the channels in a persistent slip band (PSB)
b) dislocation dense regions (DRs) such as walls in a (PSB) or veins in the matrix structure.

2. The volume fraction dv of the mesoscopic elements with critical flow stresses within the interval $(\tau_c, \tau_c + d\tau_c)$ is given by

$$dv = \varphi(\tau_c)d\tau_c \quad ,$$

(1)

where $\varphi(\tau_c)$ is the probability density function of the critical flow stresses.

3. Each mesoscopic element has an elastic-microplastic- ideal plastic stress- strain behaviour (see Fig.1). The stress- strain relation $\tau^M(\gamma^M)$ starting from the τ^M-γ^M-origin is called the virgin curve.

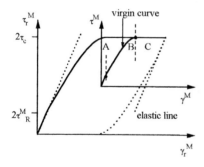

Fig. 1. Notations for description of the stress- strain relations of a mesoscopic element

After reversal of loading we assume that the stress- strain behaviour obeys the Masing- rule

$$\tau_r^M(\gamma_r^M, \tau_c) = 2\tau^M\left(\frac{\gamma_r^M}{2}, \tau_c\right)$$

(2)

where τ_r^M and γ_r^M are measured from the compression tip of the loop.

4. The deformation of the individual grain proceeds in such a way that all mesoscopic volume elements undergo the same shear strain $\gamma^M_r = \gamma_r$, where γ_r is the resolved shear strain in the grain. The total shear stress in an individual grain is given by

$$\tau_r(\gamma_r) = \int_0^\infty d\tau_c \, \varphi(\tau_c)\tau_r^M(\gamma_r, \tau_c)$$

(3)

The condition $\gamma^M_r = \gamma_r$ corresponds to a parallel circuit of all mesoscopic elements in a grain.

The Virgin Curves of the Mesoscopic Elements

Because of the different microscopic glide mechanisms the mechanical response of the PRs and DRs has to be distinguished. Consequently, the relation between the critical flow stresses τ_c and parameters which characterize the microstructure will be different for PRs and DRs, too.
To get information about these relations we calculate the virgin curves $\tau^M(\gamma^M)$ following from the glide mechanisms illustrated schematically in Fig.2.

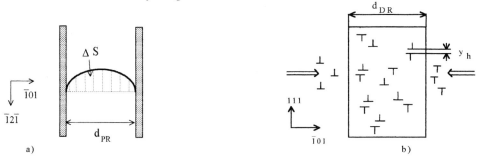

Fig. 2. Glide mechanisms in the mesoscopic material elements PR (a) and DR (b)

Dislocation Poor Regions. We assume that the microplastic deformation of a PR is produced by the undercritical bowing- out of screw dislocations (Fig. 2a).
The quantitative description of this process starts from the Orowan- equation $\gamma_p = b\rho_s\,\Delta S/d_{PR}$, where γ_p is the plastic shear strain, b the strength of the Burgers vector, ρ_s the screw dislocation density, ΔS the area covered by the dislocation during the glide motion and d_{PR} the distance of two neighbouring DRs measured in slip direction.
The inverse function of the virgin curve $\tau^{PR}(\gamma^{PR})$ is given by

$$\gamma^{PR}(\tau^{PR}) = \frac{\tau^{PR}}{G} + b\rho_s\frac{\Delta S(\tau^{PR})}{d_{PR}}, \tag{4}$$

with G as the shear modulus.
Assuming a circular curvature and using the line energy approximation with the line tension $T_s=\alpha Gb^2/2$ ($\alpha=1,5$) the function ΔS can be determined easily (Kocks et al., 1975).
The critical flow stress

$$\tau_c(d_{PR}) = \tau_R^{PR} + \frac{2T_s}{bd_{PR}} \tag{5}$$

will be attained in the case that the bowing- out of screw dislocations becomes overcriticaly. τ_R^{PR} denotes a friction stress.
Calculated virgin curves of PRs with different extensions d_{PR} are illustrated in Fig.3a.

Dislocation Dense Regions. The microplastic strain of the DRs is due to the polarisation of the edge dislocation dipoles and mobile edge dislocations penetrating into these regions (Fig.2 b).
Starting from the Orowan-equation for the plastic shear strain the virgin curve of the DRs is given by

$$\gamma^{DR}(\tau^{DR}) = \frac{\tau^{DR}}{G} + b\rho_d\left(\frac{\Delta x_v(\tau^{DR},y_h)+\Delta x_i(\tau^{DR},y_h)}{2}\right) + b\rho_m\,\bar{l}\,(\tau^{DR}) \tag{6}$$

with the dipole density ρ_d and the density of mobile edge dislocations ρ_m. The symbols Δx_v and Δx_i in (6) describe the displacement of the edge dislocations under the stress τ^{DR} in a vacancy and an interstitial dipole of the dipole width y_h. The functions $\Delta x_v(\tau^{DR}, y_h)$ and $\Delta x_i(\tau^{DR}, y_h)$ can be determined considering the equilibrium of forces acting on the dipole dislocations, assuming a quasistatic shift of the edge dislocations.

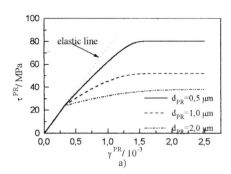
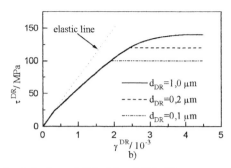

Fig. 3. Calculated virgin curves for the mesoscopic deformation behaviour of PRs (a) and DRs (b); as constants and parameters are used : G=75 GPa, v=1/3, b=0,25 nm, $\rho_s=10^{13}$ m^{-2}, $\rho_d=10^{15}$ m^{-2}, $\rho_m=10^{13}$ m^{-2}, $y_h=7$nm, $\tau_R^{PR}=\tau_R^{DR}=\tau_R=24$ MPa; The friction stress τ_R is taken from the macroscopic hysteresis loop.

The length $\bar{l}$ in (6) is the mean free path of mobile dislocations in the DRs. It follows from the interaction between the mobile dislocations and the dislocation dipoles. The mean free path $\bar{l}$ is given by

$$\bar{l}(\tau^{DR}) = \frac{1}{\rho_d \ q(\tau^{DR})} \ , \tag{7}$$

where q acts as a stress dependent capture length of the moving dislocations.
Computer simulations of the interaction between a mobile dislocation and a dislocation dipole (Neumann, 1971) show, that several reactions are possible: passing, decomposition, dipole change and equilibrium. In the case of neglecting the annihilation process the only dislocation reaction that restricts the path of a mobile dislocation is the equilibrium. In this case the capture length q can be found from Neumann plots (Neumann, 1971) as

$$q(\tau^{DR}) = \left(\frac{y_h}{0,163}\right) \ln\left(\frac{0,76Gb}{8\pi(1-v) \ y_h \ (\tau^{DR}-\tau_R^{DR})}\right) \ , \tag{8}$$

where τ_R^{DR} denotes the friction stress in the DRs and v is Poisson's ratio.
Ideal plastic deformation is assumed to set in when the mean free path $\bar{l}$ becomes equal to the half of the distance d_{DR} (yield criterion).
For the critical flow stress follow from the yield criterion with (7) and (8)

$$\tau_c(d_{DR}) = \tau_R^{DR} + \frac{0,76Gb}{8\pi(1-v) \ y_h} \exp\left(-\frac{0,326}{y_h \ \rho_d \ d_{DR}}\right) \ . \tag{9}$$

Typical virgin curves for the DRs are shown in Fig. 3b.

Hysteresis Branch of an Individual Grain

Knowing the virgin curves $\tau^M(\gamma^M)$ for the PRs and DRs in the activated slip system, the hysteresis branch τ_r (γ_r) for this glide system can be calculated with (2) and (3) provided the distribution function for the critical flow stresses $\varphi(\tau_c)$ has been determined for the considered grain.
The essential point of the modelling procedure supposed in this work is that the density function $\varphi(\tau_c)$ is correlated with the distribution frequency of the extensions of the PRs and DRs in glide direction. For nickel polycrystals cyclically deformed at room temperature distribution functions of the extensions d_{PR} and d_{DR} of PRs and DRs were measured in an extensive experimental work (Holste et al. , 1995) and the corresponding $\varphi(\tau_c)$ were calculated with (5) and (9).

For a given grain orientation we get the stress- strain relation $\sigma^i_r(\varepsilon^i_r)$ of an individual grain (i) in axial co-ordinates from

$$\varepsilon^i_r(\tau_r) = \mu^i\gamma^i_r(\tau^i_r) + \tau^i_r(\frac{1}{\mu^i E^i} - \frac{\mu^i}{G}) \ and \ \tau^i_r = \sigma^i_r \mu^i, \tag{10}$$

with the Schmid-factor μ^i and Young's modulus E^i.

THE MACROSCOPIC HYSTERESIS LOOP OF A POLYCRYSTAL

TEM-investigations of the saturation microstructure show that the dislocation arrangements in polycrystalline nickel can be classified into a few different structure types (Fig. 4).

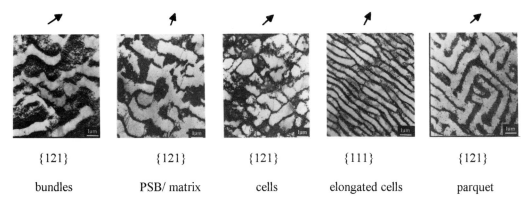

{121}	{121}	{121}	{111}	{121}
bundles	PSB/ matrix	cells	elongated cells	parquet

Fig. 4 . Types of dislocation microstructures in cyclic saturation of nickel at room temperature; the arrows mark the glide directions

Basically we assume that the observed structure reflects the local deformation behaviour resulting from the intrinsic properties of the polycrystalline aggregates (texture, etc.) and from interaction processes between the grains. All grains with the same structure type were considered as a grain group (j). Each grain of a grain group (j) is supposed to be characterized by the same stress-strain relation $\sigma^j_r(\varepsilon^j_r)$.
The orientations of all individual grains in representative macroscopic volumes of the investigated specimens have not yet been mapped. We therefore use in (10) a mean Schmid-factor $\mu_s=0,446$ and the isotropic Young's modulus E=200 GPa (nickel) for all grains.
The macroscopic hysteresis loop is obtained by averaging the hysteresis loops of the grain groups (j), taking into account the volume fractions f^j of a grain group. Supposing a homogeneous axial total strain $\varepsilon^j_r = \varepsilon_r$ on the grain size scale, the macroscopic hysteresis branch $\sigma_r(\varepsilon_r)$ is given by

$$\sigma_r(\varepsilon_r) = \sum_j f^j\sigma^j_r(\varepsilon_r) \ . \tag{11}$$

The volume fractions f^j depend characteristically on the macroscopic plastic strain amplitude ε_{pa}.
In Fig. 5 hysteresis branches are represented for grain groups with different structure types observed in a specimen cyclically deformed at $\varepsilon_{pa}= 1,35 \ 10^{-3}$. For calculating the macroscopic hysteresis branch also shown in Fig.5 the volume fractions f^j in (11) are taken as fitting parameters, because sufficient experimental data are not available, up to now. As can be seen, a good agreement can be obtained between the calculated and the experimentally determined branches.

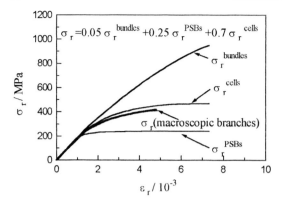

Fig. 5. Hysteresis branches for a plastic strain amplitude $\varepsilon_{pa} = 1,35 \cdot 10^{-3}$; $\sigma_r^{bundles}$, σ_r^{cells} and σ_r^{PSB} mark the grain
hysteresis loops with a bundle , cell and PSB structure, respectively; the macroscopic hysteresis branch
calculated from (11) coincides at the given stress-strain resolution with the experimentally determined
branch

RESULTS AND CONCLUSIONS

The results show that a surprising good agreement can be attained between a predicted and an experimentally
observed loop branch up to plastic strains about $2*10^{-3}$, measured from the tip of the hysteresis loop. For plastic
strains greater than $2*10^{-3}$ significant deviations occur, which point to the fact that the elastic-plastic transition
during the loading cycle is finished and the plastic deformation becomes homogeneous on a mesoscopic scale.
Because the volume fractions f are used as free fitting parameters the agreement between the calculated and the
experimentally observed hysteresis loops is not surprising. In a future work the modelling procedure has to be
proofed by taking into account experimentally determined values of f.

REFERENCES

Estrin, Y. (1991). A versatile unified constitutive model based on dislocation density evolution. In:High
    temperature constitutive modelling, ASME, Atlanta.
Holste, C. and Burmeister, H.-J. (1980). Long -Range Stresses in Cyclic Deformation. Phys. stat. sol. (a), 57,
    269-280.
Holste, C., Mecke, K. and Kleinert, W. (1990). Heterogeneous plastic deformation in fatigued pure f.c.c. metals.
    Physical Research, Akademie-Verlag Berlin, 14, 305-313.
Holste, C., Gürth, R. and Schwab, A.(1995). Wechselverformungsverhalten und Mikrostruktur von Nickel,
    Abschlußbericht (Teil B) eines von der DFG geförderten Forschungsvorhabens.
Kocks, U.F., Argon, A.S., Ashby, M.F. (1975). Thermodynamics and Kinetics of slip. In: Progress in materials
    science (B. Chalmers, J.W. Christian, T.B. Massalski, ed.), Vol.19, pp.48-53, Pergamon Press, Oxford, New
    York, Toronto, Sydney, Braunschweig.
Mughrabi, H. (1983). Dislocation wall and cell structures and long range internal stresses in deformed metal
    crystals. Acta metall., 31, 1367-1379.
Neumann, P. D. (1971). The interaction between dislocations and dislocation dipoles. Acta metall., 19 ,
    1233-1241.
Polak, J. and Klesnil, M. (1982). The hysteresis loop, a statistical theory. Fat. Engng. Mater. Strukt. , 5, 19-32.

LOW CYCLE FATIGUE AND INTERNAL STRESS MEASUREMENTS OF COPPER

K.-T. RIE and H. WITTKE

Institut für Oberflächentechnik und plasmatechnische Werkstoffentwicklung,
Technische Universität Braunschweig, Germany

ABSTRACT

By means of stress relaxation tests for cylically deformed copper at room temperature internal stresses were measured. Cyclic stress-strain curves of total and internal stress amplitude vs. plastic strain amplitude were established. Hysteresis-loops of internal stress versus plastic strain are described with a new proposed relation. These hysteresis-loops show non-Masing behaviour. A microstructural interpretation of this behaviour is given.

KEYWORDS

Internal and effective stresses; Masing and non-Masing behaviour; dislocation structure.

INTERNAL STRESS MEASUREMENTS BY STRESS RELAXATION TESTS

Internal stresses can be obtained experimentally by stress relaxation tests, stress dip tests and strain rate change tests. In this work stress relaxation tests at room temperature are performed using polycrystalline copper which was cycled to approximated saturation in uniaxial push-pull tests in the range of Low Cycle Fatigue (LCF) prior to the relaxation tests. The LCF-tests were performed as single step tests (SSTs) and two step tests (2STs) which were equivalent to the SSTs. Two different strain rates were used. In agreement with the method of Tsou and Quesnel (1982) we adopted the stress value after 30 minutes of relaxation as the internal stress.

Figure 1 shows hysteresis-loops with both the applied or total stress σ and the internal stress σ_i plotted vs. the plastic strain ϵ_p. The plastic strain is calculated according to the well known equation

$$\epsilon_p = \epsilon - \sigma/E \qquad (1)$$

with the total strain ϵ and the Young's Modulus E (here: $E=116$ GPa).

By considering different hysteresis-loops cyclic stress-strain-curves (css-curves) can be determined. The css-curves for the amplitudes of total and internal stress vs. the amplitude of plastic strain are shown in Fig. 2.

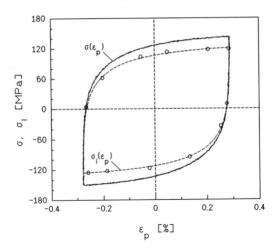

Fig. 1.   Hysteresis-loops: total and internal stress vs. plastic strain.
               $\Delta\epsilon/2 = 0.4\ \%$, $\dot{\epsilon} = 10^{-3}\ s^{-1}$;
               o: experimental data of stress relaxation tests,
               ------:   calculation analogous to eq. (2).

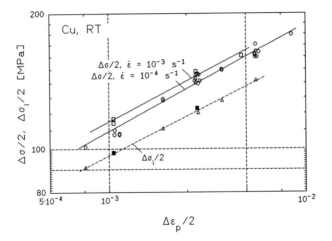

Fig. 2.   Cyclic stress-strain curves: amplitudes of applied and internal stress vs.
               amplitude of plastic strain.
               □:   data of SSTs with $\dot{\epsilon} = 10^{-3}\ s^{-1}$;
               o: data of SSTs and 2STs with $\dot{\epsilon} = 10^{-4}\ s^{-1}$;
               ■,▲:   data of stress relaxation tests after SSTs with
                      $\dot{\epsilon} = 10^{-3}\ s^{-1}$ or $\dot{\epsilon} = 10^{-4}\ s^{-1}$, respectivly.

It can be seen that the decreasement of strain rate leads to lower amplitudes of the total stress. In contrast, the amplitudes of the internal stress are nearly independent on the strain rate. This behaviour was also observed for aluminium by Tsou and Quesnel (1982) and was postulated for copper e.g. in the constitutive equations of Hatanaka and Ishimoto (1991). With the same amplitude of the plastic strain, also the shape of the σ_i-ϵ_p-hysteresis-loop is assumed to be independent on strain rate.

Another remarkable result is that the different css-curves in Fig. 2 are nearly parallel.

ANALYTICAL DESCRIPTION OF HYSTERESIS-LOOPS

The dotted line which fits the experimental points of the σ_i-ϵ_p-hysteresis-loop shown in Fig. 1 was calculated by one of two new proposed relations. These relations will be explained in the following. At first, some basic facts about the shape of the stress-strain hysteresis-loop will be described: After the transformation of the σ_i-ϵ_p-hysteresis-loop into a relative coordinate system with the origin in its lower tip, the ascending branch of the hysteresis-loop can be approximated e.g. by a power law. The power law relation corresponds in a double-logarithmic plot to a straight line. Especially in the range of medium and high strain amplitudes a nonlinear curve can be observed (Holste et al., 1987; Rie et al., 1992).

A better description can be achieved by a stretched exponential function

$$\sigma_r = A_G \cdot [1 - \exp(-(\epsilon_{pr}/\delta_G)^{\alpha_G})] \tag{2}$$

with the constants A_G, δ_G and α_G. The importance of this relation can be emphasized by our observation that the cyclic hardening curve - i.e. strain range $\Delta\sigma$ vs. cycle number N or $\Delta\sigma$ vs. $\epsilon_{p,cum}=2N\cdot\Delta\epsilon_p$, respectivly - can be well described with the aid of a stretched exponential function.

Alternativly another approach is suggested: For the consideration of this nonlinear behaviour, the log-log-curve can be approximated by a second-order polynomial. This proposal leads to the relation

$$\sigma_r = C_q \cdot \exp(-\kappa_q[\ln(\epsilon_{pr}/\delta_E)]^2) \tag{3}$$

with the constants C_q, κ_q and δ_E.

It can be concluded that for values higher then $\epsilon_{pr} \approx 2 \cdot 10^{-5}$ a very exact description of the stress-strain path was achieved with both relations.

The value of the relative plastic strain ϵ_{pr} may be calculated by means of the relative part of the total strain ϵ_r and the stress σ_r similar to the method given by eq. (1). In this case the value for E has to be determined by the slope of the point of the stress-strain-hysteresis-loop which corresponds to the tip of the σ_i-ϵ_p-hysteresis-loop.

The relations given in equations (2) and (3) are used in this work to describe σ_i-ϵ_p-hysteresis-loops. Therefore, in eq. (2) and eq. (3) we replace the relative stress σ_r by the relative internal stress σ_{ir}. For the dotted-line hysteresis-loop in Fig. 1 which is calculated by eq. (2) the values of the parameter are $A_G=260$ MPa, $\delta_G=0.039$ % and $\alpha_G=0.412$. A good agreement between experiment and calculation can be seen. When this hysteresis-loop is described with the aid of eq. (3) the parameters $C_q=256.3$ MPa, $\kappa_q=0.0282$ and $\delta_E=1.777$ % are found. The good agreement between experiment and calculation in this case is demonstrated by the hysteresis-loops in Fig. 3, which have been calculated with the aid of eq. (3).

In Fig. 3 a) σ_i-ϵ_p-hysteresis-loops for a small, a medium and a large strain amplitude of the LCF range are shown. Because the proposed relation (eq. (3)) is very appropriate to fit the experimental points, this applied relation is used in the following for verifying Masing or non-Masing-behaviour.

a)   b)

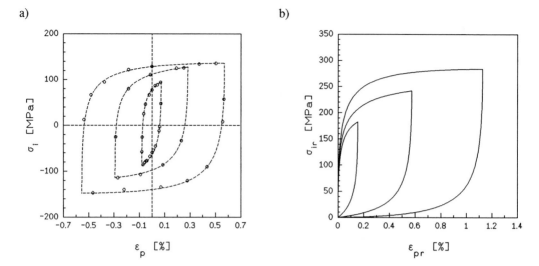

Fig. 3.   Copper, room temperature, a) hysteresis-loops of internal stress vs. plastic
               strain;
               ∘ : experimental values; ------ : calculated curve, calculation by eq. (3);
               b) hysteresis-loops of relativ internal stress vs. relativ plastic strain (without
               experimental values).
               Parameters of the former performed single step tests: strain rate $\dot\epsilon = 10^{-4} \ s^{-1}$;
               strain amplitudes $\Delta\epsilon/2$: 0.16 %, 0.4 % and 0.7%.

NON-MASING BEHAVIOUR OF THE MATERIAL

A material shows Masing behaviour when the ascending (descending) branches of hysteresis-loops
of different strain amplitudes in a relative coordinate system with the origin in the lower (upper) tip
follow a common curve. In the other case we have non-Masing behaviour. Recent results (Bayerlein
et al., 1987; Schubert, 1989) indicate that the stress-strain hysteresis-loops of copper show non-
Masing behaviour in SSTs at room temperature.

To investigate the Masing or non-Masing behaviour of the σ_i-ϵ_p-hysteresis-loops from Fig. 3 a) the
hysteresis-loops are presented in Fig. 3 b) in relative coordinates. Non-Masing behaviour can be seen
clearly.

Cyclic loading in SSTs in the range of LCF leads to the formation of a characteristic dislocation
arrangement or dislocation structure which depends on strain amplitude (see e.g. Feltner and Laird,
1962). It is assumed that these different dislocation structures are responsible for the non-Masing-
behaviour (Schubert, 1989; Christ and Mughrabi, 1992; Rie et al., 1992). A modell based on Orowan
mechanism is proposed e.g. by Schubert (1989) which correlates the elastic stress range with the
reciprocal value of the mean wall distance d_m of the dislocation cells or a similar parameter of the
dislocation structure.

The effective stress $\sigma_e = \sigma - \sigma_i$ is that fraction of the total stress causing dislocations to move at a
specific velocity. The internal stress can be defined as the stress needed to balance the dislocation
configuration at a net zero value of the plastic strain rate (Tsou and Quesnel, 1982). In agreement

with this definition, the dislocation arrangement or the dislocation structure may be represented by characteristic values of the internal stress path instead of the effective or total stress path.

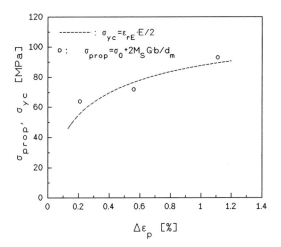

Fig. 4.   Cyclic propotional limits, σ_{prop} and σ_{yc}, in dependence of plastic strain range. The values of σ_{prop} are given by Schubert (1989); the values of σ_{yc} are determined with the help of the σ_i-ϵ_p-hysteresis-loops and described by the dotted line fit-function.

The effective stresses may also contribute to the non-Masing behaviour. But in agreement with the above mentioned model the main reason of the non-Masing behaviour is concerned with the internal stresses. Therefore, in this paper, the verification of the Masing or non-Masing behaviour is performed by means of the internal stresses instead of the total stresses usually adopted.

Now the relation between non-Masing behaviour and microstructural behaviour will be quantified with the model of Schubert (1989). In this model a microstructure dependent cyclic proportional limit

$$\sigma_{prop} = \sigma_0 + 2 \cdot M_S \cdot G \cdot b / d_m \tag{4}$$

is proposed (σ_0: lattice friction stress; M_S: Sachs-factor; G: shear modulus; b: absolut value of Burgers vector). The calculated values of σ_{prop} show good agreement with a cyclic proportional limit which is defined by a strain offset of 0.01 %. But this strain offset is arbitrary and has no physical meaning. Therefore, in the case of our σ_{ir}-ϵ_{pr}-hysteresis-loops, another way is choosen: At first, a hypothetic hysteresis-loop σ_{ir} vs. ϵ_r is constructed with the given values of σ_{ir}, ϵ_{pr} and the relation

$$\epsilon_r = \epsilon_{pr} + \sigma_{ir} / E \tag{5}$$

analogous to eq. (1). In the next step, the second deviation of a half branch of this hysteresis-loop, $d^2\sigma_{ir}/d\epsilon_r^2$ vs. ϵ_r, is constructed. The ϵ_r-value of the extrem point of this second deviation is called here $\epsilon_{r,ex}$. Alternativly to the definition given by eq. (4) a cyclic proportional limit σ_{yc} is defined now as

$$\sigma_{yc} = \epsilon_{r,ex} \cdot E / 2 \tag{6}$$

(compare Polák and Klesnil, 1982; Polák *et al.*, 1982). By this definition an uniquely applicable and physically better justified cyclic proportional limit is found.

With the values of d_m and σ_{prop} given by Schubert (1989) or Rie *et al.* (1992), respectivly, the good agreement between the two cyclic proportional limits, σ_{prop} and σ_{yc}, can be seen in Fig. 4.

SUMMARY

-   Analytical relations are proposed which give good description of hysteresis-loops σ_r vs. ϵ_{pr} and σ_{ir} vs. ϵ_{pr}.
-   The hysteresis-loops σ_{ir} vs. ϵ_{pr} show non-Masing behaviour
-   The non-Masing behaviour of copper is explained by a microstructural model.

ACKNOWLEDGEMENT

The presented work was supported by the Deutsche Forschungsgemeinschaft (DFG) in the frame of the SFB 319 "Stoffgesetze für das inelastische Verhalten metallischer Werkstoffe".

REFERENCES

Bayerlein, M.; Christ, H.-J.; Mughrabi, H. (1987): A Critical Evaluation of the Incremental Step Test. In: *Low Cycle Fatigue and Elasto-Plastic Behaviour of Materials* (K.-T. Rie, Ed.), Elsevier Applied Science, pp. 149-154.

Christ, H.-J.; Mughrabi, H. (1992): Microstructure and Fatigue; In: *Low Cycle Fatigue and Elasto-Plastic Behaviour of Materials - 3* (K.-T. Rie, Ed.), Elsevier Applied Science, pp. 56-69.

Feltner, C. E.; Laird, C. (1962): Cyclic Stress-Strain Response of F.C.C. Metalls and Alloys - II: Dislocations Structures and Mechanisms; *Acta Metallurgica, Vol. 15*, pp. 1633-1653.

Hatanaka, K.; Ishimoto, Y. (1991): A Numerical Analysis of Cyclic Stress-Strain-Response in Terms of Dislocation Motion in Copper; In: *Residual Stresses - III* (H. Fujiwara, T. Abe, K. Tanaka, Eds.), Elsevier Applied Science, pp. 549-554.

Holste, C.; Kleinert, W; Fischer, W. (1987): Deformation Stages in a Stabilized Loading Cycle of Fatigued FCC Metals (I); *Cryst. Res. Technol. 22*, pp. 419-427.

Polák, J.; Klesnil, M. (1982): The Hysteresis Loop: 1. A Statistical Theory. *Fatigue of Engineering Materials and Structures, Vol. 5, No. 1*, pp. 19-32.

Polák, J.; Klesnil, M.; Helešic, J. (1982): The Hysteresis Loop: 2. An Analysis of the Loop Shape. *Fatigue of Engineering Materials and Structures, Vol. 5, No. 1*, pp. 33-44.

Rie, K.-T.; Wittke, H.; Schubert, R. (1992): The ΔJ-Integral and the Relation between Deformation Behaviour and Microstructure in the LCF-Range; In: *Low Cycle Fatigue and Elasto-Plastic Behaviour of Materials - 3* (K.-T. Rie, Ed.), Elsevier Applied Science, pp. 514-520.

Schubert, R. (1989): Verformungsverhalten und Rißwachstum bei Low Cycle Fatigue; In: *Fortschrittsber. VDI, Reihe 18, No. 73*, VDI Verlag, Düsseldorf.

Tsou, J. C.; Quesnel, D. J. (1982): Internal Stress Measurements during the Saturation Fatigue of Polycrystalline Aluminium; *Mat. Sci. Eng., 56*, pp. 289-299.

CYCLIC RESPONSE OF DUPLEX STAINLESS STEELS: AN INTRINSIC TWO-PHASE DESCRIPTION

L. LLANES, A. MATEO and M. ANGLADA

Departamento de Ciencia de los Materiales e Ingeniería Metalúrgica, ETSEIB,
Universitat Politècnica de Catalunya, 08028 Barcelona, Spain

ABSTRACT

In this work a two-phase description of the cyclic behavior of duplex stainless steels (DSS) is discussed. Based on previous experimental work, the cyclic stress-strain response of DSS is classified as consisting of three regimes, depending upon the imposed strain. At low and high plastic strain amplitudes the measured response is amenable of being described in terms of single-phase behaviors. In the intermediate stage such simple approach is not suitable and therefore physically-based tools are recalled in order to validate estimations for the intrinsic response of each phase. These approaches include documentation on thermal aging and substructural evolution within the individual constituents.

KEYWORDS

Two-phase microstructure, cyclic deformation, physically-based description, duplex stainless steel.

INTRODUCTION

There is a large number of technically important materials which are characterized for having a dual or duplex microstructure, i.e. consisting of two phases present in relatively large amounts. These microstructures generally offer useful combination of mechanical properties, mainly as a consequence of the synergistic interaction between the individual responses of their constituents.

Duplex stainless steels (DSS) are a good example of these two-phase alloys. By adjusting the chemical composition in stainless steels, a duplex (austenite plus ferrite) microstructure may be produced which is stable at room temperature. Because their naturally strong interface between the phases, they may be considered as an ideal composite, and therefore as a model system for studying the intrinsic behavior of two-phase alloys.

DSS have been studied extensively in terms of their superplastic, tensile and stress corrosion properties (e.g. Solomon and Devine Jr., 1983). However, reports on the fatigue behavior of DSS are relatively scarce, specially within the cyclic deformation field. Recently, the authors have addressed this issue through a well-documented study on the cyclic behavior and the associated substructure evolution of DSS (Mateo et al., 1995a; Llanes et al., 1995). Based on these results, it is the aim of this work to provide a physical description of the cyclic stress-strain response of DSS.

87

CYCLIC RESPONSE AND SUBSTRUCTURE EVOLUTION OF DSS

The cyclic behavior of DSS was investigated in an AISI-329 type DSS (38vol.% γ - 62vol.% α). Fatigue testing was conducted under symmetrical uniaxial push-pull mode, total strain control and constant total strain rate. The individual substructure evolution in each constitutive phase, as a function of applied strain amplitude, was carefully documented through extensive transmission electron microscopy (TEM) studies. A detailed description of the experimental procedure used for determining the hardening/softening behavior, the cyclic stress-strain response as well as for the TEM studies has been given elsewhere (Mateo *et al.*, 1995a).

For the selected material, it was found that there is a tidy correlation between the cyclic response, in terms of both hardening-softening behavior and the cyclic stress-strain curve (CSSC), and the substructure evolution observed in each of the constitutive phases. In general, the cyclic behavior of DSS may be described in terms of three different regimes with increasing plastic strain amplitude (ε_{pl}). The measured CSSC, Fig. 1, is a nice example of such qualitative response. At low ε_{pl}, stage I, the deformation mechanisms are mainly associated with planar glide of dislocations within the austenite which is the phase that carries a large amount of the imposed strain. This initial stage is characterized, within the CSSC, by a slow rise of the saturation stress with increasing strain amplitude. At large ε_{pl} deformation is, on the other hand, concentrated within the ferritic matrix, as a result of a clear substructural evolution from vein into wall and cell structures. The development of wall structure enhances strain localization and therefore the cyclic strain hardening rate is also low in this regime III. At ε_{pl} in between, stage II, substructural changes are observed in both phases and therefore a mixed "ferritic/austenitic-like" behavior is recalled. The intermediate stage is characterized, within the CSSC, by a cyclic strain hardening rate higher than those found at small and large ε_{pl}. Hence, from a description viewpoint, a simple mixture-law based on the observed response of austenite in the first stage and ferrite in the third one may not be used. It is this intermediate regime the one where: 1) a larger interaction between the two phases is found, and 2) an intrinsic description, involving the individual strain accommodation within each phase, is more difficult to establish.

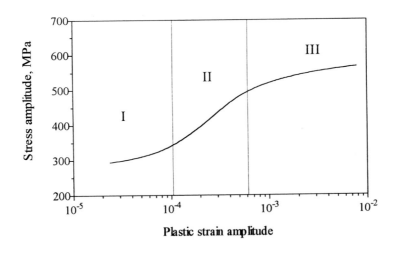

Fig. 1. CSSC of the DSS studied.

INTRINSIC BEHAVIOR OF THE INDIVIDUAL CONSTITUENTS

Based on the substructural features observed at small and large ε_{pl}, the response measured in stages I and III may be described in terms of that previously reported for single-phase austenitic (Zhong *et al.*, 1990; Li, 1991) and single-phase ferritic (Magnin and Driver, 1979; Anglada *et al.*, 1987) stainless steels respectively. Hence, the estimation of the cyclic response of DSS, within these regimes, may be established using cyclic stress-strain data from comparative single-phase materials. Mateo *et al.* (1995a, 1995b), taking into account the volume fraction parameter and considering the data available from the literature, have shown a relatively good agreement when such a comparison is made.

In order to speculate about the "intrinsic" contribution of each phase in the intermediate regime, additional experimental information seems to be required. Concerning the austenitic phase such data may be obtained from the cyclic response of DSS after being thermally embrittled (Llanes *et al.*, 1995). DSS are susceptible to severe embrittlement, usually known as "475 °C embrittlement", when exposed to the so-called intermediate temperature range (250-500 °C). It is well-known that such thermal aging promotes an increase in the tensile properties of the material together with a decrease in ductility and toughness (e.g. Chopra and Chung, 1988; Iturgoyen *et al.*, 1994). The embrittlement is the result of several microstructural changes that proceed within the ferritic constituent.

In studying the aging effects on the cyclic response of DSS, Llanes *et al.* (1995) have reported pronounced changes, when compared to that of unaged material, at ε_{pl} corresponding to the intermediate stage. As shown in Fig. 2, the stress level and the cyclic strain hardening rate are much higher, in stage II, for the aged material (at 475 °C for 200 hours) than for the unaged one. Furthermore, a detailed TEM study indicated that ferrite is plastically passive and austenite is the only phase that carries a substantial amount of plastic deformation within this stage. This is completely different from what it is observed in the unaged material. Therefore, the intrinsic response of austenite could be estimated from that measured for the aged material. However, this is not obvious and special considerations must be taken.

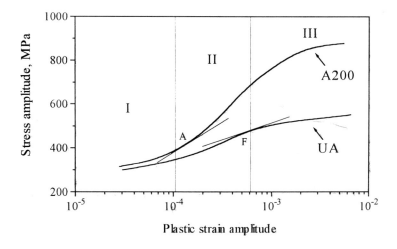

Fig. 2. CSSC of the unaged (UA) and aged (A200) DSS studied.

It is evident that the plastically deforming austenite, within the intermediate stage, is different in both materials. In the unaged DSS, austenite deforms within a matrix that is also plastically deforming. On the other hand, in the aged material austenite deforms under constraining conditions, those imposed for the surrounding brittle matrix. Therefore, a simple estimation based on volume fraction, for a given ε_{pl}, does not seem to be valid. Hence, a more physical approach must be used in order to define the appropriate individual cyclic strain hardening rates.

Figure 3 is a micrograph showing planar arrays of dislocations in deformed austenitic phase (intermediate regime) of unaged DSS. In this phase, with increasing applied strain amplitude there is not a real substructural evolution but rather structural changes of a still planar glide mode. On the other hand, in the aged material the structural features observed in austenite during stage II are mainly correlated to a wavy glide mode, i.e. cell structure (Fig. 4). In this material, a substructural scenario characterized by planar glide, similar to that observed in unaged material, is only found during stage I and in the begining of stage II (about A in Fig. 2). Hence, it is proposed that the slope in the transition between these stages is the valid one for estimating the intrinsic cyclic hardening rate of austenite in the unaged material during the intermediate regime.

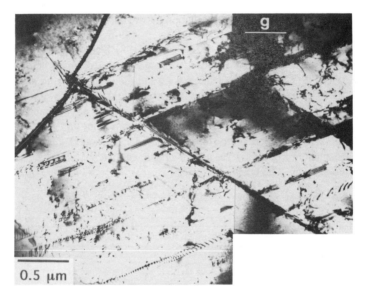

Fig. 3. Unaged DSS, austenite: planar glide mode in the intermediate stage, $\mathbf{g} = [1\bar{1}1]$.

As ferrite concerns, there is not a similar feasible treatment for inhibiting austenite's plastic activity at low and intermediate ε_{pl}. Therefore, the intrinsic response of ferrite within stage II must be estimated considering physical observations exclusively. In this second regime, ferrite substructural evolution involves the development of low density and interconnected loop patches into vein and wall structures (Fig. 5). As the material goes into the final regime, evolution into cell structure is also observed. The physical meaning of these findings is that the intrinsic cyclic strain hardening rate of the ferritic phase becomes lower and lower, with increasing imposed strain amplitude, as a consequence of strain localization. Thus, it is proposed that a slope associated with the transition regime between stage II and stage III (about F in Fig. 2) may be a suitable one for describing the intrinsic cyclic hardening rate of ferrite in the unaged material during the intermediate stage.

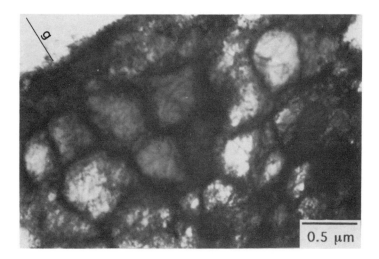

Fig. 4. Aged DSS, austenite: wavy slip mode in the intermediate regime, $\mathbf{g}$ = [200].

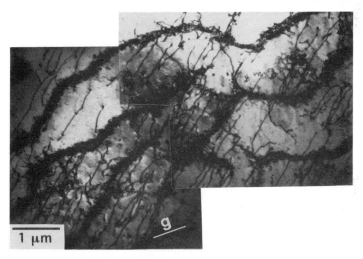

Fig. 5. Unaged DSS, ferrite: developing dislocation-free channels in stage II, $\mathbf{g}$ = [011].

SUMMARY: A TWO-PHASE DESCRIPTION

The ideas above introduced now allow us to describe the whole cyclic stress-strain response of DSS from a physical viewpoint. The austenitic phase, initially softer than the surrounding ferritic matrix, accommodates most of the imposed plastic strain during stage I. In this first regime, the behavior may then be expressed as "austenitic-like" and the cyclic strain hardening rate with it associated is qualitatively similar, after considering the volume fraction parameter, to that displayed by single-phase austenitic stainless steels.

As strain is increased austenite cyclically hardens at a rate similar to that measured during the transition from stage I to stage II in the corresponding aged DSS. This promotes ferrite's plastic activation and indicates the beginning of the intermediate stage. In this regime, the substructural development observed within ferrite implies strain localization and therefore a decreasing cyclic hardening rate. Such behavior may be correlated to that exhibited during the transition from stage II to III in the unaged material. Once plastic strain starts to get accommodated in the ferritic matrix high order substructural evolution is inhibited in austenite, phase which then only experiences substructural changes, within a planar glide mode, at a constant cyclic strain hardening rate.

Finally, in stage III ferrite is the dominant phase and it translates, with increasing imposed strain, into a further evolution to cell structure. In this final regime a good agreement is also found when comparing the behavior observed to that previously reported for single-phase ferritic stainless steels.

Hence, physically-based approaches are shown to be useful tools in order to describe the cyclic response of two-phase alloys, mainly in the regime where the individual behaviors of the constituents are strongly dependent upon each other. Because the model character of DSS, the behavior here described, as well as the approaches followed, are expected to be general for other nature composites where both phases are able to sustain plastic deformation simultaneously.

ACKNOWLEDGMENTS

The support of the European Coal and Steel Community (Contract N° 7210-MA/940) and the Spanish "Comisión Interministerial de Ciencia y Tecnología" (Grant N° MT-1497) is gratefully acknowledged.

REFERENCES

Anglada, M., M. Nasarre and J. A. Planell (1987). Mechanical behavior of a 29Cr-4Mo-2Ni superferritic stainless steel. In: *Stainless Steel '87*, pp. 474-80, The Institute of Metals, Brookfield.

Chopra, O. K. and H. M. Chung (1988). Aging degradation of cast stainless steels: effects on mechanical properties. In: *Environmental Degradation of Materials in Nuclear Power System-Water Reactors* (G. J. Theus and J. R. Weeks, eds.) pp. 737-48, TMS, Warrandale.

Iturgoyen, L., A. Mateo, L. Llanes and M. Anglada (1994). Thermal embrittlement at intermediate temperatures of AISI 329 duplex stainless steel. In: *Materials for Advanced Power Engineering* (D. Coutsouradis *et al.*, eds.) pp. 505-14, Kluwer Academic Publishers, Dordrecht.

Li, Y. -F. (1991). Cyclic response - electrochemical interaction in mono- and polycrystalline AISI 316L stainless steel. Ph. D. Thesis, University of Pennsylvania, Philadelphia.

Llanes, L., A. Mateo, L. Iturgoyen and M. Anglada (1995). Aging effects on the cyclic deformation mechanisms of a duplex stainless steel, submitted for publication.

Magnin, T. and J. H. Driver (1979). The influence of strain rate on the low cycle fatigue properties of single crystals and polycrystals of two ferritic alloys. *Mater. Sci. Eng., 39*, 175-85.

Mateo, A., L. Llanes, L. Iturgoyen and M. Anglada (1995a). Cyclic stress-strain response and dislocation substructure evolution of a ferrite-austenite stainless steel. *Acta metall. mater.*, in press.

Mateo, A., L. Llanes, J. L. Palomino and M. Anglada (1995b). Effect of material parameters on the cyclic behavior of unaged and aged duplex stainless steel, to be submitted.

Solomon, H. D. and T. D. Devine Jr. (1983). Duplex stainless steels - A tale of two phases. In: *Duplex Stainless Steels* (R. A. Lula, ed.) pp. 693-756, ASM, Ohio.

Zhong, C., N. Y. Jin, X. Zhou, E. Meng and X. Chen (1990). Cyclic deformation of AISI-310 stainless steel- I: cyclic stress-strain response. *Acta metall. mater., 38*, 2135-40.

DISTRIBUTION OF THE INTERNAL CRITICAL STRESSES IN THE TWO-PHASE STRUCTURE OF A FATIGUED DUPLEX STAINLESS STEEL

J. POLÁK[1][2], F. FARDOUN[2] and S. DEGALLAIX[2]

[1]Institute of the Physics of Materials, Academy of Sciences of the Czech Republic,
Žižkova 22, 616 62 Brno, Czech Republic
[2]Laboratoire de Mécanique de Lille, URA CNRS 1441, Ecole Centrale de Lille,
BP 48, Cité Scientifique, F-59651 Villeneuve d'Ascq Cedex, France

ABSTRACT

An austenitic-ferritic stainless steel was cyclically strained with constant strain rate and different strain amplitudes. The hysteresis loops were digitally recorded and analysed in terms of the first and second derivatives. Statistical theory of the hysteresis loop was applied in order to obtain probability density distribution of the internal critical stresses. The probability density function has two peaks, which correspond to the cyclic deformation of both phases. The position of these peaks characterises the internal critical stresses in both phases and depends very slightly on the strain amplitude.

KEYWORDS

Duplex steel, internal stress, effective stress, hysteresis loop, probability density function.

INTRODUCTION

Cyclic stress-strain response reflects the internal structure of the material. Therefore, in a homogeneous crystalline material, the hysteresis loop and the cyclic stress-strain curve are determined by the dislocation structure, the distribution and the type of the obstacles to the dislocation motion. The shape of the hysteresis loop can yield an important information concerning the internal structure of the material and its resistance to dislocation motion (Polák, 1991, Christ, 1991).

From the numerous descriptions of the cyclic plastic straining of crystalline materials, Masing model (Masing, 1923) and its generalisation using statistical theory of the hysteresis loop (Polák and Klesnil 1980, 1982, Burmeister, 1981) proved to be in a good agreement with the experimental data. The application of this model to the analysis of the hysteresis loop allows to obtain information on the distribution of the internal critical stresses and also on the effective stress in cyclically strained material.

In this contribution a two-phase material, austenitic-ferritic stainless steel has been cyclically deformed and its hysteresis loop was analysed in order to obtain information on the cyclic strength of individual components of the duplex structure.

ANALYSIS

According to the generalised statistical theory of the hysteresis loop (Polák and Klesnil, 1982), a crystalline material can be separated into an assembly of individual microvolumes each of which is ideally plastic when its internal critical stress σ_{ic} is reached. These microvolumes are arranged in such a way that compatibility conditions are fulfilled. A simple arrangement which approximates reasonably well the actual one is the arrangement in parallel. The actual state of the material is represented by the probability density function $f(\sigma_{ic})$ of these microvolumes classified according to their internal critical stresses. The stress σ on the hysteresis loop is the sum of two components, i. e. the internal stress σ_I and the effective stress σ_E according to the eq.

$$\sigma = \sigma_I + \sigma_E \tag{1}$$

The internal stress is equal to the integral of the internal stresses carried over the individual microvolumes of the material and depends thus on the probability density function $f(\sigma_{ic})$. The effective stress is equal to the integral of the effective stresses over all plasticised microvolumes. Since in a homogeneous material the effective stress is the same in all plasticised microvolumes, the macroscopic effective stress is equal to the product of the saturated microscopic effective stress σ_{es} and the fraction of the plasticised volumina. Detailed considerations give the possibility to determine the probability density function $f(\sigma_{ic})$ and the effective saturated stress σ_{es} from the shape of the hysteresis loop. Recently Polák et al. (Polák et al., 1995) discussed the methods of effective stress evaluation from the hysteresis loop. They have concluded that effective stress can be evaluated from the plot of the second derivative vs. the relative strain and corresponds to the minimum of the second derivative. The probability density function can be evaluated from the second derivative (Polák, 1991)

$$f\left(\frac{\varepsilon_r E}{2} - \sigma_{es}\right) = -\frac{2}{E^2}\frac{\partial^2 \sigma_r}{\partial \varepsilon_r^2} \qquad \text{for} \qquad 2\sigma_{es}/E \le \varepsilon_r \le 2\varepsilon_a \tag{2}$$

where ε_r and σ_r are relative coordinates of the tensile or compressive half-loop and E is the appropriate effective elastic modulus.

EXPERIMENTAL

An austenitic-ferritic stainless steel of type SAF 2205 (Sandvik Steel) alloyed with 0.18 wt.% nitrogen was supplied in the form of hot-rolled 25 mm diameter bars. They were solution-treated and machined into cylindrical specimens of 10 mm in diameter and 12.5 mm in gage length. The steel contains approximately equal fractions of austenite and ferrite. The small austenitic islands are elongated in the direction of the rolling and are surrounded by ferrite. The average dimension of the austenitic islands as well as of the neighbouring ferritic grains in transversal direction is about 10 to 15 µm.

The specimens were tested in a 100 kN computer controlled INSTRON servohydraulic testing machine. The axial strain was measured and controlled with an extensometer over the gauge length of 10 mm. The constant strain rate 4×10^{-3} s^{-1} was chosen. The stress and strain were digitally recorded with 500 to 1200 points for each hysteresis loop. Both tensile and compressive half-loops were analysed. Raw data were smoothed either by fitting second order polynomial to successive 5 experimental points or by fitting high degree polynomial to the whole half-loop. Both procedures yielded results in a good agreement. The first and second derivatives were thus obtained with good precision. The effective modulus E_{eff} was evaluated from the first derivative at the point where the second derivative reaches a relative minimum.

RESULTS

Figure 1a shows the hysteresis loops in relative coordinates and Fig. 1b the analysis of the tensile half-loop in terms of the first and second derivative divided by appropriate effective elastic modulus for the low strain amplitude 4×10^{-3}. The relative strain multiplied by the effective elastic modulus is shown on x-axis. The initial drop of the first and second derivative corresponds to the plastic strain relaxation upon unloading under the effect of the effective stress. The position of the relative minimum of the second derivative corresponds to the effective stress in a homogeneous material (Polák *et al.*, 1995). Later, the second derivative approximates the probability density function. For duplex steel two maxima, corresponding to the cyclic deformation of both phases, are the very characteristic feature of this plot.

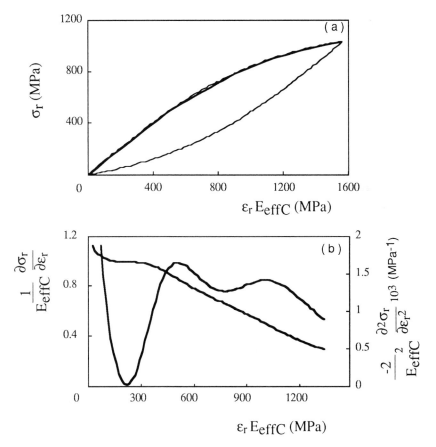

Fig. 1. Analysis of the hysteresis loop, $\varepsilon_a = 4 \times 10^{-3}$.

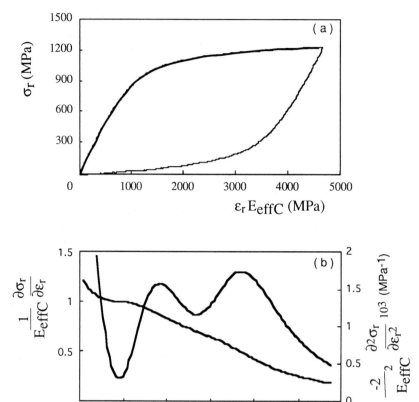

Fig. 2. Analysis of the hysteresis loop, $\varepsilon_a = 1.25 \times 10^{-2}$.

Table 1. Characteristic quantities derived from the analysis of the
hysteresis loops with different strain amplitudes.

$\varepsilon_a \cdot 10^3$	σ_{min} (MPa)	$f_{peak1} \cdot 10^3$ probability	σ_{peak1} (MPa)	$f_{peak2} \cdot 10^3$ probability	σ_{peak2} (MPa)	E_{effC} (GPa)
0.4	228	1.75	536	1.44	1026	194.1
0.7	242	1.54	528	1.76	1023	196.7
1.25	233	1.57	503	1.67	1045	186.4

Figures 2a and 2b show the same quantities for the loop run with the higher strain amplitude 1.25×10^{-2}. The general shape of the initial portion of the loop as well as of the first and second derivative is very close to that derived from the loop run with the low strain amplitude. The reproducibility was very good. The second peak becomes more pronounced when the strain amplitude increases.

The experimental data allow to determine the position of the relative minima and maxima on the plot of the second derivative. They are shown in Table 1 for three strain amplitudes. In a homogeneous material σ_{min} corresponds to the effective stress but in a duplex structure consisting of approximately equal fractions of two phases, it characterises only an average effective stress. The positions of the first peak, σ_{peak1}, correspond to the austenite and that of the second peak, σ_{peak2}, to the ferrite. The probability density f_1 and f_2 of both peaks and the effective elastic modulus in compression, E_{effC}, are also included.

DISCUSSION

The experimental data and the results of the analysis of the hysteresis loop of the austenitic-ferritic stainless steel yield reproducibly the probability density distribution of the internal critical stresses and information about the effective stress. For all strain amplitudes and different numbers of cycles both first and second derivatives decrease immediately after the stress reversal and second derivative reaches a minimum. With increasing relative strain, two well separated maxima are apparent on the plot of the second derivative, which can be identified with the deformation of the austenitic and ferritic phases respectively.

Due to the composite structure of the material, consisting of two components with different lattices, the analysis is less straightforward than in homogeneous material. Since the effective stress differs appreciably in a FCC and in a BCC metal, the minimum of the second derivative corresponds neither to the effective stress in austenite nor to that in ferrite and is expected to lie in an interval between these two extremes. Preliminary data from austenitic and ferritic materials show that the effective stresses are about 150 MPa in austenite and about 300 MPa in ferrite (Polák et al., 1995). The values of σ_{min} in Table 1 are in the middle of the expected interval.

The internal critical stresses of both phases of the microvolumes with the highest frequency can be evaluated from the position of both peaks in the plot of the second derivative only if the respective effective stresses are known. Since the effective stress decreases quickly upon unloading but is built up only slowly in reloading, eq. (2) cannot be used for the evaluation of these internal stresses. In a better approximation the internal critical stress can be estimated from the relative strain multiplied by effective elastic modulus and from the appropriate effective stress using relation

$$\sigma_{ic} = (\varepsilon_r E_{eff} - \sigma_E)/2 \tag{3}$$

Adopting the values of the effective stresses for austenite and ferrite shown above, the internal critical stresses corresponding to the microvolumes of the austenitic and ferritic phases with the highest frequency were evaluated and are plotted vs. strain amplitude in Fig. 3. Both values are nearly independent of the strain amplitude, only a slight tendency to hardening of the ferrite and to softening of the austenite with increasing strain amplitude is apparent. This behaviour can be understood in terms of the dislocation structures observed in the grains of ferrite and austenite in the duplex steel (Kruml et al., 1995). Appreciable strain localisation and subsequent softening is observed at higher strain amplitudes only in the austenite while higher amplitudes in the ferrite result in formation of cellular structure which is harder than the vein matrix structure interlaced with PSBs.

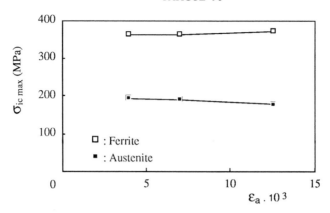

Fig. 3. Internal critical stresses of the most frequent microvolumes of austenite and ferrite.

CONCLUSIONS

Experimental study of the hysteresis loop in cyclic straining of an austenitic-ferritic stainless steel and subsequent analysis led to the following conclusions:

(i) Hysteresis loop shape reflects the duplex structure of the steel and pronounced maxima and minima appear on the second derivative. The two maxima correspond to the cyclic straining of austenite and ferrite.

(ii) The internal critical stress corresponding to these maxima can be estimated only if effective stress component is evaluated.

(iii) The internal critical stresses of the most frequent microvolumes of both phases are nearly independent of the strain amplitude.

REFERENCES

Burmeister, H.J; and C. Holste (1981). Change of the activation area during cyclic deformation II. Quantitative interpretation with a model of heterogeneous plastic deformation. *Phys. Stat Sol.*, (a) **64**, 611-629.

Christ, H.J. (1991), *Wechselverformung von Metallen*. Springer-Verlag, Berlin.

Kruml, T., J. Polák and S. Degallaix (1995). Dislocation microstructures in fatigued duplex steels. In: *Proc. Int. Conf. EUROMAT 95*, Vol. II., Associazione Italiana di Metalurgia, Milano, pp. 23-28..

Masing, G. (1923). On Heyn's hardening theory of metals due to inner elastic stresses. *Wissenschaftliche Veroffentlichungen aus dem Siemens-Konzern*, **3**, 231-239.

Polák, J. and M. Klesnil (1980). Statistická interpretace hysterézní smyčky. *Kovové Mater.* **18**, 329-344.

Polák, J. and M. Klesnil (1982). The hysteresis loop. 1. A statistical theory. *Fatigue Engng Mater. Struct.* **5**, 19-32.

Polák, J. (1991). *Cyclic plasticity and low cycle fatigue life of metals*. Elsevier, Amsterdam.

Polák, J., F. Fardoun and S. Degallaix (1995). Internal and effective stress concept in cyclic plasticity. Submitted for publication.

CYCLIC DEFORMATION BEHAVIOR OF COPPER BICRYSTALS

Y. M. Hu and Z. G. Wang

State Key Laboratory for Fatigue and Fracture of Materials, Institute of Metal Research,
Academia Sinica, Shenyang 110015, P. R. C

ABSTRACT

Cyclic deformation was applied to several kinds of copper bicrystals with different geometries in a plastic strain range of approximately 1.66×10^{-4} to 9.1×10^{-3}. Experimental results show that the bicrystals exhibit a saturation stress plateau in their cyclic stress-strain curves, and that the cyclic hardening and saturation behavior of the bicrystals is very closely related to the orientations of the component crystals and the grain boundary structure of the bicrystals. Surface observations show that activation of secondary slip due to grain boundary (GB) constraints and different degrees of dislocation interaction may be responsible for the above results.

KEYWORDS

Cyclic deformation; copper; bicrystals; hardening; saturation.

INTRODUCTION

Cyclic deformation of copper single crystals has been widely studied in order to understand the fatigue deformation mechanism in f.c.c metals (Basinski and Basinski, 1992). Copper single crystals oriented for single glide have been demonstrated to show a saturation stress plateau in their cyclic stress-strain curves in a wide strain range (Mughrabi, 1978). Copper polycrystals are considered not to exhibit a plateau under conventional strain control. However, Whether a plateau exists or not in polycrystals is still a debatable question. Bicrystals (Hook and Hirth, 1967a,b; Lim and Raj, 1985; Paider et al,1986; Sittner and Paider, 1989) have been widely used to study the role of GB on the plastic deformation of polycrystals and alloys. However, fatigue deformation of bicrystals is relatively unstudied.

In this paper, we will report the cyclic hardening and saturation behavior of copper bicrystals with different geometries and their surface slip characteristics.

EXPERIMENTAL PROCEDURES

The bicrystals used in this study were obtained by a diffusion bonding technique. Details of specimen preparation are described in a previous work (Hu et al,1995).

Three kinds of copper bicrystals designated as [$\bar{1}$35]/[$\bar{1}$35], [$\bar{1}$35]/[$\bar{2}$35], and [$\bar{2}$35]/[$\bar{2}$35] were prepared to investigated the effects of orientations of component crystals of the bicrystals on the cyclic hardening and the CSS curve features of the bicrystals. The [$\bar{1}$35]/[$\bar{1}$35] and the [$\bar{2}$35]/[$\bar{2}$35] bicrystals are iso-axial symmetrical bicrystals (Paider,1987), the orientation of grain 2 (G2) for these two bicrystals can be obtained by making an 180° rotation of grain 1(G1) around [501] and [1 $\bar{1}$ 1], respectively. The [$\bar{1}$35]/[$\bar{2}$35] bicrystal is a non-isoaxial bicrystal, the crystallographic geometry of this bicrystal is illustrated in Fig.1. The locations of the component crystals of these bicrystals in the stereograghic triangle are shown in Fig.2. The [$\bar{2}$35] crystal is located close to the 001- $\bar{1}$11 side.

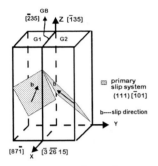

Fig.1 Geometry of the [$\bar{1}$35]/[$\bar{2}$35] bicrystal

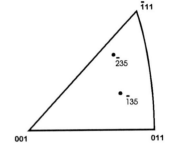

Fig.2 Orientation of the component of crystals of the bicrystals

Another two kinds of [$\bar{1}$35] iso-axial non-symmetrical bicrystals, which are noted briefly as [$\bar{1}$35]z and[$\bar{1}$35]x were prepared to investigated the effects of GB structure on the cyclic hardening and saturation behavior of the bicrystals. The orientation of G2 for these two bicrystals can be obtained by making an 180° rotation of G1 around [$\bar{1}$35] and [$\bar{3}$ $\bar{2}$6 15], respectively. These two bicrystals were cycled only at one plastic strain amplitude of 1.32×10^{-3}.

The dimension of the fatigue specimens is illustrated in Fig.3. Push-pull fatigue tests were performed on a Schenck servo-hydraulic testing machine at room temperature in air under plastic strain control. Specimen surfaces were observed after saturation of hardening by optical and scanning electron microscopy. For comparison, fatigue tests were also performed on a [$\bar{1}$35] single crystal.

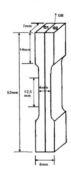

Fig. 3 Dimension of the fatigue specimens

RESULTS

Cyclic hardening

Typical cyclic hardening curves of the crystals at $\gamma_{pl} = 1.32 \times 10^{-3}$ are illustrated in Fig.4. It can be seen from Fig.4a that the curve of the [$\bar{1}$35]/[$\bar{1}$35] bicrystal is similar to that of the [$\bar{1}$35] single crystal . The [$\bar{2}$35]/[$\bar{2}$35] bicrystal shows more rapid cyclic hardening behavior, while the cyclic hardening of the [$\bar{1}$35]/[$\bar{2}$35] bicrystal is between that of the [$\bar{1}$35]/[$\bar{1}$35] and [$\bar{2}$35]/[$\bar{2}$35] bicrystals. From Fig.4b, it is also can be found that the two iso-axial non-symmetrical bicrystals [$\bar{1}$35]z and [$\bar{1}$35]x show more rapid cyclic hardening compared with the iso-axial symmetrical bicrystal [$\bar{1}$35]/[$\bar{1}$35].

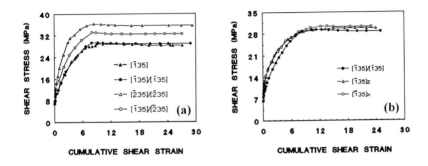

Fig.4 Cyclic hardening curves for the single crystal and the bicrystals

Cyclic saturation

All of the tested crystals show a saturation stress plateau in their CSS curves in the range of γ_{pl} from $\sim 1.66 \times 10^{-4}$ to $\sim 5.0 \times 10^{-3}$ (see Fig.5). But the plateau stresses increase in the order [$\bar{1}$35], [$\bar{1}$35]/[$\bar{1}$35], [$\bar{1}$35]/[$\bar{2}$35] and [$\bar{2}$35]/[$\bar{2}$35] from ~ 28.2 to ~ 36MPa. The plateau stress of the non-isoaxial bicrystal [$\bar{1}$35]/[$\bar{2}$35], which is approximately 32.5 MPa, seems to be the mean of those of the other two symmetrical bicrystals, which is ~ 29MPa, ~ 36MPa, respectively.

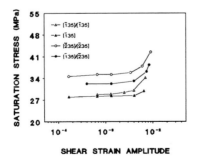

Fig.5 CSS curves for the copper single crystal and the bicrystals

From Table 1, we can seen that the saturation stresses (τ_s) of the four copper crystals at a plastic strain amplitude of 1.32×10^{-3} increase in the order [$\bar{1}35$], [$\bar{1}35$]/[$\bar{1}35$] , [$\bar{1}35$]z, [$\bar{1}35$]x from 28.2MPa to 30.2 MPa. It shows that the non-symmetrical GB exhibit larger effects on the cyclic hardening and saturation behavior of the bicrystals because the elastic and plastic strains are not compatible at interface.

Table 1 Cyclic hardening and saturation test data for the [$\bar{1}35$] iso-axial
bicrystals and the [$\bar{1}35$] single crystal.

Specimen No.	γ_{pl}	$\gamma_{pl,cum}$	τ_{max} (MPa)	τ_s (MPa)	τ_{max}/τ_s
[$\bar{1}35$]	1.32×10^{-3}	27.156	28.92	28.21	1.03
[$\bar{1}35$]/[$\bar{1}35$]	1.32×10^{-3}	28.9	29.42	28.90	1.02
[$\bar{1}35$]z	1.32×10^{-3}	25.3	30.01	29.8	1.01
[$\bar{1}35$]x	1.32×10^{-3}	27.4	30.53	30.2	1.01

<u>Surface observation</u>

No secondary slip traces were observed for the [$\bar{1}35$] single crystal until a plastic strain amplitude of 7.01×10^{-3} was adopted. While, slip characteristics for the bicrystals are quite different from those of the single crystal.

Under lower γ_{pl} ($\leq1.32\times10^{-3}$), only primary slip B4 ((111) [$\bar{1}01$]) was observed in the [$\bar{1}35$]/[$\bar{1}35$] specimen surface. Secondary slip A3 (($\bar{1}11$)[101]) was activated in the vicinity of the GB as the plastic strain amplitude increased to 4.08×10^{-3} (see Fig.6a). Deformation bands(DBs), which is frequently appeared in double-slip-oriented copper bicrystals, were observed at γ_{pl} = 7.01×10^{-3} (see Fig. 6b).

For the [$\bar{2}35$]/[$\bar{2}35$] bicrystal, secondary slip C1(($\bar{1}$ $\bar{1}1$)[011]) was activated at all strain amplitudes (see Fig.6c). Deformation bands (see Fig.6d) were observed at $\gamma_{pl} \geq 1.32\times10^{-3}$ which is lower than that for the[$\bar{1}35$]/[$\bar{1}35$] bicrystal.

For the non-isoaxial bicrystal, only primary slip was observed at γ_{pl} = 4×10^{-4}. Deformation bands and secondary slip were also observed at $\gamma_{pl} \geq 1.32\times10^{-3}$ for this kind of bicrystal.

For the iso-axial non-symmetrical bicrystals [$\bar{1}35$]z and [$\bar{1}35$]x, the elastic and plastic strains are incompatible at GB plane, secondary slip traces A3 (see Fig.7) were activated in the vicinity of GB due to the incompatible internal stress, which arises from the requirement of elastic and plastic strain compatibility at GB plane. This is the cause why the iso-axial non-symmetrical bicrystals show more rapid cyclic hardening and higher saturation stresses compared with the iso-axial symmetrical bicrystal [$\bar{1}35$]/[$\bar{1}35$].

DISCUSSIONS

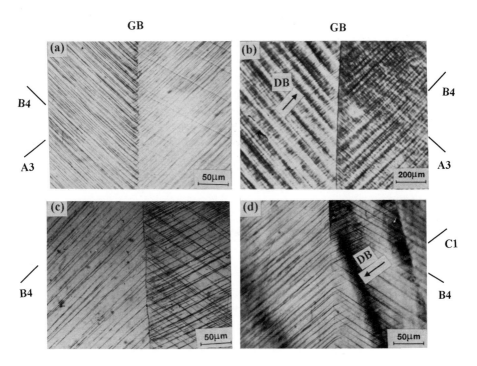

Fig.6 Slip patterns on the specimen surfaces at saturation stage. (a) [$\bar{1}$35]/[$\bar{1}$35], γ_{pl} = 4.08 $\times 10^{-3}$; (b) [$\bar{1}$35]/[$\bar{1}$35], γ_{pl} = 7.01 $\times 10^{-3}$; (c) [$\bar{2}$35]/[$\bar{2}$35], γ_{pl} =6.63$\times 10^{-4}$; (d) [$\bar{2}$35]/[$\bar{2}$35], γ_{pl} = 3.4$\times 10^{-3}$.

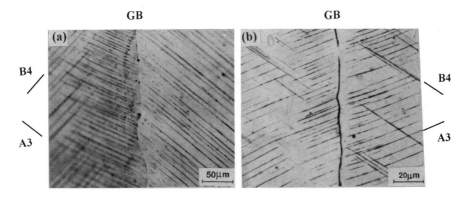

Fig.7 Slip morphologies at saturation stage near the GB for the [$\bar{1}$35] iso-axial non-symmetrical bicrystals, γ_{pl} =1.32$\times 10^{-3}$. (a) [$\bar{1}$35]z ; (b) [$\bar{1}$35]x.

Based on the surface observations, we can state that the operation of secondary slip due to the constraints of GB leads to the more rapid cyclic hardening and the higher saturation stresses for the copper bicrystals. It has been reported (Gong *et al*, 1995) that the dislocation interaction between the operative slip systems B4 and C1 which gives Lomer-Cottrell lock is stronger than that between the B4 and A3. Different degrees of dislocation interaction between primary slip system and secondary slip system may be responsible for the differences in the cyclic hardening and saturation behavior for the double-slip-oriented copper single crystals. In this study, it seems that the stronger dislocation interaction between the operative slip systems B4 and C1 may be also responsible for the more rapid cyclic hardening behavior and higher saturation stress for the [$\bar{2}$35]/[$\bar{2}$35] iso-axial symmetrical bicrystal, and the weak dislocation interaction between the operative slip systems B4 and A3 may be responsible for the similarity of cyclic hardening and saturation behavior between the [$\bar{1}$35]/[$\bar{1}$35] iso-axial symmetrical bicrystal and the [$\bar{1}$35] single crystal.

CONCLUSIONS

(1) All the bicrystals investigated exhibit a plateau in their CSS curves although secondary slip is strongly activated. It can be stated that activation of secondary slip results in the rapid cyclic hardening and higher saturation stresses for the bicrystals.

(2) Cyclic hardening and saturation behavior of the copper bicrystals is closely related the orientations of the component crystals of the bicrystals because different secondary slip will be activated, leading to different degrees of dislocation interaction.

(3) Cyclic hardening and saturation behavior of copper bicrystals is also related to the GB structure because different slip deformation in the vicinity of GB may occur.

REFERENCES

Basinski, Z. S. and S. J. Basinski (1992). Fundamental aspects of low amplitude cyclic deformation in face-centered cubic crystals. *Progress in Materials Science, 36*, 89-147.

Mughrabi, H. (1978). The cyclic hardening and saturation of copper single crystals. *Mater. Sci. Eng. 33*, 207-223.

Hook, R. E. and J. P. Hirth (1967). The deformation behavior of isoaxial bicrystals of Fe-3%Si. *Acta Metall. , 15*, 535-551.

Hook, R. E. and J. P. Hirth (1967). The deformation behavior of non-isoaxial bicrystals of Fe-3%Si. *Acta Metall. , 15*, 1089-1110.

Lim, L. C. and R. Raj (1985). Continuity of slip screw and mixed crystal dislocation across bicrystals of nickel. *Acta Metall. , 33* , 1577-1583.

Paider, V., P. P. Pal and S. Kadeckova (1986). Plastic deformation of symmetrical bicrystals having Σ3 coincidence twin boundary. *Acta Metall. , 34*, 2277-2289.

Sittner, P. and V. Paider (1989). Observation and interpretation of grain boundary compatibility effects in Fe-3.3wt%Si bicrystals. *Acta Metall. , 37*, 1717-1726.

Hu, Y. M., Z. G. Wang and G. Y. Li (1995). Cyclic deformation of a co-axial copper bicrystal. *Scripta Metall. et Mater., 33*, in print.

Paider, V. (1987). A classification of symmetrical grain boundaries. *Acta Metall. , 35*, 2035-2048.

Gong, B., Z. G. Wang and Y. W. Zhang (1995). Cyclic deformation behavior of Cu single crystals for double slip. *Mat. Sci. Eng., A194*, 171-178.

EFFECT OF SINTERING TEMPERATURE ON CYCLIC PROPERTIES OF SELECTED P/M STEELS.

Jerzy KALETA * , Artur MIKOŁAJCZYK and Jacek ŻEBRACKI
Institute of Material Science and Applied Mechanics,
Technical University of Wrocław
Wybrzeże Wyspiańskiego 27, 50-370 Wrocław, Poland.

ABSTRACT

Cyclic properties of two P/M steels with identical chemical compositions and different sintering temperatures (1120 and 1280^oC) are compared. Specimens were tested in one–directional sinusoidal tension–compression ($R = -1$, $f = 25$ Hz) over a wide range of HCF lives and in the fatigue limit region.
The tests were carried out under stress amplitude control conditions. It is shown that higher temperature of sintering results in higher strength properties in the finite endurance region with plastic properties being an exception. The sintering temperature was found to have only a negligible effect on the fatigue limit value. The present method of investigation is critically assessed including specifically the reverse magnetostriction (Villari effect).

KEYWORDS

Fatigue; cyclic properties; P/M steels; sintering temperature; Villari effect.

INTRODUCTION

The last twenty years have seen great advances in the production of sintered alloy steel parts intended for long–term cyclic load service. The automotive, engineering, aircraft and electrical industries are the most manifest examples of the trend. Static properties of sintered steels are usually close to those found in conventional steels. Fatigue properties, however, fail to repeat this pattern. They are highly dependent upon porosity which in turn is affected by pressing and sintering conditions. Alloying additions, temperature of sintering and type of heat treatment are other factors affecting fatigue properties.
Evaluation of the effect of chemical composition and manufacturing process parameters on fatigue strength, understanding the phenomena underlying fatigue failure in sintered materials and, finally, reliable life prediction techniques have become recently a most challenging research objective.
The present study was aimed at determining the effect of sintering temperature–one of the most important manufacturing parameters – on fatigue properties of P/M steels.

---

*Address correspondence to author. Fax: +48 71 21 12 35; e–mail: kaleta@immt.pwr.wroc.pl

EXPERIMENTAL DETAILS

Material specimen

Two P/M steels of the same composition but differing in sintering temperature were tested. Sintered steels were fabricated from iron powder with additions of $Mo(0.5\%)$, $Ni(4\%)$, $Cu(1.5\%)$. The sintering process was conducted at 1120 °C (Distaloy A) or at 1280 °C (Distaloy B) for 30 minutes under an atmosphere consisting of 70% N_2 and 30% H_2. The specimens had a rectangular cross-section 6 × 8 mm.

Measurement method, loading, apparatus

The specimens were subjected to alternating tension–compression ($R = -1$) under stress amplitude control and ambient temperature conditions. A sinusoidal loading spectrum was applied. The investigations were made within a broad range of life–times encompassing HCF and fatigue limit regions. The tests were conducted on a hydraulic MTS 810 pulser ($f = 25\ Hz$) connected via a system of A/D and D/A converters with a PC computer. The measurement set–up is shown in Fig. 1.

The cyclic yield limit is especially difficult to determine in sintered steels (Kaleta *et al.*, 1990, 1991b, Kaleta and Żebracki, 1994). Other properties such as the plastic strain ε_{pl} and energy ΔW corresponding to the endurance limit level are equally difficult to measure. This is largely due to their magnitudes being roughly by two orders smaller than for conventional steels (Kaleta and Piotrowski, 1992b). This was the main reason the authors supplemented conventional measuring methods with a technique based on the coupled magneto–mechanical Villari effect (Kaleta and Żebracki, 1993, 1994).

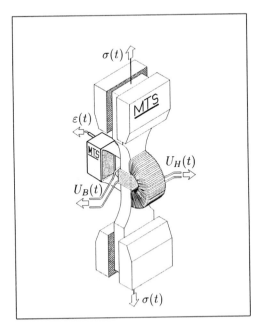

Fig. 1.   Measurement set–up

In the on–line fatigue tests the hysteresis energy ΔW and plastic strain ε_{pl} were determined (Kaleta et al., 1991a).

Alongside recorded were two voltages U_B and U_H induced in two different coils placed within the magnetic field produced by a cyclically loaded specimen. It can be easily demonstrated that voltages U_B and U_H so measured are proportional to the first derivative of magnetic induction dB/dt and magnetizing force dH/dt, respectively. The values of B and H were determined by integrating and scaling signals U_B and U_H.

Having recorded both mechanical ($\sigma(t)$ and $\varepsilon(t)$) and magnetic signals ($U_B(t)$ and $U_H(t)$) we were able to get not only typical $\sigma - \varepsilon$ plots but also a large variety of combined mechanical–magnetic characteristics sensitive to the load level (Kaleta and Żebracki, 1993, 1994). From earlier investigations it follows that for example the $U_B - \varepsilon$ loop or the magnetic energy ΔM (in the $B - H$ coordinate system) exhibit high sensivity to the first plastic deformation and fatigue limit.

DISCUSSION

In Figs 2 and 3 the Manson–Coffin law for the two materials is shown. The $2N_t$ value stands for the number of half–cycles for which intersection of the functions representing an elastic and plastic portions of the deformation occurs. Cyclic properties of the steels are summarized in Table 1.

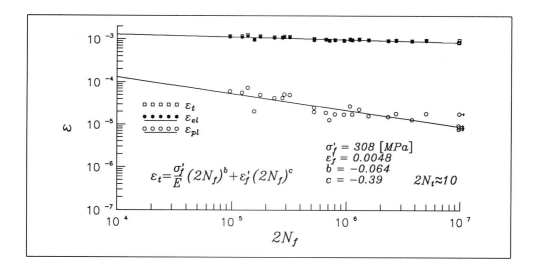

Fig. 2.   Plot of $\varepsilon_t(2N_f)$, $\varepsilon_{el}(2N_f)$ and $\varepsilon_{pl}(2N_f)$ for Distaloy A.

The presented data show the endurance limit for Distaloy A to be $\sigma_f = 110$ MPa and the associated strain: $\varepsilon_f =\sim 1 \cdot 10^{-3}$. For Distaloy B these values are $\sigma_f = 105$ MPa and $\varepsilon_f =\sim 1 \cdot 10^{-3}$, i.e. the difference is not significant.

The plots $\varepsilon_{pl}(2N_f)$ are clearly different which attested by values ε'_f and c. The higher sintering temperature resulted in lower plastic properties. It is also to be noted that the presented values of parameters b, c, n' are significantly different from those quoted in the literature for conventional steels (Kaleta et al., 1991b, Kaleta and Żebracki, 1994).

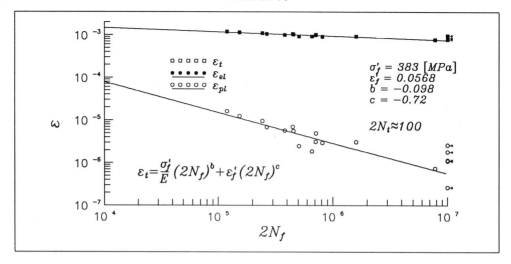

Fig. 3. Plot of $\varepsilon_t(2N_f)$, $\varepsilon_{el}(2N_f)$ and $\varepsilon_{pl}(2N_f)$ for Distaloy B.

An assessment of plastic properties can be also carried out based on the analysis of the cyclic deformation curve $\sigma_a(\varepsilon_{apl})$ (Fig. 4). It must be emphasized that for the same stress amplitude, say $\sigma_a = 110$ MPa, the plastic strain value for Distaloy A is by an order of magnitude larger than for Distaloy B.

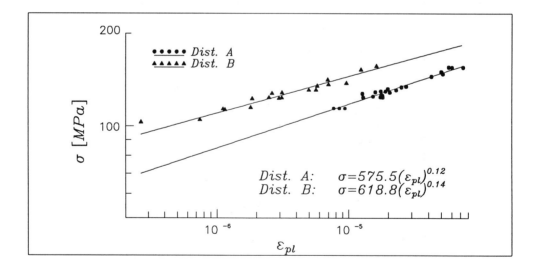

Fig. 4. Plot of stress amplitude σ_a dependence on amplitude of plastic deformation ε_{pl}.

Substantial differences in plastic properties are also seen in the $\Delta W(2N_f)$ plot (Fig. 5). The low running plot for Distaloy B is a clear manifestation of an adverse effect of higher sintering temperatures on plastic properties. In paper (Dudziński *et al.*, 1993) the authors have shown that variation in cyclic properties of the two steels resulting from different sintering temperatures can

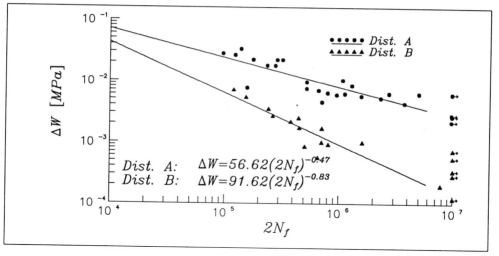

Fig. 5. Plot of energy of plastic deformation ΔW versus $2N_f$.

be actually attributed to considerable changes in the microstructure following the two manufacturing processes.

The one way of presenting how the Villari phenomenon is affected by the load level is shown in Fig.6 . The magnetic loop energy ΔM is a highly sensitive characteristics in the elastic range ($\sigma \leq \sigma_{cpl}$) as can be judged from the step portion of the plot in Fig. 6. A region of saturation is observed on the plot on attaining some marked plastic deformation in a specimen.

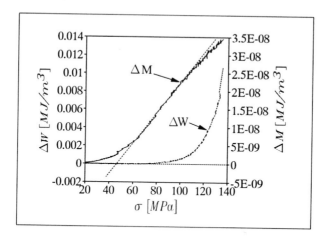

Fig. 6: Functions magnetic energy ΔM and mechanical energy ΔW vs. stress amplitude σ_a
(Distaloy A).

A plot of $\Delta W(\sigma)$ is also presented in Fig.6. The ΔW value is clearly seen to be zero within the elastic range ($\sigma \leq \sigma_{cpl}$) and to rise sharply in the elastic–plastic region. The two plots are a convincing demonstration of the fact that magnetic and mechanical phenomena are fully complementary in this case. It is further seen that the Villari effect may be of great prognostic value in the elastic range. The plot of $\Delta M(\sigma)$ (Fig.6) is characterized by three distinct intervals. The first one cores the range $0 < P_I \leq \sim 40 MPa$, in the second one good linearity is observed up to the value $\sigma_a = \sigma_f$. In the third interval $P_{III} > \sigma_f$ a saturated condition is attained. For large plastic deformations (not an easy task in sintered steel) even a drop in values of ΔM in the third interval is observed.

Table 1. Cyclic properties of sintered steels.

σ'_f [MPa]	ε'_f	b	c	k'_f	n'
Distaloy A					
308	0.0048	-0.064	-0.39	575.5	0.12
Distaloy B					
383	0.0568	-0.098	-0.72	618.8	0.14

CONCLUSIONS

- Higher temperatures of sintering result in higher strength characteristics in the finite fatigue strength region. The endurance limit, however, seems to remain unaffected.

- Higher temperatures of sintering act to decrease plastic properties of P/M steels considerably.

- The adopted experimental methodology proved fully applicable in testing sintered steels.

Acknowledgement

The authors would like to thank Dr Wojciech Myszka (TU Wroclaw) for their valued discussion and computing assistance. This work was partly sponsored by Polish Committee of Scientific Research, grant **7 T07A 019 09**.

REFERENCES

Dudziński, W., J. Kaleta, P. Kotowski, L. Zbroniec and J. Żebracki (1993). Fatigue properties and microstructure of alloyed sintered steel KA: *Report No. 7, Inst. of Materials Science and Appl. Mechanics, TU Wroclaw.*

Kaleta, J., A. Piotrowski and H. Harig (1990). Cyclic stress–strain response and fatigue limit of selected sintered steels. *Proc. 8th ECF, Torino*, 445–448.

Kaleta, J., R. Błotny and H. Harig (1991a). Energy stored in a specimen under fatigue limit loading conditions. *J. of Testing and Evaluation*, <u>19</u>, 325–333.

Kaleta, J., A. Piotrowski and H. Harig (1991b). Cyclic properties and hysteresis energy accumulation in selected sintered steels. *Proc. FEFG/ICF, Singapore*, 570–575.

Kaleta, J. and A. Piotrowski (1992b). Hysteresis energy and fatigue life of selected sintered steels. *Proc. 3rd ICLCFEPBM, Berlin*, 473–478.

Kaleta, J. and J. Żebracki (1993). Use of the Villari effect to monitor fatigue properties of E355–CC steel. *Proc. 5th ICFFT*: Montreal: EMAS.

Kaleta, J. and J. Żebracki (1994). Use of the Villari effect in fatigue of sintered steel. *Proc. 10th Int. Conf. on Exp. Mechanics, Lisbon*, 1241–1246.

EFFECT OF LOADING RAMP ON STRESS-STRAIN RESPONSE AND FATIGUE LIFE

P. LUKÁŠ and L. KUNZ

Institute of Physical Metallurgy, CAS, Žižkova 22, Brno (Czech Republic)

S. KONG, B. WEISS and R. STICKLER

University of Vienna, Währingerstrasse 42, Vienna (Austria)

ABSTRACT

The cyclic stress-strain response and the fatigue life of polycrystalline copper were shown to depend on the length and on the mode of the start-up ramp in stress-controlled tests. Basically it holds that the shorter the ramp the higher the saturated plastic strain amplitude and the shorter the fatigue life.

KEY WORDS

Loading ramp; stress-strain response; fatigue life.

INTRODUCTION

Every fatigue test is characterised by the mode of control (stress or strain control), wave shape, stress or strain cycle asymmetry, frequency of cycling, etc. In fact, all these parameters characterise steady state cycling, i.e. the cycling under the desired values of stresses (stress amplitude and mean stress in the stress-controlled tests) or strains (strain amplitude and mean strain in the strain-controlled tests) are reached and kept constant. Several papers dealing with the effect of ramp loading on the fatigue behaviour have been published in recent years. Their results are often controversial. For example, some of them show that the ramp loading shifts the cyclic stress-strain curve (CSSC) towards lower stress values (Wang et al., 1988a; Hessler et al., 1991), while others show just the opposite (Llanes et al., 1993; Kong et al., 1993). Copper single crystals cycled in strain control at zero mean strain are known to exhibit an asymmetric stress behaviour, namely the stresses in compression are a few per cent larger than those in tension (Ma et al., 1990). The consequence of this asymmetry behaviour is that the specimens cycled in stress control at zero mean stress must exhibit a tensile mean strain, manifesting itself as tensile creep strain (Eckert et al., 1987; Peters et al., 1989). In turn, the cyclic creep affects the CSSC (Lukáš et al., 1989). The high sensitivity of the CSSC to the cyclic creep deformation together with the fact that a creep deformation occurs even at zero external mean stress, indicate that the explanation of the controversial results mentioned above is to be sought just in the degree of cyclic creep deformation during the start-up period.

EXPERIMENTAL PROCEDURES

Dumbbell-shaped specimens (gauge length, 32 mm; cross-section, 6 mm x 12 mm) were machined from a cold-rolled copper. After annealing the grain size was 35 μm, while the yield stress was 17 MPa and the ultimate tensile stress 213 MPa. The tests were performed using a Shimadzu 10 kN servohydraulic system in stress control with the final stress amplitude σ_a = 63 MPa. The cycling was sinusoidal, with a frequency of 0.05 Hz during ramping and 3 Hz after the ramp was completed. Two modes of start-up procedure were used. In Mode I, the load was manually increased step-wise at every transition from negative to positive load by the same increment. Thus, every full cycle is symmetrical. In the Mode II ramping, the load was increased step-wise at every zero transition, by the same increment, from tension to compression as well as from compression to tension. Thus, none of the full cycles is symmetric. In the case of non-zero mean stress, the desired value of the mean stress was set up first. Then, either the Mode I or Mode II procedure was applied to reach gradually the desired values of the stress amplitude. In all cases, the loading started in tension. The fatigue life tests were performed in the following way. First, the specimens were ramp-loaded in Shimadzu machine to the final stress amplitude σ_a = 63 MPa with the frequency 0.05 Hz. Then the frequency was raised to 3 Hz. After 10^5 cycles were completed the cycling was continued in a Schenck machine at a frequency of 43 Hz to failure.

RESULTS AND DISCUSSION

From the continuous stress-strain records it is possible to determine two quantities, namely the plastic strain amplitude ε_{ap}, as the half-width of the elapsed hysteresis loops, and the creep strain ε_{cr}, as the

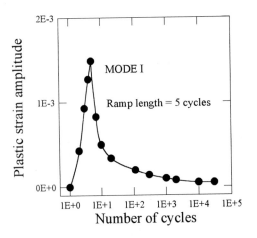
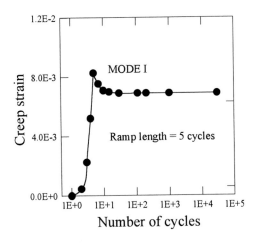

Fig.1.   Plastic strain amplitude in dependence on number of cycles for symmetrical cycling.

Fig.2.   Creep strain in dependence on number of cycles for symmetrical loading.

shift of the centre of the hysteresis loop with respect to the original zero value. Both these quantities can be plotted in dependence on the number of cycles. This is shown in Fig.1 (ε_{ap} vs. N) and Fig. 2 (ε_{cr} vs. N). The plastic strain amplitude (Fig. 1) naturally increases during the ramp. During the following

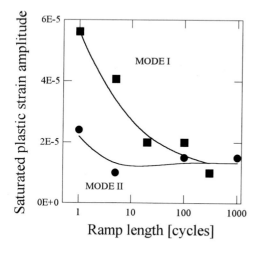

Fig.3.   Effect of loading mode and ramp length on the saturated plastic strain amplitude in the case of symmetrical cycling.

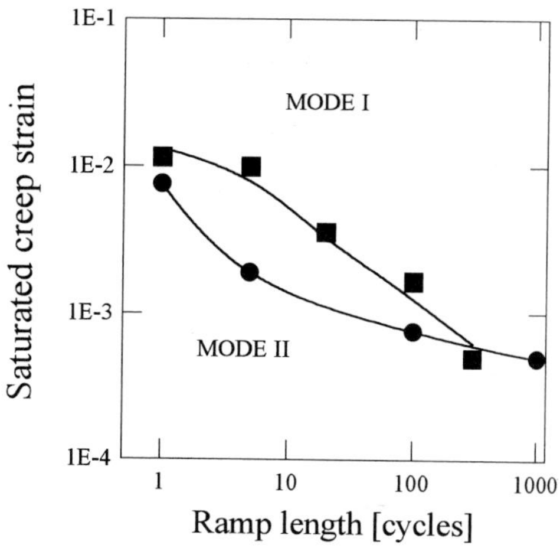

Fig.4.   Effect of loading mode and ramp length on the saturated creep strain in the case of symmetrical loading

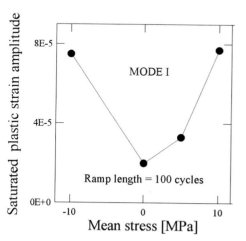

Fig.5.   Effect of mean stress on the saturated plastic strain amplitude

constant stress amplitude cycling, ε_{ap} decreases (cyclic hardening). It is important to note that there is always quite a reasonable saturation of ε_{ap}.

The creep strain (Fig. 2) increases with the number of cycles during the ramp. After the ramp is completed, the creep strain in this specific case again decreases - the specimen gets shorter again. In the case of non-zero mean stresses, the creep strain increases also after the end of the ramp. In all cases, the creep strain rate, i.e. the value of $d\varepsilon_{cr}/dN$, decreases (after the ramp is completed) with the number of cycles. This makes it possible to define (for studied range of mean stress) the saturated creep strain as the creep strain corresponding to a high number of cycles, e.g. 10^5 cycles. The saturated values of the plastic strain amplitude and of the creep strain will be denoted in the following as $\varepsilon_{ap,sat}$ and $\varepsilon_{cr,sat}$ respectively.

Both the saturated plastic strain amplitude and the creep strain in saturation depend on the ramp length and on the mode of ramping. This is shown in Figs 3 and 4 for the case of symmetrical cycling. Figure 3 shows that the value of $\varepsilon_{ap, \, sat}$ decreases with increasing ramp length. Figure 4 shows that (1) the creep strain decreases with the ramp length, and (2) the creep strain for Mode I ramping is generally higher than that for Mode II ramping. The same holds for the plastic strain amplitude (see Fig. 3). This suggests a correlation between $\varepsilon_{ap, \, sat}$ and $\varepsilon_{cr,sat}$. This point was pursued in more detail by the experiments with non-zero mean stresses. Some of the results are presented in the following diagrams. Figure 5 shows the dependence of the saturated plastic strain amplitude on the mean stress. It can be seen that the value of $\varepsilon_{ap,sat}$ increases with increasing magnitude of the mean stress, both tensile and compressive. The corresponding values of the creep strain in saturation are presented in Fig 6. As expected, it can be seen that the creep strain increases with the magnitude of the mean stress. The

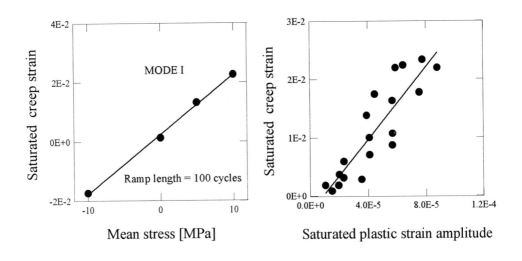

Fig.6. Correlation between the saturated   Fig.7. Correlation between the saturated
        plastic strain amplitude and the creep   plastic strain amplitude and the creep
        strain in saturation.   strain in saturation.

correlation between the saturated plastic strain amplitude and the absolute value of the cyclic creep strain in saturation is presented for all the tests performed in Fig. 7.

The TEM observation confirmed that the dislocation structure is not perfectly homogeneous; there are always differences between grains. Basically, two types of structure were observed, namely a vein structure and a loose cell structure. The vein structure was observed to occupy the majority of grains in specimens representing the lower part of Fig. 7, i.e. the specimens characterised by low values of $\varepsilon_{ap,sat}$ and $\varepsilon_{cr,sat}$. The loose cell structure is typical for specimens with high values of the $\varepsilon_{ap,sat}$ and $\varepsilon_{creep,sat}$, i.e. the specimens falling into the upper part of Fig. 7. In no cases were well-developed ladder-like persistent slip bands (PSBs) observed.

Table 1.

Specimen No.	Ramp mode	Ramp length (cycles)	Fatigue life (cycles)
1	I	1	5.24 E+06
2	I	100	1.24 E+07
3	II	1	5.93 E+06
4	II	100	1.43 E+07

The results on the effect of the ramp length on the fatigue life are presented in the table 1. It can be clearly seen that the shorter ramp results in a shorter fatigue life.

These basic experimental findings of this paper can be interpreted in such a way that there is a synergetic effect between the cyclic slip and the monotonic slip. The motion of dislocations is easier under the combined action of the cyclic and the monotonic driving forces than under the isolated action of only one of these driving forces. The higher primary slip activity triggers the action of the secondary slip systems. This is why the dislocation structure in specimens with higher values of $\varepsilon_{ap,sat}$ and $\varepsilon_{creep,sat}$ exhibits the loose cell structure as the typical structure. As already mentioned, we have not observed any well-developed, ladder-like or cell-like PSBs. It does not mean that there are none. But it means that their density is very low. Thus, the plastic strain is born here mainly by the matrix structure. The effect of ramping on the cyclic stress-strain response (Wang et al., 1988b) and on the crack nucleation and propagation (Liang et al., 1989) has been related mainly to the role of the PSBs. The present results indicate that the effect of ramping in this low amplitude region can be explained on the basis of the matrix dislocation structures and on the corresponding matrix values of the cyclic and monotonic plastic strains.

CONCLUSIONS

(1)   The mode of the ramp with which the nominally symmetrical stress-controlled fatigue tests are started affects both the stress-strain response and the fatigue life.

(2)   Basically it holds that the shorter the ramp the higher the saturated plastic strain amplitude, the higher the saturated creep strain and the shorter the fatigue life.

(3)   The application of a small mean stress leads to an increase in both $\varepsilon_{ap,sat}$ and $\varepsilon_{creep,sat}$.

ACKNOWLEDGEMENTS

The authors wish to thank the Grant Agency of the Czech Republic for its support under the Grant No. 106/93/0098

REFERENCES

Eckert, R., C. Laird and J. Bassani (1987). *Mater. Sci. Eng.*, 91, 81.

Hessler, W., B.Weiss, R. Stickler, P. Lukáš and L. Kunz (1991). *Mater. Sci. Eng.*, A145, L1.

Kong, S., B. Weiss, R. Stickler, L. Kunz and P. Lukáš (1993). In *Fatigue 93*, eds. J.-P. Bailon and J. I. Dickson, EMAS, Montreal, p. 77.

Liang F. L. and C. Laird (1989). *Mater. Sci. Eng.*, A117, 103.

Llanes, L. and C. Laird (1993). *Fatigue Fract.Engng.Mater.Struct.*, 16, 165.

Lukáš, P., L. Kunz, B. Weiss, R. Stickler and W. Hessler (1989). *Mater. Sci. Eng.*, A118 , L1.

Ma, B. T., Z. C. Wang, A. L. Radin and C. Laird (1990). *Mater. Sci. Eng.*, A129, 197.

Peters, K. F., S. Radin, A. Radin and C. Laird (1989). *Mater. Sci. Eng.* A110, 115.

Wang, Z. and C. Laird (1988a). *Mater. Sci. Eng.*, A100, 57.

Wang, Z. and C. Laird (1988b). *Mater. Sci. Eng.*, A101, L1.

PREDICTING NOTCH FATIGUE LIFETIMES

H. Y. Ahmad[1], M. P. Clode[1] and J. R. Yates[2]

1 Mechanical Engineering Department, King's College London, Strand London WC2R 2LS, UK.
2 SIRIUS, Department of Mechanical and Process Engineering, University of Sheffield, Mappin Street, Sheffield, S1 3JD, UK.

ABSTRACT

An elastic-plastic model was used to predict the fatigue lifetime of notched specimens. The model is based on experimental short and long crack growth results from un-notched specimens and the distribution of strain ahead of the notch root. Different notch shapes and geometries were tested in this investigation. A comparison was made between predicted and experimental data and good agreement was found.

KEYWORDS

Fatigue lifetime, notches, notch plastic zone size, short crack, notch size effect.

INTRODUCTION

Fatigue crack initiation at notch root requires some plasticity. It follows that the initiated crack is contained within the influence of the notch plastic stress-strain field and is called a "short crack". Thus, elastic theory such as Linear Elastic Fracture Mechanics (LEFM) can neither characterise the initiation stage nor predict the behaviour of the short crack. When the crack becomes long enough to propagate beyond the notch effective zone as a result of generating its own plasticity at the crack tip, LEFM, may correlate its growth rate successfully (Hammouda et al 1980). However, in this paper the elastic-plastic model developed by (Ahmad et al 1994a) has been used to predict the fatigue lives of specimens with different notch shapes and geometry.

EXPERIMENTAL PROCEDURE

The material used in this work was Q1N (HY80) steel which is widely used in marine and shiphull construction. All fatigue tests were performed under displacement control at a frequency of 0.5 Hz. Notched fatigue tests were performed on 12mm diameter round bars with circumferential notches. The three chosen profile were of semi-circular, V-notch 35°and 55° and root radius 0.8 and 0.2 mm respectively. The notches were of 1mm and 0.5mm depth.

Tests were carried out at nominal strain amplitudes between 0.17% and 0.25% . Notch root strains were calculated by a strain energy method proposed by (Glinka 1985). Crack growth in notched specimens was monitored by d.c. potential difference using four pairs of probes around the notch and an established crack length calibration curve (Ahmad et al 1994b).

MODELLING AND LIFETIME PREDICTION

(Ahmad et al 1994a) have presented an elastic-plastic model based on fracture mechanics for predicting fatigue crack growth rates in notched specimens. The model consists two parts. The first part, which describes the growth rate of short cracks which are either fully or partially submerged in a notch tip plastic zone. In this regime growth is assumed to occur by a stage I mode II (Ahmad et al 1994b) process in which the crack grows from its initial length a_i to the crack size (transition crack length a_t) corresponding to the notch plastic zone size Δ_n (i.e. $\Delta_n = a_t$). The growth rate expression proposed by (Ahmad et al 1994a) is

$$\frac{da}{dN} = C1 \, \varepsilon_a^{\ m} \, (\Delta n - (a + e)) \tag{1}$$

where C_1 and m are the material properties, da/dN is in mm/cycle, a is the crack length, e is a notch contribution to crack length proposed by (Smith et al 1977, 1978), ε_a is the strain amplitude ahead of the notch and Δ_n is the notch plastic zone. The notch plastic zone can be determined from finite element analysis.

The second part, which describes long crack growth. This is the final region of accelerating crack growth from a_t to failure a_f. In this region the generalised crack growth equation includes threshold and high strain terms was proposed by (Ahmad et al 1994a) as

$$da/dN = C2 (\sqrt{a} \ \varepsilon_a)^m + H a \ \Delta\varepsilon_p^{1+2n} - T \qquad (2)$$

where $C2$ is a constant which is function of the material properties and the notch correction term for the stress intensity factor proposed by (Yates 1991). The high strain term was proposed by (Tomkins 1968, 1971) and used (Ahmad et al 1994a) to describing the long crack growth at notches as

$$\frac{da}{dN} = H a \ \Delta\varepsilon_p^{(1+2n)} \qquad (3)$$

hence,

$$H = \frac{\pi^2}{8} \left(\frac{k}{(1-R) \ \sigma UTS}\right)^2 \frac{1}{(1+2n)} \qquad (4)$$

where H is a constant, when the material and the load ratio are fixed, n and k are the cyclic strain hardening exponents, σ_{UTS} is the cyclic ultimate tensile stress, and R is a load ratio.

The fatigue crack growth threshold T term was included in Paris Erdogan power law $da/dN=C(\Delta K)^m$, and was determined as 5.49×10^{-10} m/c (Ahamd et al 1994a).

Fatigue Lifetime Assessment

Fatigue lifetime assessment is based on an integration procedure which sums damage in both regimes, with a demarcation from short to long crack growth occurring at the notch plastic zone boundary.

The short crack regime equation (2) is integrated and summed from an initial crack a_i in a notched specimen, to the transition crack length a_t using the following expressions:

$$Ns = \frac{1}{C1 \ \varepsilon_a^m} \int_{a_i}^{a_t} \frac{da}{\Delta n - a} \qquad (5)$$

The integration of the above equation is given by (Ahmad 1991). Life assessment in the long crack regime involves integrating equation (8) from a_t to the final crack length a_f. The integration is evaluated numerically by substitution of suitable values of a, n, ε_a and ε_p to get N_l. Thus, the total life, N_t, is simply $N_t = N_s + N_l$.

RESULTS AND DISCUSSION

Experimental results giving the number of fatigue cycles to a given crack size versus strain amplitude for cracks in notched specimens, are plotted in Figs (1-3). These figures compare experimentally determined notched specimens total fatigue life data with the predicted propagation life curves obtained from equation (5) and the numerical integration of equation (2). The limits of the integration in equations (5) were taken to be the initial crack length a_i which taken to be the peak to trough surface roughness measurement RT of 0.5mm, to transition crack length a_t which is taken as plastic zone size Δ_n. The limits of integration of equation (2) were taken to be the transition crack length a_t to failure a_f, where a_f was taken to be the radius of the net section of the specimen.

Fig(1) shows the comparison of the predicted life time for two different shapes of notch, semi-circular notch 1mm depth, and V-shape 55° notch 1mm depth and root radius $\rho = 0.2$mm. The predicted life times were plotted with experimental data obtained from push-pull fatigue tests. Also in this figure, the effect of Δ_n on the lifetime for the two difference notches can be noticed.

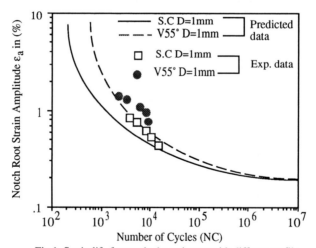

Fig 1. Strain life for notched specimens with differnt profile.

Figs (2-3) illustrate the effect of the notch depth on the lifetime. Fig 2 shows the predicted lifetime curve (solid line) for a semi-circular 1mm deep specimen and (dotted line) for a semi-circular 0.5mm deep specimen, compared with the appropriate notch experimental results. When the notch depth is bigger the contribution to the crack length 'e' will be bigger which will affect the size of the short and long crack growth zone, which in turn controls the lifetime of the notched specimens. For example the greater the notch depth, the larger the notch contribution of the crack, the longer the effective crack length, the faster the crack growth rate, so the shorter the fatigue lifetime. In all cases (Figs (2-3)) the larger notch leads to a shorter fatigue lifetime. Good agreement was found from the model with experimental data.

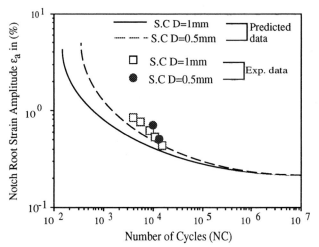

Fig 2. Strain life for notched specimens with differnt depth.

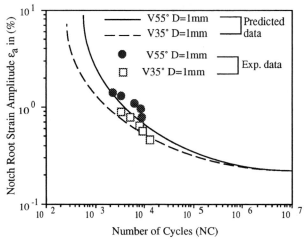

Fig 3. Strain life for V notched specimens with differnt profile.

CONCLUSIONS

A study has been made of fatigue lifetime prediction of different notch profiles in structural steel subjected to fully reversed tension-compression loading. An elastic-plastic model was used which is based on experimental short and long crack growth results from un-notched specimens and the distribution of strain ahead of the notch root. Good agreement was found between predicted and experimental data . It was found that the greater the notch depth, the larger the notch contribution to the crack length, the faster the crack growth rate, so the shorter the fatigue lifetime.

REFERENCES

Ahmad, H. Y. (1991). Fatigue crack growth at notches. Ph.D. Thesis, The University of Sheffield, UK.

Ahmad, H. Y. and J. R. Yates (1994a). An elastic-plastic model for fatigue crack growth at notches. Fatigue Fract. Engng Mater. Struct., 17, 651-660.

Ahmad, H. Y. and J. R. Yates (1994b). The influence of notch plasticity on short fatigue crack behaviour. Fatigue Fract. Engng Mater. Struct., 17, 235-242.

Glinka, G. (1985). Energy density approach to calculation of inelastic strain-stress near notches and cracks. Engng Fract. Mech., 22, 485-508.

Hammouda, M. M. and K.J. Miller (1980). Prediction of fatigue lifetime of notched members. Fatigue Engng Mater. Struct., 2, 377-386.

Smith, R. A and K. J. Miller (1977). Fatigue cracks at notches. Int. J. Mech. Sci., 19, 11-22.

Smith, R. A and K. J. Miller (1978). Prediction of fatigue regimes in notched components. Int. J. Mech. Sci., 20, 201-206.

Tomkins, B (1968). Fatgiue crack propagation-an analysis. Phil. Mag., 18, 1041-1066.

Tomkins, B (1977). Fatgiue failure in high strength metals. Phil. Mag., 23, 687-703.

Yates, R. J. (1991) A simple approximation for the stress intensity factor of a crack at notch. J. Strain Anal. Engng Design, 26, 9-13.

THE EFFECTS OF HIPPING, NOTCHES AND NOTCHES CONTAINING DEFECTS ON THE FATIGUE BEHAVIOR OF 319 CAST ALUMINUM ALLOY

A.A. Dabayeh, R.X. Xu and T.H. Topper

Deptartment of Civil Engineering, University of Waterloo, Waterloo, Ont., N2L 3G1, Canada

ABSTRACT

This paper reports the results of fatigue tests of a cast 319 aluminum alloy and the prediction of the fatigue lives by a crack growth model for short and long cracks. Fatigue tests were performed on smooth specimens in the as cast, hipped (subjected to very high pressure at an elevated temperature after casting), and notched conditions. Fatigue tests were also performed on a second set of notched specimens in which a small artificial flaw equal in size to a large casting pore was machined into the notch root. The hipped material showed an improvement of almost 100 % in fatigue strength over that of as cast material. Notched specimens and notched specimens with a drilled flaw at the notch root showed a reduction in fatigue strength of 25 % and 40 % from the as cast material fatigue strength respectively. The effect of casting defects on the fatigue life behavior of cast Al 319 alloy was modeled using a modified fracture mechanics crack growth model. The crack growth model used a numerical integration of the stress intensity factor solution for a crack growing from a notch and for a crack growing from a flaw at a notch root to predict the fatigue life. The model predictions were in good agreement with the experimental results.

KEYWORDS

fatigue, crack growth, cast aluminum, flaw, notch, prediction.

INTRODUCTION

Based on investigations of the effect of casting defects on the fatigue behavior of cast metals, it has been shown by various investigators [Evans et al.; 1956, Heuler et al.; 1992, Güngör and Edwards, 1993; Couper et al., 1990; Stanzl-Tschegg et al., 1993; Skalleurd et al., 1993] that: 1- Almost all cracks in cast aluminums and steels initiate from casting defects. 2- Surface and near surface defects initiate cracks sooner than defects embedded in the specimen. 3- An as cast surface causes a reduction in fatigue strength of about 20% when compared to polished surfaces.

In studying the influence of porosity and casting defects on fatigue crack growth Gerard and Koss [1991] found that the presence of porosity enhances short crack propagation adjacent to the pores as a result of localized, pore-induced plasticity. Their observations showed that once the crack extends beyond the plastic zone of an isolated pore it decelerates to propagation rates similar to those observed for fully dense material. This deceleration arises from a decrease in the stress concentrating effect of the defect as crack length increases.

Ting and Lawrence [1993] modeled the effect of casting defects on the fatigue life of as cast and smooth specimens of a 319 cast aluminum alloy. Fatigue life was assumed to be the sum of a crack nucleation life and a crack propagation life that included the growth of short and long cracks. They quantified the fatigue initiating defect depth "D" by measuring the square root of its projected area, T, onto the plane normal to the applied stress direction. The defect depth was then estimated based on the known T and the aspect ratio AR which is defined as the crack depth divided by half-crack length on the surface as follows:

$$D = T \sqrt{AR} \qquad (1)$$

The model then equated a pore to a notch having a diameter equal to D. They estimated the crack nucleation life for a crack initiating at a notch using the local-strain approach. They also estimated the crack propagation life by integrating the paris power law from a specified nucleated crack length, L_t, (as defined by Dowling, 1993), to the final crack length at failure. The model results indicated that nucleation life for a crack emanating from the casting defects was negligible and that the effect of defect size on fatigue life was more severe at the low stress ranges than that at high stress ranges. These results on cast aluminum alloys are for smooth components, however no studies have previously been made of the effect of a casting defect located at a notch root.

MATERIAL AND EXPERIMENTAL METHODS

The material used in this study is 319 cast aluminum alloy in a high gas condition. The high gas material had the number and size of the gas pores artificially increased by the addition of a potato to the molten metal. The chemical composition and mechanical properties of the material are given in table 1. The smooth specimens were prepared in accordance with ASTM standard E606 for constant amplitude low-cycle fatigue tests. The preparation included hand polishing of the gauge section in the loading axis direction with emery paper of grades 400 and 600. Final polishing was done using a metal polish which left a highly reflective surface which aided crack observations in crack growth tests.

All tests were carried out in a laboratory environment at room temperature (23°c) using a uniaxial , closed-loop, servo-controlled electrohydraulic testing machine. For crack growth rate measurements a K-decreasing procedure was used that minimized load interaction effects.

Table 1.Chemical composition (% by weight) and mechanical properties

alloy	Si	Cu	Cr	Mn	Mg	Fe	Ni
319	5.9	3.4	0.07	0.38	0.25	0.91	0.07

Material	0.2 % Yield Stress (MPa)	Ultimate Tensile Stress (MPa)	Elongation (%)
Al 319 High gas	145	193	1.08

A traveling microscope with a magnification of 900X was mounted on the test machine facing the specimen. A vernier with an accuracy of 0.0127 mm was attached to the microscope to measure changes in crack length. A first program, for measuring the intrinsic crack growth rates, involved testing at a constant amplitude with a high R-ratio (R>0.75). In a second program a block loading sequence was used to again measure the intrinsic crack growth rates of the three cast aluminum alloys. Each block consisted of one underload followed by 500 small cycles at a high R-ratio (R>0.75).

EXPERIMENTAL RESULTS

Crack growth rates were measured under constant and variable amplitude loading. Constant amplitude crack growth data plotted in terms of crack growth rate versus the effective stress intensity are shown in Fig. 1. An R-ratio of 0.8 was used to obtain crack growth rates in terms of the effective stress intensity. The effective threshold stress intensity factor range obtained was 1.5 MPa $\sqrt{m}$. An underload program was also applied to obtain the effective stress intensity versus crack growth rate curve. Underload stress levels were -82.8 and +41.4 MPa. The number of small cycles between underloads was 500. The crack growth rate, $(da/dn)_t$, for the block loading history was determined, then the crack growth rate for the underload cycles alone was determined during the block test by inserting a group of underload cycles and measuring their growth rate, $(da/dn)_{U.L.}$, between measurements of block growth rates. The crack growth rate for the small cycles in the block test, $(da/dn)_s$, was obtained by subtracting the growth per block due to the underload cycle, $(da/dn)_{U.L.}$, from the total growth per block, $(da/dn)_t$, and divided by the number of cycles in the block as follows:

$$(da/dn)_s = \frac{501 \ (da/dn)_t - (da/dn)_{U.L.}}{500} \qquad (2)$$

FATIGUE TESTS

Strain-life constant amplitude fatigue tests at R=-1 were conducted on smooth 319 cast aluminum alloy in the as cast and the hipped conditions. Hipping is a process in which the cast material is subjected to a high pressure at high temperature and then slowly cooled to eliminate flaws. Also strain life constant amplitude fatigue tests, R=-1, were conducted for 319 cast aluminum alloy specimens having a 3 mm hole and stress life constant amplitude fatigue tests for notched 319 cast aluminum specimens having an artificial flaw of 0.6 mm diameter drilled at the notch root of a 6 mm diameter semi-circular edge notch. The position and shape of the artificial flaw drilled is shown in Fig. 2.

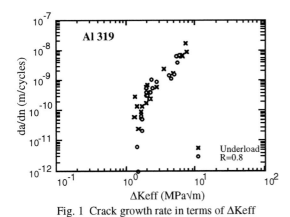

Fig. 1 Crack growth rate in terms of $\Delta Keff$
for Al 319 high gas alloy

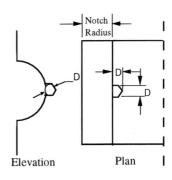

Fig. 2 Position and shape of the
artificial flaw drilled

FRACTURE MECHANICS ASSESSMENT

Fatigue life prediction was done based on a fracture mechanics model. The components of the model are described below:

1- Local Stresses and Strains

For this study both smooth or semi-circular notched components are used. Local stresses and strains were calculated using neuber's rule

$$Kt = \sqrt{K_\sigma . K_\varepsilon} \qquad\qquad (3)$$

However, since we are dealing with cracks in notches the theoretical stress concentration factor K_t for the surface is replaced by the elastic stress concentration factor K_p [Topper and El-Haddad, 1981], for a crack in a notched geometry. Equation 3, after being modified by using K_p instead of K_t, is then solved together with the material cyclic stress-strain equation to obtain the local stresses "σ" and the local strains "ε" in the vicinity of the crack tip as the crack grows in the notch field.

2- Crack Opening stress

A crack opening stress equation for constant amplitude loading [Duquesnay, 1991] was used to calculate the portion of the cycle which is effective in driving the crack in constant amplitude fatigue. The equation takes the following form:

$$S_{op} = \theta\ \sigma_{max}\ \{1 - (\ \frac{\sigma_{max}}{\sigma_y}\ )^2\ \} + \beta\ \sigma_{min} \qquad (4)$$

where σ_{max} and σ_{min} are the local maximum and minimum stresses, σ_y is the cyclic yield stress and θ and β are material constants which were determined from crack opening stress measurements on a 2024-T351 aluminum alloy [Dabayeh, 1994]. The values of θ and β were found to be 0.55 and 0.2 respectively and were assumed to be the same for all aluminum alloys. Some crack opening stress measurements were made for the cast aluminum alloy and were found to agree with the values of θ and β found before [Dabayeh, 1994].

3- K-Solutions

The K-solutions used in this study are strain based intensity factor equations for plastically strained short cracks. They describe the strain field ahead of the crack when the crack length is very small. The strain based intensity factor replaces the conventional stress intensity factor which does not apply to small cracks. The strain based K-solution used has the following form:

$$\Delta K = F\ \Delta\varepsilon\ E\ Q_\varepsilon\ \sqrt{\pi\ l} \qquad (5)$$

where   F is a geometric factor that accounts for crack front shape and finite specimen size.
        $\Delta\varepsilon$ is the local strain range.
        E is Young's modulus.
        Q_ε is a surface strain concentration factor [Abdel-Raouf et al., 1991] that accounts for short crack behavior . The surface strain concentration factor can be expressed as [Abdel-Raouf et al., 1991]:

$$Q_\varepsilon = \frac{\Delta\varepsilon}{\Delta e} = 1 + 5.3\ e^{-(\alpha l/D)} \qquad (6)$$

where   D is the grain size in the crack growth direction.
        α is a factor that accounts for the ease of cross slip in the material.
The surface strain concentration factor, Q_ε, decays rapidly with crack length and equation (5) converges with the long crack strain intensity factor equation.

The irregular shape of a shrinkage cavity was simplified to a spherical cavity. The pore in smooth specimens was equated to an edge notch having a diameter equal to that of the simplified defect or flaw. The elastic stress concentration factor K_p for a crack emanating from a notch was based on an equation proposed by Topper and El-Haddad [1981]. This equation takes the following form for circular notches:

$$K_p = \frac{K_t}{2}\ \sqrt{\frac{c/2}{c/2 + 1}}\ (1 + \frac{c/2}{c/2 + 1}\ ) \qquad (7)$$

where
        c   is the notch radius.
        l   is the crack length measured from the edge of the notch.

For large cracks equation 7 should be replaced by $\sqrt{\frac{1 + c}{1}}$ which assumes that the effective length of a crack is equal to the actual crack length, l, plus the notch depth. The stress concentration factor K_p accounts for the increase in crack tip stress due to the notch. It decreases from an initial maximum value of "K_t" to a value of $\sqrt{\frac{l+c}{l}}$ as the crack passes outside the field of influence of the notch. Exact solutions for the elastic stress concentration factor solution, K_p, for a flaw in a notch are not available. An approximate stress intensity factor solution for a defect at the notch root has been calculated using available stress intensity factor solutions for individual notches and spherical cavities. The stress intensity factor solution for a penny shaped crack emanating from an ellipsoidal cavity in an infinite solid body by Murakami et al [1982] was used in combination with the stress intensity factor solution for embedded semi-elliptical cracks in finite notched plates under tension by Neuman and Raju [1981]. This approximation is valid only when the ratio of the notch size to the cavity size is large. In this case the stress gradient from a notch decreases slowly from a spherical cavity so that the stress applied to a flaw can be considered to be uniform over much of the fatigue life.

FATIGUE LIFE PREDICTION

Fatigue life prediction is carried out by a numerical integration along the reference closure free crack growth curve between initial and final crack lengths l_0 and l_f respectively as:

$$N_f = \int_{l_0}^{l_f} \frac{da}{f(\Delta K_{eff})} \qquad (8)$$

Basinski and Basinski [1985] and Hunsche and Neumann [1986] defined cracks as singularities in a crystal related to surface intrusions and extrusions deeper than 3mm developed in persistent slip bands. The value of l_0 was assumed to be 3mm while l_f was determined by the onset of fast fracture. The crack growth stages for a crack around a flaw at the notch root, assumed in the model, are shown schematically in Fig. 3 and are summarized in the following five stages:

1- Fast crack growth of the crack around the flaw to a semi-circular crack. i.e No crack growth calculations were done and the crack around the flaw was assumed to jump to a semi-circular crack. **2-** The growth of a semi-circular crack under the effect of the flaw and notch stress fields until the crack leaves the flaw stress field. **3-** The growth of a semi-elliptical crack, which has different crack growth rates in the deep and surface directions, under the effect of the notch stress field until the crack reaches the specimen edges. **4-** Fast crack growth from a semi-elliptical crack to a through crack. **5-** Growth of a through crack under the notch stress field till fracture.

COMPARISON OF MODEL CALCULATIONS WITH EXPERIMENTAL RESULTS

The crack growth model requires the calibration of the coefficient (α/D) in equation (6). The procedure followed for calibration was to give values to (α/D) that forced the fatigue life prediction to agree with the fatigue lives of smooth specimens tested under R=-1 at a constant amplitude loading. After calibration the calibrated model was used to predict the fatigue life for the various geometries. Fig. 4 shows the predicted fatigue lives together with the experimental results for Al 319. The model prediction agrees well with the experimental results.

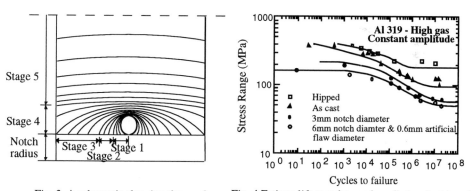

Fig. 3 A schematic showing the crack growth stages

Fig. 4 Fatigue life test data and prediction for hipped, as cast, notched and notched with artificial flaw Al 319

DISCUSSION

It has been possible through the use of a simple fracture mechanics model to predict the experimentally observed fatigue lives of 319 cast aluminum alloy. The analysis has shown that a flaw in a smooth cast component can be quantified by a circular notch having the same diameter as the flaws. It also shows that for the case of a large notch size to flaw size ratio the individual stress intensity factor solutions for a crack emanating from a flaw and a crack emanating from a notch can be combined to solve for the stress intensity factor of a crack emanating from a flaw at a notch root. However it is suggested that for the case of a small notch size to flaw size ratio a more accurate solution may be necessary. The fatigue test results showed that the hipping process results in significantly improved constant amplitude fatigue strength. A decrease in the maximum pore size from 0.9 mm to 0.2 mm results in an improvement of about 100 % from the as cast fatigue strength. The fatigue test results also showed a drop in fatigue strength from the smooth as cast speciemns to the 3 mm diameter notched specimens and a further reduction due to the 0.6 mm diameter artificial flaw at a 6 mm diameter semi-circular edge notch.

CONCLUSIONS

From this work we can draw the following conclusions:
1- Reducing the defect size by the hipping process increases the constant amplitude fatigue strength of the 319 cast aluminum alloy.
2- Notches reduced the constant amplitude, R=-1, as cast fatigue strength and the introduction of an artificial flaw at the notch root resulted in a further reduction.
3- A simple fracture mechanics model, which takes crack closure and short crack behavior into account, gave a satisfactory fatigue life predictions for smooth and notched cast components.

REFERENCES

Abdel-Raouf, H., T.H. Topper and A. Plumtree (1991). A short fatigue crack model based on the nature of the free surface and its microstructure. Scr. Met. Mat., 25, 597-602.

Basinski, Z.S. and S.J. Basinski (1985). Low amplitude fatigue of copper single crystals. Acta Metall., 33, 1307.

Couper, M.J., A.E. Nesson and J.R. Griffiths (1990). Casting defects and the fatigue behavior of an aluminum casting alloy. Fatigue Fract. Engng Mater. Struct., 13, 213-227.

Dabayeh, A (1994). "Changes in crack opening stress after overloads in a 2024-T351 aluminum alloy" M.A.Sc. thesis, University of Waterloo, Waterloo, Ontario, Canada.

Dowling, N.E (1979). Fatigue at notches and the local strain and fracture mechanics approaches. ASTM STP, 677, 247-273.

Duquesnay, D.L (1991). Fatigue damage accumulation in metals subjected to high mean stress and overload cycles. Ph.D. Thesis, University of Waterloo, Waterloo, Ontario, Canada.

Evans, E.B., L.J. Elbert, and C.W. Briggs (1956). Fatigue properties of cast and comparable wrought steels. ASTM proc., 56, 1-32.

Gerard, D.A. and D.A. Koss (1991). The influence of porosity on short fatigue crack growth at large strain amplitudes. Int. J Fatigue, 13, 345-352.

Güngör, S. and L. Edwards (1993). Effect of surface texture on fatigue life in a squeeze-cast 6082 aluminum alloy. Fatigue Fract. Engng Mater. Struct., 16, 391-403.

Heuler, P, C. Berger, and J. Motz (1992). Fatigue behavior of steel castings near-surface defects. Fatigue Fract. Engng Mater. Struct., 16, 115-136.

Hunsche, A. and P. Neumann (1986). Quantitative measurement of persistent slip band profiles and crack initiation. Acta Metall., 34, 207.

Murakami, Y., T. Norikura and T. Yasuda, T (1982). Stress intensity factors for a penny-shaped crack emanating from an ellipsoidal cavity. Trans. Japan Soc. Mech. Engrs., 48, No. 436, 1558-1565.

Neuman, J.C., Jr. and I.S. Raju (1981). Stress intensity factor equations for cracks in three dimensional finite bodies. NASA technical memorandom 83200, 1-49.

Skallerud, B., T. Iveland and G. Härkegård (1993). Fatigue life assessment of aluminum alloys with casting defects. Engineering Fracture Mechanics , 44, 857-874.

Stanzl-Tschegg, S.E., H.R. Mayer, E.K. Tscheg and A. Beste (1993). In-service loading of AISi11 aluminum cast alloy in the very high cycle regime. Int. J Fatigue, 15, 311-316.

Ting, J.C. and Jr. F.V. Lawrence (1993). Modeling the long-life fatigue behavior of a cast aluminum alloy. Fatigue Fract. Engng Mater. Struct., 16, 631-647.

Topper, T.H. and M.H. El Haddad (1981). Fatigue strength prediction of notches based on fracture mechanics. Proceedings of the international symposium on fatigue thresholds, Fatigue Thresholds Stockholm, Sweeden, 2, 777-798.

THE d*-CONCEPT - A MODEL FOR DESCRIPTION AND PROGNOSIS OF STRENGTH-PHENOMENA OF HIGHLY INHOMOGENEOUS STRAINING AT CYCLIC LOADING

S. SÄHN and V.B. PHAM

Technische Universität Dresden, Institut für Festkörpermechanik, Germany

ABSTRACT

The endurance limit of sharp notched specimens, the threshold of small cracks and the evolution of cyclic damage ahead of sharp notches were studied. The results are analysed in terms of an integral straining parameter ΔB_{V_m} that is derived by averaging the calculated local stress and strain distribution.
The investigations indicate that this straining parameter is well suited to describe the phenomena of the fatigue behaviour striking in endurance limit range as well as in the finite life region as damage evolution ahead of sharp notches till crack initiation and fatigue crack growth of small cracks. The effects of crack length, notch root radius and material behaviour are realistically described. In particular the phenomena of crack closure of steel METASAFE 900 was studied. The influence of crack closure to the parameter ΔB_{V_m} was discussed.

KEYWORDS

Fatigue, crack initiation, crack propagation, notches, small cracks.

Introduction

The state of straining at cracks and sharp notches is highly inhomogeneous and multiaxial. The assessment of this state of straining demands knowledge of
- a qualified equivalent straining ΔB_V (equivalent stress or strain) according to uniaxial state of stress
- and equivalent straining parameter ΔB_{V_m} for comparison of largely homogenous with inhomogeneous strainings.

Both of them are dependent on material and state of material. Limit states at cyclic loading are also influenced by multiaxiality of stress state like critical states at quasistatic loading. In this contribution the knowledge of a suitable cyclic equivalent straining ΔB_V is assumed.
Object here is a straining parameter for notches and cracks derived from local strainings. Different strength phenomena at cyclic loading can be described with that.

Such straining parameters at cyclic loading ΔB_{V_m} are found from local cyclic stresses and strains. That is a mean equivalent cyclic straining over a region that considerably depends on the microstructure of the material too (Neuber, 1958; Sähn et al., 1987; Sähn and Göldner, 1993)

$$\Delta B_{V_m} = \frac{1}{V^*} \int\limits_{(V^*)} \Delta B_V \, dV \overset{resp.}{=} \frac{1}{d^*} \int\limits_{(d^*)} \Delta B_V(r) \, dr \qquad (1)$$

Here are V^*- characteristic volume; d^*- characteristic length (Neuber: "Ersatzstrukturlänge"). V^* and d^* depend on the microstructure of material.
In principle we get the same straining parameter at sharp notches and cracks with $\Delta B_V(l^*)$. l^* is the distance from notch ground resp. the crack tip. l^* depends also on the microstructure of materials.

Strength Phenomena at Linear-elastic Material Behaviour

The endurance limit of differently notched specimens and the threshold values of small cracks can be derived with linear-elastic material law. The straining parameter is expressed for specimens or constructions with cracks according to the normal stress hypothesis by equ. (1) with exact stress distribution on cracks (Sähn et al., 1987)

$$\Delta B_{Vm} = \Delta \sigma_{1m} = \Delta \sigma_n \sqrt{1 + (2a)/d^*} = \sqrt{2/(\pi d^*)} \, \Delta K \sqrt{1 + d^*/(2a)} \tag{2}$$

with $\Delta K = \Delta \sigma_n \sqrt{\pi a}$; a - crack length

For sharp notches under normal stresses perpendicular to the notch we get

$$\Delta B_{Vm} = \Delta \sigma_{1m} = 2 \Delta \sigma_n \sqrt{a_K/\varrho} \cdot Y/\sqrt{1 + (2d^*)/\varrho} = \sqrt{2/(\pi d^*)} \, (\Delta K^K)/\sqrt{1 + \varrho/(2d^*)} \tag{3}$$

with $\Delta K^K = \Delta \sigma_n \sqrt{\pi a_K} \, Y$; a_K - notch length; ρ - radius of curvature; Y - geometry function.

Without statistical size effects we get from equ. (2) and (3) with $\Delta \sigma_{1m}^D = const$ for the endurance limit:

$$\Delta \sigma_{1m}^D = \Delta \sigma_D = \Delta \sigma_n^{DR} \sqrt{1 + (2a)/d^*} = \sqrt{2/(\pi d^*)} \, \Delta K_{th}^{Kl} \sqrt{1 + d^*/(2a)} = \sqrt{2/\pi d^*} \, \Delta K_{th} \tag{4}$$

and

$$\Delta \sigma_{1m}^D = 2 \Delta \sigma_{DK} \sqrt{a_K/\varrho} \, Y/\sqrt{1 + (2d^*)/\varrho} = \sqrt{2/(\pi d^*)} \, \Delta K_D^K/\sqrt{1 + \varrho/(2d^*)} \tag{5}$$

Here are $\Delta \sigma_D$ - endurance limit of unnotched specimens, $\Delta \sigma_n^{DR}$ - nominal endurance limit of specimens with cracks, ΔK_{th}^{Kl} - threshold value of small cracks, ΔK_{th} - threshold value of macro cracks, $\Delta \sigma_{DK}$ - nominal endurance limit of specimens with notches, ΔK_D^K - endurance limit of ΔK^K at notches.

From equ. (4) follows

○ for the threshold values of small cracks $\Delta K_{th}^{Kl}/\Delta K_{th} = 1/\sqrt{1 + d^*/(2a)}$ (4.1)

○ and for the endurance limit of specimens with cracks $\Delta \sigma_n^{DR}/\Delta \sigma_D = 1/\sqrt{1 + (2a)/d^*}$ (4.2)

Equ. (4.1) as well as equ. (4.2) (modified Kitagawa-diagram) are confirmed by many experiments (Tanaka et al., 1981). d^* in equ. (4.1) can be determined by experiments of threshold values of differently small cracks. In this case d^* is influenced by different crack closure at small cracks compared with macro cracks.

With the definition

○ of concentration factor $\alpha_K = \Delta \sigma_{max}/\Delta \sigma_n$

○ and of fatigue strength reduction factor $\beta_K = \Delta \sigma_D/\Delta \sigma_{DK} = \left(\Delta \sigma_D/\Delta \sigma_{1m}^D \right) \cdot \left(\Delta \sigma_{1m}^D/\Delta \sigma_{DK} \right)$

we get for sharp notches $\alpha_K = 2 \sqrt{a_K/\varrho} \, Y$ and $\beta_K = \left(\Delta \sigma_D/\Delta \sigma_{1m}^D \right) \cdot 2 \sqrt{a_K/\varrho} \, Y/\sqrt{1 + (2d^*)/\varrho}$

For the calculation of the endurance limit of sharp notched specimens there follows

$$\alpha_K/\beta_K = \left(\Delta \sigma_{1m}^D/\Delta \sigma_D \right) \sqrt{1 + (2d^*)/\varrho} \tag{6}$$

with $\Delta \sigma_{1m}^D/\Delta \sigma_D$ - describing the statistical size effect. If the failure probability is a Weibull-law then we get from comparison of the endurance limits of a plane notched specimen (with thickness B and radius of curvature ρ) with an unnotched specimen (with diameter d_0 and length l_0 : Sähn and Pyttel, 1991):

$$\Delta \sigma_{1m}^D/\Delta \sigma_D \approx [(\pi d_0 l_0)/(B \varrho)]^{1/m} \tag{7}$$

with m - Weibull exponent.

Equation (6) corresponds with a relation of Lukaš / Klesnil without statistical size effects (Lukaš ,1982) with

$\Delta \sigma_{1m}^D = \Delta \sigma_D$: $\alpha_K/\beta_K = \sqrt{1 + 4{,}5 \, (a_c/\varrho)}$ (6.1)

with $d^* = 2{,}25 a_c$; a_c - critical crack length.

Crack Initiation on Sharp Notches

The crack initiation at technical notches can be derived with local amplitudes of strain in notch ground ϵ_{Va} dependent on mean stress. At sharp notched specimens $\Delta B_{Vm} = \Delta \epsilon_{Vm} = \Delta \epsilon_{1m}$ resp. equ.(1) is a suitable straining parameter to describe the crack initiation (initiation of cracks with length of few μm, Sähn et al., 1994):

$$\left(\Delta e_{1m} - \Delta e_D\right)^n N_A = K_1\left(1 - \varepsilon_{mSt}/\varepsilon_c\right) \tag{8}$$

Here we have added linearly the damage by cyclic and static loading.
The local strain amplitudes are calculated with a FE-program and with the material law of Ramberg-Osgood:

$$\varepsilon_a = \sigma_a/E + \left(\sigma_a/K'\right)^{1/n'} \tag{9}$$

with n'$=0,165$; E$=2,0.10^5$ MPa; K'$=1610$ MPa for the steel METASAFE 900.

- Experiments without overloads on plane notched bending specimens out of METASAFE 900 are described by equ.(8) with material constants n$=1,8$; $\Delta\epsilon_D=0.004$; $K_1=0,4$; d$^*=50\mu$m and with N_A-Number of incipient crack cycles with an initiation crack of $a_A=5\mu$m. Figure 1 shows the geometry of the specimen

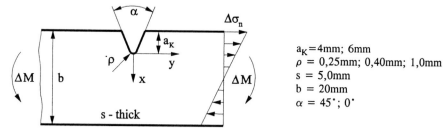

$a_K=4$mm; 6mm
$\rho = 0,25$mm; 0,40mm; 1,0mm
s $= 5,0$mm
b $= 20$mm
$\alpha = 45°; 0°$

Fig. 1:   Plane notched bending specimens

- The influence of overloads at cyclic loading near endurance limit is essentially determined by local mean stress and local damage through quasistatic strain ϵ_{mSt} and local strain oscillation range $\Delta\epsilon_{Vm}$. The overloads leave residual stresses that lead to change of endurance limit $\Delta\epsilon_D=\Delta\sigma_D/E$. The assessment of $\Delta\sigma_D$ is possible for steels with the Gerber-parabel:

$$\Delta\sigma_D/(2\sigma_w) = \sqrt{1 - \sigma_m/R_m} \tag{10}$$

Here are $\sigma_w=\sigma_{-1}$ - endurance limit under alternating loading, σ_m - mean stress, R_m - ultimate tensile strength. The damage by overloads is determined by $\epsilon_{mSt}/\epsilon_c$ with ϵ_c - critical strain dependent on multiaxiality of local stress state.

The local strains and stresses are determined with a FE-program under the condition of plane stress and plane strain for different hardening material law. Fig. 2 shows the local stress σ_y for combined hardening materials under plane strain state for a loading which leads to $N_A \approx 10^4$. The combined hardening describes the cyclic material law according to equ.:(9).
In Fig.2b is shown that the tension overloads in the phase of crack initiation ($a_A \leq 10\mu$m) have not a positive influence. But the region ahead will essentially positive influenced by pressure residual stresses after tension overloads (Fig.2c). This is the reason for the as a rule bigger lifetime after tension overloads.
Fig.3 shows the local strain oscillation range $\Delta\epsilon_y$ before and after an overload. The difference between both of them is small.

Propagation of Small Cracks

On the premises of small scale yielding we can describe the propagation of small cracks with the straining parameter:

$$\Delta B_{Vm} = \sqrt{2/(\pi d_0^*)} \, \Delta K \sqrt{1 + d_0^*/(2a)} \tag{11}$$

$$da/dN = C\left(\Delta K \sqrt{1 + d_0^*/(2a)} - \Delta K_{th}\right)^m \tag{12}$$

with   C, m - material parameter for describing propagation of macro cracks at R$=$const.

d_0^* - free parameter to adaptation on propagation of small cracks.

The parameter d_0^* is influenced by micro-support action, the different crack closure at small and macro cracks, and by different plastic deformations. The crack propagation is determined by cyclic local straining field and global by parameters out of the nominal strain field:

$$\Delta K_\varepsilon = \Delta \varepsilon_n E \sqrt{\pi a}\, Y$$

$$resp. \quad \Delta K_\gamma = \Delta \gamma\, G \sqrt{\pi a}\, Y$$

with   E - Young modulus, G - Shear modulus and $\Delta\gamma$- cyclic shear strain.

Experiments (Tokaji et al., 1986) to propagation of small cracks at $\Delta\sigma_n$=const are described satisfactorily by equ. (12) (Sähn, 1991). For instance: For fine grain C-steel S10C we get $C = 9,8.10^{-11}$, m=2,32, $d_0^* = 180\mu m$, $\Delta K_{th} = 8,07$ MPa$\sqrt{m}$, R=-1. Here are da/dN in m/LC and ΔK in MPa$\sqrt{m}$.

A conservative describing of small surface cracks is reached by equ.(12) with $\Delta K = \Delta\gamma G \sqrt{(\pi a)}Y$ at steel En8 (R_e=278 MPa, R_m=540 MPa) with C=1,37.10^{-10}, m=2,25, ΔK_{th}=5,3 MPa$\sqrt{m}$, d_0^*=200μm, at R=-1 (Küster, 1995).

Important portions of life time are taken into account with equ.(12). Also a formal loading sequence effect on life time is described with it (Sähn, 1991). In comparison with the Miner hypothesis two step loading experiments lead
- to a bigger life time at increasing loading ($\Sigma(N_i/N_{fi}) > 1$)
- and to a smaller life time at decreasing loading ($\Sigma(N_i/N_{fi}) < 1$).

Real loading effects, which are important at overloads and random tests, must be considered extra. Overloads influence the local residual stresses ahead of the crack tip and crack closure.

The continuum mechanical initiation of crack closure is dependent on a realistic modeling:

- First we take a straight crack without roughness (or profile) under bending loading. The opening and the closure of the crack here for the present are viewed without crack propagation. Fig.4 shows a crack opening profile of a crack with the crack length a=1mm at σ_{nmax}=133 MPa resp. K_{max}=8 MPa$\sqrt{m}$ and the crack opening profile at unloading until crack closure at σ_n=-1,9 MPa for elastic plastic material law (METASAFE). Table 1 shows the closure stress σ_{cl} for different lengths and different loadings. Tendentiously the influence on propagation of small cracks by crack closure dependent on crack length is discribed with d_0^* =const. in equ.(12).

Table 1:   Crack closure stress σ_{cl} for different crack lengths a and different K_{max}

K_{max} [MPa$\sqrt{m}$]	crack length a [mm]	0.075	0.1	0.5	1.0	5.0
4	σ_{nmax} [MPa]	233	202	92.5	66.5	30.2
	σ_{cl} [MPa]	-5.8	-2.0	-0.5	-0.5	0.0
8	σ_{nmax} [MPa]	466	404	185	133	60.3
	σ_{cl} [MPa]	-69.6	-56	-4.0	-1.9	-0.7
9.5	σ_{nmax} [MPa]	550	477	216	157	71.1
	σ_{cl} [MPa]	-139	-73	-6.4	-2.8	0.0

The first model gives crack closure only at pressure stresses, which essentially depend on crack length and on K_{max}. Practically it gives crack closure also already at tension stress at unloading. So we have to take an other model to describe this fact.

- Secondly we take a straight smooth crack, which is growing about Δa=25μm at σ_{nmax} for $\Delta\sigma_n$=210 MPa and R=0. Table 2 shows the beginning and the end of crack closure at unloading and of crack opening at reloading dependent on produced crack length.

Table 2:   Crack closure and crack opening stress at bending on notched specimens ($a_K = 4$mm, $\rho = 0.25$mm) with $\Delta\sigma_n = 210$ MPa, $R = 0$ for METASAFE 900 at crack propagation with $\Delta a = 25\mu$m.

crack length a [mm]	σ_n [MPa]			
	begining of crack closure	crack complete closed	begining of crack opening	crack complete opened
0.025	75	55	70	80
0.100	70	15	25	75
0.200	65	30	35	70
0.300	60	20	30	65
0.400	50	10	20	60
0.500	50	10	15	55

Out of plasticity-induced crack closure the crack closure will be influenced by roughness of crack surfaces and by the oxidation. Therefore the crack closure phenomena should be examined by calculation with continuum mechanics and by purposeful experiments.

ACKNOWLEDGMENT

The financial support of this research by the Deutsche Forschungsgemeinschaft is gratefully acknowledged.

REFERENCES

Neuber, H. (1958). *Kerbspannungslehre*, Springer-Verlag, Berlin, Göttingen, Heidelberg.

Sähn, S.; Göldner, H. *et al.* (1987). Beanspruchungsparameter für Risse und Kerben bei statischer und zyklischer Belastung. *Technische Mechanik* **8**, S. 5-17.

Sähn, S.; Göldner, H. (1993). *Bruch- und Beurteilungskriterien in der Festigkeitslehre*. Fachbuchverlag Leipzig, Köln.

Lukaš, P. (1982). Ermüdungsrißausbreitung in gekerbten Körpern. VI.Symposium Verformung und Bruch. Magdeburg.

Tanaka, K.; Nakai, Y.; Yamashita, M. (1981). Fatigue growth of small cracks. *Intern. J. of Fracture*, **17**, pp. 519-533.

Sähn, S.; Pyttel, Th. (1991). Berechnung der Kerbwirkungszahlen und Kurzrißverhalten bei zyklischer Belastung. 17. MPA-Seminar. Stuttgart.

Sähn, S.; Schaper, M. *et al.* (1994). Schädigungsentwicklung an scharfen Kerben bei zyklischer Belastung. 20. MPA-Seminar. Stuttgart.

Tokaji, K.; Ogawa, T.; Harada, Y. (1986). The growth of small cracks in a low carbon steel, the effect of microstructure limitations of linear elastic fracture mechanics. *Fatigue Fracture Engng. Mater. Struct.* **9**, pp. 205-217.

Sähn, S. (1991). Festigkeitsverhalten von Bauteilen mit kleinen Rissen und Kerben bei zyklischer Belastung. *Konstruktion* **43**, S. 9-16.

Küster, M. (1995). Einfluß von Oberflächendefekten auf die Ausbreitung kurzer Ermüdungsrisse. Großer Beleg, TU Dresden, produced on University of Sheffield.

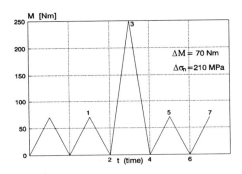

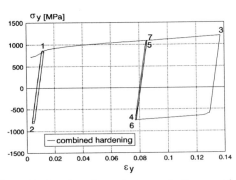

Figure 2a Loading dependent on time with
          overload

Figure 2b Local strains and stresses in the ground
          of a notch for plane strain state
          (combined hardening) at specimen
          geometry $\rho=0,25$mm, $a_K=4$mm,
          b=20mm, $\alpha=45°$, s=5mm

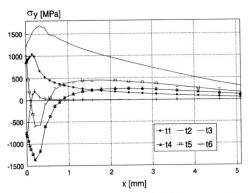

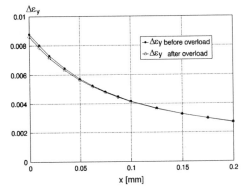

Figure 2c Stress component σ_y in the ligament of
          notch dependent on time

Figure 3 Strain oscillation range $\Delta\epsilon_y$ before and
         after an overload for plane strain.

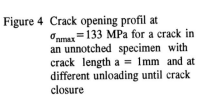

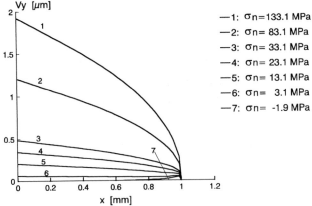

Figure 4 Crack opening profil at
         $\sigma_{nmax}=133$ MPa for a crack in
         an unnotched specimen with
         crack length a = 1mm and at
         different unloading until crack
         closure

—1: $\sigma_n=133.1$ MPa
—2: $\sigma_n=83.1$ MPa
—3: $\sigma_n=33.1$ MPa
—4: $\sigma_n=23.1$ MPa
—5: $\sigma_n=13.1$ MPa
—6: $\sigma_n=3.1$ MPa
—7: $\sigma_n=-1.9$ MPa

LOW-CYCLE FATIGUE STRENGTH
OF NOTCHED COMPONENTS FOR LOW-CARBON STEELS

AKIHIRO MORIMOTO, ATSUSHI SAITO and EIICHI MATSUMOTO

Department of Mechanical Engineering, Kinki University,
3-4-1 Kowakae Higashi-osaka Osaka, 577 Japan

ABSTRACT

The estimation of low-cycle fatigue life for the notched components is one of the most important technology for the structural materials. In this study, the low-cycle fatigue tests were carried out using the smooth round bar specimens and the notched plate specimens. The materials used are the hot-rolled low-carbon steel sheets with several strength. From the experimental results and analyses, the effect of material strength on low-cycle fatigue strength and the estimating method of low-cycle fatigue strength of notched components have been clarified.

KEYWORDS

Low-cycle fatigue; low-carbon steel; notch; stress-strain; fatigue strength reduction factor.

INTRODUCTION

The notched parts, such as a bolt hole and a welding part, exist in the members of large-scale structures. The fatigue cracks commonly initiate at these sites by repeated loadings, such as wind, earthquakes and traffic (Yamada *et al.*,1993).

In this study, the low-cycle fatigue tests were carried out using the smooth round bar specimens and the notched plate specimens. Then the effect of material strength on low-cycle fatigue strength and the estimating method on low-cycle fatigue strength of notched components were discussed.

EXPERIMENTS

Materials

The materials used are the three kinds of hot-rolled low-carbon steel sheets (SM400A, SM400B and SM490A in Japanese standard). The chemical compositions and the mechanical properties of them are shown in Tables 1 and 2.

Specimens

Table 1. Chemical compositions of materials (Wt %).

Materials	C	Si	Mn	P	S
SM400A	0.19	0.16	0.8	0.01	0.004
SM400B	0.13	0.23	1.07	0.02	0.006
SM490A	0.15	0.35	1.44	0.02	0.006

Table 2. Mechanical properties of materials.

Materials	U.Y.S. (MPa)	L.Y.S. (MPa)	T.S. (MPa)	EL. (%)	R.A. (%)	C.I.T. (J/cm^2)
SM400A	292	273	451	44.1	68.2	154
SM400B	256	256	430	43.0	78.0	279
SM490A	311	307	527	36.6	72.8	199

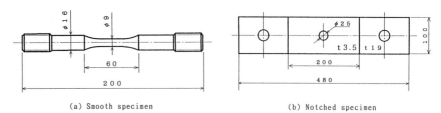

(a) Smooth specimen   (b) Notched specimen

Fig.1. Configuration of specimens.

Figures 1(a) and (b) show the fatigue specimens with dimensions of the smooth round bar and the notched plate, respectively. The axes of specimens are the same direction as the rolling direction. The smooth round bar specimen has a diameter of 9 mm and a parallel part of 30 mm. The notched plate specimen is the thickness of 3.5 mm, the width of 100 mm and has a center hole with a diameter of 25 mm (stress concentration factor K_t is 2.43).

<u>Low-cycle Fatigue Testing Method for the Smooth Round Bar Specimens</u>

The testing machine used is an electro-hydrauric servo testing instrument. The low-cycle fatigue experiment of the smooth round bar specimens was conducted under the axial strain-controlled conditions with a triangular wave form ($\dot{\varepsilon} = 2 \times 10^{-3}/s$). The fully reversed strain-controlled test (R = εmin/εmax=-1.0) and the repeated strain controlled test (R = εmin/εmax=0.7) were conducted. The length of strain controlled part is 25 mm.

<u>Low-cycle Fatigue Testing Method for the Notched Plate Specimens</u>

The low-cycle fatigue experiment of the notched plate specimens was conducted under the axial load-controlled conditions with a repeated sinusoidal wave form (6 Hz, R =Pmin/Pmax=0.1) and the strains at the root of notch were measured during the tests.

RESULTS AND DISCUSSIONS

<u>Results for the Smooth Round Bar Specimens</u>

The low-cycle fatigue test results of the smooth round bar specimens are shown in Fig.2. N_f is the cycle number to failure, when the maximum tensile stress has decreased to 75% of the stable maximum tensile stress. The following findings are obtained from this figure.
1)Each material has the same low-cycle fatigue strength, and the low-cycle fatigue life

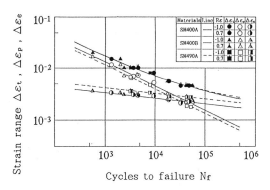

Fig.2. Results of low-cycle fatigue test
        for smooth specimens.

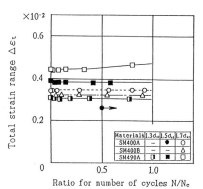

Fig.3. Strain variation at the root
        of notch with respect to
        number of cycles.

prediction by the similar equation to the universal slopes equation is practically applicable
(Manson, 1968).
(a) SM400A, SM400B

$$\Delta \varepsilon_t = \Delta \varepsilon_e + \Delta \varepsilon_p = 0.00754 N_f^{-0.112} + 0.314 N_f^{-0.458} \qquad (1)$$

(b) SM490A

$$\Delta \varepsilon_t = \Delta \varepsilon_e + \Delta \varepsilon_p = 0.00897 N_f^{-0.112} + 0.259 N_f^{-0.458} \qquad (2)$$

2) The effect of the mean strain on the low-cycle fatigue strength has not been recognized
   from the test results. The ratio of the mean strain to the fracture ductility ($\varepsilon_m/\varepsilon_f$) is
   0.01-0.02 in this test.

Results for the Notched Plate Specimens

In the members of large-structures, there are some possibilities of changes of cross sections.
The fatigue cracks commonly initiate at these sites.   Designing the civil engineering
structures, the allowable stress is increased proportionally when the wind load or the seismic
load is acting (in case of the wind load : 25% increase, in case of the seismic load : 70%
increase) (JSCE, 1980). Considering the increased rate of allowable stress on the actual
design, the low-cycle fatigue experiment was carried out on the assumption that the nominal
stress at the notched hole part was 1.3, 1.5 and 1.7 times as many as the allowable stress (
SM400A and SM400B : σ_{al}=137MPa, SM490A : σ_{al}=186MPa). Figures 3 and 4 show the total strain
range $\Delta \varepsilon_t$ variation and the mean strain ε_m variation with respect to number of cycles at the
root of notch, respectively.

As the strain ratio R = $\varepsilon_{min}/\varepsilon_{max}$ is about 0.7 at the root of notch, the effect of the mean
strain on the low-cycle fatigue life is neglected ($\varepsilon_m/\varepsilon_f$=0.01-0.02, ε_m: mean strain, ε_f:
fracture ductility) (JSME, 1983). Figure 5 shows the relationship between the total strain
range $\Delta \varepsilon_t$ at the root of notch and the number of cycles to cracking Nc (Iida et al., 1971),
corresponding to 1 mm of the crack length. The following findings are obtained from these
figures.
1) The total strain range $\Delta \varepsilon_t$ and the mean strain ε_m at the root of notch show little variation
   through the almost whole life except the beginning.
2) If the total strain range at the root of notch is equal to the total strain range of the
   smooth round bar specimen, the fatigue crack initiation life of the notched specimen is equal
   to the fatigue life of the smooth round bar specimen.

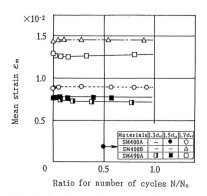

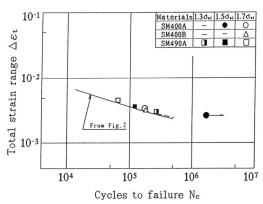

Fig. 4. Strain variation at the root of notch with respect to number of cycles.

Fig. 5. Results of low-cycle fatigue test for notched specimens.

3) The fatigue strength reduction factor K_f of the notched specimen, K_f is defined as (the total strain range $\Delta \varepsilon_{t1}$ on the fatigue life N for the smooth round bar specimen)/(the nominal total strain range $\Delta \varepsilon_{t2}$ on the fatigue life N for the notched specimen), is equal to the elastic-plastic strain concentration factor K_ε (3.1-3.4).

4) If the increased rate of the allowable stress is the same, the elastic-plastic strain concentration factor becomes to be almost the same regardless of the material strength. The higher strength material shows the higher total strain range at the root of notch. So, the higher strength material shows the lower low-cycle fatigue strength.

ESTIMATING METHOD OF LOW-CYCLE FATIGUE LIFE FOR NOTCHED COMPONENTS

The estimating method of the strain at the root of notch and the prediction method of the low-cycle fatigue life for the notched components are mentioned in this section.

The Strain Estimation at the Root of Notch

The Stowell-Hardrath-Ohman formula shown in Eq. (3) is adapted to estimate the strain at the root of notch (JSME, 1983).

$$K_\sigma = \frac{K_\varepsilon}{K_\varepsilon - K_t + 1} \tag{3}$$

where K_σ is the elastic-plastic stress concentration factor, K_ε is the elastic-plastic strain concentration factor and K_t is the elastic stress concentration factor.

The Neuber formula and so forth are proposed in place of Eq. (3) (JSME, 1983, Nakamura *et al.*, 1978). As mentioned later, the calculated strain by Eq. (3) had good accuracy. So, Eq. (3) is adapted.

Equation (4) is formed from Eqs. (1) and (2).

$$\Delta \varepsilon_e = A N_f^{-B} , \quad \Delta \varepsilon_p = C N_f^{-D} \tag{4}$$

Equation (5) is formed from Eq. (4).

$$\Delta \varepsilon_e = a \Delta \varepsilon_p^{b} \tag{5}$$

where $a = A x (1/C)^{\frac{B}{D}}$ and $b = B/D$.

When $\Delta \sigma$ is the nominal stress range, $\Delta \varepsilon$ is the nominal strain range, $\Delta \varepsilon_t$ is the total strain range at the root of notch, $\Delta \varepsilon_p$ is the plastic strain range at the root of notch and $\Delta \varepsilon_e$ is

Table 3. Material constants.

Materials	Low-cycle fatigue test						Tensile test	
	A	B	C	D	a	b	a	b
SM400A	0.00754	0.112	0.314	0.458	0.01	0.245	0.00292	0.156
SM400B	0.00754	0.112	0.314	0.458	0.01	0.245	0.00268	0.156
SM490A	0.00897	0.112	0.259	0.458	0.0125	0.245	0.00347	0.156

the elastic strain range at the root of notch. The strain estimation at the root of notch is classified into the following two cases.

The Case of the Nominal Strain in the Elastic Region.   Equation (6) shows the relationship between the nominal stress range and the nominal strain range. Equation (7) shows the relationship between the elastic strain range and the plastic strain range at the root of notch. Equation (8) is obtained from Eqs. (3), (6) and (7).

$$\Delta \sigma = E \Delta \varepsilon \tag{6}$$

$$K_\sigma \frac{\Delta \sigma}{E} = a \left(K_\varepsilon \Delta \varepsilon - K_\sigma \frac{\Delta \sigma}{E} \right)^b \tag{7}$$

$$\left(\frac{K_\varepsilon - K_t + 1}{K_\varepsilon} \right)^{1-b} (K_\varepsilon - K_t)^b = \frac{1}{a} \Delta \varepsilon^{1-b} = \frac{1}{a} \left(\frac{\Delta \sigma}{E} \right)^{1-b} \tag{8}$$

where E is Young's modulus.

The Case of the Nominal Strain in the Plastic Region.   Equation (9) shows the relationship between the nominal elastic strain range and the nominal plastic strain range. Equation (10) shows the relationship between the elastic strain range and the plastic strain range at the root of notch. Equations (11) and (12) are obtained from Eqs. (3), (9) and (10).

$$\Delta \sigma = a E \left(\Delta \varepsilon - \frac{\Delta \sigma}{E} \right)^b \tag{9}$$

$$K_\sigma \frac{\Delta \sigma}{E} = a \left(K_\varepsilon \Delta \varepsilon - K_\sigma \frac{\Delta \sigma}{E} \right)^b \tag{10}$$

$$\Delta \varepsilon = \left(\frac{\Delta \sigma}{a E} \right)^{\frac{1}{b}} + \frac{\Delta \sigma}{E} \tag{11}$$

$$\left(\frac{K_\varepsilon - K_t}{K_\varepsilon - K_t + 1} \right) \left[\left(\frac{K_\varepsilon^{1-b}}{K_\varepsilon - K_t + 1} \right)^{\frac{1}{b}} - 1 \right]^{-1} = \left(\frac{\Delta \sigma^{1-b}}{a E^{1-b}} \right)^{\frac{1}{b}} \tag{12}$$

In the case of actual design, the nominal stress range is often given. Using Eqs. (8) and (12), K is calculated by a repeated calculation. Equations (6) and (11) give the nominal strain range. The total strain range at the root of notch is equal to $K_\varepsilon \Delta \varepsilon$. So, the fatigue life of the notched components is estimated from this value and the low-cycle fatigue test results of the smooth round bar specimens. As the loading method is from the tension side in this test, the mean strain is produced at a quarter of the first cycle. The elastic-plastic strain concentration factor K_ε of the mean strain is estimated by the similar equations from the static stress-strain curve.

Table 3 shows the material constants. Figure 6 shows the estimated strain at the root of notch by this method. From these results, the estimated results are thought to have enough accuracy.

The Predicting Method of the Low-cycle Fatigue Life for the Notched Components

Using the estimated results of the strain at the root of notch and the results of the low-cycle fatigue life for the smooth round bar specimens, Figure 7 shows the predicted results of the low-cycle fatigue life for the notched specimens including another researcher's results

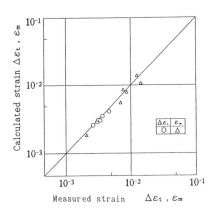

Fig.6. Estimated strain at the
root of notch.

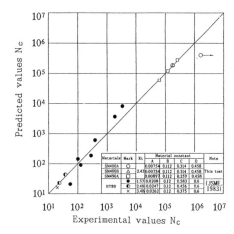

Fig.7. Relationship between the predicted
fatigue life and the experimental
fatigue life for the notched specimens.

(JSME,1983). It is recognized that the predicted low-cycle fatigue life has enough accuracy.

More,the prediction of the fatigue life is the future study in the case of the sharp notch and
the yielding for all sectional area.

CONCLUSIONS

The low-cycle fatigue tests with the low-carbon steels were carried out using the smooth round
bar specimens and the notched plate specimens. The obtained results are summarized as follows.
1) Each material has the same low-cycle fatigue strength, and the low-cycle fatigue life
prediction by the similar equation to the universal slopes equation is practically
applicable.
2) The fatigue strength reduction factor K_f of the notched specimens is equal to the elastic-
plastic strain concentration factor K_{ε}.
3) If the increased rate of the allowable stress is same, the higher strength material shows
the lower low-cycle fatigue strength.
4) The estimating method of the strain range at the root of notch and the predicting method of
the low-cycle fatigue life for the notched components have been proposed. The estimated
results had enough accuracy.

REFERENCES

Iida,K.,Urabe,Y. and Ando,Y. (1971), Low-cycle fatigue strength reduction factor for a mild
steel. *Journal of the Society of Naval Architects of Japan*, 311-320, Japan (in Japanese)
Manson,S.S. (1968). A simple procedure for estimating high-temperature low-cycle fatigue.
Experimental Mechanics, 349-355, USA
Nakamura,H.,Tsunenari,T. and Koe,S. (1978), A study on fatigue strength of notched specimen.
Journal of the Society of Materials Science, Japan, 773-779, Japan (in Japanese)
The Japan Society of Civil Engineers (1980), *Civil Engineering Handbook*, Japan (in Japanese)
The Japan Society of Mechanical Engineers (1983), *Designing Materials for The Fatigue
Strength of Metals IV*, Japan (in Japanese)
Yamada,K.,Sakai,Y.,Yamada,S. and Kondo,A. (1993). Analysis of welded details prone to fatigue
cracking. *Proceedings of The Forth East Asia-Pacific Conference on Structural Engineering
and Construction*, 1333-1338, Korea

THE INFLUENCE OF NOTCHES ON THE FATIGUE LIMIT OF A HARD BAINITIC STEEL

H. Bomas, T. Linkewitz, P. Mayr
Stiftung Institut für Werkstofftechnik, Badgasteiner Straße 3, D-28359 Bremen

F. Jablonski, R. Kienzler, K. Kutschan
Fachbereich Produktionstechnik der Universität Bremen, Badgasteiner Straße 1, D-28359 Bremen

M. Bacher-Höchst, F. Mühleder, M. Seitter, D. Wicke
Robert Bosch GmbH, Postfach 30 02 40, D-70442 Stuttgart

ABSTRACT

In this work, the influence of notches and loading conditions on the fatigue limit of a high purity high carbon steel (SAE 52 100, DIN 100 Cr 6) in a bainitic condition is presented. Since the material behaves elastic at the fatigue limit the influence of notches can be completely described by size effect and multiaxality.

A method based on the weakest link model and the Dang Van criterion for high cycle multiaxial fatigue is shown to predict the fatigue limit. The basic data of this method are the fatigue limits under torsion and under rotating bending of smooth specimens of similar geometry. Additionally, their fracture probability distribution has to be known. With these data the fatigue limits of bodies of any geometry under different zero mean stress proportional loading conditions can be calculated. An elastic finite element analysis is necessary. The predicted fatigue limits of the examined specimens and loading conditions are in good agreement with their measured values.

KEYWORDS

Fatigue limit, notch, size effect, weakest link model, multiaxiality, Dang Van criterion, finite element analysis

INTRODUCTION

Fatigue-limit design of hard steels has still some uncertainties concerning the transferability of specimen data to component behaviour. The aim of the presented investigation is to apply a transfer concept to a high strength steel, which enables the user to predict the fatigue limit of notched parts under different loadings from the measured fatigue limit of a smooth specimen.

It has already been shown that the weakest link model can be used to describe the influence of notch geometry on the fatigue limit of the hard steel which is also presented in this investigation (Bomas *et al.*, 1995). A characteristic of this steel is that no local plasticity could be determined under loading conditions near the fatigue limit.

A basic value of the weakest link model is the stress integral which describes the stressed volume (Böhm and Heckel, 1982). Since the stress integral depends on the field of a local stress parameter, an equivalent stress has to be defined for multiaxial cases. A first attempt has been made using the von-Mises criterion for describing the equivalent stress (Bomas et al., 1995). The prediction of the fatigue limit with the von Mises criterion was successful for smooth specimens and notched specimens under torsion. The fatigue limit was overestimated for notched specimens with biaxial hydrostatic notch ground stresses. In this publication the Dang Van criterion will be applied, which considers the damaging effect of hydrostatic stresses (Dang Van et al., 1989).

DANG VAN CRITERION

The Dang Van criterion is a high-cycle fatigue criterion which can be applied for any multiaxial stress-time history. If it is applied for proportional loading, it can be described by two stress parameters (Flavenot and Skalli, 1989): τ_a, the maximum shear stress amplitude, and the maximum hydrostatic stress, p_{max}, which depends on the principal stresses:

$$p_{max} = \frac{1}{3} \cdot max(\sigma_I + \sigma_{II} + \sigma_{III}) \tag{1}$$

The Dang Van criterion describes the regions of failure and endurance in the τ_a-p_{max} plane which are separated from each other by the following straight line:

$$\tau_a + \alpha \, p_{max} = \beta \tag{2}$$

The parameters α and β can be determined by the measurement of the fatigue limit under two loading conditions.

WEAKEST LINK MODEL

The weakest link model was developed by Weibull (1959) for the strength of brittle materials and later on transfered to fatigue behaviour by other authors. It is based on the thought, that fracture is caused by material or surface imperfections which are distributed in the stressed volume. Thus, the fracture probability near the fatigue limit is not only a function of the load level but also of the size of the stressed volume.

The key value of the weakest link model according to Böhm and Heckel (1982) is the so-called stress-integral I which can be interpreted as the stressed volume :

$$I = \int_V \left[\frac{\sigma(x,y,z)}{\sigma_{max}} \right]^m dV \tag{3}$$

In the fatigue limit region the fracture probability in this model obeys a Weibull distribution and depends on the maximum local stress and the stressed volume:

$$P = 1 - e^{-\frac{I}{I_0} \left(\frac{\sigma_{max}}{\sigma_0} \right)^m} \tag{4}$$

m is called the Weibull shape parameter. From equation 3, it follows for a given fracture probability that the expression

$$I \cdot \sigma_{\max}^m \qquad (5)$$

is a constant. If the maximum local stress σ_{FL0} at the fatigue limit of a reference specimen with the stress integral I_0 is known, the maximum local stress σ_{FL} at the fatigue limit of any body with the stress integral I can be calculated:

$$\sigma_{FL} = \sigma_{FL0} \cdot \left(\frac{I_0}{I}\right)^{\frac{1}{m}} \qquad (6)$$

If the reference specimen is smooth and the regarded body is notched and has the stress integral I_n, the following relation between the stress concentration factor K_t and the fatigue notch factor K_f is valid:

$$K_f = K_t \cdot \left(\frac{I_n}{I_0}\right)^{\frac{1}{m}} \qquad (7)$$

Table 1: Examined notch geometries

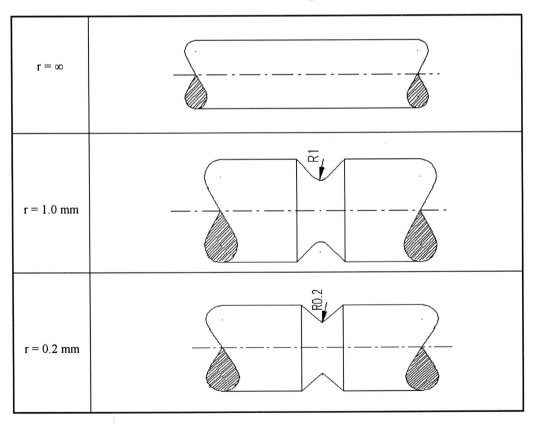

RESULTS

The experimental investigations were carried out on the bearing steel SAE 52 100 in a high purity bainitic condition (DIN: 100 Cr 6). Smooth and notched specimens with notch radii of either 1 or 0.2 mm (Table 1) were manufactured. Due to the final grinding process, compressive residual stresses of about -400 N/mm² were induced in a surface region of about 20 μm thickness.

The cyclic stress-strain curve was measured at the smooth specimens under push-pull loading. It obeys the stress-strain relation after Ramberg and Osgood (1943) with k' = 5 722 N/mm² and n' = 0.13.

The described specimens (Table 1) were used to determine the fatigue limits unter push-pull, rotating bending and torsion loading with a stress ratio of R = -1. The load varied sinusoidal with frequencies from 25 Hz to 100 Hz depending on the test machine used. The fatigue tests were carried out up to 5· 10⁶ cycles. Fig. 1 gives an overview of the experimental values.

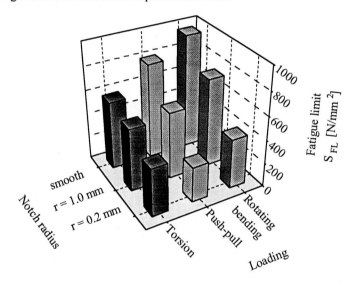

Fig. 1: Nominal fatigue limits in dependence of loading and notch radius

The smooth specimen under rotating bending and under torsion were taken to quantify the Dang Van criterion. Their fatigue limits are $\sigma_{FL,RB}$ = 967 N/mm² and τ_{FL} = 576 N/mm². These values fixed two points on the straight line characterized by equation 2, and the Dang Van diagram for the examined steel can be established (Fig. 2). Now, it is possible to express the parameters α and β in terms of these fatigue limits:

$$\alpha = 3 \cdot \left(\frac{\tau_{FL}}{\sigma_{FL}^{RB}} - \frac{1}{2} \right) \tag{8}$$

$$\beta = \tau_{FL} \tag{9}$$

By comparing a uniaxial stress condition with a multiaxial one, the equivalent stress after Dang Van, σ_{DV}, can be calculated from the principal stresses:

$$\sigma_{DV} = \sigma_I + \left(1 - \frac{\sigma_{FL}^{RB}}{2 \cdot \tau_{FL}}\right) \cdot \sigma_{II} + \left(1 - \frac{\sigma_{FL}^{RB}}{\tau_{FL}}\right) \cdot \sigma_{III} \qquad (10)$$

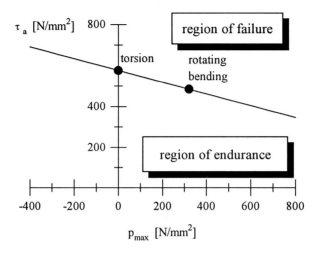

Fig. 2: Dang Van diagram of the examined steel

The local stresses of the examined specimens were determined with a finite element program. The stress integrals were calculated by replacing the stresses in equation 3 by the equivalent stress of equation 10.

The smooth specimens under rotating bending and under torsion were also taken as reference specimens for the weakest link model. Their mean stress integral and Weibull shape parameter are $I_0 = 16$ mm³ and $m = 21$. Equation 6 was used to calculate the theoretical local fatigue limits σ_{FL} of the specimens.

Fig. 3 shows the experimental local equivalent stress amplitudes of the fatigue limit together with their 10 and 90 % confidence intervals and the values calculated from equation 6 (straight line) versus the stress integrals of the specimens. The agreement of the experimental fatigue limits with the calculated ones is very good.

SUMMARY

In this work, the influence of typical component characteristica like the size of the stressed volume, stress gradients and proportional multiaxiality caused by notches and loading conditions on the fatigue limit of a high purity high carbon steel (SAE 52 100, DIN 100 Cr 6) in a bainitic condition is presented. Near the fatigue limit of this steel no local plasticity is observed. It is possible to predict the influence of notches and loading on the fatigue limit of this steel with a statistic size effect model based on the weakest link theory and the Dang Van criterion.

ACKNOWLEDGEMENTS

The authors wish to thank the „Arbeitsgemeinschaft industrieller Forschungsvereinigungen" and the „Bundesministerium für Wirtschaft" for financial support of this investigation. Further thanks go to B. Bölitz for careful experimental work.

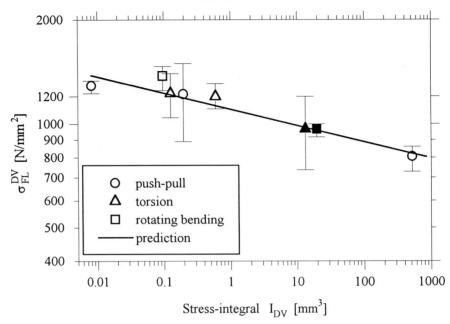

Fig. 3: Equivalent local stress amplitudes (Dang Van) at the fatigue limit as a function of stress-integral
with 10 and 90 % fracture probability interval

REFERENCES

Böhm, J., and K. Heckel (1982). Die Vorhersage der Dauerschwingfestigkeit unter Berücksichtigung des statistischen Größeneinflusses. *Zeitschrift für Werkstofftechnik*, 13, pp. 120-128

Bomas, H., T. Linkewitz, P. Mayr, F. Jablonski, R. Kienzler, K. Kutschan, M. Bacher-Höchst, F. Mühleder, M. Seitter and D. Wicke (1995). Experimental and numerical assessment of the influence of size, stress gradient and multiaxiality on the fatigue limit of a hard bainitic steel. In: *Fatigue Design 95*, (G. Marquis and J.Solin, ed.), pp. 321-332, VTT, Espoo

Dang Van, K., B. Griveau and O. Message (1989). On a new multiaxial fatigue criterion: Theory and application. In: *Biaxial and Multiaxial Fatigue*, (M. W. Brown and K. J. Miller, ed.), EGF 3, pp. 479-498, London

Flavenot, J. F., and N. Skalli (1989). A comparison of multiaxial fatigue criteria incorporating residual stress effects. In: *Biaxial and Multiaxial Fatigue*, EGF 3, (M. W. Brown and K. J. Miller, ed.), pp. 437-457, Mechanical Engineering Publications, London

Ramberg, W., and W. R. Osgood (1943). Description of stress-strain curves by three parameters. Technical report No 902, NACA

Weibull, W. (1959). Zur Abhängigkeit der Festigkeit von der Probengröße. *Ingenieur-Archiv*, pp. 360-362

IDENTIFICATION OF CYCLIC HARDENING AND $\sigma - \varepsilon$ HYSTERESIS IN OFHC COPPER

Jerzy KALETA and Grażyna ZIĘTEK
Institute of Material Science and Applied Mechanics,
Technical University of Wrocław
Wybrzeże Wyspiańskiego 27, 50-370 Wrocław, Poland.

ABSTRACT

The paper deals with an analitycal representation of cyclic hardening. The concept set forth by Lubliner (Lubliner *et al.*,1993) is applied to an elastic-plastic model. The model incorporates function $K(\varepsilon_{pl})$ understood as a hardening parameter. The model is shown to be equally well applicable to material softening in the process and its form enables simple identification procedures to be used. Experimental results are presented for OFHC copper being in the definite hardening region. The tests were carried out on cylindrical specimens subjected to uniaxial tension–compression ($R = -1$) under plastic strain amplitude control conditions. The tests were done on an MTS–810 pulser controlled via PC computer. The adopted loading ranges made it possible to get wide life variety throughout the LCF region.

KEYWORDS

Fatigue; cyclic hardening; analytic model; OFHC copper.

INTRODUCTION

The problem is part of a broader project aimed at analysing hardening and fatigue properties of face–centered cubic metals based on energy criteria. Many metals and alloys undergo hardening when loaded in a cyclic manner. Such behaviour is observed in copper, brass, aluminium, nickel and many others. Working out analytical models for cyclic hardening is still a vital problem with both theoretical and practical implications. To describe the efect of nonlinear cyclic softening or hardening different models, elastic-plastic (Caulk and Naghdi, 1978, Ohno, 1982, Eisenberg, 1976) as well as elastic-viscoplastic medias (Chaboche,1989, Lemaitre *et al.*, 1985, Naghdi and Nikkel, 1982) are proposed in literature. Given a model of hardening, it is possible then to describe all quantities associated with anelastic behaviour of a material including transient states. For a fixed load level $\varepsilon_{ap} = const$ one can determine for example $\sigma(N), \Delta W(N), W(N)$, an arc of the loop $\sigma(\varepsilon)$ and others. With these quantities known it is in turn possible to work out new energy and strain theories of the fatigue process and to verify the existing ones.

ANALYTIC MODEL OF HARDENING

At the beginning stage of cyclic loading (up to about 2000 cycles) copper shows softening that is

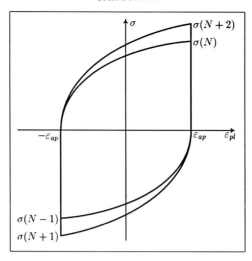

Fig. 1.   The hysteresis loop form versus the number of reversals.

replaced subsequently by hardening (Fig.1) which accompanies the deformation process until failure. To account for this behaviour an elastic-plastic model was adopted along with assumptions set forth by Lubliner.

$$\frac{d\varepsilon_{pl}}{d\sigma} = F(\sigma, \varepsilon_{pl}, K_i(\varepsilon_{pl}))\dot{\sigma} \tag{1}$$

It is assumed in the formula that $K_i(\varepsilon_{pl})$ are certain material function dependent on plastic strain. An additional assumption states that unloading conforms to the linear elastic model.
Function F was taken as:

$$F(\sigma, \varepsilon_{pl}, K_i) = \frac{\langle(\sigma - K_1(\varepsilon_{pl}))sign(\sigma - K_1(\varepsilon_{pl}))\rangle^2}{K_2(\varepsilon_{pl})} \cdot sign(\sigma - K_1(\varepsilon_{pl})) \tag{2}$$

where

$$< x > = \begin{cases} 0 & \text{for} \quad x \leq 0 \\ x & \text{for} \quad x > 0 \end{cases} \tag{3}$$

Function $K_1(\varepsilon_{pl})$ stands for plastic strain-dependent cyclic yield limit. In copper however this limit is to quickly approach zero with the number of cycles rising, hence in what follows:

$$K_1(\varepsilon_{pl}) \equiv 0. \tag{4}$$

Function $K_2(\varepsilon_{pl})$ can be considered a hardening parameter. It was assumed to be linear, namely:

$$K_2(\varepsilon_{pl}) = K(\varepsilon_{pl}) = A\varepsilon_{pl} + B(N) \tag{5}$$

It is seen from (5) that $B(N)$ is a function of the number of reversals and A is constant for each N. Variation of function $K_2(N) = K(N)$ with the number of reversals is shown in Fig.2. It is further seen that for a tensile cycle we get:

$$K(\varepsilon_{pl}) = \frac{\Delta K}{2\varepsilon_{ap}}\varepsilon_{pl} + K(0) + N\Delta K - \frac{\Delta K}{2} \tag{6}$$

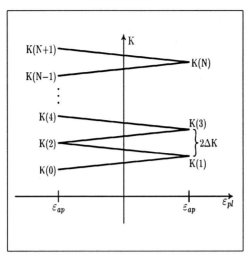

Fig. 2.   Variation of function K due to the number of reversals.

For a compressive cycle in turn expression $K(\varepsilon_{pl})$ will take the form:

$$K(\varepsilon_{pl}) = -\frac{\Delta K}{2\varepsilon_{ap}}\varepsilon_{pl} + K(0) + N\Delta K - \frac{\Delta K}{2} \tag{7}$$

Graphical interpretation of ΔK is shown in Fig. 2. As a result we get:

$$A = \frac{\Delta K}{2\varepsilon_{ap}}, \qquad B(N) = K(0) + N\Delta K - \frac{\Delta K}{2} \tag{8}$$

With these assumptions set forth it is possible to determine a relationship between plastic strain and stress. For the tension case:

$$\frac{\sigma^3}{3} = \frac{A}{2}((\varepsilon_{pl})^2 - (\varepsilon_{ap})^2) + B(\varepsilon_{pl} + \varepsilon_{ap}) \tag{9}$$

and for compression:

$$\frac{(-\sigma)^3}{3} = \frac{A}{2}((\varepsilon_{pl})^2 - (\varepsilon_{ap})^2) - B(\varepsilon_{pl} - \varepsilon_{ap}) \tag{10}$$

This in turn allows determination of a relationship between the plastic strain amplitude and the associated stress:

$$\frac{\sigma_a^3(N)}{3} = 2B\varepsilon_{ap} = 2(K(0) + \Delta K - \frac{\Delta K}{2})\varepsilon_{ap}, \tag{11}$$

and a change in the amplitude over one half-cycle

$$\frac{\sigma_a^3(N+1) - \sigma_a^3(N)}{3} = 2\Delta K \varepsilon_{ap},\tag{12}$$

The model so constructed admits of describing hardening or softening depending on the sign of ΔK. If $\Delta K > 0$, then material hardens. If in turn $\Delta K < 0$, then the softening process is taking place. In general, ΔK can vary with the number of cycles which means that the model is capable of accounting for a sequence of softening followed by hardening. It is a clear advantage of the model since many real materials exhibit such mixed behaviour.

In the following part of the paper the model is used to represent the hardening stage in copper, i.e. $\Delta K = const > 0$.

EXPERIMENTAL DETAILS

The investigation was carried out on a OFHC copper. Mechanical properties of the copper are listed in Table 1. The specimens were cylindrical in shape, of 8 mm diameter and 25 mm gauge

Table 1. Mechanical properties of the OFHC copper

Yield strength 0.2% (MPa)	σ_{UTS} (MPa)	E (MPa)	$k'(MPa)$	n'
~ 50	210	$1.1 \cdot 10^5$	467	0.193

length. They were subjected to alternating tension–compression ($R = -1$) under plastic strain amplitude control conditions. The tests were conducted on a hydraulic pulser MTS 810 connected via a system of A/D and D/A converters with an PC computer. The adopted loading ranges made it possible to get wide life variety throughout the LCF region. The measured quantities included: total strain ε_a, elastic strain ε_e and plastic strain ε_{pl}, stress σ_a, hysteresis loop area ΔW and stress/strain phase shift angle φ. The hysteresis loop area and plastic strain were determined with accuracy better than $\delta\Delta W < 7.5 \times 10^{-4}$ and $\delta\varepsilon_p < 1.2 \times 10^{-6}$, respectively. To measure the hysteresis loop area and plastic strain a modified version of the dynamic hysteresis loop method (Kaleta et $al.$,1991) was used. It consisted in expanding the stress $\sigma(t)$ and strain $\varepsilon(t)$ signals in Fourier series. The hysteresis loop area ΔW and plastic strain ε_p values were than expressed as purely analytical functions of respective harmonic components. The tests were carried out under ambient temperature conditions.

IDENTIFICATION AND DISCUSSION

Through a different grouping of terms eq. (12) can be brought into the form:

$$\sigma_a^3(N) = 6(N - N_0)\Delta K \varepsilon_{ap},\tag{13}$$

where N_0 is a certain preliminary number of reversals e.g. the number of reversals starting from which material gets harder. The above formula was taken to determine ΔK using the regression analysis. The $K(0)$ value, i.e. the K value corresponding to the number of reversals equal to N_0

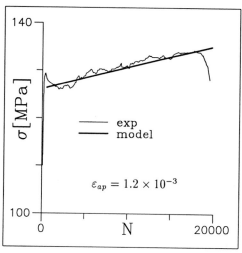

Fig. 3.   Variation of stress amplitude due to the number of cycles.

was found from (11). In Fig.3 presented are two relationship $\sigma_a(N)$ - as received experimentallyand from the identification procedure. Fig.4 gives a comparison of the actual hysteresis loop and that generated by the model with the parameters determined earlier.

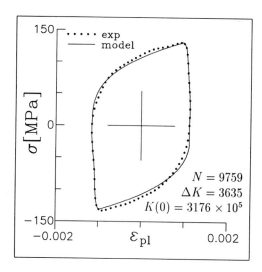

Fig. 4.   The hysteresis loop derived from the model and from the experiment.

CONCLUSIONS

1. The proposed model is a good representation of the hardening stage in OFHC coppper.

2. By assuming ΔK to be variable it is possible to account for both hardening and softening and the transition from one condition to the other.

3. The form of the model admits of using simple identification procedures.

Acknowledgement

The authors would like to thank Dr Wojciech Myszka and Andrzej Korabik (M.Sc.) (TU Wrocław) for their valued discussion and computing assistance.

REFERENCES

Caulk D. A. and Naghdi P. M. (1978). On the hardening response in small deformation of metals. *Journal of Applied Mechanics*, 45, 755–764.

Chaboche J. L. (1989). Constitutive equations for cyclic plasticity and cyclic viscoplasticity. *International Journal of Plasticity*, 5, 247–302.

Eisenberg M. A. (1976). A generalization of plastic flow theory with application to cyclic hardening and softening phenomena. *Journal of Engineering Materials and Technology*, 98, 221–228.

Kaleta J., Błotny R. and Harig H. (1991). Energy stored in a specimen under fatigue limit loading conditions. *J. of Testing and Evaluation*, 19, 325–333.

Lemaitre J., Chaboche J. L and Germain P. (1985). *Méchanique des matériaux solides*, Bordas, Paris.

Lubliner J. (1974). On loading, yield and quasi–yield hypersurfaces in plasticity theory. *Int. J. Solids Structures*, 10, 1011–1016.

Lubliner J., Taylor R. L. and Auricchio F. (1993). A new model of generalized plasticity and its numerical implementation. *Int. J. Solids Structures*, 30, No. 22, 3171–3184.

Naghdi P. M. and Nikkel D. J. Jr.(1984). Calculations for uniaxial stress and strain cycling in plasticity. *Journal of Applied Mechanics*, 51, 487–493.

Ohno N. (1982). A constitutive model of cyclic plasticity with a Nonhardening Strain Region. *Journal of Applied Mechanics*, 49, 721–727.

FATIGUE LIFE UNDER AXIAL LOADING AND PLANE BENDING CONDITIONS - EXPERIMENT AND CALCULATIONS RESULT

J. SZALA and S. MROZIŃSKI

University of Technology and Agriculture in Bydgoszcz
Department of Machine Design

ABSTRACT

The results of low - cycle fatigue investigations of 45 steel in bending and axial load conditions were presented. The cyclic properties determined in both load conditions were used in fatigue evaluations of structural mambers life. The results obtained in research were presented in fatigue curves form. The possibility of life prediction under bending from cyclic data obtained in axial load tests was verified.

KEYWORDS

fatigue life, low cycle fatigue, notched structural member, material testing

INRODUCTION

Problems presented in this paper relate to fatigue life calculations with strain method described in papers (Neuber, 1961; Tucker, 1972). It is assumed in this method that the distribution of strains in the notch area can be modelled by specimen under axial loading. The above problem is illustrated in Fig. 1. This assumption does not take into account the occurrence of strong stress and strain gradients, which influence on fatigue life. It was supported by many experimental works. It is difficult to analyse the results of tests with notched specimens. The basic problem consists in the accurate determination of stress and strain in small notch area. The influence of distribution of stress on fatigue life can be estimated during comparative test conducted under axial loading and plane bending.

The main aim of our work was the comparative analysis of 45 steel fatigue life obtained under axial loading and plane bending. In addition we estimated the appraisal of modelling the phenomena in the notch using smooth specimen under axial loading.

Quantitative influence of stress and strain gradients on fatigue life were estimated on the grounds of results of tests with specimen of different sizes under plane bending. A diagramic approach to this

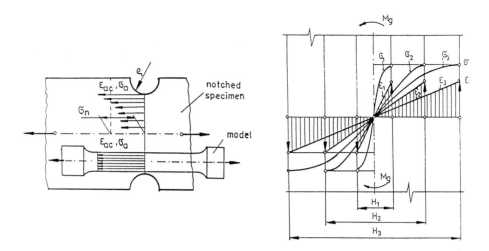

Fig. 1. Modelling of material behaviour in   Fig. 2. Methodology of experimental tests
            notch area

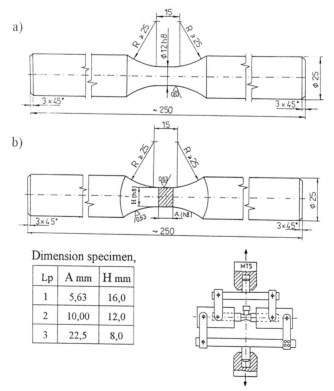

Fig. 3. Specimen used during fatigue test: a) specimen for testing under axial loading, b)
          specimen and equipment for of plane bending

problem is presented in Fig. 2. As it shows, for the same levels of total strain in extreme fibres we obtain significantly different distributions of stress and strain. During the test the specimens with heights of H=8mm, H=12 mm , H=16 mm were used. A constant parametre for different heights of specimens was ratio the moment of inertia /0,5H. The specimens were made of 45- steel. The specimen applied during low - cycle fatigue tests are presented in Fig. 3. Fatigue investigations were performed by means of an "MTS" strength test machine under constant total strain ($\varepsilon_{ac}=$ const). These fatigue tests were carried out for five levels ε_{ac} = 0.02, 0.01, 0.008, 0.005, 0.0035.

RESULTS

Test results are shown on fatigue curves in Fig. 4. The strain - life dependence was described by the Morrow's (Morrow, 1965) relation in the format presented in the figure. The values of coefficients and exponents were set in the table. On the ground of analysis of fatigue curves for axial loading and bending we can find that there are significant differences of fatigue life. For the same total strains these differences amount to 30 % for low strain levels and up to 300 % for large values of the total strains. Moreover, the influence of specimen height on fatigue life was found (influence of distribution of strain in specimen under bending).

Diagram	Dimension specimen	Kind loading	Coefficients and exponents Morrow's realation			
			b	σ_f' MPa	c	ε_f'
1	H=16 mm	bending	-0,1972	3293	-0,5518	0,5174
2	H=12 mm		-0,207	3335	-0,5384	0,4221
3	H=8 mm		-0,2018	3102	-0,5121	0,3229
4	Φ 12 mm	axial	-0,1466	1672	-0,4354	0,1713

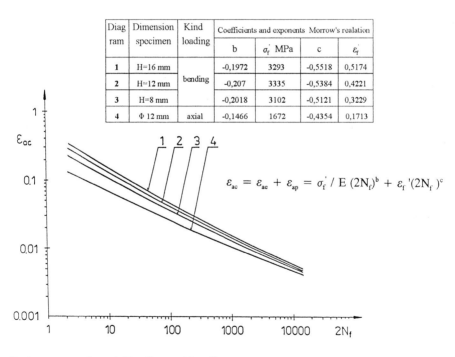

$$\varepsilon_{ac} = \varepsilon_{ae} + \varepsilon_{ap} = \sigma_f' / E \, (2N_f)^b + \varepsilon_f' (2N_f)^c$$

Fig. 4. Fatigue curves for axial loading and bending

The results of experimental work were used for fatigue life calculation. They were done for notched specimen used during experimetal tests. Obtained calculation results are presented in Fig. 5. It is possible to state that the calculation results are in the save range of fatigue life on the base of diagram position. The possition of the fatigue curves is different according to material data source. It indicates that modelling strain distribution in notch zone by bended specimen influences decrease in results of calculations and tests dispersion.

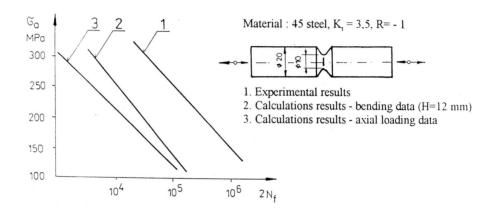

Material : 45 steel, K_t = 3,5, R= - 1

1. Experimental results
2. Calculations results - bending data (H=12 mm)
3. Calculations results - axial loading data

Fig. 5. The results of tests and calculation of notched specimens fatigue life.

However, the positive influence on approximation of tests and calculations results is reduced by difficulties occurred during low-cycle bending tests. For this reason we can find in literature the attempts to estimate the material properties for bending from data determined in axial loading conditions (Manson and Muralidharan, 1987; Megahed, 1990). It indicates that the attention is directed on the research area particularly . A verification of the proposed methodology is realized in the paper. The fatigue life under bending was determined from the cyclic properties obtained in axial load conditions. The results of verification were shown on the Fig. 6 in the fatigue plots form.

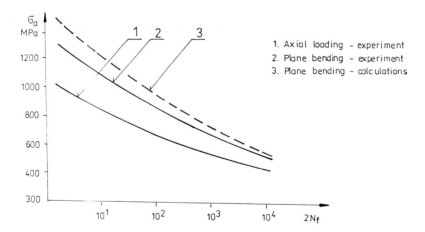

1. Axial loading – experiment
2. Plane bending – experiment
3. Plane bending – calculations

Fig. 6. The fatigue properties obtained in tests and calculations for axial load conditions and bending.

On the basis of the obtained results we can state that fatigue properties for bending can be calculated from fatigue data acquired in the axial load condition tests. The detailed analysis of the methodology described in papers (Manson and Muralidharan, 1987; Megahed, 1990) shows that the prediction of cyclic properties under bending is related to adapted stress distribution model for bended specimen in plastic - elastic strain area.

SUMMARY

The low-cycle fatigue tests results obtained in plane bending load conditions show on the influence of stress gradient on fatigue life. The comparison analysis of fatigue life obtained in tests and evaluations shows essential diferences. The modelling of local stress and strain in the notch root by smooth specimen under cyclic axial loads is doubtful. This model do not take into account two essential factors. First one is the stress distribution in notch zone (strong strain gradient), the second one the scale effect. The modelling of strain in the notch zone by bended specimen influences on the reduction of scattering of tests and computations results. It possible to predict the material properties under bending load predict on the base of cyclic properties obtained in axial tests. The results gained in this way are only approximate to the results gained in the tests. The differences arise from the adapted model of stress distribution in fatigue live calculations

REFERENCES

Manson S.S., Muralidharan U (1987). Fatigue life prediction in bending from axial fatigue information. Fatigue Fracture Engineering. Materials Structural., Vol. 9, No.5, 357-372.

Megahed M.M (1990.). Prediction of bending fatigue behaviour by the reference stress approach. Fatigue Fracture Engineering Materials Structurall ,Vol. 13, No 4, 361-374.

Morrow J.D (1965). Internal friction, damping and cyclic plasticity: Cyclic plastic strain energy and fatigue of metals. ASTM STP 378.

Neuber H (1961).Theory of stress concentration for shear strained prismatical bodies with arbitrary nonlinear stress-strain low. J. of Aplied Mech. Vol 28 Ser. E No. 4, 544 -550.

Tucker L.E (1972). A procedure for designing against fatigue failure of notched parts, society of automotive engineers. Inc, SAE Paper No. 720265 New York.

MICROMECHANISMS IN FATIGUE

THE CRACK GROWTH IN THE VICINITY OF THE TRIPLE JUNCTION IN A COPPER TRICRYSTAL UNDER CYCLIC LOADING

WEIPING JIA , SHOUXIN LI , ZHONGGUANG WANG and GUANGYI LI

State Key Laboratory for Fatigue and Fracture of Materials ,
Institute of Metal Research , Acdemia Sinica , Shenyang , 110015 , China

ABSTRACT

The effects of the grain boundary triple junction (TJ) on crack growth under high-cycle and low amplitude fatigue have been investigated in a copper tricrystal . Changing the position of the cutting notch , we got four types of specimen . When the crack transferred across GB 4mm away from the TJ , the effect of the TJ on crack growth can be ignored . In this case , the crack grew in a nominal mode I manner in double-slip grain and was deflected from its path in mono-slip grain no matter which grain the notch was located in . When the crack grew in the vicinity of the TJ , the range of suppression effects of the TJ on crack growth rates was much wider (about 4 times) than that of GB . Once the crack tip reached the GB located in this area , the GB would split and the crack would grow accelerately to the TJ . The overall effects of the TJ on crack growth rates would be the competing outcome of these two factors .

KEYWORDS

Tricrystal , grain boundary , triple junction , fatigue crack growth .

INTRODUCTION

It is well known that the grain boundary (GB) has an important effect on mechanical behaviors of polycrystals. However, the effect of grain boundary triple junction (TJ) on these behaviors is the least understood.

Recently , intergranular cracking has been found to be important at lower amplitudes , at which persistent slip bands (PSBs) operate , due to the stress concentrations caused by slip bands impingement against GBs (Figueroa et al. , 1983, Mughrubi et al. , 1983). Unfortunately , the effect of the grain boundary TJ on crack growth under high-cycle and low amplitudes fatigue is not widely investigated . In this work, copper tricrystal specimens were tested under high-cycle fatigue to examine the effect of the TJ on the process of crystallographic crack growth in the vicinity of the TJ. To make a comparison, we also designed that the cracks transferred across the GB 4mm apart from the boundary TJ (the distance that the effect of the TJ can be ignored).

EXPERIMENTAL PROCEDURES

The copper tricrystal of 99.999% purity was grown by the Bridgman method. Orientations of grains were determined by Laue technique, as shown in Fig.1. The specimens have a gauge length of 16mm and a cross sectional area of 10mm×2.5mm. All specimens have the same grain orientations. The grain boundary I-II is parallel to the loading axis and designed to be not in line with the central line of the specimen since we expected that the crack grow across the GB (or TJ) at a distance which is less than the half of the width of the specimen. To make a comparison, the notch with 1.5 - 2mm depth was spark cut as Fig.2, so we got 4 types of specimen. After careful mechanical and electrical polishing, the specimens were annealed at 800°C for 1h.

Crack growth testing was performed at room temperature in a servo-controlled testing machine, 500KN Shimadazu, under load control. The cyclic ratio of the minimum stress to the maximum stress was about 0.1, and the test frequency was 30 Hz. The crack growth rate da/dN was determined through the crack length periodically measured by means of a traveling optical microscope with a 90× zoom lens installed on the machine.

RESULTS AND DISCUSSION

From Fig.1, we can see that in grain I mono-slip was dominant, while in grain II double-slip was preferred. Fig. 3 shows that the slip morphologies are different between these grains. In grain II, two groups of slip bands appeared but were not as well developed as in grain I in which one group of slip bands with a coarser appearance emerged. In grain III, the primary slip bands were similar as in grain I because both orientations of grains were mono-slip, however, in grain III, the primary slip bands in the vicinity of the boundary TJ were not well developed as in the rest region, but the secondary slip traces were emerged. It must be the most severe area of incompatibility of deformation.

Across GB 4mm away from the boundary TJ

Li *et al.* (1994) found that in a fatigue Al bicrystal, a growing crack approaching the grain boundary was deflected in the mono-slip component, while in the multi-slip component, the crack grew in a nominal mode I manner. In this work, we found that in grain I (mono-slip) the growing crack approaching the grain boundary was deflected and after the crack entered the grain II (double-slip), it grew in a nominal mode I manner (see Fig. 4(a)). Conversely, in grain II the growing crack approaching the grain boundary grew in a nominal mode I manner and after the crack entered the grain I, it was deflected from its path (see Fig. 4(c)). In both cases, the crack initiated at notch root along primary slip bands, then grew in turn in the directions of primary and secondary slip bands.

The variations for the crack growth rates *vs.* distance between the crack tip and GB for the two situations are shown in Fig. 5(a) and (c). The crack growth rates did not drop until the crack tip approached very close to GB at a distance of about 0.15mm.
In both cases, no GB splitting was found.

These results may shed light on the mechanisms of fatigue crack transition from stage I to stage II in pure metal polycrystals. The stage I crack initiation and growth occur preferentially along a mono-slip grain in which the slip system is as expected from the maximum resolved shear stress. Very few adjacent grains have the similar mono-slip system in which the crack growth occurs without severe deflection . After the crack has passed two or three grains , it will most probably meet an adjacent grain in which the double-slip (or multi-slip)system is dominant . At this time ,the crack transferred across GB will grow in a nominal mode I manner , i.e. the crack growth transition from stage I to stage II occurs .

In the vicinity of the boundary TJ

The crack growth rates in specimen d (Fig. 2(d)) in which the notch located at grain II and pointed to the boundary TJ are shown in Fig. 5(d) . It clearly shows that when the crack tip was about 0.6 mm (about 4 times as large as that in the case of the crack approaching GB) away from the TJ , the growth rates began to drop . Fig. 5(b) shows the crack growth rates in specimen b (Fig. 2(b)) .In this specimen , since the distance between notch root and the TJ is about 0.5 mm that is within the range of effect zone of the TJ on crack growth as mention above , the crack initiation was time consuming and the crack growth rates were small even if the stress amplitude was increased .

However , in both cases , once the crack reached the TJ , the crack growth rates obviously increased The reason is as following : After the crack tip reached the nearest GB which close to the TJ , the GB would split and the crack would propagate along this GB to the TJ accelerately . When the crack tip passed the TJ , the crack would develop along another GB for a while and then turn into one of the grain interior along its primary slip trace , as shown in Fig. 4(b) and (d) . It is obvious that the internal stress and strain incompatibility are much greater at the boundary TJ than at GB , after numerous cycles fatigued , particularly in the low amplitudes and high cycle fatigue , the microcrack may be initiated at and near the boundary TJ along GBs .

CONCLUSIONS

The effects of the TJ on the crack growth are two folds :
1 .Owing to the strain incompatibility-induced internal stress in the vicinity of the TJ are greater than that near the GB which is far away from the boundary TJ , the range of suppression effects of the TJ on crack growth rate is much wider (about 4 times) than that of GB .

2 . Since the strain incompatibility in the vicinity of the TJ is much severe than that at other area , once the crack tip reached the GB located in this area , the GB may split and the crack will grow accelerately to the boundary TJ .

The overall effects of the TJ on crack growth rates would be the competing outcome of these two factors .

ACKNOWLEDGMENTS

This work was financially supported by NAMCC under Grand No . 5929100 and NNSFC under Grand No . 19392300-4 .

REFERENCES

Figueroa , J.C. and C. Laird (1983) . Crack initiation mechanisms in copper polycrystals cycled under
constant strain amplitudes and in step tests . Mater. Sci. Eng. , 60, 45-58 .

Mughrabi , H. , R. Wang , K. Differt and U. Essmann (1983) , In fatigue Mechanisms: Advances in
Quantitative Measurement of Physical Damage (edited by G.C.Shih and J.W. Provan), Mont.
Gabriel , Cadana , P.139 Mantinus Nijhoff , The Hague .

Shouxin Li , Mei Chen , Tianyi Zhang , Lizhi Sun and Zhongguang Wang (1994) . The crack growth
toward the grain boundary of a fatigued aluminum bicrystal . Scrita Metall. Mater. , 31, 897-902 .

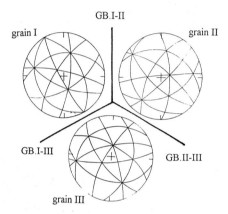

Fig. 1. Orientations of grains in the tricrystal

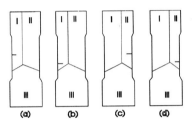

Fig. 2. The specimens for test which show
the position of spark cutting notch

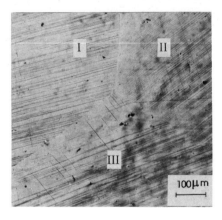

Fig. 3. Slip traces in the vicinity of the boundary triple junction

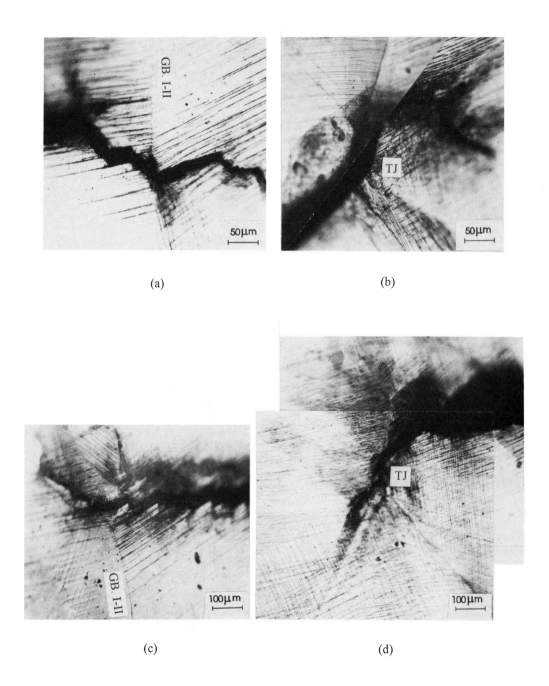

Fig. 4. Patterns of fatigue crack growth in the four specimens shown in Fig. 2.

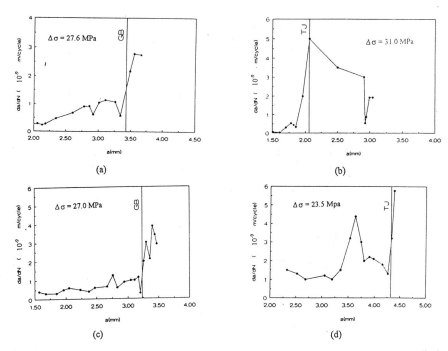

(a)   (b)

(c)   (d)

Fig. 5. Viariations of crack growth rates for the four specimens under constant applied
stress amplitudes

EFFECT OF THE GRAIN SIZE ON LOW CYCLE FATIGUE OF COPPER POLYCRYSTALS

L. GOTTE, J.MENDEZ, P. VILLECHAISE & P. VIOLAN
Laboratoire de Mécanique et de Physique des Matériaux
URA 863 / CNRS
ENSMA BP 109
86960 FUTUROSCOPE Cedex(France)
tel.: 49-49-82-09
Fax: 49-49-82-38

ABSTRACT

In order to analyze the effect of grain size on cyclic behaviour and cracking mechanisms of polycrystalline copper, Low Cycle Fatigue experiments have been performed on three different mean grain sizes of 40 µm, 150 µm and 300 µm. The cyclic response of coarse grained specimens shows a much more pronounced cyclic hardening and higher saturation stresses compared to the fine grained ones. This behaviour can be related to the <111> texture which favours multiple slip activity. It has been shown that fatigue life reduction which occurs when the grain size increases is mainly related to differences in crack growth kinetics.

KEYWORDS

Copper polycrystals, Low Cycle Fatigue, microstructure, environment, texture, crack initiation, crack propagation.

INTRODUCTION

Many investigations on the cyclic behaviour of metallic materials were focused on grain size effects(Johnson *et al.*, 1970; Mughrabi *et al.*, 1988; Christ *et al.*, 1988). These different authors have particularly studied the influence of this parameter through the dislocation structure evolution and the slip features. In copper, one of the most investigated FCC polycrystalline material, it has appeared recently that the intrinsic role of grain size remains insufficiently investigated since in most of the studies conducted on this question, the role of the texture exhibited by copper specimens has been neglected. Nevertheless, it is clear in this material that fatigue life is reduced when the grain size increases. This detrimental effect remains poorly analyzed since it is still unclear whether grain size effects are related only to crack initiation stage, only to crack growth kinetics or to both damage stages. The aim of the present work was to bring quantitative data about grain size effects on Low Cycle Fatigue(LCF) damage processes in polycrystalline copper. Moreover, in order to understand the intrinsic effect of grain dimensions, the effect of environment will be also analyzed by conducting comparative LCF tests in air and in vacuum.

EXPERIMENTAL DETAILS

Fatigue tests were carried out on cylindrical specimens mechanically and electrolytic ally polished after machining from cylindrical bars. Different annealing treatments were performed to obtain three different mean grain sizes 40 µm(FG microstructure), 150 µm (MG microstructure) and 300 µm(LG microstructure). Table and Figure 1 give respectively the experimental annealing conditions and the resulting microstructural features. Concurrently, texture measurements were performed by Electron Back Scattering Pattern technique on different sections perpendicular to the specimen axis. About one hundred grains were indexed for the MG and LG structure which allows to establish the corresponding pole figures. A classical <111> texture was found for both structures. Push-Pull fatigue tests in plastic strain control mode were conducted under two plastic strain amplitudes of $\pm 2.10^{-3}$ and $\pm 6.10^{-4}$, with a constant plastic strain rate of $2.10^{-3}s^{-1}$. The experiments were periodically interrupted for SEM examinations in order to characterize qualitatively and quantitatively surface deformation features (extrusions- intrusions, homogeneous or heterogeneous deformation distribution...) and fatigue damage(surface crack density, main crack length evolution).

RESULTS AND DISCUSSION

Cyclic behaviour - Lifetime

Fig.2 and Table 2 give the significant data of the LCF fatigue tests conducted on different copper polycrystals.
For each plastic strain amplitude, the cyclic stress curve exhibits a rapid hardening, followed by a saturation stage. In some cases a secondary hardening is also observed. As it can be seen in Fig.2 this secondary hardening stage is often revealed in vacuum when cycling is prolonged above the fatigue lifetime in air. This behaviour is in agreement with previous works on polycrystalline copper(Mendez et al.(1988), Wang & Mughrabi(1984)) which have shown that cyclic behavior is not directly influenced by the environment. Delaying crack initiation and crack growth in vacuum, allows dislocations structure transformations and cell formation which is accompagnied by an increase of the stress amplitude. Fig.2 also shows significant differences in cyclic hardening and fatigue lifetime associated with grain size evolution. For the both strain amplitude investigated, the coarser the grain is, the more pronounced the hardening rate and the higher the saturation stress. These results are in accordance with those of Liang and Laird(1989 a) and Llanes et al.(1993). The last authors assumed that the strong cyclic hardening and the high stress values observed in the coarse grained materials is directly associated with a large fraction of grains exhibiting multiple slip orientations. This behaviour is, therefore, also correlated to the very pronounced <111><100> fiber texture measured on our polycrystals.
Concerning the fatigue resistance it is clear that an increase of the grain size causes a decrease of the fatigue lifetime in copper in agreement with previous results(Liang & Laird(1989 a), Christ et al.(1988) and Lukas& Kunz(1987)) obtained in air.
It is interesting to note that the grain size effect on the lifetime reductions is weakly influenced by the level of plastic strain amplitude. This is illustrated in Fig.3 in which the results obtained in this work and those previously published by Christ et al.(1988) have been plotted. For the FG and the MG structure, the reduction factor($N_{f,FG}/ N_{f, MG}$) is approximately 2(2.5 at 6.10^{-4}, 2 at 2.10^{-3}). It must be noted that this factor is very close to the one obtained in vacuum($N_{f,FG}/ N_{f, MG}$ = 1.8 at 6.10^{-4}). The effect of the environment on fatigue damage which is clearly significant in the fatigue range investigated in this work appears therefore to be weakly influenced by the grain size.

Damage evolution

An analysis of the surface deformation features has been made by characterizing the morphology and the distribution of slip lines with extrusions + intrusions pairs. Concurrently, the fatigue damage has been studied by determining the microcracks nucleation sites or the length and the density of cracks for different fractions of the fatigue life.
For the three grain sized materials, grain boundaries(GB) constitute the preferential initiation sites in accordance with previous observations in LCF tests(Mendez(1984), Christ et al.(1988), Mughrabi & Wang(1988), Liang & Laird(1989b)).
 For the FG structure at the low plastic strain amplitude, two types of GB's nucleation have been observed. The first one concerns steps growth located at the boundary of two neighboring grains whose configuration leads to deformation incompatibilities as it has been previously noted by Kim and

Laird(1978). The second one comes from the impingement of PSB's against a grain boundary. At $\Delta\varepsilon_p/2 = \pm 2.10^{-3}$, the first nucleation mechanism seems to be predominent.

For the MG and LG structure, only the first type of GB's crack initiation process is observed, whatever the plastic strain amplitude is since GB's cracks are observed to occur before PSB's are formed.

Under vacuum, the sites for crack initiation are also the grain boundaries. However, due to the significant delay for crack initiation in vacuum, the surface roughness in the grains adjacent to the initiation sites is much higher in this environment than in air.

Therefore, grain size and environment do not modifie crack initiation sites which are clearly the grain boundaries in air as in vacuum, the intragranular damage remaining negligeable. Moreover, intergranular damage mechanisms also remains predominent in crack growth stages, specially for low plastic strain amplitudes. This predominance of intergranular cracking appears therefore directly related to the detrimental effect in fatigue resistance of grain size increase.

We have determined surface cracks density(ρ) and surface microcrack length distribution for different number of cycles and at failure. Whatever the structure is, an increase of the total cracks density at failure with the plastic strain amplitude has been noted(Fig.4a). Moreover, the longest secondary cracks have a length clearly higher in the MG than in the FG material, even though in the first one the total density is lower. Three categories of cracks have been considered according to their length. Classes I and II correspond respectively to newborn cracks or to cracks which have just propagated in the same GB segment or in the neighboring boundaries but which still remains located near the surface. Classe III cracks corresponds to longer cracks which can be considered as starting propagating into the specimen bulk. The limits of the different classes of length for the FG and the MG structures are indicated in Table 3; these limits being defined in relation with the microstructural dimensions(GB segments), they are somewhat different from a microstructure to the other.

Fig.4b shows clearly that the secondary cracks of type III are about five times more numerous in the MG than in the FG structure. Such a difference in distribution would already indicate that increasing the grain size accelerates crack growth.

Crack growth kinetics

Measurements of surface crack length, performed at different fractions of the fatigue life allowed to determine the number of cycles to initiate the fatal crack, leading to the specimen failure, and the initial dimension L_0 of this crack. It also allowed to establish the evolution of its surface length L with the number of cycles N.

The L-N data corresponding to the FG and MG structures at $\Delta\varepsilon_p/2 = \pm 6.10^{-4}$ and 2.10^{-3} in air, are reported in Fig.5.

In all the investigated conditions, the number of cycles to initiate the main crack is very low compared to the total fatigue lifetime, consequently, fatigue resistance appears mainly associated with crack propagation.

Therefore it could not be concluded, at least for the strain amplitudes investigated in this work, that there is a significant influence of the grain size on the number of cycles to initiation. In contrast, it is clear that the initial crack length L_0 of the main crack, directly associated to the microstructural features, increases with the grain size(Fig.5a). Moreover, Fig.5b shows that the crack growth rate, which increases with $\Delta\varepsilon_p/2$ for a given microstructure, is enhanced by the grain size whatever the strain amplitude is. In previous work(Ghammouri(1990)) it has been shown that in a FG copper, the length of the main crack can be expressed as a function of the initial grain size, the number of cycles N and the mechanical parameters by the relation

$$L = L_0 \exp(A \ (\Delta\varepsilon_p/2)^\alpha \ N) \qquad (1)$$

When the number of cycles to crack initiation No can be neglected as it is the case for the tests conditions considered here, the coefficients A and α can be directly related to those of the Coffin-Manson law $\Delta\varepsilon_p/2 = CpN_f^{-m}$ with $\alpha = 1/m$ and $A = \ln(L_f/L_0)/Cp^{1/m}$, where L_f is the surface length of the crack at failure; for our specimens L_f is approximatively 5000 μm.

For the FG and MG specimens, a good correlation between experimental data and calculation is obtained with $L_{FG}(\mu m) = 33 \exp(84.5 \ (\Delta\varepsilon_p/2)^{1.98} \ N)$ and $L_{MG} (\mu m) = 50 \exp(50.3 \ (\Delta\varepsilon_p/2)^{1.78} \ N)$.

Lo = 33 μm for the FG structure and Lo = 50 μm for the MG structure are in agreement with microstructural dimensions.(Fig.5a)

Relation(1) gives an evolution of the crack growth rate expressed by

$$dL/dN = A \ (\Delta\varepsilon_p/2)^{\alpha} \ L \quad (2)$$

This relation allows to calculate the ratio between the crack growth rates corresponding to the different grain sized copper polycrystals. It appears that the ratio $dL/dN_{MG}/dL/dN_{FG}$ is approximately 2.6 for 6.10^{-4} and 2 for 2.10^{-3} . So very close to the inverse of the corresponding fatigue lifetimes.

It appears therefore clearly that the detrimental effect of grain size is mainly associated with the crack growth stage. However it can be expected that for coarser grains the increase of the L_0 could also play an important role in the fatigue resistance decrease.

CONCLUSION

- Increasing the grain size produces a detrimental effect of the fatigue resistance in air of the polycrystalline copper. This effect is weakly influenced by the environment since it is of the same order than the one observed in vacuum.
- In the L.C.F range crack initiation stage constitute a small part of the fatigue life which is therefore mainly associated with the crack growth stage.
- By establishing experimental data and modelling of crack growth, it has been shown that the fatigue life reduction observed when the grain size increases is due to higher initial crack length but mainly to higher crack growth rates.

REFERENCES

Christ, Mughrabi, Wittig-Link, (1988), *in Proc.Int.Coll.Basic Mechanisms in Fatigue of Metals*, 83-92, Elsevier Amsterdam.

Ghammouri, (1990), *Doctorate Thesis*, University of Poitiers(France)

Johnson, Feltner, (1970), *Metall.Trans.*, **1**, 1161

Kim, Laird, (1978), *Act.Metall.Mater.*, **26**, 777

Liang, Laird,(1989 a), *Mater.Sci.Engng.*, **A 117**, 95,

Liang, Laird,(1989 b), *Mater.Sci.Engng.*, **A 117**, 83-93

Llanes, Rollet, Laird, Bassani, (1993), *Act.Metall.Mater.*, **41**, N°9, 2667

Lukas, Kunz, (1985), *Mater.Sci.Engng.*, **74**, L1-L5

Mendez, (1984), *Doctorate Thesis*, University of Poitiers(France)

Mendez, Violan, (1988), *in Basic Questions in Fatigue*, Volume II, ASTM STP 924 Philadelphia, 196-210

Mughrabi, Wang, (1984), *Mater.Sci.Engng.*, **63**, 147-163

                    (1988),*in Proc.Int.Coll.Basic Mechanisms in Fatigue of Metals*, 1-13, Elsevier Amsterdam.

Villechaise, (1991), *Doctorate Thesis,* University of Poitiers(France)

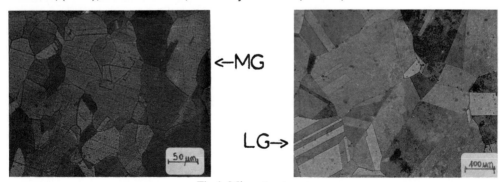

Fig.1: Microstructures

Table 1: Annealing Conditions

FG	MG	LG
360°C, 2H30, Vacuum	780°C,2H30, Vacuum	885°C, 2H30, Vacuum

Table 2: Tests conditions-Lifetime

Microstructure	Amplitude Environment	Cycles to failure (10^3)	Manson-Coffin Coefficients	Kinetics coefficients (AIR)
FG	610^{-4} Air	125	m=0.506 Cp=0.23	A=84.5
	Vacuum	880	m=0.415 Cp=0.19	α=1.98 Lo=33μm
	210^{-3} Air Vacuum	12 53		
MG	610^{-4} Air	51	m=0.561 Cp=0.26	A=50.3
	Vacuum	496		α=1.78 Lo=50μm
	210^{-3} Air	6		
LG	610^{-4} Air	12.3*	*coalescence, early failure	

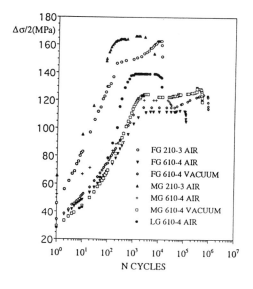

Fig.2: Cyclic strengthening curves

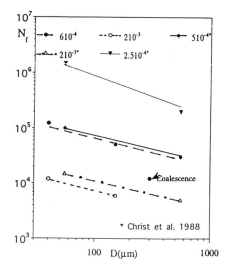

Fig.3 : Grain size effect in constant plastic strain amplitude

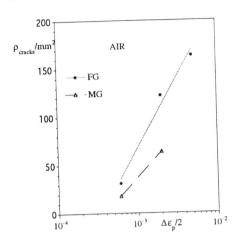

a- Total crack density

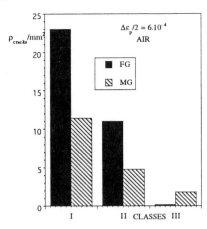

b- Cracks distribution
in function of their length

Fig.4: Damage distribution

Table 3: Surface damage characteristics at failure in air

Total cracks density (cracks/mm2)		Classes of length		
$\Delta\varepsilon_p/2 =$ 6.10^{-4}	2.10^{-3}	I	II	III L(μm)
FG 31.3	121.9	L≤ 40	40 < L≤100	L> 100
MG 18	64	L≤ 60	60 < L≤120	L> 120

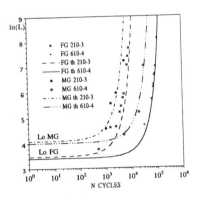

a- Microstructural effect

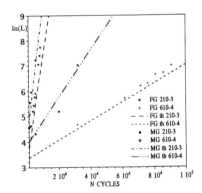

b-Grain size effect on
crack growth rate

Fig.5: Main crack evolution

Cyclic Behavior and Fatigue Life of TiNiCu Shape Memory Alloy

Toshio SAKUMA[*1], Uichi IWATA[*1] and Yuji KIMURA[*2]

*1 Central Research Institute of Electric Power Industry
    11-1, Iwato-Kita 2-Chome, Komae-shi, Tokyo 201, Japan
*2 Department of Chemical Engineering, Kogakuin University
    1-24-2, Nishishinjuku, Shinjuku-ku, Tokyo 160, Japan

ABSTRACT

The heat engine using the shape memory alloy as working elements is one of the useful direct conversion methods of low grade heat. Authors have proposed the reciprocating type heat engine using shape memory alloy wires. However, in order to undertake reliable engineering designs and optimum material selections, it is imperative that cyclic behaviors of shape memory alloy are well understood. This paper describes the recovery stress, reversible shape memory effect and residual strain in relation to number of cycles. This paper also describes the relationship between the fatigue lives and the maximum stress and strain. Furthermore, from the observations of fractured surface, the fracture origin is investigated and the cause of failure is discussed.

KEYWORDS

Shape Memory Alloy; TiNiCu Alloy; Cyclic Behavior; Fatigue Life; Fatigue Fracture Morphologies.

INTRODUCTION

Applications of shape memory alloy to various industries have been studied and some shape memory alloys are put into practical uses. The heat engine using the shape memory alloys as working elements is one of the useful direct conversion methods of low grade heat. Therefore, various types of heat engine using shape memory alloys are proposed and studied(Ginell *et al.*, 1979). Authors have proposed the reciprocating type heat engine using shape memory alloy wires(Mizutani *et al.*, 1993). However, in order to undertake reliable engineering designs and optimum material selections, it is imperative that cyclic behaviors and fatigue lives of shape memory alloy are well understood. For the TiNi alloy, cyclic behaviors and fatigue lives undergoing repeated shape memory effect or pseudoelasticity are studied(Tobushi *et al.*, 1991a,b). These cyclic characteristics have been investigated under conditions of heating-cooling cycles at constant strain or cyclic deformations at constant temperature. However, in case the shape memory alloy is used as working elements of engine, constancies of strains and temperatures are not maintained. Therefore, the investigations of cyclic characteristics are required under conditions of varying strains and temperatures.

In this paper, the cyclic behaviors and the fatigue lives of shape memory alloy are investigated experimentally. And, the cyclic behaviors of recovery stresses, two-way strains with reversible shape memory effect and residual strains are indicated in relation to cyclic number for various strains and heating temperatures. Furthermore, the fractured surfaces are observed by SEM and the causes of failure are also discussed.

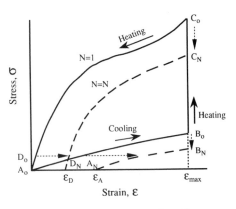

Fig.1 Stress-strain diagram of cyclic tests.

EXPERIMENTAL PROCEDURE

The cyclic tests are that the specimen at cooling(parent phase) is elongated to a given strain at constant rate by a pulse motor and the specimen is heated at a given strain up to a given temperature by hot water, thereafter the specimen is recovered at constant rate by a pulse motor maintaining a given temperature. Stress-strain diagram of cyclic tests is shown in Fig.1. The shape memory wire used is TiNiCu alloy. The copper addition changes the martensite structure to the orthorhombic structure and reduces the thermal hysteresis to 10-15K. The cooling temperature is 293K constant, and the heating temperature is 323K to 363K for the tansformation temperature A_f of 313K and 343K to 363K for the transformation temperature A_f of 333K. And the variations of strain are from 1% to 7%.

CYCLIC BEHAVIORS

The principle of engine is that the shape memory wire at cooling is elongated by the recovery stress generated at heating the deformed wire, causing the continuous reciprocating motion by repeating heating and cooling alternatively. Therefore, in order to undertake the optimum design of the heat engine using shape memory alloy, the cyclic behaviors of recovery stress σ_R, two-way strain ε_A and residual strain ε_D shown in Fig.1 are necessary to be investigated.

Recovery Stress

Fig.2 shows the recovery stresses σ_R in relation to the cyclic number N for various heating temperature T_H. The recovery stress is largely dependant upon the heating temperature and increases with the increase of heating temperature. However, the recovery stress varies with the number of cycles. In case the heating temperature T_H is lower than about 340K, the recovery stresses σ_R rarely varies with the number of cycles. But, in case the heating temperature T_H becomes higher than about 340K, the recovery stress increases until cyclic numbers become around 100, and then decreases

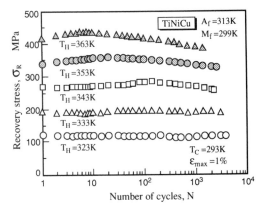

Fig.2 Variations of recovery stress with the cyclic number for various heating temperature.

with the increase of cyclic numbers. The recovery stress also varies with the strain. Fig.3 shows the variations of the recovery stress with the cyclic numbers for various strains. In case the strain is smaller than 5%, the recovery stresses are almost same and do not vary with the cyclic numbers. However, in case the strain is larger than 5%,the recovery stresses decrease largely until the cyclic numbers reach to about 50.

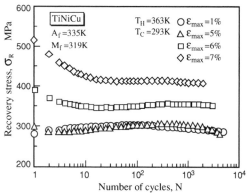

Fig.3 Variations of recovery stress with the cyclic number for various strains.

Reversible Shape Memory Effect

The wire specimen of TiNiCu shape memory alloy is deformed and is heated under constant strain up to higher than transformation temperature A_f, and then, when it is cooled, the reversible shape memory effect is exhibited, that is, it is elongated under unloading. This effect is convenient for engine because the deformation works at cooling are decreased.

Fig.4 shows the two-way strain ratio($\varepsilon_A/\varepsilon_{max}$)in relation to the number of cycles N for various strains ε_{max}. It is recognized that the two-way strain ratios largely increase within about 10-50 cycles and they become close to almost constant values larger than 50 cycles. In case the strain is below 5%, as the strain ε_{max} becomes large, the two-way strain ratio increases. On the contrary, in case the strain is larger than 5%, the same decreases. These variations of two-way strain are caused by internal stress due to irreversible defect such as dislocation. This defect increases with the increase of strain. Accordingly, the two-way strain are considered to be increased with the increase of strain because the internal stress increases with the increase of defect. However, when the strain ε_{max} is larger than 5 %, the plastic deformation is partially generated in structures and internal stress is decreased. Thus, in case the strain ε_{max} becomes larger than 5 %, the two-way strain is considered to be decreased.

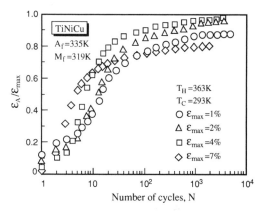

Fig.4 Variations of two-way strain ratio with cyclic numbers for various strains.

Residual Strain

For deciding the strain amplitude of engine, it is important to appreciate the variations of residual strain with cyclic numbers.

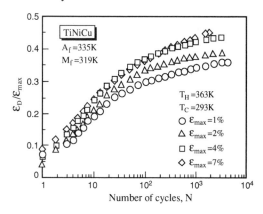

Fig.5 Variations of residual strain ratio with cyclic numbers for various strains.

Fig.5 shows the variations of residual strain ratio ($\varepsilon_D/\varepsilon_{max}$) with cyclic numbers for various strain ε_{max}. The same tendencies as two-way strain ratio is recognized, that is, the residual strain ratios largely increase within about 100 cycles and they become close to almost constant values beyond 100 cycles. In case the strain is below 4%, as the strain ε_{max} becomes large, the residual strain ratio increases. However, in case the strain is 4 % or more, the residual strain ratios are almost same as those of 4%. These increases of residual strain are considered to be caused by irreversible deformation due to dislocation in structures.

FATIGUE LIFE

Cyclic Numbers to Failure

The recovery stress is largely dependant upon the heating temperature, and varies with heating temperature even in a same strain. Fig.6 shows the relashionship between maximum stress(recovery stress) and cyclic numbers to failere for various heating temperatures and strains. The larger the maximum stress is, the cyclic numbers to failure becomes longer. Therefore, even in the same maximum strain ε_{max}, the cyclic numbers to failure decrease with the increase of heating temperature. In

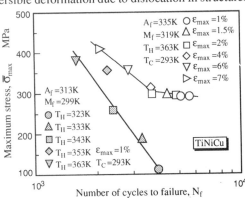

Fig.6 Relashonship between maximum stress and cyclic numbers to failure for various strains and heating temperatures.

case of the same heating temperature, maximum stresses rarely vary with the increase of maximum strains up to 5% although they vary with maximum strains larger than 5% as shown in Fig.3. However, as shown in Fig.10, the cyclic numbers to failure decrease with the increase of maximum strain.

Size of Stable Fatigue Fractured Region on the Fractured Surface

Fatigue fractured surfaces are investigated by SEM on specimens which have different diameters and are cyclically deformed under conditions of different strains and heating temperatures. It is observed that the

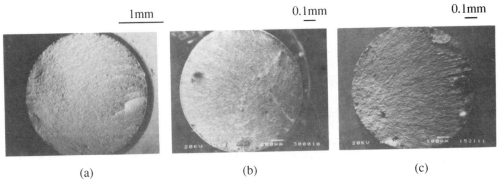

Fig.7 Sizes of stable fatigue fracture region on the fractured surface,
(a) D=3mm, T_H =363K,T_C =293K, ε=1%, $\dot{\varepsilon}$=0.5%/s, N_f =4965, d_{max} /D=0.342,
(b) D=1mm, T_H =363K,T_C =293K, ε=2%, $\dot{\varepsilon}$=0.5%/s, N_f =5962, d_{max} /D=not clear,
(c) D=1mm, T_H =323K,T_C =293K, ε=1%, $\dot{\varepsilon}$=0.05%/s, N_f =6283, d_{max} /D=0.299.

size of stable fatigue crack growth region (d_{max}/D) on the fracutred surface depends upon diameters and testing conditions of specimens. Fraction values of specimen deformed under 1 % strain condition, shown in Fig.7(a), are larger than that of under 2% strain shown in Fig.7(b). And, fractions under the same strain condition have the dependency upon the diameter as show in Fig.7(a) and (c). The specimen with larger diameter indicates relatively larger fraction of stable fatigue fracture region.

The Striation Spacing on the Fatigue Fracutred Region

The relashonship of the striation spacings (S) with the diameters of specimen and the test conditions is investigated. Fig.8 shows the striations observed on the fatigue fractured surface. As a result, no decisive relationship is established between striation spacing and diameter of specimen, strain value and heating temperature. However, it is observed that there are certain relationships betweeen the striation spacing and the roughness of fractured surface, that is, the amount of plastic deformation during the crack growth process is related to fatigue life. In the specimens of relatively short fatigue life, comparatively large striation spacings and plastic deformations on the fatigue fractured surface are observed.

10μ m 1μ m 5μ m

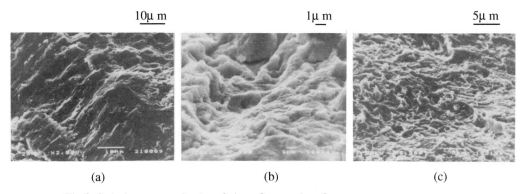

(a)   (b)   (c)

Fig.8 Striations recognized on fatigue fractured surfaces,
   (a) D=3mm, ε=1%, ε=0.5%/s, T_H =363K,T_C =293K, N_f =4965, S=1.46μ m,
   (b) D=1mm, ε=2%, ε=0.5%/s., T_H =363K,T_C =293K, N_f =5962, S=1μ m,
   (c) D=1mm, ε=1%, ε=0.05%/s, T_H =323K,T_C =293K, N_f =6283, S=0.726μ m.

10μ m 10μ m

The Origin of Fatigue Crack

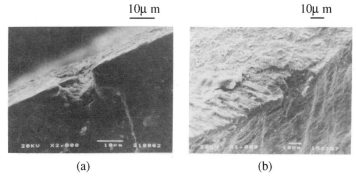

The crack origin site are also examined in detail by SEM. In almost all cases, fatigue cracks are initiated from the pitlike groove on the speci-men surface as shown in Fig.9. Also, the larger the defects are, which are found in the fatigue crack origin region on the surface of specimen, the fatigue lives

(a)   (b)

Fig.9 The origin site of fatigue crack,(a)D=3mm, ε=1%,ε=0.5%/s, T_H=363K,T_C=293K,N_f =4965,Deeper pitlike groove, (b)D=1mm, ε=1%, ε=0.05%/s, T_H=323K,T_C=293K, N_f =6283, Shallow pitlike groove.

come shorter.

Total Output Work to Failure

For the reciprocating engine using shape memory alloy, total output work is more essential rather than fatigue life. Fig.10 shows the total output works W_T to failure and fatigue lives in relation to strain ε_{max}. Works W_T are obtained by the difference between summation of works done during shape memory effect and that of works required for deformation to a given strain ε_{max}.

The output work per cycle increases with the increase of strain ε_{max}. However, fatigue lives decrease with the increase of strain ε_{max}. Therefore, the total output works become maximum in the visinity of about 5%.

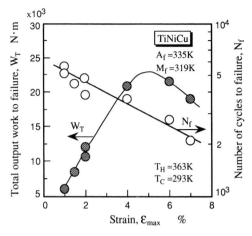

Fig.10 Variations of total output works to failure and fatigue lives with strain.

CONCLUSION

On the cyclic behaviors and fatigue lives of TiNiCu shape memory alloy, cyclic tests are carried out under conditions of varying strains and temperatures. And, cyclic behaviors of recovery stresses, two-way strains and residual strains are discussed in views of maximum strain and heating temperature. Furthermore, on the fatigue lives, the number of cycles to failure is estimated on the base of the maximum stress and strain for various strains and heating temperatures. Also, observations of fractured surfaces by SEM, fatigue lives and fracture origin are discussed. As a result, it is clarified that in order to get as much strain energy (total output work) as possible possessed of TiNiCu shape memory alloy, it is necessary to determine optimum conditios of strain and heating temperature as shown in Fig.10.

REFERENCES

Ginell, W. S., J. L. Mcnichols and J. S. Cory (1979). Nitinol Heat Engines for low-grade thermal energy conversion. *Mech. Eng.*, 101-5, 28-33.

Mizutani, H., T. Sakuma, K. Wada, U. Iwata and K. Shiroyama (1993). New System of Solid Phase Heat Engine Using Shape Memory Alloy. *ICOPE-93*, 1, 41-46.

Tobushi, H., Y. Ohashi, T. Hori and H. Yamamoto (1991a). Cyclic Deformation of TiNi Shape Memory Alloy Helical springs. *Trans. Jpn. Soc. Mech. Eng.*, 57-533, Ser. A, 121-126.

Tobushi, H., H. Iwanaga, K. Tanaka, T. Hori and T. Sawada (1991b). Cyclic Deformation of TiNi Shape Memory Alloys. *Trans. Jpn. Soc. Mech. Eng.*, 57-533 A, 651-658.

Fatigue Crack Growth Behaviour of Titanium Single Crystals

K. TAKASHIMA*, Y. MINE*, S. ANDO*, H. TONDA*, Y. HIGO** and P. BOWEN***

*Department of Materials Science and Resource Engineering, Kumamoto University
Kurokami, Kumamoto 860, Japan*

**Precision and Intelligence Laboratory, Tokyo Institute of Technology,
Nagatsuta, Yokohama 227, Japan*

****School of Metallurgy and Materials/IRC in Materials for High Performance Applications,
The University of Birmingham, Edgbaston, Birmingham B15 2TT, UK*

ABSTRACT

The fatigue crack growth behaviour of α-titanium single crystals has been investigated in laboratory air at room temperature. Two types of SEN specimens **A** and **B** with different notch orientations were prepared by a strain-annealing technique. The notch plane and direction in the **A**-specimen were close to $(\bar{1}2\bar{1}0)$ and $[10\bar{1}0]$, and those in the **B**-specimen were close to $(1\bar{1}00)$ and $[0001]$, respectively. Both fatigue crack growth resistance and ΔK_{th} were higher for the **A**-specimen than for the **B**-specimen. The crack in the **A**-specimen propagated parallel to $(01\bar{1}0)$ $[2\bar{1}\bar{1}0]$ and two sets of prismatic slip traces $((1\bar{1}00)$ $[11\bar{2}0]$ and $(10\bar{1}0)$ $[1\bar{2}10]$) were found near to the crack tip of the **A**-specimen. Therefore, the crack in the **A**-specimen is deduced to extend by alternating shear on two intersecting prismatic slip systems ($(1\bar{1}00)$ $[11\bar{2}0]$ and $(10\bar{1}0)$ $[1\bar{2}10]$). The fatigue crack propagated along the $[0001]$ direction in the **B**-specimen. Although very fine traces due to twinning were observed near to the crack surface, large slip bands were not found. Ridges and valleys which are deduced to be associated with twin interfaces were observed on the fatigue surface. This result suggests that twinning occurs at the crack tip if prismatic slip is difficult to activate, and that the crack may grow by the separation of twin interfaces.

KEYWORDS

Fatigue crack growth, Titanium, Single crystal, Slip system, Mechanical twinning

INTRODUCTION

Fatigue crack growth in crystalline metals and alloys are fundamentally based on cyclic plastic deformation processes at the crack tip. The crack growth increment for each cyclic loading (da/dN) in such materials often occurs in the range of 10^{-6} to 10^{-3} mm, and this extension is usually smaller than the grain size of such materials. Therefore, it is extremely important to know the fatigue crack growth mechanisms that may occur in single crystals to promote increased understanding of the intrinsic fatigue crack growth resistance of materials. There have been several investigations into the fatigue crack growth mechanisms for FCC and BCC crystals. For FCC crystals, one fatigue crack growth model that is accepted generally is based on the "unzipping" (alternate shear) of two slip planes (e.g., (111) and

($11\bar{1}$) if the crack growth direction is [$\bar{1}\bar{1}0$]). This model agrees well with experimental results (Neumann, 1969, Pelloux, 1969, 1970, Bowles and Broek, 1972). A similar model for the "unzipping" (alternate shear) of slip planes by cyclic loading has been also presented for BCC crystals (Fukui *et al.*, 1977, Nunomura and Fukui, 1977). However, there have been few studies on fatigue crack growth mechanisms in HCP crystals (Ward-Close and Beevers, 1980). This may be partly due to the fact that the slip systems of HCP crystals have not yet been identified precisely. In addition, mechanical twinning may occur in HCP crystals to accommodate its deformation, since slip systems in HCP metals are limited. This may make it difficult to clarify fatigue crack growth mechanisms in HCP metals. Among HCP crystals, α-titanium and its alloys have been applied to aerospace structures and engine components because of their excellent specific strength and corrosion resistance. Therefore, it is important to evaluate the fatigue mechanism of such materials. In this present study, the fatigue crack growth behaviour in α-titanium single crystals with different orientations has been investigated to elucidate crack growth mechanisms that can occur.

EXPERIMENTAL PROCEDURE

The starting material used in this study was a commercially pure titanium rolled plate of 4 mm in thickness. The oxygen content of the plate was 0.043 wt%. Coarse grain specimens were prepared by a strain-annealing technique. Specimens with dimensions of 20 mm in width and 150 mm in length were cut from the plate. A plastic strain of 7% was applied to the plate by tensile loading, and then the specimen was heated to a temperature of 1200 °C and held for 10.8 ks. Subsequently, the specimen was furnace cooled to a temperature of 850 °C, and held for 86.4 ks, and then furnace cooled to room temperature. Coarse grained plate with a grain size of approximately 30 mm was obtained by this method. The orientation of each grain was determined by the back-reflection X-ray Laue method. Two types of SEN specimens **A** and **B** with different notch orientations were then cut from this coarse grained plate. The size of the SEN specimen was 7 mm in width and 22 mm in length. A through thickness notch of 1.75 mm was introduced in the centre of the specimen by use of a diamond saw to give an initial notch to width ratio, a_0/W, equal to 0.25. The orientation of each specimen is shown in Fig. 1, where NP, ND and NN denote the notch plane, notch direction and the normal to the notch plane, respectively. The notch plane and direction of the **A**-specimen were close to ($\bar{1}2\bar{1}0$) and [$10\bar{1}0$], and those of the **B**-specimen were close to ($1\bar{1}00$) and [0001], respectively. In the **A**-specimen, prismatic slip is expected to be activated easily, and first order pyramidal slip is expected to be activated in the **B**-specimen. The specimens were annealed at a temperature of 650 °C for 36 ks to eliminate any effect of residual strains during machining, and then chemically polished to remove the surface oxide layer formed during heat treatments.

Fatigue crack growth tests were carried out at room temperature in air under three point bending. All fatigue tests were conducted at a frequency of 10 Hz and a stress ratio, R (the ratio of the minimum to maximum stress intensity factor applied over the fatigue cycle) of 0.1 using an Shimadzu servo-hydraulic testing machine. The crack length was monitored using the electrical potential difference method. After the crack started to grow, the load range was decreased gradually until the crack growth rate approached the near threshold region (defined here as less than 10^{-8} mm/cycle). Crack growth rates were then measured under a conditions of increasing ΔK with crack extension, i.e. under a constant cyclic loading range.

RESULTS

Fatigue crack growth resistance curves in the **A**- and **B**-specimens as a function of applied stress intensity factor range ΔK are plotted in Fig. 2. The fatigue crack growth rate of the **B**-specimen is higher than that in the **A**-specimen at a given value of ΔK; and ΔK_{th} for the **A**-specimen is higher than that of the **B**-specimen. A definite dependence of crystallographic orientation on fatigue crack growth behaviour is thus observed.

Figure 3 shows the crack growth profile of the **A**-specimen. Two sets of slip traces were visible near the crack tip. One slip direction is inclined at approximately 30° to the notch direction and the other is at approximately right angles to the notch direction. Therefore, the slip directions are parallel to [$11\bar{2}0$] and [$1\bar{2}10$] and so these slip traces are identified to be prismatic slip on ($1\bar{1}00$) [$11\bar{2}0$] and ($10\bar{1}0$) [$1\bar{2}10$]. A crack initiated along the [$11\bar{2}0$] direction and then changed its growth direction from [$11\bar{2}0$] to [$2\bar{1}\bar{1}0$] after approximately 0.4 mm of crack extension from the notch root (see Fig. 3). Figure 5 shows a scanning electron micrograph of the fatigue surface of the **A**-specimen. Striation-like equispaced markings were observed on the fatigue surface of the **A**-specimen, and these markings were aligned parallel to [0001]. Figure 4 shows the crack growth profile of the **B**-specimen. The fatigue crack propagated along [0001] in the **B**-specimen. Although very fine traces, which are deduced to correlate with twinning, have been observed using Nomarski interference techniques near the crack surface, the large slip bands (seen for example in Fig. 3) were not found. Figure 6 shows a scanning electron micrograph of the fatigue surface of the **B**-specimen. The fatigue surface is composed of regularly spaced ridges and valleys, and some markings were observed on each ridge. The peaks of the ridges were parallel to [$11\bar{2}0$]. Some grooves parallel to [0001] were also visible on the ridges and valleys.

DISCUSSION

In α-titanium crystals, the prismatic slip systems (of the type $\{1\bar{1}00\}<11\bar{2}0>$) have been considered as the principal deformation mode (Partridge, 1967), and such slip systems are easy to activate for the **A**-specimen. Therefore, prismatic slip (of the type ($1\bar{1}00$) [$11\bar{2}0$] and ($10\bar{1}0$) [$1\bar{2}10$]) occurred at the crack tip in the **A**-specimen as shown in Fig. 3. Also, striation-like markings which were parallel to [0001] were observed on the fatigue surface as shown in Fig. 5, and the crack propagates largely parallel to the [$2\bar{1}\bar{1}0$] direction. This suggests that the crack propagates by alternating shear on two intersecting prismatic slip systems. Thus, the fatigue crack will grow by a cyclic shear process if prismatic slip is favoured at the crack tip. However, the spacing between the striation-like markings was approximately 1 μm and this is not coincident with the macroscopic crack growth rate. Indeed, they were approximately between 10^2 and 10^3 times faster than that predicted from the crack growth rates measured by the electrical potential difference method. This suggests strongly that these "striations" are formed by a discontinuous crack growth process. Other less distinct markings along [$2\bar{1}\bar{1}3$] were also observed on the fatigue surface (see Fig. 5). This direction is consistent with that of first or second order pyramidal slip. However, it is unlikely that crack growth can be associated with this deformation in this specimen.

The fatigue crack propagated along [0001] in the **B**-specimen. Although very fine traces due to twinning have been observed near the crack surface, large slip bands were not found. A "sawtooth" zigzag pattern is observed on the fatigue surface. This pattern is composed of ridges and valleys and they seem to have a crystallographic relationship as suggested by Fig. 6. Each pattern is parallel to [$11\bar{2}0$] and a groove along [0001] is observed. Fine markings parallel to [0001] were also visible on the fatigue surface. Thus first order pyramidal slip seems to be active in this specimen. However, it is difficult to produce such a pattern by a slip process alone. In the orientation of this specimen, twin systems (on planes of the type $\{10\bar{1}1\}$ and $\{10\bar{1}2\}$ for this orientation) are also expected to be active. The resolved shear stress acting here on the $\{10\bar{1}2\}$ twinning plane is higher than that on the $\{10\bar{1}1\}$ twinning plane, so that twin deformation on the $\{10\bar{1}2\}$ twinning plane is expected to occur in the **B**-specimen. If two sets of $(1\bar{1}02)$ and $(\bar{1}102)$ twins occur at the crack tip, then a "sawtooth" type of fatigue surface might be considered to form as a result of the separation of twin interfaces by cyclic loading. The fine markings on the ridges may be due to prismatic slip traces, but the detail of the formation process is unclear. This result suggests that twinning may occur at the crack tip if prismatic slip is difficult to activate, and that the crack may grow by the separation of twin interfaces. However, these slip systems and their flow stresses are known to depend critically on the content of interstitials such as O, C, N, H and other impurities. Thus, further investigation is required to quantify if these crack growth mechanisms occur generally under cyclic loading in α-titanium single crystals.

CONCLUSION

The fatigue crack growth behaviour of α-titanium single crystals has been investigated in laboratory air at room temperature. Two types of SEN specimens **A** and **B** with different notch orientations were prepared and the notch plane and direction of the **A**-specimen were close to $(\bar{1}2\bar{1}0)$ and $[10\bar{1}0]$, and those of the **B**-specimen were close to $(1\bar{1}00)$ and $[0001]$, respectively. Both fatigue crack growth resistance and ΔK_{th} were higher in the **A**-specimen than those in the **B**-specimen, and a definite dependence of crystallographic orientation on fatigue crack growth was observed. The crack in the **A**-specimen propagates parallel to $(01\bar{1}0)$ $[2\bar{1}\bar{1}0]$ and two sets of prismatic slip traces ($(1\bar{1}00)$ $[11\bar{2}0]$ and $(10\bar{1}0)$ $[1\bar{2}10]$) were found near to the crack tip of the **A**-specimen. Therefore, the crack in the **A**-specimen is deduced to extend by alternating shear on two intersecting prismatic slip systems ($(1\bar{1}00)$ $[11\bar{2}0]$ and $(10\bar{1}0)$ $[1\bar{2}10]$). The fatigue crack propagated along $[0001]$ in the **B**-specimen. Although very fine traces due to twinning were observed near the crack surface, large slip bands were not found. Ridges and valleys which may be associated with twin interfaces were observed on the fatigue surface. This result suggests that twinning may occur at the crack tip if prismatic slip is difficult to activate, and that the crack may grow by the separation of twin interfaces.

REFERENCES

Bowles, C. Q. and D. Broek (1972). On the formation of fatigue striations, *Int. J. Fract. Mech.*, **8**, 75-85.

Fukui, Y., Y. Higo and S. Nunomura (1977). Crystallographic dependence of fatigue crack propagation in bcc metals, *J. Jpn. Inst. Metals,* **41**, 400-405.

Neumann, P. (1969). Coarse slip model of fatigue, *Acta Met.*, **17**, 1219-1225.

Nunomura, S. and Y. Fukui (1977). On the transcrystal fatigue crack propagation mechanism of bcc metals, *J. Jpn. Inst. Metals,* **41**, 405-412.

Partridge, P. G. (1967). The crystallography and deformation mode of hexagonal close-packed metals, *Metallurgical Review*, **12**, 169-194.

Pelloux, R. M. N. (1969). Mechanisms of formation of ductile fatigue striations, *Trans. ASM*, **62**, 281-285.

Pelloux, R. M. N. (1970). Crack extension by alternating shear, *Eng. Fract. Mech.*, **1**, 697-704.

Ward-Close, C. M. and C. J. Beevers (1980). The influence of grain orientation on the mode and rate of fatigue crack growth in α-titanium, *Met. Trans.*, **11A**, 1007-1017.

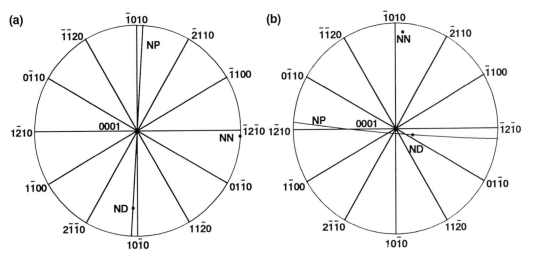

Fig. 1 Orientations of **A**- and **B**-specimens shown in (a) and (b) respectively.
NP: notch plane, ND: notch direction, NN: normal to notch plane.

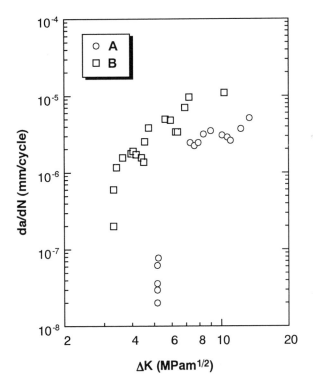

Fig. 2 Fatigue crack growth resistance curves of **A**- and **B**-specimens.

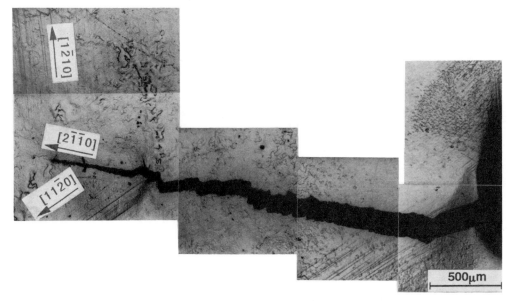

Fig. 3 Crack profile of **A**-specimen.

Fig. 4 Crack profile of **B**-specimen.

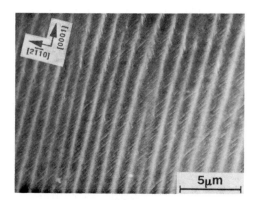

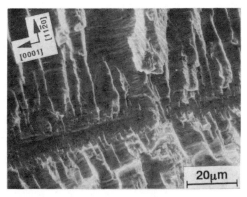

Fig. 5 Scanning electron micrograph of
fatigue surface of **A**-specimen.

Fig. 6 Scanning electron micrograph of
fatigue surface of **B**-specimen.

MICROMECHANISMS OF FATIGUE IN TITANIUM
SINGLE CRYSTALS

TAN XIAOLI and GU HAICHENG

Research Institute for Strength of Metals
Xi' an Jiaotong University, Xi' an, 710049 China

ABSTRACT

High purity titanium single crystals with different orientations were used to study the fatigue mechanisms of HCP metals. Reversed push-pull cyclic tests were conducted and cyclic stress-strain curves (CSSCs) were determined using multiple step test method. The CSSCs can be classified into three groups according to the crystal orientations. Persistent slip bands, cyclic twins and micro-cracks were found on the surfaces of fatigued specimens by SEM observation. TEM analyses on the thin foils sliced from fatigued specimens show the dislocation patterns, twin structures and stacking faults related to crystal orientations. Synergetic concept is expected to interpret the macroscopic, mesoscopic and microscopic phenomena corresponding to fatigue process in titanium.

KEYWORDS

Fatigue mechanism; titanium; single crystal; dislocation; twin; stacking fault; crack; synergetics

INTRODUCTION

Numerous investigations on fatigue mechanism have been conducted on metals and alloys which posses FCC and BCC crystal structure, whereas the information on HCP metals is still limited although the hexagonal metals, such as Be, Mg, Co, Zr, Ti, *etc.*, play an important role in modern science and technology. Titanium is one among some twenty hexagonal metals and has found wide application in various fields. It is now established that slip with $\langle a \rangle$ Burgers vector in titanium occurs on $\{10\bar{1}0\}$, $\{10\bar{1}1\}$ and (0001) planes. Also, the first-order pyramidal planes $\{10\bar{1}1\}$ and the second-order pyramidal planes $\{11\bar{2}2\}$ with $\langle c+a \rangle$ Burgers vector are available for slip (Numakura *et al.*, 1986). Among these, prismatic slip is principally chosen as the deformation mode.

Six different twinning planes, *i.e.* $\{10\bar{1}1\}$, $\{10\bar{1}2\}$, $\{11\bar{2}1\}$, $\{11\bar{2}2\}$, $\{11\bar{2}3\}$ and $\{11\bar{2}4\}$, have been reported in titanium (Mullins *et al.* , 1981; Ishiyama *et al.* , 1990). The complex slip and twinning behavior makes it difficult to clarify the micromechanism especially under cyclic loading. To the authors' knowledge, research on the fatigue behavior of titanium single crystals has not been carried out until most recent years (Gu *et al.* , 1994). The present study provides information on the fatigue mechanism of titanium single crystals.

EXPERIMENTAL PROCEDURE

Strain-annealing method was employed to grow single crystals from floting zone-melting refined titanium (Gu *et al.* , 1994). Specimens with a rectangularly shaped section were prepared with a programmable spark erosion machine under good cooling conditions. After careful grinding, the specimens were chemically polished so as to provide contrast of slip bands and twins on a smooth background. The orientations of the single crystal specimens were determined using the X-ray Laue method.

Reversed push-pull cyclic tests were conducted with a length of gauge section monitored. The cyclic stress-strain curves (CSSCs) were determined using a multiple step test method. The crystallographic factors of slip bands and twins were determined using the trace analysis technique and were checked with the Schmid factor calculation. Minute specimens sliced from a single crystal with [0001] direction parallel to the axis were fatigued in the chamber of a scanning electron microscope (SEM) for *in situ* observation   (Tan *et al.* , 1994). The substructure of fatigued specimens was examined on thin foils with a transmission electron microscope (TEM).

RESULTS AND DISCUSSION

Cyclic Stress-Strain Curves

The CSSCs of crystals with different orientations can be classified into three types, as qualitatively illustrated in Fig. 1. Curve A, exhibiting a plateau, represents those of single crystals oriented near the $10\bar{1}0$-$11\bar{2}0$ border of the unit triangle of stereographic projection. In these crystals prismatic slip is favourable. Curve B stands for those of crystals oriented in the middle of the unit triangle. In this group of crystals cross slip and/or twinning will occur in addition to prismatic slip. The curve exhibits a stage at which the cyclic stress increases slightly with increment of the strain amplitude. Curve C   characterizes those of crystals oriented near the 0001-$10\bar{1}0$ or 0001-$11\bar{2}0$ border of the unit triangle. In these crystals double prismatic slip and twinning are geometrically favoured. Thus the curve is steep. The evidence indicates that the crystal orientation has a strong influence on the CSSC, not only on its stress level but also on its shape.

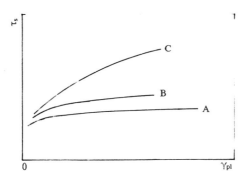

Fig. 1.   CSSCs of crystals with different orientations.

SEM Observation

Slip on prismatic plane along ⟨a⟩ direction is also the most principal deformation mode during cycling of titanium single crystals, and this was justified by SEM observation on the surfaces of fatigued specimens. Persistent slip bands appeared on the specimen surfaces are shown in Fig. 2 (a). However, ⟨a⟩ slip will be forbidden in those single crystals oriented near 0001 pole of the unit triangle due to the very low value of the Schmid factor. Observation on the minute specimens sliced from the [0001] oriented single crystal found that the cyclic strain was accommodated almost entirely by cyclic twins, as shown in Fig. 2(b).

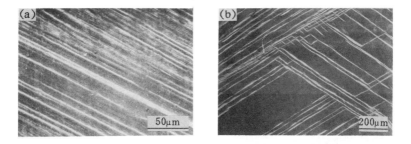

Fig. 2.   Cyclic deformation appearances on the surface of titanium single
crystals: (a) persistent slip bands, (b) cyclic twin bands.

Fatigue crack initiation in titanium single crystals is found to be related to persistent slip bands or cyclic twins (Tan *et al.*, 1995a). In single crystals with favorable orientations for prismatic slip, $\{10\bar{1}0\}$ ⟨$11\bar{2}0$⟩ slip system was activated. Cracks usually occurred in localized slip bands (Fig. 3 (a)). Although twin boundaries are the plane imperfection with the lowest energy, their roles in crack initiation have long been recognized. In the single crystal oriented for difficult glide, it was found that microcracks initiated at the midrib of cyclic twin bands just like midrib in martensite plate of high carbon steel (Fig. 3(b)). On the basis of crystallographic analysis it is suggested that the cyclic twin bands consist of $\{10\,\overline{1}2\}$ and $\{11\,\overline{2}2\}$ types of twins. Actually the microcracks along the midrib are the results of the impingement of these two sets of twins upon each other (Tan *et al.*, 1994).

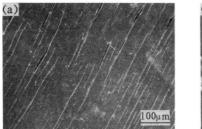 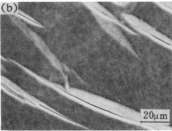

Fig. 3.   Fatigue crack initiation in titanium single crystals: (a) cracks along
slip bands, (b) cracks along the midrib of cyclic twin bands.

TEM Analysis

Thin foils sliced along particular planes according to the predetermined orientations from fatigued single crystals were examined with TEM. Dislocations, twins, and stacking faults are found to interact on each other in a complicated manner.

Dislocation pattern is found to be sensitive to the crystal orientation. In single crystals oriented for single slip, dislocation walls or channels are prone to form, whereas in crystals oriented for double or multiple slip, equiaxial dislocation cells are presented, as shown in Fig. 4.

 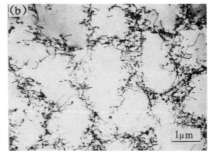

Fig. 4.   Dislocation structures in fatigued titanium single
crystals: (a) wall structure, (b) equiaxial cells.

Cyclic twins were found in most of fatigued crystals. Various fine structures can be observed in cyclic twins under high magnifications. Figure 5(a) shows some small second-order twins arranged regularly in a host twin. An intriguing phenomenon is that a stack of stacking faults formed in a $\{11\overline{2}1\}$ twin (Fig. 5 (b)). Titanium is known to have a high stacking fault energy. The formation of a ribbon of stacking fault during $\{11\overline{2}1\}$ twinning is suggested to be expressed as (Tan $et\ al.$, 1995b).

$$\frac{a}{3}[11\overline{2}0]_{(1\overline{1}00)} \longrightarrow \frac{a}{9}[\overline{1}\,\overline{1}20]_{(\overline{1}100)} + \frac{2a}{9}[\overline{1}\,\overline{1}20]_{(\overline{1}100)} - 2b_t \tag{1}$$

Where b_t is the Burgers vector of a unit $\{11\overline{2}1\}$ twin dislocation. This leads to the conclusion that the energy barrier during the formation of stacking faults may be overcome by twinning process in titanium. The interaction between coexisting twin and dislocation walls was also found(Fig. 6).

(a) 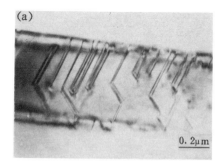 (b)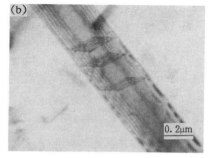

Fig. 5.   Fine structures in cyclic twins:(a) second-order twins, (b) stacking faults.

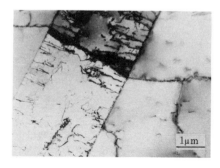

Fig. 6.   Interaction between cyclic twin and dislocation wall.

Fatigue is a highly irreversible process far from thermodynamic equilibrium. There is no doubt that a specimen undergoing fatigue in an open system, which exchanges energy with environment. The transfer of energy in the form of work done on the specimen by the applied testing stresses and the dissipation of energy within the specimen leads to the evolution of self-organizing structure. From microscopic viewpoint, the cyclic deformation is attributed to the movement and interaction of dislocations. As a result, wall or cell structures are formed. Twinning is another important deformation mode in titanium, especially in crystals where slip is unfavorable. And dislocation loops, stacking faults, as well as second-order twins are associated with twinning. Furthermore, the evidence shows that microcracks are formed along persistent slip bands and cyclic twin bands. Thus, the collective effects and mutual interactions of dislocations, twins, stacking faults and microcracks render the fatigue damage into more complex and more difficult problems. Synergetics deals with the systems composed of many subsystems and studies the joint action of these subsystems. Despite the complexity, it is expected that the synergetics conception will shed a high-light on modeling fatigue damage at all macroscopic, mesoscopic and microscopic levels.

CONCLUSIONS

The CSSCs of titanium crystals are dependent on orientations. In crystals where single prismatic slip is favourable, the CSSCs present a plateau stage, whereas those of double slip oriented crystals are steep. Persistent slip bands and cyclic twin bands are observed to accommodate the cyclic deformation, consequently fatigue microcracks are found to initiate along slip bands and twin bands. Dislocation walls and cells are observed in fatigued crystals. Fine structures of cyclic twins are revealed as: accumulating dislocation loops, regularly arranged second-order twins, and stacking faults. During the fatigue process of titanium single crystals, interactions among dislocations, twins, stacking faults, microcracks, *etc.*, should be emphasized.

ACKNOWLEDGEMENTS

The authors are grateful for support from the National Natural Science Foundation of China.

REFERENCES

Gu Haicheng, Guo Huifang, Zhang Shufen and C. Laird (1994). Orientation dependence of cyclic deformation in high purity titanium single crystals, *Mat. Sci. Eng.*, **A188**, 23-36.

Ishiyama S., S. Hanada and O. Izumi (1990). Orientation dependence of twinning in commercially pure titanium, *J. Japan Inst. Metals*, **54**, 976-984.

Mullins S. and B. M. Patchett (1981). Deformation microstructures in titanium sheet metal, *Metall. Trans. A*, **12A**, 853-863.

Numakura H., Y. Minonishi and M. Koiwa (1986). $\langle\bar{1}\bar{1}23\rangle\{10\bar{1}1\}$ slip in titanium polycrystals at room temperature, *Scr. Metall.*, **20**, 1581-1586.

Tan Xiaoli, Gu Haicheng, Zhang Shufen and C. Laird (1994). Loading mode dependence of deformation microstructure in a high purity titanium single crystal oriented for difficult glide, *Mat, Sci. Eng.*, **A189**, 77-84.

Tan Xiaoli and Gu Haicheng (1995a). Fatigue crack initiation in high purity titanium crystals, *Int. J. Fatigue*, in press.

Tan Xiaoli and Gu Haicheng (1995b). Stacking faults in fatigued titanium single crystals, *Scr. Metall. Mater.*, in press.

CYCLIC DEFORMATION BEHAVIOUR OF DIFFERENT CRYSTALLOGRAPHIC ORIENTATION AL SINGLE CRYSTAL

Xia,Y.B. (State Key Laboratory for Fatigue and Fracture of Materials,
Institute of Metal Research, Academia Sinica, Shenyang   110015   P.R. China)

ABSTRACT

The symmetrical push–pull fatigue tests at constant strain amplitudes of 6×10^{-4} and 2×10^{-3} were carried out with different crystallographic orientation (CO) aluminium single crystals (ASC) at ambient temperature and in air. cyclic stress response, σ, internal friction, Q^{-1} and ultrasonic attenuation, $\Delta\alpha$ were measured and the stress–strain hysteresis loops for selected cycles were recorded. The friction stress, σ_f and back stress, σ_b were obtained from the loops. The results indicate that the first of all, the σ, Q^{-1}, $\Delta\alpha$, σ_f and σ_b change repidly and then keep constant basically in cyclic saturation stage for all crystals used in present paper. The effect of the CO on the parameters above is as follows: either the value of the σ, $\Delta\alpha$, σ_f and σ_b or their verying rate for single–slip (SS) ASC are smaller than that for double–slip (DS) and multi–slip (MS) ASC, while the Q^{-1} of the former is larger than that of the two latters. The hysteresis loop shaps of different CO ASC have distinction obviously. The SEM observation on the surfaces of the specimens fatigued shows that the slip bands of SS ASC are wider than that of MS ASC, and DS ASC are intervened between SS and MS ASC.

KEYWORDS

Crystallographic orientaion (CO); aluminium single crystal (ASC); single–slip (SS); double–slip(DS); multi–slip(MS).

INTRODUCTION

The investigation of material fatigue behaviour had more time than a century. The research of metal fatigue mechanism also had several decades up to now, but this problem is far away from full understanding of its lows and natures to date. The study of cyclic deformation behaviour of single crystals is very necessary to understanding the fatigue mechanism of malerials. The previous works were mostly concentrated on copper single crystals (Laird et al., 1986, Mughrali, 1978, chong et al., 1981) and a few on ASC. A few results of studied the cyclic stress response and the surface slip morphology of Al and Al alloy single crystals (Abel et al., 1979, Vorren et al., 1988) and studied the cyclic deformation behaviour of Al single crystals (Alden et al., 1961) were reported and are scattered and disputed.

The authors had reported the cyclic deformation behaviour of coarse grained pure alumiuium polycrystals (Xia et al., 1990, Liu et al. 1989). The paper investigated and compared the effect of the CO on the σ, Q^{-1}, $\Delta\alpha$, σ_f, and σ_b of ASC during cyclic deformation.

EXPERIMENTAL

The specimens of $8 \times 8 \times 70$mm of ASC which were grown using 99.999% Al by the Bridgeman technigue were cut out utilizing electro−spark. The tensile axis orientation of the specimens was checked by X ray back reflection method.

Cyclic deformation tests were carried out in push−pull mode under strain control at ambient temperature in air on Fatigue Internal Friction Instrument (Zhu et al., 1988) for the strain amplitude of 6×10^{-4} and on 5KN shimadzu 4825 fatigue machine for the strain amplitude of 2×10^{-3}. All specimens were cycled to 3000 cycles. For preselected clyces, the stress−strain hysteresis loops were recorded by X−Y recorder. The stress σ, Q^{-1} and $\Delta\alpha$ during cycling was measured.

The σ_f and σ_b were calculated from the hysteresis loops. Finally, the slip morphologies on fatigue specimen surface were observed in S360 SEM.

RESULTS

Cyclic stress response

The stress response curves of different CO ASC were presented in Fig.1. The hardening behaviour of these crystals shows many similarity. The crystals harden rapidly in first few cycles and reach saturation, finally, secondary cycle hardening occurs. All of the crystals undergo stress overshooting and presoftening between initial hardening and saturation, i.e. the σ exceeds firstly the saturation stress σ_{sa}, then decreases and approaches the σ_{sa}. The effect of the CO on initial hardening rate and the σ_{sa} value is obvious. The σ_{sa} of DS and MS ASC are about twice of

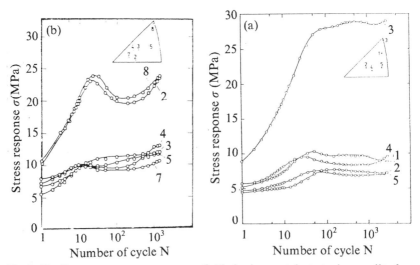

Fig.1 Cyclic stress response curves of Al single crystal at strain amplitude of 6×10^{-4}(a) and 2×10^{-3}(b).

SS one and the formers have obviouser stress overshooting and presoftening.

Internal friction Q^{-1} during cycling

Fig.2 shows variation of the Q^{-1} vs cycles under strain amplitude of 6×10^{-4} at room temperature for different CO ASC. The Q^{-1} decreases rapidly with cycles in first. After reaching a minimum value (valley), the Q^{-1} keeps fundamentally constant in cyclic saturation stage. The reducing rate of the Q^{-1} of MS ASC is more large than that of SS ASC and the Q^{-1} value of the former is only a half of the one of the latter at steady state.

Ultrasonic attenuation $\Delta\alpha$ during cycle

Fig.3 is the curve of the $\Delta\alpha$ at maximum σ of each cycle vs cyclic number, N, the $\Delta\alpha$ increases with increasing the N at early stage of fatigue life. As comparing with Fig.1(a) it can be found that the $\Delta\alpha$ starts to decrease previous to the beginning of cyclic saturation state. For MS ASC, the $\Delta\alpha$ increases again in secondary cyclic hardening stage.

Friction stress and back stress

After method employed (Kuhlman–Wilsdorf et al.,1979), the friction stress, σ_f and back stress, σ_b were obtained from the hysteresis loops as follows:

$$\sigma_f = \frac{(\sigma_e + \sigma_s)}{2} \tag{1}$$

$$\sigma_b = \frac{(\sigma_e - \sigma_s)}{2} \tag{2}$$

where the σ_e is maximum flow stress and the σ_s is yield stress of hysteresis loop. Fig.4 is the curves of the σ_b vs cumulative plastic strain, ε_{cum}. Their shape is many similar to the one of the $\sigma-N$ curves (in Fig.1). The effect of CO on the σ_b is obvious. The σ_b value of the crystals which

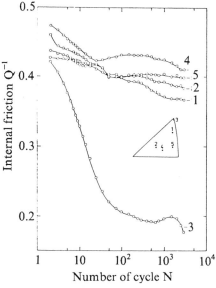

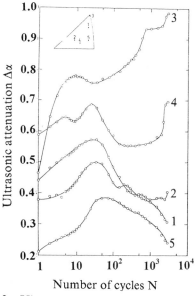

Fig.2 Variation of internal friction Q^{-1} of Al single crystals during cyclic deformation at strain amplitade of 6×10^{-4}

Fig.3 Ultrasonic attenuation $\Delta\alpha$ vs cycle number N curvest of Al single crystals at strain amplitude of 6×10^{-4}.

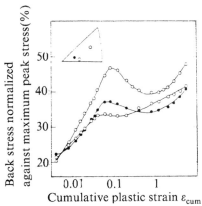

Fig.4 The back stress during cyclic deformation of Al single crystals at strain amplitude of 2×10^{-3}.

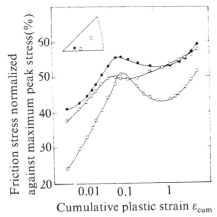

Fig.5 The friction stress during cyclic deformation for different slip orientation crystals at strain amplitude of 2×10^{-3} and room temperature.

CO is near 123 (expressing by coreless circle) is low at cyclic saturation state. The σ_b increases again in secondary cyclic hardening stage. The increasing rate of the σ_b of the DS ASC near 012 or 013 (indicating by dark dots) is larger than that of SS ASC.

Fig.5 shows a plat of the σ_f vs the ε_{cum}. The variation of the σ_f and the σ_b with the ε_{cum} has similar rule, but the σ_f value is just in reverse with σ_b. The effect of CO on the σ_f is also obvioys. The overshoot of the σ_f is larger and its value is lower for DS and MS ASC than those for SS one.

Stress–strain hysteresis loop and slip morphology

The three examples of the hysteresis loops and the slip band morphology of SS, DS and MS ASC are shown in Fig.6(a), (b) and (c),respectively. Accompanied by the increase of stress, the hysteresis loops of the SS ASC become progressively more pointed (from a to b, c, d and e) in Fig.6(a), this change rate is relaxation. The width of the slip bands of this specimen cycled to 3000 cycles is rather more wide. For MS ASC, conspicuous increase of σ is accompanied by the change of the loop shape (from a to b, c and d in Fig.6(c)) in initial hardening stage. The loops develop from e to f during cyclic presoftening stage and in secondary cyclic hardening stage the loops become gradually more pointed (from g to h in Fig.6(c)). The space of the slip bands for this crystal are rather fine. It is noticeable that the hysteresis loops of MS ASC have two yield. Fig.6(b) shows the variation of the loops with cycling and the slip morphology of DS ASC, the specimen also cycled for 3000 cycles.

DISCUSSIONS

It is well known, σ is an applied force which is needed for dislocation to overcome resistances. The resistance are caused by interaction between dislications, dislocation and defect (e.g. solute atoms, vacuities and precipitations etc.), dislocation and boundary (such as grain boundaries, phase boundaries, etc.) and so on. The resistance is mainly interaction among dislocations for high pure metallic single crystals. In general, the hardening rate of DS and MS ASC is larger than that of SS ASC.

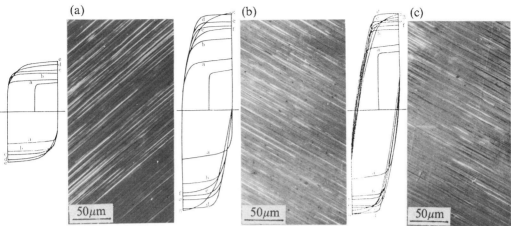

Fig.6 Changes in the shape of hysteresis loops during
cyclic deformation and photographs on the sur—
face of specimen cycled for 3000 cycles at strain
amplitude of 2×10^{-3}. Keys next to them corre—
spond to cycle numbers as follows:
(a) a,1; b,5; c,20; d,500; e,1400 for mono—slip
crystal
(b) a,1; b,5; c,10, d,20; e,50; f,200; g, 1400 for
double slip crystal.
(c) a,1;b,5;c,10;d,20;e,50;f,200;g,1000;h,1400 for
multi—slip crystal.

Q^{-1} is the sum of energy loss in one cycle, it results from dislocation motion and depends on the
quanlity, the resistance sufered and the moving distance of mobile dislocations.

$\Delta\alpha$ reflects instantaneous case of dislocation vibration aroud itself, because of high frequency
and small amplitude of ultrasonic wave, Although, all the $\Delta\alpha$ and the Q^{-1} relate to structures and
motion of dislocations, but there are difference between them. By TEM observation(Fei, 1991),
the maximum of $\Delta\alpha$ corresponds to parallel dislocation tangle bands and an amount of isolate
dislocations between the bands, while the maximum of σ and the valley of Q^{-1} value correspond
to embryonic dislocation cells. Therefore, when $\Delta\alpha$ has achieved amximum, σ contnues to in—
crease and Q^{-1} decrease yet.
In subsequent cycle, the dislocation tangle increase and transform into embryonic dislocation
cells. The moveable dislocation segments shorten and $\Delta\alpha$ begins to decrease. At the beginning of
saturation state and the Q^{-1} valley, embryonic dislocation cells have been formed, while, dislocations only bow from cell wealls and are sucked up by the walls, but hardly transmit the walls. In
the case the change only thins walls. increase the dislocation density of walls and orientation difference between cells, while cell size keep to constant (Xia et al. 1990). σ is inversely propor-
tioned to cell size, consequently, often arriving saturation state, σ unchanges and Q^{-1} transforms
from decreases to stable. The density and the lengeth of dislocation segments in cells lessen and
shorten, respectively. These make the $\Delta\alpha$ to be smaller.

CONCLUSIONS

The variation of Q^{-1} and σ with N is synchronous basically, and the direction of their change is opposed each ohter. The Q^{-1} valley corresponds to the beginning of saturation stage. The decrease rate and valley value of Q^{-1} of MS ASC are higher and lower, respectively, than those of SS ASC.

The variation of $\Delta\alpha$ with N for different CO ASC are not alike because of different mechanisms of dislocation structure and motion.

The loop shape of three kinds of crystals (SS ,DS and MS ASC) has large disparity. For SS ASC, the loop shape approximates rectangular, the σ_s approaches the σ_e and the σ_f is larger than the σ_b. For MS ASC, the σ_s is more little than the σ_e (the σ_s is equal to zero for DS ASC and is a negative value for MS ASC), the σ_f is less than the σ_b. The value and the change of σ_b and σ_f are related to dislocation structure and its change in materials during cyclic deformation.

REFERENCES

Laird, C., Charslcy, P. and Mughrabi, H.(1986). Low eneray dislocation structure produced by cyclic deformation. Mater. Sci. Eng., 81, 433.

Mughrabi, H.(1978). The cyclic hardening and saturation behaviour of copper single crystals. Mater. Sci.Eng., 33, 207.

Chong, Alex S. and laird, C. (1981). Mechanisms of fatigue hardening in copper single crystals: the effects of strain amplitude and orientation. Mater.Sci.Eng., 51, 111.

Abel, A., Wilkelm. M. and Gerold, V.(1979). Low–cycle fatigue of single crystals of a Cu–Al alloys. Mater.Sci.Eng., 37, 187.

Vorren, O. and Ryum, N. (1988). Cyclic deformation of Al single crystals: effect of the crystallographic orientaton. Acta Metall., 36, 1443.

Alden, T.H., Backofen, W.A.(1961). The formation of fatigue cracks in aluminum single crystals. Acta metall., 9, 352.

Xia, Y.B., Wang, Z.G. and Wang,R.H. (1990). Secondary hardening and dislocation evolution in low cycle fatigue of polycrystalline aluminium. phys.stat.sol.(a), 120, 125.

Wei Liu, Zhongguang Wang and Yuebo Xia (1989). Crystallographic characteristics of slip band continuity across grain boundaries in multicrystalline aluminium during cyclic deformation. phys.stat. sol.(a), 114, 177.

Zhu Zhengang, Zhou Xing and Fei Guangtao (1988). Fatigue internal friction instrument. Chinese Journal of seientific instrlument, 9, 396.

Kuhlman–Wilsdorf, D. and Laird, C.(1979). Dislocation behavior in fatigue II friction stress and back stress as inferred from an analysis of hysteresis loops. Mater.Sci.Eng., 37, 111.

Fei Guangtao (1991).The relation between the variation of internal friction, ultrasonic attenuation and dislocation configuration in aluminium and dilute aluminium under the early stage of fatigue cycling. Master' Gegres Treatise, Institute of solid state physics, Academcia sinica. P.R. China.

LOW-CYCLE IMPACT FATIGUE BEHAVIOURS AND DISLOCATION STRUCTURES OF BRASS

Pingsheng Yang and Meng Zhang

Institute of Materials Science, Nanchang University,
Nanchang 330047,P.R.China

ABSTRACT

In this work, low-cycle impact fatigue(LCIF) tests were carried out and the alterations of dislocation structures during the process of fatigue were investigated by using TEM analysis on polycrystal annealed brass. The results showed that under LCIF loads a continuous hardening but not saturation hardening appeared, and the cyclic hardening capacity was about 2 times as great as that in LCF. The LCIF life of brass conform to the Coffin-Manson law too. The transition life is very high. That indicates that the material is ductile in fatigue and is insensitive to strain rate. The TEM analysis of dislocation structures well interpreted the cyclic hardening behaviours in LCF and LCIF from microscope angle.

KEYWORDS

Brass; low-cycle fatigue; dislocation structure; cyclic hardening; strain rate

Brass is a fcc metallic alloy with a lower stacking fault energy and a series of studies have been conducted on the fatigue mechanism of the material. However, few reports on low cyclic impact fatigue behaviours and dislocation structures of brass have published. Obviously, it is important to clarify those problems in understanding the strain rate effects on such metals. The study will also contribute to the investigations on kinematics and kinetices of dislocation slip.

EXPERIMENTAL METHOD AND MATERIALS

The low-cycle impact fatigue tests(LCIF) were conducted by using the push-pull impact fatigue apparatus designed by the authors, in which the loading assembly is actually a combination of a Hopkinson's pressure bar and an extension bar. The strain rates in the specimens may attain 400 s^{-1}.

The testing principles and the determination method of stress-strain are identical with those described by the authors(Yang et al.,1994). In order to clarify the effects of strain rate on low-cycle fatigue behaviour of materials, ordinary low-cycle fatigue(LCF) was conducted in parallel by using an Instron 1342 electronic-hydraulic testing machine with a trigonal wave function and strain-controlled at 0.2 Hz frequency.

The material used in this investigation was H62 brass (Wt-%: 60.5-63.5 Cu, bal. Zn). The specimens were 580℃ annealed for 1 hr, the mechanical properties are as follows: yield stress, σ y = 150 Mpa; ultimate tensile strength, σ b = 370Mpa.

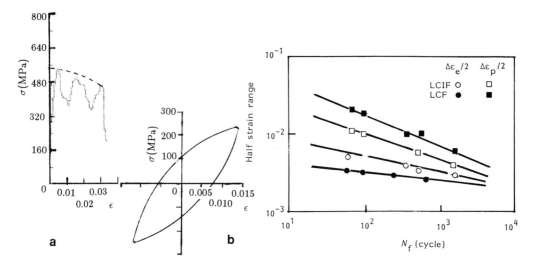

Fig.1.   Hysteresis loop:
         (a) LCIF; (b) LCF

Fig.2.   Relationship between half strain
         range $\Delta\varepsilon_e/2$, $\Delta\varepsilon_p/2$ and cycles
         to failure for brass

EXPERIMENTAL RESULTS AND DISCUSSION

Hysteresis Loop

In LCIF tests as well as in LCF tests, the hysteresis loop were recorded at specified cycles in order to study cyclic stress-strain behaviours and their changes. Typical hysteresis loops for LCIF and LCF are shown in Fig.1. In the stress-strain loops of LCIF, there are periodic stress falls forming an attached oscillation, as shown by the solid line. This does not belong in behaviour of materials, so a smoothing procedure was undertaken, as shown by the dotted line. Some differences in the shape of LCIF and LCF loops are considered to be related to the loading wave function, which are trapezoidal in LCIF and trigonal in LCF.

In this investigation the strain rate in LCIF tests was 100--300 s^{-1}, which was 10^4 times higher than that in LCF tests and 10^6 times higher than in the static tension tests. At this high rate of strain, the elastic limit and flow stress were obviously raised compared with those in LCF and static tension tests: the so-called 'overstress'. The flow stress at 2.0% permanent deformation in LCIF tests was 300 Mpa, that is $(\sigma_D - \sigma_S)/\sigma_S = 1.0$. Comparing with other experimental results by the authors (Yang et al.,1994;Yang and Zhou,1994), the overstress of brass is equal roughly to that of austenitic stainless steel but lower than that of mild steel $((\sigma_D - \sigma_S)/\sigma_S = 1.6)$ and higher than those of aluminium (0.25) duralumin (0.1) and medium-carbon alloy steel (0). It is found that the overstress is related to more the crystal lattice category and microstructures than strength of materials.

Cyclic Hardening

In both LCIF and LCF tests a cyclic hardening appeared (Fig. 3). However, there are some differences in the cyclic hardening curves: (1) In LCF test a fast hardening occurred at first in several percentage of N_f then cyclic stability appeared, but in LCIF tests a continuous cyclic hardening appeared. (2) In LCIF tests the cyclic hardening capacity was twice as strong as that in LCF tests.

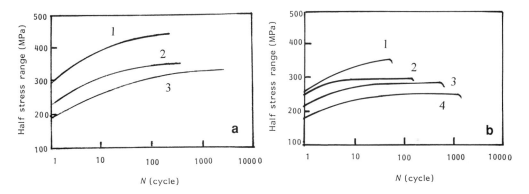

Fig.3   Cyclic hardening curves of brass:
(a) LCIF; 1). $\Delta\varepsilon_T / 2 = 0.0092$; 2). $\Delta\varepsilon_T / 2 = 0.0090$; 3). $\Delta\varepsilon_T / 2 = 0.0070$;
(b) LCF; 1). $\Delta\varepsilon_T / 2 = 0.0244$; 2). $\Delta\varepsilon_T / 2 = 0.0129$; 3). $\Delta\varepsilon_T / 2 = 0.0127$;
4). $\Delta\varepsilon_T / 2 = 0.0084$

It is generally considered that the cyclic hardening behaviours of matalic materials are related to their stress-strain behaviours in tension test. With a lower stacking fault energy, brass may appear three-stages hardening in deformation even at atmospheric temperature.The flow curves of metalic crystals are significantly influenced by strain rates and temperature: along with the increase in strain rate stage I and stage II become longer and the stress (strain) at which stage III begins, τ_3, (γ_3) is raised.

Accordingly the cyclic hardening features of brass in LCIF can be explained as follows: because of γ_3 increases obviously under high strain rates, but the cumulated strain of LCIF tests in this study had not come up to the γ_3 value yet, so a continuous cyclic hardening but not cyclic saturation occurred. Also as result of the obvious increase in τ_3, the cyclic hardening capacity in LCIF tests may be very great. Furthermore the TEM analysis on fatigue dislocation structures had explained the differences in cyclic hardening behaviours between LCF and LCIF.

Alterations of Dislocation Structures

Cyclic hardening(softening) behaviours result from hardening factors and softening factors in cyclic deformation. The former is related to the dislocation multiplication and the interactions of dislocation with second phase particles, point defects and other dislocations, while the latter to the formation of the low energy dislocation structures such as band and cell. As well known, the variations in dislocation arrangement conform to the universal principle of that free-energy tend towards lower. Cyclic deformation in push-pull straining provides ideal conditions for achieving low energy dislocation structures because large cumulative strains give rise to high dislocation densities and to-and-fro dislocation motions enhance entrapment probabilities. The observations on the dislocation configurations by authors confirm that also, but there are some differences in evolutions of dislocation structures between LCIF and LCF.

Under the egual life condition the evolution processes of dislocation structures were analysed in LCIF (N_f= 1500 $\Delta \varepsilon_e/2 = 0.003$, $\Delta \varepsilon_p/2 = 0.004$) and   LCF (N_f= 1600 $\Delta \varepsilon_e/2 = 0.0024$, $\Delta \varepsilon_p/2 = 0.006$). The TEM analysis shows that the dislocation structures of not-fatigued specimen were a few scattered dislocation lines and loops (Fig.4). In LCIF tests after 1% N_f cycles dislocation density increased, more dislocation loop and dipole appeared (Fig.5). After 10% N_f cycles dislocation density increased further, the dislocation lines become shorter and transformed into dislocation tangle. In failed specimen the substructures consisted of dislocation loops, dislocation braid and tangles. At some fields, the tangle has evolved into a embryonic form of dislocation cell, but no cell appeared (Fig.6).

In LCF tests, after 10% N_f cycles the tangles constituted the embryonic cell (Fig.7) which is similar to that of failed specimen in LCIF, besides there were the deformation bands(Fig.8). In the failed specimen dislocation cell and subcrystal appeared (Fig.9). In other words, in LCIF process the hardening factors: dislocation multiplication and formation of high energy structures played a dominant role, so the cyclic hardening occured from beginning to end. While in LCF at the initial stage of fatigue a rapid multiplication of dislocations led to cyclic hardening, but soon afterwards, the softening substructures shch as cell and deformation band appeared, but balanced the hardening factors, therefore the material came into cyclic stability.

Fatigue Life

By plotting $\Delta \varepsilon_e/2$, $\Delta \varepsilon_p/2$ at 50%N_f against the cycles to failure (Fig.2), it was found that their relations in both LCF and LCIF all conformed to the Coffin-Manson law. There were similar circumstances to that of other experimental materials: the $\Delta \varepsilon_e/2$ – N_f curve of LCIF was above

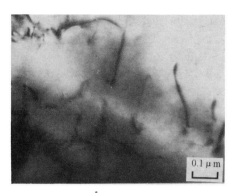

Fig.4.   The substructure of not fatigue
         specimen of brass

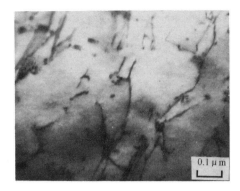

Fig .5.   The substructure in brass after
          1%N_f during LCIF

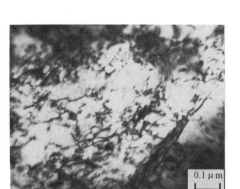

Fig.6.   The substructure in brass after
         failure in LCIF

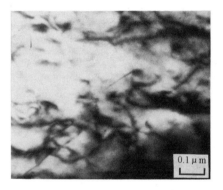

Fig.7.   The Substructure in brass after
         10%N_f during LCF

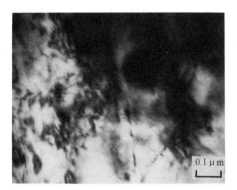

Fig.8.   The substructure in brass after
         10%N_f during LCF

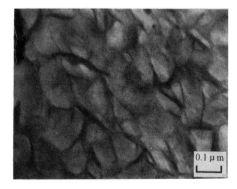

Fig.9.   The substructure in brass after
         faillure in LCF

that of LCF, which accords with the phenomenon of overstress; by contrast, the $\Delta \varepsilon_p/2 - N_f$ curve of LCIF laid below that of LCF, showing that cyclic deformation at high strain rates causes heavier fatigue damage. The transition life of LCIF inevitably shifts left. As shown in the figure, the transition life shifts left from about 6×10^4 cycles for LCF to about 1.5×10^4 cycles for LCIF. However, the magnitude of shift was less than that of mild steel and medium-carbon alloy steel obviously but near to austenitic stainless steel and aluminium, and the transition lifes of brass were the highest in experimented materials either in LCF or LCIF. As well known, the transition life characterizes the ductility of materials in fatigue, so we consider whether it is related to those that brass, aluminium and austenitic steel were well used at low temperature.

CONCLUSIONS

1. Under LCIF loads the flow stress of brass shows a moderate sensitivity to strain rate.
2. In LCIF tests for brass a continuous cyclic hardening appears but the hardening capacity was much stronger as compared with LCF. This hardening behaviours have been well explained by the TEM analysis on the fatigue dislocation structures.
3. The fatigue life in LCIF conform to the Conffin-Manson law; but the transition life of brass was very high. That indicates a big dutility and insensitivity to strain rate in fatigue.

ACKNOWLEDGEMENT

The research was supported by the National Natural Science Foundation of China.

REFERENCES

Yang, P.S. and Q. F. Zhang (1994). Push-pull impact fatigue testing machine and experimental procedures . Journal of Nanchang University (Natural Science), 18, 41-48.

Yang, P.S., X. N. Liao and H. J. Zhou (1994). High strain-rate low-cycle impact fatigue of a medium-carbon alloy steel. Int. J. Fatigue, 16, 327-330.

Yang, P.S. and H.J. Zhou (1994). Low -cycle impact fatigue of mild steel and austenitic stainless steel. Int. J. Fatigue, 16, 567-570.

MEAN STRESS SENSITIVITY OF THE HCF STRENGTH IN TIMETAL 1100: MICROSTRUCTURAL EFFECTS

J. LINDEMANN, A. BERG and L. WAGNER

Materials Technology, Technical University of Brandenburg at Cottbus
03013 Cottbus, Germany

ABSTRACT

The HCF strengths of fully lamellar as well as duplex (α_p phase in lamellar matrix) microstructures in the near α titanium alloy TIMETAL 1100 were evaluated in fully reversed (R = −1) and tension-zero-tension (R = 0.1) loading. The prior β grain size and α_p volume fraction were varied in the fully lamellar and duplex microstructures, respectively. While the variation in HCF strengths among the various microstructures was not very pronounced at R = −1, the fully lamellar microstructures were by far superior to the duplex structures at R = 0.1. The best overall performance was found for the fine grained fully lamellar microstructure.

KEYWORDS

Fully lamellar microstructures; duplex microstructures; fatigue crack nucleation; mean stress sensitivity.

INTRODUCTION

It has long been recognized that ($\alpha+\beta$) titanium alloys such as Ti-6Al-4V tend to exhibit a drastic decrease in the endurance limit in terms of stress amplitude if the fatigue loading is changed from fully reversed (R = −1) to tension-zero-tension (R = 0) loading [1, 2]. While the reason for this poor fatigue performance at R = 0 is not yet fully understood, there is experimental evidence that this so-called anomalous mean stress effect associated with equiaxed or duplex microstructures does not occur in β annealed fully lamellar microstructures [3]. Unfortunately, fully lamellar microstructures usually exhibit low ductility values due to the large packet and β grain sizes. However, fine prior β grains can easily be produced in TIMETAL 1100 by conventional thermomechanical treatments since in this alloy having a silicide solvus temperature above the β transus temperature, β annealing is possible with the presence of large volume fractions of silicides which inhibit β grain growth [4 - 6].

EXPERIMENTAL PROCEDURE

The TIMETAL 1100 material was received from Timet (USA). The chemical composition of the alloy and details of the thermomechanical treatments to produce fully lamellar as well as duplex microstructures are reported elsewhere in these proceedings [7]. Crystallographic texture measurements were performed by means of a computer controlled texture goniometer using Cu-Kα radiation. The random intensities were determined by measurements on hot isostatically pressed powder specimens. Tensile tests were conducted on cylindrical specimens (gage length 20 mm, diameter 4 mm) at ambient and elevated (600 °C) temperatures in air. The initial strain rate was $\dot{e}$ = 8.3 x 10^{-4} s^{-1}. Fatigue tests were performed on electrolytically polished hour glass shaped specimens in rotating beam loading (R = −1) and in axial tension-zero-tension (R = 0.1) loading by means of a servohydraulic testing machine. Tests were performed at a frequency of 60 s^{-1} in air.

EXPERIMENTAL RESULTS AND DISCUSSION

Annealing the alloy TIMETAL 1100 at 1060 °C and 1012 °C resulted in fully lamellar structures (Fig. 1) with prior β grain sizes of 450 (Fig. 1a: LC) and 160 μm (Fig. 1b: LF), respectively. For both duplex microstructures D60 (Fig. 1c) and D20 (Fig. 1d), the α_p size is about 15 μm. While the width of the α lamellae is about 1 μm for both, the length of the α lamellae and the colony sizes are larger in D20 (Fig. 1d) as opposed to D60 (Fig. 1c) due to the coarser prior β grains.

Examples of the (0002) pole figures of the lamellar and duplex microstructures are shown in Fig. 2. The pole figure of LF (Fig. 2a) indicates a comparatively weak texture which was also measured for LC. The pole figure of D20 is shown in Fig. 2b indicating a mixed B/T type of texture. No change in this pole figure was measured for D60, i.e., both lamellar α and α_p phase must be similarly textured.

Table 2 - Tensile properties of TIMETAL 1100

a) lamellar

Microstructure	$\sigma_{0.2}$ (MPa)	UTS (MPa)	e_u (%)	El (%)	σ_F (MPa)	ε_F = ln A$_0$/A$_F$	E (GPa)
LC	955	1065	6.1	6.2	1170	0.10	116
LF	935	1040	4.8	11.1	1233	0.22	116

b) duplex

Microstructure	$\sigma_{0.2}$ (MPa)	UTS (MPa)	e_u (%)	El (%)	σ_F (MPa)	ε_F = ln A$_0$/A$_F$	E (GPa)
D20-TD	965	1032	3.3	9.2	1341	0.32	113
D20-RD	982	1060	5.4	14.5	1344	0.29	109
D60-TD	953	994	3.2	11.4	1320	0.36	113
D60-RD	935	1010	6.2	13.2	1259	0.25	106

Tensile test results are listed in Table 2 comparing the data of the various microstructures. Yield stress values are similar between LC and LF, while LF is markedly superior to LC regarding tensile elongation to fracture El and ductility ε_F. Similarly, yield stress values are very similar between D20 and D60, while D20 is somewhat superior with respect to tensile strength.

The S-N curves are shown in Fig. 3 comparing the fatigue results at R = −1 and R = 0.1 of the fully lamellar microstructures (Fig. 3a) and the duplex microstructures as loaded in TD (Fig. 3b) and RD (Fig. 3c). The endurance limits of the various microstructures and loading directions differ by only about 60 MPa at R = −1 with a highest value of σ_a (10^7) = 550 MPa for LF (Fig. 3a) and a lowest value of 490 MPa for D60 loaded in TD (Fig. 3b). As can also be seen in Fig. 3, the variation in HCF strengths is by far greater at R = 0.1 with a highest value of σ_{max} (10^7) = 900 MPa for LC (Fig. 3a) and a lowest value of 600 MPa for D20/D60 loaded in TD (Fig. 3b). The HCF strengths at R = −1 and R = 0.1 are summarized in Fig. 4. While loading the duplex microstructures in RD as opposed to TD results in somewhat higher fatigue strengths at R = 0.1, the performance of the fully lamellar structures at R = 0.1 is by far the best. Thus, the results of earlier work on Ti-6Al-4V [3] are confirmed in the present study. In [3], the anomalous mean stress effect observed in duplex as opposed to lamellar microstructures was explained by easy crack nucleation through dislocations piling up against α_p phase boundaries at low maximum stresses at which single slip can be activated. Nevertheless, duplex structures were recommended in Ti-6Al-4V since the fully lamellar structures, though being superior with regard to HCF at R = 0.1, always possess poor LCF characteristics due to inevitable coarse β grain sizes in that alloy. On the contrary and as also shown in previous and parallel work [6 - 9] on TIMETAL 1100, fully lamellar microstructures which combine excellent LCF, HCF and creep strengths can easily be produced in TIMETAL 1100 without suffering the typical drawbacks of the duplex structures, such as directionality of mechanical properties and anomalous mean stress sensitivity of the fatigue strength.

ACKNOWLEDGEMENTS

This work was supported by the Deutsche Forschungsgemeinschaft.

REFERENCES

1.   R. K. Steele and A. J. McEvily, Eng. Frac. Mech., 8 (1976) 31.

2.   R. K. Steele and A. J. McEvily, Fracture Mechanics and Technology, Ed. G. C. Sih and C. L. Chow, Sijthoff and Nordhoff, Netherlands (1977) 33.

3.   S. Adachi, L. Wagner and G. Lütjering, Fatigue of Engg. Mat. and Struct., IMechE (1986) 67.

4.   M. Peters et. al., Microstructure/Property Relationships in Titanium Alloys and Titanium Aluminides (Y. W. Kim and R. R. Boyer, eds.) TMS, Warrendale, Pa (1991).

5.   D. Weinem, J. Kumpfert, M. Peters and W. A. Kaysser, Processing Window of the Near-α-Titanium Alloy TIMETAL 1100 to Produce a Fine Grained β-Structure, Mat. Sci. and Engg. (1995).

6.   A. Styczynski, L. Wagner, C. Müller and H. E. Exner, Microstructure/Property Relationships of Titanium Alloys (S. Ankem and S. A. Hall, eds.) TMS-AIME (1994) 83 - 90.

7.   A. Berg, J. Lindemann and L. Wagner, Fatigue '96 DVM (1996).

8.   A. Berg, J. Kiese and L. Wagner, Light-Weight Alloys for Aerospace Applications III (E. W. Lee, K. V. Jata, N. J. Kim and W. E. Frazier, eds.) TMS-AIME (1995).

9.   J. Lindemann, A. Styczynski and L. Wagner, Light-Weight Alloys for Aerospace Applications III (E. W. Lee, K. V. Jata, N. J. Kim and W. E. Frazier, eds.) TMS-AIME (1995).

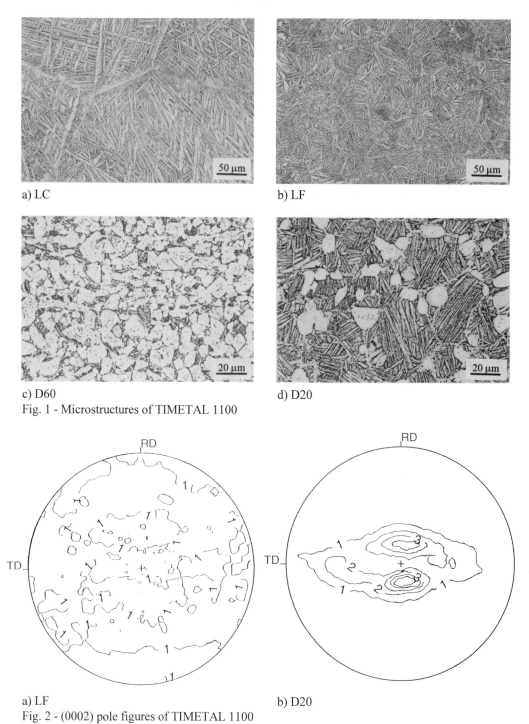

a) LC

b) LF

c) D60

d) D20

Fig. 1 - Microstructures of TIMETAL 1100

a) LF

b) D20

Fig. 2 - (0002) pole figures of TIMETAL 1100

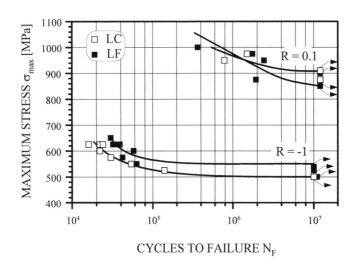

a) Lamellar microstructures

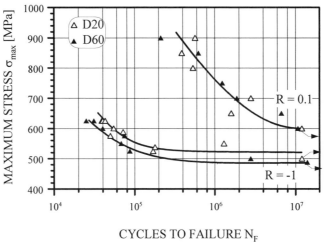

b) Duplex microstructures (TD)

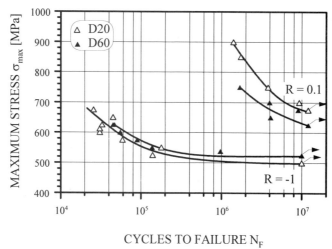

c) Duplex microstructures (RD)

Fig.3 - Fatigue behavior of the various conditions in TIMETAL 1100

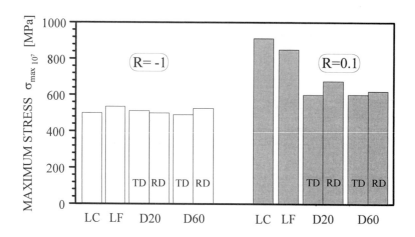

Fig.4 - HCF-strengths of the various conditions in TIMETAL 1100

FATIGUE LIFE OF DUPLEX STAINLESS STEELS:
INFLUENCE OF AGEING AND ENVIRONMENT

A. MATEO[1], P. VIOLAN[2], L. LLANES[1], J. MÉNDEZ[2] and M. ANGLADA[1]

1.- Dept. Ciencia de los Materiales e Ingeniería Metalúrgica
ETSEIB, Univ. Politècnica de Catalunya, 08028 Barcelona, Spain
2.- Laboratoire de Mécanique et de Physique des Matériaux
(URA N°863/CNRS), ENSMA, 86960 Futuroscope Cedex, France

ABSTRACT

The influence of two extrinsic factors, namely ageing and environment, on the fatigue behaviour of an UNS S31803 duplex stainless steel is studied. The cyclic stress-strain response for both unaged and aged (200 hours at 475°C) steel is initially determined. Then, constant-amplitude tests to failure, at a selected plastic strain amplitude, are performed in both laboratory air and in vacuum. The results, in terms of fatigue life and for the chosen amplitude, indicate that ageing increases the fatigue life and tests performed in vacuum last ten times longer than those conducted in air. These experimental findings are discussed in terms of ageing effects on the plastic activity of the ferritic matrix under different experimental conditions.

KEYWORDS

Duplex Stainless Steel (DSS); low-cycle fatigue (LCF); fatigue life; ageing embrittlement; environmental conditions.

INTRODUCTION

Duplex (austenite plus ferrite) stainless steels (DSS) are excellent materials for components in the oil, chemical and power industries. This is due to their combination of high strength and toughness together with high resistance to corrosion, which result from their dual microstructure (Solomon and Devine Jr., 1983). However, one of the main problems which appears along the service life of DSS is the severe embrittlement that they suffer when exposed to the intermediate temperature range (250°C-500°C). This phenomenon, attributed mainly to the spinodal decomposition of the ferritic phase, conditions the industrial applicability of these materials (*e.g.* Chopra and Chung, 1988; Pumphrey and Akhurst, 1990; Iturgoyen *et al.*, 1994).

Many of the industrial applications for which DSS are potential candidates involve cyclic loading. However, fundamental knowledge about fatigue of DSS, under both unaged and aged conditions, is relatively scarce. In this work the fatigue life of a DSS has been investigated and discussed as a function of ageing conditions and environmental factors. In doing so, the cyclic stress-strain curves (CSSC) for both unaged and aged steel were initially determined. Then, from the measured CSSCs one plastic strain amplitude was selected in order to perform constant-amplitude tests and compare the number of cycles to reach failure in both air and vacuum.

EXPERIMENTAL PROCEDURE

The DSS studied in this work was of type UNS S31803. After an annealing treatment (at 1050°C for one hour), the final microstructure consisted of 45% austenite on a ferritic matrix. Table 1 gives its chemical composition, as well as that of the constitutive phases.

Table 1. - Chemical composition of the UNS S31803 steel studied and partitioning of the alloying elements between the constitutive phases, (wt.%).

	C	N	Cr	Ni	Mo	Mn	Si
UNS-S31803	0.025	0.13	22.00	5.50	3.00	1.60	0.30
Phase α	-	-	27.53	2.86	4.82	1.46	0.61
Phase γ	-	-	20.42	6.31	1.19	1.79	0.05

Two types of cylindrical specimens were used, both of diameter 6 mm, but with different gauge lengths (25 and 10 mm), because of the distinct gripping fixtures available in the two laboratories involved in this research. In both cases, one set of specimens was tested in the annealed or unaged condition and another after ageing treatment at 475°C for 200 hours. The CSSCs were determined performing incremental-step tests in an Instron servohydraulic machine. These tests were carried out under fully reversed total strain control in laboratory air. Constant amplitude tests were conducted to failure in both air and vacuum (pressure lower than 10^{-3} Pa), in an Instron electromechanical machine under symmetrical uniaxial push-pull mode and plastic strain control. After the fatigue tests, the specimens surfaces were observed by scanning electron microscopy (SEM).

EXPERIMENTAL RESULTS

The cyclic stress-strain response, for unaged and aged materials, is shown in Fig. 1. This figure includes saturation or final (in the cases where a saturation behaviour was not reached) stresses and plastic strain amplitudes (ε_{pl}) from incremental-step tests. Stress-strain response was determined following an experimental procedure similar to that reported by Llanes et al. (1995). The CSSCs allow to define three stages in the two cases, but at different stress levels for each condition. While the behaviour is relatively independent of ageing within the first stage (at the lowest ε_{pl}), a slope change is observed at about $\varepsilon_{pl} = 2 \times 10^{-4}$ and the two curves start to depart from each other. The slope change is more pronounced for the aged condition, i.e. the cyclic strain hardening rate is larger for the material after the ageing treatment. The end of this intermediate range of ε_{pl} is determined by a second change of slope, at values of $\varepsilon_{pl} \approx 10^{-3}$. The third stage exhibits a low cyclic hardening rate for both conditions.

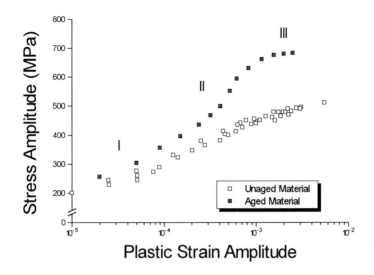

Fig. 1. CSSCs of the studied DSS.

Considering the found cyclic stress-strain responses, a ε_{pl} of 6×10^{-4} was selected in order to perform several constant-amplitude tests to failure. This amplitude is within the second stage of the CSSCs, regime where the differences between the responses of unaged and aged material are more pronounced. Hence, significative differences between the number of cycles to failure of the material in both conditions are expected at the chosen ε_{pl}. Taking into account fatigue life's differences, an amplitude within stage III could have been suitable too. However, because of the large values of ε_{pl} in this regime, the fatigue lives were expected to be too short and extrinsic effects, *i.e.* these due to environment, possibly difficult to appreciate. Being the study of such influence on fatigue life one of the aims of these work, a ε_{pl} within the intermediate regime was rather chosen.

Figure 2 shows the cyclic hardening-softening curves corresponding to the tests performed in vacuum for both unaged and aged materials. Results obtained in air are only indicated by giving the stress fall due to the last period of growth of the fatal crack. Before that period the cyclic behaviours were identical in both environments. Both aged and unaged materials are characterised by a short period of primary hardening followed by a softening stage. It can be noted that for the aged DSS failure is reached in air (32000 cycles) before the end of this softening period. A saturation stage is only obtained after about 10^5 cycles. In contrast, for the unaged condition, saturation is nearly reached at failure in air. The aged material is therefore characterised by a more pronounced softening. The stress amplitudes obtained under plastic strain control are in good agreement with the corresponding values in Fig. 1, obtained under total strain control condition. The aged material exhibits a stress amplitude response about 150 to 100 MPa higher that in the unaged one.

The results, in terms of fatigue life, are shown in Table 2. For comparison purposes, results obtained on an austenitic stainless steel are also presented (Méndez *et al.*, 1993). Several aspects may be noticed:
 - first, ageing increases the fatigue life in both air and vacuum by a factor of approximately two.

- second, tests performed in vacuum lasted about ten times longer than those conducted in air.
- third, the effect of environment in DSS material is higher than on austenitic stainless steel (17Cr-12Ni).

Table 2. Fatigue lives (number of cycles to failure) of the investigated DSS and 316L austenitic SS (Méndez et al., 1993) under applied $\varepsilon_{pl} = 6 \times 10^{-4}$, in both air and vacuum.

	Air	Vacuum
Unaged	18000	198000
Aged (200 hours at 475°C)	32000	415000
316L Austenitic SS	150000	530000

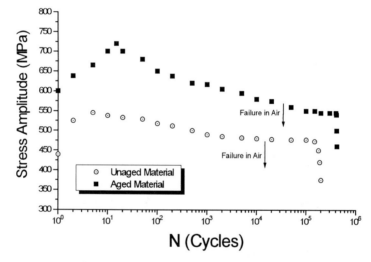

Fig. 2. Hardening-softening response of the studied DSS.

Strain localisation and crack nucleation features were investigated through SEM. Fig. 3 shows the typical aspect of the surface damage in vacuum tested samples. In the aged material (Fig. 3b) deformation appears highly localised in the austenitic phase (in dark). Intense slip bands with marked extrusions are observed. Crack initiation usually takes place within these slip bands, leading to the formation of numerous surface microcracks. It may be seen that such cracks have a great difficulty to grow through the ferritic phase resulting in crack branching or crack arrest at the α/γ interface. On the other hand, only very few cracks form in the unaged material and plastic deformation seems to be concentrated in the ferritic phase (Fig. 3a).

DISCUSSION

The differences on cyclic behaviour observed in Fig. 1 must be related to changes on the active cyclic deformation mechanisms for each material condition and for each interval of ε_{pl}, as has been discussed by the authors, for an AISI-329 type DSS, elsewhere (Mateo et al., 1995; Llanes et al., 1995). The much higher stress values found for the aged steel compared with those measured for the unaged material

(Iturgoyen, 1994) may be associated with the overall yield strength increment produced by the ferrite embrittlement due to spinodal decomposition of this phase.

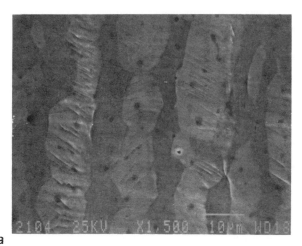

a

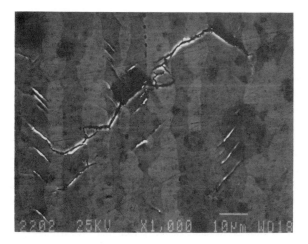

Fig. 3. Surface damage in vacuum tested samples. a) Unaged material; b) Aged material

The increase in fatigue life for aged materials at applied $\varepsilon_{pl} = 6\times10^{-4}$, as compared to that of unaged DSS, should also be related to embrittlement from spinodal decomposition of ferrite. Surface damage features seem to indicate that, at the studied ε_{pl}, crack nucleation is associated with the austenitic phase in the aged material and with the ferritic matrix in the unaged one. Moreover, the afterwards local crack growth scenario is also different for both materials depending upon the ductility level of the phase in which the crack must then propagate. In the aged materials the brittle ferrite matrix is not able to carry out plastic deformation and therefore cracks experience a noticeable difficulty to propagate through it. On the other hand, in the unaged DSS austenite is also able to accommodate plastic deformation and facilitates crack extension. These ideas are in agreement with previous reports on the observation of a higher fatigue crack propagation threshold for aged DSS than for the unaged material (Iturgoyen et al., 1993). Therefore, increasing fatigue life with ageing, at intermediate ε_{pl}, must be associated with the passive plastic role of ferrite within the stages of nucleation and initial crack growth beyond the austenitic grains.

The much more pronounced environmental effect observed in DSS, as compared to that previously reported in austenitic SS, is very noticeable. The fact that such relative effect seems to be independent of the plastic role of ferrite suggests that the presence of this phase alone must be, at least partly, responsible for it. However, the physical reason for such an extrinsic degradation is not clear and further work in attempting to answer such a query is currently in progress.

ACKNOWLEDGEMENT

The research reported in this paper was funded by the European Coal and Steel Commission (contract 7210-MA/940), the Spanish *Comisión Interministerial de Ciencia y Tecnología* (contract MT-1497) and the Spanish *Ministerio de Educación y Ciencia* and the French *Ministère des Affaires Étrangères* under the Picasso Programme (joint actions 95102 and HF94-245 respectively). We gratefully acknowledge this support.

REFERENCES

Chopra, O. K. and H. M. Chung (1988). Aging degradation of cast stainless steels: effects on mechanical properties. In: *Environmental Degradation of Materials in Nuclear Power System-Water Reactors*, (G. J. Theus and J. R. Weeks, eds.), pp. 737-748. TMS, Warrendale.

Iturgoyen, L., J. Alcalá and M. Anglada (1993). The influence of ageing at intermediate temperatures on the fatigue properties of a duplex stainless steel. In: *Proc. Fifth Int. Conf. on Fatigue and Fatigue Thresholds*, (J. -P. Bailon and J. I. Dickson, eds.), pp. 669-674. EMAS, London.

Iturgoyen, L. (1994). Ph.D. Thesis, Universitat Politècnica de Catalunya, Barcelona.

Iturgoyen, L., A. Mateo, L. Llanes and M. Anglada (1994). Thermal embrittlement at intermediate temperatures of AISI 329 duplex stainless steel. In: *Materials for Advanced Power Engineering*, (D. Coutsouradis et al., eds.), Vol. I, pp. 505-514. Kluwer Academic Publishers, Dordretch.

Llanes, L., A. Mateo, L. Iturgoyen and M. Anglada (1995). Aging effects on the cyclic deformation mechanisms of a duplex stainless steel, submitted for publication.

Mateo, A., L. Llanes, L. Iturgoyen and M. Anglada (1995). Cyclic stress-strain response and dislocation substructure evolution of a ferrite-austenite stainless steel. *Acta metall. mater*, in press.

Méndez, J., P. Villechaise, P. Violan and J. Delafond (1993). Environment and deformation interactions on the fatigue resistance of a 316L Stainless Steel treated by dynamical ion mixing. In: *Corrosion-Deformation Interactions, CDI'92*, (T. Magnin and J. M. Gras, eds.), pp. 741-754. Les Editions de Physique, Les Ulis.

Pumphrey, P. H. and K. N. Akhurst (1990). Aging kinetics of CF3 cast stainless steel in temperature range 300-400°C. *Mater. Sci. Technol.*, 6, 211-219.

Solomon, H. D. and T. D. Devine Jr. (1983). Duplex stainless steels - A tale of two phases. In: *Duplex Stainless Steels*, (R. A. Lula, ed.) pp. 693-756. ASM, Ohio.

Fracture Mode Transition in Low-Temperature Low-Cycle Fatigue of Pure Iron

Desheng XIA,	Graduate Student of Keio University,
	3-14-1 Hiyoshi, Kohoku-ku, Yokohama 223, Japan
Jun KOMOTORI,	Keio University
Kaoru KUWANO,	All Nippon Airways, Co. Ltd. Japan
Masao SHIMIZU,	Keio University

ABSTRACT

A commercial grade pure iron was used to investigate the low cycle fatigue behaviors at low temperature, with special focus on the role of deformation twins in the fracture process of the low cycle fatigue. The fracture surfaces were observed using scanning electron microscope to clarify the fracture mechanisms. It was found that a fracture mode transition occurs from surface to internal fracture mode with increase in plastic strain range ($\Delta\varepsilon_p$) at the low temperature. In a small $\Delta\varepsilon_p$ regime, a surface crack results in the final fracture of specimen. In a large $\Delta\varepsilon_p$ regime, however, a crack initiated inside the material leads to the final fracture. In this mode, three types of internal crack initiation sites have been observed: internal crack originates at (i) the intersection of deformation twin and grain boundary, (ii) the intersection of deformation twins and (iii) the intersection of deformation twin and inclusion. The applicability of Manson-Coffin law to the low-temperature low-cycle fatigue was also discussed.

KEYWORDS

Low cycle fatigue; deformation twin; fracture mode transition; low temperature; Manson-Coffin law; pure iron.

1. INTRODUCTION

It is well known that Manson-Coffin law can be applied to many materials in the low cycle fatigue at room temperature. However, few studies have been reported on the low cycle fatigue at low temperature such as that in the ductile-to-brittle transition range of the material.

At low temperatures, some ductile metals become brittle (notably those having body-centered cubic structure), and reveal cleavage fracture when tension load is applied. Hull (1960) and Honda (1961) have reported that the cleavage fracture of silicon-iron single crystal results from a microcrack formed at the intersection of deformation twins. In this study, a commercial grade pure iron was used to investigate the low cycle fatigue behavior at low temperature, with special focus on the role of deformation twins in the fracture process of the low cycle fatigue. The applicability of Manson-Coffin law to low-temperature low-cycle fatigue is also discussed.

2. MATERIAL AND EXPERIMENTAL PROCEDURES

The material was a commercial grade pure iron with the chemical composition as given in table 1. It was annealed in a vacuum furnace at 1200°C for 4 hours to obtain the microstructure of ferrite single phase with average grain size of 400μm, and was machined into hourglass shape specimen with minimum diameter of 6mm. After electropolishing the specimen surface, strain controlled low cycle fatigue tests were performed under push-pull loading condition at -140°C and at room temperature (R.T.). The testing temperature was controlled within ±0.2°C by using liquid nitrogen. Observation of the fracture surface was also carried out using a Scanning Electron Microscope (SEM).

3. RESULTS AND DISCUSSION

3.1 Fracture Mode Transition in Low Cycle Fatigue at Low Temperature

Before fatigue test, static tension tests were carried out at the temperatures ranging from R.T. to -150°C to clarify the effect of testing temperature on the value of reduction in area (R.A.) and fracture mechanism. It is found that: (i) rapid decrease in the value of R.A. occurs from -70°C (Fig. 1), and (ii) at the testing temperature lower than -90°C, serrations appear in the stress-strain curve and a large number of deformation twins can be observed as shown in Fig. 2. In order to investigate the effect of deformation twins on fatigue fracture mechanism and on fatigue life properties, low cycle fatigue tests were performed at -140°C and at room temperature (R.T.) . Fig. 3 shows the results of the fatigue tests. It should be noted that the fatigue lives decrease with the decrease in temperature and this behavior becomes remarkable in the relatively large plastic strain range ($\Delta\varepsilon_p$). To clarify the reason for this, observation of the fracture surface was performed and the fracture origin was specified based on the result of examination of the characteristics of river patterns.

As a result, two types of fracture modes were observed depending on the level of $\Delta\varepsilon_p$ as described below:

1) In a small $\Delta\varepsilon_p$ regime, a fatigue crack initiated at the surface of specimen leads to the final fracture (Surface Fracture Mode). In this case, fatigue striations can be observed on the fracture surface (Fig.4).

2) In a large $\Delta\varepsilon_p$ regime, however, a fatigue crack initiated from the inside of specimen leads to the final fracture (Internal Fracture Mode). In this mode, three types of internal crack initiation sites can be observed as follows:

  (a) Internal crack initiation at the intersection of deformation twin and grain boundary: Fig. 5 shows the typical feature of this type of fracture origin. Dimple patterns were observed at the crack initiation site (arrow mark). In this case, it is deduced that the intersection of deformation twin and grain boundary leads to the formation of microvoids inside the material due to the stress concentration. These microvoids result in the final cleavage fracture of the specimen.

  (b) Internal crack initiation at the intersection of deformation twins: Fig. 6 shows the typical feature of this type of fracture origin. Three deformation twins (A, B and C) which intersect each other can be observed at the crack initiation site (arrow mark). In this case, it is considered that stress concentration at the intersection site leads to the final cleavage fracture of the specimen.

  (c) Internal crack initiation at the intersection of deformation twin and inclusion: Fig. 7 shows the typical feature of this type of fracture origin. It should be noted that the deformation twin is stopped at the inclusion (arrow mark), and the final cleavage fracture starts at the tip of the deformation twin.

From the results of observation, it is clear that in the internal fracture mode, deformation twins strongly affect the initiation of internal cracks at -140°C.

3.2 Fracture Mode Transition and Manson-Coffin Relation

It has been reported by the authors (Shimada *et al.*, 1987; Komotori *et al.*, 1991, 1993a, 1993b) that, in an extremely low cycle fatigue regime where specimen fails in internal fracture mode, the final fracture occurs at a strain cycle count less than that expected from Manson-Coffin law for ordinary low cycle fatigue regime. In such a situation, the fatigue life is determined by the competition between two failure limit lines corresponding to the surface and internal fracture modes respectively. Fig. 8 shows the schematic diagram that explains this concept.

Fig. 9 shows the results of low cycle fatigue tests at -140°C and R.T.. In this figure, the solid and hollow marks show the results for the specimens fractured in internal and surface fracture modes, respectively. It should be noted that at room temperature, where fracture mode transition does not occur, the fatigue lives can be expressed by a single Manson-Coffin line over the wide range of $\Delta\varepsilon p$. At -140°C, however, there exist two failure limit lines corresponding to the surface and internal fracture modes respectively. These behaviors can be clearly explained by introducing the concept of "The Competition of Two Failure Limit Lines" as shown in Fig. 8.

4. CONCLUSIONS

A study was made to clarify the fracture mechanism of commercial grade pure iron in low cycle fatigue at low temperature, with special focus on the role of deformation twins in the initiation of microcrack which leads to the final fracture of the specimen. The applicability of Manson-Coffin relation to low-temperature low-cycle fatigue is also discussed in relation to the fracture mode transition behavior. The results are summarized as follows:

(1) A transition of fracture mode occurs from surface to internal fracture mode with increase in plastic strain range ($\Delta\varepsilon p$) at -140°C. In a small $\Delta\varepsilon p$ regime, the initiation and propagation of a surface crack leads to the final fracture of the specimen. In a large $\Delta\varepsilon p$ regime, however, a crack initiated inside the material leads to the final fracture.

(2) Deformation twins strongly affect the initiation of internal crack at -140°C. Three types of internal crack initiation sites have been observed: internal crack originates at (i) the intersection of deformation twin and grain boundary, (ii) the intersection of deformation twins and (iii) the intersection of deformation twin and inclusion.

(3) At room temperature, where fracture mode transition does not occur, fatigue lives can be expressed by a single Manson-Coffin line. At -140°C, however, there exist two failure limit lines corresponding to the surface and internal fracture modes respectively.

REFERENCES

Honda, R. (1961). Cleavage fracture in single crystals of silicon iron. *J. Phys. Soc. Japan*, **16**, 1309-

1321.

Hull, D. (1960). Twinning and fracture of single crystals of 3% silicon iron. *Acta Met.* **8**, 11-18.

Komotori, J., M. Yokoyama and M. Shimizu (1991). Microstructual effect on extremely low cycle fatigue of dual phase steel. *Int. Conf. on Mechanical Behavior of Materials (ICM6)*, **2**, 517-523.

Komotori, J., T. Adachi and M. Shimizu (1993a). Effect of spheroidal graphites on characteristic of low cycle fatigue life of ductile cast iron. *Int. Conf. on Mechanical Behavior of Ductile Cast Iron and other Cast Metals*, 79-84.

Komotori, J., T. Adachi and M. Shimizu (1993b). Evaluation of internal fatigue damage in extremely low cycle fatigue through the measurement of local material density. *Asian Pacific Conference on Fracture and Strength '93*, 169-173.

Shimada, K., J. Komotori and M. Shimizu (1987). Fracture mode transition and damage in extremely low cycle fatigue. *Int. Conf. on Low Cycle Fatigue and Elasto-Plastic Behavior of Materials*, 680-686.

Table 1 Chemical composition

C	Si	Mn	P	S	Cu	Ni	Cr
0.03	0.01	0.06	0.006	0.004	0.01	0.01	0.01

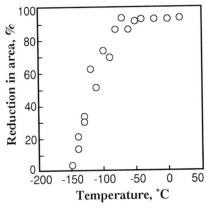

Fig. 1 Relationship between temperature and ductility

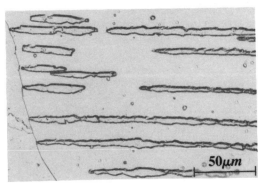

Fig.2 Deformation twins observed at longitudinal section (tension at -140°C)

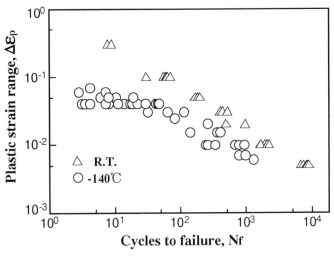

Fig.3 Results of low cycle fatigue test

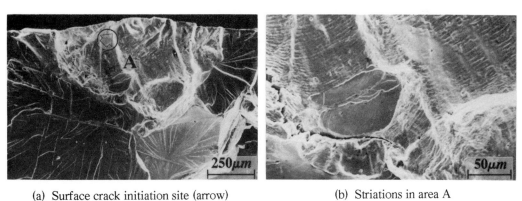

(a) Surface crack initiation site (arrow)   (b) Striations in area A

Fig. 4 Typical feature of surface fracture mode(-140°C, $\Delta \varepsilon_p$=0.02, Nf=118)

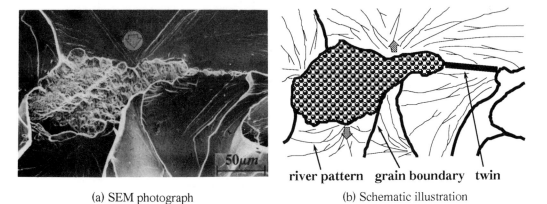

river pattern   grain boundary   twin

(a) SEM photograph   (b) Schematic illustration

Fig. 5 Internal crack initiation site with intersection of deformation twin and grain boundary
(-140°C, $\Delta \varepsilon_p$=0.05, Nf=19)

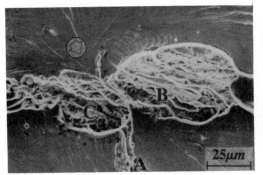
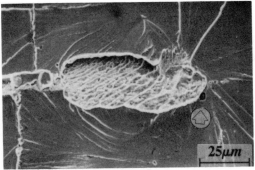

Fig. 6 Internal crack initiation site with
intersection of three deformation twins
(-140˚C, Δεp=0.06, Nf=7)

Fig. 7 Internal crack initiation site with
intersection of deformation twin and
inclusions (-140˚C, Δεp=0.05, Nf=6)

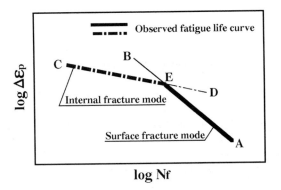

Fig. 8 Schematic diagram illustrating the concept of
"the competition of two failure limit lines"

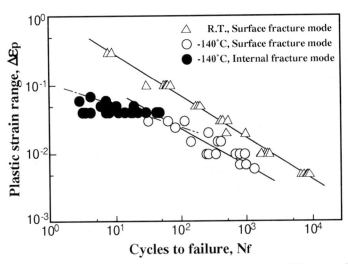

Fig. 9 Effect of fracture mode transition on low cycle fatigue life properties

Effect of Manganese Content on Existence of Non-propagating Micro-cracks of Extra-low-carbon Steels

Shin-ichi NISHIDA*, Nobusuke HATTORI* and Shin-ichi MATTORI**

* Faculty of Science and Engineering, Saga University,
   Honjo 1, Saga, 840, Japan
** Saga University (Graduate school)

ABSTRACT

The fatigue properties of the extra-low-carbon steels have been evaluated in this test. In addition, the effect of manganese content on existence of non-propagating micro-cracks of the materials have been also investigated using three kinds of extra-low carbon steels ($0.0022 \sim 0.0025\%$C), whose manganese content only is changed by 0.28%(material A), 1.58%(material B) and 2.35%(material C), respectively. The fatigue test had been performed by the Ono-type rotating bending fatigue testing machine. In this study, it has been confirmed that the fatigue limits of material A, B and C were 137MPa, 170MPa and 200MPa, respectively. That is, the fatigue limit of the materials increases with increasing the manganese content. This result is considered due to the effect of solid solution strengthening based on the difference of manganese content. In addition, the fatigue micro-cracks of all specimens of materials A and B initiate at the ferritic grain boundaries or in its neighborhood within 10% of each total fatigue life ratio. On the other hand, those of material C initiate within 5 %. Furthermore, the non-propagating micro-cracks have been observed in the specimen of material A and B subjected to the stress amplitude of fatigue limit by 1×10^7 cycles, but not in the specimen of material C.

KEYWORDS

Extra-low-carbon steels; Manganese content; Non-propagating micro-cracks;Fatigue properties

INTRODUCTION

In order to decrease manufacturing cost, it is necessary to improve deep drawing ability of the materials used. It has been tried to minimize the carbon content or nitrogen one in the materials, because the manufacturing cost closely depends on these kinds of content with keeping the strength of materials at higher level. The new materials have been recently developed by making the interstitial atoms, such as carbon or nitrogen, free from the material and adding Mn, P and Si for compensating the decrease of the strength of material. Though the materials have been widely utilized more and more for various kinds of machine components and structures, there are few reports about fatigue properties about these kinds of materials.

The authors mainly try to evaluate the fatigue properties of the extra-low-carbon steels.   In addition, the effect of manganese content on existence of non-propagating micro-cracks of the materials have been also investigated using three kinds of extra-low carbon steels.

EXPERIMENTAL PROCEDURE

The materials used in this test are three kinds of extra-low carbon steels(0.0022~0.0025%C), whose manganese content only is changed by 0.28% (material A), 1.58% (material B) and 2.35% (material C), respectively.

Table 1 lists their chemical composition.

Fig. 1 shows the shape and dimensions of fatigue specimen, which was cut out by coinciding the rolling direction with the specimen axis and by making a partial shallow notch at its rolling surface side.   This shallow notch exists for limiting fatigue damaged part and does not affect for its fatigue strength at all (Nisitani and Hasuo, 1978).   Fatigue crack initiation behavior of each specimen was observed by the successive taken replica method in the circumferential direction at the specimen's surface.

All of specimens are machined and then after polished with emery paper.   After electro-polished, they are annealed in vacuum at the temperature of 600℃ for 1 hour.   The fatigue test had been performed by the Ono-type rotating bending fatigue testing machine of 15 N·m capacity.

RESULTS AND DISCUSSION

Fig. 2 shows the S-N curves on the three kinds of extra-low carbon steels.   The fatigue limit by 1

Table 1   Chemical composition.   mass%

Materials	C	Si	Mn	P	S	Al	Ti	N
A	0.0022	0.012	0.82	0.05	0.006	0.021	0.040	0.0024
B	0.0025	0.012	1.58	0.05	0.006	0.021	0.040	0.0022
C	0.0022	0.011	2.35	0.05	0.006	0.022	0.041	0.0031

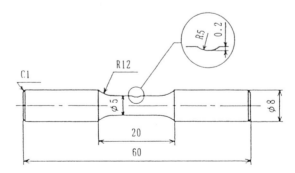

Fig.1.   Shape and dimensions of specimens.

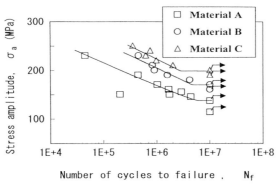

Fig.2.   S-N curves .

$\times 10^7$ cycles of each material is (σ_{wo})$_A$=137MPa, (σ_{wo})$_B$=170MPa , and (σ_{wo})$_C$=200MPa, respectively.   The fatigue limit of these materials, shown in this figure, increases with an increase in the manganese content, and the fatigue limit of material C is higher than that of material A by 63MPa.   In addition, the number of cycles to failure of material C becomes considerably longer than that of material A or B under the same stress amplitude over the fatigue limit, that is, the S-N curve of material C shifts above as compared with those of material A and B.   It should be noted that the manganese is effective to improve fatigue strength of extra-low carbon steels.   The high strength of material C is considered to be due to the solid solution strengthening by manganese in matrix.

Fig. 3 shows the representative successive observation results obtained by replica method for fatigue crack initiation behavior of material A tested under a stress amplitude of 170 MPa, which is higher than fatigue limit of this material by 33MPa.   The fatigue cracks initiate in grain boundary or in its neighborhood within about 10 % of each total fatigue life, then propagate with increasing the cyclic number.

Fig. 4 shows the representative successive observation results for material B under the stress amplitude of 210 MPa.   As shown in this figure, the fatigue crack initiation behavior of material B is essentially the same as that of material A, and the micro-cracks also initiate within about 10 % of the total fatigue life.

Fig. 5 shows the representative successive observation results for material C under the stress amplitude of 230 MPa.   The fatigue crack initiation behavior of this material is the same as that of the other two materials.   However, the cyclic ratio of fatigue crack initiation of this material is about 5 % of the total fatigue life, and that is slightly faster than those of the other two materials.

That is, it is recognized that there is scarcely effective of manganese on fatigue strength for three materials.   In contrast, there is no effect of manganese on the fatigue crack initiation behavior for these materials.

Fig. 6 shows the representative successive observation results in material A subjected to the stress amplitude of fatigue limit by 1×10^7 cycles.   As shown in this figure, it seems that the tip of fatigue crack is growing slightly at this stage, yet.   As shown in Fig.7, however, the value of fatigue crack propagation from 0.5×10^7 cycles to 1×10^7 cycles of material A is only 5 μ m. Also, it has been reported by Nisitani et al. (1994) that the growth of fatigue crack under the

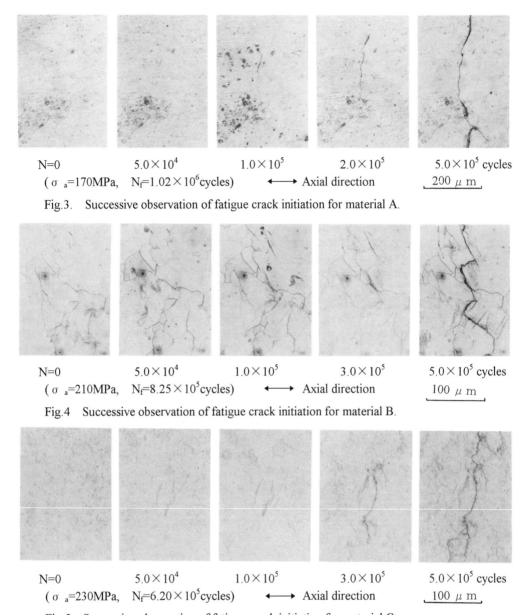

N=0 5.0×10^4 1.0×10^5 2.0×10^5 5.0×10^5 cycles
(σ_a=170MPa, N_f=1.02$\times 10^6$cycles) $\longleftrightarrow$ Axial direction |___200 μ m___|

Fig.3.   Successive observation of fatigue crack initiation for material A.

N=0 5.0×10^4 1.0×10^5 3.0×10^5 5.0×10^5 cycles
(σ_a=210MPa, N_f=8.25$\times 10^5$cycles) $\longleftrightarrow$ Axial direction |___100 μ m___|

Fig.4   Successive observation of fatigue crack initiation for material B.

N=0 5.0×10^4 1.0×10^5 3.0×10^5 5.0×10^5 cycles
(σ_a=230MPa, N_f=6.20$\times 10^5$cycles) $\longleftrightarrow$ Axial direction |___100 μ m___|

Fig.5   Successive observation of fatigue crack initiation for material C.

stress amplitude of fatigue limit on the same extra-low carbon IF (Interstitial-Free) steels had been not observed until 8×10^7 cycles after 1×10^7 cycles. Therefore, it is concluded that the non-propagating micro-cracks exist in material A.

Fig. 8 shows the results of material B subjected to the stress amplitude of fatigue limit by 1×10^7 cycles. As shown in this figure, this result is essentially the same as that of material A.

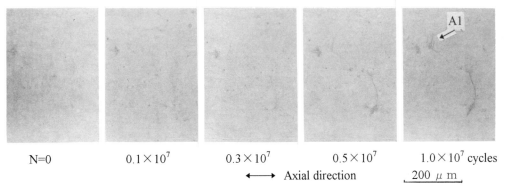

N=0 0.1×10^7 0.3×10^7 0.5×10^7 1.0×10^7 cycles

⟷ Axial direction   200 μ m

Fig.6   Specimen's surface state of material A subjected to the stress amplitude of fatigue limit
(σ_{w0}=137MPa) by 1×10^7 cycles.

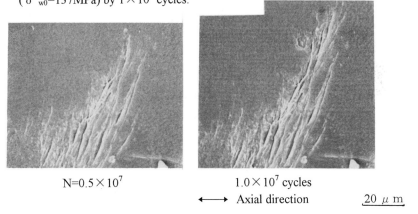

N=0.5×10^7 1.0×10^7 cycles

⟷ Axial direction   20 μ m

Fig.7   Detail of A1 in Fig.6.

Fig. 9 shows the results of material C subjected to the stress amplitude of fatigue limit by 1×10^7
cycles.   Though there appeared only some damaged parts such as slip bands, the non-propagating
micro-cracks have not been recognized.   Namely, the non-propagating micro-cracks are observed
in the specimen's surfaces of material A and B, while the non-propagating micro-cracks are not
observed in the case of material C.   These results are different from that obtained for material A
and B.   Manganese is easy to segregated, and carbon is attracted to the segregated manganese.
Therefore, the carbon become poor at the rest of regions in matrix.   As a result, the action of
carbon in material decreases with an increse in manganese content.   Thus, the non-propagating
micro-cracks dose not exist in the material C which contains much manganese.   From the above
results, the manganese which is contained in this kind of material, i.e., extra-low carbon steels
effects on not only fatigue crack propagation but also fatigue crack initiation.

CONCLUSIONS

The fatigue tests have been performed to investigate the effect of manganese on fatigue strength
and fatigue crack initiation characteristics of extra-low carbon steels.   The main results obtained
in this study are as follows;

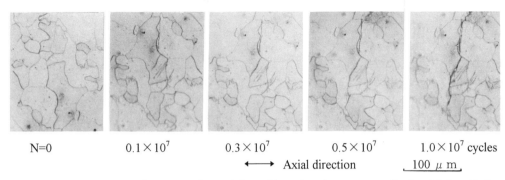

N=0 0.1×10^7 0.3×10^7 0.5×10^7 1.0×10^7 cycles

◄─────► Axial direction   100 μ m

Fig.8   Specimen's surface state of material B subjected to the stress amplitude of fatigue limit
         (σ_{w0}=170MPa) by 1×10^7 cycles.

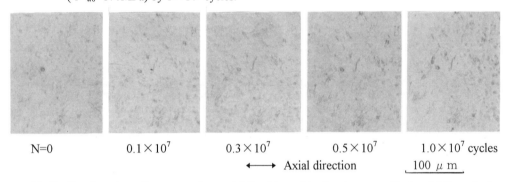

N=0 0.1×10^7 0.3×10^7 0.5×10^7 1.0×10^7 cycles

◄─────► Axial direction   100 μ m

Fig.9   Specimen's surface state of material C subjected to the stress amplitude of fatigue limit
         (σ_{w0}=200MPa) by 1×10^7 cycles.

(1) The fatigue limit of this kind of material increases with an increase in manganese content in its matrix.   The fatigue limit of material C (Mn;2.35%) becomes larger than that of material A (Mn;0.28%) by 1.5 times.   This result is considered due to the effect of solid solution strengthening based on the difference of manganese content.

(2) The fatigue micro-cracks of all specimens of materials A and B initiate at the ferritic grain boundaries or in its neighborhood within 10% of each total fatigue life ratio.   On the other hand, those of material C initiate within 5 %.   That is, it is considered that the manganese in this kind of material also effects the behavior of fatigue crack initiation.

(3) The non-propagating micro-cracks have been observed in the specimen of material A and B subjected to the stress amplitude of fatigue limit by 1×10^7 cycles, but not in the specimen of material C.   It is considered that the action of carbon in material is strongly restricted with an increase in manganese content.

REFERENCES

H. Nisitani, Y. Hasuo (1978).   Effect of mean stress on axial fatigue process in quenched and
     tempered carbon. *JSME (in Japanese)*, **44**, No.377, 1-7.
H. Nisitani, T. Yakushiji, M. Kage, K. Ogata, s. Ohta (1994).   Effect of strain ageing on the
     non-propagation of fatigue crack of low-carbon steel. *JSME (in Japanese)*, No.948-2, 50-53.

MICROSTRUCTURAL INVESTIGATIONS OF FATIGUED TEMPERED 12% CR-STEEL X 18 CrMoVNb 12 1

R. GERSINSKA

Federal Office for Radiation Protection, Postfach 10 01 49, 38201 Salzgitter

ABSTRACT

The low cycle fatigue properties of a tempered 12% Cr-steel X 18 CrMoVNb 12 1 have been investigated at different temperatures, strain rates and strain amplitudes and correlated with the microstructure. The influence of the temper treatment has been examined. Plastic deformation results in an extrusion formation on the surface and in a cellular dislocation structure in the interior of the specimens. The new microstructure increases the mean-free-path length of dislocations and explains softening. Different strengthening mechanisms are discussed. The main contributions to saturation stresses come from the interaction of moving dislocations with dislocations and subcell boundaries. The contribution of precipitates is small but they influence the subcell formation.

KEYWORDS

Microstructure, tempered steel, fatigue properties, strengthening mechanisms, dislocations

INTRODUCTION

The tempered steel X 18 CrMoVNb 12 1 combines a high resistance against pore swelling and a low tendency to radiation-induced He-embrittlement with high thermic conductivity and good restistance against water and liquid metal corrosion (Böhm and Hauck, 1967; Wassilew et al., 1983; Walters et al., 1987). Therefore, the steel was selected as possible structural material for the first wall and as building material for future nuclear fusion reactors (Harries, 1986). In this reactors, the material is apart from other stresses periodically exposed to repeated thermal stresses leading to complex time-dependencies of temperature, stress and strain. Such loadings can be simulated by isothermic, strain-controlled fatigue experiments. Investigations of the fatigue behaviour of the X 18 CrMoVNb 12 1 material are therefore of particular importance. In this report, microstructural changes during isothermic fatigue are investigated to gain information on the specific behaviour of this material. A detailed description of the work is presented by Gersinska (1992).

EXPERIMENTAL

The investigated material is a X 18 CrMoVNb 12 1 steel. Its chemical composition is listed in table 1.

Table 1. Chemical composition of X 18 CrMoVNb 12 1, charge 53645.

Element	C	Cr	Ni	Mo	V	Nb	Si	Mn	S	P	B	N	Zr	Al	Co	Cu
Weight - %	0.13	10.6	0.87	0.77	0.22	0.16	0.37	0.82	0.004	0.005	0.0085	0.02	0.053	0.054	0.01	0.015

The heat treatment of the as-received material consists of hardening (1075°C/30'), temper treatment (750°C/2h) and air cooling. Low cycle fatigue tests were perfored between room temperature and 650°C by Scheibe and Schmitt (1991). The total strain amplitudes varied between 0.5% and 1.5%. The fatigue tests were conducted with a triangular waveform at a strain rate of $3 \cdot 10^{-3}$ s^{-1}. Different experiments were stopped at definded number of cycles (1%, 5%, 25%, 50%, 75% N_f = number of cycles to fracture) to investigate the microstructural development at the surface and in the interior of specimens. The influence of the temper temperature on the cyclic strain behaviour was examined in various specimens cycled at 250°C, 350°C and 450°C with a total strain amplitude of 0.75%. Scanning electron microscope (SEM) analysis of the fracture surfaces and of the external surfaces of the specimens were perfored in a Philips SEM 505. A Jeol JEM-2000 FX transmission electron microscope (TEM), operating at 200 kV and equipped with a tilting specimen stage and an energy-dispersive X-Ray analyser (EDX) were used for the examination of metallic foils.

RESULTS and DISCUSSION

The fatigue behaviour of the X 18 CrMoVNb 12 1 steel is generally marked by cyclic softening (Fig. 1) expressed by a decreasing stress amplitude with an increasing number of cycles. Higher strain amplitudes are causing higher stresses and higher temperatures in turn lower stresses.

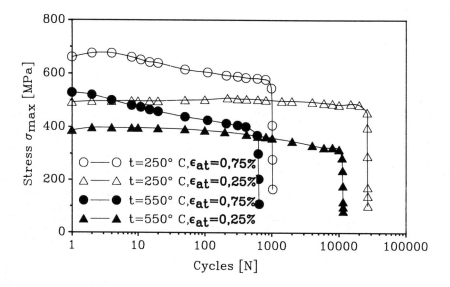

Fig. 1. Stress as a function of number of cycles

SEM investigations have been performed to investigate the development of cracks (fig. 2). After 1% N_f the first slip lines are visible. There angle to the load is about 45°. After 5% N_f slip bands and after 25 % N_f the first microcracks have been observed on the extrusions and intrusions. The cracks start always at the surface of the specimens and are transcrystalline in character.

TEM investigations of the microstructure were performed on as-received material, fatigued specimens and specimens of interrupted experiments. The as-received material is in a fully tempered state. Its microstructure consists of a typical lath structure with chromium rich $M_{23}C_6$ carbides on prior austenite grain boundaries and lath boundaries (fig. 3). In all fatigued specimens a cellular dislocation structure has been observed (fig. 4). The dislocation densities of tempered material and fatigued specimens have been measured. The dislocation density within the cells is smaller than between the cells and smaller than the density of the tempered material ($1,8 \cdot 10^{14}$ m^{-2}). The dislocation densities of cells vary between $2 \cdot 10^{13}$ m^{-2} (650°C) and $1 \cdot 10^{14}$ m^{-2} (20°C), dependent on temperature. The cell size has been measured in the as-received material and in all fatigued specimens.

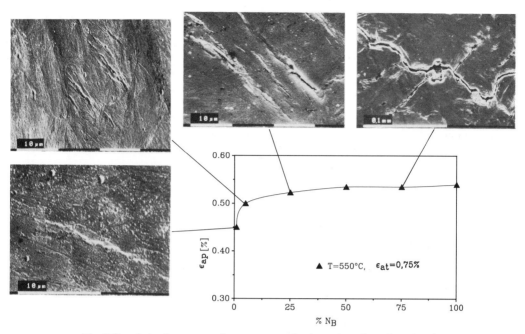

Fig. 2 Correlation between surface structure, plastic strain and number of cycles

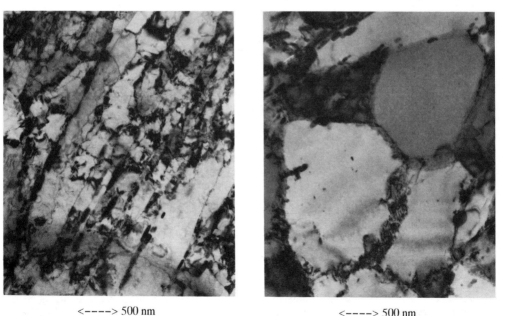

<----> 500 nm
Fig. 3   Microstructure of tempered material

<----> 500 nm
Fig. 4   Microstructure of fatigued specimen
(T=550° C, ε_{at}=0.75%)

Fig. 5 is showing a summary of all determined cell sizes of specimens cycled at $20°C \leq T \leq 650°$ with $0.25\% \leq \varepsilon_{at} \leq 0.75\%$ until fracture. The demonstrated relationship between the measured cell sizes and the saturation stress amplitudes includes a dependence on temperature as well as on strain. All data may, however, be described by a single curve.

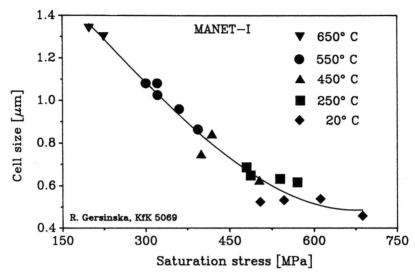

Fig. 5   Relationship between the measured cell size and the saturation stress
         amplitude and various individual contributions

With cyclic strain, the examined material shows a transition from initially a lath-type structure to a type of cellular dislocation structure, i.e. wavy deformation takes place and is guided by thermally activated slip processes. With rising temperature, the mean cell diameter is increasing. The influence of carbides inhibiting the dislocation movement is decreasing with increasing temperature. At higher temperatures, carbides are frequently visible in the cell interior. This may be explained by thermally activated slip processes enabling a by-passing of carbides. The cell walls are becoming thinner and are more and more resembling sub-grain boundaries. With rising strain amplitude, the cell diameter is decreasing. Stronger deformations require a higher dislocation density. This will lead to a decrease in cell size since more cell walls are formed than at lower strain amplitudes. The formation of cells has been investigated in interrupted specimens (fig. 6). After 1% N_f the first cells are visible. After 25% N_f the number of cells are nearly constant and only changes in cell size have been measured. The increase in cell size in the interior of specimen correlates with the reduction of measured stress at constant total strain amplitude.

On several specimens cycled at different temperatures (250°C and 550°C) and strain amplitudes, EDX-analyses of the $M_{23}C_6$ carbides were performed. A change in their composition (table 2) and configuration due to deformation is not observed. The mean distance and mean size of precipitates was determined by imaging system analysis. The mean precipitate size is D=0.1μm and the mean distance is l=0.38μm.

Table 2. Result from the EDX-analysis of extraction replicas.

Element	Fe	Cr	Mo
Weight - %	26.2±2.8	60.0±3.8	13.8±4.9

Cyclic experiments were run at various strain rates ($\varepsilon = 3 \cdot 10^{-3} s^{-1}$, $\varepsilon = 3 \cdot 10^{-4} s^{-1}$ and $\varepsilon = 3 \cdot 10^{-5} s^{-1}$) at different temperatures (20°C≤T≤650°C) with a total strain of ε_{at}=0.75%. A variation in strain rate results in only minor changes of the cyclic deformation behaviour in the investigated material. Fatigue damage is predominant also at reduced strain rates. Lower stress amplitudes are measured at higher temperatures (T>450°C) only. This is a sign of creep fatigue which is not of essential significance for the cyclic deformation behaviour.

Cyclic deformation curves of specimens tempered with T_A=600°C are showing the typical softening characteristics. A lower temper temperature of 600°C effects a greater strength in the material in comparison to 750°C. This is due to smaller martensite lath distances (0.28 μm) and higher dislocation density ($2.7\pm0.5\cdot10^{14} m^{-2}$). Microstructural investigations reveal the typical development of cells.

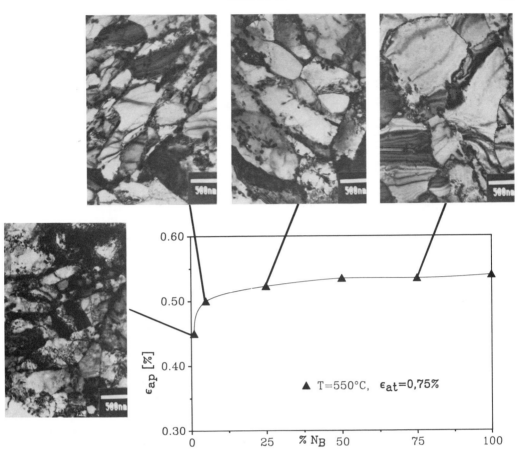

Fig. 6   Correlation between microstructure, plastic strain amplitude and number of cycles

The elastic interaction of dislocations with slip hindrances results in different strengthening mechanisms which contribute to the flow stress of the tempered steel. Since the interaction between slip dislocations and precipitates are reversely proportional to the particle distance, no contributions to strengthening are expected from this interaction, due to the large average distances between particles (0.38 μm). The interaction between slip dislocations and foreign atoms contributes to the so-called mixed crystal strengthening. It is proportional to the concentration c of foreign atoms (at%) and a material-dependent exponent n: $\sigma_{MK} \sim G(T) \cdot c^n$, $G(T)$ is the temperature-dependent shear modulus. This stress contributions are small (about 50 to 100 MPa (Leslie, 1972)) compared with the measured saturation stresses. The contibution of interactions between dislocations depends on the dislocation density (Holt, 1970): $\sigma_V = M \cdot \alpha_1 \cdot G(T) \cdot b \cdot \sqrt{\rho}$, M=2.75 is the Taylor factor, b the Burgers vector ($2.48 \cdot 10^{-10}$ m) and ρ the dislocation density. The constant $\alpha_1 = 0.3$ is a function in the arrangement of dislocation. Moving dislocations must overcome the stress fields of other dislocations. This contributions to saturation stresses have been estimated with the help of the measured dislocation densities. The contribution of the interaction of dislocations with subcell boundaries can be estimated with the help of the empirical Hall-Petch-relationship (Bradley and Polonis, 1988): $\sigma_{GF} = M \cdot \alpha_2 \cdot G(T) \cdot b \cdot d^m$. Hereby is m a material-dependent exponent and α_2 a factor of proportionality that may result from the measured flowstress dependence on sub-grain size d. After subtracting the realistically estimated proportions of mixed crystal strengthening and dislocation strengthening from saturation stress amplitudes and considering the temperatur-dependent shear modulus $G(T)$, the relation between the remaining contribution from the dislocation cell boundaries σ_{GF} and the measured cell size d is described by $\sigma_{GF} = 1.65 \cdot 10^{-7} \cdot G(T) \cdot d^{-0.7}$. The tensile yield strength and the expected saturation stresses of fatigued specimens have been calculated as sum of the different contributing parts and compared with the experimental data. In Fig. 7, the experimental data (triangles) and calculated saturation stress amplitudes (continuous line) are shown at $\varepsilon_{at} = 0.75\%$ and different temperatures.

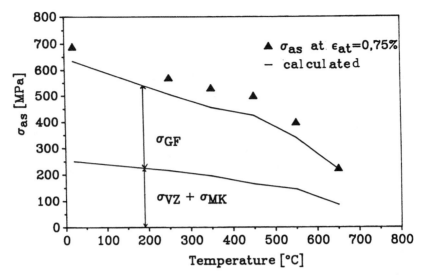

Fig. 7   Temperature dependence of saturation stress amplitude and various individual contributions

The dislocation contribution σ_{VZ} depends on the dislocation density in the cell interior which is not influenced by the total strain amplitude ε_{at}. The fractions of alloy atoms σ_{MK} remain unchanged during deformation. The sum of these fractions is also shown in Fig. 7. The fraction σ_{GF} in cyclic deformed material is based on the interaction between dislocations and cell boundaries. Microstructural investigations at higher temperatures or smaller strain amplitudes show a considerable increase in cell size which explains the lesser saturation stress amplitudes. Creep influenced fatiguing, expected to occur at T>550°C, is also reducing stress amplitudes.

In conclusion it may be said that the strengthening contribution of precipitates and of mixed crystal strengthening to the measured saturation stress is small. However, the influence of precipitates on cell formation should not be neglected since precipitates may inhibit slip dislocations, facilitate cell formation and possibly influence the cell size. Dislocations and cell boundaries are greatly contributing to strengthening. Softening measured during cyclic deformation is due to the reduced dislocation density in the cell interior and the growing cell diameters. The newly formed microstructure is providing slip dislocations a larger mean-free-path length. On the basis of the discussed considerations, a correlation is possible between quantitatively processed microstructural data and the mechanical properties as yield strenth and saturation stress at cyclic deformation.

This study was financed by a doctorate stipendium from the Karlsruhe Research Center. Thanks are expressed for the extended confidence.

REFERENCES

Böhm, H. and Hauck, H. (1967),(1969). Journal of Nucl. Materials 21, 112-113, 29, 184-190.
Bradley, E. R. and Polonis, D. H. (1988). Acta Metall. 36, 393-402
Gersinska, R. (1992). Research Center Karlsruhe, KfK 5069, ISSN 0303-4003.
Harries, D. R. (1986). Radiat. Eff. 101, 3-19.
Holt, D., (1970). J. of Appl Phys. V41, 8, 3197-3201
Leslie, W. C. (1972). Met. Trans., 3, 5-26.
Scheibe, W. and Schmitt, R. (1991). Proceedings of the Workshop on Fatigue of Fusion Reactor Candidate Materials, Vevey, Switzerland.
Walters, G. P., Mazey, D. J., Murphy, S. M., Hanks, W., Bolster, D. E., Kurcook, D., Sowden, B. C. and Kimber, R. I. (1987). Harwell Report AERE R 12620.
Wassilew, C., Ehrlich, K. and Anderko, K. (1983). Brigthon, G. B., BNES, Vol. 1, 161-164.

THE INFLUENCE OF NITROGEN AND GRAIN SIZE ON THE CYCLIC DEFORMATION BEHAVIOUR OF AUSTENITIC STAINLESS STEEL

M. Nyström, U. Lindstedt*, B. Karlsson* and J-O. Nilsson

AB Sandvik Steel, R&D Centre, S-811 81 Sandviken, Sweden
*Department of Engineering Metals, Chalmers University of Technology, S-412 96 Göteborg, Sweden

ABSTRACT

The present study concerns the fatigue behaviour of two austenitic stainless steels (base composition corresponding to 316L) with two different nitrogen levels, 0.14 and 0.29% respectively. The influence of grain size on the mechanical properties has been investigated. The grain size dependence of the monotonic as well as the cyclic stress strain curves was shown to follow a Hall-Petch type relationship. The influence of the grain size was found to be lower for the cyclic than for the monotonic yield stresses. The lower grain size sensitivity of the cyclic yield stresses can be explained by the successive breakdown of the planar slip mode during cyclic straining. For both static and cyclic deformation modes the dependence on the grain size increases with higher nitrogen content. Dislocation structures of the various conditions are studied with transmission electron microscopy and related to the mechanical behaviour of the two materials.

KEYWORDS

Austenitic stainless steel; slip bands; grain size effects; Hall-Petch relation; monotonic and cyclic straining; nitrogen.

INTRODUCTION

Austenitic stainless steels are used in a wide variety of applications because of their good formability and mechanical properties combined with excellent corrosion resistance. The austenitic stainless steels constitute the most widely used type of stainless steel. The influence of the microstructure on the mechanical behaviour of stainless steels is therefore of great interest. It has been shown that the strength of austenitic stainless steel can be raised by solid solution hardening, *e.g.* by adding nitrogen. The addition of nitrogen leads to increased yield stress and better fatigue properties (Mullner *et al.,* 1993, Degallaix *et al.*, 1986, Vogt *et al.*, 1984). Furthermore, a grain size dependence of the monotonic yield stress, following the Hall-Petch model, has been observed (Norström, 1977, Pickering, 1976). However, available information about the cyclic deformation properties at ambient temperature is scarce. The aim of this work, therefore, is to evaluate the grain-size dependence of the monotonic and cyclic stress-strain behaviour of austenitic stainless steel. Two different nitrogen levels were selected in order to promote different slip character of the dislocations and hence also differences in dislocation activity at the grain boundaries. In a companion paper the nucleation and growth behaviour of short surface cracks has been assessed for the steels used in this study (Lindstedt *et al.*, 1995).

MATERIALS

The two steels used in this study are fully austenitic stainless steels, where one is the commercial grade 316LN with 0.14%N (denoted LN) and the other an experimental alloy with 0.29%N, denoted HN. The chemical compositions (Table 1) are similar except for higher contents of nitrogen and chromium in the HN. Each steel was studied at three different grain sizes (Table 2). The present variation in chromium content has a minor influence on the mechanical properties (Pickering, 1976).

Table 1.   Chemical composition (weight-%).

Material	C	Si	Mn	P	S	Cr	Ni	Mo	Cu	N
LN	0.024	0.50	1.16	0.025	0.001	17.4	10.01	2.76	0.42	0.140
HN	0.027	0.47	0.99	0.017	0.012	21.6	9.83	2.64	0.04	0.294

EXPERIMENTAL

Short cylindrical test bars with a gauge diameter of 5 mm and gauge length of 15 mm were machined from 12 mm thick plates. The surfaces of the reduced section of the specimens were ground and polished, the last polishing step with 1 μm diamond paste. The low cycle fatigue (lcf) tests were carried out in push-pull mode and the tests were controlled by the total strain. The strain wave-form was sinusoidal with a mean strain rate of $5 \cdot 10^{-3}$ s^{-1}. The tensile specimens were cylindrical test bars with gauge lengths of 30 mm. The strain rate employed was $3 \cdot 10^{-4}$ s^{-1}. Thin foils were produced by electropolishing in 15% perchloric acid in ethanol and examined in a JEOL 2000-FX transmission electron microscope (TEM).

RESULTS AND DISCUSSION

<u>Monotonic yield stress</u>

The results from the tensile tests are shown in Table 2. It is clear that the increased nitrogen content in the HN steel raises the yield stress by approximately 30% irrespective of grain size. Also the ultimate tensile strength increases with increasing nitrogen content while the uniform elongation decreases. The yield stress is approximately linearly dependent on $\lambda^{-1/2}$ for both materials, where λ is the mean intercept grain size, Fig. 1. This means that the materials follow the classic Hall-Petch type behaviour (Petch, 1953), that can be written as

$$R_{p0.2}^m = \sigma_o^m + k_y^m \lambda^{-1/2} \tag{1}$$

where the symbols σ_o^m and k_y^m have been chosen to identify the case of monotonic loading. The parameter σ_0^m can be interpreted as a friction stress in the lattice influenced by interstitially dissolved nitrogen atoms. The difference in solid solution effects of nitrogen in the two materials thus reflects

Table 2. Monotonic and cyclic mechanical data.

Material	λ [μm]	$R^m_{p0.2}$ [MPa]	R_m [MPa]	A_g [%]	$R^c_{p0.2}$ [MPa]	σ^m_0 [MPa]	k^m_y *	σ^c_0 [MPa]	k^c_y *
	43	311	658	62	289				
LN	78	292	637	65	270	240	14.9	236	10.7
	130	283	630	68.5	268				
	47	404	768	52.5	338				
HN	80	371	741	55	313	286	24.8	263	15.6
	185	346	722	60	302				

- $R^m_{p0.2}$ and $R^c_{p0.2}$ evaluated using 0.2% off-set strain.
- σ^m_0, k^m_y from Eq. 1; σ^c_0, k^c_y from Eq. 2. *(k in $MPa\sqrt{mm}$)

differences in σ^m_o. The mechanism behind the effect of nitrogen on lattice friction has earlier been discussed (Grujicic *et al.*, 1988). Lattice straining and short range order have been proposed but neither of these mechanisms is free from objections.

The physical reason for the stronger grain size dependence of the yield stress in high-nitrogen steels is likely to be due to a more pronounced planar slip mode of the dislocations. Such planar glide tends to extend the glide length of individual dislocations; the probability thus increases for single dislocations to interact with grain boundaries. On the contrary, in materials with wavy slip character of the dislocations there is a reduction of the efficiency of grain boundaries. Thus a shift from wavy slip to planar glide would make grain boundary hardening more effective, which results in higher k^m_y-values. Although both materials in the present study are likely to deform essentially by planar glide, the increase in nitrogen content from 0.14 to 0.29% evidently enhances this deformation mode.

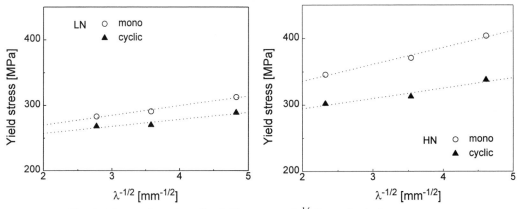

Fig. 1. Monotonic and cyclic yield stress vs. $\lambda^{-\frac{1}{2}}$ (λ=grain size).

Virgin material was found to be virtually dislocation-free as shown by TEM of the grip sections of tensile specimens. Supporting evidence of planar slip in tensile specimens deformed to a total strain of 0.8% was found in both the LN and HN steels, examples of which are given in Figs. 2 and 3. Although planar slip may seem to be more pronounced in the HN steel when comparing Figs. 2 and 3 no significant difference in the degree of planarity could be assessed when a large number of grains were qualitatively examined in the TEM. The most likely reason for this is that planar slip is so pronounced already in material with 0.14% nitrogen that the moderate tensile deformation (0.8%) is insufficient to produce differences in the dislocation configurations between LN and HN steels. This is consistent with previous observations of dislocation configurations in a highly alloyed 27%Cr-31%Ni austenitic stainless steel with nitrogen concentrations ranging from 0.04% to 0.35% (Nilsson and Thorvaldsson, 1985). These authors found planar slip already for a nitrogen concentration of 0.11% after tensile deformation at ambient temperature to a total strain of 1.5-2%. In this context it should be mentioned that Fig. 3 is an illustrative but rare example of dislocation motion in an fcc alloy since dislocations in all four {111} slip planes have become visible.

Fig. 2 Planar slip in LN steel (λ=80μm) after tensile deformation to 0.8% total strain. Beam direction ⟨011⟩. TEM.

Fig. 3 Planar slip in coarse-grained HN steel (λ=185μm) after tensile testing to 0.8% strain. Beam direction ⟨011⟩. TEM.

<u>Cyclic yield stress</u>

The cyclic yield stresses of LN and HN for different grain sizes are presented in Fig. 1 and Table 2. The cyclic yield stresses correspond to 0.2% off-set strain in the cyclic stress-strain curves (taken from multi-step test). As Fig. 1 indicates, the cyclic, like the monotonic, yield stresses exhibit a grain-size dependence of Hall-Petch type:

$$R_{p0.2}^c = \sigma_0^c + k_y^c \lambda^{-1/2} \qquad (2)$$

Evaluated values of σ_0^c and k_y^c for the two materials are shown in Table 2. Comparing the σ_0-values in Table 2 for cyclic and monotonic straining reveals higher friction stresses - as represented by σ_0 - for the high-nitrogen steel in both loading modes. For the low-nitrogen steel LN the friction stress is similar for monotonic and cyclic loading conditions. Besides planar glide the relatively low nitrogen level is expected to favour also tangling of the dislocations upon both monotonic and cyclic loading (Vogt *et al.*, 1993), which would indicate similar friction stresses in both modes. However, in the case of the high-nitrogen steel a more marked softening follows upon cycling, as expressed by a decrease of the friction stress from σ_0^m=286 MPa to σ_0^c=263 MPa. Such a softening following upon increased nitrogen content has earlier been observed (Degallaix *et al.*, 1988). If short range order is assumed to determine the hardening, a cyclic deformation mode would create a more pronounced break down of this order and thus a lowered friction stress.

It is evident that the k_y-factors for cyclic yield stresses are smaller than those for monotonic yielding (Table 2), indicating a smaller grain size dependence in cycling loading. For both monotonic and cyclic loading, however, k_y is larger for the HN steel with its higher nitrogen content. It is expected that less pronounced planar glide in cyclic loading, mostly so in the low-nitrogen steel LN, makes the influence of grain boundaries on plastification weaker, which in turn would lower the k_y-values. These arguments are supported by observations on outer surfaces of cyclically strained specimens, where SEM imaging revealed clear traces of well defined planar glide in the high-nitrogen steel HN. The LN steel, on the other hand, exhibit only a few slip bands of very weak character.

The influence of nitrogen concentration and grain size during cyclic loading were studied microstructurally in the LN material of grain size 78 μm and HN material of grain sizes 80 and 185 μm using TEM. These qualitative analyses failed to reveal significant differences in dislocation configurations although a large number of specimens were carefully examined. In all three material conditions areas showing clear signs of planar slip could be observed. However, it was also possible to find other areas in the same specimens where dislocation cells were predominant. Two important conclusions can be drawn from these examinations. Firstly, cross-slip of dislocations is so difficult already for a nitrogen concentration of 0.14% that cyclic loading is unable to entirely obliterate the planar dislocation slip character observed in tensile specimens. Secondly, the dislocation configurations vary considerably from one grain to another, presumably reflecting small-scale inhomogeneities in stress state. Figs. 4 and 5 illustrate these conclusions, Fig. 4 showing planar dislocation arrays in cyclically strained specimens of the LN steel, and Fig. 5 showing dislocation cell networks in cyclically strained fine-grained HN steel. Although the slip characters were very similar, one significant difference could be observed. When comparing fine-grained and coarse-grained HN-steels cell formation was found to be much more pronounced in the fine-grained steel, indicating promoted dislocation cross-slip.

CONCLUSIONS

1. The monotonic yield stress increases with decreasing grain size according to a Hall-Petch relationship. Increased nitrogen content results both in higher friction stress and increased grain size dependence. The larger k_y-factor at higher nitrogen contents is explained by more pronounced planar slip.

2. The cyclic yield stress also follows a Hall-Petch type relation but with a smaller k_y-factor than for the monotonic yield stress, indicating less grain size sensitivity. Again the k_y-factor raises with increasing nitrogen content. The smaller grain size sensitivity is associated with a gradual breakdown of the planar dislocation glide during cyclic straining.

3. Cross-slip of dislocations is difficult already for a nitrogen concentration of 0.14%. The dislocation configurations for both materials vary considerably from one grain to another, presumably reflecting small-scale inhomogeneities in stress state.

Fig 4 Planar arrays of dislocations in LN steel   Fig 5 Dislocation cells formed during cyclic
       (λ=80µm) after cyclic loading. Electron   loading in HN steel (λ=78µm). TEM.
       beam parallel to $\langle 001 \rangle$. TEM.

REFERENCES

Degallaix, S., J. Foct and A. Hendry (1986). Mechanical Behaviour of High-nitrogen Stainless Steels, *Materials Science and Technology,* **2**, 946-950.

Degallaix, S., G. Degallaix and J. Foct (1988). Influence of Nitrogen Solutes and Precipitates on Low Cycle Fatigue of 316L Stainless Steels, *ASTM, STP 942,* 798-811.

Grujicic, M., J.-O. Nilsson, W. S. Owen and T. Thorvaldsson (1988). Basic Deformation Mechanisms in Nitrogen Strengthened Stable Austenitic Stainless Steels, In: *High Nitrogen Steels* (J. Foct, Ed.) 151-158.

Lindstedt, U., B. Karlsson and M. Nyström (1995). The Influence of Nitrogen and Grain Size on the Nucleation and Growth of Surface Cracks in Austenitic Stainless Steel, *This conference*

Mullner, P., C. Solenthaler, P. Uggowitzer and M. O. Speidel (1993). On the Effect of Nitrogen on the Dislocation Structure of Austenitic Stainless Steel, *Materials Science and Engineering,* **A164**, 164-169.

Nilsson, J.-O. and T. Thorvaldsson (1985) The Influence of Nitrogen on Microstructure and Strength of a High-alloy Austenitic Stainless Steel, *Scand. J. Metallurgy,* **15**, 83-89.

Norström, L.-Å. (1977). The Influence of Nitrogen and Grain Size on Yield Strength in Type 316L Austenitic Stainless Steel, *Metal Science,* **11**, 208-212.

Petch, N. J. (1953). The Cleavage Strength of Polycrystals, *Journal of the Iron and Steel Institute,* **174**, 25-28.

Pickering, F. B. (1976). Physical Metallurgy of Stainless Steel Developments, *International Metal Review,* **21**, 227-268.

Vogt, J.-B., T. Magnin and J. Foct (1993) Effective Stresses and Microstructure in Cyclically Deformed 316L Austenitic Stainless Steel: Effect of Temperature and Nitrogen Content, *Fatigue and Fracture of Engineering Materials and Structures,* **16**, 555-564 .

Vogt, J.-B., S. Degallaix and J. Foct (1984). Low Cycle Fatigue Life Enhancement of 316L Stainless Steel by Nitrogen Alloying, *International Journal of Fatigue,* **6**, 211-215.

Effect of Spheroidal Graphite on Low Cycle Fatigue Properties of Ferritic Ductile Cast Iron

Shoji harada, Yoshihito Kuroshima, Yoshihiro Fukushima

Department of Mechanical Engineering, Kyushu Institute of Technolgy, 1-1 Sensuicho, Tobataku, Kitakyushu 804 JAPAN

and

Takahiro Ueda

Kyushu Works, Hitach Metals Co. Ltd., 35 Nagahama, Kandacho, Miyakogun, Fukuoka 800-03 JAPAN

ABSTRACT

The fatigue processes of four kinds of ferritic ductile cast iron (FDI) having different spheroidal graphite (SG) morphologies are examined. The specimens employed are a high nodule counts (HNC), a low nodule counts (LNC) FDI and U, L materials sliced from the upper and lower sides of a Y-shaped block respectively. Through experiments and computer simulation, it is finally shown that two patterns of dominant fatigue mechanism, ie., primary growth of a single crack and joining type of growth of plural subcracks observed in HNC, L and LNC, U respectively, are mainly caused by the difference of 3D size distribution of SG.

KEYWORD

Low-cycle fatigue (LCF); Ductile cast iron; Distribution of spheroidal graphite; Fatigue mechanism.

INTRODUCTION

Ductile cast iron (DI) or SG cast iron is being regarded as one of the low cost advanced materials and is widely applied to automobile parts. However, the low strength reliability of DI, mostly being attributed to microstructural fluctuation, still limits its application. To ensure the reliability strength evaluation based on the microstructure-oriented quantitative analysis is indispensable. In this regard, the present authors have already carried out a series of studies mainly related to the LCF properties of DI (Harada, S. et al., 1992, 1993, 1995).

According to the previous reports (Harada, S. et al., 1995), difference of nodule counts and slicing location in Y-shaped block in ferritic ductile cast iron (FDI) induced difference in fatigue mechanism of initiation, propagation and coalescence of subcracks.

The primary interest of the present paper is to clarify how the morphology of the spheroidal graphite (SG) affects the fatigue process. To extract the SG effect explicitly four kinds of specimens with different SG morphology were LCF-tested and the fatigue mechanisms were carefully observed by metallographic and fractographic observations. Then, computer simulation technique was applied to actually demonstrate whether the fatigue behaviors observed in four kinds of specimen can be reproduced by computer simulation only using SG information as an input data.

MATERIALS AND EXPERIMENTAL PROCEDURES

The materials tested are two kinds of FDI, i.e., HNC, LNC (FCD400) and FCD370. The materials HNC, LNC have nodule counts of 229.6/mm^2 and 84.7/mm^2 respectively. The chemical compositions and the SG morphology-related parameters of these materials are tabulated in Table 1 and Table 2 respectively. Rather low nodularity found in Table 2 comes from the difference of its definition (Harada, S. et al., 1993). The L, U specimens were sliced from the lower and upper portion of the FCD370 Y-shaped block of 32×40×250mm^3 in size. Figure 1 shows the distribution of the SG morphology-related parameters in the thickness direction of Y-shaped block. The slicing locations of L, U specimens are also shown in the figure. An hour-glass-shaped fatigue specimen with a minimum diameter of 8mm was machined from each material. Cyclic stress-controlled fatigue tests were conducted in a electro-hydraulic servo-controlled fatigue testing machine at a speed of 0.1 to 0.5Hz. Fatigue process was successively observed on each specimen surface with the aid of a plastic replication technique. In fractographic observation, a special attention was paid to examine internaly located SG-originated preceding subcracks by carefully tracing striations formed around SG.

EXPERIMENTAL RESULTS AND DISCUSSIONS

Fatigue Life and Mechanism

Figure 2 shows the S-N diagram of each material. Clearly the materials HNC, LNC show a reversing trend of fatigue life around Nf≈10^3, while the materials L, U show no discernible difference of fatigue life. These results mean that the fatigue life is not simply determined by nodule counts and that the distribution of nodule counts observed in the Y-shaped block (or actual element) has only a slight effect on the fatigue life.

According to the results of successive observation of microscopic fatigue process on specimen surface, it was found that two types of dominant fatigue mechanism, namely, dominant growth of a single leading crack and growth by coalescence of subcracks were observed. In particular, the coalescence type of crack growth was observed at the later fatigue stage when rather large subcracks initiated separately grew up to a length of about 1mm and linked up each other. The former type was seen in L and HNC specimens and the latter in U and LNC specimens. Growth curve of a leading crack in each specimen reflected those crack growth behaviors. Figures 3(a) and (b) show the crack growth curves of the L, U specimens and HNC, LNC specimens respectively. In case of U, LNC specimens, the crack growth curves imply rather step-like pattern, which corresponded to the joining type of crack growth. While in case of L, HNC specimens, the crack growth curves are rather smooth as shown in Figure 3 (a), in comparison with other specimens. The step-like pattern and joining pattern found in main crack growth correspond to the coalescence type of crack growth and the dominant growth of the single crack respectively.

The results of fractographic observations revealed two characteristic features of crack growth behavior. The first point is concerned with an initiation of preceding crack at internally located SG. Figure 4(a) depicts an example of initiation and growth of subcracks at internally located SG. Figure 4(b) implies a highly magnified view at the point A in Figure 4(a). The arrow marks in Figure 4(a) show the directions of local crack growth, free from the direction of main crack growth, determined by higher magnification of striation. Furthermore, a closer observation around SG in Figure 4(b) reveals that a discrepancy is formed at SG/matrix interface and the striation spacing indicated by an arrow is fairly small, comparing with the predicted value from surface crack growth rate. Judging from these fractographic informations, it should be noted that subcracks or preceding cracks might have been formed by debonding of SG/matrix interface before arrival of the main crack. The formation of these internal subcracks might have accelerated the main crack growth when joining each other.

The second point of the features of crack growth behavior is related to the role of large-sized SG at the initial stage of crack growth. Even in L, HNC specimens where the dominant growth of single crack was observed, subcracks showed independent growth until they grew up to $300\mu m$. Therefore, two types of governing mechanism of main crack growth might have been induced by the difference of crack growth from $300\mu m$ to $500\mu m$. Taking into account of the results of fractographic observation, Regarding the crack growth behavior at this stage, the results of fractographic observation showed that the fatigue failure origins were mostly located at the microshrinkages in case of the crack growth pattern of single crack governing type, while they were located both at the microshrinkages and large-sized SG.

Computer simulation of the effect of SG on the fatigue mechanism

The results of the successive observation of the fatigue process on specimen surface and fractographs suggested that the large-sized SG played an important role to control the main fatigue process. This trend was checked by computer simulation of fatigue process. Figure 5 shows the 3D size distribution of SG in each specimen, evaluated by converting 2D information of SG distribution in terms of a statistical approach developed by Saltykov (Kuroshima et al., 1990). Using this as an input data and assuming coalescence condition of subcracks, the computer simulation was done. The results are shown in Figures 6(a) and (b) for HNC and LNC specimen respectively. The SG sized distribution-induced difference of fatigue processes are well simulated.

Conclusions

The experimental results showed that SG primarily controls the fatigue process of FDI with different SG morphologies. The difference of the fatigue process mainly induced by the SG size distribution was well demonstrated by computer simulation.

References

Harada, S., Y. Akiniwa and T. Ueda (1992). The effect of microstructure on the low-cycle fatigue behavior of ductile cast iron. LCF-3, 124-129.

Harada, S., T. Ueda, Y. Akiniwa and M. Yano (1993). Small crack growth in ductile cast iron with different microstructure. Strength of Ductile Cast Iron '93, H. Nisitani, S.Harada,T.Kobayashi ed. JSME-MMD, Kitakyushu, 91-96.

Harada, S.,Y. Kuroshima, Y. Fukushima. and T.Ueda.(1995). The effect of nodule counts on the low-cycle fatigue properties of ferritic ductile cast iron. PVP95, Vol. 306, 247-252.

Kuroshima, Y., M. Shimizu and K. Kawasaki (1990). Relationbetween critical size of inclusion responsible for fatigue failure and inclusion size distribution. Proc. KSME/JSME Joint Conf, Fracture and Strength '90. 73-78.

Table 1 Chemical compsition, wt%

	C	Si	Mn	P	S	Mg
HNC	3.61	2.61	0.41	0.02	0.01	0.03
LNC	3.74	2.58	0.45	0.02	0.00	0.03
L, U	3.76	2.21	0.19	0.017	0.016	0.038

Table 2 SG morphology-related parameters

	n_g (/mm^2)	f_g (%)	D_m (μm)	h_g (%)
HNC	229.6	7.5	21.4	64.3
LNC	84.7	10.8	36.3	63.9
L	181.0	6.8	18.7	58.0
U	130.0	7.4	21.8	53.8

n_g : nodule count in a unit area, f_g : volume fraction of graphite,
D_m : mean diameter of graphite grain, h_g : nodularity

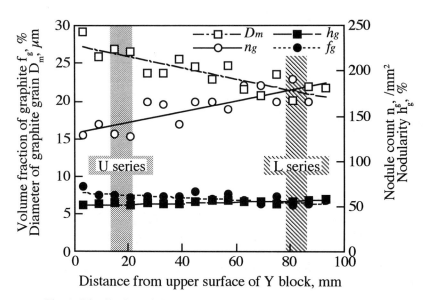

Fig. 1. Distribution of the SGmorphology-related parameters in the
thickness direction of a Y-shaped block.

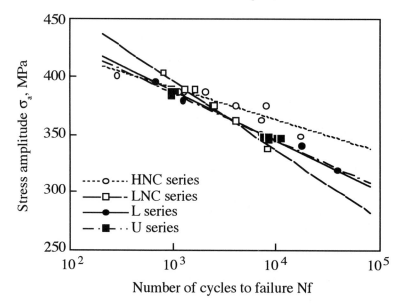

Fig. 2. S-N curves of the four kinds materials.

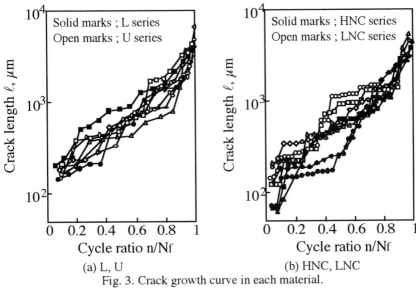

(a) L, U   (b) HNC, LNC

Fig. 3. Crack growth curve in each material.

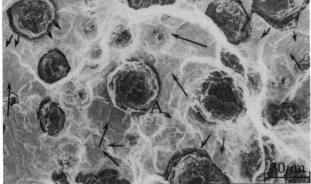

(a) Initiation of preceding crack originated at internally locaated SG

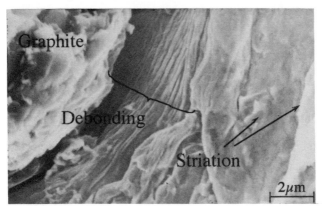

(b) Debonding and striation formation at SG/matrix interface

Fig. 4. An example of preceding crack originated at internally located SG.

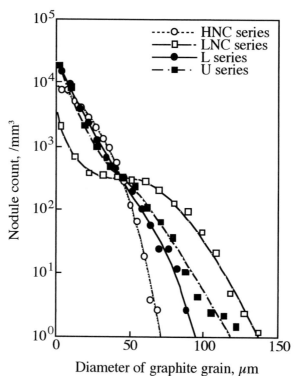

Fig. 5. 3D SG size distribution of each material.

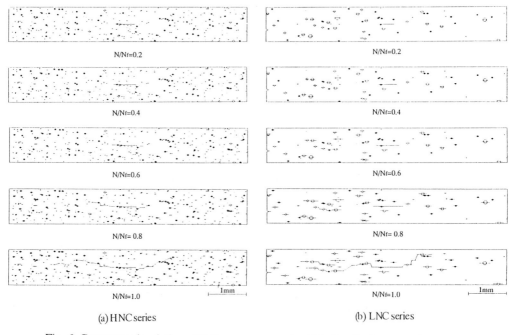

(a) HNC series   (b) LNC series

Fig. 6. Computer simulation of fatigue process using SG distribution as an input data.

FATIGUE STRENGTH OF AUSTEMPERED SPHEROIDAL GRAPHITE CAST IRON UNDER TEMPERATURE ENVIRONMENT

Shigeru DOI *, Kazumichi SHIMIZU**
Masao TAKAHARA***, and Katsuyuki SATO*

*   Oita University, 700 Dannoharu, Oita 870-11, Japan
Fax 0975-69-7003   Tel 0975-69-3311
**   Oita National College of Tech, 1666 maki , Oita 870-11, Japan
***   Isuzu Motor Co.LTD, 3-25-1 Tanomachi Kawasaki-ku
Kawasaki - shi   210, Japan

ABSTRACT

As one of the fatigue strength dominating factors of austempered spheroidal cast iron (ADI), the effects of temperature environment on high cycle fatigue were investigated. The temperatures of using ADI are desirably at the Ms transformation point or less. In this study, therefore, a series of long life rotating bending tests were carried out at 453 K that is below the Ms transformation temperature.

The resultant S-N curves are upward convex, indicating that no fatigue limit exists in the long life region. In the region of 10^6 cycles ore more, fish eye type fractures were observed in all the test pieces. The positions from which fish eyes occurred were about 200 μm from the surfaces in all the stress levels.

KEYWORDS

High Cycle Fatigue; Austempered spheroidal graphite cast iron; Rotating Bending Fatigue; Fish Eye; S-N Curve; Ms Transformation.

INTRODUCTION

The diversity of austempered spheroidal graphite cast iron (ADI) has been expected as one of new materials. The authors (Doi et al.,1990) have performed a series of studies regarding the high cycle fatigue of ADI, and reported the results of fractography (Doi.,1994) and mechanism (Doi et al.,1994) of combined fracture as basic data. The mechanism of crack initiation and propagation, however, varies with relations between the properties of graphite and substrate structures, and is provided with many unknown portions from the quantitative point of view. Suzuki et al. (Suzuki et al.,1985) carried out a series of fatigue tests by changing the substrate structure from ferrite to bainite to clarify their fatigue strength characteristics. The authors (Doi et al.,1994) prepared test pieces containing local martensite formed by the work hardening of ADI surfaces and those in which martensite was removed by electro-polishing and the resultant substrate consisted of retained austenite and bainite. The fatigue strength characteristics of these two kinds of test

245

pieces were compared in detail.

As part of the studies to investigate the effects of environmental factors on fatigue strength, this paper paid attention to the temperature range equal to or less than Ms transformation, and the high cycle fatigue behavior in this range was compared with that obtained at room temperature. An actual example of thermal environment differing from the above-mentioned temperature atmosphere is gears coming into contact with each other for a long time. In this case, the thermal environment of this type is formed in the contacting portions of the gears. The material change in the contacting portions as well as the thermal environment is said to decrease fatigue strength. In this connection, because the decrease in fatigue strength includes contact fatigue, it is not clear whether or not the decrease in fatigue strength stems from generated heat alone. However, there is no research report discussing fatigue strength at the Ms transformation point or less, so it seems important for the authors to investigate high cycle fatigue strength under temperature environment.

EXPERIMENTAL METHODS

The test pieces were made of the austempered ADI material (hereinafter, referred to as HADI material). Its chemical composition, mechanical properties and structural characteristics are shown in Table 1. The grip of the test piece was cut perpendicular to its axis, and the cross section was polished, etched and subjected to oxidation baking for observation by an optical microscope. Heat treatment and the resulted metal structure on ADI is shown in Fig. 1. Figure 2 gives the shape of the uniform-gauge test piece with 7 mm diameter parallel portion. The special furnace was made to keep low temperature atmosphere, and the temperature sensor was fixed, as shown in the right side of Fig. 2, about 2 mm away from the parallel portion of the test piece to control the temperature atmosphere. Figure 3 shows temperature-time curves. After reaching the set temperature, the test piece was held at the temperature for about 10 hr to avoid the formation of temperature gradient toward the center of the test piece or temperature dispersion, followed by the start of an experiment. The testing temperature was 453 K. Fatigue limit is defined by stress amplitude that could bear 10^8 cycle repetition on the basis of the results of experiments at room temperature.

The chucking rods and thermostatic chamber were additionally provided for the Shimadzu-type rotating bending tester (capacity: 98 Nm, number of revolution: 3400 rpm), allowing its capacity to be nearly equal to that of a small tester with a maximum capacity of 14.7 Nm.

RESULTS OF EXPERIMENTS

Figure 4 shows S-N curves. The results of small-size rotating bending tests (diameter of the parallel portion: 6 mm) at room temperature are also shown (hereinafter referred to as A and B materials). In Fig. 4, the number of testing points for the HADI material is small, but rupture life becomes long on the low stress side, leading to extremely slow-gradient S-N curve. In the A and B materials at room temperature, horizontally crimped portions exist between the high and low stress sides, indicating that crack propagation stops temporarily. At 453 K, however, temperature dependence appears and rupture takes place after 10^6 cycle or more. This may be because martensite formed at the end of a crack in the initial stage of fatigue thermally diffuses, and thereby because a kind of blunting temporarily occurs at the end of the crack.

Table 1 Chemical composition, mechanical propertie and S.G. related values (Wt %)

C	Si	Mn	P	S	Cr	Mg	Cu
3.6	2.2	0.3	0.0	0.01	——	0.0	0.7

Tensile strength σ_B MPa	Yield strength σ_Y MPa	Elongation δ %	Vickers hardness Hv	hg % 1)	ng % 2)	dg μ m 3)
1017	813	10	358	64.9	246	15

1)hg: nodularity 2)ng: graphite nodule count 3)dg: mean diameter of S.G. (S.G. : spheroidal graphite)

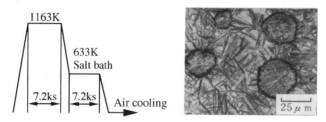

Fig.1 Heat treatment on ADI and microstructures

Fatigue fracture, therefore, does not originate from the integration of cracks distributed over the surface on the low stress side. As shown in Fig. 5, fish eye type fractures can be observed in all the test pieces after 10^6 cycles or more. This tendency differs to a slight extent from that indicated by the experimental results (Doi et al.,1994) obtained by using electro-polished ADI materials, but may stem from the relaxation of substrate's sensitivity for initial cracks on the surfaces of the test pieces.

DISCUSSIONS

S-N Curves

A variety of phenomena taking place under thermal environment have not been fully understood. The authors, therefore, performed the rotating bending tests at lower temperatures to investigate the effects of temperature atmosphere on fatigue life.

The S-N curve shown in Fig. 6 is upward convex because of clear influence of the temperature environment. Taking dispersion into account, this S-N curve could be shown by a straight line. Although temperature was controlled not to cause temperature gradient, the test piece was sensitive to internal defects, and fish eye type fracture appeared in the early stage of its life, indicating the tendency of non-existence of fatigue limit. The reason is that the fracture mechanism on high stress side differs from that on low stress side; namely, the mechanism of crack propagation sometimes depends on the integration of cracks distributed over the surface and subsequent growth, whereas internal defect type fracture

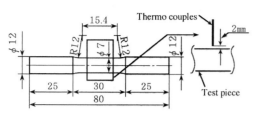

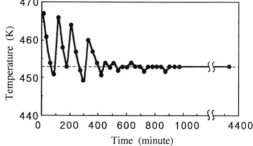

Fig.2 Dimension and profile of tested specimen

Fig.3 Temperature - Time curves

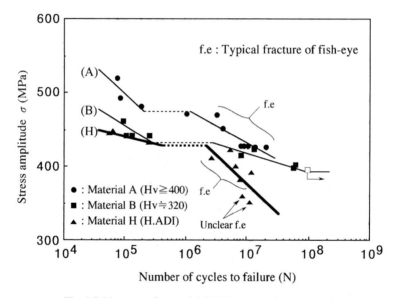

Fig.4 S-N curve of material HADI comparing materials A and B

originates, in addition to the stress concentration to internal defects, from degraded materials due to bubbles contained in the substrate structures around micro defects remaining in the course of melting and solidification. The two-step crimps observed in the S-N curves of the A and B materials result from, in addition mainly to the above-mentioned factor, the difference in the thickness of the worked layer of the surface. At a surface temperature of about 453 K, on the other hand, the main cause on the low stress side is internal defect type fracture. There is no document showing this factor clearly, but the material of the test piece may change as if it were martempered (423 - 473 K). One of the causes of delayed surface crack propagation is considered to be the structural change of substrate at the end of the crack due to martempering at around 453 K. Temperatures adopted in the experiments are equal to or less than the Ms transformation point, but holding time is as long as 10 hr, so that the amount of austenite possibly decreases;

accordingly, part of the test piece was cut into two parts - one is for heating test and the other for test at room temperature, and the amounts of retained austenite of each part were measured by using X-rays. The results showed that the amounts of retained austenite were about 40% and about 36.5% before and after heating, respectively, indicating slight decrease. Also, the states of graphite after heating were observed with an optical microscope, but no remarkable change in the substrate structure was found. Crack propagation from surface-distributed cracks is not remarkable on the low stress side, but internal defect initiated fracture tends to occur. This tendency is also found in the fracture mechanism of the buff-finished materials reported by the authors (Doi et al.,1994) in which fish eye initiated fracture took place in the long life region.

From the above, it was made clear that the fatigue behavior of the ADI material at room temperature or at Ms transformation temperature or less 453 K in our experiments showed the upward-convex S-N curves unlike two-step-crimped ones.

CONCLUSIONS

The fatigue behavior of the ADI material at the Ms transformation point or less showed further complicated

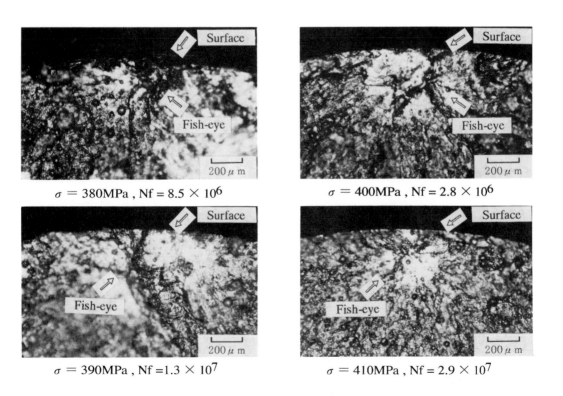

$\sigma = 380$MPa , Nf = 8.5×10^6

$\sigma = 400$MPa , Nf = 2.8×10^6

$\sigma = 390$MPa , Nf =1.3×10^7

$\sigma = 410$MPa , Nf = 2.9×10^7

Fig.5 Typical fish eye pattern of material HADI

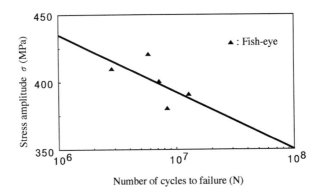

Fig.6 Magnified S-N curve of HADI in fish-eye region

fraction factors in addition to the factors indicated by the two-step-crimp fracture mechanism. The experimental results can be summarized as follows:
1) Fatigue fracture on the long life side is all fish eye type.
2) Crack initiation and propagation may partly stem from the material change in the retained austenite of the substrate. The inhibition of this material change may lead to the S-N curves different from those for two-step-crimp fracture due to crack initiation from the surface followed by propagation and fracture.

REFERENCES

S. Doi et al.: Trans of The Japanese Society of Mechanical Engineers, 56-531, A(1990), 2205-2209.
S. Doi : Trans of The Japanese Society of Mechanical Engineers, 60-570, A(1994), 331-336.
S. Doi et al.: Trans of The Japanese Society of Mechanical Engineers, 60-574, A(1994), 1315-1318.
H. Suzuki et al.: Trans of The Japanese Society of Mechanical Engineers, 51-464, A(1985), 1224.
S. Doi et al.: Trans of The Japanese Society of Mechanical Engineers, 60-575, A(1994), 1493-1497.

INITIATION AND PROPAGATION OF SHORT CRACKS

INITIATION AND PROPAGATION OF SHORT FATIGUE CRACKS

K. J. MILLER

Director, Structural Integrity Research Institute of the
University of Sheffield (SIRIUS)
Mappin Street, Sheffield, S1 3JD, UK.

ABSTRACT

This paper concentrates on the mechanics of microstructure-influenced fatigue crack growth. A brief historical review is followed by a description of the boundaries of applicability of the three fundamentally different fatigue fracture mechanics equations which are based on the physics of fatigue crack growth processes.

From knowledge of the influence of microstructure on fatigue crack growth, eg: crystallographic orientation, grain size and second phase structures, the paper concludes with a discussion of the effects, on microstructural fracture mechanics, of temperature, corrosive environments, surface finish and shot peening recently determined from experiments in the author's laboratories.

KEYWORDS

Initiation; Short Fatigue Cracks; Grain Size; Microstructural Mechanics; Fatigue Limits; Surface Treatments; Environment.

INTRODUCTION, DEFINITIONS AND REFERENCES

All defects can cause fatigue cracks to grow, irrespective of their size, shape and orientation, if a sufficiently high cyclic stress is applied. The problem for the engineer is to select the most appropriate form of the fracture mechanics equations and to be aware of the limitations or boundary conditions of applicability of these equations when describing crack extension.

It is unfortunate that two words in fatigue literature can mislead the engineer. The first is "initiation" and the second is "short"; as distinct from "long" cracks. Reports on the fatigue of metals frequently discuss experimental findings in terms of the number of cycles to initiate a fatigue crack followed by the number of cycles to propagate the crack; hence initiation is thought to be distinct from propagation. The words "long" and "short" can only refer to a distinction in crack size and regrettably current literature frequently refers to the anomalous behaviour of short fatigue cracks. As will be seen in this paper, "short" cracks require high stresses for their propagation while "long" cracks can grow at both low and high stress levels with apparently unrelated (ie: anomalous) behaviour.

Cracks subjected to low stress levels have a threshold or fatigue limit which is frequently determined by recording cracks sporadically growing at speeds as low as 10^{-10} to 10^{-11} m/cycle. If a crack starts to grow from a surface notch or crack-like defect as small as 1 to 10 microns, then it may take between one million and ten million cycles to propagate to a length of 100 microns. In the present paper the initiation phase is considered to be of zero duration, and is replaced by a slow crack propagation phase best described in terms of microstructural fracture mechanics.

Space does not permit a large number of references, and so readers are requested to examine the list of past papers provided at the end of this presentation for a more extensive source of reference material.

HISTORICAL REVIEW

The first attempt to consider very slow crack growth during "initiation" was made by Zachariah who started his research in 1971; see [1] and Fig 1(a). Experimental data was collected for both constant amplitude and two-level cumulative damage, reversed cyclic torsion tests. Using hollow cylindrical specimens shear type cracks dominated throughout the lifetime until close to failure; this simplified the analysis. In Fig 1(a) the initial crack length a_0 can be equated to the depth of very shallow machining grooves on the smooth, polished surfaces while the failure crack length a_f was assumed constant for all stress levels; ie: 1mm which was the specimen wall thickness. Over the range of 10^2 to 10^7 cycles to failure N_0 was considered to be zero (ie: the initiation phase does not exist) and the number of cycles to failure when a_f was attained was noted for all tests. Therefore the only unknowns in Fig 1(a) were the transition crack length a_t and the slope α which was considered to be negative; a not unreasonable assumption, in that the initial cracks would develop to shorter lengths at lower stress-strain ranges. In the two level, strain range controlled tests those cracks generated in the first phase were considered to continue propagation immediately the second level phase began. To avoid residual stress effects in the high-to-low series of experiments an intermediate stress relief heat-treatment was introduced. This treatment was not considered necessary for the low-to-high strain level test series. All data was analysed by computer to obtain the optimum values of α and a_t and Fig 2 indicates the model depicted in Fig 1(a) was delightfully accurate in mathematical terms. This figure also shows that the Palmgren-Miner, linear accumulation of damage hypothesis, can be both dangerously inaccurate and also over-conservative should the early stages of damage, here considered to be slow crack propagation, be neglected in an analysis.

Subsequently Ibrahim conducted a re-analysis (see reference [2]) in order to achieve a more realistic physical interpretation of the accumulation of damage in terms of fatigue crack growth. He considered that the slope α could be positive ($\alpha > 90°$) on the basis that the transition to rapid crack growth would occur at shorter transition crack lengths at the higher stress levels. Figure 1[b] indicates this later model while Fig 1(c) is a composite of both viewpoints, both of which are consistent with current views that there is a transition zone for cracks transferring to a continuum mechanics (non-microstructural) description of crack growth; this intermediate phase being larger at low stress levels and indeed may be of infinite duration at very low stress levels, thereby explaining why fatigue cracks can exist below the fatigue limit of steels.

Since that time, much research has been done, eg: [3], [4] and [5] that substantiates the hypothesis of Fig 1(c); in particular the exact determination of the very small crack lengths at transition points via the constant monitoring of the growth of microstructurally short cracks using various techniques including the recently developed acoustic microscope [6]. Of particular importance is the effect of microstructure, the environment and surface state on the transition zone shown in Fig 1(c). These factors are the subject of the present paper.

FATIGUE CRACK GROWTH PROCESSES

Fortunately only two basic forms of crack growth mechanisms exist irrespective of the uniaxial, biaxial or multiaxial loading system employed [7]. These were defined by Forsyth [8] as Stage I growth and Stage II growth, see Fig 3. The former occurs on a plane of maximum shear strain, the latter on a plane whose normal is in the direction of the maximum tensile stress. Depending on the stress-state and its intensity, either Stage I or Stage II cracking will dominate the fatigue lifetime of specimens, components and structures.

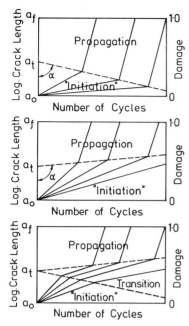

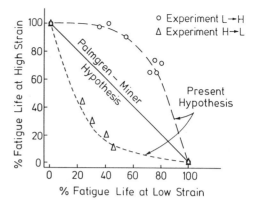

Fig. 1 The "initiation" period considered as
slow short-crack propagation at
both low and high stress levels.
(a) Zachariah hypothesis
(b) Ibrahim hypothesis
(c) Current view

Fig. 2 Typical life-fraction results by
the Zachariah hypothesis [1].
Experimental data and theoretical
predictions for 2-strain range levels,
cumulative damage tests, where
N_f (low) = 1000 cycles and
N_f (high) = 400,000 cycles

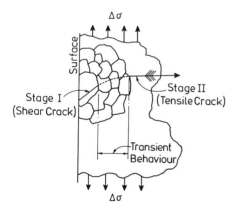

Fig. 3 Stage I, Transient and Stage II
crack growth paths as observed in
a uniaxial push-pull fatigue test.

Fig. 4 Transient crack growth planes in
the microstructure-dominated
phase of fatigue lifetime

Figure 3 schematically illustrates the crack growth path in 2-dimensions for a uniaxial fatigue test. The
initial Stage I shear crack will usually start at a surface stress concentration, for example, a triple grain-
boundary point, and grow relatively quickly in a transgranular manner, across one grain. The associated crack tip plasticity will be in the form of a persistent slip band; this taking some time to fully

develop and stabilize. The crack increases in size from approximately zero microns to the depth of the largest surface grain; a grain which provides the longest slip planes and whose crystallographic directions are suitably orientated in relation to the direction of the maximum applied shear stress.

Further growth of the crack is now difficult and the aforementioned transition zone begins. In this sense the first grain boundary can be regarded as a primary barrier since beyond this barrier the crack has to grow across the several grains that surround the first grain. Consequently the different slip plane orientations, the different crack growth directions and the different lengths of crack paths through to the next grain boundary barrier, all of which are associated with the different near-neighbour grains, cause the crack to have a 3-dimensional tortuous profile, that can be both transgranular and intergranular in nature as the crack front attempts to maintain a regular, compatible and low-energy profile. Fig 4 illustrates the tortuous 3-Dimensional nature of this very slow crack growth period which is the transition phase illustrated in Fig 1(c).

Should the applied stress level not be sufficient to propagate cracks on such unfavourably inclined planes as depicted in Fig 4, and in directions not synonymous with the maximum shear (or tensile) stresses, then the crack will be arrested. It follows that the fatigue resistance of a metal can be equated to the transition phase in terms of the highest stress level which will not allow the initial crack to proceed beyond the dominant microstructural barrier; represented in Fig 3 as a grain boundary. Other barriers to the continuation of crack growth can be twin planes within a metal crystal and, at a larger scale, strings of pearlite separating the ferrite matrix of a medium carbon steel [9].

Should the stress level be sufficient to propagate the Stage I crack of Fig 3 across the first barrier then, due to the increased crack tip plasticity associated with the increasing crack length, the next series of similar barriers in the crack path will not be of sufficient strength to cause crack arrest although decelerations and accelerations in the crack speed will occur as, respectively, the crack approaches and then passes through successive barriers.

Eventually the transition zone will cease when the crack is sufficiently long to permit opening of the crack front, the development of mutually perpendicular shear ears at the crack tip, leading to the growth of a Stage II (tensile) crack. At this point the microstructure has limited influence and crack growth can be described by continuum mechanics.

FATIGUE FRACTURE MECHANICS

Before, during and after the transition zone the description of fatigue crack growth will depend on the current crack length, the stress state and the stress level. Obviously the slow speed of a 3-D meandering crack within the transition zone will be difficult to quantify, however an excellent shear crack growth model exists [10], to which readers are referred for a description of the transition zone. Additionally boundary conditions can be formulated from experimental studies that include the quantification of the effects of microstructure.

Figure 5 illustrates the three different forms of fracture mechanics applicable to the fatigue process. For long cracks propagating under low applied stress levels, linear elastic fracture mechanics (LEFM) provides an adequate description of fatigue crack growth behaviour, but when the initial crack size is physically small (eg 0.5 mm) the stress level for propagation is necessarily high and the assumptions that were invoked for a LEFM analysis may be inappropriate. Consequently an EPFM (elastic-plastic fracture mechanics) description of crack growth is necessary. Taking da/dN as crack speed, $\Delta\gamma$ as the applied shear strain range, a as crack depth, β, B and D as experimentally determined material parameters, then

$$da/dN = B\Delta\gamma^\beta a - D \tag{1}$$

Equation (1) has to be modified when the microstructure influences growth as is reflected by the two lower curves of Fig 5 i.e. when the initial crack size is smaller than the spacing between the dominant microstructural barriers. Microstructural Fracture Mechanics (MFM) equations have the form

$$da / dN = A\Delta\gamma^{\alpha}(d - a)\qquad(2)$$

where d is the barrier spacing and A and α are material parameters. The transition zone X-X in Fig 5 is

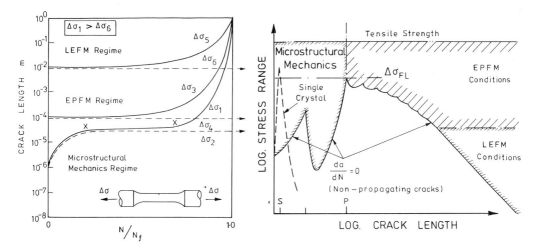

Fig. 5 The three zones of elastic (LEFM),
plastic (EPFM) and microstructural
(MFM) fracture mechanics in relation
to the stress levels that are close to the
propagation limit for the given initial
crack lengths

Fig. 6 The fatigue limit, expressed as zero
crack growth rate, schematically
shown as a function of stress range
and crack size.

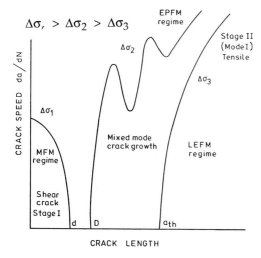

Fig. 7 Fatigue crack growth rates in the
MFM, EPFM and LEFM zones of
behaviour along with their respective
thresholds d, D and a_{th} at stress

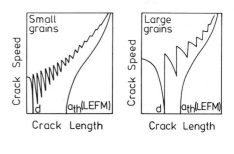

Fig. 8 The effect of grain size
on crack growth rate
in the MFM zone at a
stress range close to
the fatigue limit.

bounded by equations (1) and (2). The story of the fatigue of metals (as opposed to the fatigue of structures and LEFM) is directly concerned with the difficulty a crack experiences when attempting to propagate through the dominant microstructural barrier of a given metallic texture. The initial crack growth condition can be determined from equation (2). At first the crack speed is fast (d >> a ~ o) but when the crack approaches the barrier (a ~ d) the crack growth rate tends to zero..

From Fig 5 it is clear that each initial crack size, over the range of one micron to one metre, (each described by different equations) must have its own fatigue limit given by da/dN = 0. Fig 6 shows the 3 different fatigue limits associated with MFM, EPFM and LEFM; see [11] and [12]. Of some importance is that the fatigue limit of a steel is not associated with LEFM parameters and particularly the frequently quoted effective threshold condition $\Delta K_{eff,th}$. This latter parameter is not associated with material texture. Indeed the assumption of limited crack-tip plasticity, necessary for LEFM analyses, does not hold when the stress-range $\Delta\sigma$ exceeds approximately 0.7 times the cyclic yield stress, σ_{cy}. Certainly above σ_{cy} the crack tip plastic zone will be extensive and even "long" cracks subjected to these stress levels can not be described by LEFM, instead EPFM models of the form introduced by Dugdale [13] and Tomkins [14] are required. This is the reason for the apparent anomalous behaviour of long cracks as will be seen in Fig 7.

MICROSTRUCTURAL INFLUENCES

Figure 7 shows the threshold conditions D and d given respectively by Eqs (1) and (2). At stress-strain levels below that associated with d and just above that associated with D, the microstructure plays an important role. The third threshold condition given by a_{th} in Fig 7 can be determined from long crack-low stress experiments using LEFM type equations of the form

$$da/dN = C\left(\Delta\gamma\sqrt{\pi a}\right)^n \tag{3}$$

This latter equation will not be referred to further in this paper but it should be noted that in some materials the threshold states given by d, D and a_{th} (determined respectively from Eqs. (2)(1) and(3)) may be of similar dimensions.

Grain Size Effects

It has long been known that the fatigue limit of metals can be increased by reducing the grain size. The reverse may apply when considering the fatigue limit of a metallic structure; see [15].

To increase the fatigue resistance of a polycrystalline metal it is required to have as many barriers to crack growth as possible particularly so during the early stages of propagation, i.e. to have the maximum number of grains per unit volume when the initial crack size is of a length less than the grain dimensions This is because the number of cycles to failure, N_f, is simply the integration of the appropriate crack growth laws. For a given stress range, the smaller the crack length and the more barriers, then the more decelerations and the slower the average growth rate, see Fig 8, and an extended lifetime. Furthermore the smaller the grain size, the higher the yield stress and the smaller the crack-tip plastic zone size. Briefly Eq. (2) indicates the necessity to decrease the value of d in order to decrease da/dN. Two recent publications [16] [17] should be consulted.

Inclusions; the effect of size, shape and orientation

Should an inclusion, of a certain size, shape and orientation, span the transition stage then Stage II crack growth will not be delayed and lifetime will be dramatically reduced. In this respect it is not the size of the inclusion itself which is important but rather the ratio of the inclusion size to the grain size, or to be more precise - the distance between the major microstructural barriers to crack growth. Should inclusions of a given size be embedded in a coarse grained material then little effect on lifetime may be recorded, but the same inclusions in a fine grained material can eliminate the transition phase. It is for this reason that inclusions are of greater consequence in high strength steels because these steels invariably have a very small grain size.

The orientation of an elongated inclusion is also important. Should the major axis of such inclusions be in the same direction of crack growth then a weak link in the transition path between Stage I and Stage II fatigue crack growth may result. Conversely if the minor axis is in the crack growth direction then a weak ductile inclusion could act as a barrier. Further details on the important effects of inclusion geometry can be read in reference [12].

Second-phase structures

Second phases, particularly in a continuous form within a given single phase matrix, can substantially increase the fatigue resistance of a metal if they are of very high strength. For example pearlite, in a matrix of ferrite grains is a lamellar structure of ferrite and cementite (the second phase) and this constituent is responsible for the increased fatigue strength. Such a composite microstructure provides a very strong barrier to crack growth [9], far more effective than the ferrite grain boundaries themselves. Indeed it is not difficult to find long non-propagating cracks across several ferrite grains of a medium carbon steel at stress ranges below its fatigue limit.

A whole range of heat treatments can provide various forms of mixed phase microstructures which can lead to an improved fatigue resistance. This is always associated with the early crack growth in the weaker phase being slowed down or arrested by the stronger phase which necessarily should be of a near continuous form in order to act as arrestors or deflectors of the crack growth path.

Grain Orientation and Plasticity

The fatigue resistance of a metal is related to the difficulty of transferring crack-tip plastic displacements from one grain to the next. Persistent slip bands of a prescribed orientation must necessarily stop at a grain boundary. However if a metal has an fcc. structure (e.g. Aluminium) each grain has 4 slip planes each with 3 slip directions giving 12 slip systems. The spread of plasticity in fcc. metals is therefore easier than in bcc. or hcp structures. For this reason polycrystalline aluminium has no defined fatigue limit.

Two possibilities exist for continued crack growth when the crack reaches the grain boundary. First the crystallographic mis-orientation with the next grain is minimal or alternatively the crack tip stress field induces a high local stress concentration inside the second grain which produces a new crack which propagates backwards to meet the first crack.

Noting an initial comment that cracks initiate at stress concentrations such as triple grain-boundary points on a surface, one technique to increase fatigue resistance of a metal is to remove grain boundaries, i.e. use single crystals. It follows that there are 3 fundamentally different fatigue limits to be considered by engineers, see Fig 6, namely the polycrystalline material threshold, d, the continuum mechanics threshold, ΔK_{th}, and the single crystal threshold; the subject of a future paper.

ENVIRONMENT EFFECTS

The major contributors to the lowering of the fatigue resistance are aggressive environments and increases in temperature. In this paper the effect of each will be considered separately, however it must be remembered that should high temperatures and aggressive environments be simultaneously involved the synergetic effect can be devastating [18].

Aggressive Environments

The effect of various environments on long crack growth is well documented. Much less is reported on physically small cracks and yet this topic is far more important. A comparison of fatigue endurance

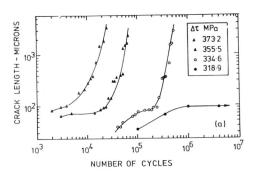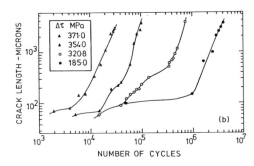

Fig. 9 The effect of shear stress range on fatigue crack growth and lifetime at room temperature
(a) in an in-air environment
(b) in a 0.6 M, NaCl solution

curves of metals tested in air and in an aggressive environment indicates that the fatigue limit of a steel can be eliminated although at higher stress levels little difference on lifetime may result. The reason is that the fatigue limit arises from the strength of microstructural barriers that provide the transition in short crack growth behaviour. If the deleterious environment is applied at the instant a crack reaches the barrier, then the transitional phase can be removed. Fig 9 shows that, in air, the barrier is effective in arresting a 80 - 100 micron crack (fatigue limit) at a shear stress range of 318 MPa. However in a salt-water solution [19] the barrier is easily overcome at this stress level and even at a stress level as low as 185 MPa because of the synergism between cyclic plasticity and environmental attack.

In mathematical terms the environment need only be applied for that period of time that permits the crack to grow between the two threshold states given by d and a_{th} obtained from, respectively, equations (2) and (3), that is the linking of MFM with LEFM.

For a more detailed appreciation of the effect of environment, readers are referred to reference [20].

Temperature

As temperature increases, time-dependent effects in both deformation and fracture processes occur which cause the cyclic lifetime to decrease. In the recent past, many researchers have investigated this problem in experiments which controlled the stress-strain response of a specimen, ie: a deformation approach was adopted. Strain Range Partitioning involving various components of creep, plasticity and elastic follow-up behaviour was introduced as was a Frequency-Modified Lifetime approach. In recent times however, attention has been focused on the effect of temperature on microstructural short-crack behaviour.

In [21] and [22] the authors consider the effect of temperature on the transition behaviour of cracks. For a given stress level in one-hour hold-time tests, in which a 316 AISI Stainless Steel was subjected to reversed bending, the lifetime is reduced primarily (but not only) due to an acceleration of the initial period prior to Stage II crack growth. Additionally the behaviour of long compressive hold-times could be compared to tensile hold times. In total, more than three thousand small cracks were examined. Small oxide patches on the surface produced many cracks and these could span grain boundaries. Additionally the sustained high temperature which created a lower yield stress could introduce time dependent (ie: creep) deformation processes. All of these effects led to the creation of longer, and an increased number of, surface cracks that rapidly coalesced to form a much earlier Stage II crack growth phase.

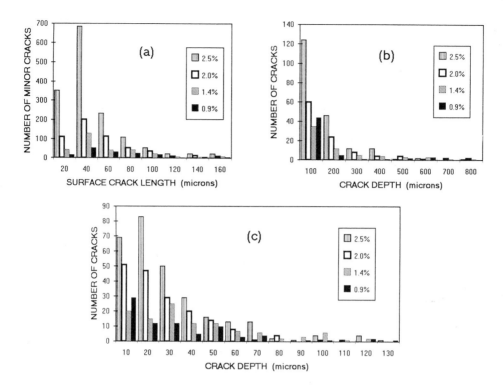

Fig. 10 The effect of strain range on the distribution of surface crack dimensions for specimens
      subjected to one-hour hold times
      (a) surface crack lengths of minor cracks on the tensile hold side
      (b) crack depths on the tensile hold side
      (c) crack depths on the compressive hold side

Figure 10 provides some information on the distribution of both minor and major cracks taken from
that study for four different strain range conditions. It can be seen that although only one deep crack
eventually causes failure, a great number of surface cracks are created. The major conclusion of this
study is that the higher the temperature, the longer the hold time then the sooner the transition phase is
by-passed.

SURFACE EFFECTS

Since most fatigue cracks initiate at a free surface, then the pre-fatigue surface condition is very important. This section considers the effect of the surface profile on transition crack growth and the effect of
surface treatments that attempt to prolong the initial crack growth period.

Surface Finish

Consider three different surface profiles generated by rough machining, grinding and polishing which
leave, respectively, surface micronotches of 25, 5 and 1 micron depth. Also consider the probability

of these micronotches creating or being equated to a surface crack that crosses a dominant microstructural barrier in two different materials, one with an average grain size of 50 μm and one with 5 μm grains. Let us say that the probability of finding a single surface grain that is twice the average grain size is the same for both materials, ie the grain which is the most likely to be the initiation site.

Simple calculations show that the transition phase is immediately eliminated should the fine-grained material have a rough surface finish. In this case LEFM parameters can be invoked to predict the lifetime of components manufactured to such stupid conditions. More complex probability calculations involving MFM are required to assess the effect of grinding on the lifetime of the finer grained material. Interestingly the effect of polishing will not have any effect on the lifetime of the coarse grained material and perhaps only a slight effect on the fine grained material.

It follows that Weibull statistics should be considered from the viewpoint of the Stage I to Stage II transition phase and involve experimental studies that examine the distribution of microstructural barriers to crack growth in relation to the MFM parameters given in Eq. (2). Some recent work [23] on this topic should be examined.

Surface Treatments

Shot-peening and cladding are two processes frequently introduced to improve a metals resistance to surface cracking.

Shot peening can greatly distort the surface texture and the previously planar path for crack growth at the initiation point. Additionally residual compressive stresses are introduced by shot peening but these can be eliminated at high temperatures and by cyclic stress relaxation processes. The most significant effect is considered to be, [20], the creation of a large number of distorted plastically deformed zones within the near surface grains and these new zones will probably extend to well beyond the depth of the original transition zone and the compressive residual stress field, ie. into the zone where the complementary tensile residual stress field should exist; see Fig 11 taken from [20]. It follows that the shot-peening process effectively introduces a greatly increased number of barriers of various orientations into the near-surface grains. This has two effects (i) they cause crack tip plastic zones to follow distorted orientations thereby adding to the already tortuous nature of the crack growth

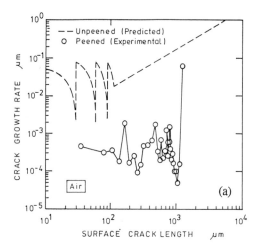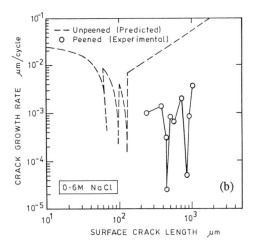

Fig. 11 Influence of shot-peening on short fatigue crack growth in a high strength steel
    (a) $\Delta\tau$ = 1050 MPa (AIR); N_f unpeened = 114,000 & N_f peened = 1.4 x 10^6 cycles
    (b) $\Delta\tau$ = 820 MPa (0.6 M NaCl) N_f unpeened = 107,000 & N_f peened = 270,000 cycles

path in the transition zone, and (ii) reduce the size of the crack tip plastic zone due to the increased yield stress of the work-hardened material.

Regretfully space does not permit a lengthy discussion on surface treatments such as nitriding, carburizing and cladding. When considering these treatments a distinction needs to be made between relocation of the zone of crack initiation and the effect of the introduction of a different microstructure. Frequently it is the different microstructure that is the major parameter affecting lifetime (crack growth mechanics). Cladding is usually introduced to resist other processes of deterioration leading to crack growth, ie. wear, erosion and surface corrosion, rather than the fatigue process itself [24].

CONCLUSIONS

This paper has tried to portray the transition-zone between different crack-growth processes as the important period of the fatigue lifetime of engineering materials. In this context fatigue damage is related to crack growth and the fatigue crack initiation period is regarded as non-existent, being replaced by a microstructural crack propagation phase which necessarily must be quantified by a Microstructural Fracture Mechanics (MFM) approach and not Linear Elastic Fracture Mechanics (LEFM).

Initiation and physically small crack growth behaviour of a metal is seen to be distinct from the crack growth behaviour of an engineering structure, even though the structure may be made from the same metal. Such a structure can be analysed by continuum damage mechanics by assuming that it already contains a defect significantly larger than the microstructural dimensions discussed in this paper.

REFERENCES / BIBLIOGRAPHY

[1]   K. J. Miller and K. P. Zachariah (1977). Cumulative damage laws for fatigue crack initiation and stage I propagation. *J Strain Analysis*, 12 (4), 262-270.

[2]   M. F. E. Ibrahim and K. J. Miller (1980). Determination of fatigue crack initiation life. *Fatigue of Engng. Mater. and Struct.* 2, 351-360.

[3]   The Behaviour of Short Fatigue Cracks (1986). Eds. K. J. Miller and E. R. de los Rios. EGF (ESIS) Publication 1. Mechanical Engineering Publications (Instn. Mech. Engs) London, 560 pages.

[4]   International Journal: Fatigue and Fracture of Engineering Materials and Structures. *Special Issue "Short fatigue cracks"* (1991) 14 (2/3) 143-372.

[5]   Short Fatigue Cracks (1992). Eds K. J. Miller and E. R. de los Rios. ESIS Publication 13. Mechanical Engineering Publications (Inst. Mech. Engs) London, 507 pages.

[6]   J. Z. Pan, P. J. Jenkins, E. R. de los Rios and K. J. Miller (1993). Detecting small fatigue cracks by acoustic microscopy. *Fatigue Fract. Engng. Mater. Struct.*. 16 (12) 1329-1337.

[7]   M. W. Brown and K. J. Miller, (1979). Initiation and growth of cracks in biaxial fatigue. *Fatigue Fract. Engng. Mater. Struct.* 1 (231-246).

[8]   P. J. E. Forsyth (1961). A two stage process of fatigue crack propagation, in Proc. Symp. on *"Crack Propagation"*. The College of Aeronautics, Cranfield. HMSO. London 76-94.

[9]   K. J. Miller, H. J. Mohamed, M. W. Brown and E. R. de los Rios (1986)., in *Small Fatigue Cracks* Eds. R. O. Ritchie and J. Lankford. The Metallurgical Society USA. 639-656.

[10] A. Navarro and E. R. de los Rios (1988). Short and long fatigue crack growth: a unified model. *Phil. Mag.* 57, 15-36.

[11] K. J. Miller (1991). Metal fatigue - past, current and future. *The John Player Lecture,* Proc. Inst. Mech. Engs. London, 205 (C5), 291-304.

[12] K. J. Miller (1993). Materials science perspective of metal fatigue resistance. *Materials Science and Technology* 9, 453-462.

[13] D. S. Dugdale (1960). Yielding in steel sheets containing slits. *J. Mech. Phys. Solids* 8, 100-104.

[14] B. Tomkins (1968). Fatigue crack propagation - an analysis. *Philos. Mag.* 18, 1041-1066.

[15] K. J. Miller (1993). The two thresholds of fatigue behaviour. *Fatigue Fract. Engng. Mater. Struct.* 16, 931-939.

[16] A. Turnbull and E. R. de los Rios. (1995). The effect of grain size on fatigue crack growth in an aluminium magnesium alloy. *Fatigue Fract. Engng. Mater. Struct.* 18, 1355-1366.

[17] A. Turnbull and E. R. de los Rios (1995). The effect of grain size on the fatigue of commercially pure aluminium. *Fatigue Fract. Engng. Mater. Struct.* To be published.

[18] J. D. Atkinson, Z. Chen and J. Yu (1995). A predictive model for corrosion fatigue crack growth rates in RPV steels exposed to PWR environments. in *"Fatigue and Crack Growth: Environmental Effects, Modeling Studies, and Design Considerations"*. Ed:- S. Yukawa. ASME Conference 1995. PVP - Vol 306, 3-18.

[19] R. Akid and K. J. Miller (1990). The initiation and growth of short fatigue cracks in an aqueous saline environment, in *Environment Assisted Fatigue*. Eds. P. Scott and R. A. Cottis. EGF (ESIS) Publication 7. Mechanical Engineering Publications (Instn. Mech. Engs.) London. 415-434.

[20] K. J. Miller and R. Akid (1996). The application of microstructural fracture mechanics to various metal surfaces. *Proc. Royal Soc.* To be published.

[21] N. Gao, M. W. Brown and K. J. Miller (1995). Crack growth morphology and microstructural changes in 316 stainless steel under creep-fatigue cycling. *Fatigue Fract. Engng. Mater. Struct..* 18. (December issue). To be published.

[22] N. Gao, M. W. Brown and K. J. Miller (1995). Short crack coalescence and growth in 316 stainless steel subjected to cyclic and time dependent deformation. *Fatigue Fract. Engng. Mater. Struct.* 18. (December issue). To be published.

[23] Y. Bergengren, M. Larsson and A. Melander (1995). The influence of machining defects and inclusions on the fatigue properties of a hardened spring steel. *Fatigue Fract. Engng. Mater. Struct.* 18, 1071-1087.

[24] A. A. Rakitsky, E. R. de los Rios and K. J. Miller (1994). Fatigue resistance of a medium carbon steel with a wear resistant thermal spray coating. *Fatigue Fract. Engng. Mater. Struct.* 17, 563-570.

STATISTICAL EVALUATION OF FATIGUE LIFE CONSIDERING SCATTER OF SMALL CRACK GROWTH RATE

N. KAWAGOISHI[*1], M. GOTO[*2], H. NISITANI[*3], T. TOYOHIRO[*4] and H. TANAKA[*1]

*1 Faculty of Engineering, Kagoshima University, Kagoshima 890, Japan
*2 Faculty of Engineering, Oita University, Oita 870-11, Japan
*3 Faculty of Engineering, Kyushu Sangyou University, Fukuoka 812, Japan
*4 Miyakonojo National College of Technology, Miyakonojo 885, Japan

ABSTRACT

The distribution characteristics of the growth rates of small crack and the relationship between the distribution of the growth rates of small crack and the one of fatigue lives were investigated through the small-crack growth law, $dl/dN = C_1 \sigma_a^n l$. Although the crack growth rate in a plain specimen fluctuates markedly due to the influence of the microstructure when a crack length is comparable with the size of microstructure, a scatter of the crack growth rates among the specimens can be evaluated through the distribution of the crack growth rates in the region where the growth behavior is mainly controlled by a mechanical parameter. Furthermore, the distribution of fatigue lives in plain specimens can be predicted through the statistical properties of the crack growth rates which are controlled by a mechanical parameter.

KEYWORDS

Fatigue; statistical property; life prediction; small crack; crack growth rate; small crack growth law.

INTRODUCTION

Fatigue properties exhibit a large scatter in general. Therefore, it is important to know the statistical properties and physical meanings of fatigue behaviors in reliable designs of machines and structures.
Since fatigue process is divided into the crack initiation process and its growth process, statistical analyses are necessary in each process. However, the growth process is more significant in service, because fatigue life is mainly controlled by the growth life of cracks, especially small cracks (Kawagoishi et al., 1992).

In the present study, rotating bending fatigue tests of two kinds of carbon steels were carried out on plain specimens and specimens with two small blind holes in order to investigate the distribution characteristics of the growth rates of small crack and the relationship between the distribution of the growth rates of small crack and the one of fatigue lives were investigated through the small-crack growth law, $dl/dN = C_1 \sigma_a^n l$ (Nisitani, 1981).

EXPERIMENTAL PROCEDURES

Materials used are rolled round bars of 0.15% and 0.42% carbon steels. Their chemical compositions (wt%) are 0.15C, 0.15Si, 0.41Mn, 0.014P, 0.008S and 0.42C, 0.27Si, 0.81Mn, 0.013P, 0.016S (remainder Fe), respectively. These materials were heat treated as shown in Table 1 to get a ferritic-pearlitic structure (F-P steel) and a ferritic-martensitic structure (F-M steel). Their mechanical properties are 377 MPa yield stress, 659 MPa tensile strength and 54 % reduction of area in F-P steel and 413 MPa, 650 MPa and 60 %, respectively in F-M steel.

Figure 1 shows the shape and dimensions of specimens. Fatigue tests were carried out using plain specimens and specimens with two small blind holes at the center of plain specimen. After machining, the specimens of F-P steel were re-annealed for 1 hr at 600°C to remove the residual stress. Furthermore, all the specimens were electro-polished to remove about 20μm from the surface layer in order to make the observation of the surface state easier. The crack length is measured by replication method and defined as the length in the circumferential direction along specimen surface. The distributions of fatigue lives and crack growth rates were examined using about fifteen specimens under each stress level. The machine used is the Ono type rotating bending fatigue testing machine (capacity 15N·m, frequency about 50Hz).

RESULTS AND DISCUSSION

First, the evaluation method of the growth rate of a small crack, its statistical treatment and the relationship between the distribution of the crack growth rates and the one of fatigue lives are investigated based on the results of a ferritic-pearlitic steel.

Figure 2 shows the crack growth curves for holed specimens. Although the relationship between ln$l$ and N among specimens shows a wide scatter, the l vs N relation in each specimen is approximated by a

Table 1. Heat treatment

Material	Structure	Heat treatment
S15C	Ferritic-pearlitic	840 °C 1h, F.C.
S45C	Ferritic-martensitic	900 °C 1h, F.C., 790 °C 0.5h, W.Q.

F.C.: Furnace cooled, W.Q.: Water quenched

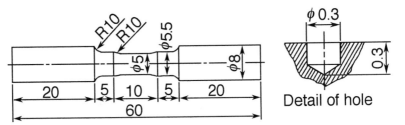

Fig. 1. Shape and dimensions of specimens

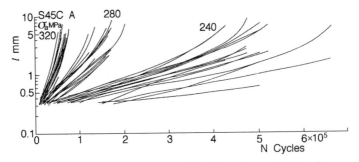

Fig. 2. Crack growth curves of holed specimen (F-P steel)

straight line under a constant stress level when the crack length is smaller than 1~2 mm. That is, the crack growth rate is nearly proportional to the crack length, and the crack growth rates can be evaluated by the small-crack growth law, Eq.(1) in this material (Nisitani et al., 1985).

$$dl/dN = C_1 \sigma_a^n l \tag{1}$$

Therefore, the crack growth rate for each specimen can be expressed by the following equation under a constant stress level.

$$dl/dN = C_2 l \tag{2}$$

Consequently, the distribution of crack growth rates can be evaluated by the distribution of constant C_2 in Eq.(2).

It is reported that the distribution of fatigue data fits to the log-normal distribution or the Weibull distribution in many studies on the statistical analysis of fatigue data (Goto, 1993).
Figure 3 shows the distribution of constant C_2 in Eq.(2) plotted on the log-normal probability paper. The distribution of constant C_2 fits to the log-normal distribution.
The results of Fig.3 and Eq.(2) suggest that the distribution of crack growth lives may be estimated by Eq.(3).

$$N_{l_1 \to l_2} = \frac{1}{C_2} \ln \frac{l_2}{l_1} \tag{3}$$

where, $N_{l_1 \to l_2}$ is the growth life of crack from l_1 to l_2
Figure 4 shows the estimated distribution of crack growth lives by Eq.(3) and the measured one. As seen from Fig.4, the distribution of crack growth lives can be estimated by the one of constant C_2, that is, Eq.(3).

In the second phase, we consider about the relationship between the distribution of the growth rates of small crack and the one of fatigue lives in plain specimens, because the crack growth behavior in a plain specimen is strongly influenced by microstructure (Kawagoishi et al., 1993).

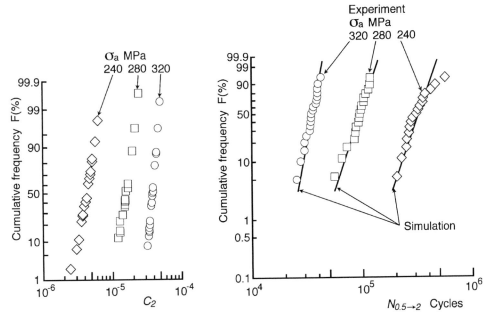

Fig. 3. Distribution of C_2
(log-normal probability, F-P steel)

Fig. 4. Estimated distribution of crack growth lives
(F-P steel)

Figure 5 shows S-N curve for plain specimens. Fatigue lives also exhibit a wide scatter.
Figure 6 shows the Weibull plots of fatigue lives. The distribution of fatigue lives follows the Weibull distribution.
Figure 7 shows the relationship between $\ln l$ and N/N_f for plain specimens, where N_f is the number of cycles to failure. Most of fatigue life is occupied by the growth life of a small crack and the growth life of crack from 50μm to 1mm, $N_{0.05\to1}$, where Eq.(2) holds, is roughly 50% of the fatigue life N_f similar to the results in many metals (Nisitani et al., 1992). That is,

$$N_f = 2N_{0.05\to1} \tag{4}$$

From Eqs.(3) and (4), the following equation is obtained.

$$N_f = (2/C_2)\ln(1/0.05) = 6/C_2 \tag{5}$$

This relation means that the distribution of fatigue lives can be estimated by the one of crack growth rates, i.e. C_2 in holed specimens, and the reverse is also true.

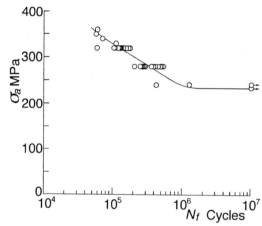

Fig. 5. S-N curve of plain specimens
(F-P steel)

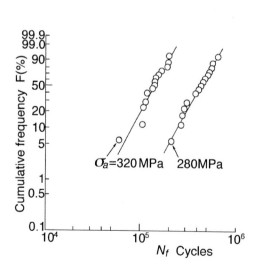

Fig. 6. Distribution of fatigue lives
(Weibull distribution, F-P steel)

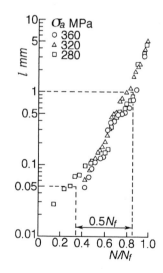

Fig. 7. $\ln l$ and N/N_f relation for plain specimen
(F-P steel)

Figure 8 shows the Weibull plots of N_f estimated by Eq.(5) considering the distribution of C_2 shown in Fig.3. In this figure, the measured distribution of N_f is also shown by symbols. The estimated distribution of N_f is in good agreement with the measured one.

Next, the validity of the statistical evaluation method of the fatigue lives stated above is confirmed in a ferritic-martensitic steel.
Figure 9 shows the distribution of constant C_2 plotted on the log-normal probability paper.
Figure 10 shows the Weibull plots of N_f estimated by Eq.(5) considering the distribution of C_2 shown in Fig.9. The estimated distribution of N_f is in good agreement with the measured one. These results suggest that the distribution of crack growth rates can be also estimated by the one of fatigue lives which is accumulated widely as a data base, and this is very convenient in practice.

CONCLUSIONS

Rotating bending fatigue tests were performed on plain specimens and specimens with two small blind holes of two kinds of carbon steels in order to investigate the fatigue life statistically considering a scatter of small crack growth rates. The crack growth rates are expressed by the small-crack growth law, $dl/dN = C_1 \sigma_a^n \, l$. Furthermore, the growth law is re-written to the relation of $dl/dN = C_2 \, l$ and the relationship between the fatigue life and the constant C_2 is expressed by the equation, $C_2 \, N_f = 6$, for plain specimens of 5mm diameter when a stress level is constant. Consequently, the distribution of fatigue lives can be estimated by the one of crack growth rates, and the reverse is also true.

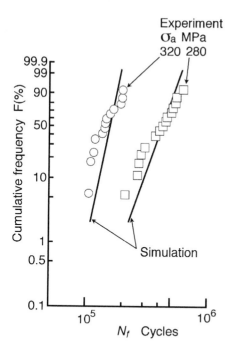

Fig. 8. Estimated distribution of fatigue lives
(F-P steel)

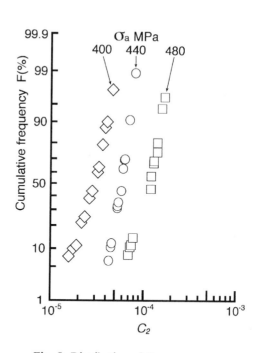

Fig. 9. Distribution of C_2
(log-normal probability, F-M steel)

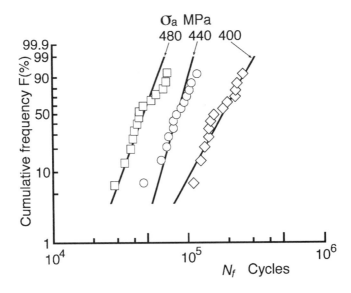

Fig. 10. Estimated distribution of fatigue lives (F-M steel)

REFERENCES

Kawagoishi, N., Nisitani, H. and Toyohiro, T. (1992). Minimum fatigue crack length for the application of small-crack growth law. JSME Int. Ser. I, 35, 234-240.

Nisitani, H. (1981). Unifying treatment of fatigue crack growth in small, large and non-propagating cracks. Mechanics of Fatigue-ASME AMD, 47, 151-166.

Nisitani, H. and Kawagoishi, N. (1985). Characteristics of fatigue crack growth in quenched and tempered 0.45 % carbon steel. Trans. JSME, A, 51, 1668-1676.

Goto, M. (1993). Scatter in small crack propagation and fatigue behaviour in carbon steel. Fatigue Eng. Mater. Struct., 16, 795-810.

Kawagoishi, N., Nisitani, H. and Toyohiro, T. (1993). Effect of microstructure on the characteristics of small fatigue crack growth in dual-phase steels. JSME Int. Ser. I, 36, 126-133.

Nisitani, H., Goto, M. and Kawagoishi, N. (1992). A small-crack growth law and its related phenomena. Eng. Frac. Mech., 41, 499-513.

PREDICTION OF NON-PROPAGATING FATIGUE CRACK BEHAVIOUR IN ALUMINUM ALLOYS

A. Plumtree and A. Varvani-Farahani

Department of Mechanical Engineering
University of Waterloo, Waterloo, ON Canada N2L 3G1

ABSTRACT

The non-propagating behaviour of short fatigue cracks in an aluminum 2024-T351 alloy plate was investigated with respect to the three major directions. The largest number of cracks was observed in the longitudinal direction which initiated at cracked Al_7Cu_2Fe particles. By comparison, non-propagating cracks formed in the matrix of the transverse and short-transverse specimens. The profile of the cracks was found to be independent of direction in spite of the large differences in grain size. A model based on surface strain redistribution and the development of crack closure was applied to successfully predict the endurance limit and existence of non-propagating cracks.

KEYWORDS

Non-propagating fatigue cracks, crack initiation stress, strain distribution factor.

INTRODUCTION

The effect of microstructure on the fatigue crack behaviour at low stress levels in the vicinity of the endurance limit cannot be ignored, particularly in understanding the initiation stress and behaviour of non-propagating fatigue cracks. A short fatigue crack growth model based on surface strain redistribution and the development of crack closure was proposed by Abdel-Raouf, *et al (1991, 1992)*. This model incorporated the grain size to explain the existence and length of non-propagating fatigue cracks. The present work has been carried out on non-propagating cracks in an Al 2024-T351 alloy to assess the applicability of the proposed model.

EXPERIMENTAL PROCEDURE

A 2024 aluminum alloy was received in the form of a 76mm thick plate which had been solution heat treated at 495°C, then stretched to about 1.5 percent strain at the mill and naturally aged (T351 condition). The average grain sizes in the rolling or longitudinal (L-), transverse (T-) and short transverse (S-) directions were approximately 415µm, 158µm and 34µm respectively.

For cyclic testing, flat waisted specimens 76 x 29 x 2 mm were machined in six different directions - LT, TL, TS, ST, SL and LS. The first letter indicates the specimen orientation with respect to the rolling direction and the second gives the thickness and hence crack depth direction, again with respect to the rolling direction. After machining, the specimens were mechanically polished to a fine finish (0.25µm diamond). The tests were performed using a two tonne Amsler Vibrophore Fatigue Testing Machine type 10HF422 operating under stress control at a frequency of 90-95 Hz. All specimens were cycled under a constant amplitude stress (minimum to maximum stress ratio, R = -1).

Following cyclic testing the specimens were carefully examined and if any cracks were present, their lengths were measured using a film replication technique and direct observation by optical microscope. Different specimens possessed lengths and number of cracks which were recorded. Crack depths and profiles were measured using a Confocal Scanning Laser Microscope (CSLM) and also by following the crack length during removal of successive surface layers.

RESULTS

Considering the endurance limit stresses after 10^7 cycles, the LT, LS and ST specimens possessed the lowest values (138 MPa). The TL and TS specimens possessed intermediate values (141 and 146 MPa) whereas the highest endurance limit (148 MPa) was recorded with the SL specimens.

Microscopic observation revealed that the initiation stress at which a crack was first detected after 10^7 cycles was dependent on specimen direction. In all cases, the crack initiation stress was lower than the endurance limit. The initiation stresses for the LT, TL, TS, ST, SL and LS specimens were 107, 131, 134, 131, 121 and 117 MPa respectively. In the LT and LS specimens surface cracks initiated at Al_7Cu_2 Fe particles, while in the other specimens nucleation occurred within the matrix. Fig. 1 shows typical cracked particles and accompanying non-propagating

Fig. 1. Cracked particles with non-propagating
fatigue cracks. Stress axis vertical.
LT specimen. Magnification x 1000.

cracks in a longitudinal specimen (LT). Grain boundaries were the major barriers to the growth of all non-propagating cracks. The length of most of these cracks was generally less than one-grain and only a few cracks were as long as, or exceeded, two grains. These longer cracks resulted from the coalescence of smaller cracks.

Different specimen directions displayed various crack populations and dimensions. In the LT and LS directions the number of cracks was highest due to the cracks in the particles lying normal to the loading axis. In addition to the greater possibility of fatigue crack initiation, subsequent coalescence of fatigue microcracks was more likely. Microscopic observation revealed that the number of cracks in the LS and LT directions was 25 and 60 respectively. The LS and LT specimens possessed non-propagating crack lengths ranging from 2 to 435 µm and 4 to 70µm respectively as the stress level was increased from the initiation stress to the endurance limit, which, despite the presence of cracked particles, was just below that for the other specimens. The ST and SL specimens contained 13 to 24 and 3 to 10 cracks with lengths of 2 to 17µm and 4 to 67µm respectively as the stress level was increased from the initiation level to the endurance limit. The fewest cracks were observed in the TS and TL specimens. These were found to have 6 to 8 and 3 to 4 cracks with a range of crack lengths of 5 to 77µm and 5 to 72µm respectively. It was observed that the larger surface grains possessed the greatest potential for crack initiation and it is interesting to note that all the grains containing cracks had a higher surface area than the average. Because of less constraint, these larger surface grains facilitated persistent slip band and subsequent extrusion formation.

Using CSLM, the aspect ratio (crack depth, a, to total surface crack length, 2c) was found to be constant at 0.42 and independent of specimen direction for cracks within the first surface grain. This result was verified by successively removing a known surface layer thickness and measuring the corresponding crack length. Both surface layer removal and CSLM techniques showed that the first stage of fatigue crack growth occurred along an inclined plane of 40°-50° to the axis of the specimen.

DISCUSSION

The behaviour of non-propagating fatigue cracks based on surface strain redistribution and the development of crack closure related to the microstructure of the material has been described by Abdel-Raouf *et al (1991, 1992)*. This model also has the potential to explain the existence of non-propagating cracks and their depths at different stress levels below the endurance limit.

Because of increasing constraint, the local strain range, $\Delta\varepsilon$, decays with depth from the surface eventually reaching the nominal strain range, Δe, at a depth of about five grain diameters. The local strain range variation with crack depth can be expressed through the strain distribution factor $Q_\varepsilon (=\Delta\varepsilon/\Delta e)$ by

$$Q_\varepsilon = 1 + q \exp(-\gamma a) \tag{1}$$

where q is a constant (=5.3), γ is the reciprocal of the grain size (D) and a is the crack depth.

The intrinsic nominal stress range (ΔS_{ith}) may be calculated using

$$\Delta S_{ith} = \frac{\Delta K_{ith}}{Q_\varepsilon (\pi a)^{1/2}} \tag{2}$$

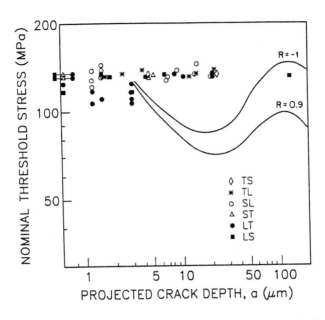

Fig. 2. Variation of stress amplitude with crack depth

where ΔK_{ith} is the intrinsic stress intensity factor range at a stress ratio (R=minimum to maximum stress) of 0.9 when closure is absent and F is the crack shape factor.

Applying a closure factor, H_{cl} *(Abdel-Raouf et al, 1991)*, the threshold stress range (ΔS_{th}) may be determined using

$$\Delta S_{th} = \Delta S_{ith} \, H_{cl} \qquad (3)$$

The values of threshold stress range at stress ratios of R=0.9 (ΔK_{ith}) and R=-1 were calculated using equations (2) and (3) and plotted against crack depth in Fig. 2. For a minimum crack depth of 3 µm the nominal threshold stress (128 MPa) for a stress ratio of R=-1 (D=34 µm) is approximately the same as that experimentally observed when cracked particles were not instrumental in fatigue crack initiation (121 to 134 MPa).

Experimentally, the endurance limits for the different LT, LS, ST, SL, TL and TS specimens were 138, 138, 138, 148, 141 and 146 MPa respectively. According to the model, the endurance stress should be 146 MPa for R=-1, corresponding to a crack depth of 135 µm (about four grain diameters). The experimental results for all the specimens are in very good agreement with those predicted, provided the smallest average grain size (i.e. 34µm) is used in the model. This indicates that the first grain boundary encountered by a persistent slip band crack controls the strain redistribution.

CONCLUSIONS

1)   Since numerous cracked elongated particles were favourably oriented for fatigue crack initiation in the L-direction, these specimens contained the greatest number of non-propagating cracks. Although fatigue cracks initiated at the lowest observed stress amplitudes, the corresponding endurance limits were only slightly lower than those for the

S- and T- direction specimens.

2)   Although grain boundaries were the major barriers to growth, the aspect ratio of the non-propagating cracks was independent of grain size which varied with specimen direction.

3)   A non-propagating short fatigue crack model based on strain redistribution and crack closure development was used to successfully predict the initiation stress, endurance limit and length of non-propagating cracks. When applying this model, the smallest average grain size should be considered.

ACKNOWLEDGEMENTS

This research has been supported by the Natural Sciences and Engineering Research Council of Canada through Grant 0002770. A.V-F wishes to thank the Ministry of Culture and Higher Education of Iran for a scholarship. The authors would like to express their thanks to Marlene Dolson and Dianne Hause for typing the manuscript.

REFERENCES

1.   Abdel-Raouf, H., Topper T.H. and Plumtree, A., (1992), A Model for the Fatigue Limit and Short Crack Behaviour Related to Surface Strain Redistribution, *Fatigue Fracture Engineering Material Struct.* **15**, 895-909.
2.   Abdel-Raouf, H., Topper, T.H. and Plumtree, A.,(1991), A Short Fatigue Crack Model Based on the Nature of the Free Surface and Its Microstructure, *Scripta Metall. et Mat.*, **25**, 597-602.

LOW CYCLE FATIGUE OF ZIRCALOY-4

Achour BELOUCIF and Jacques STOLARZ

Ecole Nationale Supérieure des Mines, Centre SMS, URA CNRS 1884
158, cours Fauriel
42023 SAINT-ETIENNE CEDEX 2
FRANCE

ABSTRACT

Low cycle tension-compression fatigue tests under plastic strain control were performed on annealed and recrystallized Zircaloy-4. The fatigue damage is decribed in terms of evolution of surface microcracks. First microcracks appear at about 25% of fatigue life e.g. much later than in b.c.c. and f.c.c. alloys. A full description of microcrack evolution before volumic crack growth is given for both heat treatments and it is related to the alloy's microstructure and texture. The results are compared with low cycle fatigue studies on cubic alloys.

KEYWORDS

Zircaloy-4; low cycle fatigue; surface microcracking.

INTRODUCTION

Power fluctuations in pressurized water reactors (PWR) cause cyclic strains in the Zircaloy-4 cladding of the fuel rods. The complete analysis of damage of Zr-4 claddings must take into account environmental parameters, in particular corrosion-deformation interactions (CDI) due mainly to the presence of products of nuclear fission. In order to understand complex CDI processes it is necessary to have knowledge about elementary mechanisms of mechanical damage by cyclic straining. However, the fatigue studies of Zircaloys concerned mainly interactions between fatigue and creep (O'Donnel and Langer, 1964, Bocek *et al.*, 1986) at elevated temperatures (350°C and higher) and damage mechanisms in iodine containing environments (Nakatsuka and Hayashi, 1981, Schuster and Lemaignan, 1989). On the other hand, no systematic study of low cycle fatigue damage in an inert environment has been undertaken. Such studies allow to obtain informations about elementary damage processes and to understand the material damage under more complex conditions. In particular, recent studies (Magnin and Bataille, 1994) show that systematic investigations of surface damage during very early stages of low cycle fatigue tests can give access to macroscopic parameters of the fatigue process like the Coffin-Manson relation. Up to now, studies of this kind were undertaken only on cubic alloys, mainly stainless steels. Therefore, it seems very interesting to analyse the surface damage processes in a hexagonal alloy like Zircaloy-4 and to compare the results with those obtained on b.c.c. and f.c.c. metals. On the other hand, the knowledge of elementary damage mechanisms through cyclic straining in an inert environment is supposed to constitue a basis for a better understanding of corrosion-deformation interactions in PWR.

EXPERIMENTAL METHOD

Experiments were carried out on an industrial Zircaloy-4 (supplied by FRAMATOME) containing 1.5% Sn, 0.20% Fe and 0.11% Cr (wt. %).

Smooth fatigue specimens (ϕ 6 mm) were given an annealing treatment at 505°C during one hour, followed for a part of samples by a recrystallizing treatment at 700°C during two hours under argon. Corresponding microstructures are shown in Figure 1.

Compared to the annealed alloy with its elongated grains (Fig.1a), the recrystallized structure is composed of equiaxed grains of 20-25 μm diameter (Fig.1b).

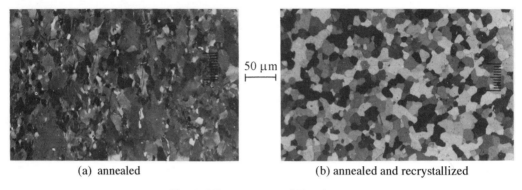

50 μm

(a) annealed   (b) annealed and recrystallized

Fig. 1. Microstructure of Zircaloy-4

Both annealed and recrystallized alloys exhibit a strong texture with (1 0 -1 1) planes perpendicular and basal (0 0 0 2) planes parallel to the rod axis. This texture is much more pronounced in the annealed alloy (Fig.2).

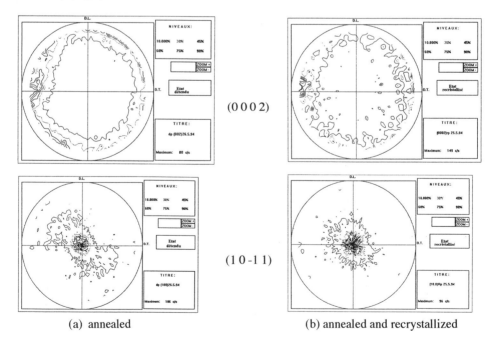

(0 0 0 2)

(1 0 -1 1)

(a) annealed   (b) annealed and recrystallized

Fig. 2. Crystallographic texture of Zircaloy-4

Table 1. Mechanical characteristics of Zircaloy-4

	Re app. (MPa)	Re 0.2 (MPa)	Rm (MPa)	A %	Z %
Annealed	400	510	675	15.6	23
Recrystallized	250	360	500	16	26

The fatigue tests were performed in tension-compression under symmetrical plastic strain control ($\Delta\varepsilon_p/2 = \pm 2 \cdot 10^{-3}$), at constant strain rate ($d\varepsilon/dt = 10^{-3}$ s^{-1}) using a servohydraulic machine. All tests were carried out under vacuum ($< 10^{-3}$ Pa) because of a very important decrease of fatigue resistance of Zircaloy-4 in ambient air.

At first, cyclic consolidation curves were registered and fatigue life N_r determined for both heat treatments. After that, another samples were given various number of fatigue cycles between 10 and 95% N_r. Surface microcrack initiation and evolution was then observed with a scanning electron microscope.

RESULTS

Fatigue life and crack propagation mode

Figure 3 shows cyclic consolidation curves (evolution of the tension peak stress) for both annealed and recrystallized Zircaloy-4. Fatigue life was determined as corresponding to a decrease of the tension peak stress of 10%.

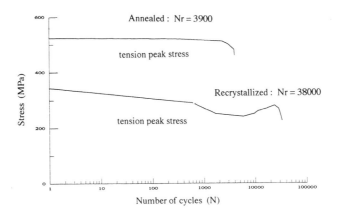

Fig.3. Evolution of peak stresses during low cycle fatigue test

Under given testing conditions, fatigue life of the annealed Zircaloy-4 is of about 3900 cycles. A slight softening is observed during the whole test. For the recrystallized alloy fatigue life is of about 34000 cycles. Until 3000 cycles a softening occurs in a similar way as in the annealed Zircaloy. Between 3000 and 7000 cycles, peak stress stabilizes, after what a slight hardening takes place.

Initiation and growth of surface microcracks

Low cycle fatigue studies of b.c.c. and f.c.c. metals and alloys show that superficial microcracking can occur during the major part of lifetime of a sample. For example, in stainless steels, first microcracks appeare already at about 10% N_r (Magnin et al., 1985).

According to the classification proposed for cubic alloys (Magnin *et al.*, 1985), four different types of microcracks are considered. Their length is related to the mean grain diameter (about 25 μm for the recrystallized Zircaloy-4). For the annealed alloy, 25 μm represents the average grain width (Fig.1a). Therefore, for Zircaloy-4, four types of microcracks have following dimensions:

$$\text{Type I:} \qquad l_I < 25 \ \mu m$$
$$\text{Type II:} \qquad 25 \ \mu m < l_{II} < 50 \ \mu m$$
$$\text{Type III:} \qquad 50 \ \mu m < l_{III} < 250 \ \mu m$$
$$\text{Type IV:} \qquad l_{IV} > 250 \ \mu m$$

The length of type I microcracks does not exceed the average grain diameter. They can be created inside the grains or at grain boundaries. Type II and III microcracks are formed through coalescence of type I microcracks in adjacent grains or through their growth at the surface. Volumic propagation becomes possible for type IV cracks.

Surface SEM observations after initial stages of fatigue tests (10 to 30% N_r) show that for both heat treatments first type I microcracks nucleate at intersections of slip lines/bands with sample surface near to 25%N_r e.g. much later, in terms of N_{nucl}/N_r, compared to cubic alloys (Magnin *et al.*, 1985). After initiation, type I surface microcrack density d increases very rapidly. In the annealed alloy, d_I reaches about 70 per mm^2 after 30%N_r. Type II crack formation through coalescence and/or growth of smaller ones occurs almost instantaneously after initiation.

Global evolution of microcrack densities is presented in figure 4. In spite of the difference between peak stress levels, one observes similar evolution of microcracking for both heat treatments:
-      type I crack initiation at 25%N_r
-      rapid coalescence and growth of type I microcracks
-      creation of type III cracks at about 45%N_r.
The main difference between the two alloys is those of crack densities. For instance, maximal value of type I crack density reaches 125 per mm^2 for annealed Zicaloy and only about 70 per mm^2 after recrystallization (Fig.4).

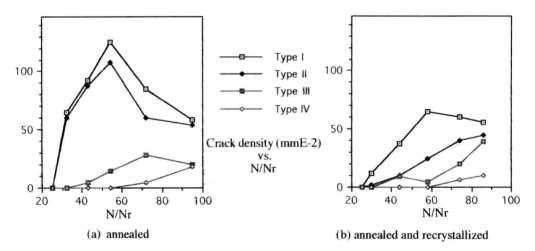

(a) annealed   (b) annealed and recrystallized

Fig. 4. Evolution of densities of surface microcracks in low cycle fatigue tests

Surface distribution of microcracks is not isotropic. Figure 5a shows that a regular microcrack network with overall orientations perpendicular vs. sample axis can be observed in the annealed Zircaloy-4. Although this tendency is less pronounced after recrystallizing (Fig. 5b), general orientation of longer cracks (type III and IV) remains unchanged. The propagation mode is mainly transgranular, with some intergranular crack segments in the recrystallized alloy.

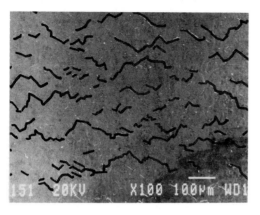

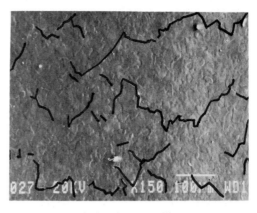

(a) annealed   (b) annealed and recrystallized

Fig. 5. Surface microcracking in Zircaloy-4 at 86% N_r

DISCUSSION AND CONCLUSION

The aim of this work was to study damage mechanisms in low cycle fatigue under plastic strain control of Zircaloy-4 through analyzing surface microcrack evolution at different stages of the fatigue test. Zircaloy-4 has been investigated after annealing and recrystallization.

Following mechanisms of surface fatgue damage have been found:
- the type I (shorter from the mean grain diameter) microcrack initiation occurs for both alloys after a number of cycles which corresponds to about 25 per cent of fatigue life;
- a quasi instantaneaous evolution of type I to type II cracks is observed for both heat treatments;
- the initiation and propagation of surface microcracks are mainly transgranular;
- type III and type IV cracks are oriented perpendicularly vs. sample axis and form a regular network over the whole surface.

The reason for a creation of an anisotropic network of cracks in Zircaloy seems to be related to the crystallographic texture of the alloy with (0002) planes parallel to the sample axis. In this way, first type I microcracks which nucleate at slip lines have necessarily almost perpendicular orientations vs. sample axis. Since propagation mode is mainly transgranular, after grain boundary crossing, microcracks propagate along another slip system with an orientation close to the first one.
In the recrystallized Zircaloy, with its equiaxed grains and less texture (Fig.2b), type I microcracks can have more random orientation, according to the orientation of slip systems vs. sample axis. Moreover, as propagation can, in this case, have a mixed character, microcrack growth and coalescence can occur in accord to the local orientation of grain boundaries and slip systems in adjacent grains which leads to a more isotropic microcrack distribution than in the annealed Zircaloy.

It is interesting to compare evolutions of microcracks in Zircaloy-4 and in a 316L stainless steel strained under plastic strain control (Magnin *et al.*, 1985). Figure 6 represents densities of type I and type II microcracks in Zircaloy-4 (annealed and recrystallized) and in 316L SS tested under plastic strain control at the same amplitude of plastic strain.

The main difference between both curves concerns the microcrack initiation which takes place at about 10-15% N_r for the 316L and only at 25% N_r for the Zircaloy-4. This difference is not due to peak stress levels in both alloys. For given testing conditions, the maximal peak stress in tension reaches 150 MPa in 316L against 350 in recrystallized and 550 MPa in the annealed Zircaloy-4. In spite of this, crack initiation occurs much earlier (in terms of N/Nr) in the stainless steel than in the

Zircaloy. On the other hand, N_{init}/N_r ratios are the same for both annealed and recrystallized Zircaloy. Further studies on other h.c. metals and alloys will be necessary to determine if the result obtained is a particularity of the hexagonal structure.

On the other hand, it is interesting that microcrack densities measured during low cycle fatigue tests of 316L and of Zircaloy-4 are very similar to each other in spite of all differences of microstructure and mechanical parameters of testing.

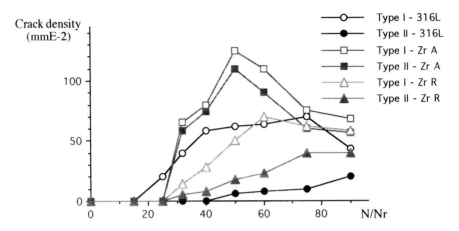

Fig. 6. Comparison of microcracks evolution in 316L stainless steel and in Zircaloy-4

Acklowledgements: Thanks are owed to Mr. J. Joseph, from Société FRAGEMA, Lyon - France, for the provision of material and helpful discussions

REFERENCES

Bocek M., Alvarez-Armas I., Armas A.F. and Petersen C. (1986). Low cycle deformation behaviour of Zircaloy-4 at elevated temperatures. *Z. Werkstofftech.*, **17**, 317-327.
O'Donnel W.J. and Langer B.F. (1964). Fatigue design basis for Zircaloy components. *Nuclear Science and Engineering*, **20**, 1-12.
Magnin T. and Bataille A. (1994). Comparison of the surface cracking process in uniaxial and multiaxial fatigue. *Proceedings: Fourth International Conference on Biaxial/Multiaxial Fatigue.* Paris, France May 31 - June 3, 1994.
Magnin T., Coudreuse L. and Lardon J.M. (1985). A quantitative approach to fatigue damage evolution in FCC and BCC stainless steels. *Scripta Metallurgica*, **19**, 1487.
Nakatsuka M. and Hayashi Y. (1981). Effect of iodine on fatigue properties of Zircaloy-2. Fuel cladding under cyclic tensile stress. *Journal of Nuclear Materials*, **18**, 785-792.
Schuster I. and Lemaignan C. (1989). Characterization of Zircaloy corrosion fatigue phenomena in an iodine environment. *Journal of Nuclear Materials*, **166**, 348-363.

BEHAVIOUR OF SMALL CRACKS IN A 2017-T4 ALUMINUM ALLOY

M. GOTO[1], N. KAWAGOISHI[2], H. NISITANI[3] and H. MIYAGAWA[1]

[1]:Department of Mechanical Engineering, Oita University, Oita 870-11, Japan
[2]:Department of Mechanical Engineering, Kagoshima University, Kagoshima 890, Japan
[3]:Faculty of Engineering, Kyushu-Sangyo University, Fukuoka 812, Japan

ABSTRACT

Since fatigue life of a plain specimen is controlled by the propagation life of a small crack, it is crucial to estimate the growth behaviour of a small crack for design of machines and structures. In this study, using an aluminum alloy 2017-T4, rotating bending fatigue tests of plain specimens were carried out. The initiation and propagation behaviour of a small crack was clarified based on the successive observation of surface.

KEYWORDS

Fatigue; crack propagation; small-crack growth law; crack length distribution; plain specimen

INTRODUCTION

Since fatigue life of a plain specimen is controlled by the propagation life of a small crack, it is crucial to estimate the growth behaviour of a small crack for design of machines and structures. The growth rate of small cracks cannot be treated by the Linear Elastic Fracture Mechanics, but it can be determined by a term $\sigma_a^n l$, where σ_a is the stress amplitude and l is the crack length (Nisitani *et al.*,1987). On the other hand, aluminum alloy is widely used as structural material. Most of studies for Al alloy have related to the long crack propagation in which the small scale yielding condition holds. However, the studies concerning the small crack initiation and propagation behaviour are not abundant.

In this study, using an aluminum alloy 2017-T4, rotating bending fatigue tests of plain specimens were carried out. The initiation and propagation behaviour was observed by the plastic replica technic. In order to study the change in crack length distribution due to the stress cycling, the behaviour of all the cracks beyond 10 μ m initiated within the specific region whose area is 10-20 mm^2 were examined. The crack length data was analyzed by assuming the Weibull distribution.

EXPERIMENTAL PROCEDURES

The material was an age-hardened Al alloy 2017-T4 in the form of rolled bar about 20 mm in diameter. The chemical composition (wt %) was 3.99 Cu, 0.52 Si, 0.18 Fe, 0.45 Mn, 0.61 Mg, 0.01 Cr, 0.04 Zn, 0.02 Ti, remainder aluminum. The mechanical properties were 303 MPa proof stress, 464 MPa ultimate tensile strength and 43.7 % reduction of area.

All the tests were carried out using a rotating bending fatigue machine operating at 3,000 rev/min. Specimens used in the tests were plain specimens with 8 mm in diameter. Before testing, all the specimens were electropolished to remove about 25 μ m from the surface layer in order to facilitate observations of the surface state. The observations of fatigue damage on the specimen surface and the measurement of crack length were made via plastic replicas using optical microscope at a magnification of $\times 400$. The crack length, l, is the length along the circumferential direction on the specimen surface. The stress value referred to is that of the nominal stress amplitude, σ_a, at the minimum cross section.

EXPERIMENTAL RESULTS AND DISCUSSION

Crack initiation and propagation behaviour

Figure 1 shows the S-N curve. The fatigue strength at 10^7 cycles was 118 MPa. In the present section,

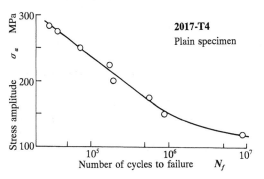

Fig.1   S-N curve of plain specimen

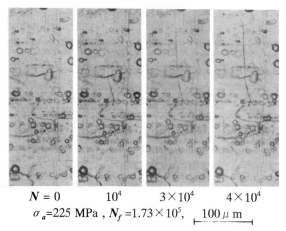

$N = 0$ 10^4 3×10^4 4×10^4
σ_a=225 MPa , N_f =1.73$\times 10^5$,   100μ m

Fig.2   Change in surface states around a major crack

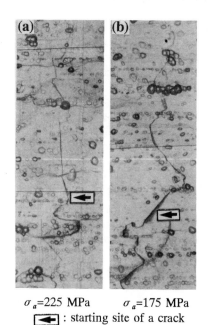

σ_a=225 MPa σ_a=175 MPa
⬅ : starting site of a crack
100μ m

Fig.3   Crack growth path;
(a) σ_a=225 MPa, and
(b) σ_a=175 MPa

in order to investigate the physical basis of fatigue damage, the behaviour of major crack which led to the final fracture of a specimen was examined at each stress amplitude.

Figure 2 shows the change in surface state around the initiation site of a major crack. The stress amplitude is 225 MPa. At the early stages of cycling, a shear microcrack initiates within a grain. The dimension of a crack is extremely small when compared to the grain size. After initiation, a shear crack propagates up to the grain boundaries. A change in propagation direction occurs at grain boundaries. When the contiguous grains have the suitable slip plane, a crack continues growing with shear mode in the grains. However, a crack propagates along the grain boundaries, when the slip direction of the contiguous grains is inconvenient for shear crack growth.

Figure 3 shows the fatigue crack growth path. (a) is at 225 MPa and (b) is at 175 MPa. There are significant differences in crack growth path between at 225 MPa and 175 MPa, i.e. the growth path at a low stress amplitude exhibits remarkable large zigzag pattern when compare to a large stress amplitude. At a large stress amplitude, although the microstructure affects strongly the growth behaviour of extremely small crack (l < 0.1 mm), the effect of microstructure decreases gradually as a crack grows longer and the tensile mode controls growth behaviour instead of shear mode. At a small stress amplitude, the growth path of comparatively large cracks is also affected by the microstructure. The shear mode propagation was observed frequently in large cracks in excess of 1 mm. At a small stress amplitude, it may be difficult to turn into the tensile mode from shear mode because of small driving force in crack growth. Moreover, the large zigzag growth pass tends to suppress an increase in the driving force because of the crack surface roughness. At a low stress amplitude, thus, a large crack propagates with shear mode and exhibits large zigzag growth pass.

Figure 4(a) shows the crack propagation curve (ln l versus N relation). When the stress amplitude exceeds 225 MPa, each relation in the range l > 0.1 mm can be approximated by a straight line. However, the shape of growth curves at 175 MPa exhibits the upward convex type. Figure 4(b) shows the ln l versus N/N_f, relative number of cycles, relation. The relation at the stress above 225 MPa can be represented approximately by a straight line independent of stress amplitude, whereas the relation at 175 MPa can be shown by a convex curve. This indicates that the crack growth characteristics depends on the magnitude of stress amplitude.

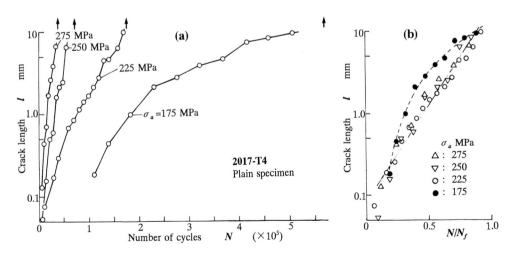

Fig.4   Crack growth data; (a) ln l vs N relation, and (b) ln l vs N/N_f relation

Small crack growth law of 2017-T4

It has been widely known that the growth rate of a long crack in which the condition of small scale yielding is satisfied can be treated by the Linear Elastic Fracture Mechanics. On the other hand, the results of fatigue tests for various carbon steels and low alloy steels have suggested that the growth rate of a small crack has been determined uniquely by the generic term $\sigma_a^n l$ (Nisitani *et al.*, 1992; Goto *et al.*, 1994), not by the stress intensity factor range, ΔK. Figure 5 shows the *dl/dN* versus ΔK relation ($\Delta K = \Delta \sigma \sqrt{\pi l}$). ΔK is effective parameter for determining the growth rate of long cracks, however *dl/dN* of small crack cannot be determined uniquely by ΔK.

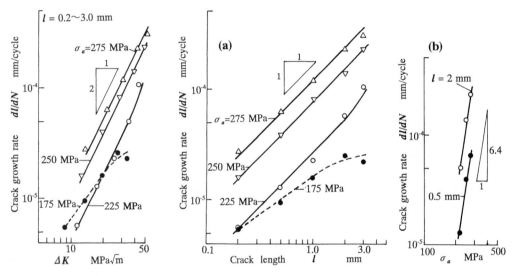

Fig.5 *dl/dN* vs ΔK relation   Fig.6 Dependency of *dl/dN* on ; (a) l, and (b) σ_a

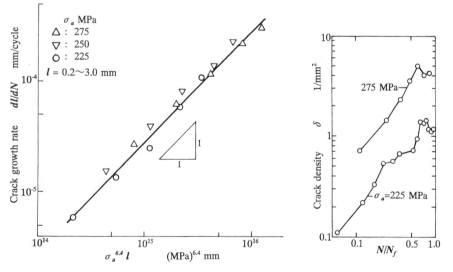

Fig.7 *dl/dN* vs $\sigma_a^n l$ relation ($n \doteqdot 6.4$)   Fig.8 δ vs N/N_f relation

Figure 6(a) shows the dependency of dl/dN on crack length. For every constant stress amplitude excepting 175 MPa, a straight line can be drawn approximately. The slope is about unity. Accordingly, dl/dN is proportional to l. Figure 6(b) shows the dependency of dl/dN on stress amplitude. In the high stress amplitude above 225 MPa, the dependency is expressed by the relation $dl/dN \propto \sigma_a^n$. The value of n is constant and about 6.4 in this case. Putting the results of Fig.6(a) and (b) together, we can obtain the growth law of a small crack, i.e.

$$dl/dN = C \sigma_a^n l \qquad \text{(where } n \text{ is approximately 6.4)} \qquad (1)$$

Figure 7 shows the dl/dN versus $\sigma_a^n l$ relation. It is found that the growth rate of small crack under high stress amplitude ($\sigma_a > 225$ MPa) is determined uniquely by $\sigma_a^n l$, not by ΔK. When the stress amplitude is relatively small ($\sigma_a \ll 225$ MPa), the dl/dN of small crack cannot be treated by both the ΔK and $\sigma_a^n l$. This closely relates with that the growth behaviour of crack under such a small stress is liable to be affected by the microstructure.

Crack length distribution

In order to study the change in crack length distribution due to the stress cycling, the behaviour of all the cracks beyond 10 μm initiated within the specific region whose area is 10-20 mm^2 were examined. To estimate the scatter characteristics quantitatively, the distribution properties for crack growth data must be analyzed. In what follows, distribution studies of each data set were performed using the three-parameter Weibull distribution function $F(x)$ expressed by the following equation:

$$F(x) = 1 - \exp\left[-\left(\frac{x - \eta}{\gamma}\right)^m\right] \qquad (2)$$

Here, the three constants m, η and γ are the shape parameter, scale parameter and location parameter, respectively. Determination of the three Weibull parameters is made by a correlation factor method, i.e. the correlation coefficient is the maximum value determined by a linear regression technique.

Figure 8 shows the relation between the crack density given by the mean value of the number of

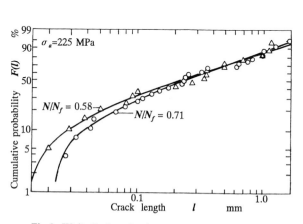

Fig.9 Weibull plots for the crack length distribution

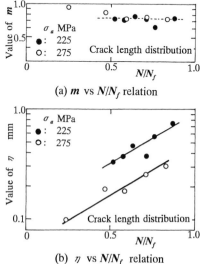

(a) m vs N/N_f relation

(b) η vs N/N_f relation

Fig.10 The value of Weibull parameters

cracks per square mm, δ, and the relative number of cycles. The value of δ increases with an increase in N/N_f in the range $N/N_f<0.7$, then it exhibits decreasing trend with further increasing N/N_f. The decreasing tendency of δ at the later stages of cycling results from the coalescence of cracks.

Figure 9 shows the crack length distribution data at various stages of fatigue life represented on the Weibull probability paper. The curved lines in the figure show those distribution functions. The experimental results are in good agreement with the distribution functions. The distributions of crack length are represented by the three-parameter Weibull distribution, as shown for other materials (Goto, 1990).

Figure 10(a) shows the relation between the shape parameter, m, of the crack length distribution and the relative number of cycles, N/N_f, at each stress amplitude. The values of m are 0.7 to 0.8 independent of stress amplitude and N/N_f. Figure 10(b) shows the $\ln \eta$, logarithm of scale parameter, versus N/N_f relation. The relation is approximated by a straight line when the stress amplitude is constant. Thus, we can evaluate the crack length distribution at arbitrary cycle ratio. However, the relations shown in Fig.10 may be hold in the range $\sigma_a>225$ MPa, because the crack growth characteristics at $\sigma_a=175$ MPa was different from at $\sigma_a>225$ MPa as shown previously.

CONCLUSIONS

Rotating bending fatigue tests of a 2017-T4 Al alloy were performed and the successive observations of surface were made to clarify the initiation and propagation behaviour of a small crack. At early stages of cycling, a small shear crack initiated within the grain. After initiation, a crack propagates under the influence of the microstructure. However, when the stress amplitude exceeds 225 MPa, the effect of microstructure on the crack growth behaviour decreases as crack grows longer. In such a case, the crack growth is controlled mainly by the tensile mode. On the other hand, at the stress amplitude far below 225 MPa, the shear mode propagation dominates even for long cracks in excess of 1 mm. The crack growth rate, dl/dN, at a high stress was determined by $\sigma_a^n l$, whereas the dl/dN at a low stress could not be evaluated by the mechanical terms ($\sigma_a^n l$ and ΔK) because of the strong microstructure dependency. The length distribution of all the cracks beyond 10 μ m initiated within the region whose area is 10-20 mm^2 was represented by the Weibull distribution. The value of shape parameter for crack length distribution was 0.7 to 0.8 independent of N/N_f and σ_a.

ACKNOWLEDGEMENTS

The financial supports of the Grant-in-Aid for Scientific Research from Oita University are greatly appreciated.

REFERENCES

H. Nisitani and M. Goto (1987). A small crack growth law and its application to the evaluation of fatigue life. *The Behaviour of Short Fatigue Cracks* (Edited by K.J.Miller and E.R. de los Rios), EGF **1**, MEP,pp.461-478.
H. Nisitani, M. Goto and N. Kawagoishi (1992) A small-crack growth law and its related phenomena.*Eng. Fract. Mech.* **41**, pp.499-513.
M. Goto and H. Nisitani (1994), Fatigue life prediction of heat-treated carbon steels and low alloy steels based on a small crack growth law. *Fatigue Eng. Mater. Struct.* **16**, pp.171-185.
M. Goto (1990), Statistical investigation of the behaviour of microcracks in carbon steels. *Fatigue Fract. Mech.* **14**, pp.833-845.

AN INVESTIGATION OF THE GROWTH OF SMALL CORNER CRACKS FROM SMALL FLAWS IN 6061-T651 ALUMINUM

R.L. Carlson, D.S. Dancila and G.A. Kardomateas

School of Aerospace Engineering, Georgia Institute of Technology,
Atlanta, Georgia 30332-0150, U.S.A.

ABSTRACT

A three-point-bending fatigue test procedure designed to investigate the growth of small corner cracks is described. The results of experiments in which cracks emanate from very small corner notches are presented. An intermittent growth-arrest behavior was observed and abrupt jumps were of the order of the transverse grain size. It is hypothesized that the arrest periods are the result of a grain boundary pinning mechanism at one of the three to five grains along a wavy crack front.

KEYWORDS

small fatigue cracks; grow-arrest behavior; flaws.

INTRODUCTION

About twenty years ago, Pearson (1975) discovered that the growth of small fatigue cracks did not obey laws of growth which are applicable for long cracks. Since that time small fatigue crack growth has been the subject of many investigations. An excellent summary of salient features of the behavior has been included in a recent paper by Halliday, Poole and Bowen(1995). Many of the studies which have been reported have focused on the growth of `natural' cracks; i.e., cracks which were initiated on the surfaces of highly polished specimens. Often, the surface `thumbnail' cracks which developed emanated from cracks in brittle intermetallic inclusions. These inclusion cracks could perhaps be identified as the `equivalent initial flaws' which form the basis for the durability analyses proposed by Manning, Yang and Welch (1992).

Structural components in service generally do not have highly polished surfaces and often mechanical flaws such as nicks or gouges are introduced during manufacturing or maintenance operations. Although these flaws can be very small, they can, if located in regions of high stress, become sites for the initiation of fatigue cracks. Gangloff et al (1992) and Pickard, Brown and Hicks (1983) have described small fatigue crack test results which were obtained from specimens with flaws in the form of small notches. Results were presented for edge and corner notched specimens which were tested in tension. It may also be noted that recommendations for notch preparation are presented in the American Society for Testing and Materials Test Method E 647 (1993).

The objective of this paper is to present test data on the growth of small fatigue cracks which have been initiated at small corner notches. In contrast to the tensile loading state used by Pickard, Brown and Hicks (1983), the specimens used were loaded in three point bending as shown in Fig. 1. By virtue of the orientation of the specimen, the maximum tensile stress is developed at the middle of the gauge section, at the top corner. Note that the region of high tensile stress is isolated so that the specimen could be used to develop `natural' corner cracks at a reasonably predictable location. The results presented here, however, involve crack growth from a small corner notch at the high tensile stress location.

EXPERIMENTAL PROGRAM

Test specimens were machined from 16 mm diameter bar stock of the aluminum alloy 6061-T651. The 0.2 percent yield strength of the material was 283 MPa and the ultimate strength was 293 MPa. The average transverse grain size was 200 microns. This corresponds to the plane on which crack propagation occurs.

The gage section of the test specimen had 10.2 mm square cross section. Corner notches with a 60 degree included angle were introduced by use of a digitally controlled slitting saw. The faces adjacent to the notched corner were then metallographically polished by use of five grades of abrasive papers ranging from 600 to 2000 grit. Final polishing was done by use of 3 and 1 micron diamond paste, respectively. Polishing enhances the optical visibility of the fatigue cracks. Polishing <u>after</u> notching also provides a potential for obtaining very small notches; i.e., removal of surface layers reduces the original notch depth. By this procedure it may be possible to produce notch depths which are of the order of some of the larger inclusions which have been observed to be crack initiation sites. Note also that since the loading state is bending, the possibility of eccentric loading that could occur for subsequent tensile loading is not a problem.

Tests were conducted on an Instron servo-hydraulic testing machine which applied sinusoidal loading at 10 Hz. For the tests reported the load ratio R was 0.0625. An optical measurement system was used to monitor crack growth. It consisted of a telemicroscope with an optical resolution of 2.5 μm and a working range of 0.55 to 1.7m. The crack image was viewed by use of a video camera and a monitor. A three axis translation mount and a digital position readout provided a crack growth increment of the order of 10 μm. The crack image was enhanced by use of a thin paint film applied prior to each measurement. The paint had an alcohol solvent. During crack initiation and extension, the paint membrane film was locally distorted and indicated the location of the highly strained surface adjacent to the crack tip. Crack lengths were measured at regular intervals to provide plots of crack length versus loading cycles.

Cracks emanating from the notches were introduced by applying loads which gave a nominal maximum stress of 0.9 of the yield strength. Small cracks were initiated after about 200,000 cycles. After crack initiation, the load was reduced by 50 percent. No further growth was observed during an additional 100,000 cycles. The load was then increased by 10 percent and crack growth resumed. This resulted in a maximum load that was about 0.5 of the yield strength. In one experiment growth of a moderately long crack was monitored to establish the near threshold region. In a second experiment measurements of the growth of a small crack were obtained.

Limitations on the use of the stress intensity factor as a correlation parameter have been identified by a number of investigators (Miller, 1982; Lankford and Davidson, 1983; Chan and Lankford, 1988; Navarro and de Los Rios, 1988). Since it provides a means of comparing long and small crack growth data, however, it has been used here. Its usefulness as a parameter for design is another issue. Computation of the stress intensity factor used is based on results for a corner crack in a bar with a rectangular cross-section in the NASA/FLAGRO computer program(1989). The test specimens had square cross-sections and results for equal bending moments were superimposed to provide the test bending moment about the neutral axis (the cross-section diagonal).

The long crack growth data provided a relatively smooth crack length versus cycles curve which made it possible to construct the near threshold, log - log plot of growth rate versus range of stress intensity factor depicted in Fig. 2. Data for crack growth during both a decreasing load and a constant load were obtained. The straight line through the data points can be represented by the equation

$$\frac{da}{dN} = 10^{-8}\left(\frac{\Delta K}{6.7}\right)^{28} \tag{1}$$

From the steepness of the slope it is clear that a small error in the determination of the stress intensity factor in a design calculation could result in a large error in the computed growth rate.

For the range of growth rates and stress intensity factors in Fig. 2 it is of interest to consider a non-distorted Cartesian plot of eq.(1) The plot in Fig. 3 illustrates nicely how the growth rate goes to zero as ΔK goes to zero. The additional data shown are discussed in a subsequent section.

Data for small crack growth are presented in Fig. 4. The repeating arrest of growth pattern observed by other investigators is clearly present. Loading was discontinued at intervals and the specimen was removed to measure crack growth on the second face of the specimen. The two types of data points represent measurements from the two faces. Note that growth on side A initially lagged behind that for side B. Also, the crack paths were not always normal to the specimen axis and some branching was observed. One branch eventually became dominant.

Miller (1982) has presented a qualitative description of crack growth history by using crack length on a left-hand ordinate and the sizes of microstructural features on a right-hand ordinate. This provides a convenient perspective for comparing crack length with such features as inclusions and grain size. If, for the corner crack, it is assumed that the arc of the crack front is centered at the crack corner, the number of grains - on the average - along a crack front for a given crack length can be determined from the equation

$$g = \frac{\pi a}{2D} \tag{2}$$

where a is the crack length (or radius to the crack front from the corner), g is the number of grains along the crack front and D is the transverse grain size. Equation (2) has been used to determine the scale of the right-hand ordinate of Fig. 4. Comparisons of the two ordinates, then, indicates the number of grains encountered - on the average - for a given crack length.

DISCUSSION

When a crack front encounters a small number of grains, the effects of grain boundaries and grain orientations can be expected to influence the manner in which the crack extends. The references cited previously have discussed these effects and their affect on the use of the stress intensity factor as a correlation parameter. A comparison of values on the ordinate scales of Fig. 4 provides insight into the microstructural features encountered by the crack front. In addition to the crack surface being nonplanar it is likely that not all of the grains along a crack front are arranged in an orderly row. For example, an advancing crack would not pass all of the grain boundaries at the same time. Thus, when a crack has advanced part of the way through a grain, the grain boundary of an adjacent grain could be behind and pin the overall crack advance until the crack pushed through the boundary. This hypothesis requires that even during periods during which no measurable growth is occurring, there is crack extension across some of the grains along the crack front. With this sequence of events in mind, it is of interest to observe that the jumps or steps in growth are of the order of the grain size. For the initial portion of the growth the abrupt growth steps are about one third of the grain size.

The deviation from planar, uniform growth could promote obstruction to closure conditions. This conjecture is supported by Halliday, Poole and Bowen(1995) who found evidence for substantial levels of closure obstruction in small 'thumb-nail' fatigue crack experiments.

The number of grains along the crack front ranges from three to less than five during the step type growth. It may be noted that for this alloy a short, through edge crack in a 5 mm thick sheet would encounter about 25 grains. Also, the number of grains along the front of a "thumb-nail" crack would be about twice those for a small corner crack of the same depth. Since the stress intensity factor is insensitive to these details, it cannot be expected to account for behaviors which reflect these differences.

An examination of the initial growth features in Figure 4 would indicate that an elaborate scheme for computing growth rates is not warranted. Often, in fact, growth rate data are simply represented by clusters of discrete data points on log - log plots of growth rate versus range of stress intensity factor. Nevertheless, continuing growth is occurring and a growth trend curve is indicated. A simple method for representing the growth rates has been adopted. A trend curve has been developed by connecting successive inner corners of the steps. The rates, so determined, are indicated in Figure 3 for both faces of the corner. The small crack growth curve is to the left of the near threshold curve.

CONCLUSIONS

1. A relatively simple test procedure has been developed for investigating small corner fatigue cracks under three point bending. It could be used to initiate natural corner fatigue cracks. Here it was used to initiate small corner cracks from small notches.

2. The distortion of log - log plots for near threshold data can be eliminated by the use of Cartesian coordinate plots.

3. The jumps during growth and arrest behavior for small corner cracks are of the order of the grain size for the aluminum alloy 6061-T651 tested.

4. Using the stress intensity factor as a comparison parameter indicates that for small R, small corner cracks in the 6061-T651 can grow below threshold.

5. Intermittent growth is probably the result of an uneven advance along the crack front. It is hypothesized that the arrest period is the result of a pinning mechanism at a grain boundary of one of the 3 or 4 grains along the crack front.

6. The fatigue crack growth behavior of small and short cracks of the same length and with the same grain size can be expected to be different because of the large difference in the number of grains along their crack fronts. Also, small 'thumb-nail' cracks can be expected to have about twice as many grains along their fronts as corner cracks. Since the stress intensity factor is insensitive to these details, it cannot be expected to account for behaviors which reflect these differences.

ACKNOWLEDGMENTS

The results presented were obtained as a part of a program sponsored by the Warner Robins Air Logistics Center, Robins AFB under Contract No. F09603-91-G-0096-0013. The authors are grateful for the interest and support provided by the Project Monitor, Mr. Gary Chamberlain. Bob Cummings of the AF Corrosion and Materials Engineering Branch at Robins AFB performed the metallographic work.

REFERENCES

American Society for Testing and Materials Proposed Small Cracks Appendix to Test Method E 647 (1993).

Chan, K.S. and Lankford, J. (1988). The role of microstructural dissimilitude in fatigue and fracture of small cracks. Acta Metall., 36, 193-206.

Gangloff, R.P. et al (1992). Direct current electrical potential measurement of the growth of small cracks, in Small Crack Test Methods (J.M. Larsen and J.E. Allison, Eds.), ASTM STP 1149, 116-168.

Halliday, M.D., Poole, P. and Bowen, P. (1995). In-situ SEM measurements of crack closure for small fatigue cracks in aluminum 2024-T351, Accepted for publication in J. Fatigue Fracture Engng. Mater. Struct.

Lankford, J. and Davidson, D.L. (1983). Near-threshold crack tip strain and crack opening for large and small fatigue cracks. In: Fatigue Crack Growth Threshold Concepts (D. Davidson and S. Suresh, Eds.), 447-463.

Manning, S.D., Yang, J.N. and Welch, K.M. (1992). Aircraft structural maintenance scheduling based on risk and individual aircraft tracking. In: Theoretical Concepts and Numerical Analysis of Fatigue (A.F. Blom and C. J. Beevers, Eds.), 401-420.

Miller, K.J. (1982). The short crack problem, J. Fatigue Fracture Engng. Mater. Struct., 5, 223-232.

NASA/FLAGRO (1989). Fatigue crack growth computer program.

Navarro, A. and de Los Rios, E.R. (1988). Short and long fatigue crack growth: a unified model, Phil. Mag., 57, 15-36.

Pearson, S. (1975). Investigation of fatigue cracks in commercial Al alloys and subsequent propagation of very short fatigue cracks, Engng. Fract. Mech., 7, 235-247.

Pickard, A.C., Brown, C.W. and Hicks, M.A. (1983). The development of advanced specimen testing and analysis techniques applied to fracture mechanics lifing of gas turbine components. In: Advances in Life Prediction Methods (D.A. Woodward and J.R. Whitehead), 173-178.

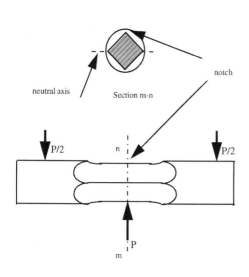

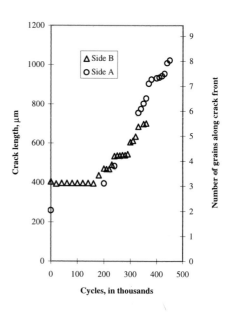

Figure 1. Corner crack specimen

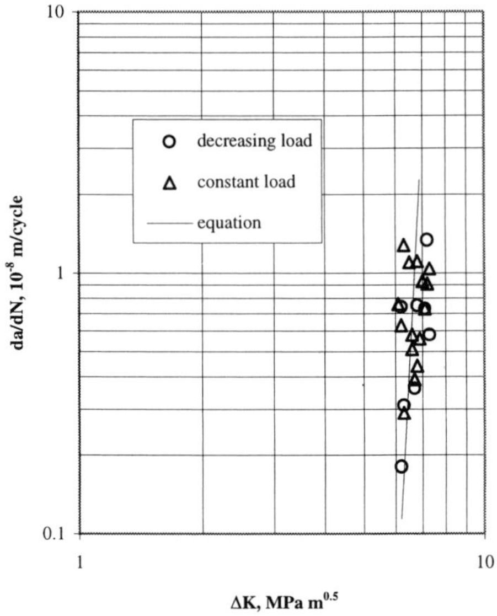

Figure 2. Log da/dN vs. log ΔK.

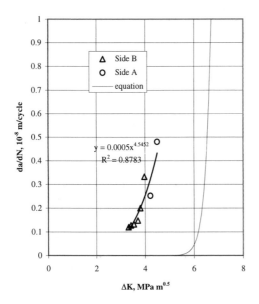

Figure 3. da/dN vs. ΔK.

Figure 4. Crack extension and number of grains vs. cycles.

THE EFFECT OF PERIODIC COMPRESSIVE OVERLOADS ON THE CRACK SURFACE ROUGHNESS AND CRACK GROWTH RATE OF SHORT FATIGUE CRACKS IN 1045 STEEL

A. Varvani-Farahani and T. H. Topper

Engineering Department, University of Waterloo,
Waterloo, Ontario, N2L 3G1, Canada

ABSTRACT

An examination of the growth of short fatigue cracks under constant amplitude loading and during load histories having periodic compressive overloads of various magnitudes revealed that the fracture surface near the crack tip and the crack growth rate changed dramatically with the magnitude of the compressive overload. The height of the surface irregularities reduced as the compressive overload increased progressively flattened fracture surface asperities near the crack tip. This resulted in a reduced crack closure stress and a higher crack growth rate.

Confocal Scanning Laser Microscopy (CSLM) image processing of the fracture surface in an area immediately behind the fatigue crack tip was used to measure the height of asperities for constant amplitude loading (R=-1 and S_{ampl}=138 MPa) and for periodic compressive overloads of -300, -360, and -430 MPa (followed by 50 small fatigue cycles). Asperity height decreased from 28 μm in constant amplitude loading to 18, 13, and 8 μm for -300, -360, and -430 MPa overloads, respectively. The crack growth rate increased as asperity height decreased. A complementary investigation using Scanning Electron Microscope (SEM) revealed compression-induced abrasion marks (due to compressive overloads of -430 MPa). The abrasion marks corresponded to the region close to the crack tip location when the compressive overloads, were applied.

Fracture surface profiles obtained using CSLM showed that a surface crack initiates and propagates into the specimen a distance of 175-320 μm on a plane inclined to the applied stress. The plane of crack growth then shifts to the plane perpendicular to the applied stress. It is noteworthy that asperity height (obtained by CSLM) during the inclined crack growth (Stage I) is less than for the horizontal (Stage II) crack growth. A careful observation of the fracture surface of the specimens using SEM verified that fracture surface asperity height during Stage I fatigue crack growth was much less than that in Stage II growth.

KEYWORDS

Short fatigue cracks; compressive overload; fracture surface asperity; asperity height; Stage I-Mode II

INTRODUCTION

Numerous investigations have shown that, for both long and short fatigue cracks, the application of compressive overloads of near yield stress magnitude leads to an acceleration in crack growth rates. Some of the clearest evidence of the deleterious effects of compressive overloads on crack growth resistance are found in the work of Topper and Yu (1985). They showed that the near-threshold fatigue crack growth rates measured using a loading spectrum with periodic compressive overloads are always higher than those for constant amplitude fatigue. In this paper, a new technique (CSLM) provides a quantitative study of the relationship of fracture surface asperity characteristics, under constant amplitude and periodic compressive overloading conditions, to crack growth rate and crack closure stress for short fatigue cracks in SAE 1045 Steel.

EXPERIMENTAL PROCEDURE

Material

The material examined in this investigation was SAE 1045 Steel in the form of 2.5 in diameter bar stock with the following chemical composition (W%): 0.46 C, 0.17 Si, 0.81 Mn, 0.027 P, 0.023 S, and remainder Fe. This material is a medium carbon heat treatable steel which is widely used in the automotive industry. The microstructure of SAE 1045 Steel after a final polishing showed pearlitic-ferritic features containing up to 30 μm long sulfide inclusions in the extrusion direction. The modulus of elasticity is 206000 MPa, the cyclic yield stress is 448 MPa, and the fatigue limit stress (R=-1) is 300 MPa.

Specimen Design

Fig 1 shows a uniaxial fatigue specimen designed to aid in the observation of short fatigue cracks. The specimen has a flat gauge area to facilitate microscopic observations, and to facilitate breaking of a cracked specimen in liquid nitrogen. A small notch (r=0.5 mm) was machined in one side of the gauge area. This provides a high stress concentration which determines the crack initiation location.

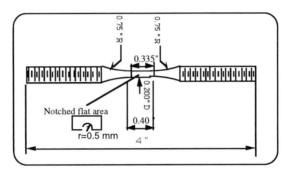

Fig.1 Uniaxial Fatigue Specimen Geometry.

Short Fatigue Crack Tests

Short crack growth tests were carried out for four different load histories: (a) constant amplitude loading with a stress level of the order of the fatigue limit for a 0.5mm notched specimen (138 MPa), (b) a periodic compressive overload -300 MPa followed by n small cycles (n=50, 100, 500, 1000 cycles), (c) a periodic compressive overload of -360 MPa followed by n small cycles (n=50, 100, 500, 1000 cycles), and (d) a periodic compressive overload of -430 MPa followed by 50 constant amplitude small cycles.

Crack growth rates of short fatigue cracks were monitored by measuring the crack length as the number of cycles increased using an optical microscope and JAVA system (Jandel Video Analysis Software) with an accuracy of 1μm.

Fracture Surface Observations

Fracture surfaces of the short fatigue cracks were observed after fatigue tested specimens were broken in liquid nitrogen (-200 °C). Breaking specimens in liquid nitrogen provided a brittle fracture and no plasticity was observed on the fracture surfaces.

Confocal Scanning Laser Microscope (CSLM)

In this study, the fracture surface and the variation in the height of asperities on the fracture surface were observed using a confocal scanning laser microscope. The laser beam source and the power used were Ar-Kr 488nm and 0.25 mW, respectively. The apparatus and technique were described in previous work (Varvani-Farahani, 1994).

First, the laser beam was centered on the area of the fracture surface adjoining the crack tip by direct observation through an attached optical microscope. The fracture surface of an area of 1mm^2 at the crack tip was scanned by the laser beam (through an aperture of 0.42) and reflected to the detector. In order to make a three dimensional profile of fracture surface asperities, a piezo-electric stage, was used. The piezo-electric stage controls the distance between specimen and microscope. This provides successive images of level contours of the asperities from their peaks to their valleys. Each image provided a matrix of 512 rows and 512 columns. Note that the row numbers increase along the thickness of specimen and the column numbers increase along the crack length and reach 512 at the crack tip. Every component of this matrix corresponds to a height value of the scanned fracture surface area. All images were combined to create a configuration of the fracture surface profile. Taking different slices through this profile and determining an average value of fracture surface asperity height in each slice, revealed that the asperity height is dramatically influenced by the magnitude of the compressive overload.

Fractography of Fracture Surface Roughness

First, the fracture surface of the specimens was gold plated and then fractographic examinations was carried out using a scanning electron microscope (SEM). Fracture surface roughness and the fracture paths of short fatigue cracks was studied. The maximum shear plane and the roughness of this plane during Stage I crack growth (Mode II) was determined. The height of plastically crushed asperities after the application of compressive overloads was determined for each magnitude of overload.

RESULTS

Short Fatigue Crack Growth Rate

Propagation rates of fatigue cracks, one millimeter in length, in notched SAE 1045 Steel specimens which were subjected to periodic compressive overloads superimposed on R=-1 cycles, were investigated. A comparison of specimens tested under constant amplitude and variable loading conditions showed that fatigue crack growth is drastically influenced by the magnitude of overloads, and the number of small fatigue cycles following the overload.

The results indicated that the fatigue crack growth rates obtained from a loading spectrum with periodic compressive overloads were always higher than for constant amplitude loading. Fig 2 shows crack length versus number of cycle plots for cracks for various magnitudes of the compressive overloads (-300, -360, -430 MPa). The number of small cycles following an overload was 50 for all tests. These results show that as the magnitude of compressive overloads increases, the growth rate of short fatigue cracks increases significantly. It was also observed that when the number of small cycles between overloads (n) was increased, the growth rate of short fatigue cracks decreased. Fig 3 illustrates short crack growth behavior following a compressive overload of -360 MPa.

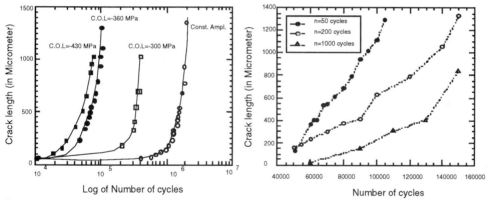

Fig.2 Crack Growth Behavior under Constant and Periodic Compressive Overloads of -300, -360, -430 MPa.

Fig.3 Crack Growth Behavior as the Number of Cycles between Compressive Overloads changes.

Figs 4a-b illustrate a trend of increasing crack growth rate as periodic compressive overloads of -300 and -360 MPa were followed by 200 and 500 small fatigue cycles. These figures also show that as the magnitude of compressive overload increases, from -300 MPa to -360 MPa, the crack growth rate increases. The variations in crack growth rate, as crack length increases, are associated with grain boundaries which act as microstructural barriers to crack growth. When a crack reaches a grain boundary the crack growth rate drops and after it passes the grain boundary the crack growth rate increases.

Crack Path Morphology

An SEM examination of the short fatigue cracks revealed the crack path morphology after the crack turned from Stage I to Stage II crack growth. The nature of this deflection between Stage I and Stage II growth appeared to be crystallographic and involved extensive slip band cracking.

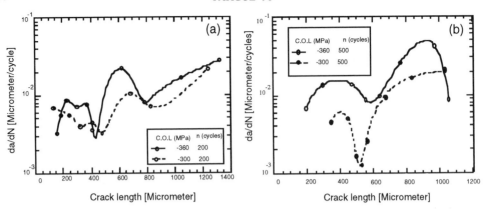

Fig.4 Comparison of Short Fatigue Crack Growth Rates under two magnitude of Compressive Overloads
of -300 MPa and -360 MPa and different Number of Cycles between Overloads.

Fig 5 illustrates the crack path of a short fatigue crack for SAE 1045 Steel. This morphology indicates the rough faces of the crack which results in roughness-induced crack closure.

Aspect Ratio and Crack Depth Profile

The ratio of crack depth to half crack length (a/c) was obtained in SAE 1045 Steel for a few short fatigue cracks with the length of 50 μm using CSLM. The technique used in this measurement is described in a previous paper (Varvani-Farahani *et al*., 1995). Aspect ratios (a/c) of cracks which experienced fatigue loading just above the fatigue limit stress were found to be approximately 0.8. Using the CSLM technique, the crack depth profile (in Stage I crack growth) was found to be semi-elliptical in shape.

Fracture Surface Roughness

An investigation of the fracture surface of short fatigue cracks using CSLM image processing and the SEM system revealed the height variation of fracture surface asperities along the crack front and specimen thickness for different loading conditions.

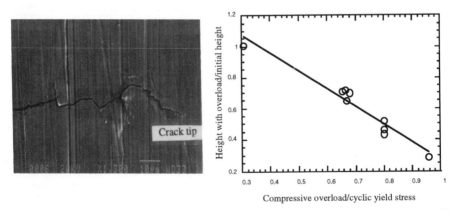

Fig.5 SEM Micrograph of a Surface Crack showing   Fig.6 The Change of the Average Height of Asperities VS.
the Irregular Path of a Crack.   Compressive Overload Stress.

The average height of asperities in an area immediately behind the crack tip was measured using CSLM image processing of the fracture surface for various levels of constant amplitude and periodic compressive overloading. Fig 6 shows a trend of decreasing asperity height as the magnitude of the compressive overload increases.
A linear relation of normalized asperity height versus normalized compressive overload level can be expressed as:

$$\frac{h_{ol}}{h_0} = 1.41 - 1.13\left(\frac{\sigma_{ol}}{\sigma_{ys}}\right) \tag{1}$$

where σ_{ol}, ,and σ_{ys} are the compressive overload stress and the cyclic yield stress, respectively. The initial asperity height (h_0) was measured for a constant amplitude loading condition. The height of a plastically crushed asperity after compressive overloading is denoted as h_{ol} in the above linear regression.

The decrease of height as the magnitude of the compressive overload increases, indicates that compressive overloads flatten the fracture surface asperities. A complementary investigation using a SEM showed compression-induced abrasion marks in an area immediately behind the crack tip. Fig 7a is a scanning electron micrograph of an asperity immediately behind the crack tip showing the area flattened due to a compressive overload of -430 MPa followed by n=50 small fatigue cycles. This is compared to the height of an asperity in constant amplitude loading (R=-1) and a stress amplitude of 138 MPa in Fig 7b.

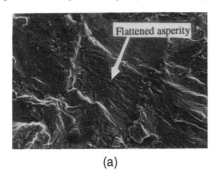
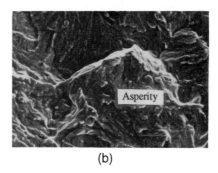

(a)   (b)

Fig.7 A Scanning Electron Micrograph of a Fracture Surface Asperity: a) under Periodic Compressive Overload of -430 MPa and b) under Constant Amplitude Loading Condition.

Using CSLM, the decrease in the asperity height (fracture surface roughness) under constant amplitude loading was verified when the superimposed compressive overloads of -300, -360, and -430 MPa are applied. The height of fracture surface irregularities near the crack tip decreased (from 28, to 18, 13, and 8 µm, respectively) as compressive overload stresses increased (from constant amplitude loading to -300, -360, and -430 MPa, respectively). The overloads resulted in significant increases in crack growth rates.

Fracture surface profiles obtained using CSLM showed that a surface crack initiates and propagates into the specimen a distance of 175-320 µm on a plane inclined to the plane perpendicular to the applied stress. The plane of crack growth then shifts to the plane perpendicular to the direction of the applied stress. It is noteworthy that the height of the asperities (obtained by CSLM) during inclined crack growth (Stage I) is less than for horizontal crack growth (Stage II). This may partly explain why short fatigue cracks (Stage I) have a low closure stress. A careful observation of the fracture surface of the specimen using a SEM verified that the fracture surface roughness height during Stage I fatigue crack growth was much less than that in Stage II crack growth.

DISCUSSION

Results of this study show that the crack growth rate of short fatigue cracks in SAE 1045 Steel increases dramatically as the magnitude of the applied compressive overload increases. A compressive overload led to a flattening of roughness asperities and therefore a reduction in closure stress. In this regard, Henkener et al., (1990) reported that the crack growth rate increased as the compressive overload increased. The increased compressive overload led to an increase in the range of the effective stress intensity factor. Herman et al., (1989) and Hertzberg et al., (1992) also showed that a low closure stress (due to compressive loads) is associated with the crushing of asperities in the crack wake.

The crack growth behavior of short fatigue cracks under three periodic compressive overloads (Fig.2) shows that the increase in crack growth rate as the overload increases from -300 MPa to -360 MPa is much higher than when the overload increases from -360 to -430 MPa. For a 1mm increase in crack length in this figure, a test with a -300MPa overload requires 8000 blocks (50*8000=400,000 cycles) while the tests with -360 and -430 MPa overloads required about 1800 and 1500 blocks, respectively. In this regard, Kemper et al., (1989) and Tack and Beevers (1990) observed a similar saturation effect in which increases in compressive overload beyond a certain level did not result in additional increases in crack growth rate.

Fig 3 illustrates the influence of the number of small cycles following an intermittent overload of -360 MPa. A higher crack growth rate is associated with a smaller number of cycles between two compressive overloads. Pompetzki et al., (1990) investigated the effects of overloads on fatigue damage in smooth specimens of SAE 1045 Steel. They conducted three sets of fatigue tests, in which the number of cycles between the intermittent overload was varied. The results showed that fatigue resistance given by life curves decreased as the number of small cycles between overloads decreased. They also reported that compressive overloads decreased the crack opening stress and increased the effective stress and that as cycling at the lower stress level continued, the crack opening stress increased and the effective stress and damage per cycle decreased.

In this study, SEM fractography and a CSLM investigation of the crack fracture surface indicated that Stage I growth of short cracks in SAE 1045 Steel occurred in an inclined plane. The height of asperities on this plane (Stage I) was found to be less than that for Stage II. Hence, the first stage of short fatigue crack growth has a lower closure stress. This agrees with the results of Minakawa et al., (1984). They showed that the closure level developed from zero closure as a function of the crack length for a Mod 9Cr-1Mo steel. The opening stress intensity factor rapidly built up in the first 200 µm of crack growth.

CONCLUSIONS

1. The height of surface irregularities is reduced as the magnitude of a periodic compressive overload increases due to plastic crushing of asperities near the crack tip. This reduces the crack closure stress and results in a higher crack growth rate.
2. The fracture surface roughness profile obtained by CSLM reveals that the asperity height is much less for the inclined part of the crack surface (Stage I crack growth) than for the horizontal part of the crack surface (Stage II crack growth). Therefore, Stage I crack growth should have less closure.
3. Observations in the SEM showed abrasion marks and asperities plastically crushed due to compressive overloads.
4. The number of small fatigue cycles between two intermittent compressive overloads influenced the crack growth rate. The crack growth rate increased as the number of small cycles between overloads decreased.

REFERENCES

Henkener, J. A., T. D. Scheumann, and A.F. Grandt (1990), Fatigue crack growth behavior of a peakaged Al-2.6LI-0.09ZR alloy. In: *Proc. 4th Int. Conf. on Fatigue and Fracture Thresholds* (H. Kitagawa and T. Tanaka, eds), Vol.II, pp. 957-62. MCEP, Honolulu.

Herman, W. A., R. W. Hertzberg, and R. Jaccard (1989), Prediction and simulation of fatigue crack growth under conditions of low crack closure. In: *Advances in fracture Research 7th Int. Conf. on Fracture*, (K. Salama, K. Ravi-Chanadar, D. M. R. Taplin, and P. Rama Rao, eds), Vol. 2, pp.1417-26, Pergomon Press, Houston.

Hertzberg, R. W., W. A. Herman, T. Clark, and R. Jaccard (1992), Simulation of short crack and other low closure loading conditions utilizing constant K_{max} ΔK-decreasing fatigue crack growth procedures. In: *Small-crack Test Methods,* (J.M. Larsen and J. E. Allison, eds), pp. 197-220, ASTM STP 1149, USA.

Kemper, H., B. Weiss, and R. Stickler (1989), An alternative presentation of the effect of the stress-ratio on the fatigue threshold. *Engineering Fracture Mechanics*, **32**, 591-600

Minakawa, K., H. Nakamura, and J. Mc Evilly (1984), On the development of crack closure with crack advance in a ferritic Steel. *Scripta Metallurgica*, **18**, 1371-74

Pompetzki, M. A., T. H. Topper, and D. L. Duquesnay (1990), The effect of compressive underloads and tensile overloads on fatigue damage accumulation in SAE 1045 Steel. *Int. J. Fatigue*, **12**, 207-213

Tack, A. J. and C. J. Beevers (1990), The influence of compressive loading on fatigue crack propagation in three aerospace bearing Steels. In *Proc. 4th Int. Conf. on Fracture and Fatigue Thresholds*, (H. Kitagawa and T. Tanaka, eds), Vol.II, pp.1179-84 , MCEP Ltd, Honolulu,.

Topper, T. H. and M. T Yu (1985), The effect of overloads on threshold and crack closure. *Int. J. fatigue*, **7**, 159-164

Varvani-Farahani, A. (1994), Non-propagating fatigue crack behavior in different orientations of rolled Al 2024-T351 alloy.. MASc Thesis, University of Waterloo, Waterloo, Ont., N2L 3G1, Canada.

Varvani-Farahani, A., T. H. Topper, and A. Plumtree.Confocal scanning laser microscopy measurements of the growth and morphology of microstructurally short cracks in Al 2024-T351 alloy. (submitted to *Fatigue and Fracture of Engineering Materials and Structures*-July 1995).

Crack Initiation and Small Fatigue Crack Growth Behaviour of Beta Ti-15V-3Cr-3Al-3Sn Alloy

Kazuaki SHIOZAWA and Hiroki MATSUSHITA

Department of Mechanical Systems Engineering
Faculty of Engineering, Toyama University
3190 Gofuku, Toyama 930, JAPAN

ABSTRACT

Fatigue strength, crack initiation and small crack growth behavior in beta titanium alloy, Ti-15V-3Cr-3Al-3Sn, were investigated using smooth specimen subjected to axial fully reversed loading in air at room temperature and in saline solution. Two types of microstructure were prepared by the heat treatment; one is homogeneous beta phase obtained by solution treatment (ST) and another is the microstructure having precipitated alpha phase in beta phase obtained with solution treatment followed by aging treatment (STA). From the fatigue tests, same fatigue resistance was obtained both in air and in saline solution. In S-N diagram on the STA alloy; a significant change in the slope of S-N curve was observed, accompanying the transition from surface to subsurface crack initiation. Surface fatigue crack initiated in (prior) beta grain at below 10% of fatigue life and period of small fatigue crack growth was about 90% of fatigue life. The small fatigue crack growth rate of ST and STA alloy was determined uniquely by the term of $\sigma_a/\sigma_{0.2}$ ($\sigma_{0.2}$:proof stress).

KEYWORDS

Fatigue; Corrosion fatigue; Crack initiation; Small crack; Crack growth rate; Subsurface crack initiation; Beta-titanium alloy; Ti-15V-3Cr-3Al-3Sn.

INTRODUCTION

The high strength-to-density ratio of titanium alloy makes it a very attractive design choice in engineering structure and components, as automobile, train, aircraft and aerospace, requiring high specific strength, reliability, energy-efficient and weight-saving. Recently, beta-phase titanium alloy contained beta stabilizing elements have attracted special interest, because it is possible to cold work easily and to get high tensile properties after solution treatment and aging. The few studies performed on the beta titanium alloy revealed about the fatigue behavior (Tanaka *et al.*, 1992, Bian *et al.*, 1995), although alpha + beta alloy, such as Ti-6Al-4V, has been studied by many researchers (for example, JSMS Committee of Fatigue, 1994).

To clarify the fatigue strength, crack initiation and small crack growth behavior in beta titanium alloy, the most common beta Ti-15V-3Cr-3Al-3Sn alloy was employed in this study and investigated using smooth specimen subjected to axial fatigue in air and in saline solution at room temperature. Two types of microstructure were prepared by the heat treatment; one is homogeneous beta phase obtained by solution treatment and another is the microstructure having precipitated alpha phase in beta phase obtained with solution treatment followed by aging treatment for 24 hours. Fatigue strength and crack growth behavior of the beta titanium alloy were compared with experimental results obtained from the specimen of Ti-6Al-4V alloy tested under the same condition.

EXPERIMENTAL PROCEDURE

Testing Materials

Materials tested in this experiment were beta titanium alloy of Ti-15V-3Cr-3Al-3Sn (named as Ti-15-3) taken from hot-drawn bar of 30 mm diameter and alpha + beta titanium alloy of Ti-6Al-4V (named as Ti-6-4) taken from a 26 mm diameter bar of hot-drawn. The chemical composition of the material is shown in Table 1. The Ti-15-3 alloy was examined in two heat-treated conditions. One was only solutionized under 1073K, 0.5 hour, water quench and then the microstructure was a homogeneous beta single phase with grain size of about 52.4 μ m (Ti-15-3ST). Another was solutionized under the same condition as the Ti-15-3ST and then aged at 783K, 24 hour, air cool (Ti-15-3STA). The microstructure obtained was consisted of Widmanstatten α -plates in an aged β matrix having prior β -grain size of about 52.4 μ m. Ti-6-4 annealed at 1173K, 2 hour, air cool had an equiaxed alpha microstructure of about 4.2 μ m diameter. Mechanical properties of three kinds of materials heat treated are listed in Table 2.

Table 1 Chemical composition of materials used (wt%) .

Material	Al	V	Cr	Sn	Fe	O	C	N	H	Ti
Ti-15V-3Cr-3Sn-3Al	3.5	14.8	3.4	3.0	0.15	0.12	0.005	0.01	0.001	Bal.
Ti-6Al-4V	6.55	4.15	—	—	0.26	0.187	0.005	0.004	0.0038	Bal.

Table 2 Mechanical properties of materials used.

Material	0.2% proof stress $\sigma_{0.2}$ (MPa)	Tensile strength σ_B (MPa)	Elongation δ (%)	Reduction of area ψ (%)	Vickers hardness Hv
Ti-15V-3Cr-3Sn-3Al ST	750	775	28.0	66.3	259
Ti-15V-3Cr-3Sn-3Al STA	1156	1261	12.0	23.1	376
Ti-6Al-4V	924	1019	17.8	41.8	30

Specimen used for fatigue test was machined from the drawn bar heat treated as smooth round-bar with a gauge diameter of 5 mm and gauge length 6 mm. Specimen surface was mechanically polished with emery-paper up to grade #2000 and subsequently electropolished about 15 μ m thick.

Testing Method

Fatigue test was performed in air at room temperature and in 3.0%NaCl aqueous solution environment by using push-pull electric-hydraulic servo machine which operated at 20Hz. The fully reversed sinusoidal stress wave form was employed, that is the stress ratio $R(=\sigma_{min}/\sigma_{max})$=-1. In corrosion fatigue test, the saline solution controlled at $298 \pm 1K$ was continuously circulated in a plastic reservoir through the tank at a flow rate of about 30ml/min. Following the immersion of a specimen in the saline solution for one hour under unloading and free corrosion condition, corrosion fatigue test was started.

Initiation and propagation of crack on the specimen surface were measured by the optical microscope on the replica which was taken from the surface interrupted at various fractions of fatigue life.

EXPERIMENTAL RESULTS AND DISCUSSION

S-N Curve

Experimental results obtained by the fatigue tests in air and in saline solution were summarized in Fig.1. From this figure, any difference of fatigue strength between testing environments can not be observed for three kinds of materials. Therefore, it implies that these materials have good corrosion fatigue resistance in saline solution. An obvious endurance limit was seen for both Ti-15-3ST and Ti-6-4 as reported in the literature. But special interest was observed in S-N diagram on the aged alloy, Ti-15-3STA. The first part of the curve up to about 5x10^6 cycles looks similar to the curves of Ti-15-3ST and

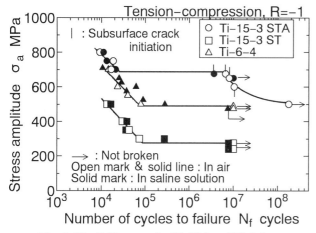

Fig. 1 The S-N curves for Ti-15-3 and Ti-6-4
under stress ratio of -1 in air and 3%NaCl
aqueous solution.

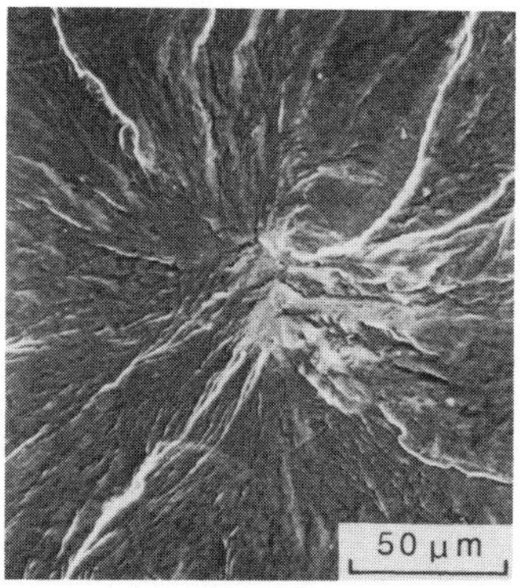

Fig. 2 A typical example of a subsurface crack initiation site
in Ti-15-3STA.

Ti-6-4, however, after the curve flattens out there was a sharp drop in the fatigue strength. This behavior
was observed not only in air but also in saline solution, and the sharp drop occurred at same stress
amplitude and number of stress cycles. This drop in the fatigue strength is associated with a change in
the mechanism of fatigue crack initiation which changes from surface initiation for cyclic lives, $N \leqq$
5×10^6, to subsurface initiation for $N > 5 \times 10^6$.

Observation on Fracture Surface

Figure 2 provides a typical example of a subsurface crack initiation site in Ti-15-3STA observed by
SEM. At high stress amplitude level, crack initiated at specimen surface due to cracking by crystal slip.
On the other hand, fatigue crack origin was located at subsurface site at low stress amplitude whereafter
the fatigue crack grew radially, normal to the tensile stress axis until it reaches the specimen surface, at
which point it has the circular form. A facet inclined to the tensile axis was observed at crack origin,
which is the order of 30 to 50 μ m in size and located at 60 to 600 μ m below the surface. And no defect
at this site was observed even at much higher magnification, such as a silicide or a microvoid. The
subsurface crack initiation behavior was reported in Ti-6Al-4V alloy by some researchers (Ruppen *et
al.*,1979, Atrens *et al.*, 1983, Adachi *et al.*,1985), when cycles to failure was greater than 10^6 and axial
loading condition was under $R \geqq 0$.   Since a subsurface cracking mode was observed under the R=-1
(fully reversed axial loading) in Ti-15-3STA having Widmanstatten structure in this study, this mode is
of fundamental as well as practical interest. The mechanisms of subsurface crack initiation and transition
of initiation site from surface to subsurface are not clear at this point. But it is emphasized from this
results that subsurface crack initiation mode occurs under any testing conditions,   and can not be
explained only by means of the mean stress effect and also the development of residual compression
stresses in surface regions.

Small Fatigue Crack Growth Behavior

Crack initiation and propagation behavior of three kinds of specimens were investigated under the
fatigue test at high stress amplitude level, that is the condition of surface crack initiation. Fig.3 shows a
typical example of the surface small fatigue crack observed by SEM. Surface crack initiation site is in the
prior beta grain for Ti-15-3ST and Ti-15-3STA, and in the equiaxed alpha grain for Ti-6-4.

Optical micrograph of the crack is shown in Fig.4, when test was interrupted and the plastic replica was
taken from the specimen surface. It can be seen from this figure that the small surface crack growth was
affected by the microstructure. The orientation of crack growth path in Ti-15-3ST changes through a

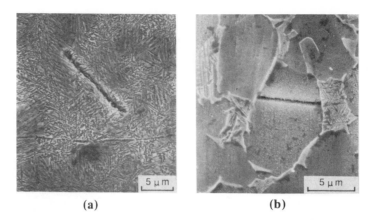

(a)   **(b)**

Fig. 3 A typical example of the surface crack initiation site observed by SEM; (a) Ti-15-3STA and (b) Ti-6-4. Specimen surface was etched after interrupting the fatigue test.

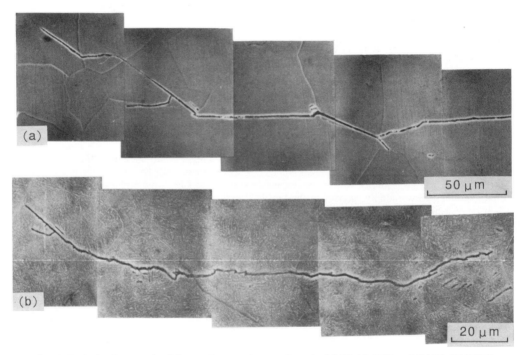

Fig. 4 Optical micrograph of the surface crack growth path; (a) Ti-15-3ST and (b) Ti-15-3STA.

large angle at the grain boundaries and is straight in the grain (Fig.4a). For Ti-15-3STA having Widmanstatten or acicular structure, crack growth keeps straight on and is not so strongly affected by colony and prior beta grain boundary (Fig.4b).

The experimental relationship between surface crack length, c, and the relative fatigue cycles, N/N_f is presented in Fig.5 for three kinds of specimens. The fatigue crack initiated at below $N/N_f=0.1$ for all specimens. This value obtained is smaller than that of steels and aluminum alloys reported in the literature. It is also evident that crack growth life up to 1mm is more than 90% on fatigue life. Therefore, consideration of small fatigue crack growth behavior is important for a titanium alloy. Crack initiation and growth behavior in corrosion fatigue were same as those in air.

Small Fatigue Crack Growth Law

Figure 6 shows the *dc/dN* vs. ΔK diagram, where ΔK was calculated by the method proposed by Raju-Newman considering the aspect ratio of crack geometry. It can be seen that no differences of *dc/dN* between materials is observed at high ΔK region, but at below a ΔK of about 15MPam$^{1/2}$ Ti-15-3ST and Ti-15-3STA show higher crack growth rates as compared with Ti-6-4.

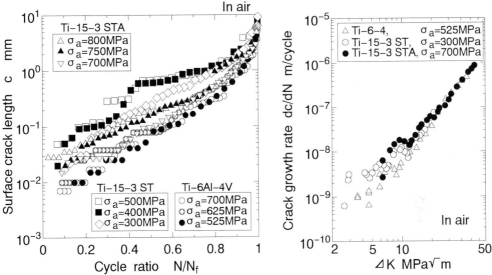

Fig. 5 Experimental relationship between surface fatigue crack length and relative fatigue cycles for Ti-15-3ST, Ti-15-3STA and Ti-6-4 tested in air.

Fig. 6 *dc/dN* vs. ΔK diagram for Ti-15-3ST, Ti-15-3STA and Ti-6-4 tested in air.

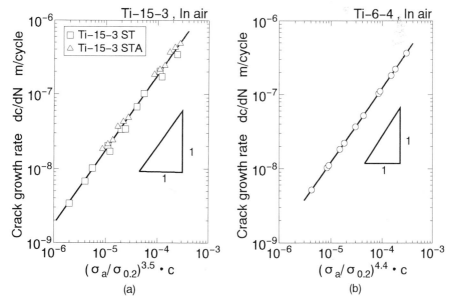

Fig. 7 Small fatigue crack growth law of (a) Ti-15-3 and (b) Ti-6-4.

Because the ($\sigma_a/\sigma_{0.2}$) was above 0.5 in this experiment, the condition of small scale yielding was not satisfied at a tip of small fatigue crack. Also the dependency of applied stress amplitude on crack growth rate was found in the relationship between dc/dN and ΔK. Therefore, small surface fatigue crack growth rate was evaluated in this study by application of the small crack growth law proposed by Nishitani *et al.*(1987). Fig.7 shows the relationship between dc/dN and ($\sigma_a/\sigma_{0.2}$)$^{3.5} \cdot c$ for two kinds of Ti-15-3 alloys. It reveals an excellent linear relationship and dc/dN can be expressed by following equation;

$$dc/dN = 1.81 \times 10^{-3} (\sigma_a/\sigma_{0.2})^{3.5} \cdot c. \tag{1}$$

It is of great interest from this result that growth rate of a small fatigue crack in two kinds of microstructure on Ti-15-3 alloy can be represented uniquely with a term of ($\sigma_a/\sigma_{0.2}$) considering the effect of material properties as proof stress. On the other hand, the small fatigue crack growth rate for Ti-6-4 was presented as follows;

$$dc/dN = 1.20 \times 10^{-3} (\sigma_a/\sigma_{0.2})^{4.4} \cdot c. \tag{2}$$

CONCLUSIONS

(1) A change in fatigue crack initiation mechanism has been observed for beta Ti-15V-3Cr-3Al-3Sn alloy having Widmanstatten structure (STA) under the stress ratio R=-1. Fatigue crack initiated at the surface for short cyclic lives, i.e. $N \leq 5 \times 10^6$ cycles, in contrast, for long cyclic lives and lower stress amplitude, i.e. $N > 5 \times 10^6$ cycles, subsurface fatigue crack initiation was observed.

(2) Subsurface crack initiation was observed not only in air but also in saline solution at same stress amplitude and number of stress cycles. A significant decrease in fatigue strength, after flattening of the fatigue curve, was associated with the transition of crack initiation mechanisms.

(3) In Ti-15V-3Cr-3Al-3Sn and Ti-6Al-4V, the S-N curve in saline solution was same as that in air. Therefore, these materials have a good corrosion fatigue resistance in saline solution.

(4) A surface fatigue crack initiated at below 10% of fatigue life and a small fatigue crack growth life was more than 90%. This behaviour did not depend on the condition of heat treatment and testing environment.

(5) Grain boundaries act as barriers against small surface fatigue crack growth and the orientation of crack growth path changed through a large angle at the grain boundaries for the Ti-15V-3Cr-3Al-3Sn solutionized (ST) and Ti-6Al-4V. But small surface crack in Ti-15V-3Cr-3Al-3Sn-STA grows along a continuous path through the grain and/or colony boundary.

(6) Small fatigue crack growth rate of Ti-15V-3Cr-3Al-3Sn-ST and -STA was determined uniquely by the term of ($\sigma_a/\sigma_{0.2}$) and was expressed by a small-crack growth law written with Eq.(1).

REFERENCES

Ruppen,J., P.Bhowal, D.Eylon and J.McEvily (1979). On the process of subsurface fatigue crack initiation in Ti-6Al-4V. *ASTM-STP 675* (J.T.Fong, Ed.), pp.47-68.

Atrens,A., W.Hoffelner, T.W.Duering and J.E.Allison (1983). Subsurface crack initiation in high cycle fatigue in Ti6Al4V and in a typical martensitic stainless steel. *Scripta Metallu.*, **17**, 601-606.

Adachi,S., L.Wagner and G.Lutjering (1985). Influence of microstructure and mean stress on fatigue strength of Ti-6Al-4V. *Proc. 5th Int. Conf. on Titanium Sci. and Tech.*, **4**, 2139-2146.

Nishitani,H. and M.Goto (1987). A small crack growth law and its application to the evaluation of fatigue life. *The Behaviour of Short Fatigue Cracks* (K.J.Miller and E.R.de los Rios, Ed.), pp.461-478.

Tanaka,S., H.Nishitani, S.Nishida, W.Fujisaki, T.Teranishi and M.Honda (1992). Initiation and propagation of cracks in Ti-15V-3Cr-3Sn-3Al alloy. *Trans. Jpn. Soc. Mech. Eng.*, **58**, 2268-2273.

JSMS, Committee of Fatigue (1994), *MIcrostructure, Mechanical Properties and Fatigue Strength of ($\alpha + \beta$)Titanium Alloy*.

Bian,J., K.Tokaji, M.Nakajima and T.Ogawa (1995). Effect of microstructure on fatigue behaviour of Ti-15Mo-5Zr-3Al alloy. *J. Soc. Mat. Sci., Japan*, **44**, 933-938.

Effect of Environment (vacuum, inert gas, nitrogen and air) on the Chemical Composition of the Plastic Zone at the Crack Tip of Titanium.

T.H. Myeong, R. Iguchi and Y. Higo

Precision and Intelligence Laboratory, Tokyo Institute of Technology, 4259 Nagatsuta, Midori-Ku, Yokohama, Kanagawa 226, Japan

ABSTRACT

Fatigue crack growth tests were performed on titanium in vacuum($\sim 10^{-7}$Torr), inert gases (He, Ar), nitrogen and air (R.H.=20%). Fracture surface morphologies were different from each other, even where these environments were mild. Microcracks, which were parallel to striations, were observed on the fracture surfaces, and their frequency increased as the environment became more active (He < N_2 < Air). It is of interest to note that no striation was observed on the fracture surface tested in Ar. These results suggest that environment has some effects on the deformation behaviour in the plastic zone at the crack tip. In order to consider this suggestion, the concentration of nitrogen in the plastic zone around a crack surface tested in nitrogen was analysed using electron probe micro analysis (EPMA). The concentration of nitrogen in the plastic zone, especially in the cyclic plastic zone, increased significantly. The result indicates that nitrogen may be adsorbed on the fresh surface produced at the crack tip during loading and diffuse into the cyclic plastic zone with cyclic dislocation movement. Considering all the results, it is thought that atoms of other environmental gases including argon have some effects on the chemical composition of the cyclic plastic zone as well as nitrogen.

KEYWORDS

titanium; fatigue crack growth tests; fracture surface morphology; crack tip; plastic zone; mild environment; chemical composition analysis.

INTRODUCTION

Fatigue crack growth requires cyclic plastic deformation at the crack tip. Therefore, it is important to understand the plastic deformation behaviour at the crack tip during fatigue tests. A number of studies to explain fatigue crack growth behaviour have been reported, but most of them were conducted in laboratory air. Air consists of many kinds of gases such as nitrogen, oxygen, argon, water vapor, etc. However, few studies have be done on the adsorption and absorption behaviour of the gases during fatigue cycling, and their contributions to the plastic deformation behaviour is unknown. Many studies of the effects of environment on fatigue crack growth rate have been performed. It is well known that mild environments such as inert gas, nitrogen and air do not have any strong effect on fatigue crack growth rate in Paris regime (Ritchie, 1977). Fatigue crack growth rate, however, does not give information about fatigue crack growth mechanism directly because it is only a value which represents the average growth distance per plastic deformation cycle at a certain stress intensity range (ΔK). When a fatigue test is conducted in an environment, the gas existing in the environment would be adsorbed on the fresh surface produced at the crack tip during loading.

The objectives of this study is to gain a better understanding the effects of environmental gases on deformation behaviour in the plastic zone at a crack through the use of fractography and a chemical composition analysis.

EXPERIMENTAL PROCEDURES

The material used was a hot-rolled pure titanium polycrystalline plate of 37mm thickness. Compact test (CT) specimens with 42mm width and 6mm thickness were removed from the material. Fatigue tests were performed under sinusoidal loading at a frequency of 10Hz and stress ratio of 0.1. The fatigue tests were carried out at room temperature (approximately 24℃), in an environment of either vacuum($\sim 10^{-7}$Torr), dry and high purity inert gases (Ar, He), a nitrogen gas (N), or air (R.H.=20%). After the fatigue tests, fracture surface observations were carried out using a scanning electron microscope (SEM). A specimen tested in nitrogen was sectioned at the middle of the specimen thickness and the concentration of nitrogen was analysed using an electron probe micro analysis (EPMA) at every 10μ m from the crack surface to a depth of 700μ m in the direction perpendicular to the crack surface (Lengauer et al., 1992)

RESULTS AND DISCUSSION

Fracture Surface Observation

A fracture surface tested in each environment was observed at the positions where values of ΔK were 15, 20 and 30 MPa m$^{1/2}$. No obvious difference was found in the surface morphology at a fuction of ΔK in each environment. The fractographs observed at a ΔK of 30MPa m$^{1/2}$ were shown in Fig. 1.

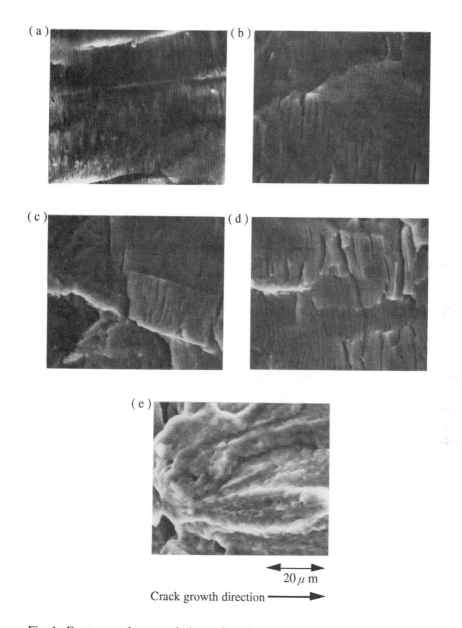

20 μ m

Crack growth direction ⟶

Fig. 1. Fracture surface morphology of specimens tested in (a) vacuum,
(b) helium, (c) nitrogen, (d) air, and (e) argon (ΔK = 30 MPam$^{1/2}$).

In Fig. 1 except (e), microcracks which are parallel to the striations were observed and the frequency of them increased with the environment becoming active (He<N$_2$ <Air). However,   no fatigue striation was observed on the fracture surface tested in argon (Fig. 1(e)). These results suggest that environment has some effects on the deformation behaviour in the plastic zone at the crack tip during fatigue crack growth. In order to consider this suggestion, the concentration of nitrogen in the plastic zone around a crack surface which was tested in nitrogen was analysed using electron probe micro analysis (EPMA).

Chemical Composition Analysis

 The relative concentration of nitrogen is shown in Fig. 2 as a function of the distance from the fracture surface.

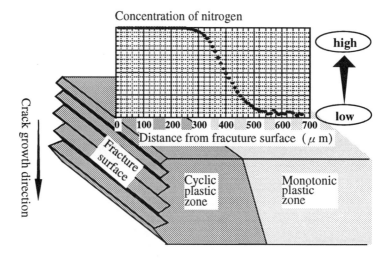

Fig. 2. Chemical composition of N in the plastic zone around the crack
surface which was tested in N$_2$(Δ K = 25 MPa m$^{1/2}$).

The concentration change in N in the plastic zone around the crack surface was found. In a region near the fracture surface, the concentration of nitrogen was high. From 260 μ m the concentration of nitrogen decreased gradually with the distance from the fracture surface and became constant at approximately 550 μ m. It is clearly demonstrated that environment have some effects on the chemical composition of the plastic zone at the crack tip during fatigue crack growth.

The analysed position was at a Δ K of 25 MPa m$^{1/2}$, where the cyclic plastic zone size under plain strain condition, r_p, was approximately 200μ m. This was calculated using the equation (1) (M.Klesnil et al., 1980) .

$$r_p = \frac{1}{\pi}\left(\frac{\Delta K_I}{2\sigma_y}\right)^2 \cdot \cos^2\frac{\theta}{2}\left[(1-2v)^2 + 3\sin^2\frac{\theta}{2}\right] \tag{1}$$

where σ_y is the monotonic yield stress of this material (450MPa) (Shimojo, 1992), v is the Poisson's ratio and θ is $\pi/2$. A comparison of the calculated value with the analysed result shows that the size of the high concentration region of nitrogen almost agreed with the cyclic plastic zone size.

A possible mechanism of N invasion into the plastic zone during fatigue crack growth is schematically shown in Fig. 3.

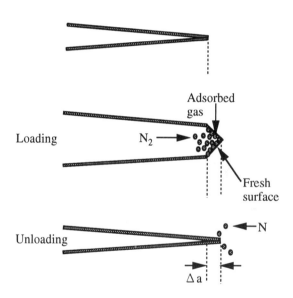

Fig. 3. Schematic showing the invasion mechanism of N
into the plastic zone

During loading fresh surfaces are produced at the crack tip. Nitrogen molecules in environment are adsorbed on the fresh surfaces. During unloading the adsorbed nitrogen invades into the plastic zone during dislocation movement.

Considering the difference in fracture surface morphologies in the environments and the chemical composition analysis, it can be thought that atoms or molecules of other environmental gases including argon have some effects on the chemical composition in the cyclic plastic zone as well as nitrogen.

CONCLUSION

Fatigue crack growth tests were performed on titanium in vacuum($\sim 10^{-7}$Torr), inert gases (He, Ar), nitrogen and air (R.H.=20%). The results were obtained as follows.

1) Even in mild environments, the fracture surface morphologies, including microcrack formation, were different from each other.
2) The concentration of nitrogen in the plastic zone around a crack surface which was tested in nitrogen increased significantly. The size of the increased region was almost in accordance with the cyclic plastic zone size. It is suggested that nitrogen may be adsorbed on the fresh surface produced at the crack tip during loading and diffuse into the cyclic plastic zone with cyclic dislocation movement.
3) It is thought that atoms of other environmental gases including argon have some effects on the chemical composition in the cyclic plastic zone as well as nitrogen.

REFERENCES

M.Klesnil and P.Lukas (1980). Materials Science Monographs, 7. Fatigue of Metallic Materials (Elsevier), pp 88-101.
M.Shimojo (1992). PhD thesis, Tokyo Institute of Technology, Japan.
R.O.Ritchie (1977). Influence of microstructure on near-threshold fatigue-crack propagation in ultra-high strength steel. Metal Science, 11, 368-381.
W.Lengauer, J.Bauer, A.Guillou, D.Ansel, J.P.Bars, M.Bohn, E.Etchessahar, J.Debuigne and P.Ettmayer (1992). WDS-EPMA nitrogen profile determination in TiN/Ti diffusion couples using homotypic standard materials. Mikrochimica. Acta, 107, 303-310.

INITIATION OF SHORT FATIGUE CRACKS IN RAILWAY AXLES

V. GROS, C. PRIOUL and Ph. BOMPARD

Ecole Centrale Paris, Laboratoire MSS/MAT, CNRS URA 850
Grande Voie des Vignes, 92295 Chatenay Malabry, France.

ABSTRACT

Bending fatigue tests were carried out using both smooth and notched specimens of a normalised medium carbon steel used for railway axles. The process of early crack development was observed by the replication method. This technique allows an accurate description of both crack initiation and short crack behaviour.

Crack initiation on smooth specimens occurs along the extrusion/intrusion bands in ferrite and the initial short cracks propagate in an irregular manner, due to periodical arrest on pearlite colonies. When crossing the pearlite the cracks propagate in the ferritic inter-lamellae spacing but microcracks grow preferentially in ferrite and generally bypass the pearlite colonies. Short cracks propagate much faster than long ones and below the threshold ΔK_{th} measured for long cracks. The different scales of the problem, and especially the limit length between short and long cracks, are reported in a Kitagawa diagram. In notched specimens, the crack initiation stage characterized by persistent slip band formation is no more observed.

KEYWORDS

Crack initiation, short fatigue cracks, medium carbon steel, Kitagawa's diagram,

INTRODUCTION

The railway axle is one of the main security pieces of trains. It is made of normalised medium carbon steel (0.4% C), and subjected to rotary bending. Fatigue is a potential cause for axle fracture. The few cracks, which can be detected during periodic non destructive testings, are initiated on notches resulting from the impact of ballast on the axle body. Therefore, in order to adapt the survey periodicity of axles to new service conditions, it is of great importance to accurately evaluate the initiation and propagation stages from notches.

As shown in Fig. 1 a real impact leads to a large perturbation of the fracture surface, with a high material hardening under the notch, an important surface relief (and so a large stress concentration) and even an adiabatic shear band related to the high speed impact. Due to the complexity of the local stress field around the notch, this first approach will not consider the consequences of adiabatic shear band formation and will concentrate mainly on the initiation stage.

For this purpose, the fatigue crack initiation has been studied first using bending tests on smooth specimens. The processes of crack initiation and propagation have been followed using a plastic replica method. The evolution of the cracks during the initiation, growth and coalescence stages, has been investigated by successive observations of the specimen surface. These measurements allow to determine the different scales of the problem according to Miller's terminology (Miller, 1993) and to define the limit between short and long fatigue cracks on a Kitagawa's diagram. In a second stage, notched specimens have been tested in order to determine the effect on fatigue life of both residual compressive stresses and stress concentrations in the notch.

Fig. 1 : Metallographic profile of a real impact

MATERIAL AND EXPERIMENTAL PROCEDURE

The material tested is a normalised medium carbon steel in the form of rolled bars about 65 mm in diameter. The chemical composition (wt. %) is reported in Table 1. The micro-structure is ferrito-pearlitic with lamellar pearlite and the mean grain size is about 20 μm.

Table 1 : Chemical composition (Weight %)

	C	Mn	Cr	Ni	Ti	Cu	Si	P	S
%	0,41	0,76	0,09	0,08	0,01	0,19	0,23	0,01	0,02

The mechanical properties (yield stress σ_{ys}, ultimate tensile strength σ_u, fatigue limit σ_w) are reported in Table 2.

Table 2 : Mechanical properties

σ_{ys} (MPa)	σ_u (MPa)	σ_w (MPa)	A %
350	600	270	25

The specimens have been machined from the bars after a 160 minute normalizing heat treatment at 860°C. Fatigue smooth or notched specimens consist of parallelepipedic bars of 90 mm in length, 5.7 mm in width and 4 mm in thickness. After machining, the smooth specimens are polished using emery papers and then electropolished to facilitate the observations of small crack growth. A light nital etching is performed in order to reveal the different phases and to characterize the interaction between microstructure and small crack behaviour. Notched specimens have undergone impact by a cone-shaped projectile in order to simulate impacted railway axles. These specimens are mechanically polished and etched with nital. Fatigue tests have been performed with a frequency of 25 Hz on a Schenck PWO machine applying a pure bending loading. The observations of crack initiation and short crack propagation have been made via cellulose acetate replicas which are taken from the surface of the specimens during fatigue tests and observed by optical microscopy. The evolution of the surface crack length with the number of cycles was quantified by examining the replicas in a reverse order.

RESULTS AND DISCUSSION

Smooth specimens

The micrographic observations of the replicas confirm that the initial surface damage results from the emergence of persistant slip bands in ferrite grains. The fatigue cracks initiate in these grains and grow along extrusions/intrusions bands (Fig. 2a). In the early stage of the propagation, fatigue cracks propagate in an irregular manner, growing preferentially in ferrite or along ferrite/pearlite boundaries. This discontinuity of the propagation is due to periodical crack arrest on pearlite colonies which acts as the first microstructural barrier to crack propagation. These observations agree with those of Togaji in a similar S45C steel (Tokaji *et al.*, 1988). In the second stage, the cracks change their growth direction and propagate perpendicularly to the principal stress direction. Figure 2b shows that this propagation of the crack through pearlite is carried out preferentially into the ferrite phase along the cementite lamellae which act as strong barriers to crack propagation.

 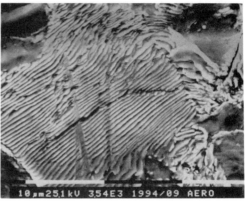

a : crack initiation in ferrite   *b : crack propagation in pearlite*

Fig. 2 : Crack paths at the surface of the specimen

During the first stage of crack propagation, a large discontinuity of the crack growth rate is measured (Fig. 3). This figure also shows the linking of the secondary cracks and their periodical arrests. If we

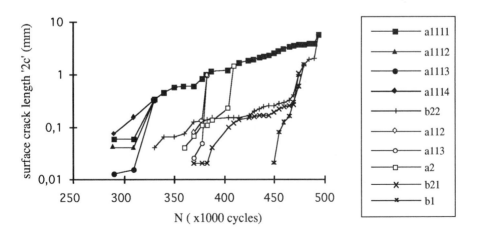

Fig. 3 : Growth data of cracks leading to final failure

plot the evolution of crack growth rate versus crack length (curve not presented), the smaller distance between two successive minima is about 20 μm, which corresponds to the mean grain size. So, the first microstructural barrier to crack propagation in this material appears to be the grain boundary. The fracture of the specimen occurs by coalescence of the secondary cracks which are lined up, labelled a_1 and b_1 in Fig. 3. The first crack a_1, results from the coalescence of two cracks a_{11} and a_{12}, each of them having a filiation a_{11i} and a_{12j} etc ...

Several non-propagating microcracks are observed, with a maximum size of about 350 μm (but most of them are smaller than the grain size). The crack front shape has being previously determined as been elliptical (Gros, 1994), and the relationship between crack length measured at the surface (2c) and crack depth (a) is 'a = 0.33(2c)'. The propagation becomes more regular as the crack grows and the effect of crack arrest on grain boundary vanishes once the crack exceeds a specific length which is a caracteristic of the material. This surface length is about 1 mm which corresponds to about 0.33 mm in depth.

The different scales of the problem can be summarized on a Kitagawa diagram (Fig. 4) where d_0 is the grain size, d is the limit between short and long cracks and l is the length from which we can apply the Linear Elastic Fracture Mechanic (LEFM). If the stress intensity factor corresponding to the fatigue limit (270 MPa) is calculated for d = 330 μm a value very close to 7 MPa.√m is obtained, which is the value of the threshold stress intensity factor determined for long cracks. So, these experimental observations of the fatigue crack initiation stage are in very good agreement with the results obtained in the propagation stage.

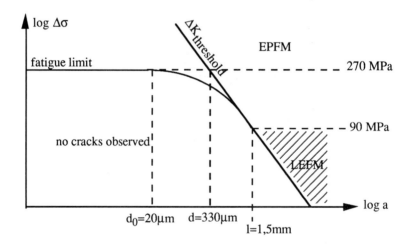

Fig. 4 : Kitagawa diagram of the material

The influence of the applied stress on short crack growth behaviour is reported in Fig. 5 which shows the evolution of the main crack (which leeds to the final failure) as a function of the cycle ratio N/Nf. Whatever the stress level, the data fall upon the same curve. This figure also shows that more than 70% of the fatigue life is devoted to the initiation and growth of small cracks (shorter than 1 mm). The main difference beetwen the three curves seems to be in the initiation stage since the first crack appears at 10%, 40% and 60% of the fatigue life for 350, 340, and 305 MPa respectively.

Using the Newman's relation to calculate the stress intensity factor (Newman et al., 1981) it is possible to plot the evolution of the crack growth rate versus the stress intensity factor (Fig. 6). The comparison between short and long crack growth rates confirms that short cracks propagate much faster than long ones. Furthermore, short cracks can propagate below the threshold ΔK_{th} measured for long cracks.

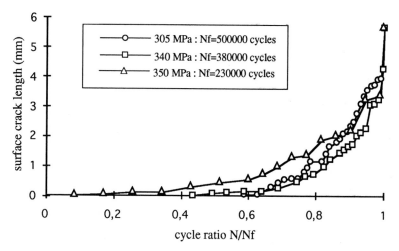

Fig. 5: Influence of the applied stress on the fatigue crack growth

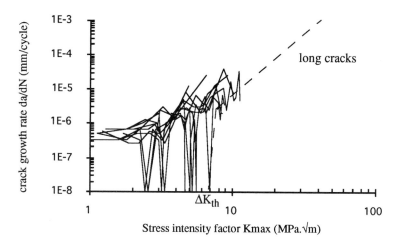

Fig. 6 : Crack growth rate versus stress intensity factor

Notched specimens

Fatigue tests on notched specimens show that the presence of an initial defect, like a notch, suppresses the crack initiation stage characterized by persistent slip bands formation. The fracture of notched specimens usually results from the coalescence of three major cracks which initiate on the edge and at the bottom of the notch (Fig. 7a). These initial cracks, which are not aligned, connect together to form a main crack. Outside the notch, this main crack reveals some branching.

These initiation sites correspond to the three initial zones of maximum axial stress, as confirmed by 3D finite element calculations (Fig. 7b) of the notch effect, on a quarter of a specimen. Nevertheless, no simulation of the residual stresses resulting from the impact has been introduced.

Since a reduction of the fatigue life by a factor of two (N_f = 200000 cycles for σ_{max} = 340 MPa) is measured, the effect of impact appears to be highly detrimental. So, the compressive stress field due to the impact does not compensate the stress concentration effect. Nevertheless, it must be noticed that the stress concentration effect (K_t) cannot alone explain this fatigue life reduction in impacted specimens. Indeed, the application of the Neuber's rule with either K_t or K_f leads to about 50000

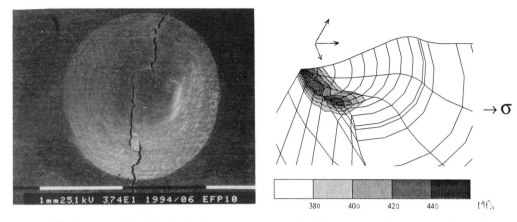

a : Coalescence of the three initial cracks   b : Maximum axial stresses at the notch

Fig. 7 : Crack propagation in a notched specimen

cycles to failure, which is much less than the experimental result. Consequently compressive stresses must be introduced in a complete fatigue life modelling.

If the crack length versus the cycle ratio is plotted for notched specimens, it appears that the curve (not reported) can also be superimposed on the curves obtained for smooth specimens (Fig. 5).

CONCLUSIONS

In smooth specimens of normalized ferrito-pearlitic steels, crack initiation occurs along the extrusion/intrusion bands in ferrite and the cracks propagate in an irregular manner, due to periodical arrest on pearlite. The microcracks grow preferentially in ferrite and when crossing the pearlite, the crack propagates in the ferritic inter-lamellae spacing. The comparison between crack growth rates of short and long cracks shows that short cracks propagate much faster than long ones and below the threshold ΔK_{th} measured for long cracks. The initiation mechanisms are different in notched specimens. These qualitative observations are not sufficient to explain the detrimental effect introduced by a notch, on the fatigue life of a component. Finite element calculations are needed to simulate impacts and to quantify residual stress fields.

This experimental study on smooth and notched specimens must be confirmed by traction-compression tests since the stress gradient in bending tests is probably too steep to be representative of the real stress gradient near the surface of railway axles.

Aknowledgements : The authors would like to thank the laboratories division of the S.N.C.F. (French railway) for their financial support and scientific contribution to this work.

REFERENCES

Gros, V. (1994). Rapport interne ECP-SNCF.

Kitagawa, H. and S. Takahashi (1976). Applicability of fracture mechanics to very small cracks or the cracks in the early stage. *Proc. 2nd Int. Conf. Mech. Behaviour of Materials, Boston*, 627-631.

Miller, K. J. (1993). Materials science perspective of metal fatigue resistance. *Materials Science and Technology*, Vol. 9, 453-462.

Newman, J. C. and I. S. Raju (1981). An empirical stress-intensity factor equation for the surface crack. *Engineering Fracture Mechanics*, Vol. 15, n° 1-2, 185-192.

Tokaji, K., T. Ogawa and S. Osako (1988). The growth of microstructurally small fatigue cracks in a ferritic-pearlitic steel. *Fatigue Fract. Engng. Mater. Struct.*, Vol. 11, n° 5, 331-342.

FATIGUE CRACK INITIATION AND PROPAGATION BEHAVIOR IN PEARLITE STRUCTURES

C.URASHIMA* and S.NISHIDA**

*Yawata R & D Laboratory, Nippon Steel Corporation,
Tobihata 1-1, Kitakyusyu 804, Japan
**Faculty of Science & Engineering, Saga University,
Honjyo-Machi 1 Saga 840, Japan

ABSTRACT

Fatigue processes in pearlite structures have been investigated, and several experiments were performed to improve the fatigue strength and fatigue limit ratio of the rail with full pearlite structure. From the results of metallurgical investigations, following results were obtained.

(1) Fatigue crack initiated from inter-lamella slip between cementite and ferrite in pearlite structure. Therefore, the fatigue strength of pearlitic steels is affected by the lamella spacing of pearlite related to the easiness of slip under loading of cyclic stress.

(2) Initiated fatigue crack is arrested by pearlite brock boundary. Therefore, the fatigue limit ratio is affected by the size of pearlite blocks.

(3) In order to obtain higher fatigue strength and higher fatigue limit ratio, it is desirable to reduce the lamella spacing and block size of pearlite.

KEYWORDS

Fatigue, Pearlite structure, Crack initiation, Crack propagation, Fatigue limit, Fatigue limit ratio

INTRODUCTION

In recent years, the speed of passenger trains and the load of freight trains have been increased to raise the efficiency of railway transport. In this connection, the occurrence of fatigue failure to rails has been sometimes observed(Besuner, 1978, Sugino et al., 1991) .

There have been many studies on fatigue of pearlitic steels(Gray III et al., 1985, Ishii et al., 1983, Sato, 1979). However, there is scarcely any study on such fundamental points as the initiation of fatigue crack and propagation behavior of small crack. The purpose of this study is to improve the fatigue strength and fatigue limit ratio of rail steels with full pearlite structure. In this study, the fatigue process of pearlitic steels is first investigated in detail and the factors of pearlite structure in the initiation and propagation of fatigue crack are clarified. Then, the effects of the lamella spacing and block size of pearlite structure on fatigue limit and fatigue limit ratio were investigated. Measures to improve the fatigue strength of pearlitic steels are considered on the basis of the results of these investigations.

EXPERIMENTAL METHOD

The materials used for test were taken from the heads of JIS 60k standard carbon steel in rails and AREA 136lb low-alloy Cr-V rails. The former was used for the microcsopic observation of fatigue process and its chemical composition and mechanical properties are shown in Table 1. The latter was used for to investigate the effect of the factors of pearlite structure on fatigue strength. The chemical composition of this rail steel are shown in Table 2, and the heat treatment conditions for changing the factors of pearlite structure and obtained mechanical properties are shown in Table 3. In this heat treatment, two kinds of pearlite lamella spacing and five kinds of block size were obtain. The optical microstrucure of its shows in Fig.1.

Table 1 Chemical composition and mechanical properties of material used
for the microscopic observation of fatigue process.

Chemical composition, (mass,%)					Mechanical Properties			
C	Si	Mn	P	S	P.S,MPa	T.S,MPa	El,%	R.A,%
0.78	0.23	0.86	0.013	0.005	534.1	872.2	14.0	24.0

Table 2 Chemical composition of material used for to investigate the
effect of the factors of pearlite structure on fatigue strength   (mass,%)

C	Si	Mn	P	S	Cr	V
0.75	0.28	1.27	0.015	0.010	0.76	0.12

Table 3 Heat treatment conditions and mechanical properties for to inveastigate
the effect of the factors of pearlite structure on fatigue strength

Symb.	Heat treatment conditions	Mechanical properties				Pearlite block size,No.
		P.S MPa	T.S MPa	El %	R.A %	
CVA	1250°C,30min 650°C,1hr in salt bath → A.C.	639.0	1038.8	13.6	20.6	2
CVB	1000°C,30min 650°C,1hr in salt bath → A.C.	632.1	1093.7	15.3	29.4	4
CVC	850°C,30min 650°C,1hr in salt bath → A.C.	551.7	1041.7	17.0	46.1	6
CV1	1000°C,30min 550°C,1hr in salt bath → A.C.	1177.0	1504.3	13.8	11.2	8
CV3	850°C,30min 550°C,1hr in salt bath → A.C.	1079.0	1387.7	21.4	45.4	10

The shape and size of fatigue specimen is shown in Fig.2. A dull shallow notch was given to part of the specimen surface in order to observe the fatigue process by limit the initiation point of a fatigue crack. The specimens were subjected to stress-relief treatment after polishing with emery paper, and were then electrically polished before they were used in fatigue test. The fatigue test was performed at a cyclic speed of 42Hz using 14.7Nm Ono-type rotating bending fatigue test machine. The fatigue process of some specimens were sampled as replica using AC film every constant number of cycles under two kinds of cyclic stress, and the fatigue process was investigated.

Fig. 1   Optical microstructures of each heat treatment specimen

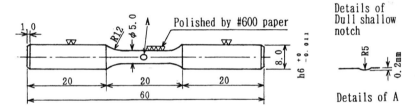

Fig. 2   Shape and size of fatigue test specimen

RESULTS AND DISCUSSION

Observation of Fatigue Process

The fatigue test was performed under cyclic stress of about 30MPa higher than the fatigue limit.   In the very initial stage of number of cycles, slip occures within a pearlite colony that has a constant inclination relative to the loading direction of stress, and a fatigue crack is initiated by this slip.   Figure 3 shows an enlarged picture of the initiation point of the fatigue crack.   Fatigue crack initiation is characterized by the facet that is initiated by the inter-lamella slip of pearlite structure.   This feature is different from the knowledge that in the case of static tension; the slip crossing a lamella is the initiation point of failure(Takahashi et al., 1978).   It might be thought that the inter-lamella slip of pearlite that serves as the initiation point of a fatigue crack occurs along the slip plane of ferrite.   Macroscopically, a small crack initiated by inter-lamella slip propagates vertically with respect to the stress axis while changing its propagation direction at the pearlite block boundary sometimes along lamella, sometimes across lamella. Figure 4 shows a non-propagating fatigue crack observed in a specimen that indicates the fatigue limit. The fatigue crack is arrested at the pearlite block boundary.   From this, it might be thought that the propagation of small fatigue crack is geartly affected by pearlite block size.

The macroscopic fatigue crack propagation rates of rail steels with full pearlite structure are a little fast than those of general steels when the comparison is made using the same ΔK value.   One of the feature

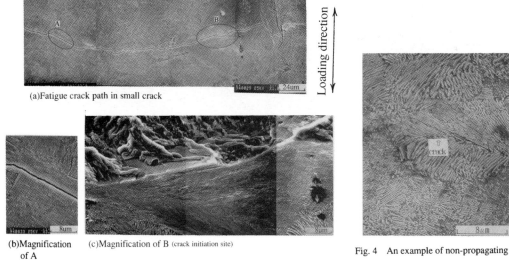

(a)Fatigue crack path in small crack

(b)Magnification   (c)Magnification of B (crack initiation site)
of A

Fig. 3   Enlarged picture of the initiation point of fatigue crack
(6a=401.8MPa, N=19.06×10^4 cycles)

Fig. 4   An example of non-propagating
fatigue crack
(6a=372.4MPa, N=10^7 cycles)

of this crack propagation is that no traces of slip are observed near the main faituge crack.   Figure 5 shows an enlarged picture of part of the features of crack propagation in the large-crack propagation region(ΔK=25MPa $\sqrt{m}$).   No slip that acrosses a pearlite lamella is observed near the crack.   Another feature of pearlitic steel is that you can not observe the striation that is one of tyipical pattern of fatigue fracture as illustrated in Fig. 6.   Only pearlite lamella themselves are observed on the fracture surface of the fatigue specimen.   The reason why no striations are observed on the fatigue fracture surface may be

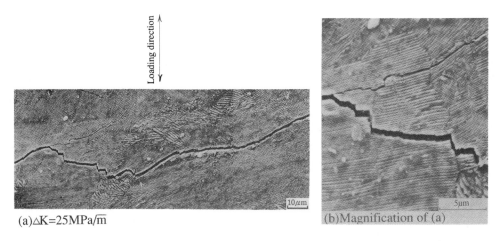

(a)ΔK=25MPa/$\overline{m}$ (b)Magnification of (a)

Fig. 5   An example of fatigue crack propagation behavior in large crack(ΔK=25MPa$\sqrt{m}$)

that hard cementite and soft ferrite form layers at very narrow intervals, with the result that the slip at the front of the crack is hindered by hard cementite.

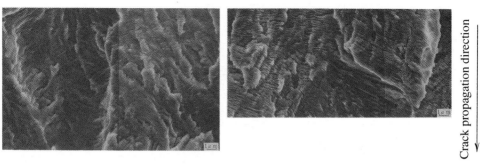

Crack propagation direction

Fig. 6   An example of fatigue fracture surface (ΔK=30MPa√m)

Improvement of Fatigue limit and Fatigue limit ratio

Table 4 and Figure 7 show the results of the rotating bending fatigue test in which of specimens with two kinds of pearlite lamella spacing and five kinds of block size.   Figure 8 and Figure 9 show the relationship between fatigue limit and pearlite lamella spacing and block size respectively.   Fatigue strength is affected by the pearlite lamella spacing; the specimen with smaller lamella spacing shows higher fatigue limit.   It might be thought that this is because of the occurrence of the inter-lamella slip that initiates a fatigue crack becomes difficult by reducing the lamella spacing.

Table 4   Fatigue test results of each heat treatment specimen

Symb.	Pearlite block size No.	Pearlite lamellar spacing,μm	Tensile strength T.S,MPa	Fatigue limit 6wb,MPa	Fatigue strength ratio
CVA	2	0.152	1038.8	402	0.387
CVB	4	0.125	1093.7	470	0.430
CVC	6	0.133	1041.7	451	0.433
CV1	8	0.050	1504.3	735	0.489
CV3	10	0.048	1387.7	588	0.424

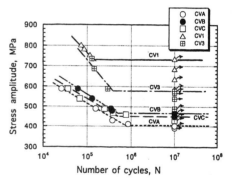

Fig. 7   S-N curves of each heat treatment specimen

There is a relatively good correlation between the fatigue limit ratio and the pearlite block size.   The fatigue limit ratio improves with decreasing block size.   It is considered that the reason why the fatigue limit ratio is affected by block size is related to arrest resistance of the initiated fatigue crack.   However, an optimum size exists.   It might be thought that this is because of that the initiation site of faitigue crack changes from inter-lamella slip of pearlite to inter-face of pearlite block boundary.

From the above results, it was clarified that the improvement of the fatigue strength and fatigue limit ratio of rail steels can be achieved by reducing the pearlite lamella spacing and controlling the block size.

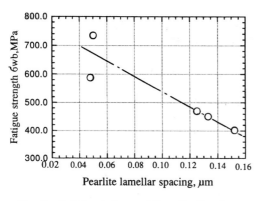

Fig. 8   Relation between fatigue limit and
         pearlite lamella spacing

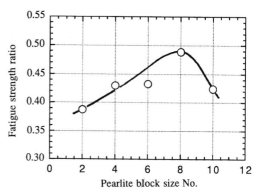

Fig. 9   Relation between fatigue limit ratio and
         pearlite block size

CONCLUSIONS

To improve the fatigue strength and fatigue limit ratio of rail steels with full pearlite structure, the fatigue process was microscopically observed and the effect of the factors of pearlite structure on fatigue properties was investigated. The following results were obtained.

(1) The fatigue crack initiated from inter-lamella slip between cementite and ferrite in pearlite structure. Initiated fatigue crack was arrested, and its propagation rate was delayed at the pearlite block boundary. Therefore, the pearlite lamella spacing affects fatigue strength with respect to inter-lamella slip initiation limit stress, and the block size influences the fatigue limit ratio with relative to the arrest resistance of a fatigue crack propagation.

(2) The fatigue crack in full pearlite structure does not propagate by the striation forming mechanism based on slip at the tip of the crack. It propagates as cutting lamellae of pearlite or being accompanied by inter-lamella slip.   Therefore, striation observes in general steels are not observed on fracture surfaces of pearlite structure.

(3)   The improvement of the fatigue strength and fatigue limit ratio of rail steels with full pearlite structure can be achieved by reducing the pearlite lamella spacing and controlling the block size.

REFERENCES

P.M.Besuner (1978).   Fracture mechanics analysis of rails with shell-initiated transverse cracks. ASTM STP, 644, 303-329.

G.T.Gray III, A.W.Thompson and J.C.Williams (1985).   Influence of microstructure on fatigue crack initiation in fully pearlitic steels. Metallurgical Transactions A, 16A, 753-760.

H.Ishii and T.Sasaki (1983).   The effect of microstructure and environment on the fatigue of eutectoid steels. Journal of the Japanese Society for Strength and Fracture of Materials, 17, 65-77.

T.Sato (1979).   The effect of the structure on the impact and fatigue properties of weld-affected zone in a high carbon steel. Journal of the Japan Institute of Metals, 43, 908-917.

K.Sugino, H.Kageyama, C.Urashima and A.Kikuchi (1991).   Metallurgical improvement of rail for the reduction of rail-wheel contact fatigue failures. Wear, 144, 319-328.

T.Takahashi, M.Nagumo and Y.Asano (1978).   Microstructures dominating the ductility of eutectoid pearlite steels. Journal of the Japan Institute of Metals, 44, 708-715.

THE INFLUENCE OF NITROGEN AND GRAIN SIZE ON THE NUCLEATION AND GROWTH OF SURFACE CRACKS IN AUSTENITIC STAINLESS STEEL

U. Lindstedt, B. Karlsson and M. Nyström*

Department of Engineering Metals
Chalmers University of Technology
S-412 96 Göteborg, Sweden

*AB Sandvik Steel
R&D Centre
S-811 81 Sandviken, Sweden

ABSTRACT

The evolution of short fatigue cracks during low cycle fatigue testing in austenitic stainless steel with base composition resembling 316L has been studied. The influence of the nitrogen content and grain size was investigated. Two steels were studied: one steel with 0.14% nitrogen at grain sizes 43 and 130 μm respectively and one with 0.29% nitrogen at grain size 47 μm. Surface cracks begin to nucleate at an early stage during cyclic straining. In the low nitrogen steel the cracks appear earlier. In both steels there is a saturation in the crack density appearing after about half of the life time. The crack growth rate is faster in the low nitrogen steel leading to shorter total life time. The growth character of surface cracks is more related to slip bands in the high nitrogen steel.

KEYWORDS

Austenitic stainless steel; high nitrogen steel; grain size dependence; low cycle fatigue; short fatigue cracks; replication technique.

INTRODUCTION

Austenitic stainless steels are used in a wide variety of applications because of their good formability and mechanical properties combined with excellent corrosion resistance. The austenitic stainless steels are the most widely used type of stainless steel, often in constructions subjected to dynamic loads. The influence of the microstructure on the fatigue properties of stainless steels is therefore of great interest.

An important factor determining the fatigue life of a material is the initiation and growth of short cracks. In homogenous materials short cracks are expected to initiate at the surface, due to formation of extrusions and intrusions on the surface. In a microstructure containing stiff particles, the stresses and strains are higher at particles intersected by a free surface than at particles within the matrix (Levin and Karlsson, 1993). This observation further confirms that the surface runs the bigger risk for fatigue crack nucleation. It has been established that surface cracks often initiate in the early stages of the fatigue life (Chang, 1986) and that successive growth and linking of cracks occur until final failure. The growth behaviour of short cracks deviates from the behaviour of long cracks. For short cracks, an initial rapid growth rate at very low ΔK's is often followed by a deceleration as the crack grows. Finally, at a certain size, the crack growth of the short crack approaches the behaviour of a long crack. The deceleration at low ΔK's has been attributed to development of crack closure and to interaction between the crack and grain boundaries (Suresh, 1991).

The aim of this work is to study the nucleation and growth of short surface cracks during cyclic straining until final fracture. In a companion paper (Nyström et al., 1995) the influence of the nitrogen content and grain size on the cyclic deformation response of austenitic stainless steel is studied.

MATERIALS

The two materials used in this study are fully austenitic stainless steels, where one is the commercial grade 316LN named LN with 0.14%N and the other an experimental alloy denoted HN with 0.29%N. The chemical compositions are shown in Tab. 1. The LN steel was studied at two different grain sizes, λ=43 and λ=130 μm respectively, and the HN steel at λ=47 μm (λ=mean intercept grain size).

Table 1. Chemical composition (weight-%).

Material	C	Si	Mn	P	S	Cr	Ni	Mo	Cu	N
LN	0.024	0.50	1.16	0.025	0.001	17.4	10.01	2.76	0.42	0.140
HN	0.027	0.47	0.99	0.017	0.012	21.6	9.83	2.64	0.04	0.294

EXPERIMENTAL

Short cylindrical test bars with a gauge diameter of 5 mm and gauge length of 15 mm were machined from 12 mm thick plates. The surfaces of the reduced section of the specimens were ground and polished, the last polishing step with 1 μm diamond paste. The low cycle fatigue (lcf) tests were performed at room temperature in a servo-hydraulic testing machine (Instron 8500). The tests were controlled by the total strain ($\Delta\varepsilon_t$) and the machine was running in the push-pull mode. The strain wave form was sinusoidal with a mean strain rate of $5 \cdot 10^{-3}$ s^{-1}. Very good reproducibility and low scatter in the results were observed during the cyclic tests, indicating very good uniformity in the grain structure.

Nucleation and growth of surface cracks were studied by a replication technique. During the lcf testing, the tests were successively stopped to produce replications of the outer surfaces of the specimens. At each preselected number of strain cycles the specimen was left in tension load to ensure open cracks during replication. Later evaluation of replicated cracks in an optical microscope allowed cracks with lengths >5 μm to be recorded. $\Delta\varepsilon_t/2=0.01$ was selected as a comparative total strain amplitude and it corresponds to a life time of a few thousand cycles.

RESULTS AND DISCUSSION

The surface cracks were measured by individual lengths and number per unit area by optical microscopy of the replicas. According to (Rubinstein, 1986) the stress intensity factor can be raised if two neighbouring cracks are favourably positioned. Thus, closely placed cracks are likely to interfere and be transferred into one single crack. This feature of two interacting cracks was taken into account at the recording of "effective" lengths and densities of surface cracks.

The results of the crack measurements for the two steels at the smallest grain size are given in Figs. 1a and 1b. Nucleation of cracks takes place in the early stages of the total life time (Fig. 1a). This nucleation is in progress until a saturation plateau is reached at about half of the life time. It can be noted that the initiation of surface cracks starts earlier in the low nitrogen steel LN and that the crack density in this steel reaches a higher level (Fig. 1a). The growth rate of the cracks is also higher in the LN steel (Fig. 1b). Studying the data points in Fig. 1b, there appears to be an indication of intermediate saturation of the mean crack length in both steels corresponding to a dip in the growth rate of the surface cracks. This dip, as said earlier, can be caused by crack closure and/or grain boundary-crack interactions.

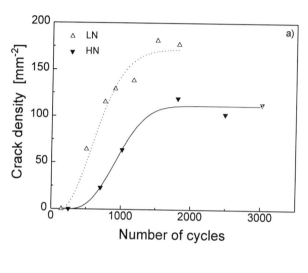

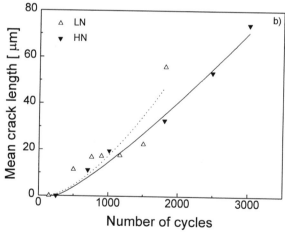

Fig. 1.   a) Surface crack density on specimen surfaces during lcf testing at $\Delta\varepsilon_t/2=0.01$. HN ($\lambda=47$ μm) and LN ($\lambda=43$ μm). b) Surface crack growth during lcf testing at $\Delta\varepsilon_t/2=0.01$. HN ($\lambda=47$ μm) and LN ($\lambda=43$ μm). The number of cycles to failure for the two specimens was N_f(HN)=3200 and N_f(LN)=1900 respectively.

The arithmetic mean crack length a_{mean} of the short surface cracks as a function of the number of cycles, Fig. 1b, can, if the dip described above is not considered, approximately be described by an exponential expression:

$$a_{mean} = k_1 \cdot (N - N_0)^m \tag{1}$$

where a_{mean} is the visible mean crack length on the surface, N the number of cycles and N_0 the number of cycles before cracks appear at the surface. Such a growth behaviour of short cracks has been seen earlier (Vasek and Polák, 1991) although it can not be expected to occur for all materials (Evertsson, 1993; Levin and Karlsson, 1993).

The nucleation of surface cracks as seen in Fig. 1a can be approximated with an Avrami type of equation describing the density of cracks N_A as a function of the number of strain cycles N:

$$N_A = k_2 \cdot \left(1 - exp\left(-k_3 \cdot (N - N_0)^n\right)\right) \tag{2}$$

There is a marked difference in the crack density N_A between the two steels at all cycle numbers (Fig. 1a). On the other hand, the difference in mean crack length at a given number of cycles is less clear (Fig. 1b), except for the fact that the LN steel fails at a lower cycle number. Studies of the replicas reveal a large scatter in crack size for the LN steel and this steel has a lot more small surface cracks (<10 μm) than the HN steel. This probably has to do with the more intense nucleation of cracks, *cf.* Fig. 1a. The higher amount of small cracks in the LN steel keeps the mean crack length on a relatively low level. In Fig. 2 the standard deviation is added to the mean crack length and Eq. (1) is used to describe the growth. Here it is seen that $a_{mean}+\sigma$ grows much faster in the LN steel than in the HN steel. Thus, in the LN steel a few surface cracks that form early grow fast and determine the life to failure. In the HN steel, such large cracks are not so readily formed.

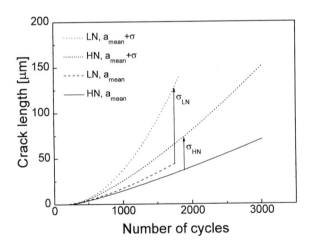

Fig. 2.   Surface crack growth during lcf testing at $\Delta\varepsilon_t/2=0.01$. HN ($\lambda$=47 μm) and LN ($\lambda$=43 μm). Note the much larger standard deviations of the crack lengths for the LN steel, partly explaining the shorter fatigue life of this steel.

Replication studies of the LN steel at the coarser grain size show the same principal behaviour in the evolution of the mean surface crack length and surface crack density as for the steels discussed above.

During low cycle fatigue testing, the stress is always lower in the LN steel compared to the HN steel at comparative strain levels. This deformation response together with the crack growth behaviour presented in Fig. 2 clearly show that the short cracks in the LN steel have a lower crack growth resistance than those in the high nitrogen steel HN. The reason for this can be sensed from the SEM micrographs in Fig. 3, showing the smooth surfaces of the lcf specimens. The growth of short cracks in the high nitrogen steel HN is accompanied by opening of slip bands adjacent to the cracks. This mechanism can be compared with crack branching and it consumes energy, thus slowing down the propagation rate. The SEM studies further show that the HN steel has intense slip band formation on the surface, Fig. 3b. The surface of the LN steel, in contrast, is sparsely covered with slip bands, Fig. 3a. Nitrogen, as said earlier, promotes planar slip and it is obvious that the difference in nitrogen content between the LN and HN steels has an effect on protruding slip bands on the surface. Due to these surface features, a difference in the propagation of short surface cracks in the two steels is observed. When a surface crack propagates in the HN steel it jumps between slip bands on the surface and the crack gets a jagged shape. In the LN steel the surface cracks have a more smooth character.

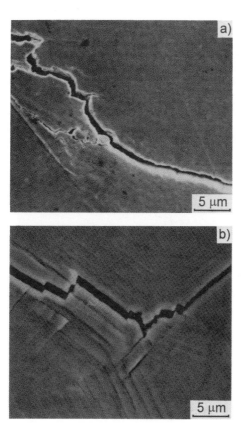

Fig. 3.   SEM micrographs showing the different growth characteristics of short cracks in the a) LN and b) HN steels. Note the opening of slip bands close to the main crack in the HN steel.

CONCLUSIONS

The present study concerns austenitic stainless steels with two different nitrogen levels (0.14 and 0.29 weight-% respectively). One of the materials was a commercial grade 316LN, and the other one an experimental alloy with similar base composition. The nucleation and growth of short surface cracks are described. The following main conclusions can be drawn:

1. Short cracks begin to nucleate at the surface at the early stages of cyclic straining. In the low nitrogen steel cracks appear earlier. In both materials a saturation in the total number of cracks is reached at about half of the life time. The growth rate of appearing cracks is larger for the low nitrogen steel, leading to a more rapid reaching of critical crack sizes and consequently shorter life time of the steel.

2. The formation of slip bands on the surface is more intense in the high nitrogen steel. The character of the growing cracks is strongly related to these slip bands in the high nitrogen steel while the low nitrogen steel develops smoother cracks. This difference in slip band formation and growth behaviour of surface cracks in the two materials can partly explain the shorter fatigue life of the low nitrogen steel.

ACKNOWLEDGEMENTS

This work was financially supported by the Swedish Research Council for Engineering Sciences. Dr Peter Sotkovszki is acknowledged for assisting during metallographic studies.

REFERENCES

Chang, N. S. (1986). Fatigue Crack Initiation and Detection of Surface Damage in an Age Hardening Stainless Steel, Dissertation, Wayne State University, Detroit, Michigan.

Evertsson, M. (1993). Fatigue of Duplex Stainless Steels, Diploma Work, Chalmers University of Technology, Göteborg.

Levin, M. and B. Karlsson (1993). Crack Initiation and Growth During Low-Cycle Fatigue of Discontinuously Reinforced Metal-Matrix Composites. *International Journal of Fatigue*, **15**, 377-387.

Nyström, M., U. Lindstedt, B. Karlsson and J.-O. Nilsson (1995). The Influence of Nitrogen and Grain Size on the Cyclic Deformation Behaviour of Austenitic Stainless Steel. *This conference*.

Rubinstein, A. A. (1986). Macrocrack-Microdefect Interaction. *Journal of Applied Mechanics*, **53**, 505-510.

Suresh, S. (1991). *Fatigue of Materials*, Press syndicate of the University of Cambridge, Cambridge.

Vasek, A. and J. Polák (1991). Low Cycle Fatigue Damage Accumulation in Armco-Iron. *Fatigue and Fracture of Engineering Materials and Structures*, **14**, 193-204.

Fatigue crack propagation
in Spheroidal Graphite cast Iron (SGI)

Y.NADOT*   N.RANGANATHAN*   J.MENDEZ*   and   A.S.BERANGER**

* Laboratoire de Mécanique et de Physique des Matériaux - URA CNRS 863
ENSMA - Site du Futuroscope - BP 109
86960 FUTUROSCOPE cedex FRANCE
tel : (33) 49 49 82 18   fax : (33) 49 49 82 38

** RENAULT SA
Direction de la recherche - Service 60152
860 Quai de Stalingrad   92109 BOULOGNE BILLANCOURT cedex FRANCE

ABSTRACT

SGI materials contain different types of heterogeneities in the microstructure which could explain important scatter in fatigue lives. Observations on fracture surfaces tend to show that surface defects are more dangerous than internal defects. This could be attributed to environmental effects on crack propagation. In order to verify this point, fatigue crack growth behaviour was analysed in air and in vacuum. Results show a high sensitivity of SGI to environment : crack growth rate is much higher in air than in vacuum even at high ΔK levels.

KEYWORDS

crack propagation ; nodular graphite iron ; environment.

INTRODUCTION

Spheroidal Graphite cast Iron is a very good material for industrial applications : mechanical properties are similar to those of commonly used steels and castability is better than for low carbon steels. This is the reason why they are used in the car industry for engine and safety components (ex : crankshafts and suspension arms). However, due to casting process, microstructure heterogeneities are found both in industrial and laboratory castings. Many studies showed that fatigue behaviour depends on the characteristics of these defects (nature, size and distribution) and specially on their position, inside the material or at the specimen surface. These two types of cracks do not propagate under the same environmental conditions. The role of this factor must be therefore elucidated and crack propagation laws in air and vacuum established.

In the first part of this paper the different types of defects at the origin of fatigue failure in SGI materials are described. A general correlation between the defect position and fatigue life is discussed. The second part deals with crack growth behaviour in air and vacuum. Environmental effects are discussed through the differences in crack growth rates, crack closure and fracture mechanisms on a ferritic nodular cast iron.

TYPICAL MICROSTRUCTURE OF SGI

<u>Microstructure of the as cast material</u> :

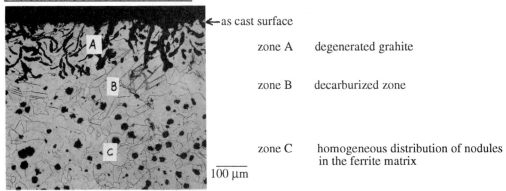

←as cast surface

zone A   degenerated grahite

zone B   decarburized zone

zone C   homogeneous distribution of nodules
             in the ferrite matrix

100 μm

fig. 1 : evolution of the microstructure from
the surface to the bulk in the as cast material

The microstructure illustrated in fig. 1 is similar from one casting to another, however other defects can also be found in different casting :

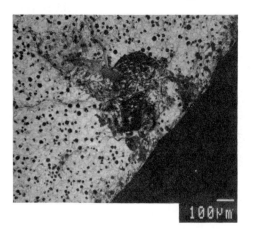

<u>near cast surface defect</u> : dross defect produced by components that take part in the casting process ; it can be 2 mm deep from the casting surface.

fig. 2 : SEM image of a dross defect on a fracture surface using backscattered electrons

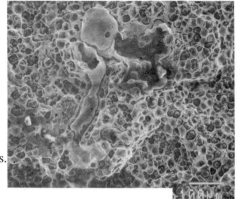

<u>internal defects</u> : shrinkages are found inside the matrix. The biggest (Ø 2-3 mm) are in the center of the sample (the last part cooled inside the mould). Many microshrinkages (less than 500 μm) are observed distributed in the matrix in all the castings.

fig. 3 : microshrinkage on a fracture surface

RELATION STRUCTURE / FATIGUE LIFE

Under cyclic loading, heterogeneities in the microstructure of the as cast material lead to local stress concentration which give rise to crack initiation. In a recent study in collaboration with Renault SA concerning the endurance limit of a ferritic nodular cast iron (Nadot et al., 1995) we have shown that lifetime is dependent on the type of defect :
- for the same size range, surface defects give rise to a lifetime always lower than internal defects
- for the same type of defect (surface or internal), size is the parameter controlling lifetime.

Illustration of those conclusions are given schematically in the following.

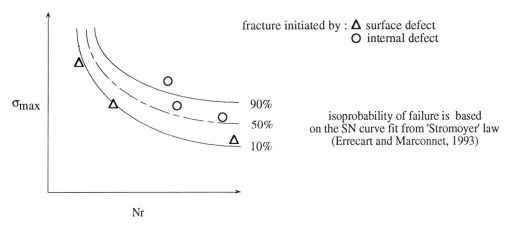

fig. 4 : lifetime dependence with the type of defect ; R = -1 ; AIR

Fig. 4 illustrates the fact that surface defects are more dangerous than internal defects : when the main crack is initiated from a surface defect, fatigue life is lower than the 50% SN curve. In contrast the internal defects that cause the formation of the fatal crack are associated with a fatigue life higher than the 50% SN curve.
Internal crack develops a circular front from the center of the specimen so that the crack mainly propagates in inert atmosphere.

These observations allow to point out different factors controlling fatigue life in such a material :

- enviromental effects on crack initiation and crack propagation
- defect size and position

In the following, this study will be focussed on the effect of environment on crack growth behaviour.

CRACK GROWTH BEHAVIOUR in a SGI

material and experimental details
In the present study, SGI has been provided by Renault SA. The matrix contains 95% ferrite and 5% pearlite, the volumic fraction of nodules is 10% with a ferrite grain size of 30 μm and a graphite size of 15 μm. Mechanical properties are : E=180 GPa, σ_y=380 MPa and UTS=510 MPa. The material tested here was taken from the center of a casting block : cast surface was 5 mm machined in order to eliminate surface heterogeneities so that the present results concern the bulk matrix. Only a few microshrinkages are observed that do not modify long crack behaviour. Tests were conducted at 35

Hz on a servohydraulic machine. CT 75 (B=12 mm) specimens were used and crack length was mesured on polished surface using a travelling microscope. Crack closure was measured by the mean of a back-face gage and COD extensometer using the differential compliance method. Two environments were studied at room temperature : ambient air and high vacuum (4.10^{-4} Pa).

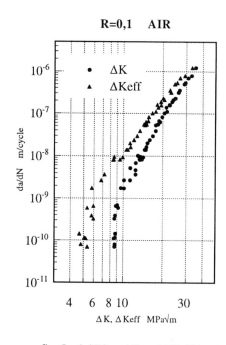

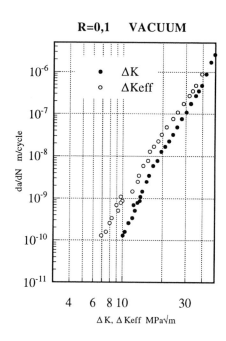

fig. 5 : da/dN vs ΔK and ΔKeff in air   fig. 6 : da/dN vs ΔK and ΔKeff in vacuum

general features
Fatigue crack growth rate curves in air and vacuum are presented in fig. 5 and 6. The effect of crack closure is readily seen by comparing the da/dN vs ΔK and the da/dN vs ΔKeff curves . The effective curve in air shows two different parts : a linear portion described by the mean of the Paris law and the low growth rate regime where crack propagation decreases rapidly with ΔKeff (threshold regime). In contrast the effective curve in vacuum is linear in all the domain explored.

crack closure
It is noticeable that for this material crack closure in vacuum remains significant at high ΔK levels (fig. 6). This effect is more pronounced on SGI than on ferritic steels (Taylor, 1985) ; that tends to prove that there is a special effect induced by nodules present in the ferritic matrix as mentionned by Ogawa (1987). Attention is given on the fact that the present results are strictly normed by ASTM procedure (ASTM, 1987) in order to ensure that load history can not influence the rate of closure induced by roughness and plasticity. For high crack growth rates (da/dN > 10^{-8} m/cycle in air) closure is similar in air and vacuum at the same ΔK. However, for low ΔK levels corresponding to growth rates between 10^{-8} and 10^{-9} m/cycle in air, crack closure becomes higher in air than in vacuum. We will see in the following that environment plays a major role in this growth rate range. Near the threshold regime, a consequent oxide layer is visible on the fracture surface. However crack closure is not observed to change ; that tends to prove that this oxide layer does not induce significant mechanical effects on crack closure.

environmental effects

The main characteristics of the curves of fig. 5 and 6 consists in their different growth rates. For $\Delta Keff < 10$ MPa$\sqrt{m}$ crack growth rate in air is one order of magnitude faster than in vacuum. This difference is considerable but of the same order as in other metallic materials. For $\Delta K = 30$MPa$\sqrt{m}$ the difference in growth rates between air and vacuum is still significant : half an order of magnitude which is a high value compared to metallic materials in general (steels, titanium and aluminium alloys). Environment has also a strong effect on the threshold ΔK level ; low growth rate regime in air shows a discontinuity : crack propagation decreases rapidly for little changes in ΔK. This phenomenon is not observed in vacuum. All these point shows that crack growth rate in ferritic SGI is very sensitive to environment through all the domain studied.

cracking mechanisms

As it can be seen on figures 7 and 10, cracking mechanisms at high ΔK level are similar in air and vacuum : cleavage-like is found on a fully transgranular fracture. At low ΔK levels (da/dN $< 10^{-8}$ m/cycle) micromechanisms of crack propagation are different. In air, transgranular fracture mode is still predominent but intergranular fracture also occurs (30% on the fracture surface - see fig. 8). In fig. 11 it is shown that only a few intergranular zones exist in vacuum. That clearly shows that intergranular failure is due to an embrittlement induced by ambient air. The increase of crack closure discussed before (at the same growth rate) appears therefore related to the appearance of intergranular fracture. In fig. 9 and 12, we can see that the influence of environment is not only related to this intergranular embrittlement. Transgranular microfracture mechanisms are also different in air and vacuum : fig. 9 shows brittle fracture surfaces in air with crystallographic facets oriented in the crack growth direction. In contrast, fig. 12 reveals that in vacuum microfracture processes are characterized by fine striation-like marks perpendicular to the crack propagation ; fracture surface appears therefore to be much more ductile than in air.

CONCLUSION

SGI materials exhibit different types of inhomogeneities associated with the casting process. According to the nature, the size and the position of these heterogeneities in the fatigue samples, failure takes place near the surface or in the bulk. The important scatter observed in fatigue lifetime of SGI materials seems to be related to such crack initiation sites position. For the same size, surface defects are more damaging than internal defects.

The effect of the environment on fatigue crack propagation of SGI materials has been characterized by comparing crack growth rates in air and vacuum. It has been shown that crack propagation behaviour is strongly affected by the ambient air in all the domain of ΔK explored. The effect is more pronounced at high ΔK level compared to other metallic materials.

This strong influence of environment on the fatigue crack propagation of SGI materials can explain partially the negative role played by surface defects compared with internal defects.

ACKNOWLEDGEMENTS

The financial and technical support of Renault SA for this study is greatfully acknowledged.

REFERENCES

ASTM E647-86a

Errecart R. and E. Marconnet (1993). publication interne. Renault SA

Nadot, Y., J. Mendez, N. Ranganathan and A.S. Beranger (1995). Fatigue damage in Spheroidal Graphite cast Iron. To be published

Ogawa, T. and H. Kobayashi (1987). Near-threshold fatigue crack growth and crack closure in nodular cast iron. *Fat. Fract. Engng. Mater and Struct.* vol. 10, No. 4, pp. 273-280

Taylor, D. (1985). *A compendium of fatigue thresholds and growth rates* . EMAS. United Kingdom.

crack propagation

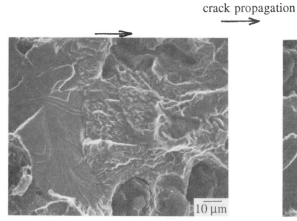

fig. 7 : cleavage-like in air
$\Delta Keff=18$ MPa$\sqrt{m}$ da/dN=10^{-7} m/cycle

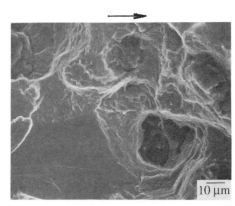

fig. 10 : cleavage-like in vacuum
$\Delta Keff=24$ MPa$\sqrt{m}$ da/dN=10^{-7} m/cycle

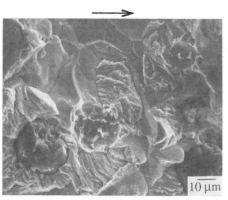

fig. 8 : mixed mode in air
$\Delta Keff=8$ MPa$\sqrt{m}$ da/dN=$4\ 10^{-9}$ m/cycle

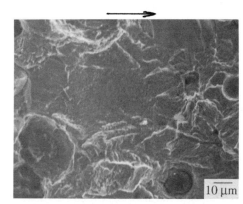

fig. 11 : transgranular mode in vacuum
$\Delta Keff=10$ MPa$\sqrt{m}$ da/dN=10^{-9} m/cycle

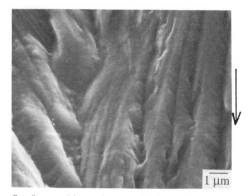

fig. 9 : cracking micromechanisms in air
$\Delta Keff=8$ MPa$\sqrt{m}$

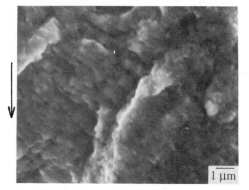

fig. 12 : cracking micromechanisms in
vacuum $\Delta Keff=8$ MPa$\sqrt{m}$

ANALYSIS OF SMALL FATIGUE CRACKS IN HSLA-80 STEEL

R. C. McCLUNG and T. Y. TORNG

Southwest Research Institute
P. O. Drawer 28510
San Antonio, TX 78228-0510 USA

ABSTRACT

Fatigue crack growth rate tests are performed with both large and small cracks in HSLA-80 steel. Small cracks grow at ΔK levels below the large crack threshold. The large and small crack data are rationalized with engineering models which account for plasticity-induced crack closure. Probabilistic crack growth models are developed to describe the increased scatter in the small crack data.

KEYWORDS

fatigue crack growth, small cracks, HSLA-80 steel, crack closure, probabilistic analysis

INTRODUCTION

"Small" fatigue cracks are sometimes observed to grow faster than traditional "large" cracks at the same nominal value of the cyclic driving force, ΔK. In some cases, these "small" cracks are small in comparison to characteristic microstructural dimensions, such as grain size. When the crack grows beyond this microstructural scale (e.g., 5-10 grain diameters), the growth rates often merge with the trends of the large crack data. In other cases, "small" cracks are small in comparison to characteristic mechanical dimensions, such as notch or crack-tip plastic zones, because local or nominal stresses are relatively large and strict small-scale yielding conditions are violated. Both types of small cracks sometimes grow at nominal applied ΔK values less than the large-crack threshold value, ΔK_{th}.

Small-crack behavior can be especially important for structural life prediction, since cracks are often small for a significant portion of the total life. The general trend to replace traditional safe-life design methods (e.g., strain-life analysis) with damage tolerance methods (i.e., crack growth analysis) for highly stressed fatigue-critical components further increases the significance of the small-crack problem. However, robust engineering methods to analyze small-crack behavior for life prediction purposes are not generally mature. Although fatigue life management techniques based on large-crack behavior are relatively mature, it is not always clear how these data and techniques can be applied to small cracks. Are the same correlating parameters valid? Is the large crack threshold valid? And what about the large scatter commonly observed in small-crack behavior? Can this scatter be addressed in the life prediction task, and if so, how?

This short paper summarizes research into small-crack behavior in an HSLA-80 steel of particular interest for ship and submarine applications. The growth rates of large and small cracks are critically compared. Engineering models are developed to correlate small-crack and large-crack growth rates, and a probabilistic model is developed to treat the unique scatter in the small-crack regime.

EXPERIMENTAL PROCEDURE

All experiments were conducted with HSLA 80 steel. Details of material chemistry, heat treatment, and microstructure are given elsewhere (Davidson, Chan, and McClung, 1996). The as-received microstructure consisted mostly of fine polygonal and acicular ferrite grains with sulfide inclusions. Grain sizes ranged from 1 to 10 µm, with an average diameter around 2 to 4 µm. The 0.2% offset yield strength and ultimate strength were approximately 88 ksi (607 MPa) and 99 ksi (683 MPa).

Small-crack tests were performed with two different specimen designs. Beams of square cross-section 4 x 4 mm by 52 mm long were loaded in 3-point bending under nominal zero-max load cycling. Outer fiber stresses were required to exceed yield in these specimens order to initiate cracks, so the local (outer fiber) stress ratio (R) was not zero. In order to initiate cracks more easily, hourglass specimens with 0.1875 in. (4.76 mm) minimum diameter were loaded in rotating bending ($R = -1$). Low stress machining and hand polishing techniques were used to minimize surface residual stresses.

Crack growth was recorded for several different cracks on the same specimen by periodically taking acetate replicas of the specimen surface. Cracks nearly always nucleated at inclusions, which ranged in size from 2 to 12 µm, with an average diameter around 5 µm. These inclusion sizes provided an effective lower limit to the smallest crack size studied. Tests were stopped when the largest cracks exceeded about 2 mm in surface length. Selected specimens were sectioned at the end of the test in order to determine typical aspect ratios for the surface cracks: nearly all were approximately semi-circular in shape. Secant and second-order incremental polynomial methods were used to compute da/dN from the raw a vs. N data. The nominal stress intensity factor, ΔK, was calculated from the surface half-crack length, a, and the nominal applied stress range, $\Delta \sigma$, according to the relationship $\Delta K = 1.3 \Delta \sigma (a)^{1/2}$.

FCG data for large cracks in CT specimens at $R = 0.1$ were obtained from Nussbaumer et al. (1995) for the same heat of HSLA-80 and from Todd et al. (1993) for a comparable HSLA steel.

DETERMINISTIC ANALYSIS

Crack growth data for both large and small cracks are compared on the basis of the nominal full-range ΔK in Fig. 1. Note, first of all, that the small cracks grew at nominal ΔK values significantly lower than the large crack threshold value, ΔK_{th}, around 7 MPa·m$^{1/2}$. Second, note that the large and small-crack data, which included different stress ratios and maximum stresses, were not well-correlated by this nominal ΔK.

One possible approach to correlating the large and small-crack data is based on the recommendations of ASTM Test Method E 647, "Standard Test Method for Measurement of Fatigue Crack Growth Rates." This test method instructs that $\Delta K = K_{max} - K_{min}$ when $R \geq 0$, but that ΔK should be calculated according to $\Delta K = K_{max}$ when $R < 0$ (i.e., take only the tensile portion of the stress intensity factor range). The large and small-crack data plotted according to these definitions show reasonable agreement (see Fig. 2).

Another first-order correction to the nominal ΔK to address the effects of applied stress and stress ratio can be derived from considerations of plasticity-induced crack closure. Of several different proposed closure mechanisms, the development of a plastic wake and the corresponding influence on residual stress fields is usually the most significant outside of the near-threshold regime (McClung, 1991). The closure-corrected effective stress intensity factor range, ΔK_{eff}, is related to the nominal full-range ΔK according to $\Delta K_{eff} = U \Delta K$, where U is the effective stress intensity factor range ratio, $U = (\sigma_{max} - \sigma_{open})/(\sigma_{max} - \sigma_{min})$. In this work, U was estimated analytically from the modified-Dugdale model of Newman (1984), which can be expressed as a closed-form equation giving $\sigma_{open}/\sigma_{max}$ as a function of $\sigma_{max}/\sigma_{flow}$, R, and the stress state. Here σ_{flow} is the average of the yield and ultimate strengths. The stress state is quantified by the constraint factor α, where $\alpha = 1$ for plane stress, $\alpha = 3$ for full plane strain, and intermediate values represent partial constraint. The Newman model is based on a center crack in an infinite plate. This model can be applied satisfactorily to other geometries (McClung, 1994) by reinterpreting $\sigma_{max}/\sigma_{flow}$ as K_{max}/K_{flow}, where $K_{flow} = \sigma_{flow}(\pi a)^{1/2}$.

The maximum stress for the rotating bending specimens varied from 75 to 81 ksi (520 to 560 MPa). The local stress ratio for the square beam tests was estimated as $R = -0.35$ by assuming that the maximum local stress at the outer fiber was approximately equal to the yield stress due to plastic deformation at the outer fiber, but that the stress range was still equal to the full applied (elastic) outer fiber bending stress range. The ratio of K_{max}/K_{flow} to $\sigma_{max}/\sigma_{flow}$ for the small cracks was 0.73. The maximum stresses (loads) for the large crack tests were small enough that no strong dependence of $\sigma_{open}/\sigma_{max}$ on K_{max}/K_{flow} was present. The stress state was identified as plane strain ($\alpha = 3$) for the large crack tests. The estimated stress state varied for the small-crack tests, but the differences between plane stress and plane strain closure stresses were negligible at these large maximum stresses. Calculated values of U were approximately $U = 0.8$ for large cracks, $U = 0.45$ for the rotating bending tests, and $U = 0.6$ for the square beam tests.

Large and small-crack data are compared on the basis of this closure-corrected ΔK_{eff} in Fig 3. The correlation is reasonably strong. Above about 10 MPa·m$^{1/2}$, the large and small-crack data are nearly coincident. At lower ΔK_{eff} values, the central tendencies of the small-crack data are consistent with a downward extrapolation of the Paris equation from the large crack regime. A linear regression of all large and small-crack data together gave a Paris slope of $m = 3.23$.

The ability of the simple ASTM approach to correlate these large-crack and small-crack data in a similar manner to the ΔK_{eff} method might seem to imply that closure considerations are not important. However, the ASTM approach itself makes an implicit assumption about the effective stress range. The condition that $\Delta K = K_{max}$ when $R < 0$ is consistent with the common assumption that the crack is closed when the applied stresses are compressive. This assumption that $\sigma_{open} = 0$ is often an acceptably accurate estimate for small cracks under large, fully-reversed stresses, and for large cracks (outside of the near-threshold regime) at small applied stresses, which are precisely the conditions in these tests. This ASTM correlation may break down if the closure behavior for either large or small cracks changes significantly.

In summary, these data and analyses indicate that a small-crack effect can occur in HSLA-80 steel at larger maximum stresses, even when cracks are large compared to the grain size. However, a closure-corrected ΔK_{eff} approach successfully correlates small-crack data with large-crack data, at least in the case where cyclic plastic strains are negligible. If the stress range were large enough to introduce cyclic plasticity, ΔK_{eff} may become inadequate, and an alternative elastic-plastic parameter such as ΔJ_{eff} may be required (McClung and Sehitoglu, 1991). The central tendencies of the small-crack data are adequately described by a downward extrapolation of the large-crack Paris equation, neglecting the large-crack threshold, which is not obeyed by the small-crack data.

PROBABILISTIC ANALYSIS

Although both large and small cracks exhibit some growth rate variability, the apparent scatter in microcrack growth rates in the low ΔK regime is generally much greater, as seen in Figs 1-3. This scatter can complicate an engineering life assessment, since an ordinary mean value computation could be significantly nonconservative in the important small-crack regime. Some explicit treatment of variability may be necessary.

Many stochastic FCG models have been proposed to treat this inherent variability. From an engineering perspective, the lognormal random variable (LRV) model proposed by Yang et al. (1985) is one of the most useful. This approach employs a simple mathematical model for which the analytical solution can be directly derived, and all crack growth parameters and model statistics can be estimated from limited data. The general form of the LRV model is

$$\log \left(\frac{da}{dN} \right) = \log \left[f(\Delta K, R, \alpha, etc.) \right] + \log \left[(X(t)) \right] \tag{1}$$

Here f is the user-defined FCG model, such as the Paris equation, and $X(t)$ is a non-negative stationary lognormal random process. However, the conventional LRV model assumes that variability in $\log(da/dN)$ is the same at all ΔK levels, which is not indicated by the experimental data. Therefore, a modified LRV model has been derived (Torng and McClung, 1994) in which the standard deviation of the stochastic random variable $Z(t) = \log[X(t)]$ was itself modeled as a time-dependent function, so that variability in $\log(da/dN)$ is greater in the small crack regime. Both the standard and modified LRV models are shown in Fig. 4 as applied to the combined large-crack and small-crack data correlated by ΔK_{eff}.

ACKNOWLEDGEMENTS

Support for this research was provided by the Office of Naval Research, contract N0014-91-C-0214, program officer Dr. A. K. Vasudevan. Dr. D. L. Davidson supervised the small-crack tests, which were performed by B. K. Chapa. Dr. R. J. Dexter and Prof. J. A. Todd are thanked for providing large crack data.

REFERENCES

Davidson, D. L., K. S. Chan, and R. C. McClung (1996). Cu bearing HSLA steels: The influence of microstructure on the initiation and growth of small fatigue cracks. *Metall. Mater. Trans.* (submitted).

Nussbaumer, A. C., R. J. Dexter, J. W. Fisher, and E. J. Kaufmann (1995). Propagation of very long fatigue cracks in a cellular box beam. Fracture Mechanics: 25th Volume, ASTM STP 1220, pp. 518-532.

McClung, R. C. (1991). The influence of applied stress, crack length, and stress intensity factor on crack closure. *Metall. Trans. A*, **22A**, 1559-1571.

McClung, R. C. (1994). Finite element analysis of specimen geometry effects on fatigue crack closure. *Fatigue Fract. Engng Mater. Struct.*, **17**, 861-872.

McClung, R. C. and H. Sehitoglu (1991). Characterization of fatigue crack growth in intermediate and large scale yielding. *ASME J. Engng Mater. Technol.*, **113**, 15-22.

Newman, J. C. Jr. (1984). A crack opening stress equation for fatigue crack growth. *Int. J. Fracture*, **24**, R131-R135.

Todd, J. A., L. Chen, E. Y. Yankov, and H. Tao (1993). A comparison of the near-threshold corrosion fatigue crack propagation rates in Mil S-24645 HSLA steel and its weld metal. *ASME J. Offshore Mechan. Arctic Engng*, **115**, 131-136.

Torng, T. Y. and R. C. McClung (1994). Probabilistic fatigue life prediction methods for small and large fatigue cracks. Proc. 35th SDM Conference, AIAA, pp. 1514-1524.

Yang, J. N., W. H. Hsi, and S. D. Manning (1985). Stochastic crack propagation with applications to durability and damage tolerance analysis. AFWAL-TR-85-3062.

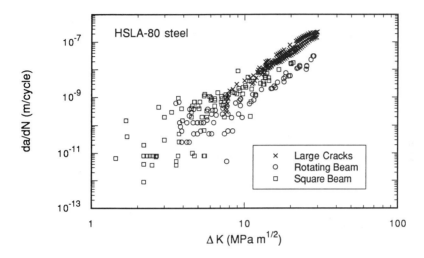

Fig. 1. Small-crack and large-crack HSLA-80 data correlated by ΔK calculated from full stress range

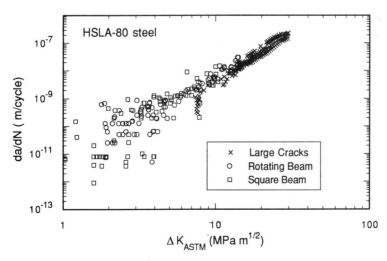

Fig. 2. Small-crack and large-crack HSLA-80 data correlated by ΔK as defined in ASTM E 647

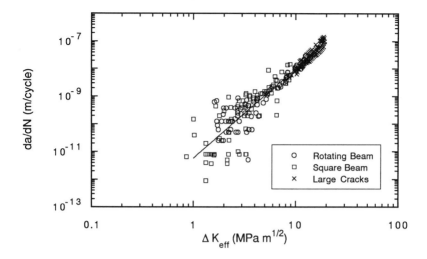

Fig. 3. Small-crack and large-crack HSLA-80 data correlated by
closure-corrected ΔK_{eff}

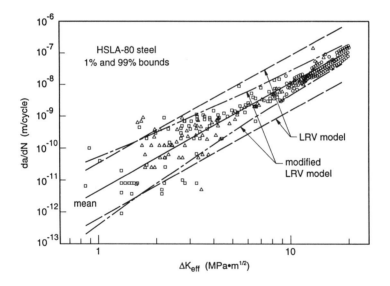

Fig. 4. Stochastic treatments of small-crack and large-crack HSLA-80 data, based on
conventional and modified lognormal random variable models

INTRINSIC AND EXTRINSIC INFLUENCES ON CRACK PROPAGATION

FINITE ELEMENT PERSPECTIVES ON THE MECHANICS OF FATIGUE CRACK CLOSURE

R. C. McCLUNG

Southwest Research Institute
P. O. Drawer 28510
San Antonio, TX 78228-0510 USA

ABSTRACT

The existence and significance of plasticity-induced fatigue crack closure is examined from the perspective of recent investigations using an elastic-plastic finite element (FE) simulation of stationary and growing fatigue cracks. The FE results provide insight into the mechanics phenomena which cause closure to occur, and the manner in which closure influences fatigue damage at the crack tip. FE studies also point towards an engineering framework for the practical implementation of closure information in design, and help to identify engineering applications in which closure may be used to predict growth rate behaviors. Some important outstanding issues are identified and discussed.

KEYWORDS

plasticity-induced fatigue crack closure, elastic-plastic finite element analysis, fatigue crack growth

INTRODUCTION

The first publication of the fatigue crack closure phenomenon by Elber (1968) was one of the more influential events in the study of fatigue crack growth (FCG). Considerable effort has been expended over the ensuing years to characterize closure and its relationship to various FCG rate behaviors. However, while many researchers and engineers have agreed that closure is a significant phenomenon with specific, quantifiable implications for FCG rates, voices of substantial dissent have regularly registered their complaints. These minority positions have argued that closure does not occur to the extent claimed by its proponents, and/or that closure descriptions are inadequate to explain observed FCG behavior. Furthermore, even though the proponents of closure are generally in the majority in the research community, the closure concept has not been integrated into most practical fatigue design and evaluation techniques practiced in industry today. The user community has not yet fully endorsed the idea of closure, perhaps either because they have not found its implementation necessary to solve their daily problems, or because the necessary implementation tools were not readily available.

The research community has many different tools available to provide windows into the closure phenomenon. Numerous experimental techniques have been developed and applied to measure closure, ranging from simple remote compliance methods to direct, high-resolution observation of the crack-tip region itself. These test methods are of first importance to establish the existence of closure ("a single test is worth a thousand theories") and identify material and loading variables which influence closure. However, these experimental techniques cannot, in general, provide substantial insight into the actual mechanics of crack closure—the near-tip stresses and strains which influence closure, or are themselves influenced by closure. Furthermore, since many experimental techniques for closure measurement are somewhat indirect, they often provide only generic empirical information about the connection between test parameters and closure levels.

Analytical tools for closure studies are able to provide this detailed information about the mechanics of crack closure and often, in turn, the functional relationship between key variables and closure response. Therefore, closure analysis is able to interpret and expand upon the experimental observations, guiding new experiments, and sometimes even serving in a validating role for the test technique itself. Furthermore, the analytical tools provide a predictive capability which is necessary for the eventual implementation of closure technology in a design setting.

Analytical closure tools fall into two major categories. Elastic-plastic finite element (FE) simulations of fatigue cracks are the most common tools, with over 75 papers in the literature. The second major category comprises simpler analytical models based on modified versions of the Dugdale crack formulation, of which the most well-known is the FASTRAN model (Newman, 1981). Most of these tools address only plasticity-induced closure, although some limited studies into roughness- and oxide-induced closure have been conducted or are in progress.

The purpose of this short paper is to provide some critical perspective on the mechanics of fatigue crack closure, and especially plasticity-induced closure, from the primary viewpoint of the FE method. We will review some of what we have learned and what we do not yet understand about both the "fundamental" and the "practical" mechanics of closure. From the fundamental perspective, we will summarize our understanding as to how closure occurs and why it influences FCG behavior. From the practical perspective, we will discuss under what conditions the closure phenomenon is important for design and what is required to facilitate this application.

FINITE ELEMENT MODEL

The FE model discussed here has evolved and been used extensively over the past ten years to study a wide range of problems. Detailed descriptions of the model are available elsewhere (McClung and Sehitoglu, 1989; McClung, Thacker, and Roy, 1991). A brief summary is given here for convenience.

FE meshes composed of four-noded linear strain elements were cycled between minimum and maximum applied stresses. Crack geometry was usually a center-cracked plate in uniform tension, but some analyses were conducted with edge-cracked plates in tension and/or bending. The mesh was very finely spaced along the crack line in the vicinity of the crack tip. At the minimum load on each cycle, while the crack was closed, the boundary conditions were changed at the crack tip to allow the crack to "grow" by one element length the next time it opened. This growth process typically continued for twenty to fifty cycles until the crack had grown relatively far beyond its original location. During each cycle, stresses and displacements along the crack surface were closely monitored and boundary conditions were appropriately changed as the crack opened or closed. The material model usually employed linear kinematic hardening with a von Mises yield surface, either with "low" hardening (H/E = 0.01, where H is the plastic modulus $d\bar{\sigma}/d\bar{\epsilon}_p$, and E is the elastic modulus) or "high" hardening (H/E = 0.07). Most analyses have been plane stress, but a limited number of plane strain studies have also been conducted.

This FE model has been validated through extensive comparison to available experimental information, including remote and local measurements of closure stresses, and local measurements of near-tip strains and displacements (McClung, 1991a; McClung and Davidson, 1991). While specific quantitative details sometimes differ (as do the results of various closure experiments), the general trends of the numerical model are consistent with test measurements. Furthermore, the calculated changes in closure levels are generally consistent with experimentally observed changes in FCG rates. It must be emphasized that the FE model itself must be constructed and exercised very carefully to preserve the validity of the results (McClung and Sehitoglu, 1989).

FUNDAMENTAL MECHANICS OF CRACK CLOSURE

Finite element results provide direct insight into the mechanics phenomena which cause plasticity-induced closure to occur. Figure 1 shows the cumulative axial plastic strains at zero load for a plane stress fatigue crack which has grown under $R = 0$ loading from left-to-right into the current field of view. The plastic strains out in front of the crack tip were induced by the most recent load cycle, as the crack tip singularity elevated the stresses ahead of the crack well into the plastic regime. Since these plastic strains are not reversed upon unloading to zero remote load, they remain as permanently stretched material while the crack tip moves ahead, leaving behind the plastic wake clearly shown in the contour plot. The material required to feed this plastic stretch comes from plastic contraction in the out-of-plane (zz) direction; the plastic strains ϵ_{zz}^P are nearly mirror images of the ϵ_{yy}^P strains.

The material remote from the crack tip, which remains elastic, constrains the plastic wake upon unloading, and induces compressive axial stresses which serve to clamp the crack closed even at some nonzero remote loads. The residual stresses at zero load corresponding to Fig. 1 are shown in Fig. 2. Note, in particular, the compressive σ_{yy} stresses which extend for an appreciable distance behind the crack tip, gradually increasing in intensity near the crack tip. As the crack is loaded remotely, these residual stresses are gradually overcome closer and closer to the crack tip, causing the crack to gradually "peel open," and this sequence has been confirmed by direct experimental measurements (McClung and Davidson, 1991).

These figures all illustrate the closure process in plane stress. However, some scholars have suggested that closure does not occur in plane strain, noting that plastic contraction in the thickness direction is not available to feed axial plastic stretch. Finite element studies (Sehitoglu and Sun, 1991; McClung, Thacker, and Roy, 1991) have shown that closure can, in fact, occur in plane strain, and that the material required to feed axial plastic stretch comes from in-plane transverse contraction (i.e., negative ϵ_{xx}^P strains). A plot of cumulative plastic strains for a plane strain fatigue crack looks a great deal like Fig. 1, except that ϵ_{xx}^P and ϵ_{zz}^P switch places, and the magnitudes of all plastic strains are scaled down by nearly a factor of five. The in-plane residual stress field for the plane strain fatigue crack is considerably more complex due to the in-plane plastic contraction. Color contour plots providing further information about the residual stress and plastic strain fields in both plane stress and plane strain are available elsewhere (McClung, Thacker, and Roy, 1991).

Further evidence of this residual plastic stretch and the associated residual stress field is available in the crack opening displacements behind the crack tip, Fig. 3 (McClung and Davidson, 1991). Here we compare COD profiles for a growing elastic-plastic fatigue crack, a stationary elastic-plastic crack (i.e., no plastic wake), and an elastic crack, all loaded to the same maximum applied stress ($0.3\sigma_0$ for the e-p cracks). The plastic stretch itself directly accounts for about 60% of the difference between the stationary and growing elastic-plastic crack COD values. The remainder of the difference can likely be attributed to the influence of near-tip residual stresses.

Given that the mechanics of the growing fatigue crack lead naturally to the development of "crack closure," the next key question is how crack closure influences the growth rates of fatigue cracks. Without invoking a specific theory of fatigue crack growth, it remains clear that the activation of fatigue damage at the crack tip which leads to crack advance is naturally linked to the stresses and plastic strains in the immediate vicinity of the crack tip.

Near-tip stresses at maximum load for stationary and growing elastic-plastic fatigue cracks are compared in Fig. 4 (McClung, 1991b). The reduction in near-tip stresses induced by the closure phenomenon is evident, although the difference is not great. It is interesting to note that the calculated plastic zone width is actually greater for the growing fatigue crack (with closure) than for the stationary crack (without closure). Since the near-tip stresses are slightly lower, slightly more plastic redistribution of the theoretical elastic crack-tip stress field is required to satisfy equilibrium.

The differences between elastic-plastic cracks with and without closure are more evident in the near-tip strain distributions, Fig. 5 (McClung and Davidson, 1991). Note that the total strains for the two elastic-plastic cracks are predominantly plastic, and it is likely these near-tip plastic strains which can be linked most directly to fatigue damage at the crack tip. Extrapolating the available numerical data even closer to the crack tip itself, the plastic strains for the growing crack with closure have been reduced by nearly a factor of two in comparison to the stationary crack without closure.

Another window into the influence of closure on near-tip plastic deformation is the reversed plastic zone (RPZ) size ahead of the crack tip, Δr_p. It is within this RPZ that the intense cyclic deformation occurs which leads directly to fatigue damage, and the size of the RPZ is one measure of its intensity. The RPZ size has even been proposed as a promising parameter (with direct physical meaning) for the correlation of FCG rates. Figure 6 (McClung, 1991b) shows calculated values of RPZ size as a fraction of the monotonic plastic zone size, r_p (e.g. the width of the plastic zone at maximum load evident in Fig. 4). Remember the famous Rice (1967) result that the RPZ size should be 1/4 of the monotonic PZ size. This simple theory agrees approximately with the calculated $\Delta r_p/r_p$ ratio for stationary elastic-plastic cracks. However, crack closure is shown to cause a significant decrease in the RPZ size, because the premature closing of the crack tip prematurely turns off the singularity at the crack tip which is driving the reversed deformation during unloading. These smaller values of the $\Delta r_p/r_p$ ratio have been confirmed by high resolution experimental measurements of plastic zone sizes (Davidson and Lankford, 1980). Clearly, crack closure can cause a significant decrease in the intensity of the reversed plastic strains at the crack tip, and this naturally leads to a corresponding decrease in the driving force for fatigue crack growth.

PRACTICAL MECHANICS OF CRACK CLOSURE

From a more pragmatic perspective, the crack closure concept is useful only to the extent that it provides some direct assistance in understanding and correlating important fatigue crack growth rate behaviors. These engineering applications of closure require a practical mechanics framework to describe when and how closure has a quantifiable impact on growth rates.

In particular, in order to employ closure in an actual design or life management setting, the engineer must be able to make a rational connection between closure behavior of a laboratory specimen and closure behavior of a crack in a real component or structure. This condition can be met either by enforcing adequate similitude between specimen and structure, or by quantitatively characterizing closure adequately in both specimen and structure. Most engineering structures are far too complex to admit the similitude approach, however, and the quantitative approach is instead required.

Characterization in the specimen is relatively straightforward, since the simplicity of common specimen geometries usually facilitates direct experimental measurement of closure and sometimes a rigorous numerical solution as well. The baseline center-cracked plate geometry, for example, has been extensively studied with FE methods, and Newman (1984) has exercised his modified Dugdale model to generate a convenient closed-form engineering equation for the normalized crack opening stress $\sigma_{open}/\sigma_{max}$ as a function of $\sigma_{max}/\sigma_{flow}$, R, and stress state. Here σ_{flow} is interpreted as the average of the yield and ultimate strengths to accommodate strain hardening. The stress state is quantified by the constraint factor α, which ranges from $\alpha = 1$ (plane stress) to $\alpha = 3$ (plane strain).

A similar characterization of closure in real structures is more difficult, however, since complex geometries of component and/or crack usually preclude direct measurement or direct analytical simulation, especially in a routine design context. However, this inability should not, by itself, be a barrier to engineering implementation of closure theory. In comparison, our inability to measure the stress intensity factor K directly in a real structure (or the impracticality of obtaining an exact closed-form solution for that K) does not prevent us from employing K effectively in engineering design and analysis. We can use K dependably because we have a practical mechanics framework to define the structural K and the laboratory K consistently, and to estimate K in the structure adequately.

A similar practical mechanics framework to correlate closure in different configurations now appears available. A normalized stress intensity factor parameter, K_{max}/K_{flow}, where $K_{flow} = \sigma_{flow}(\pi a)^{1/2}$, has been shown to correlate closure in three contrasting geometries, as shown in Fig. 7 (McClung, 1994). The correlation appears nearly exact in the small-scale yielding regime, gradually breaking down at higher loads. Furthermore, since K_{max}/K_{flow} reduces to the Newman parameter $\sigma_{max}/\sigma_{flow}$ for the center-cracked plate, the convenient Newman closure equation can be extended to a much wider range of geometries, thereby providing a simple engineering solution to many closure problems.

The skeptic—and the prudent—may still properly ask why much routine FCG design and analysis has been so successful for so many years without considering closure. The answer has several parts. In some cases, closure does not occur to a significant extent. For example, while closure does occur under plane strain conditions—lower loads, thicker sections—as discussed earlier, the extent of closure is relatively minor. Normalized crack opening stresses are typically around $\sigma_{open}/\sigma_{max} = 0.15$ to 0.20 for $R = 0$ (McClung, Thacker, and Roy, 1991) and do not change appreciably with K_{max}/K_{flow}. Further reductions in closure at higher stress ratios cannot cause a large percentage change in U, therefore. As long as baseline FCG data and structural applications share these plane strain conditions, closure is largely a non-issue. Closure can also drop out of the picture under plane stress conditions if specimen and structure are sufficiently similar (e.g., testing and application of thin sheet aluminum), assuming no other effects such as variable amplitude loading.

In other cases, design methods may contain implicit (unintentional?) closure information without an explicit closure factor. For example, the ASTM Test Method E 647 definition of ΔK stipulates that $\Delta K = K_{max}$ when $P_{min} < 0$, which is equivalent to a ΔK_{eff} approach where $\sigma_{open}/\sigma_{max} = 0$. While this closure inference is not exactly true, it is a reasonably approximation for many negative stress ratios, especially at the larger stress amplitudes sometimes encountered with fully reversed loading. The approximate equivalence of closure and ASTM approaches for the analysis of a set of large and small crack data is illustrated elsewhere in this conference (McClung and Torng, 1996). Furthermore, the ASTM E 647 limits on net-section yielding for FCG test validity correspond almost exactly to the conditions beyond which net-section yielding causes significant changes in crack closure behavior (McClung, 1994), which in turn influences growth rates. Other FCG analysis techniques, such as empirical overload retardation models or mean stress models, may also serve to substitute for more direct treatments of the closure phenomenon which actually motivates the observed behavior.

So when *is* plasticity-induced crack closure an important phenomenon with significant implications for changes in FCG rates? Changes in crack closure response with applied stress (or applied K_{max}/K_{flow}) are more pronounced in plane stress (e.g., thin sheet materials such as aluminum aircraft structures) and at lower stress ratios (compare $R = 0$ and $R = -1$ in Fig. 7). These effects are observed most clearly in the Paris regime (McClung, 1991a).

Crack closure appears to be particularly significant for elastic-plastic fatigue crack growth (EPFCG) (McClung and Sehitoglu, 1991; McClung et al., 1995), where stresses are large and often fully-reversed, and plasticity is widespread. A closure-corrected elastic-plastic parameter, ΔJ_{eff}, has proven useful in correlating EPFCG rates. Closure corrections are necessary, in particular, to relate EPFCG conditions to the conventional small-scale yielding (SSY) conditions prevailing in baseline FCG testing, so that crack growth properties defined in terms of K can be used to predict EPFCG behavior. Figure 8 (McClung et al., 1995) illustrates how ΔJ_{eff} can be used to correlate elastic-plastic and SSY ($R = 0$) FCG data. Here the Inconel 718 has a cyclic yield stress around 135 ksi, so the EPFCG tests at $R = 0$ see monotonic but not cyclic plasticity, and the $R = -1$ tests experience fully cyclic plasticity.

Closure can be an important factor in explaining crack growth behavior near notches (McClung and Sehitoglu, 1992), where crack opening levels change significantly as the crack grows out of the notch root (see Fig. 9). The decreases in crack opening stresses for short cracks at the notch root are quantitatively consistent with observed increases in FCG rates, and closure arguments are also successful in describing the different notch effects at different nominal stress ratios (McClung and Sehitoglu, 1992).

Closure also provides a practical explanation for numerous variable amplitude loading effects on FCG rates, such as retardation following an overload. These effects are more commonly treated in engineering practice with empirical models such as Wheeler or Willenborg, and these models are reasonably successful with adequate calibration on similar histories. However, the empirical models are not general and fail to describe some variable amplitude effects, such as the acceleration observed following an overload in the elastic-plastic regime (McClung and Sehitoglu, 1988). A comparison of the computed closure histories for 100% overloads with different baseline stresses (see Fig. 10) shows that opening levels first decrease and then increase following an overload, but that the decrease for large amplitude load histories is much more severe and far-ranging, causing a pronounced acceleration in FCG rates (McClung, 1992). A subsequent overload is likely to occur before the crack grows into the region of enhanced closure.

Closure is commonly cited as a primary factor in anomalous small crack behavior. While closure is different for many small cracks, some clear thinking is required. First of all, closure is not negligible for all small cracks, as is commonly claimed. High-resolution experimental studies have shown that small cracks do exhibit closure, albeit at different levels than large cracks (Lankford et al., 1984). The primary driver for this difference may not be the crack size, however, so much as the applied stress, since many small cracks are studied at relatively high stress levels. These are often mechanically-small cracks in which the crack size is small compared to the crack tip PZ size due to the load amplitude, and the general effects of stress level on closure are well-documented (McClung, 1991a). This explanation is also loosely applicable to microstructurally-small cracks which exhibit enhanced crack-tip microplasticity within a single grain.

Crack closure is slowly finding wider acceptance in applications settings, and FCG analysis tools which include closure are more commonly available. For example, the current version of NASGRO (formerly called NASA/FLAGRO), a widely used FCG code with extensive NASA support, includes some rudimentary closure algorithms to treat mean stress effects. Forthcoming versions of NASGRO will include a relatively complete strip-yield closure model suitable for the treatment of variable amplitude loading. Longer-range plans call for the incorporation of a closure-based treatment of EPFCG within NASGRO (McClung et al., 1995). These engineering implementations of closure technology are unlikely to include FE methods, but FE studies can play a key role in validating the simpler Dugdale treatments of closure which will be invoked.

Before closing, it is useful to point out briefly some important outstanding issues relative to the practical mechanics of closure in an applications setting. First of all, the FCG threshold phenomenon appears to be associated, at least in part, with changes in closure behavior. This has been confirmed by both experimental measurement (Hudak and Davidson, 1988) and FE simulation of threshold test methods (McClung, 1991c), although FE analyses have proven inadequate to simulate fully the complex mixed-mode behavior of near-threshold cracks (McClung and Davidson, 1991). However, at present this is mostly an academic understanding, and engineering methods to predict growth rates based on a characterization of near-threshold closure response are not available. Closure behavior in the near-threshold regime appears to be fundamentally different than in the Paris regime (McClung, 1991a), but it should be noted that plasticity contributions to near-threshold closure are not negligible, as is commonly assumed.

On the other hand, non-plasticity contributions to closure, such as oxide- and roughness-induced phenomena, can also be significant both near and beyond the threshold regime. Unfortunately, predictive engineering models for these other closure behaviors are far less mature than for plasticity effects, although several studies are currently in progress. Full implementation of a practical FCG methodology based on closure concepts will require substantial progress in modeling of these alternative mechanisms.

Finally, the engineering treatment of plasticity-induced closure with a FASTRAN-like approach does require some characterization of the stress state, perhaps in terms of the Newman constraint parameter α. But proper identification of the stress state is not an easy task, especially for three-dimensional crack geometries. Simple rules-of-thumb based on comparison of the crack-tip plastic zone size to the specimen thickness are available and have been used with some success in ductile fracture characterization, but may not be adequate for FCG. Newman sometimes employs α as a fitting parameter, rather than derive its value from first principles. Further work is clearly required. Since constraint effects are also a significant current issue in elastic-plastic fracture mechanics (i.e., constraint effects on tearing or cleavage), the chances of some significant progress may be enhanced.

CONCLUSIONS

The crack closure concept is not a panacea, a solution to every FCG rate effect and anomaly. But extensive FE research has clearly demonstrated the fundamental mechanics by which plasticity-induced closure occurs and by which closure influences the fatigue crack growth process. And FE studies have identified specific FCG behaviors which can be adequately explained by closure arguments. While some engineering procedures are now available to address closure in a practical applications context, further work is needed to provide the more complete set of closure insights and tools required to support a complete engineering implementation of the concept.

REFERENCES

Davidson, D. L. and J. Lankford, Jr. (1980). Plastic strain distribution at the tips of propagating fatigue cracks. *ASME J. Engng Mater. Technol.*, **98**, 24-29.

Hudak, S. J. and D. L. Davidson (1988). The dependence of crack closure on fatigue loading variables. *Mechanics of Fatigue Crack Closure, ASTM STP 982*, pp. 121-138.

Lankford, J., D. L. Davidson, and K. S. Chan (1984). The influence of crack tip plasticity in the growth of small fatigue cracks. *Metall. Trans. A*, 15A, 1579-1588.

McClung, R. C., and H. Sehitoglu (1988). Closure behavior of small cracks under high strain fatigue histories. *Mechanics of Fatigue Crack Closure, ASTM STP 982*, pp. 279-299.

McClung, R. C. and H. Sehitoglu (1989). On the finite element analysis of fatigue crack closure— 1. Basic modeling issues, and 2. Numerical results. *Engng Fract. Mech.*, **33**, 237-272.

McClung, R. C. (1991a). The influence of applied stress, crack length, and stress intensity factor on crack closure. *Metall. Trans. A*, **22A**, 1559-1571.

McClung, R. C. (1991b). Crack closure and plastic zone sizes in fatigue. *Fatigue Fract. Engng Mater. Struct.*, **14**, 455-468.

McClung, R. C. (1991c). Finite element analysis of crack closure during simulated fatigue threshold testing. *Int. J. Fract.*, **52**, 145-157.

McClung, R. C. and D. L. Davidson (1991). High resolution numerical and experimental studies of fatigue cracks. *Engng Fract. Mech.*, **39**, 113-130.

McClung, R. C. and H. Sehitoglu (1991). Characterization of fatigue crack growth in intermediate and large scale yielding. *J. Engng Mater. Technol.*, **113**, 15-22.

McClung, R. C., B. H. Thacker, and S. Roy (1991). Finite element visualization of fatigue crack closure in plane stress and plane strain. *Int. J. Fract.*, **50**, 27-49.

McClung, R. C. (1992). Finite element modeling of fatigue crack growth. *Theoretical Concepts and Numerical Analysis of Fatigue*. EMAS, pp. 153-172.

McClung, R. C. and H. Sehitoglu (1992). Closure and growth of fatigue cracks at notches. *ASME J. Engng Mater. Technol.*, **114**, 1-7.

McClung, R. C. (1994). Finite element analysis of specimen geometry effects on fatigue crack closure. *Fatigue Fract. Engng Mater. Struct.*, **17**, 861-872.

McClung, R. C., G. G. Chell, D. A. Russell, and G. E. Orient (1995). A practical methodology for elastic-plastic fatigue crack growth. 27th National Symposium on Fatigue and Fracture Mechanics, Williamsburg, Virginia, June 1995 (in review for ASTM STP).

McClung, R. C. and T. Y. Torng (1996). Analysis of small fatigue cracks in HSLA-80 steel. Proc. FATIGUE '96, Sixth International Fatigue Conference, Berlin.

Newman, J. C. Jr. (1981). A crack-closure model for predicting fatigue crack growth under aircraft spectrum loading. *Methods and Models for Predicting Fatigue Crack Growth Under Random Loading, ASTM STP 748*, pp. 53-84.

Newman, J. C. Jr. (1984). A crack opening stress equation for fatigue crack growth. *Int. J. Fracture*, **24**, R131-R135.

Rice, J. R. (1967). Mechanics of crack tip deformation and extension by fatigue. *Fatigue Crack Propagation, ASTM STP 415*, pp. 247-309.

Sehitoglu, H. and W. Sun (1991). Modeling of plane strain fatigue crack closure. *ASME J. Engng Mater. Technol.*, **113**, 31-40.

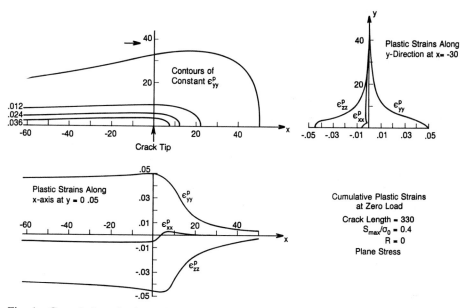

Fig. 1. Cumulative plastic strains at zero load near a growing plane stress fatigue crack

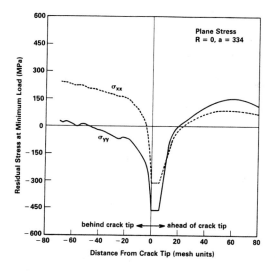

Fig. 2. Residual stresses along the crack line at minimum load for a plane stress fatigue crack
(McClung, Thacker, and Roy, 1991)

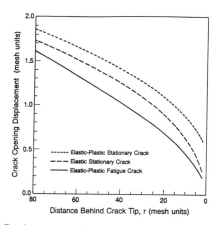

Fig. 3. Crack opening displacements for various crack types

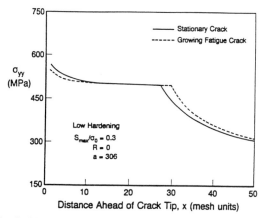

Fig. 4. Near-tip axial stress distributions at maximum load

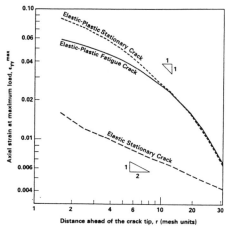

Fig. 5. Strains ahead of crack tip for various crack types (McClung and Davidson, 1991)

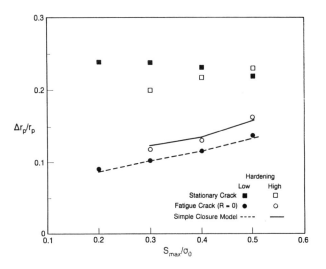

Fig. 6. Ratio of forward to reversed crack-tip plastic zone sizes
for stationary and growing fatigue cracks

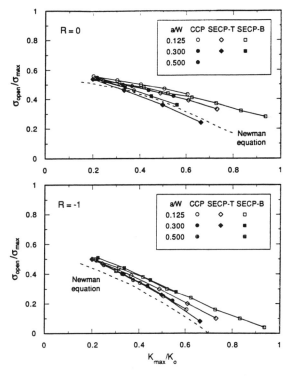

Fig. 7. Normalized crack opening stresses as a function of
normalized stress intensity factor (McClung, 1994)

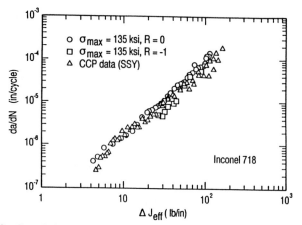

Fig. 8. Correlation of experimental EPFCG data (McClung et al., 1995)

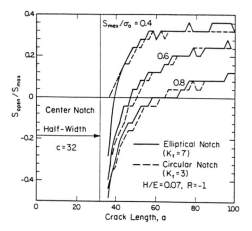

Fig. 9. Changes in crack opening levels for cracks growing out of notches

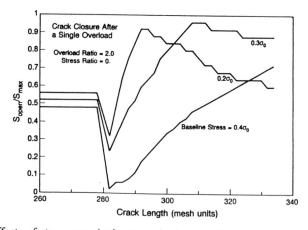

Fig. 10. Effects of stress magnitude on crack closure following a single overload cycle

ENERGY CONSIDERATIONS OF THE FATIGUE CRACK CLOSURE PROCESS

D. L. Chen*, B. Weiss and R. Stickler
Institute for Physical Chemistry - Materials Science, University of Vienna
Währinger Straße 42, A - 1090 Vienna, Austria

ABSTRACT

An energy point of view is applied to examine the closing situation of both closure-free and closure-affected cracks. It was found that the closing of the closure-free crack obeys rigorously the energy conservation law. A concept of shielding work has been proposed to satisfy the energy conservation law in the closure-affected case. The shielding work was indicated to be proportional to the shielding stress intensity range.

KEYWORDS

Fatigue, fracture mechanics, crack closure, energy principle, closure-free crack, closure-affected crack, energy conservation law.

1. INTRODUCTION

For over two decades the concept of fatigue crack closure has been utilized to explain the effects of various factors on the fatigue crack growth behavior. However, in recent years a number of authors has expressed their doubts on the conventional closure concept. For instance, it has been pointed out that "the conventionally measured closure level bears little relation to the actual situation at the crack tip", and "the use of conventional closure concept to infer the level of the effective stress intensity range is called into question" [e.g., Vecchio et al., 1986, Hertzberg et al., 1988, Garz et al., 1989]. The ASTM Task Group E24.04.04 has also indicated that the conventional closure measurements exhibited serious inconsistencies depending on the laboratory, investigator and technique used [Philips, 1989]. In particular, it has recently been reported by several authors that "there is no closure from plasticity, either from the plasticity ahead of the crack tip or from the wake" [Vasudevan et al., 1993], implying that the initial closure concept proposed by Elber [1970] was totally denied and rejected. From these observations reported in the literature it may be concluded that the conventional crack closure concept seems problematic.

In previous publications [Chen et al., 1991, Weiss et al., 1992, Chen et al., 1994a] the present authors have proposed a modified crack closure definition by considering the crack opening displacement range (ΔCOD) experienced by the crack tip as a controlling parameter for the propagation of a fatigue crack. By comparing the applied load-COD behavior of an idealized closure-free crack and that of realistic closure-affected crack, a shielding stress intensity range ΔK_{sh} has been defined, which is used to calculate the effective stress intensity range experienced actually by the crack tip, or sometimes called effective driving force for the propagation of a fatigue crack (ΔK_{eff}):

$$\Delta K_{eff} = \Delta K - \Delta K_{sh} , \tag{1}$$

where ΔK is the externally applied stress intensity range (= $K_{max} - K_{min}$). In the proposed closure concept the role of the lower portion of loading cycle below the crack opening point K_{op} has been taken into account. Hence, the underestimation of the effective driving force evaluated by the conventional closure concept, which has been pointed out by some investigators [Vecchio et al., 1986, Hertzberg et

*on leave from the State-Key Laboratory for Fatigue and Fracture of Materials, Institute of Metal Research, Chinese Academy of Sciences, Shenyang 110015, P.R.China.

al., 1988, Garz *et al.*, 1989], is expected to be solved. However, physical implication of the defined shielding stress intensity range seems not so clear. This paper is thus aimed to extend the modified crack closure considerations by examining the difference in closing processes between the closure-free and closure-affected cracks from an energy point of view.

2. LOAD-DISPLACEMENT BEHAVIOR DURING CRACK CLOSING

The application of the energy principle requires a relationship of the externally applied load versus the displacement at the load application position, rather than versus the crack opening displacement. Thus it is essential to find out such relationships for both the closure-free case and closure-affected cracks.

2.1 Load-Displacement Behavior for the Closure-Free Crack

According to Tada, Paris and Irwin [1973] the total displacement at the load application point is composed of two parts: the displacement due to the presence of a crack (i.e., u_{crack}) and the elastic displacement when the crack does not exist ($u_{no\ crack}$), namely:

$$u = u_{no\ crack} + u_{crack} . \tag{2}$$

If the distance between two points of the load application, i.e., the distance between two fixtures is assumed to be $2l$, then we have:

$$u_{no\ crack} = \frac{2l\sigma}{E}, \tag{3}$$

where σ is the externally applied stress, E is the Young's modulus. For center cracked tension (CCT) specimen the displacement due to the presence of a crack may be calculated as:

$$u_{crack} = \frac{4Ya\sigma}{E'}G_1 , \tag{4}$$

$$G_1 = \frac{a^2 - (1-\alpha)l\left(\sqrt{a^2 + l^2} - l\right)}{a\sqrt{a^2 + l^2}} , \tag{4a}$$

where a is the half-crack length, Y is a geometric correction factor [Chen *et al.*, 1994b], α is a constant and E' is an equivalent Young's modulus dependent on plane stress or plane strain. From equs (2)-(4), one can deduce a relation between the externally applied load and the displacement at the load application point as follows:

$$P = \eta u , \tag{5}$$

$$\eta = \frac{Wt}{\frac{2l}{E} + \frac{4aYG_1}{E'}} , \tag{5a}$$

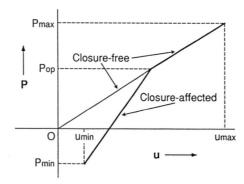

Fig.1. Response of the externally applied load vs. the displacement at the loading position for the closure-free and closure-affected cracks.

where P is the externally applied load, W is the specimen width and t is the specimen thickness. It can be seen that for the closure-free case (i.e., considered by classical fracture mechanics) the load-displacement response is characterized by a linear relation passing through the coordinate origin (u=0, P=0), as shown in Fig.1.

2.2 Load-Displacement Behavior for the Closure-Affected Crack

In a previous paper [Chen et al., 1996], a model for the fatigue crack closing process has been proposed. The impediments of crack closing are simulated by an elastic wedge, representing the asperity-, oxide-, and any artificial wedge-induced crack closure mechanisms. In this case the displacement at the load application point cannot be calculated by equ.(5) any more. Thus the superposition principle in fracture mechanics has to be applied to calculate the total displacement due to both the externally applied stress (σ) and the resistant force (P_1) produced by the wedge to the crack surfaces. Equ. (5) has given the displacement arising from the externally applied stress. The displacement at the load application point, caused by the resistant force P_1 near the crack tip, u_{p_1}, can be calculated to be:

$$u_{P_1} = \frac{8 P_1}{\pi t E'}(G_2 + G_3) , \tag{6}$$

$$G_2 = ln\left(1 + \sqrt{\frac{2 a a_1}{a^2 + l^2}}\right), \tag{6a}$$

$$G_3 = \frac{\alpha l^2}{a^2 + l^2}\sqrt{\frac{2 a a_1}{a^2 + l^2}} , \tag{6b}$$

where a_1 is a distance of the wedge to the crack tip. Using the following correlation between the applied stress and the resistant force (details of the derivation can be found elsewhere [Chen et al., 1996]):

$$P_1 = \frac{\pi}{8}\frac{\lambda}{1 + \lambda}t\,(h_1\,E' - 4Y\,\sigma\sqrt{2 a a_1}) , \tag{7}$$

$$\lambda = \frac{8 E_1 W_1}{\pi h_1 E'} , \tag{7a}$$

where E_1, h_1, W_1 are the Young's modulus, height and width of the wedge, respectively. From equs (5)-(7), and using the superposition principle, one can obtain a relation between the externally applied load and the displacement at the load application position for the closure-affected case:

$$P = \beta u - \gamma , \tag{8}$$

$$\beta = \frac{W\,t}{\frac{2l}{E} + \frac{4Y}{E'}\left[aG_1 - \frac{\lambda\sqrt{2 a a_1}}{1 + \lambda}(G_2 + G_3)\right]} , \tag{8a}$$

$$\frac{\gamma}{\beta} = \frac{\lambda\,h_1}{1 + \lambda}(G_2 + G_3) . \tag{8b}$$

In comparison with the load-displacement response at the loading position for the closure-free case (equ.(5)), it is noticed that the slope of the latter case is greater than that of the former case, i.e., $\beta > \eta$ being dependent on the parameter λ and the position of the wedge. Another explicit feature for the closure-affected case is that the load-displacement plot does not pass through the coordinate origin point, as indicated in Fig.1.

3. ENERGY CONSIDERATIONS OF THE CRACK CLOSING PROCESS

3.1 Energy Considerations for the Closure-Free Crack

The elastic strain energy (also termed as the elastic potential energy) of the system, U, and the work done by the externally applied load on the system, W, can be calculated by the following formulas:

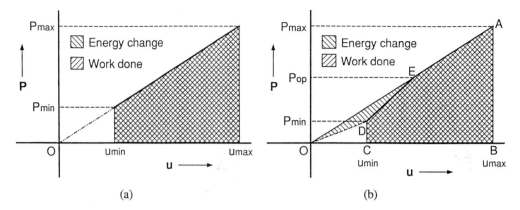

Fig.2. Elastic strain energy change and the work done by the externally applied load on the system for
(a) the closure-free crack and (b) the closure-affected crack.

$$U = \frac{1}{2}Pu ,$$ (9)

$$W = \int_{u_1}^{u_2} \bar{P}\, d\bar{u} ,$$ (10)

where u_1 is the displacement at the beginning of the process considered, u_2 is the displacement at the end
of the process.

Let us consider now an unloading process from P_{max} to P_{min}, corresponding to a variation in the
displacement from u_{max} to u_{min} (Fig. 2(a)), the decrease (or release) in the elastic strain energy, ΔU, can
be expressed as:

$$\Delta U = \frac{1}{2\eta}(P_{min}^2 - P_{max}^2) .$$ (11)

In the unloading process the externally applied load would simultaneously do the work on the system
due to the change of the displacement. It should be noted that the work done in this process is the
negative work because the displacement is reduced. The work obtained for the closure-free case is:

$$W = \int_{u_{max}}^{u_{min}} \eta u\, du = \frac{1}{2\eta}(P_{min}^2 - P_{max}^2) .$$ (12)

It can be seen from equs (11) and (12) that the release in the elastic strain energy of the system in the
process of the unloading from P_{max} to P_{min} is just equal to the negative work done by the applied load on
the system (see also Fig.2(a)). This statement expresses exactly the energy conversion law. Thus, the
energy change during the crack closure or opening *for the closure-free case* rigorously obeys the energy
conservation law.

3.2 Energy Considerations for the Closure-Affected Crack

It is apparent that for this case an identical situation to that for the closure-free case occurs when $P_{min} \geq$
P_{op}. Hence, an unloading process from P_{max} to P_{min} below P_{op} is considered only in the following. The
decrease in the elastic strain energy of the system can be derived as follows:

$$\Delta U = \frac{1}{2\beta}P_{min}^2 + \frac{\gamma}{2\beta}P_{min} - \frac{1}{2\eta}P_{max}^2 .$$ (13)

This change is shown in Fig.2(b) by an area of ABCDO(E)A. The work done by the applied load on the
system for the closure-affected crack can be calculated as:

$$W = \int_{u_{max}}^{u_{op}} \eta u \, du + \int_{u_{op}}^{u_{min}} (\beta u - \gamma) \, du = \frac{1}{2\beta} P_{min}^2 + \frac{\gamma}{2\beta} P_{op} - \frac{1}{2\eta} P_{max}^2 \, . \tag{14}$$

The value of the work is also demonstrated in Fig.2(b) by an area of ABCDEA. Obviously, in this closing process the work done by the applied load on the system (area ABCDEA) is smaller than the release in the elastic strain energy (area ABCDOA). However, in terms of energy conversion law, the energy release and the work done should be equal. Therefore, there must exist a portion of "non-work" done by the externally applied load that does not play any role on the system during unloading. We call this part of "non-work" as shielding work, W_{sh}. From equs (13) with (14) one can readily obtain an expression for W_{sh}:

$$W_{sh} = -\frac{\gamma}{2\beta}(P_{op} - P_{min}) \, . \tag{15}$$

The magnitude of W_{sh} is graphically indicated by an area of ODE in Fig.2(b). It should be noted that the defined shielding work has the same symbols (positive or negative) as that of the work done. In the tensile regime $W_{sh} > 0$ for the loading cycle and $W_{sh} < 0$ for the unloading cycle. A similar analysis of the energy change and the work done during the loading from P_{min} to P_{max} results in a positive shielding work, which has the same absolute value as given by equ. (15). Irrespective of the unloading or loading, $W_{sh} = 0$ for $P_{min} \geq P_{op}$.

By using the obtained relation of the range P_{op}-P_{min} to the shielding stress intensity range, ΔK_{sh} [Chen et al., 1996], one can derive the following expression:

$$\Delta K_{sh} = \xi W_{sh} \, , \tag{16}$$

$$\xi = \pm \frac{2Y \sqrt{\pi a}}{(G_2 + G_3) h_1 W t} \, , \tag{16a}$$

where the ($\pm$) symbols mean that for the loading cycle the symbol (+) is used, for the unloading cycle the symbol (-) is taken. It can be seen from equ.(16) that the shielding stress intensity range is proportional to the defined shielding work, depending on the height (h_1) and location (a_1) of the wedge, geometric correction factor (Y), crack length (2a), the distance of clamps (2l), the specimen width (W) and thickness (t). Equation (16) expresses a physical interpretation of the shielding stress intensity range. Therefore, the physical implication of the shielding stress intensity in a crack closure process can be understood to be: the shielding stress intensity range is equivalent to the shielding work not done by the externally applied load on the system for a given specimen material and loading condition.

CONCLUSIONS

The current status of the knowledge of fatigue crack closure effect is briefly reviewed. The process of the fatigue crack closing for both closure-free and closure-affected cases is examined by an energy point of view.

In the closure-free case the energy conservation law is rigorously obeyed, since the release in the elastic strain energy of the system is just equal to the work done by the externally applied load on the system.

For the closure-affected case the work done by the applied load on the system during unloading is found to be less than the release in the elastic strain energy, which obviously violates the energy conservation law. Thus a concept of shielding work has been proposed to fulfill the energy conservation law in the closure-affected case. The shielding work is finally indicated to be proportional to the shielding stress intensity range.

ACKNOWLEDGMENTS

This work is part of the cooperative project between the National Science Foundation of Austria (P 8032-TEC and P10160-PHY) and the National Natural Science Foundation of China (C9 and A18). The financial support received from the Fonds zur Förderung der Wissenschaftlichen Forschung, Vienna, is gratefully acknowledged. The authors would like to thank Professor Z. G. Wang, Director of the State-Key Laboratory for Fatigue and Fracture of Materials, Institute of Metal Research, Chinese Academy of Sciences, for his interest and fruitful discussion.

REFERENCES

Chen, D.L., B. Weiss and R. Stickler (1991). A new evaluation procedure for crack closure. *Inter. J. Fatigue,* 13, 327-331.

Chen, D.L., B. Weiss and R. Stickler (1994a). The effective fatigue threshold: significance of the loading cycle below the crack opening load. *Inter. J. Fatigue,* 16, 485-491.

Chen, D.L., B. Weiss and R. Stickler (1994b). A new approach for the determination of stress intensity factors for finite width plate. *Eng. Fract. Mech.,* 48, 561-571.

Chen, D.L., B. Weiss and R. Stickler (1996). A model for crack closure. *Eng. Fract. Mech.,* in print.

Elber, W. (1970). Fatigue crack closure under cyclic tension. *Eng. Fract. Mech.,* 2, 37-45.

Garz, R.E. and M.N. James (1989). Observations on evaluating fatigue crack closure from compliance traces. *Inter. J. Fatigue,* 11, 437-440.

Hertzberg, R.W., C.H. Newton and R. Jaccard (1988). Crack closure: correlation and confusion. In: *Mechanics of Fatigue Crack Closure, ASTM STP 982* (J.C. Newman Jr. and W. Elber, eds), American Society for Testing and Materials, Philadelphia, pp. 139-148.

Philips, E.P., Results of the round robin on opening-load measurement. *NASA Technical Memorandum 101601,* Langley Research Center, Hampton, Virginia, May 1989.

Tada, H., P.C. Paris and G.R. Irwin (1973). *The Stress Analysis of Cracks Handbook.* Del Research Corporation, Hellertown, Pennsylvania.

Vasudevan, A.K., K. Sadananda and N. Louat (1993). Critical evaluation of crack closure and related phenomena: I. background and experimental analysis. In: *Fatigue 93* (J.P. Bailon and J.I. Dickson, eds), Engineering Materials Advisory Services Ltd., Warley, West Midlands, U.K., Vol.I, pp.565-570.

Vecchio, R.S., J.S. Crompton and R.W. Hertzberg (1986). Anomalous aspects of crack closure. *Inter. J. Fract.,* 31, R29-R33.

Weiss, B., D.L. Chen and R. Stickler (1992). Test procedures and a new concept for near-threshold fatigue closure. In: *Theoretical Concepts and Numerical Analysis of Fatigue* (A.F. Blom and C.J. Beevers, eds), Engineering Materials Advisory Services Ltd., Warley, West Midlands, U.K., pp.173-199.

PLASTICITY-INDUCED CRACK CLOSURE UNDER PLANE STRAIN CONDITIONS IN TERMS OF DISLOCATION ARRANGEMENT

F.O. Riemelmoser and R. Pippan

Erich-Schmid-Institut für Festkörperphysik der
Österreichischen Akademie der Wissenschaften
A-8700 Leoben, Jahnstraße 12, Austria

ABSTRACT

Fatigue crack growth rate is influenced significantly by crack closure. In our investigations we want to clarify whether or not crack closure under plane strain conditions can occur. As a tool for describing crack tip plasticity we use a dislocation model. In this study some important features of the model are presented and its application to growing mode I fatigue cracks is achieved. It is then shown that crack closure under plane strain conditions is not the exception but rather the rule.

KEYWORDS

Fatigue, crack closure, plane strain, dislocation

INTRODUCTION

In the early 1970's Elber (1971) discovered experimentally that a fatigue crack can close even at high tensile loads. He held the residual strains in the wake of the crack responsable for the premature contact of the crack faces. Since this pioneer work many investigations, both experimental and analytical studies, have been carried out to explain this phenomenon. These studies have led to the present unterstanding of crack closure. It is commonly accepted that the premature contact of the crack faces under plane stress conditions can be explained by an additional wedge which is filled in the crack (Budiansky and Hutchinson, 1978). The extra material is said to come from the necking of the specimen (out-of-plane flow). However, this model is not applicable to situations under plane strain conditions. By definition out-of-plane flow is not allowed. If we take this and the constance of volume during plastic deformation into account, the assumption of an additional wedge is not reasonable. Thus, many researchers believe that crack closure under plane strain conditions can not occur. Even there are ad hoc statements (Vasudevan et al., 1994) which should demonstrate that premature crack face contact is physically impossible. In the second chapter of this paper we will furnish proof to the contrary.
It is necessary to mention that there are, of course, investigations concerning crack closure under plane strain conditions by means of numerical integration tools (FEM, BEM) as well. Unfortunately their

results are ambigious (Fleck and Newman 1988, McClung et al. 1991).Most likely these ambiguities are caused by problems of these methods in treating nonlinear fracture mechanics. It is needless to say that with FE and BE simulations the object under consideration is discretised into distinct elements. In each of these subdomains the strain and stress field obey a fixed function; it is not necessary for these functions to coincide with the real stress and strain distribution. The error which is made in doing so is small generally but at locations (like crack tips) with large stress gradients or large deformations respectively the error is often not kept within reasonable bounds. Furthermore, crack closure under plane strain conditions is an effect which takes place only in the vicinity of the crack tip. It is small wonder that there arise difficulties in treating crack closure problems under plane strain conditions by means of FEM.

CRACK CONTOUR AND DISLOCATIONS

By using a dislocation model Vasudevan *et al.* (1994) "demonstrated" that crack closure under plane strain conditions can never occur. In the sequel their ideas are discussed. To begin with let us summarize their basic conclusions:

• A dislocation which is originated at the crack tip forms a ledge in the crack face (Fig. 1).

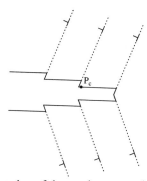

Fig. 1: Schematic representation of the crack contour due to dislocations

• The y displacement at the crack face due to dislocations is smaller than the ledge.
• The connection of point 1 and 2 forces the crack to be fully opened.

Indeed, point 1 and 2 are valid; however it will be subsequently shown that the conclusion (point 3) is not fulfilled automatically. Let us consider the y displacement of the crack face on the left side of the second ledge due to the 2nd dislocation. It is negative but it is smaller than the ledge (Vasudevan). Let us now estimate the y displacement of the crack face caused by the 2nd dislocation on the right side of its ledge, say at the point P_c? It is also negative! Otherwise there would be a discontinuity in the displacement field of the upper half plane which is not allowed except at the positions of the dislocations.

At P_c there is only one contribution to the y displacement which keeps the crack open, but there are the other dislocations which force the crack to close. It is easy to choose a distribution of dislocations which then results in a displacement at P_c which is negative in sum - in this case the crack is closed.

In the subsequent chapters it is shown that crack closure takes place inevitably in the vicinity of the crack tip - it is caused by the dislocations which are formed by cycling loading of a crack tip and which remain in the wake of the crack.

THE DISLOCATION MODEL

The fundamental aspects and inherent assumptions of the dislocation model are discussed elsewhere (Riemelmoser *et al.*, in prep.). There we concentrated our efforts to cyclic crack tip plasticity at a stationary fatigue crack (= no crack advance was allowed). In this study a growing fatigue crack is considered. We will briefly discuss the basic features of the dislocation model and give our attentions to items which are typical for growing fatigue cracks.

• The initial system is a homogeneous isotropic linear elastic medium which is cut along the negative x_1-axis.
• The strains and stresses in the initial, dislocation free system are described by the common crack tip fields. As a measure of their strength the stress intensity factor K is used.
• The dislocations originate at the crack tip only. A positive dislocation is generated if the local k at the crack tip reaches a critical value k_e (Rice and Thomson, 1974). Upon dislocation emission the crack advances by an increment equal to $|b| \cdot \cos\alpha$ (b = Burgers vector, α = angle of inclination).
• All dislocations lie on parallel slip planes (Fig. 2). The latter are inclined to the crack plane. The angle of inclination $\alpha = 70.5°$.

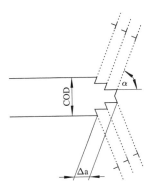

Fig. 2: Schematic representation of the dislocation arrangement, crack growth and crack tip blunting upon dislocation emission

• A dislocation does not move if the amount of the total force on it is less than friction stress τ_0.
• The dislocation configuration is symmetric to the x_1-axis; a pure mode I loading is guaranteed.
• Upon unloading the dislocations are allowed to return to and to annihilate at the crack tip.
• Free surface once created (upon dislocation emission) is never allowed to re-weld again.
• Parallel to the reduction of the applied K (in the unloading stage) the local k is decreased as well. If k_{local} becomes equal or less $-k_e$ negative dislocations are generated. Hereby "negativ" corresponds to the sign of the Burgers vector, i.e. the Burgers vector of a positive and a negative dislocation have the some magnitude but they point into opposite directions.
• Upon emission of a positive dislocation (Fig. 2) COD increases contrarily to the reduction of COD by emission of a negative dislocation.
• A positive and a negative dislocation on the same slip plane annihilate if their mutual distance becomes less than 10 times the core radius.

• The annihilation of a positive dislocation at the crack tip or the generation of a negative one, respectively causes crack closure in the very vicinty of the crack tip (Fig. 3).

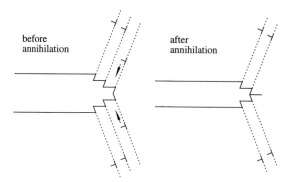

Fig.3: Unreal crack closure due to dislocation annihilation

This kind of crack closure is subsequently termed "unreal" to point out that it does not influence the crack growth rate. In our investigations the unreal crack closure was taken into consideration by a crack length correction procedure: Upon unloading the system was treated as if the crack faces were allowed to re-weld again but only in that crack growth increment which was produced during the last loading cycle. Of course, in the subsequent loading stage the crack was allowed to open to the full length again.
•Material data: Shear modulus μ=80 GPa; poison ratio ν=0.3; friction stress τ_0=μ/2000;
                critical stress intensity necessary for dislocation emission k_e= $0.4 \cdot \mu \cdot \sqrt{b}$.
Process parameter: Maximum load K_{max}=$4.0 \cdot k_e$; minimum load K_{min}: increasing.

FATIGUE CRACK GROWTH IN THE LIGHT OF THE DISLOCATION MODEL

According to our assumptions a free surface, once created, will not re-weld again. Thus the crack becomes longer by dislocation generation and annihilation. This is illustrated with Fig. 4 in which the change of the crack contour during the first and the second cycle is depicted.

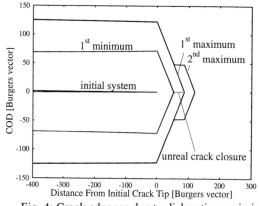

Fig. 4: Crack advance due to dislocation emission and annihilation

In Fig. 5 a sequence of pictures demonstrates the behavior of the crack upon the second unloading.

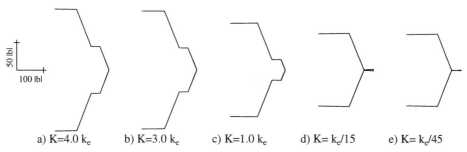

a) K=4.0 k_e b) K=3.0 k_e c) K=1.0 k_e d) K= $k_e/15$ e) K= $k_e/45$

Fig. 5: The crack upon the second unloading

Very interesting is the transition from fig. 5d to 5e. In 5d the crack is still fully opened. After the next load reduction had been carried out a further pair of dislocations returned to the crack tip. The resulting crack contour is then shown in 5e; the crack became partially closed. Crack face contact do, of course, influence the stress field in the body and the dislocation arrangement.

To overcome the difficulties which arise herewith, we have employed a K_{min} raising procedure in such a manner that K_{min} was increased whenever the crack became closed (closed in the sense of real crack closure, i.e. the point of first contact is somewhere behind the crack tip (Fig 5e)). The development of the loading range is sketched in Fig. 6.

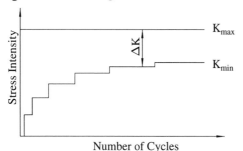

Fig. 6: ΔK reduction scheme as to prevent crack closure

DISCUSSION

The view of the growing crack (Fig. 7) clearly indicates the existence of premature crack face contact up to very large R-ratios (0.61). Furthermore we want to call the reader's attention to the contour of the crack in the neighborhood of the crack tip. Obviously it is similar in each of the pictures in Fig. 7b, though the R-ratio increases from 0.21 in the first picture to 0.61 in the last one. Hence, it appears that the crack face deformation due to the applied mean load is always compensated by dislocations, whatever the R-ratio (mean load) might be. It seems there is a "cloud" of dislocations formed by a cyclic loaded crack tip in such a way that the vicinity of the crack tip is always fully shielded independently of the applied mean load; the dislocations density in this cloud increases with larger R-ratios. It can be concluded that the shape of the crack in the vicinity of the crack tip does not depend on the loading history . The crack contour and the closure behavior is mainly influenced by elastic constants, by the activated glide system and the stress intensity necessary to generate dislocations.

The results suggest that crack closure under plane strain conditions is not only due to low R-ratios but also that it takes place in general.

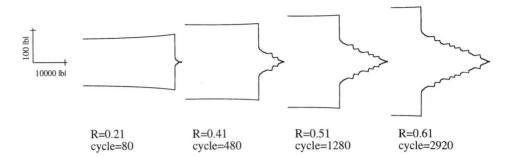

Fig.7a: The crack contour in the progress of fatigue crack growth.

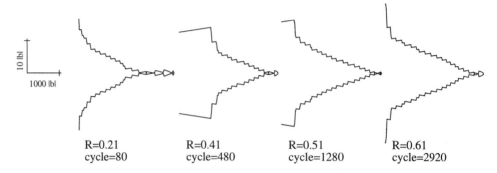

Fig.7b: The crack tip at higher magnification .

ACKNOWLEDGEMENT

This work was supported by the FFF (Fonds zur Förderung der wissenschaftlichen Forschung), Project Nr. P 10116 TEC.

REFERENCES

Budiansky, B. and J.W. Hutchinson (1978). Analysis of closure in fatigue crack growth. *J. Appl. Mech.* 45, 267-276.

Elber, W. (1971). The significance of fatigue crack closure. *ASTM STP* 486, 230-242.

Fleck, N.A. and J.C. Newman, Jr. (1988). Analysis of crack closure under plane strain conditions. *Mechanics of Fatigue Crack Closure, ASTM STP* 982, 319-341.

McClung, R.C., B.H. Thacker and S. Roy (1991). Finite element visualization of fatigue crack closure in plane stress and plane strain. *Int. J. Fract.* 50, 27-49.

Rice, J.R. and R. Thomson (1974). Ductile versus brittle behaviour of crystals. *Phil. Mag.* 29, 73-97.

Riemelmoser, F.O., R. Pippan and H.P. Stüwe (in prep.). Cyclic crack tip plasticity: A dislocation approach.

Vasudevan, A.K., K. Sadananda and N. Louat (1994). A review of crack closure, fatigue threshold and related phenomena. *Mat. Sci. Eng.* A188, 1-22.

THE INFLUENCE OF MICROSTRUCTURE ON FATIGUE CRACK PROPAGATION AND CRACK CLOSURE BEHAVIOUR IN THE THRESHOLD REGIME

TH. AUF DEM BRINKE, H.-J. CHRIST and K. DETERT

Institut für Werkstofftechnik, Universität-GH-Siegen,
D-57068 Siegen, Fed. Rep. Germany

ABSTRACT

Fatigue crack propagation behaviour and thresholds of two steels, the heat resistant austenitic Alloy 800 and the ultra high strength 18%-Nickel-maraging steel Ultrafort 6355, were investigated with particular attention to the relationship to microstructural parameters and crack closure. Inflections in the log da/dN vs. log ΔK_{eff} plots are related to an equivalence of the microstructural element size to the plastic zone size. The combination of this information with the "striation model" developed by Roven and Nes (1991) will help to predict fatigue crack propagation.

KEYWORDS

Fatigue; Fatigue crack propagation; Fatigue crack closure; Microstructure

INTRODUCTION

Fatigue crack propagation and fatigue threshold have gained great importance for a damage tolerant design. The influence of the microstructure in the near threshold regime is eminent and a frequent object of investigation. Since experiments are costly and time-consuming, there is a real need for the development of new models and the deduction of time-saving lifetime assessment procedures.

MATERIALS AND EXPERIMENTAL

The materials investigated in this study are two steels which differ both in microstructure and mechanical properties, the heat resistant austenitic steel Incoloy™ 800 (common designation Alloy 800) in sheet form of 8 mm thickness and the ultra high strength maraging steel Ultrafort™ 6355 in forged block form of 71 mm x 130 mm, respectively. The Incoloy was soft annealed at 960°C/1h/air, the Ultrafort was solution annealed at 860°C/1h/air and aged at 460°C/6h/air. The chemical composition and tensile properties of both alloys are reported in Table 1 and 2. The LCF-data listed in Table 2 are according to the Manson-Coffin relation with $\sigma_{ys}{}'$ as the cyclic yield strength.

Table 1.   Chemical composition of the alloys

Alloy	Amount in wt.-%					
Ultrafort 6355	Fe	C	Ni	Co	Mo	Ti
X 2 NiCoMo 18 12	65.59	0.008	18.99	9.83	4.86	0.34
Incoloy 800	Fe	C	Ni	Cr	Al	Ti
X 10 NiCrAlTi 32 20	47.45	0.068	30.3	20.3	0.28	0.32

Table 2.   Tensile and low cycle fatigue properties

Alloy	E	σ_{ys}	σ_{UTS}	% Elon	% Area	σ_{ys}'	ε_f'	c
	MPa	MPa	MPa	gation	Red.	MPa	(C.-M.)	(C.-M.)
Ultrafort	180000	1920	2050	4.7	36	1688	0.466	-0.924
Incoloy	196000	238	564	39.4	55.6	309	0.593	-0.611

Two specimen geometries in the L-T-orientation were used for obtaining fatigue crack propagation curves and thresholds, compact tension specimens (CT-specimens, W=50 mm, t=8 mm) on a servohydraulic and small, sharpy-like, single edge notch bend specimens (W=15 mm, t=8 mm) loaded in four-point-bending on a resonant testing machine according to ASTM E 647-88. Fatigue testing was carried out at frequencies of 20 Hz and 140-200 Hz, respectively, at 20°C in air under sinusoidal loading with load ratios R from 0,1 to 0,7. In order to get a wide spectrum of da/dN vs. ΔK values, the tests were conducted by holding the load-range constant (increasing ΔK) and by applying the load shedding procedure as recommended by Saxena with load ratio R held constant (in order to study the influence of crack closure) and with constant maximum load K_{max} (in order to get closure-free effective thresholds by elimination of the crack closure effect). The crack length was monitored throughout the test using the compliance method (CT-specimen) or an indirect potential drop method (bending-specimen) calibrated by optical measurements. Special attention was paid to crack closure effects which were expected to have a substantial influence on the fatigue crack propagation rate in the near threshold regime. The stress intensity factor for crack opening K_{op} was determined from load-displacement- (CT) and moment-backface-strain- (bending) diagrams. The start of non-linearity on the load-displacement diagrams was taken as the opening load F_{op} and the opening moment M_{op}, respectively. In order to eliminate crack closure, some specimens were precracked in cyclic compression ($F_{max} \leq 0$; $F_{min} < 0$).

RESULTS AND DISCUSSION

Fatigue Crack Propagation

The upper branch of the fatigue crack propagation curves shown in Fig. 1 and Fig. 2 represent results obtained in crack growth experiments on CT-specimens, whereas the results of the lower branch were observed on bending-specimens. Since the CT-specimens were precracked in compression, no detectable crack closure was found even at R=0,1. The length of the precracks of about 50-100 μm appeared to be in the range of the cyclic plastic zone size as mentioned by Pippan (1987). Due to high compressive loads from precracking the austenitic steel showed excessive crack acceleration during fatigue at small stress ratios, cf. Fig. 1. Throughout testing of the bending specimens a small degree of crack closure occurred. The fatigue crack propagation curves of both materials (Fig. 1 and Fig. 2) show that crack propagation can be described by a curve which exhibits four inflections dividing the plots into five regions. Fractographic studies confirmed different fracture mechanisms for these regions.

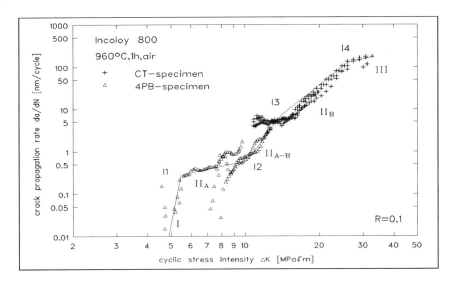

Fig. 1:   Log da/dN vs. ΔK plot for the Incoloy obtained in constant load amplitude (CT) and in decreasing ΔK(bending)-tests, respectively. Stress ratio R=0,1 was kept constant. Solid line accentuates inflections I1..I4 (see Table 3).

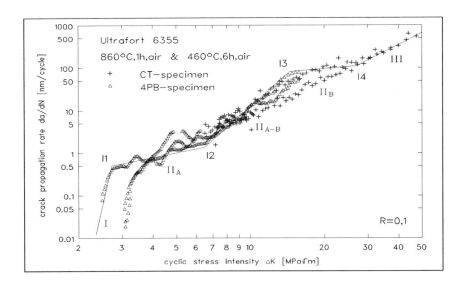

Fig. 2:   Fatigue crack propagation plot for the Ultrafort obtained in constant load amplitude (CT) and in decreasing ΔK(bending)-tests, respectively. Stress ratio R=0,1 was kept constant. Solid line accentuates inflections I1..I4 (see Table 3).

In addition to the commonly used partition of the log da/dN vs. log ΔK plot into three regions, Jono and Song (1984) recommended the further division of Region II into three subregions II_A, II_{A-B} and II_B. Clark (1986) observed on a Cobalt-free maraging steel two inflections in the near threshold regime lying beyond the ones obtained in the present paper since Clark took not the effective cyclic stress intensities. The measured values for the transition points I1..I4 are listed in Table 3. Since no detectable crack closure was found ΔK was assumed as ΔK_{eff}.

Table 3:   Measured effective cyclic stress intensities at transition points

Material	$\Delta K_{eff,I1}$ MPa$\sqrt{m}$	$\Delta K_{eff,I2}$ MPa$\sqrt{m}$	$\Delta K_{eff,I3}$ MPa$\sqrt{m}$	$\Delta K_{eff,I4}$ MPa$\sqrt{m}$
Ultrafort	3	6	15	25
Incoloy	5	8..10	14..15	25

Attention was paid in particular to the relationship between microstructural element size and crack growth rate or plastic zone size equivalence. Hornbogen and zum Gahr (1976) pointed out that for heterogenously deforming microstructures grain boundaries impede crack growth when the cyclic stress intensity is between the points where the sizes of the static and the cyclic plastic zone reach the grain size. Yoder et al. (1982) assumed that the transition from crystallographic controlled to stress controlled fatigue crack propagation depends on the relation of the maximum length of the crack path in a slip-plane s_{max} to the cyclic plastic zone size $r_{pl,c}$. Based on these ideas Detert et al. (1988) derived from the equation for the cyclic plastic zone size (Eq. 1) a relation for the cyclic stress intensity at the transition from crystallographic to stress controlled fatigue crack propagation (Eq. 2):

$$r_{pl,c} = \frac{1}{3\pi}\left(\frac{\Delta K_t}{2\sigma_{ys}}\right)^2 \tag{1}$$

with the assumption that $r_{pl,c}=s_{max}$

$$\Delta K_t = \sigma_{ys}\cdot 2\cdot\sqrt{(3\pi\cdot s_{max})} \tag{2}$$

In Table 4 the results for the effective cyclic stress intensity factors are listed which were calculated according to equation (1) taking the metallographic detected microstructural dimensions (prior austenitic grain size D_{Gp}, grain size D_G, particle size D_P, martensite lath width D_{Mb} and martensite lath length D_{Ml}) as plastic zone size.

Table 4:   Cyclic stress intensities calculated according to Eq. 2

microstructural dimensions	Ultrafort				Incoloy	
	D_{Gp}	D_P	D_{Mb}	D_{Ml}	D_G	D_P
[μm]	24	4..10	1..2	5..20	104	10
ΔK_t (calculated)						
[MPa$\sqrt{m}$]	57	23.4..37	3.7..16.6	26.2..52.3	14.4	4.5

It can be seen from Tables 3 and 4 that the prior austenitic grain size D_{Gp} of the Ultrafort doesn't affect crack growth whereas the martensite lath width D_{mb}, the martensite lath length D_{ml} and the Ti-particle size D_P can be related to the magnitude of the inflections in the log da/dN vs. log ΔK plot. In case of the Incoloy grain size D_G and particle size D_P fit to the measured inflection data. Due to scatter in microstructural dimensions the obtained transition stress intensities can not be associated exactly with the measured inflections.

In order to find reasons for the differences in fatigue crack propagation behaviour, crack length was plotted against cyclic stress intensity and regions of interest were studied in more detailled fractographic investigations by means of SEM. The crack path appears to be transgranular with only small amounts of intergranular fracture observed in region II_A. Signs of roughness induced crack closure are evident for the Incoloy and attributed to the formation of secondary cracks due to separation of fragments from the crack face and their interaction with the crack faces. The part of the coarse Ti(C,N)-particles for fatigue crack propagation is indicated by their dominating occurrence in region II_{A-B} on the fracture surface compared to the small amount of these particles on polished sections of the materials. Typical for both alloys is the formation of fine substriations and oxide layers on the fracture surface in the near threshold region. In the case of the Ultrafort the appearance of oxide layers on the fracture surface is dependent on testing frequency only above 20 Hz and linked with oxide induced crack closure. Clark (1986) suggested that the mechanism of intergranular facetting is hydrogen-assisted due to the production of hydrogen by an oxide thickening process at the crack tip. The occurrence of intergranular fracture is unrelated to any plastic zone size-grain size equivalence. For high cyclic stress intensities (region II_B and above) the typical striations form on both materials. The striation spacing is in all cases larger than the fatigue crack propagation rate per cycle. This indicates that the typical one striation per cycle model is not valid as was pointed out by Roven and Nes (1991). The authors refined the existing accumulated damage, LCF- models for fatigue crack propagation based on the striation formation. The resulting relation for the crack propagation rate leads to a good fit in the log da/dN vs. log ΔK plot when the function of the cyclic stress strain curve is established. The need to obtain the microstructural parameters controlling crack growth is still evident and a possible approach is the simple cyclic plastic zone size - microstructural element size equivalence.

Fatigue Threshold

Threshold data are displayed in Fig. 3 for the Incoloy (right) and for the Ultrafort (left). The Incoloy data show a wide scatter and even the effective thresholds for R-values smaller than 0,7 are remarkably higher than those obtained from K_{max}=const. tests. Therefore it is difficult to obtain a valid value for the critical stress ratio R_c defined as

$$R_c = \frac{K_{op}}{K_{op} + \Delta K_{th,eff}} \qquad (3)$$

according to Detert et al. (1991). When threshold conditions are met, ΔK_{th} is highly dependent on the position of the crack tip relative to microstructural obstacles. Since the Incoloy has a coarse microstructure a remarkable scatter of fatigue thresholds results. For the Ultrafort all measurements are in good accordance to literature and slightly elevated towards low R-values caused by oxide induced crack closure. It is notable that there exists a linear relationship between the effective threshold $\Delta K_{th,eff}$ and the elastic modulus as was demonstrated by Liaw et al. (1983) leading to the relationship

$$b = \left(\frac{\Delta K_{th,eff}}{E}\right)^2 \qquad (4)$$

with b as the smallest atomic spacing in lattice. Results evaluated after Eq. 4 (dashed lines in Fig. 3) are in reasonable accordance to measured data for the effective threshold $\Delta K_{th,eff}$. In contradiction to this empirical function, Yokobori et al. (1989) formulated a dislocation dynamics based function taking the microstructural dependence into account by relating the threshold to the grain size and the Burgers-vector.

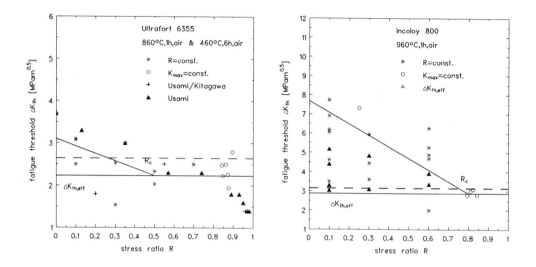

Fig. 3:   Fatigue thresholds for the maraging steel and for the austenitic steel determined in ΔK decreasing tests according to ASTM E-647. Intersection of solid lines (ΔK_{th}, $\Delta K_{th,eff}$) assigns R_c. Dashed line shows results of Eq. 4.

REFERENCES

ASTM-STP E 647-88 (1988). *Annual Book of ASTM Standards*, ASTM, Philadelphia, PA, Vol. 03.01

Clark, G. (1986). *Fatigue Fract. Engng. Mater. Struct.*, Vol. 9, No. 2, 131-142

Detert, K. Ibas, O. and R. Scheffel (1988). *Z. f. Metallkunde*, 79, 564-571

Detert, K. Ibas, O. and R. Scheffel (1991). *Z. f. Metallkunde*, 82, 225-229

Hornbogen, E. and zum Gahr, K. H. (1976). *Acta Metall.*, 24, 581-592

Jono, M. and J. Song (1984). In: *"Fatigue 1984"*, *Proc. of 2nd Int. Conf. Birmingham*, UK, (C. J. Beevers, Ed.), Vol. II, Engineering Materials Advisory Ltd., 717-726

Liaw, P. K. Leax, T. R. and W. A. Logsdon (1983). *Acta Metall.*, 31, 1581-1598

Pippan, R. (1987). *Fatigue Fract. Engng. Mater. Struct.*, Vol. 9, No. 5, 319-328

Roven, H. J. and E. Nes (1991). *Acta metall. mater.*, 39, 1735-1754

Yoder, G. R. Cooley, L. A. and T. W. Crooker (1982). *Scripta Metal.*, 16, 1021-1025

Yokobori Jr., A. T. Isogai, T. Yokobori, T. Maekawa, I. and Y. Tanabe (1989). In: *"Advances in Fracture Research"*, *Proc. of 7th Int. Conf. on Fracture*, Houston, USA, (D. M. R. Taplin, Ed.), Vol. 2, Pergamon Press, 1453-1463

A UNIFIED FRAMEWORK FOR FATIGUE CRACK GROWTH

K. Sadananda and A. K. Vasudevan[*]

Code 6323, Naval Research Laboratory, Washington D.C. 20375
[*]Code 332, Office of Naval Research, Arlington, VA 22217

ABSTRACT

We show that the two parametric approach that we have developed earlier provides a unified frame work that can explain both the short and long crack growth behavior as well as the endurance limit in a S-N curve without resorting to extraneous factors such as crack closure. In extending our analysis to growth of short cracks, we reject the current notion of lack of similitude for the short cracks, and express the similitude as a fundamental postulate that *for a given crack growth mechanism, equal crack tip driving forces result in equal crack growth rates.*.

INTRODUCTION

Fatigue damage normally involves three stages; crack nucleation, a short and a long crack growth. Long crack growth is well characterized using Linear Elastic Fracture Mechanics (LEFM). Short crack growth behavior has been considered as anomalous, since crack growth rates differ from long crack growth rates. Plasticity induced closure has been attributed as the primary source for the lack of similitude (Suresh and Ritchie, 1984). Crack closure concepts, in spite of several reservations and uncertainties, are generally taken for granted as valid concepts. We have critically examined using fundamental concepts from dislocation theory (Vasudevan et al. 1994; Louat et al. 1993) crack closure. We have shown on the basis of dislocation theory that plasticity originating from the crack tip can not contribute to its closure. If crack closure is not significant, then, one has to explain all the phenomena hitherto attributed to crack closure. To do that, we have recently developed new concepts (Vasudevan and Sadananda 1995; Sadananda and Vasudevan 1995 a, b). It involves recognition of fatigue as two parametric problem requiring two load parameters, and hence two thresholds for fatigue crack growth rather than one. Fig. 1 shows a fundamental threshold curve in terms of ΔK vs. K_{max}. The two asymptotic limiting values in the curve correspond to the two critical thresholds, ΔK^{*}_{th} and K^{*}_{max} that must be met simultaneously for a crack to grow. We now extend these concepts to short crack growth behavior and show that there are no anomalous effects, and the behavior can be adequately explained using the two parametric description. In addition, we bridge the gap between the short crack and the crack nucleation thus providing a unified frame work for the characterization of fatigue damage.

SHORT CRACK GROWTH BEHAVIOR

Behavior of short cracks has been considered as anomalous (Suresh and Ritchie 1984, Smith and Miller, 1978), since they differ from that of long cracks. Fig. 2 compares schematically the behavior

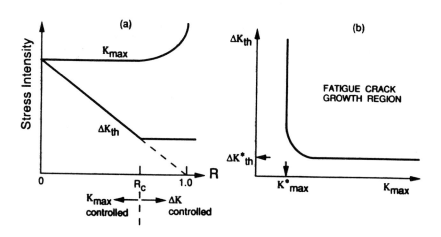

Fig. 1. (a) ΔK and K_{max} as a function of load ratio, R and (b) Fundamental threshold curve in terms of ΔK versus K_{max} showing two limiting thresholds for crack growth

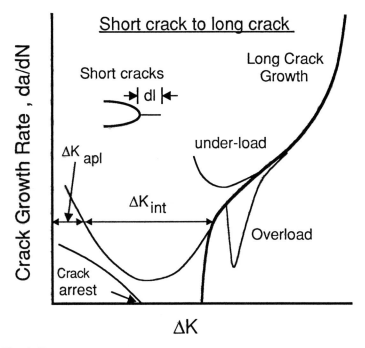

Fig. 2. Growth behavior of short cracks in relation to that of long cracks.

of short cracks and long cracks. The contrasting behavior of the two cracks can be summarized as follows: (a) There is a no uniqueness associated with short cracks. (b) Short cracks grow at stress intensities below long crack thresholds. (c) The same ΔK does not result in the same crack growth rates (lack of similitude). (d) Shorts cracks can accelerate and decelerate as ΔK increased. (e) In some cases deceleration could lead to complete crack arrest. And (f) finally, for cracks comparable to grain size, boundaries can inhibit crack growth.

Crack closure from plasticity in the wake of a crack is attributed as the primary source for the lack of similitude, between long and short cracks (Lankford 1982). The implication is that the abnormality is not with the short cracks but with long cracks, since crack closure in long cracks reduces their applied driving forces. This means that the short crack growth behavior is intrinsic, and that of long crack is not, and hence long crack growth data needs to be corrected for crack closure.

We question these generally accepted notions. Since plasticity originated from the crack cannot contribute to its closure we need to look for a different explanation for the observed differences in the growth behavior of short and long cracks. We note that irrespective of the nature and the source of short cracks, that is, whether they are mechanically, microstructurally or physically short, or are growing cracks from pre-existing notches or initiated at free surfaces, their behavior are all similar. Fig. 2, in fact, depicts the general behavior with the cracks growing at applied stress intensities below the thresholds of long cracks, and with decelerating and accelerating growth, and in some cases with complete arrest.   In this paper, we look for the generalities that contribute to deviations from the long crack growth behavior without any consideration of anomalous behavior.

THE ROLE OF INTERNAL STRESSES

First, we consider that the behavior of a long crack is fundamental with the requirement of two thresholds, ΔK^*_{th} and K^*_{max}, to be satisfied simultaneously for it to grow. Since cracks originating from the notches grow in an accentuated stress field, and further since all other cracks also behave similarly, we presume that there are always some internal stresses to aid the growth of short cracks. In the case of the notches, the internal stresses are identifiable as the notch tip stresses. In other cases, the internal stresses may not be easily identifiable. Some of these stresses may not be preexisting before fatigue tests, as in smooth specimens, but are internally generated as a result of the accumulation of fatigue damage. Internal stresses, in general, affect the K_{max} and K_{min} to the same proportion and hence the R-ratio while affecting ΔK to a lesser degree. Because of local high R, the short crack in the limit behaves similar to a long crack at high R ratio (Pippan et al.1987).

The generally accepted notion that short cracks grow at stress intensities lower than the long crack growth threshold is based on the consideration of only one threshold parameter, as in Fig. 2.   From our concepts, crack, short or long, grows only when both thresholds are met simultaneously, Fig. 1. For most metals and alloys, the cyclic threshold, ΔK^*_{th} is small and is usually of the order of 1 MPa $\sqrt{m}$ for Al alloys, while the other threshold, K^*_{max}, is generally higher, of the order of 3-5 MPa$\sqrt{m}$ depending on the microstructure and crack tip environment. Even short cracks have to meet these requirements.   When the applied stress amplitude for short cracks are high enough to provide the cyclic threshold, the augmented local K_{max} helps to meet the threshold K_{max} thus ensuring fatigue crack growth. With increase in length of the crack, the contribution from the internal stresses to K_{max} decreases, while contribution from the crack-tip stress field increases due to increase in its length.   Fig. 3 schematically shows the variation of the contributions from the internal stresses, the crack tip stresses and the total combined stresses in relation to the thresholds.   As shown in the figure, the total stress reaches a minimum, and if the minimum is greater than the threshold K^*_{max}, then short cracks decelerates and accelerates. On the other hand, if the minimum is lower than the threshold, then the deceleration leads to crack arrest.

Therefore, short crack is not different from a long crack except for the internal stress fields (whatever the sources of these stresses may be) which are deterministic, and that the two critical thresholds are the same independent of the crack size for a given material and crack tip environment. Significant test to test variations in the short crack growth data arise due to the variability in the internal stresses from specimen to specimen rather than in the limitation of the crack tip parameters. Fig. 4 is the crack growth data (Tanaka and Nakai, 1983) involving short cracks nucleated from notches, surface cracks and from corner cracks. The internal stresses (ordinate) are computed from the differences in the driving forces in the long and short cracks. The data are plotted against a factor proportional to the length of the short cracks.   It confirms that the internal stresses near the free surface crack (corner or surface cracks) are not different from the stresses ahead of a notch. Hence we conclude that all short cracks irrespective of their origin grow in internal stress fields

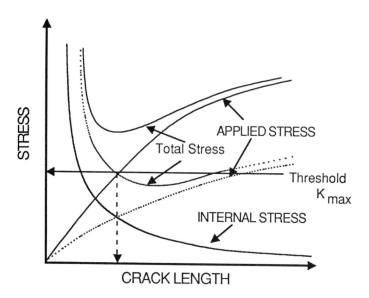

Fig. 3. Variation of internal, applied and total stresses as a function of crack length

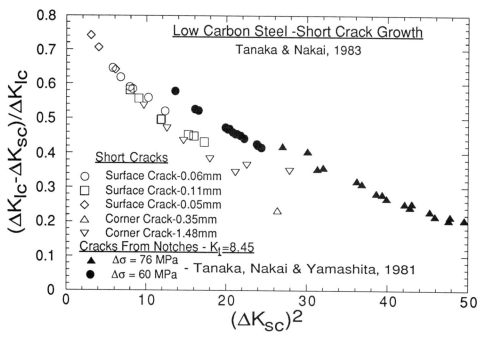

Fig. 4. Internal stresses computed from crack growth data are identical irrespective of the source of the short crack (sc) - stresses and lengths normalized with respect to long crack (lc)

which accentuate the applied stresses to provide the needed driving forces. From these results we reach an important conclusions which in fact reinforces a similitude between the long and short cracks and among various short cracks which is reformulated as follows: Similar *crack tip forces contribute to similar crack growth rates if the crack growth mechanisms are similar.*

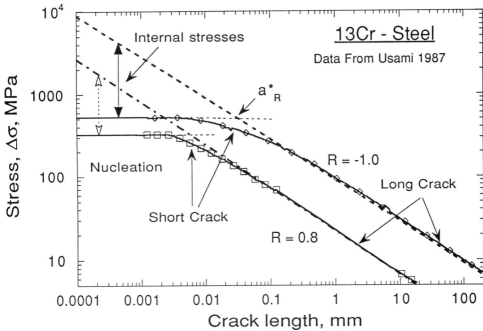

Fig. 5. Kitagawa diagram representing endurance limit and thresholds for short and long cracks.

CRACK NUCLEATION AND ENDURANCE LIMIT

We now extend the analysis to crack nucleation. Stress amplitude at endurance limit in S-N curve and the threshold for short and long cracks are combindly represented by the Kitagawa diagram using Usami's data at different R ratios (Usami 1987) as shown in Fig. 5. Linear portion at large crack lengths with slope of one half is related to square root singularity under LEFM considerations. The curves level off at different endurance limits for different R ratios. Linear portion at each R ratio intersects the horizontal portion at different crack lengths, $a*_R$. Physically critical crack length, $a*_R$, signifies that for a given R ratio, a nucleated or short crack has to grow to that size before it begins to grow as a long crack. With increase in R ratio the $a*_R$ decreases and this can be related to constant $K*_{max}$ required particularly in the K_{max} controlled conditions ($R<R_c$) with R_c of the order of 0.6. For crack lengths less than $a*_R$, cracks behave as short cracks. For them to grow, internal stresses are required. For a smooth specimen, at endurance limit, the internal stresses have to be generated in situ at, say, intrusions and extrusions, till a crack becomes self sustaining. If the stresses are normalized by the respective endurance limits and the crack lengths by $a*_R$ then all the data fall in one curve as shown in Fig. 6. The extrapolation of the linear portion in the small crack growth regime indicates the actual stress required to sustain a crack less than $a*_R$ for it to grow without the aid of any internal stresses. Since the endurance limit provides the limiting applied stress, the difference between the extrapolated linear curve and the horizontal curve is the measure of the minimum internal stresses required for a crack of length less than $a*_R$ to grow. If the internal stress gradients are larger than the minimum then acceleration and deceleration of cracks occur in relation to the long crack growth behavior. Attributing the short crack growth to some anomalous behavior is, therefore, due to lack of recognition of these fundamental two threshold requirements and the role of internal stresses in aiding or obstructing the growth. Thus the two parametric approach provides a unified frame work for characterization of fatigue crack growth.

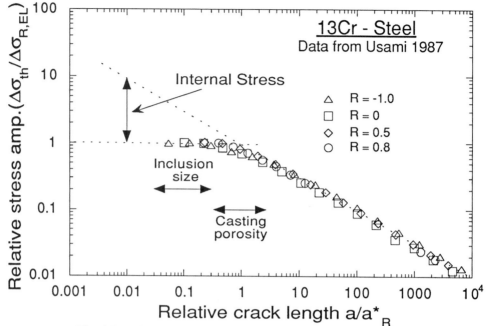

Fig. 6. Data from Fig. 5 for various R-ratios in normalized coordinates.

SUMMARY AND CONCLUSIONS

We have developed a unified frame work that accounts for the behavior of both short and long cracks as well as the endurance limits in a S-N curve. We show that there are no anomalous behaviors or the need for extrinsic factors such as crack closure. Further we show that the concept of similitude is valid, which we reformulate as: *Equal crack tip forces cause equal crack growth rates provided the governing mechanisms are equal.*

REFERENCES

Lankford, J., (1992). *Fat. Mat & Struc.* 5, .233-248.

Louat, N., Sadananda, K., Vasudevan, A. K. and Duesbery, M., (1993). *Met. Trans.*, 24A, 2225.

Pippan, R., Berger, M. and Stuwe, H. P., (1987), *Met. Trans.* 18A, .429-435.

Sadananda, K. and Vasudevan, A.K., (1995 a) , *ASTM STP-1220*, 484-501.

Sadananda, K. and A. K. Vasudevan,, 1995b, Reevaluation of Short Crack Growth behavior, 27th National Fracture Mechanics, June 1995, *ASTM-STP*, submitted for publication.

Smith, R.A. and Miller, K. J., (1978) *Int. J. Mech. Sci.*, 20, pp.201-206

Suresh, S. and Ritchie, R. (1984). *Int. Met. Rev.*, 29, 445-476.

Tanaka, K. and Nakai, Y. (1883) *Fat. Eng. Mat. & Struc.* 6, 315-327.

Usami, S. (1987)"Short Crack Fatigue Properties, in Current Research on Fatigue Cracks,- Vol. 1, Eds. T. Tanaka, T et al. Elesevire London, 119-147.

Vasudevan, A. K., Sadananda, K., and Louat, N., (1994). *Mat. Sci. Eng.* A188, 1-22. .

Vasudevan, A.K. and Sadananda, K. (1995) *Met. Trans.*, 26A, 1221-1234.

The Effect of Thickness on the Rate of Fatigue Crack Growth

H. Bao and A. J. McEvily

Department of Metallurgy, The University of Connecticut,
Storrs, CT 06269, USA

ABSTRACT

A study of the effect of thickness on the rate of fatigue crack growth has been carried out. Two specimen thicknesses (0.3 mm and 6.35 mm) were tested. The materials investigated were 1018 steel, 9Cr-1Mo steel, and the aluminum alloys 2024, 6061 and IN9021. Thickness effect was found only in the 9Cr-1Mo and IN9021 alloys, where at high ΔK level the rate of growth increased with increased thickness. The cause of this thickness effect was attributed to pronounced localized contraction on the surface at the crack tip. A high yield to tensile strength ratio was responsible for this contraction.

KEYWORDS

Fatigue, Thickness Effects, Crack Closure, Work Hardening .

INTRODUCTION

In the ASTM standard test method for the measurement of fatigue crack growth rates (E647), section 5.1.3 points out that " fatigue crack growth rate data are not always geometry independent in the strict sense since thickness effects sometimes occur. However, data on the influence of thickness are mixed. Fatigue crack growth rates over a wide range of ΔK have been reported to either increase, decrease, or remain unaffected as specimen thickness is increased".(A review of the literature on this subject may be found in reference 1.) A number of factors contribute to this inconsistency with respect to the effect of thickness, and among them residual stresses, microstructural variations, and the state of stress at the surface of a specimen compared to that in the interior may be reflected in the effect of thickness on fatigue crack growth behavior. In the present paper we shall be concerned only with the last of these items.

The ratio of the plastic zone size to the thickness of a plane specimen is an indication of the relative amounts of plane stress compared to plane strain existing at a crack front. Fatigue crack growth studies at growth rates below 10^{-3} mm/cycle are usually carried out on specimens whose thickness, B, is large enough such that the plastic zone size to thickness ratio is much less than 1.0, and this ratio decreases even further as the threshold level is approached. Therefore even though a region of plane stress always exists at the surface of the specimen, specimens are usually tested under predominantly plane strain conditions, and the more so as the threshold level is approached. Nevertheless, in view of this difference in the state of stress through the thickness of a specimen one wonders if, in the absence of any surface residual stresses or any through-thickness metallurgical variation, there is any difference in the rate of growth under plane stress as compared to plane strain conditions which might translate into a thickness effect. In order to shed more light on this question we have attempted to maximize any plane stress contribution by testing extremely thin specimens, i.e., 0.3 mm in thickness, in a variety of alloys and have compared the results obtained with those for thicker specimens, i.e., 6.35 mm in thickness.

MATERIALS, SPECIMENS and TESTS

MATERIALS: The alloys studied were 1018 steel and 9Cr-1Mo steel, and the aluminum alloys 2024-T351, 6061-T6 and IN9021-T4. The mechanical properties of the materials are given in Table 1.

Table 1.   Mechanical Properties of Materials

Materials	σ_y (MPa)	σ_u (MPa)	% elong.	σ_y/σ_u
1018	260	504	32	0.52
9Cr-1Mo	531	668	26	0.79
Al6061-T6	276	310	12	0.89
Al2024-T351	345	483	18	0.71
IN9021-T4	530	592	8	0.90

SPECIMENS: The specimens of 6.35 mm and 0.3 mm thick were fabricated in accord with the standard ASTM compact tension (CT) specimen specification [E647-88]. The dimensions for the 6.35 mm thick compact tension specimen were H=6.85 cm and W=5.72 cm. Smaller specimens were used for the 0.3 mm thick specimens, namely, H=3.05 cm and W=2.59 cm. To eliminate any microstructural variations between specimens, both 6.35 mm and 0.3 mm thick specimens were machined from the same plates, which were at least 8 mm thick. The specimen surfaces on both sides were polished to a surface finish of 1 μm by using fine emery papers and diamond paste, which facilitated the location of the crack tip in the crack growth studies and therefore measurement of the crack length.

EXPERIMENTAL SET-UP : The tests were carried out at room temperature in laboratory air (50% relative humidity) using an electro-servo hydraulic testing unit with a Digital Control System (Instron 8500). The test system was operated under a load control mode with a sinusoidal waveform. The cyclic frequency was 30 Hz for the crack propagation rate of less than 10^{-5} mm/cycle, and lower frequency (i.e. 1Hz) for higher crack growth rates. For 0.3 mm thick specimens, special guide plates and grips were designed and fabricated to prevent any possible out-of-plane movement for the 0.3 mm thick specimens without introducing frictional loading.

During the tests, the crack lengths were measured by a traveling optical microscope system, the Questar M100. With a video camera module, the image was observed on the screen of a TV monitor. The magnification used during the tests was about 1150x (resolution ≈5 μm). A stroboscopic light was also used to facilitate the clear visual observation of the crack tip during cycling.

MEASUREMENT OF CRACK CLOSURE: The crack closure measurements were taken for each ΔK level after a steady growth rate had been reached. The technique used for measuring the crack closure in this program was the Modified Elastic Compliance Method, which used a subtraction technique with the conventional compliance method. The displacement signal for the 6.35 mm thick specimens was obtained from a strain gauge glued on the backface of the specimen, whereas, for the 0.3 mm thick specimens, two strain gauges were glued one on each side surface near the back edge.

FATIGUE CRACK GROWTH PROCEDURES: Specimens of both thickness were precracked at an intermediate ΔK level with a crack growth rate on the order of 10^{-6} mm/cycle. In order to eliminate the effect of chevron notch on the crack closure, chevron notches of the 6.35 mm thick specimens were removed by an EDM procedure after the crack had propagated 2 mm beyond the chevron notch. Such a procedure was unnecessary for the 0.3 mm thick specimens since they contained no chevron notch.

The decreasing ΔK test method was used to establish da/dN vs. ΔK curves, with the decrease in ΔK in each step-down being less than 5%. At each ΔK level, the load was adjusted as crack propagated about 0.1 mm. Crack growth rate and crack closure level were measured after the crack had passed through the transient region due to the step-down of the ΔK level. The stress intensity factors were calculated in accord with ASTM Standard E399-88.

EXPERIMENTAL RESULTS

da/dN vs. ΔK: For 1018 steel as well as aluminum alloys AL2024 and AL6061, specimen thickness had no significant effect on the rate of fatigue crack growth. However, there were distinct thickness effects in the case of 9Cr-1Mo steel and the aluminum alloy IN9021, and the 0.3 mm thick specimens exhibited slower fatigue crack growth rates than did the 6.35 mm thick specimens. Fig. 1 and Fig. 2 show results for 1018 and 9Cr-1Mo steels.

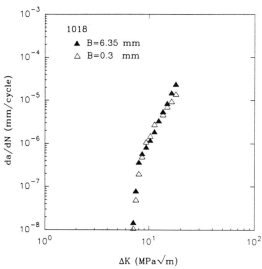

Fig.1. da/dN vs. ΔK plot of 1018 steel for both specimen thicknesses.

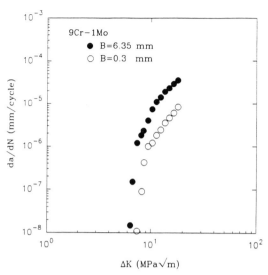

Fig. 2. da/dN vs. ΔK plot of 9Cr-1Mo steel for both specimen thicknesses.

<u>K_{op} vs. ΔK</u>:   The crack closures for 0.3 and 6.35 mm thick specimens of 1018 steel were the same, and remained practically a constant with decreasing ΔK, as shown Fig. 3. Meanwhile, the crack closure for 0.3 mm thick specimen of 9Cr-1Mo steel varied with ΔK, which was about one half of the ΔK value, and the crack closure for 6.35 mm thick specimen of 9Cr-1Mo steel was independent of ΔK (also shown in Fig. 3). The crack closure levels for 0.3 and 6.35 mm thick specimens of Al6061 alloy were the same, and decreased with decreasing ΔK. For IN9021 alloy, the crack closure level for 0.3 mm thick specimens was higher than that of 6.35 mm thick specimens, with the latter remaining almost constant. For the same material, the crack closure levels for 0.3 mm thick specimens decreased to the value of the thick specimens in the near threshold range

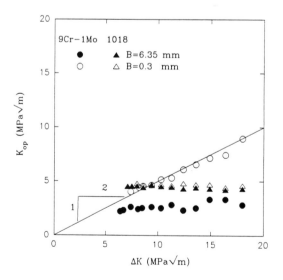

Fig.3.   K_{op} vs. ΔK plot for 9Cr-1Mo steel and 1018 steel for both specimen thicknesses.

ANALYSIS AND DISCUSSION

Why did some alloys (9Cr-1Mo and IN9021) exhibit thickness effect on the fatigue crack growth rate whereas the others did not (1018, AL2024, AL6061)? For alloys which did exhibit a thickness effect, crack closure values were higher for 0.3 mm thick specimens than for 6.35 mm thick specimens, and this crack closure reduced the driving force for fatigue crack propagation, ΔK_{eff}. Fatigue crack growth rates plotted vs. ΔK_{eff} were independent of thickness for the materials investigated. Therefore, specimen thickness was not a parameter governing the relationship between fatigue crack growth rate and ΔK_{eff}.

Where a thickness effect was found it was considered that plasticity induced closure due to lateral contraction at the surface was the most likely cause. This lateral contraction increased with increasing ΔK and caused closure to increase. For the thick specimens, the surface region was just a small fraction of the crack front and material along most of the crack front was constrained by their neighboring material against deformation in the thickness direction. Therefore, the crack opening in the thick specimens was roughness controlled, and exhibited no significant variation with increasing ΔK for either the 9Cr-1Mo steel nor the IN9021 aluminum alloy.

The lateral contraction across the crack in 6.35 mm thick specimens was measured by a profilometer. Test results showed that the 9Cr-1Mo steel specimen had more lateral contraction, or necking, across the fatigue crack than did the 1018 specimen, as illustrated in Fig. 4. Similarly, the IN9021 showed pronounced localized contraction along the fatigue crack.

Lateral contraction is a reflection of the work hardening capacity of the material, which is controlled by the ratio of the yield strength and the ultimate strength of the material. The higher the ratio, the lower the work hardening capacity of the material. As pointed out by Keeler and Backofen [2], the beginning of unstable flow and subsequent fracturing is determined by the stress ratio and material properties, and strain tends to be better distributed with lower values of the yield-to-tensile strength ratio.

In addition, strain rate sensitivity can play a role with respect to the degree of localization of a neck. For example, in low carbon steels which have a relatively high strain-rate dependence, the necks are quite gradual. In materials with a lower strain rate dependence (e.g. aluminum alloys), the necks are much sharper [3]. Further, Saxena and Chadfield [4] have shown that the strain rate sensitivity of steels drops rapidly with increase in flow stress, which is also consistent with the greater tendency for more localized contraction at the crack tip in the case of the 9Cr-1Mo steel as compared to the 1018 steel.

From Table 1 it is seen that the 1018 steel has a lower σ_y/σ_u ratio than does the 9Cr-1Mo steel. Therefore there will be a less tendency for strain localization in the 1018 steel. This situation also holds for the 2024 aluminum alloy in comparison to the IN9021 alloy, and as a consequence there is no thickness effect in the 2024 alloy. For the 6061 aluminum alloy, the σ_y/σ_u ratio is relatively high, which should lead to a thickness effect. However none was found. It is suggested that the absence of the thickness effect for the 6061 alloy is associated with its relatively low yield stress, which lead to extensive through thickness plane stress deformation and similar plasticity induced closure levels in both thicknesses as observed. The similarity in plasticity induced closure levels thereby eliminated any thickness effect.

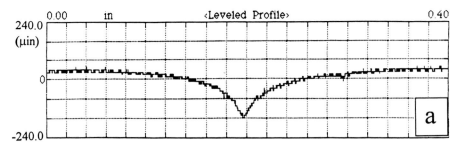

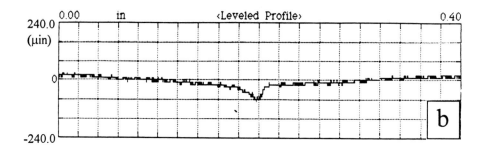

Fig.4. The profiles of fatigue crack on 6.35 mm thick specimen.
(a) 9Cr-1Mo steel, ΔK=30 MPa√m; (b) 1018 steel, ΔK=30MPa√m.

CONCLUSIONS

1) No thickness effect on the rate of crack growth was found for 1018 steel , and 2024, 6061 aluminum alloys.

2) *a.* A thickness effect was found for 9Cr-1Mo and IN9021 alloys.

   b. The thickness effect was due to pronounced surface lateral contraction at the crack tip, which increased the surface closure level and reduced ΔK_{eff}.

   c. Pronounced surface lateral contraction is promoted by a higher yield to tensile strength ration of the materials.

REFERENCE

1. Bao, H. (1994). " Thickness Effects on Fatigue Crack Growth", Ph.D Dissertation, The University of Connecticut.

2. Backofen, W. A. and Keeler, S. P.(1963) ASM Tran. Vol. 56, Mar. p. 25-48.

3. Hosford, W. M. and Caddell, R. M. (1993). Metal Forming, second ed., p. 319.

4. Saxena, A. and Chadfield, D. A. (1976). SAE paper 760209.

THE PLASTIC WAKE OF A LONG FATIGUE CRACK REVISITED

Jean-Paul Baïlon and Zhao-Xiong Tong
Department of Metallurgy and Materials Engineering, Ecole Polytechnique
PO Box 6079, Station Centre-Ville, Montreal (Quebec), Canada, H3C 3A7

ABSTRACT

The dislocation structures formed in the plastic wake of a long fatigue crack, propagating in commercially pure copper at several ΔK levels and in two environments (air and vacuum), has been characterized in detail by TEM. Combining the dependence of the dislocation cell size d on the distance Y from the crack surface with basic equations given by the theory of linear elastic fracture mechanics makes it possible to calculate the equivalent stress and strain gradients ($\overline{\sigma}_e$-vs-Y and ε_e-vs-Y) in the region nearest the crack surface ($5 \leq Y \leq 250$ μm). The microstructural and the mechanical approaches give very coherent results and even allow to estimate the friction stress σ_0 and the cyclic stress hardening coefficient σ_{e0} of the material included in the cyclic plastic zone. It is also shown that these two material parameters are influenced by the test environment (laboratory air and vacuum) and physical reasons of this influence are proposed.

KEYWORDS

Fatigue crack propagation; dislocation structures; plastic zone; stress gradient; strain gradient; cyclic stress hardening coefficient; friction stress; environment; copper.

INTRODUCTION

Since it is recognized that the mechanism of fatigue crack propagation is intimately related to the plastic deformation occurring around the crack tip, several methods have been employed to investigate the plastic strain distribution within the plastic zone [1-4]. Among them, the TEM foil technique may be the most powerful. Polycrystalline copper has been particularly studied by Lukas and co-workers [5-7], who have shown that the size of the dislocation cells is smallest within a distance $Y \leq 10$ μm from the crack surface and can be considered as constant. According to these authors, the cell size, as measured by the mean intercept distance λ, is depending on the applied ΔK according to equation 1:

$$\lambda = c/\Delta K \qquad (1)$$

By gathering the experimental results on the relationship between the size of dislocation cells formed in low cycle fatigue (LCF) specimens and measured by TEM and the saturation stress amplitude σ_s

reached in the LCF specimen, Kayali and Plumtree [8] proposed equation (2) valid for several materials, where $\sigma_s = \Delta\sigma/2$ is the saturation stress amplitude; σ_0 is the friction stress; E is Young's modulus of the material; b is the Burgers vector; A is an empirical material constant (equal to 3.6 for copper) and λ is the dislocation cell size as measured by the mean intercept distance. Equation (3) is another form of equation (2) where k' is again a material constant which can deduced from multislip single crystal data rather than from the extrapolation of a Hall-Petch plot for cells of infinite size [9]. If the characteristic dimension of dislocation cells is expressed by the equivalent diameter d, eq. (2) can be rewritten as eq. (4), with α in eq. (5) equal to 1.267 according to ASTM standard for measurement of grain size [10].

$$(\sigma_s - \sigma_0) = AEb/d \quad (2)$$

$$\sigma_s = \sigma_0 + k'/\lambda \quad (3)$$

$$(\sigma_s - \sigma_0) = \alpha AEb/d \quad (4)$$

$$d = \alpha\lambda \quad (5)$$

To our knowledge, the magnitude of the local stress and that of the local strain in the plastic zone, as well as their dependence on the distance Y from the crack tip, have never been simultaneously deduced or calculated for the same material. The objective of this paper is thus to combine a microstructural and a mechanical approach of the plastic wake associated with a long fatigue crack propagating in commercially pure copper in order to quantitatively deduce the stress and strain gradients in this zone.

THE MICROSTRUCTURAL APPROACH

In a recent paper [11], we have presented a detailed report of the dislocation structures observed by transmission electron microscopy (TEM) and formed in the plastic zone of a long fatigue crack running in commercially pure copper. We have shown how the aspect of these dislocation structures changes with the distance Y from the crack surface and we have presented the dependence of these microstructural features on the applied ΔK and on the environment (air or vacuum). For a given applied ΔK and in a given environment, the various dislocations structures formed in the plastic zone are, in order of increasing distance Y from the crack surface: i) well formed dislocation cells, ii) parallel or labyrinth walls, iii) ladder-like persistent slip bands (PSB) embedded in a matrix consisting of veins, iiii) then finally, loop patches and dislocation tangles. Figure 1 illustrates the characteristic features of the dislocation structures which are formed at different distances Y in the plastic zone according to the applied ΔK and for the two environments. The characteristic dimension of the dislocation cells formed in the region nearest the crack surface and measured by the equivalent diameter d decreases with increasing distance Y from the crack surface as shown in Fig. 2. One noticeable feature of this figure is that all the plots $log(1/d)$-vs-$log(Y)$ appear as straight lines and have nearly the same slope m', so the cell diameter d can be expressed ca be given by eq. (6)

$$\frac{1}{d} = \left(\frac{1}{d_{0,\Delta K}}\right)\Big/Y^{m'} \quad (6)$$

It is striking to note that the slope m' of the straight lines shown in Fig. 2 is nearly constant, independently the value of the applied ΔK or the type of environment. The mean value of m' is equal to 0.29 ± 0.04. This suggests that all the experimental data can be consolidated if a mechanical approach of the stress and strain gradients in the plastic zone is introduced as done in the following section.

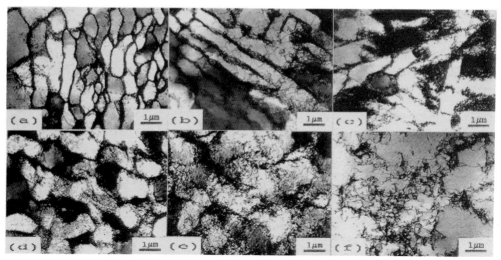

Fig. 1: Dislocation structures at different distances Y from the crack surface:
in vacuum ($da/dN = 3 \times 10^{-7}$ mm/c, $\Delta K = 7.1$ MPa√m , R=0.1) (a) $Y = 15$ μm, (b) $Y = 40$ μm,
(c) $Y = 200$ μm or in air ($da/dN = 5 \times 10^{-6}$ mm/c, $\Delta K = 5,0$ MPa√m , R=0.5) (d) $Y = 15$ μm,
(e) $Y = 170$ μm, (f) $Y = 300$ μm

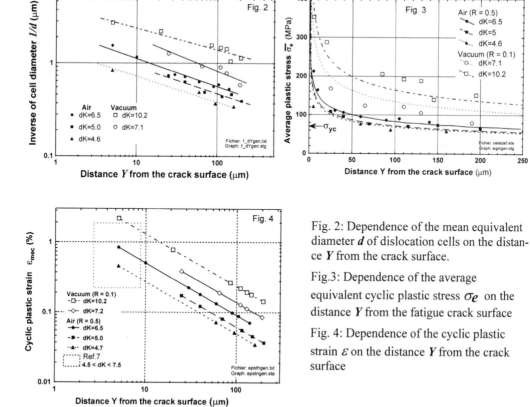

Fig. 2: Dependence of the mean equivalent diameter d of dislocation cells on the distance Y from the crack surface.

Fig.3: Dependence of the average equivalent cyclic plastic stress σ_e on the distance Y from the fatigue crack surface

Fig. 4: Dependence of the cyclic plastic strain ε on the distance Y from the crack surface

THE MECHANICAL APPROACH

McClintock [12] expressed the local stress and strain at a point **P** located at a distance r, θ (in polar coordinates) from the crack tip by equations (7), where $\Sigma_{ij}(\theta)$ and $E_{ij}(\theta)$ are function of θ; I_n is a function of the strain hardening exponent n of the material; J is Rice's integral; $m = n/(1+n)$; σ_{e0} is the stress hardening coefficient, a material constant defined by the relationship between the equivalent plastic stress σ_e and the equivalent plastic strain ε_e as given by equation (8). In the case of linear elastic fracture mechanics, Rice's integral J is related to Irwin's stress intensity factor K_I according to equation (9), where E and v are Young's modulus and Poisson's ratio of the material respectively.

$$\sigma_{ij}(r,\theta) = r^{-m}\sigma_{e0}(J/\sigma_{e0}I_n)^m \Sigma_{ij}(\theta) \qquad (7a)$$

$$\varepsilon_{ij}(r,\theta) = r^{m-1}(J/\sigma_{e0}I_n)^{1-m} E_{ij}(\theta) \qquad (7b)$$

$$\sigma_e = \sigma_{e0}\varepsilon_e^n \qquad (8)$$

$$J = \frac{(1-v^2)}{E}K_I^2 \qquad (9)$$

By combining equations (7) and (9), the stress and strain at a given point located in the plastic zone can be related to the stress intensity factor K in the case of monotonic loading or to the stress intensity range ΔK in the case of cyclic loading and one can estimate the stress and strain distributions in the plastic zone based on this approach. It must be however pointed out that, in the plastic wake of a propagating crack, a unit volume **P** of material, located at a distance Y from the crack plane, is submitted to a cyclic stress range which is continuously changing as this unit volume is swept by the plastic zone which is moving at the fatigue crack propagation rate da/dN. Furthermore, the orientation of the principal stress axis in this unit volume is also continuously changing during this sweeping process. Consequently, if the relationship between σ_s and d, as given by equation (4) and deduced from uniaxial cyclic tension in LCF tests, is assumed to be still applicable to the plastic zone of a fatigue crack propagating in a CT specimen, the stress value deduced from measurements of the dislocation cell size in the plastic zone will be some kind of an average equivalent cyclic $\overline{\sigma}_e$ stress noted.

From equation (8), this average equivalent cyclic stress $\overline{\sigma}_e$ can be expressed by equation (10), where $\overline{\Sigma}$ is now a mean value of $\Sigma_{ij}(\theta)$, averaged over θ (from θ_1 to θ_2) and $\overline{r}$ may be taken as the distance Y ($\overline{r} = Y$) perpendicular to the crack surface.

$$\overline{\sigma}_e = \overline{r}^{-m}\sigma_{e0}(J/\sigma_{e0}I_n)^m \overline{\Sigma} \qquad (10)$$

COMBINING THE TWO APPROACHES

With the assumption that $\overline{r} = Y$, by substituting equation (9) into equation (10) and replacing K with ΔK for a cycling loading, one obtains equations (11) and (12).

$$\overline{\sigma}_e = C'(\Delta K^2/Y)^m \qquad (11)$$

$$C' = \sigma_{e0}[(1-v^2)/E\sigma_{e0}I_n]^m \overline{\Sigma} \qquad (12)$$

By comparing equations (6) and (11) and recalling that $\sigma_s \propto 1/d$ according to equation (4), it can be expected that the exponents m' and $m = n/(1+n)$ should be equal. The experimental value of the cyclic strain hardening exponent n for copper studied in the present work was already obtained by Marchand [13] for the same material and was equal to 0.37; thus a m value of 0.275 is obtained. This value of m is in good agreement with the average value of $m' = 0.29 \pm 0.04$ obtained from Fig. 2. If it is now assumed that the average equivalent cyclic stress $\overline{\sigma}_e$ as given by equation (10) and reached in a unit volume P of material located in the plastic zone at a distance Y from the crack surface is equal to the cyclic saturation stress σ_s as given by equation (4), one obtains equations (13), (14) and (15):

As presented in detail in ref. [14], a least square regression analysis done through the experimental data $1/d$-vs-$(\Delta K^2/Y)^m$ gave an excellent agreement between the experimental points and the theoretical equation (13) with the correlation coefficient r equal to 0.965 and 0.935 for air and vacuum respectively.

$$\frac{1}{d} = B + C\left(\frac{\Delta K^2}{Y}\right)^m \tag{13}$$

$$B = -\frac{\sigma_0}{\alpha EAb} \tag{14}$$

$$C = \frac{C'}{aEAb} = \frac{\sigma_{e0}\left[(1-v^2)/E\sigma_{e0}I_n\right]^m \overline{\Sigma}}{\alpha EAb} \tag{15}$$

From the experimental value of B and C parameters of eq. (13), one can now easily deduce the value of the friction stress σ_0 from equation (14) and that of the C' parameter from eq. (15) by knowing the values of the other constants for copper ($\alpha = 1.267$; $E = 110$ GPa; $A = 3.6$; $b = 0.256$ nm). We obtain $\sigma_0 = 14.2$ MPa or 4.6 MPa and $C' = 2.40$ or 3.64 for air or vacuum respectively. With these values, it is now straightforward to calculate the stress gradient $\overline{\sigma}_e - vs - Y$ in the plastic zone, either from the mechanical approach as summarized by eq. (11) with the known value of the C' parameter or from the microstructural approach as given by eq. (4) with the known value of the friction stress σ_0. The stress gradient $\overline{\sigma}_e - vs - Y$ in the plastic zone is shown in figure 3, where the curves correspond to eq. (11) (the mechanical approach) and the experimental points are obtained with eq. (4) (the microstructural approach). The agreement between the two approaches can be considered as very good.

Combining equation (8) and (11) make it now possible to obtain the strain gradient ε -vs- Y which is given by equations (16) and (17).

$$\varepsilon = D\left(\frac{\Delta K^2}{Y}\right)^{1/1+n} \tag{16}$$

For LCF specimens of the same copper tested in air, Marchand [13] has already obtained a value of 1125 MPa for the stress hardening coefficient σ_{e0air} of equation (8). Since the ratio C'_{vac}/C'_{air} is equal to $(\sigma_{e0vac}/\sigma_{e0air})^{1-m}$

$$D = \left(\frac{C'}{\sigma_{e0}}\right) \tag{17}$$

according to equation (15), the value of σ_{e0vac} is thus equal to 1540 MPa. With these respective values of σ_{e0}, it is now possible to calculate the strain gradient ε -vs- Y in the plastic zone according to the environment as shown in Fig. 4. The values of the cyclic plastic strain ε varies from 0.035% (for the lowest applied ΔK and at a distance $Y \approx 200$ μm from the crack surface) up to 2.2% (for the highest applied ΔK and at distance $Y \approx 5$ μm). These values of the cyclic plastic strain are in good agreement with those already published by Lukas [7] or those proposed in our previous paper [11].

DISCUSSION

We will briefly comment on the value of the friction stress σ_0 and on that of the stress hardening coefficient σ_{e0} which both depend on the environment as shown above. The value (14.2 MPa) of the friction stress σ_0 obtained in air is higher than that obtained in vacuum (4.6 MPa). This latter value is in agreement with the 3.8 MPa value of σ_0 in copper as deduced from LCF experiments and reported by Kayali and Plumtree [8]. In LCF tests as well as in fatigue crack propagation tests carried out in an inert environment like vacuum, a unit volume of material located either into the bulk of LCF specimen or in the plastic zone of the fatigue crack is not submitted to a chemical reaction with the external environment. Thus one can expect that the values of σ_0 deduced either from LCF tests or from fatigue propagation tests carried out in vacuum to be almost the same (3.8 and 4.6 MPa respectively). On the other hand, as already proposed by Marchand et al.[15] for fatigue propagation tests carried out in a reactive environment such as laboratory air, the dissociation of the water vapor contained in laboratory air on the fresh fracture surfaces produces atomic hydrogen which is then swept in the plastic zone by the to and fro movements of dislocations. This enrichment of the plastic zone with atomic hydrogen produces an interstitial solid solution hardening of the material located in this plastic zone with a concomitant increase of the local friction stress σ_0 of that strained material. The higher value (14.2 MPa) of the friction stress σ_0 obtained for tests carried out in air is likely due to this interstitial solid solution hardening of the plastic zone by atomic hydrogen produced on the fresh fracture surfaces and swept into the plastic zone by dislocation movement.

The increase of the cyclic stress hardening coefficient σ_{e0} for the material tested in vacuum is understandable if one recalls that, in vacuum, the crack propagation rate da/dN is lower than that obtained in air for a given value of the stress intensity factor range ΔK as shown in ref. [11]. When tested in vacuum at a given ΔK, the material of the plastic zone is thus submitted before cracking to a larger number of load cycles than that sustained by the material tested in air at the same ΔK. Consequently when the test is carried out in vacuum, a secondary cyclic strain hardening of the plastic zone is expected with a concomitant increase of σ_{e0}. This secondary cyclic strain hardening has been already reported for LCF experiments on single or polycrystalline copper fatigued in vacuum after a number of applied cycles which has exceeded that for which the LCF specimen has already failed in air [16, 17]. As shown in Fig. 2, the dislocation cells, formed at a given distance Y of the crack surface in CT specimens tested in vacuum, presented some differences in their appearance with those obtained in air, their boundaries being more dense, the misorientation between cells being larger and the interior of the cells containing fewer dislocations. All these features of the dislocation structures clearly indicate that, when tested in vacuum, the material within the plastic zone has sustained a number of load cycles larger than that reached in air at cracking. Thus, the increase of the cyclic stress hardening coefficient σ_{e0} in vacuum is probably the consequence of the secondary cyclic strain hardening which occurs in vacuum.

REFERENCES

1) Stickler R., C.W. Hughes and G.R. Booker (1971). *Scanning Electron Microscopy*, 473-480.
2) Davidson D.L. and J. Lankford (1980). *Fatigue Engng. Mater. Struct.*, **3**, 289-303.
3) Chalant G. and L. Remy (1982). *Eng. Fract. Mech.*, **16**, 707-720.
4) Grosskreuts J.C.and G.G. Shaw (1967). In: Fatigue Crack Propagation, ASTM STP **415**, 226-273.
5) Lukas P., M. Klesnil and R. Fiedler (1969). *Philos. Mag.*, **20**, 799-805.
6) Lukas P. and L. Kuns (1984). *Mater. Sci. Engng.*, **62**, 149-157.
7) Lukas P.(1984) In Fatigue 84, (C.J.Beevers, A. Blom ed.), EMAS Ltd, Wharley (UK), **1**, 479-495.
8) Kayali E.S. and A. Plumtree (1982). *Metall.. Trans. A*, **13**, 1033-1041.
9) Kocks U.F.(1970). *Metall. Trans. A*, **1**, 1121-43
10) ASTM Standards E1382 (1995). Measurement of grain size, **03.01**, section 3, 899.
11) Zhao-Xiong Tong and J.P. Baïlon (1995). *Fatigue Fract. Engng. Mater Struct.*, **18**, 847-859.
12) McClintock F.A.(1968) Plastic aspects fracture (Chap. 2), In: Fracture, An Advanced Treatise, **III**, (H. Liebowitz ed.), 162
13) Marchand (1983). *Thesis of Master degree*, Dept of Metallurgical Engng., Ecole Polytechnique, Université de Montréal, 94
14) Baïlon J.-P. and Zhao-Xiong Tong (1995). Stress and strain gradients in the plastic zone of a fatigue crack, submitted for publication in *Fatigue Fract. Engng. Mater. Struct.*
15) Marchand N., J.P. Baïlon and J.I. Dickson (1988). *Metall. Trans. A*, **19**, 2575-2587.
16) Wang R. and H. Mughrabi (1984). *Mater. Sci. Eng.*, **63**, 147-163 and **65**, 235-243.
17) Mendez J., P. Violan and G. Gasc (1986). In: The Behaviour of Short Fatigue Cracks, (K.J. Miller and E.R. de Los Rios, ed.), EGF Publication, Mech. Engng. Publications, London, 145-161.

MODE II FATIGUE CRACK PROPAGATION IN A MARAGING STEEL

C. PINNA and V. DOQUET

Laboratoire de Mécanique des Solides,
CNRS URA 317, Ecole Polytechnique,
91128 Palaiseau Cedex, FRANCE

ABSTRACT

The kinetics and micromechanisms of mode II crack propagation in a maraging steel is investigated using CTS specimens loaded either by a conventional machine or a special one able to work inside a SEM. For ΔK_{II} above 18 MPa$\sqrt{m}$ decelerating mode II propagation followed by branch crack development behind the crack tip was observed. Real time measurements of crack flanks displacements in the SEM allow the determination of the effective ΔK_{II} and of the real contact surface over which friction forces act. Crack flanks wear tends to sustain mode II growth.

KEYWORDS

micromechanisms, mode II, crack, propagation, maraging, friction, wear

INTRODUCTION

Although fatigue crack growth usually occurs in mode I, shear mode propagation has been observed in some cases, especially in contact fatigue or under large amplitude torsional loading. Studies dealing with mode III fatigue crack growth have shown the large influence of friction forces acting along crack flanks on propagation (see for example Tschegg et al., 1988). The work of Smith (1988) on the threshold value of ΔK_{II} for the development of a branch-crack suggests a similar influence of friction effects for mode II. Besides, it is easier to appreciate because of the plane geometry associated with this kind of propagation allowing direct observation of crack tip mechanisms. The aim of this study is a better understanding of mode II crack propagation mechanisms with special attention paid to friction effects.

EXPERIMENTAL PROCEDURE

The material investigated is a M250 maraging steel (Z2NKD 18-8-5, $R_{0.2} = 1910$ MPa, $R_u = 1973$ MPa) used by SNECMA in turboreactors. Its chemical composition is given in Table 1.

elements	Fe	Ni	Co	Mo	Ti	Al	B
weight (%)	base	18.29	8.13	4.76	0.45	0.1	0.0024

Table 1. - composition of the maraging steel

The data supplied by SNECMA do not reveal any anisotropy as far as fatigue crack propagation is concerned. Compact Tension Shear specimens (Fig. 1) were chosen for this study. Smith (1988) has laid stress on the influence of pre-cracking techniques on further mode II propagation. According to his recommendations, pre-cracking was conducted at low ΔK_I (15 MPa$\sqrt{m}$) and low R (0.3) in order to minimize asperity size and residual stresses at the crack tip, until a pre-crack about 1 mm long was produced. Then, lines perpendicular to the pre-crack, were engraved every 50 μm behind and ahead the pre-crack tip with a microdurometer in order to measure mode II crack flanks displacements. Three shear fatigue tests were carried out at constant ΔK_{II} - 18, 24 and 30 MPa$\sqrt{m}$ respectively - with a frequency of 0.5 Hz and R = 0. K_{II} evaluations were made using the expressions obtained for CTS specimens by Smith (1988) through finite elements calculations. For a mode II crack, the maximum extension of the plastic zone can be evaluated as :

$$r_p^a = \frac{3}{2\pi}\left(\frac{K_{max}}{\sigma_y}\right)^2 , \qquad (1)$$

ahead of the tip and

$$r_p^b = \frac{4}{2\pi}\left(\frac{K_{max}}{\sigma_y}\right)^2 , \qquad (2)$$

behind. For the highest ΔK_{II} employed here, these extensions amount to 118 μm and 157 μm respectively so that Small Scale Yielding conditions prevailed for all our tests. Plastic replicas of the cracks were periodically prepared. Tests were run until mode II crack propagation gave way to a significant branch crack development. An additional test was carried out with $\Delta K_{II} = 27$ MPa$\sqrt{m}$ and ν = 0.025 Hz on a biaxial fatigue machine specially designed to work inside the chamber of a scanning electron microscope (for further details see Doquet et al., 1994). It is important to note that although observations and measurements were periodically made under vacuum, the test itself was performed in laboratory air. Microgrids with a 5 μm pitch were laid on the surface behind and ahead the crack tip allowing periodical crack flank displacement measurements with a good precision : 0.3 μm.

RESULTS

In spite of constant ΔK_{II}, deceleration of crack growth is observed, and this deceleration is discernable earlier for low ΔK_{II} than for higher values (Fig. 2). These results are quite similar to those reported by Tschegg et al. (1988) for mode III. In all cases, deceleration leads to the development of a branch-crack, 20 to 300 microns **behind** the mode II crack tip, at an an angle ranging from 58 to 75 degrees to the mode II propagation plane, that is not too far from the angle which corresponds to the maximum tangential stress $\sigma_{\theta\theta max}$ (Fig. 3). **Initial** crack propagation rates seem to be related to nominal ΔK_{II} by a Paris law but this correlation has not much interest since it represents only a short period of the propagation with no account of deceleration and, furthermore, it depends on the pre-crack lenght. The final lenght of the crack at the end of mode II propagation arise with nominal ΔK_{II} in a linear fashion for the limited number of tests carried out (Fig. 4) : for 18, 24 and 30 MPa$\sqrt{m}$ the distance covered by the crack in mode II is 120, 600 and 1300 μm respectively. During the test carried out on the machine allowing SEM observations, pictures of crack flanks were made at both K_{max} and K_{min}. That way, it was possible to check that crack flanks relative displacements at K_{max} are really pure mode II (tangential displacements), in spite of some **local** opening displacements induced by the shape of asperities on fracture surfaces. These opening displacements can essentially be observed along the pre-crack flanks, that show a ± 45° "zig-zag" profile typical of propagation under low ΔK_I and favourable to asperity-induced opening. The profile of the mode II part of the crack is much flatter and the opening displacement along this part is thus much more limited than along the pre-crack. Wear is evident from the large quantity of fretting debris that comes out from the fracture surface and has to be periodically removed to allow observation. Along the mode II part of the crack, several "steps" and many secondary cracks inclined 90 to 110° to the mean propagation line can be observed (Fig. 5). Besides, the presence, ahead of the crack tip, of longitudinal and transversal intense slip lines must be underlined. This can be related to calculations of stress and strain fields made by Hutchinson (1968) for a mode II crack, assuming perfect plasticity and stress-free crack flanks. As can be seen on Fig. 6, the "fan zones" where the shear strain is singular are oriented at about 0 and 110°. The propagation of the mode II crack thus probably follows intense slip bands. On a more quantitative point of view, one of the profiles of the amplitude of relative tangential displacements along the crack flanks measured during the test at $\Delta K_{II} = 27$ MPa$\sqrt{m}$ is shown on Fig. 7 together with the theoretical profile associated with nominal ΔK_{II}, in the framework of Linear Fracture Mechanics for a plane stress state, (that is :

$$\Delta u_{II} = \frac{8 \, \Delta K_{II}^{nom}}{E \sqrt{2 \pi}} \sqrt{r} , \tag{3}$$

where r is the distance to the crak tip). The fact that the real amplitude of mode II displacements is inferior to the amplitude evaluated with ΔK_{II} nominal has to be noted first. It clearly shows that friction forces reduce the effective crack driving force. However mode II displacements are approximatively proportional to the square root of the distance to the crack tip. This tends to confirm Deng's analytical calculation (1994) showing that friction forces do not modify the **type** of stress and strain fields singularity that is still $1/\sqrt{r}$. An effective stress intensity amplitude, ΔK_{II}^{eff}, can thus be defined and can be evaluated from the real displacement profile using equation (3) where ΔK_{II}^{eff} replaces ΔK_{II} nominal. For example, ΔK_{II}^{eff} was evaluated to 21 MPa$\sqrt{m}$ for the profile shown on Figure 8 which corresponds to an attenuation of the nominal stress intensity factor : $\Delta K_{II}^{nom} - \Delta K_{II}^{eff}$ equal to 6 MPa$\sqrt{m}$. This attenuation due to friction forces will be denoted by $K_{II}^{friction}$ in the following. An other technique was also used to determine the value of $K_{II}^{friction}$. It relies on the determination, by real time observations in the SEM, of the threshold value of nominal K_{II} at the onset of crack flanks slip, at the tip of the crack, during the loading stage. At this point, K_{II}^{nom} equals $K_{II}^{friction}$. Although much more direct than the preceeding technique, it seems less accurate since it relies on a visual impression. That is why, in the following of this study, the technique based on displacement profiles, that provides a good precision on ΔK_{II}^{eff} (± 1 MPa$\sqrt{m}$), will systematically be used to measure the decrease of ΔK_{II}^{eff} as the crack propagates, to relate it to the decreasing crack growth rate and to determine an effective threshold. But at the present date, the crack loaded on the machine allowing SEM observations has not yet shown significant deceleration nor branching, and the measurements made on replicas taken during "classical" tests do not offer a sufficient precision (± 4 MPa$\sqrt{m}$ on ΔK_{II}). The intensity of the frictional attenuation, K_f can be related to the distribution of shear friction stresses along crack flanks $\tau(x)$ using the technique of weight functions. According to Smith (1988) :

$$K_{II}^f = \int_0^a \frac{2 \tau(x)}{\sqrt{2 \pi x}} dx , \tag{4}$$

where the origin of the x axis is at the root of the notch and a is the abscissa of the tip of the crack. To apply this formula, it is necessary to determine the real contact area over which friction forces act, allowance made for asperities on crack flanks. SEM observations allow such a determination, at least at the surface of the specimen on which the portions of the crack that are closed can be distinguished from those that are open. The sum of the lengths, projected on the mean propagation plane, of segments that are closed at K_{max} was measured separately for the mode I pre-crack and the mode II part because it is evident that the different morphologies of these two parts lead to different behaviours as concerns friction. The real contact length along the pre-crack represents 45% of its total length, at the beginning of the mode II test, whereas it is only 30% for the mode II part. Ten thousand cycles later these percentages had dropped, for the same areas, to 35 and 20% respectively. This can probably be attributed to wear. There are thus two opposite influences on K_f : that of the **nominal** increase of contact surfaces due to crack growth and that of wear that reduces the **effective** contact surface.

DISCUSSION

The rather large distances over which mode II propagation takes place in the maraging steel for the moderate ΔK_{II} values employed here may seem somewhat surprising. Our opinion is that the high hardness of this material prevents the crushing of crack flanks asperities by plastic deformation thus limiting adherence phenomena and as a consequence the intensity of friction. A high hardness is however not a necessary condition to allow mode II propagation. Quite important mode II propagations were in fact observed in aluminium alloys by Otsuka et al. (1987). Other microstructural and mechanical parameters certainly favour mode II propagation as for example the capacity of a material to wear away. To explain the fact that branch-cracks develop **behind** the crack tip, it can be suggested that whereas the numerous 90° microcracks left in the wake of the growing crack are quickly unloaded when the mode II crack tip propagates away, those formed when the main crack has decelerated enough or even stopped are continuously loaded at a high level and thus enabled to grow.

CONCLUSIONS

- For ΔK_{II} values above 18 MPa $\sqrt{m}$, decelerating mode II crack propagation followed by branch-crack development behind the crack tip was observed in a maraging steel.

- At a microscopic level, the profile of mode II cracks is relatively flat with occasional "steps" at 90° to the mean line, following intense slip bands.

- Real time measurements of crack flanks displacements in a SEM allow the determination of the effective ΔK_{II} and thus of the attenuation of the nominal ΔK_{II} due to friction forces.

- The real contact surface over which friction forces act can also be evaluated through SEM observations.

- During mode II propagation, crack flanks undergo extensive wear that sustains further crack growth.

ACKNOWLEDGMENTS

The financial support of this study is provided by SNECMA and CNRS.

REFERENCES

Deng, X. (1994). A note on interface cracks with and without friction in contact zone. Transactions of the ASME, J. of Applied Mechanics, 61, 994-995.

Doquet, V., D. Caldemaison and T. Bretheau (1994). Combined tension and torsion cyclic tests inside a scanning electron microscope. Proceedings of the Fourth International Conference on Biaxial and Multiaxial Fatigue, St Germain en Laye, II, 19-26.

Hutchinson, J. W. (1968). Plastic stress and strain fields at a crack tip. J. Mech. Phys. Solids, 16, 337-347.

Otsuka, A., K. Mori and K. Tohgo (1987). Mode II fatigue crack growth in aluminium alloys. In Current Research on Fatigue cracks, Current Japanese Materials Research, 1, Elsevier Applied Science, London, 149-180.

Smith, M. C. and R. A. Smith (1988). Toward an understanding of mode II fatigue crack growth. Basic Questions in fatigue, I, ASTM STP 924, J. T. Fong and R. J. Fields, Eds., Philadelphia, 260-280.

Tschegg, E. K. and S. E. Stanzl, S. E. (1988). The significance of sliding mode crack closure on mode III fatigue crack growth. Basic Questions in fatigue, I, ASTM STP 924, J. T. Fong and R. J. Fields, Eds., Philadelphia, 214-232.

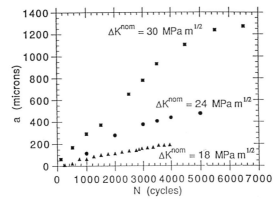

Fig. 1. Compact Tension Shear specimen.

Fig. 2. Length covered in mode II a_{II} vs number of cycle for different ΔK_{II}.

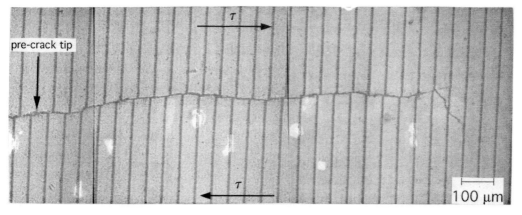

Fig. 3. Aspect of a Mode II crack with a branch crack at $K_{II}^{max} = 30$ MPa$\sqrt{m}$.

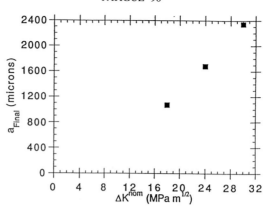

Fig. 4. Total length of the cracks at the end of mode II propagation vs ΔK_{II}.

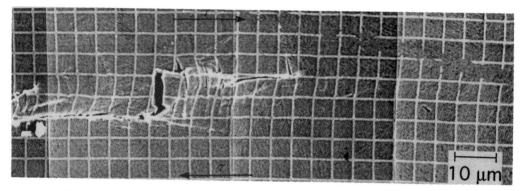

Fig. 5. Aspect of mode II crack flanks at $K_{min} = 0$ for the test at $\Delta K = 27$ MPa√m̅.

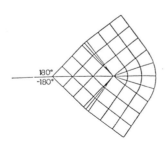

Fig. 6. Stress distribution at the tip of a shear crack for perfect plasticity.

(Hutchinson 1968)

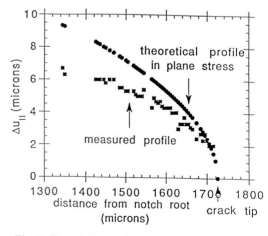

Fig. 7. One of the profiles of relative tangential displacements along the crack flanks measured for $\Delta K_{II}^{nom} = 27$ MPa√m̅.

MIXED MODE CRACK GROWTH IN A SINGLE CRYSTAL NI-BASE SUPERALLOY

R. JOHN [1], T. NICHOLAS [2], A.F. LACKEY [1], and W.J. PORTER [1]

[1] University of Dayton Research Institute,
1031 Irving Avenue, Dayton, OH 45419-0128, U.S.A., and
[2] Wright Laboratory (WL/MLLN), Materials Directorate,
Wright Patterson Air Force Base, OH 45433-7817, U.S.A.

ABSTRACT

The crack growth from machined flaws oriented near-parallel to a $\{\bar{1}44\}$ plane of a single crystal alloy subjected to mixed mode (modes I and II) loading was investigated. Disks with a middle crack were subjected to compressive fatigue loading. Various combinations of mode I and mode II loading were achieved by orienting the machined flaw at different angles to the loading axis. Under mixed mode loading with $K_I > 0$, the crack was growing even at $\Delta K_{II} \approx 0.6$ MPa$\sqrt{m}$ compared to a pure mode I threshold, $\Delta K_{I,th} \approx 10$ MPa$\sqrt{m}$. For this crystallographic orientation subjected to mixed mode fatigue loading conditions with $K_I > 0$, damage tolerance of the single crystal superalloy component could be severely overestimated if only mode I data were used in design.

KEYWORDS

Fatigue crack growth; mixed mode; mode I; mode II; single crystal; superalloy; threshold.

INTRODUCTION

Ni-base single crystal materials are used in many applications such as advanced turbine engine blades, which require high temperature, creep resistant, and high stiffness characteristics. These superalloys are primarily loaded in the <001> direction. Recently, cracking distress has been observed in some blades under certain loading conditions. Detailed fractographic investigation of the failure surfaces indicated that mode II (sliding) type cracking along the {111} plane could be the mode of crack growth in the early stages of fatigue. To aid in the understanding of this process, a program was initiated to characterize the crack growth along the {111} planes of a single crystal alloy subjected to mixed mode (combined modes I and II) loading. This paper discusses the fatigue crack growth from machined flaws which were oriented near-parallel to $\{\bar{1}44\}$. Note that the angle between {111} and $\{\bar{1}44\}$ planes is about 45°.

EXPERIMENTAL PROCEDURE

The geometry adapted for testing under mixed mode loading was the "Brazilian" <u>D</u>isk with a <u>M</u>iddle crack subjected to <u>C</u>ompressive loading, DM(C), as shown in Fig. 1.

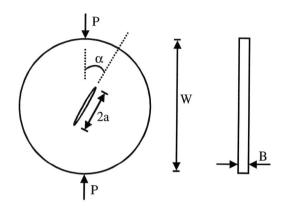

Fig. 1. Schematic of a centrally notched disk
      subjected to compression.

The DM(C) geometry has been used to investigate the mixed mode fracture of polymethacrylate (Atkinson *et al.*, 1982) and sintered carbide (Yarema *et al.*, 1984), and mixed mode fatigue crack growth in high strength steel 35NCD16 and stainless steel 316 (Louah *et al.*, 1987). The DM(C) geometry is ideal for testing under a wide range of mode I and mode II loading conditions. Mode I stress intensity factor, K_I and mode II stress intensity factor, K_{II} solutions are available (Atkinson *et al.*, 1982; Yarema *et al.*, 1984) for this geometry for 2a/W up to 0.6. Various ratios of mode II stress intensity factor to mode I stress intensity factor, K_{II}/K_I can be achieved at the crack tip by orienting the crack at different angles (α) to the loading axis (Fig. 1). Figure 2 shows the effect of flaw orientation on K_I and K_{II} for a crack length to width ratio, 2a/W = 0.30. When $\alpha = 0$, $K_{II} = 0$, resulting in pure mode I loading conditions, and when $\alpha \approx 27°$, $K_I = 0$, resulting in pure mode II loading conditions. For $\alpha > 27°$, $K_I < 0$, implying that the crack surfaces could be in contact with each other depending on the initial notch width and applied load ratio.

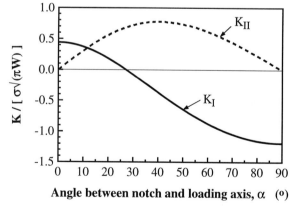

Fig. 2. Effect of flaw orientation on K_{II}
      and K_I.

$$\left(\sigma = \frac{2P}{\pi WB}, \quad \frac{2a}{W} = 0.3 \right)$$

Angle between notch and loading axis, α (°)

Figure 3 shows that for a constant load and $\alpha = 27°$, K_I decreases and K_{II} increases with increase in 2a/W. Hence, if a test is started with $\alpha = 27°$ and 2a/W = 0.3, as the crack extends in a self-similar path, the loading condition at the crack tip changes from pure mode II to mixed mode loading with K_I < 0.

Fig. 3. Variation of K_{II} and K_I with normalized crack length.

$$\left(\sigma = \frac{2P}{\pi WB}, \quad \alpha = 27° \right)$$

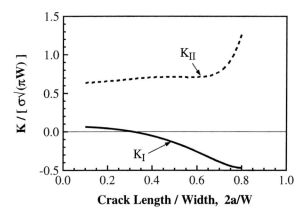

In this investigation, the specimens were prepared from cylindrical stock of a single crystal Ni-base superalloy material such that $\{\bar{1}44\}$ was near-parallel to the electrical discharge machined notch and $\{100\}$ to the surface of the disk. Typical dimensions were: diameter, W = 30.25 mm, thickness, B = 2.22 mm, and initial crack length, $2a_0$ = 9.16 mm.

The DM(C) specimens were loaded in compression using ceramic anvils. The direct current electric potential (DCEP) technique was used to monitor the crack growth. The DCEP measurements were verified with periodic optical crack length measurements. All the tests were conducted at room temperature in laboratory air at a frequency of 10-20 Hz with a stress ratio, R, of 0.1 (=K_{min}/K_{max} at crack tip). One specimen was tested at each α = 0, 16, 27 and 32°. The tests were started under constant load until crack initiation, after which the applied load was shed simulating a threshold-type decreasing K_{max} test. After the crack reached near-arrest conditions or $2a/W \approx 0.8$, all the mixed mode specimens were rotated back to α = 0° and loaded under mode I conditions with R=0.5. This ensured that the crack extended to failure with minimum contact between the crack surfaces.

RESULTS AND DISCUSSION

The tests conducted at each angle α resulted in near-self-similar crack growth with the crack extension collinear with the notch. The observed crack pattern is contrary to the expected curvilinear path followed by cracks experiencing mixed mode loading. Typically, pure mode II loading of a crack in an isotropic material would result in crack initiation at about 70° to the original crack (Hussain et al., 1974). Under continued loading, the crack will propagate towards the point of load application in the disk (Atkinson et al., 1982; Louah et al., 1987). Hence, the results imply that the combination of mixed mode loading and orientation of the initial flaw used in this study results in crack propagation along the "weak planes" contrary to that expected in an isotropic material. The DCEP solution for a center-cracked geometry was used during the tests, consistent with the observed self-similar crack propagation in the DM(C) specimens. The DCEP measurements were within 10% of the optical measurements for $0.3 \leq 2a/W \leq 0.8$.

Using the measured crack length versus cycles response, K_I, K_{II}, and the crack growth rate, da/dN were calculated as a function of crack length. Figure 4 shows the variation of K_I and K_{II} with increasing crack length for α = 16 and 32°.

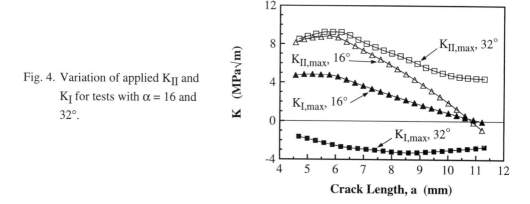

Fig. 4. Variation of applied K_{II} and K_I for tests with α = 16 and 32°.

Note that during all the tests, the load was decreasing with increasing crack length. For the test with α = 16° mixed mode conditions existed throughout most of the test with $K_{II} \approx 2K_I$. K_I was > 0 until the test was stopped and K_{II} was > 0 until very close to the end of the test. In contrast, for the test with α = 32°, K_{II} was > 0 and K_I was < 0 throughout the test. Negative K_I implies that the crack surfaces could have been in contact at least at longer crack lengths. The third test, conducted at α = 27°, was similar to that of the test at α = 32° except that K_I started at zero at the beginning of the test and became slightly negative with increase in crack length. In the fourth test with α = 0, K_{II} was always equal to 0.0, producing a pure mode I crack growth.

Figure 5 summarizes the crack growth rate data obtained in the decreasing load tests at each angle α. The range of growth rates obtained show the effect of both the magnitude and sign of K_I on the crack growth rate (da/dN) versus K_{II} range (ΔK_{II}) response of the material.

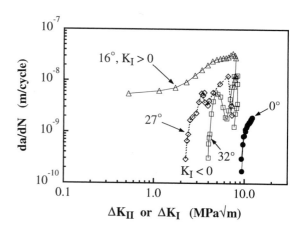

Fig. 5. Effect of flaw orientation on crack growth behavior under mixed mode loading.

For the mode I test ($\alpha = 0°$), da/dN versus ΔK_I is plotted in the figure since $\Delta K_{II} = 0$ in this test. All the three mixed mode tests ($\alpha = 16$, 27 and 32°), produced significantly lower crack growth rates as a function of ΔK_{II} compared to the mode I test. This implies that any damage tolerance based design using only mode I data can lead to overestimation of crack growth life when actual service conditions produce mixed mode loading along a potential failure plane.

A significant observation is the effect of negative K_I on the mode II crack growth rates as seen by comparing the data obtained at $\alpha = 16$, 27 and 32°. Note that negative K_I implies that frictional contact could occur between the crack surfaces during the tests. Clearly, the data for $\alpha = 27$ and 32° indicate lower growth rates than for 16° where K_I is positive. The data also show that for similar crack growth rates, ΔK_{II} for $\alpha = 32°$ is larger than that for $\alpha = 27°$. As discussed earlier, even though K_I was negative for the tests with $\alpha = 27$ and 32°, the magnitude of K_I for $\alpha = 27°$ was less than that for $\alpha = 32°$. Accordingly, the frictional resistance generated behind the crack tip during crack growth with $\alpha = 32°$ could be expected to be higher than that with $\alpha = 27°$. The additional frictional resistance could explain the slower growth rates measured for $\alpha = 32°$ compared to that for $\alpha = 27°$ as seen in Fig. 5.

The near-threshold behavior observed in tests at $\alpha = 27$ and 32° occurs even though the magnitude of ΔK_{II} is larger than that of ΔK_I. In contrast, the test conducted at $\alpha = 16°$ with $K_I > 0$ did not reveal threshold-type behavior. The crack continued to grow at a near-constant rate ($\approx 5 \times 10^{-9}$ m/cycle) until the test was stopped. At this stage, the calculated ΔK_{II} and ΔK_I were near zero implying that mixed mode loading with $K_I > 0$ and corresponding low values of ΔK_{II} could result in crack growth even under extremely low loads. This mixed mode loading condition in conjunction with high cycle fatigue (HCF) loading could easily produce significant crack extension leading to failure due to the large number of cycles under HCF. The low values of ΔK_{II} observed during this study are close to the resolved shear stress intensity factor range, ΔK_{rss} which produced similar crack growth rates as reported by Telesman and Ghosn (1989). An attempt was made to correlate all the data in Fig. 5 using $\Delta K_{eff} = \sqrt{\left(\Delta K_I\right)^2 + \left(\Delta K_{II}\right)^2}$ as the correlating parameter, but the resulting data did not collapse into a single curve or band.

Figure 6 shows the fracture surface of the specimen tested with $\alpha = 27°$. The fracture surface under mixed mode loading is distinctly different from that under mode I loading. The mixed mode fracture surface is non-planar throughout the test while the mode I fracture surface is planar.

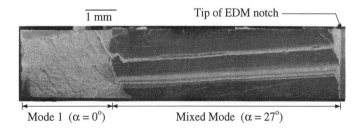

Fig. 6. Fracture surface of specimen tested at $\alpha = 27°$ followed by $\alpha = 0°$.

The mode I fracture surface is similar to that observed in the specimen tested with $\alpha = 0°$. The non-planar fracture surface was observed in the mixed mode region of all the tests.

During this study, K was calculated assuming that the crack surface is perpendicular to the sides of the disk specimen. Due to the non-planar nature of the fracture surface under mixed mode loading, the actual K calculated using the resolved stresses on the inclined surfaces can be expected to lower than the values reported in Figures 4 and 5.

CONCLUSIONS

The centrally notched disk geometry, DM(C) can be used to characterize the mixed mode crack growth behavior in single crystal superalloys using the DCEP technique to monitor self-similar crack growth. A particular flaw orientation with respect to the primary loading axis of the single crystal geometry can result in self-similar crack growth when the material is subjected to mixed mode loading, contrary to that observed in isotropic materials. Mixed mode loading with $K_I > 0$ can produce crack growth rates under very low values of ΔK_{II}. Threshold ΔK_{II} values much lower than threshold ΔK_I values were obtained along the same crystallographic plane in the same direction. These observations indicate that the damage tolerance of the single crystal superalloy component could be severely overestimated if only mode I data were used in design.

ACKNOWLEDGMENT

This research was conducted at Wright Laboratory (WL/MLLN), Materials Directorate, Wright-Patterson Air Force Base, OH 45433-7817, USA. The authors from the University of Dayton Research Institute were supported under on-site contract number F33615-94-5200 (Project Monitor: Mr. J.R. Jira). The machined specimens were provided by Pratt & Whitney, West Palm Beach, FL, USA. The authors also gratefully acknowledge the numerous technical discussions with Dr. S.E. Cunningham and Dr. D.P. DeLuca, from Pratt & Whitney.

REFERENCES

Atkinson, C., R.E. Smelser, and J. Sanchez (1982). Combined Mode Fracture via the Cracked Brazilian Disk Test. *International Journal of Fracture*, 18, 279-291.

Hussain, M.A., S.L. Pu, and J.H. Underwood, J.H. (1974). Strain Energy Release Rate For a Crack Under Combined Mode I and Mode II. *Fracture Analysis, ASTM STP 560*, pp. 2-28, American Society For Testing and Materials, Philadelphia, PA, U.S.A.

Louah, M., G. Pluvinage, and A. Bia (1987). Mixed Mode Crack Growth Using the Brasilian Disc. *Fatigue 87*, (R.O. Ritchie and E.A. Starke, Jr., ed.), pp. 969-977, Engineering Materials Advisory Services, Ltd., West Midlands, U.K.

Telesman, J. and L.J. Ghosn (1989). The Unusual Near-Threshold FCG Behavior of a Single Crystal Superalloy and the Resolved Shear Stress as the Crack Driving Force. *Engineering Fracture Mechanics*, 34, 1183-1196.

Yarema, S.Y., G.S. Ivanitskya, A.L. Maistrenko, and A.I. Zboromirskii (1984). Crack Development in a Sintered Carbide in Combined Deformation of Types I and II. *Problemy Prochnosti*, 8, 51-56.

FATIGUE CRACK GROWTH CHARACTERISTICS OF AUSTENITIC STAINLESS STEEL AT CRYOGENIC TEMPERATURE UNDER HIGH-MAGNETIC FIELD

Takayuki Suzuki and Kazumi Hirano

Mechanical Engineering Laboratory, AIST, MITI
Namiki 1-2, Tsukuba-shi, Ibaraki-ken 305, Japan

ABSTRACT

The fatigue crack growth tests of austenitic stainless steel SUS316L were performed at 293K, 77K and 4K under the non-magnetic field and at 4K under the high-magnetic field. Under the non-magnetic field the fatigue crack growth resistance increased with decreasing the test temperature. Under the high-magnetic field the fatigue crack growth resistance increased comparing with that under the non-magnetic field. The effects of the test temperature and the high-magnetic field on the fatigue crack growth resistance were discussed by the concept of martensitic transformation induced crack closure.

KEYWORDS

Superconducting applied device; cryogenic temperature; high-magnetic field; fatigue crack growth; martensitic transformation induced crack closure

INTRODUCTION

Recently superconducting applied devices such as the power generator have been developed. It is the most important and urgent to clarify the fracture characteristics at the cryogenic temperature and under the high-magnetic field in order to ensure the structural integrity of these devices for a long time. There are a few studies of the tensile properties(Fultz et al., 1984, Fultz et al., 1986, Kurita et al., 1994) and the elastic-plastic fracture toughness(Fukushima et al., 1990, Chan et al., 1990, Chan et al., 1994) at the cryogenic temperature and under the high-magnetic field, and the effect of the high-magnetic field on these properties has been clarified. However, there exists only a study(Fultz et al., 1984) of the fatigue crack growth resistance at the cryogenic temperature and under the high magnetic field, and the effect of the high-magnetic field on the fatigue crack growth characteristics has not always been clear.

In this study, the high-magnetic cryogenic temperature material testing system has been developed. And the fatigue crack growth tests of austenitic stainless steel SUS316L are conducted at the cryogenic temperature and under the high-magnetic field. The effects of the test temperature and the high-magnetic field on the fatigue crack growth characteristics are discussed.

MATERIALS AND EXPERIMENTAL PROCEDURE

Experimental Device

The developed high-magnetic cryogenic temperature material testing system(Suzuki et al, 1992) mainly consists of an electro-hydraulic material testing machine, a mini-computer, a superconducting magnet, and a liquid helium tank. Various kinds of fracture tests are also performed at the cryogenic temperature and under the high-magnetic fields using this material testing system.

The support structure inside the cryostat is mainly made of austenitic stainless steel SUS304L and some flanges are made of FRP because of their small thermal conductivity. The liquid helium level meter is equipped with the support structure. The evaporation rate of liquid helium is approximately 2.4l/hour and the liquid helium can be introduced continuously from the liquid helium tank through the transfer-tube.

The superconducting magnet used in this system is a split-type one. The maximum magnet field is 6T. It is remarkable that the magnetic field direction can be settled either parallel or perpendicular to the loading axis, and that it can easily be changed by varying the direction of the superconducting magnet to the support structure.

Materials and Specimen

The material used in this study is austenitic stainless steel SUS316L (C0.019, Si0.70, Mn1.31, P0.034, S0.006, Ni12.18, Cr17.81, Mo2.13Wt%). The 1CT specimen(thickness is 10mm) was used for the 293K and 77K tests, and the 0.4CT specimen(thickness is 4mm) was used for the 4K tests both under the non-magnetic and high-magnetic fields. The fatigue pre-cracks were introduced by ΔK decreasing tests at 77K except for the 293K tests.

Experimental Procedure

Fatigue crack growth tests were performed by the high-magnetic cryogenic material testing system. Under the non-magnetic field ΔK increasing tests were performed at 293K, 77K, and 4K at the stress ratio R=0.05 with the frequency f=10Hz. At 77K, ΔK decreasing test was also performed near the threshold at the stress ratio R=0.05 with the frequency f=10Hz. Under the high-magnetic field ΔK was kept nearly constant at the stress ratio R=0.05 with the frequency f=10Hz. And the 6T high magnetic field was imposed on the specimen parallel to the loading axis while the crack length was 7.0-9.6mm. The ΔK value increased from 24.0 to 27.8MPam$^{1/2}$ under the high-magnetic field.

The volume fraction of martensitic transformation was measured by the X-ray diffraction method on the fatigue fracture surface. The irradiated area on the specimen surface is 1x2 mm. The α' martensitic volume fraction was calculated by the method proposed by Durnin(Durnin et al., 1968).

EXPERIMENTAL RESULTS

Fatigue Crack Growth Characteristics under Non-magnetic Field

da/dN-ΔK Relationships. The da/dN-ΔK relationships under the non-magnetic field are shown in Fig.1. Fatigue crack growth resistance increases as the test temperature decreases. The fatigue crack growth resistance at 4K is larger than those at 77K and 293K. As compared with the previous data(Suzuki et al., 1993), the fatigue crack growth resistance of SUS316L is smaller than that of A286 steel at each test temperature. Especially at 77K, it is remarkable that the difference of fatigue crack growth resistance between SUS316L and A286 steel increases as ΔK decreases near the threshold. In a case of A286 steel the da/dN-ΔK relationship near the threshold is not represented by the Paris's law. While in a case of SUS316L the da/dN-ΔK relationship is represented by the Paris's law even near the threshold.

Crack Closure Behavior. The K_{open}/K_{max}-K_{max} relationships are shown in Fig.2. It is found that the K_{open}/K_{max} values at the cryogenic temperature(77K, 4K) are larger than those at 293K. As compared with the previous data(Suzuki et al., 1993), it is remarkable that the K_{open}/K_{max} values measured at 77K are nearly constant even near the threshold at 77K. In a case of A286 steel, the K_{open}/K_{max} values measured at 77K increase as K_{max} decreases near the threshold.

da/dN-ΔK_{eff} Relationships. The da/dN-ΔK_{eff} relationships are shown in Fig.3. By using ΔK_{eff} instead of ΔK, the difference of the fatigue crack growth resistance decreases among these test temperatures. Then it is found that the effect of the test temperature on the fatigue crack growth resistance is basically explained by the crack closure concept.

C_{α}-K_{max} Relationships. The martensitic volume fraction C_{α} and K_{max} relationships are shown in Fig.4. At each test temperature, C_{α} increases as K_{max} increases. The C_{α} values at the cryogenic temperature (77K, 4K) are larger than those at 293K within the limits of this test results. It is known that meta-stable austenite such as SUS316L is easily transformed into martensite by the strain-induced martensitic transformation. And it is found that the amount of the strain induced martensitic transformation depends greatly on the test temperature.

Cryogenic Fatigue Crack Growth Characteristics under High-magnetic Field

Fatigue Crack Growth Resistance. The da/dN-ΔK relationships both under the high-magnetic and non-magnetic fields are shown in Fig.5. It is found that the fatigue crack growth resistance under the high-magnetic field is larger than that under the non-magnetic field.

da/dN-ΔK_{eff} Relationships. The K_{open}/K_{max}-a relationships both under the high-magnetic and non-magnetic fields are shown in Fig.6. It is found that the K_{open}/K_{max} values increase under the high-magnetic field comparing those under the non-magnetic field.

The da/dN-ΔK_{eff} relationships both under the high-magnetic and non-magnetic fields are also shown in Fig.5. It is found that there is little difference in the da/dN-ΔK_{eff} relationships between the non-magnetic and high-magnetic fields.

C_{α}-a, C_{α}-K_{max} Relationships. The C_{α}-a relationships are shown in Fig.6. The C_{α}-K_{max} relationships are also shown in Fig.4. From these figures it is found that the C_{α} values under the high-magnetic field are larger than those under the non-magnetic field. Then it is concluded that the strain induced martensitic transformation occurs easily under the high-magnetic field.

Fractographic Observation

The SEM micrographs of fatigue fracture surface are shown in Fig.7(a)-(d). By comparing with (a), observed at 293K, and(b),(c) and (d), observed at 77K, 4K and 4K•6T, it is found that the flat and brittle fracture surface is clearly shown at fractographs(b),(c) and (d).

At 77K, such a typical fracture surface is observed near the threshold. In a case of A286 steel(Suzuki et al., 1993), the rough fracture surface is observed at 77K and K_{open}/K_{max} increases as K_{max} decreases near the threshold. It results in the roughness induced crack closure. However, in a case of SUS316L, the fracture surface is flat even near the threshold, and the roughness induced crack closure does not occur. Then the difference of fatigue crack growth characteristics near the threshold between SUS316L and A286 is explained by the roughness induced crack closure.

There is little difference between fractographs(c) and (d) and the effect of high-magnetic field on the fracture topography is not observed.

DISCUSSION

From the experimental results, it is found that the fatigue crack growth resistance increases both at the cryogenic temperature and under the high-magnetic field. It is also found that the amount of martensitic transformation increases both at the cryogenic temperature and under the high-magnetic field. The amount of martensitic transformation is consistent with the increase of fatigue crack growth resistance, and then the effect of the martensite on the fatigue crack growth resistance both at the cryogenic temperature and under the high-magnetic field is discussed as follows.

First, the effect of the transformation induced crack closure(Suresh et al., 1984) is considered. Transformation induced crack closure occurs because of the volume expansion by the martensitic transformation around the crack tip. Actually, as shown in Fig.2 and Fig.6, the K_{open}/K_{max} values increase both at the cryogenic temperature(77K, 4K) and under the high-magnetic field because of the increase of the martensite around the crack tip. And the crack growth resistance is rationalized by using the effective stress intensity factor range which takes account of the effect of martensitic transformation induced crack closure.

Second, the effect of crack kink or crack deflection is considered. It was reported, (Chan et al., 1990), that stress intensity factor might decrease because of crack kink or crack deflection when a crack propagated in martensite. However, no crack kink or crack deflection is observed by the fractographs. It is found that this mechanism does not control the fatigue crack growth characteristics.

Then it is concluded that the effects of both the test temperature and the high-magnetic field on fatigue crack growth resistance are mainly explained by the martensitic transformation induced crack closure.

CONCLUSION

Fatigue crack growth tests of austenitic stainless steel SUS316L were performed at the cryogenic temperature and under high-magnetic field by using the newly developed high-magnetic cryogenic temperature material testing system.

(1)Under the non-magnetic field the fatigue crack growth resistance increases as the test temperature decreases.
(2)Under the high-magnetic field the fatigue crack growth resistance increases comparing that under the non-magnetic field.
(3)The effects of both the test temperature and the high-magnetic field on fatigue crack growth resistance are mainly explained by the martensitic transformation induced crack closure.

ACKNOWLEDGMENTS

The authors wish to thank Prof. M. Kikuchi(Science University of Tokyo) for many valuable suggestions. The authors also wish to thank Mr. K. Nagai(Former student of Science University of Tokyo) and Mr. Y. Matsunuma(Student of Tokyo Science University) for their contribution in the experiments as the graduation study.

REFERENCES

Fultz, B., G.M.Chang, Kopa R. and J.W.Morris, Jr.(1984). Magneto-mechanical effects in 304 stainless steels, Adv. Cryo. Eng., 30, 253-262.
Fultz, B. and J.W. Morris Jr.(1986). Effects of high magnetic fields on the flow stress of 18-8 stainless steels, Acta Metall., 34, 379-384.
Fukushima, E., Kobatake S., Tanaka M. and S. Ogiwara(1988). Fracture toughness tests on 304 stainless steel in high magnetic fields at cryogenic temperature, Adv.Cryo. Eng., 34, 367- 370.
Chan, J.W., Glazer, J., Mei, Z. and J.W.Morris, Jr.(1990). 4.2K Fracture toughness of 304

stainless steel in a magnetic field, Adv.Cryo.Eng., 36B, 1299-1306.

Chan, J.W., Chu. D., Tseng. C. and J.W.Morris Jr.(1994). Cryogenic fracture behavior of 316LN in magnetic fields up to 14.6T, Adv.Cryo.Eng., 40B, 1215-1221.

Kurita, Y., Shimonosono, T.and K. Shibata(1994). Effects of an 8 tesla magnetic field on tensile deformation of stainless steels at 4K, Adv.Cryo.Eng., 40B, 1223-1229.

Suzuki, T. and K.Hirano(1992). R&D on magneto-mechanical cryogenic material testing system, Proc. of the 1992 annual meeting of JSME/MMD, 920-72, 607-608.

Durnin, J.and K.J. Ridial(1968). Determination of retained austenite in steel by X-ray diffraction, J. of the Iron and Steel Institute, January, 60-67.

Suzuki, T. and K. Hirano(1993). Cryogenic fatigue crack growth characteristics of structural material for the superconducting generator, Proc. of Asian Pacific Conference on FRACTURE and STRENGTH'93, 117-121.

Suresh, S. and R.O.Ritchie(1984). Near-Threshold Fatigue Crack Propagation:A Perspective on the Role of Crack Closure,In:Fatigue Crack Growth Threthold Concepts, The Metallugical Society of AIME, 227-61.

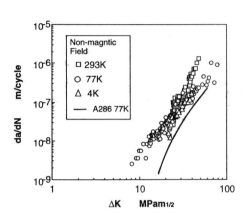

Fig.1 da/dN-ΔK relationships for SUS316L under non-magnetic field

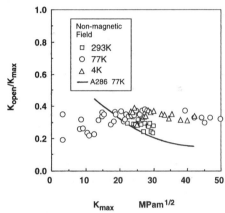

Fig.2 K_{open}/K_{max} -K_{max} relationships for SUS316L

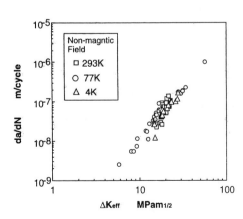

Fig.3 da/dN-ΔK_{eff} relationships for SUS316L under non-magnetic field

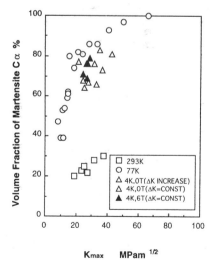

Fig.4 Comparisons of C_α-K_{max} relationships between non-magnetic and high-magnetic fields

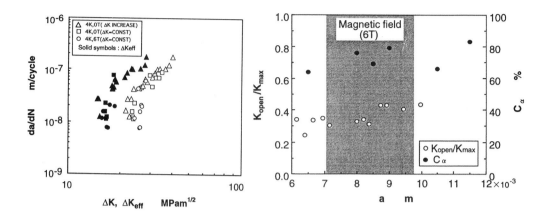

Fig.5 da/dN-ΔK and ΔK$_{eff}$ relationships for SUS316L at 4K under high-magnetic and non-magnetic fields

Fig.6 K$_{open}$/K$_{max}$ and C$_\alpha$ -a relationships under high-magnetic and non-magnetic fields

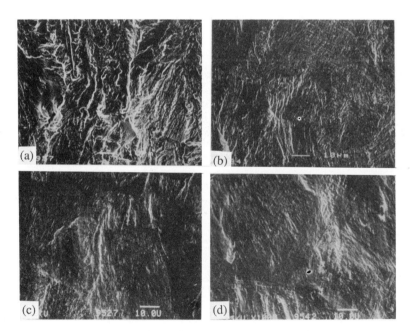

Fig.7 SEM micrographs of fatigue fracture surface
(a)293K, ΔK=22.0MPam$^{1/2}$
(b)77K, ΔK=13.7MPam$^{1/2}$
(c)4K, ΔK=18.6MPam$^{1/2}$
(d)4K•6T, ΔK=24.7MPam$^{1/2}$

DIRECT OBSERVATION OF THE CRACK TIP DEFORMATION AND THE CRACK CLOSURE

R. Pippan, Ch. Bichler, Ch. Sommitsch and O. Kolednik

Erich-Schmid-Institut für Festkörperphysik der
Österreichischen Akademie der Wissenschaften
A-8700 Leoben, Jahnstraße 12, Austria

ABSTRACT

Direct observations of the crack tip deformation during fatigue with the scanning electron microscope were performed. The crack closure behavior at the surface and the midsection of the specimen was studied. The experiments show clearly that plasticity induced crack closure occurs at the surface (plane stress region) and the midsection (plane strain region) of the specimen.

KEYWORDS

Closure measurement, crack tip deformation, plasticity induced closure, fatigue crack propagation, striations.

INTRODUCTION

Fatigue crack propagation in metals is caused by the cyclic plastic deformation of the crack tip, which is determined by the mechanical properties of the material, the stress intensity range and the contribution of crack closure (or contact shielding). It is usually assumed that near the threshold of the stress intensity range the oxide and the roughness induced closure are the most important contact shielding effects. At larger crack propagation rates the plasticity induced crack closure is the dominant contact yielding effect, since the thickness of the oxide layer and the "height" of "asperities" caused by the mismatch of the rough fracture surface are small in relation to the cyclic plastic crack tip opening displacement. But there are significant discrepancies in related literature about the existence of crack closure - especially in the plane strain case. A number of researchers have concluded that plasticity induced crack closure does not occur under plane strain conditions for a steady state growing fatigue crack. There are experimental (Lindley *et al.*, 1974, Ewalds *et al.*, 1978 and Minakawa *et al.*, 1984) and numerical (Fleck and Newman, 1988) studies which come to this conclusion. However, the opposite behavior is reported by experimental (Dawicke *et al.*, 1990 and Pitoniak *et al.*, 1974, Pippan *et al.*, 1994) and numerical (see for example Chermahini *et al.*, 1991 and Mc. Clung, 1991) investigations, too.

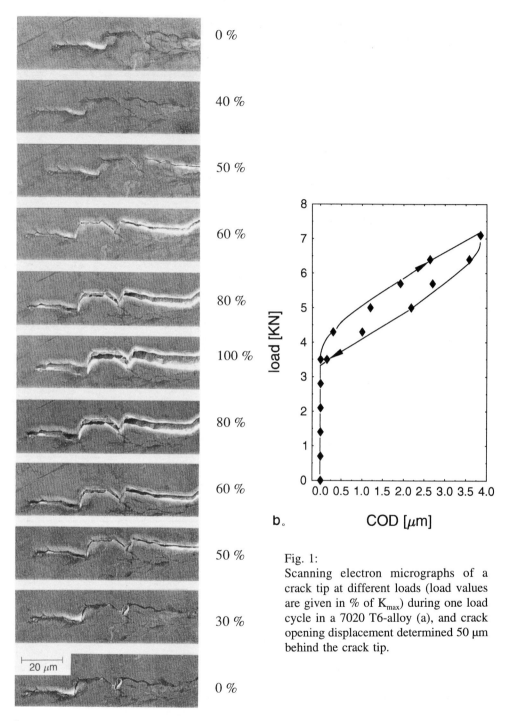

0 %

40 %

50 %

60 %

80 %

100 %

80 %

60 %

50 %

30 %

0 %

b.

a.

Fig. 1:
Scanning electron micrographs of a crack tip at different loads (load values are given in % of K_{max}) during one load cycle in a 7020 T6-alloy (a), and crack opening displacement determined 50 µm behind the crack tip.

The main problem from the theoretical point of view is to visualize the additional volume in the plane strain case which is necessary to explain the plasticity induced closure. From the experimental point of view, the measurement of crack closure - especially in the midsection of the specimen is problematic (Davidson, 1991, Allison, 1988, Chen *et. al.*, 1991, Suresh, 1991, and Hertzberg *et al.*, 1988). The purpose of this paper is to present different experimental results which indicate that plasticity induced closure occurs under both plane stress and plane strain conditions.

SURFACE OBSERVATIONS

Fig. 1a shows the scanning electron micrographs of the crack tip at different loads during one load cycle. The material was a rolled 7020 T6 alloy (LT orientation) with a yield strength of 280 MPa and ultimate tensile strength of 330 MPa. The CT-specimen with a width of 50 mm and thickness of 16 mm was directly loaded within the scanning electron microscope with a stress intensity factor range $\Delta K = 20$ MPa$\sqrt{m}$ at a stress ratio R = 0. For these experiments the small scale yielding conditions and in the midsection of the specimen the plane strain conditions are fulfilled.

In Fig. 1b the crack opening displacement measured 50 μm behind the crack tip is plotted as a function of the load. It can be seen that the crack opens between 40 and 50 % of the maximum load and closes at somewhat smaller load. It should be pointed out that closure of the crack in this case is easier to determine than during opening, since the crack opening displacement at the tip during loading is proportional to $(K-K_{op})^2$ (K_{op} is the stress intensity where the crack tip opens). During unloading the reduction of the crack tip opening displacement is proportional to $(K_{max}-K)^2$, until the crack is closed, ie. there is a sharp change of the crack tip opening displacement during closure. From Fig. 1 it is evident that roughness induced closure and the oxide induced closure are not the reasons that the crack flanks in the vicinity of the tip are closed over 40 % of the load amplitude. Therefore, the contacts of the crack flanks are caused by the plastic deformed region in the (near) wake of the crack. It is often argued that plasticity induced crack closure is only a plane stress effect and therefore occurs only in the near surface region. But we will now show that it also takes place in the midsection of the specimen.

INVESTIGATION OF THE MIDSECTION BEHAVIOR

The study of the deformation of the crack tip in the midsection of the specimens is more difficult. We will present two experiments which illustrate the plasticity induced crack closure under plane strain conditions.

Closure determination with a rising load amplitude test

Fig. 2a presents schematically the technique used to determine the closure stress intensity in the midsection of the specimen. This technique is very similar to that of Döker *et al.*, (1988).

Applications of this technique were also presented by Zhang *et al.*, (1993). The idea of this technique is simple:

> A crack can propagate when $K_{max}-\Delta K_{eff\ th}$ is larger than the closure stress intensity. If one knows $\Delta K_{eff\ th}$ one has to determine the K_{max} value where the crack begins to propagate.

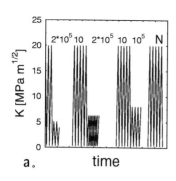

Fig. 2:   Schematic description of the loading procedure to determine the closure stress
           intensity in the midsection of a specimen (a), and the resulting SEM fractograph of
           a 7020 T6 alloy which was fatigued with $\Delta K = 20$ MPa$\sqrt{m}$ at $R = 0$.

In order to measure the closure stress intensity with this technique we interrupted the test at the
minimum load. A constant load amplitude sequence with a small ΔK range was then applied. This
load amplitude was increased in steps till we reached an amplitude where we expected that the
crack will grow. The load amplitude where the first extension of the crack occurs were determined
on the broken specimen halves in the scanning electron microscope. Fig. 2b presents the fracture
surface which was produced during a load sequence as depicted in Fig. 2b. The same material and
loading parameters were used as in the experiment which are shown in Fig. 1.

In order to distinguish at which step of the small amplitudes the crack begins to propagate, each
sequence was separated by 10 large load cycles. This causes 10 striations and allows to distinguish
on the fracture surface at which load sequence a crack extension occurred. From the fractograph it
can be seen that:
*   at the load sequence with $K_{max} = 8$ MPa$\sqrt{m}$ a crack extension of about 3 μm occurred (a
         similar behavior was also observed at $K_{max} = 7$ MPa$\sqrt{m}$ but this is not shown in Fig. 2).
*   at the load sequence with $K_{max} = 6$ MPa$\sqrt{m}$ there are only few regions where a very small
         extension of the crack was observed and at smaller K_{max} values no such marks were found
         on the fracture surface.
The effective threshold of this alloy is 0,8 MPa$\sqrt{m}$ (Pippan *et al.*, 1994), hence in the midsection
of the specimen the crack at $\Delta K_1 = 20$ MPa$\sqrt{m}$ and $R = 0$ closes or opens at a stress intensity which
is somewhat larger than 5 MPa$\sqrt{m}$.

Estimate of crack closure from the crack propagation rate

Fig. 3 presents the striation width determined in a constant ΔK test in a 7020 T6 alloy. The test was
performed on specimens which were pre-cracked with a small stress intensity range ($\Delta K = 6$
MPa$\sqrt{m}$, $R = 0.1$). Since the pre-crack after the first cycle with $\Delta K = 20$ MPa$\sqrt{m}$ does not close
(Pippan *et al.*, 1993 and Fleck and Newman, 1988) the crack flanks can contact each other only on
the newly created fracture surface (striations). Hence the closure load at the beginning of such
constant ΔK test is near the minimum load. With an increase of the crack length, the closure load
increases till it reaches a constant value. This causes a reduction of the crack growth rate from 1.2

μm/cycle to about 0.6 μm/cycle (mean values). If we assume that da/dN is proportional to ΔK^2_{eff} (or the cyclic crack tip opening displacement) the crack closure stress intensity should be 6 MPa√m, which agrees well with the results of the rising load amplitude experiments.

A similar behavior is depicted in Fig. 4. It shows the change of striation width in a constant ΔK test (79 MPa√m, R = 0) in a cold rolled austenitic steel (18 - 8) with a yield strength of 890 MPa and an ultimate tensile strength of 1010 MPa. One can clearly see that the striation width decreases with increase of the crack extension, which is caused by the rising of the closure stress intensity.

pre-crack

ΔK=

20MPa√m

10μm

(a)   (b)

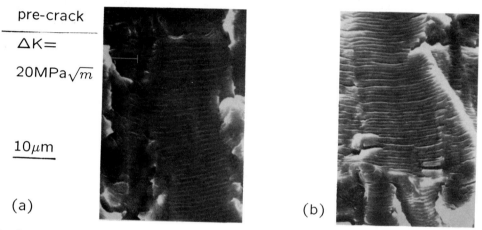

Fig. 3:   Comparison of the striation width determined in the midsection of a 7020 T6 specimen at the beginning of a ΔK constant test (no crack closure) (a) and the steady state condition, where the crack closure has reduced the propagation rate (b).

ΔK=70MPa√m

R=0

pre-crack

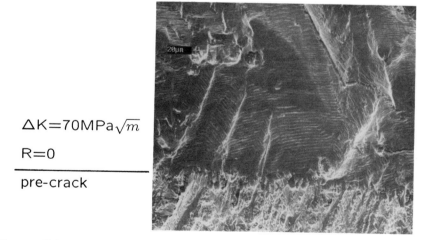

Fig. 4:   Change of the striation spacings with increasing crack extension in a constant ΔK test (70 MPa√m, R = 0) in an austenitic steel.

CONCLUSION

The different experiments indicate clearly that plasticity induced crack closure occurs at the surface (plane stress region) and in the midsection (plane strain region) of a specimen.

REFERENCES

Allison, J.E. (1988). In: *Fracture Mechanics: 18 th Symposium*. (D.T. Read and R.P. Reed. ed.), pp 913-933, ASTM STP 945, Philadelphia.

Chen, D.L., B. Weiss and R. Stickler (1991): A new evaluation procedure for crack closure. *Int. f. Fatigue*, 13, 327-331.

Chermahini, R.G. and A.F. Blom (1991). Variation of crack-opening stresses in three-dimensions: finite thickness plate. *Theor. Appl. Fract. Mech.* 15, 267-276.

Darwicke, D.S., J.C. Newman Jr. and A.F. Grandt Jr. (1990): In: *Proc. Fatigue* 90 (H. Kitagawa and T. Tanaka, ed.), pp. 1283-1288. MCEP Materials and Component Engineering Publications ltd., Birmingham.

Davidson D.L. (1991). Fatigue crack closure. *Eng. Fracture Mech.*, 21, 393-402

Döker, H. and V. Bachmann (1988). In: *Mechanics of Fatigue Crack Closure*. (J.C. Newman, Jr. and W. Elber, ed.) pp 247-249. ASTM STP 982, Philadelphia.

Elber, W. (1970). Fatigue Crack Closure Under Cyclic Tension. *Engineering Fracture Mechanics*, 2, 1970, 37-45.

Ewalds, H.L. and R.T. Furnée (1978). Crack closure measurement along the fatigue crack front of center cracked specimens. *Int. J. Frac.*, 14, R-53-R55.

Fleck, N.A. and J.C. Newman Jr. (1988). In: *Mechanics of Fatigue Crack Closure*. (J.C. Newman Jr. and W. Elber, ed.) pp 319-341. ASTM STP 982, Philadelphia.

Hertzberg, R.W., C.H. Newton and R. Jaccard (1988). In: *Mechanics of Fatigue Crack Closure* (J.C. Newman Jr. and W. Elber, ed.) pp 139-148. ASTM STP 982, Philadelphia.

Lindley, T.C. and C.E. Richards (1974). *Mater Sci Engng*, 14, 281-293

McClung, R.C. (1991). The influence of applied stress, crack length, and stress intensity factor on crack closure. *Metall. Trans.*, 22A, 1559-1571.

Minakawa, K., G. Levan and A.J. McEvily (1984). The influence of load ratio on fatigue crack growth in 7090-T6 and in 9021-T4P/M aluminum alloys. *Metall. Trans.*, 17A, 1787-1795.

Pippan, R., O. Kolednik and M. Lang (1994). A mechanism for plasticity-induced crack closure under plane strain conditions. *Fatigue Fract. Engng. Mater. Struct.*, 17, 721-726

Pippan, R., M. Lang and O. Kolednik (1993). In: *Fatigue* 93, (J.P. Bailon and J.I. Dickson, ed.), 3. Vol. pp 1819-1924. Engineering Materials Advisory Service LTD, EMAS, Warley, UK.

Pitoniak, F.J., A.F. Grandt, L.T. Montulli and P.F. Packman (1974). Fatigue crack retardation and closure in polymethylmethacryle. *Engng. Fract. Mech.*, 6, 663-670.

Suresh, S., (1991) *Fatigue of Materials*. Cambridge University Press.

Vasudevan, A.K., K. Sadananda and N. Lonat (1994). A review of crack closure, fatigue crack threshold and related phenomena. *Mater. Sci. Eng.*, A188, 1-22.

Zhang, S.J., H. Döker, H. Nowack, K. Schulte and K.H. Trautmann (1993). In:*Advanced in Fatigue Lifetime Predictive Techniques* (M.R. Mitchell and R.W. Landgraf, ed.), second Volume, pp 54-71. ASTM STP 1211, Philadelphia.

THRESHOLD VALUE FOR ENGINEERING APPLICATIONS

THRESHOLD VALUE FOR ENGINEERING APPLICATION

R. Pippan

Erich-Schmid-Institut für Festkörperphysik der
Österreichischen Akademie der Wissenschaften
A-8700 Leoben, Jahnstraße 12, Austria

ABSTRACT

Threshold of fatigue crack propagation ΔK_{th} is usually measured by reducing the load amplitude at a constant stress ratio and mode I loading. One denotes this ΔK_{th} value as long crack threshold. However the minimum crack length where the long crack threshold becomes inapplicable is usually unknown. Most engineering problems are associated with small flaws, mixed mode loading and variable amplitude loading. This discrepancy between the measurement conditions and the engineering problems is the main reason that ΔK_{th} is not widespread in engineering design.
Resistance curves (R-curves) for the threshold should allow a more reliable application of ΔK_{th} values to engineering problems. A recently proposed simple technique to determine such R-curves will be considered and applications of to mixed mode and overload effects are discussed.

KEYWORDS

Fatigue threshold, resistance curve, fatigue limit prediction, short cracks, mixed mode loading, variable amplitude.

INTRODUCTION

The engineering fatigue design codes are commonly based on S/N curves (the endurance limit as determined on smooth or notched specimens). However, in practice the engineering components contain flaws of a size between few μm and few mm. These flaws are usually taken into account by a safety factor (endurance limit / maximum stress). The stresses in such components are significantly smaller than the yield stress or the endurance limit and linear fracture mechanics should be applicable. But a damage tolerance concept based on ΔK_{th} is not commonly used in design (see for example Brook, 1984). One of the main reasons for this is that the threshold according to the ASTM E 647 procedure is applicable only to long cracks. For smaller

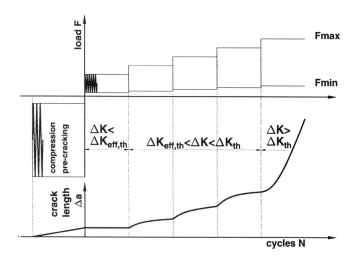

Fig. 1:   Schematic illustration of the loading procedure of a test to determine the R-curve for
               fatigue propagation threshold and the fatigue crack growth curve.

crack lengths the threshold is smaller. Hence a fatigue limit estimation of a component which
contains a small flaw based on the long crack threshold is nonconservative. Furthermore the
minimum necessary length for application of the long crack behavior is usually unknown.

Threshold for physical short cracks

Romaniv et al., (1981), Tanaka et al., (1988), Pippan et al., (1986, 1987), Pineau (1986) and others
show that the threshold of stress intensity range increases with increasing crack length until it
reaches the "constant" long crack threshold value. The increase of ΔK_{th} with increasing crack
extension can be interpreted as an R-curve (Tanaka et al., 1983, 1988, Pippan et al., 1986, 1994)
and can be used to calculate the fatigue limit of components containing small defects, notches or
short cracks. Many authors (see for example, Tanaka et al., 1983, 1988, Pippan et al., 1994, Pineau,
1986, Suresh, 1991, Mc Evily et al., 1991) explained this increase of the threshold by an increase
of the contribution of crack tip shielding (Ritchie, 1988) which leads to an increase of the crack
closure stress intensity.

Tanaka and Akinawa (1988) measured this increase of the crack closure stress intensity - which
should be equal to the change of ΔK_{th} - as a function of crack extension for short cracks. But the
measurement of the crack closure load is associated with difficulties and a relative large uncertainty,
especially in the short crack region. Suresh (1985), Pippan (1987) and Novack and Marissen (1987)
proposed a method to measure the long crack threshold and the crack growth curve on specimens
pre-cracked in cyclic compression.

It was recently shown (Pippan et al., 1994) that the above test can also determine in addition an R-
curve for the threshold. This paper presents a short introduction to this method, some examples,

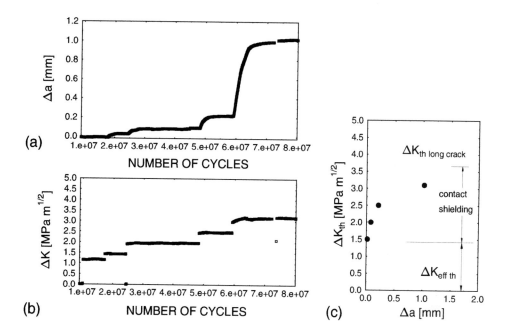

Fig. 2:   Typical experimental result of the first steps of a rising load amplitude experiment
            in a 21 vol % Al_2O_3 particles reinforced 6061 T6 alloy, (a) change of crack
            extension, (b) change of stress intensity, (c) R-curve for the threshold.

some applications in the short crack region, the limitations, and the potential application of
extension of this method to mixed mode loading and variable amplitude loading.

Remarks upon the R-curves concept

The R-curve is defined as a plot of the resistance to fracture versus crack extension. If the R-curve
is independent of the geometry- i.e. if it is a material parameter only - one can easily determine the
critical load of a component which contains a defect. The application of this concept for static
loading is described in many text books for fracture mechanics.

If one defines the threshold ΔK_{th} as the resistance of a material against fatigue crack propagation it
is also possible to use this R-curve concept for fatigue problems (Pook, 1983, Tanaka et al., 1983,
1988, and Pippan, 1986, 1994). It should be noted that the geometric independence of the "static"
R-curve is usually not controlled, even if in some cases it is evident that the R-curve depends on
the geometry or loading parameters. Hence, this point will be discussed in this paper in detail.

DETERMINATION OF R-CURVES ON PRECRACKS WITHOUT CONTACT SHIELDING

Fig. 1 shows schematically how the experiments are performed. The pre-cracks are produced in cyclic compression. The necessary load amplitude is not significantly larger than in cyclic tension. In cyclic compression a crack emanates from the notch, but then the growth rate decreases progressively until the crack completely stops propagating (Suresh, 1985, Pippan, 1987). The advantage of pre-cracking the specimen in cyclic compression is that the stress intensity where the crack closes is below zero at the beginning of the fatigue crack growth test. Hence, one can perform the threshold test by increasing the load amplitude in steps until the threshold value of a long crack is reached. If the load amplitude corresponds to a ΔK which is smaller than the effective threshold $\Delta K_{eff\,th}$ the crack does not grow. The first propagation is observed if ΔK is larger than $\Delta K_{eff\,th}$. Therefore this technique allows one to determine an upper and a lower bound for $\Delta K_{eff\,th}$. At the load steps, where the amplitudes correspond to a ΔK which is larger than $\Delta K_{eff\,th}$ and smaller than the long crack threshold, the crack starts to grow and stops where $\Delta K_{eff} = \Delta K_{eff\,th}$. Finally there is a load level where the crack does not stop. From this point on the test can be continued to measure the da/dN vs ΔK diagram for long cracks. These measurements provide an upper and a lower bound for the long crack threshold. If we plot the extension of the crack where it stops growing versus the corresponding ΔK we obtain an R-curve for the threshold of stress intensity range.

Such an example, performed on a 21 vol % Al_2O_3 particles reinforced 6061 T6 aluminum alloy (yield stress 360 MPa and fracture stress 408 MPa), is shown in Fig. 2. The CT specimen (Thickness 5 mm width W = 25 mm depth of the notch 10 mm) was prefatigued at $\Delta K = 14$ MPa $\sqrt{m}$ and R = 20 (pure compression) to initiate a crack of about 0.2 mm length measured from the notch root. The actual test was performed at R = 0.1. The crack length was monitored by DC-potential measurement and is plotted in Fig. 2a for the first few steps of the load amplitude. In Fig. 2b the change of the stress intensity range is shown. Initially the crack grew at a load amplitude which corresponded to a ΔK of 1.4 MPa $\sqrt{m}$. Therefore the effective threshold $\Delta K_{eff\,th}$ - in other words, the threshold of an open crack - should be between 1.2 and 1.4 MPa $\sqrt{m}$.

Fig. 2c shows a plot of the crack extension Δa versus the amplitude ΔK leading to crack stop, which can be seen as the locus of threshold as a function of the crack extension i.e. as a kind of R-curve for fatigue crack propagation (Pippan, 1994).

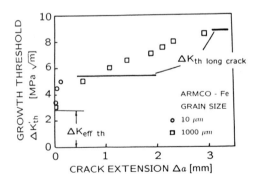

Fig. 3:   R-curve for the fatigue crack propagation threshold of ARMCO-iron for different grain sizes at R = 0.1.

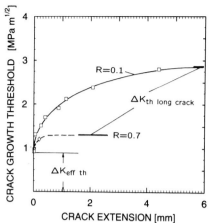

Fig. 4:   R-curve for fatigue crack propagation threshold of a 7075 alloy LT-orientation at R
            = 0.1 and 0.7

Two additional examples are presented in Fig. 3 and 4. Fig. 3 shows the influence of the grain size
at R = 0.1 in ARMCO iron (Pippan, 1991). In the fine grained material the R-curve increases faster
than in the coarse grained material. But the long crack threshold in the coarse grained material is
significantly larger. The extension of the crack where the long crack threshold is reached, is
markedly larger in the coarse grained material than in the fine grained material.

Fig. 4 compares the R-curves for the threshold at R = 0.1 and R = 0.7 of a 7075 aluminum alloy
(special heat treatment of AMAG - Ranshofen, yield stress 624 MPa and fracture stress 650 MPa,
details of this experiment were shown in Pippan *et al.*, (1994)). The effective threshold lies between
0.8 and 0.9 MPa√m and the threshold increases with the extension of the crack at both R-ratios. The
contribution of the effect of crack closure is much larger at R = 0.1 than at R = 0.7 and the
extension of the crack, where the long crack threshold is reached, is significantly larger at R = 0.1
than at R = 0.7. Note that the relatively large contribution of crack closure at R = 0.7 may be the
reason that in similar alloys at larger R-values no constant long crack threshold is observed (see for
example, Marci (1992)).

Estimation of a fatigue limit versus defect size diagram

Application of the R-curve concept to fatigue loading in analogy to static loading gives the fatigue
limit for a certain defect size. An example for such application is shown in Fig. 5. The resistance
curve is drawn into a driving force diagram (ΔK as a). The difference between the origin of the R-
curve and the driving force diagram is set equal to the chosen defect size. Driving force curves for
different stress amplitudes are indicated in Fig. 5a. The ΔK vs Δa curve which is tangential to the
R-curve gives the fatigue limit for the chosen defect size. At smaller stress amplitudes the crack
may propagate at first but it should stop growing after a certain extension, which is determined by
the intersection of the two curves.

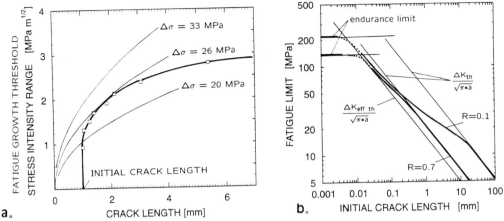

Fig. 5:   Application of the R-curve concept to the 7075 alloy (a) and the resulting fatigue
           limit versus initial crack length (defect size) diagram (b).

At larger load amplitudes the driving force is always larger than the resistance (threshold). Hence
such a component should fail.

Fig. 5b presents the fatigue limit versus defect size diagram, which was determined from the R-
curve of Fig. 4. We assumed a "component" width which is much larger than the defect size, the
driving force is given by $\Delta K = \Delta \sigma \sqrt{\pi a}$. The different limit curves are indicated in this Figure:
*   the limit curve from the R-curve,
*   the limit lines calculated from the long crack thresholds and the effective threshold,
*   and the upper bound for very small defects - the fatigue limit of smooth specimens.

The influence of the R-curve on the fatigue limit vs. initial crack size curve is evident.

IS THIS R-CURVE INDEPENDENT OF GEOMETRY AND LOADING PARAMETERS?

It is a common problem of static loading that the R-curves may depend on the geometry, but it is
often ignored. In fatigue the R-curve may depend on the geometry and (or) the loading history.
Hence one has to determine the limits for the application of a certain R-curve, or to find a lower
limit for the R-curves - which allows a conservative design.

The first question which should be discussed is: Is the measured R-curve influenced by the testing
parameters? Different studies show that the measured long crack threshold often depends on the
reduction rate (see for example, Cadman *et al.*, (1981)). Is a similar effect possible also in the rising
load amplitude test on specimens precracked in cyclic compression? There are different possible
parameters which may influence the determined R-curve and the long crack threshold value:
-       the pre cracking conditions
-       the height of the load steps
-       and the specimen size and notch depth.

It is difficult (or impossible) to create an precrack without creating residual stresses or damaging the material immediately ahead of it. This is essential since it is this region in which short crack propagation and R-curve must be studied (Suresh and Ritchie, 1984). With the exception of special simple materials, Pippan (1987), where a subsequent annealing or precracking before heat treatment can guaranty to remove such stresses without change of microstructure. In front of a pre-crack initiated in cyclic compression there is a small region with residual tensile stresses (Pippan et al., 1994). This opens the growing crack - in other words this reduces the contribution of contact shielding for small extensions of crack. It should be pointed out that this reduction leads to a "conservative" R-curve. The size of the zone with residual tensile stresses depends on the applied load amplitude during pre-cracking in compression and the number of the loading cycles. To reduce the size of this zone the used load amplitude should be as small as possible (Pippan et al., 1994).

The increase of the R-curve for ΔK_{th} is mainly caused by an increase in the contribution of contact shielding. The three most important mechanisms are the plasticity, roughness and the corrosion debris induced crack closure. In such experiments with rising load amplitude, the plasticity induced crack closure should be controlled by the actual ΔK. The stress ratio and crack extension should not be influenced significantly by the height of the load steps. The roughness induced closure is determined by the roughness of the fracture surface and the extension of the crack, which define the number of possible contacts. Since in the near threshold region the roughness is not significantly influenced by the crack growth rate this contribution should not be changed by the height of the load steps. The thickness of the corrosion layer (or debris) depends on the crack growth rate and the other closure contributions (or the contact stresses) (Suresh, et al., 1981). The thickness of the corrosion layer may grow during cyclic loading without crack propagation. This can produce an artificial high crack closure load after stopping the crack progatation. When such behavior is expected the load steps (or the increase of K_{max}) should be not to small, in order to allow the crack to propagate out of this high artificial crack closure region.

Can the specimen geometry influence the measured R-curve?

In a constant load amplitude test ΔK increases with rising crack extension, hence one can not determine that part of the R-curve where $d\Delta K_{th}/da$ is smaller than $d\Delta K/da$ in the test with constant amplitude. But in usual fracture mechanic specimens $d\Delta K/da$ is small and therefore the specimen geometry should not influence the R-curve (or the long crack threshold value).

What are the limits for the application?

Testing of the limitation of the R-curve application has not been extensive (Tanaka et al., 1983, 1988, Pippan et al., 1986).

Some simple limits exist:
- the small scale yielding conditions must be fulfilled, otherwise ΔK is meaningless
- the fatigue limit of smooth specimens - larger stress amplitudes lead to failure independent of defect size.

Note that the fatigue limit of smooth specimens of most materials is not determined by the threshold or the R-curve of a material - it is controlled by initiation of cracks or the propagation conditions of microstructural short cracks. The stress amplitudes, which are somewhat smaller than the fatigue

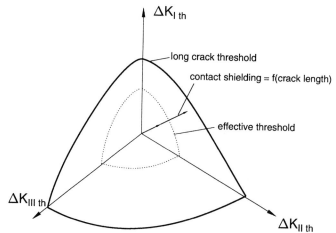

Fig. 6:   Schematic illustration of the threshold behavior under mixed mode conditions for given mean load parameters (stress ratios)

limit of smooth specimens, are critical for the application of the R-curve concept (at significantly smaller stress amplitudes there should be no problems). At such large stress amplitudes the contribution of contact shielding may be influenced by the stress amplitude in addition, and not only by the crack extension and ΔK. In such case the R-curve should only be used as rough estimate, but the effective threshold gives a safe lower limit which is independent of the R-ratio and loading history.

EXTENSION OF THIS R-CURVE TECHNIQUE TO OTHER LOADING CONDITIONS

The described method of measuring the R-curve is not limited to constant amplitude, mode I loading.

Mixed mode loading

The worst case of a defect in fatigue design is a crack without contact shielding (an open crack). This is independent of the loading mode. The threshold of crack propagation for mixed mode loading is determined by a corresponding effective threshold, which is independent of the crack extension, and a contribution of contact shielding, which should depend on the crack extension. It should be possible to describe this increase of contact shielding at the threshold by an R-curve

The expected threshold behavior is schematically depicted in Fig. 6. The effective threshold, the long crack threshold and the threshold for a certain crack extension are given by different surfaces in the ΔK space (ΔK_I, ΔK_{II}, ΔK_{III}). The $\Delta K_{eff\,th}$ surface should be independent of the crack extension and the stress ratios. But the contribution of contact shielding depends on crack extension and

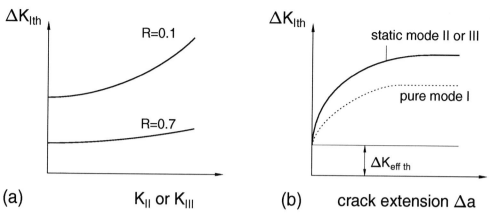

Fig. 7:   Schematic illustration of the change of the long crack threshold as a function of a
static mode II or III loading (a), expected R-curve behavior (b).

loading parameters $R_I = K_{Imin}/K_{Imax}$, $R_{II} = K_{IImin}/K_{IImax}$ and $R_{III} = K_{IIImin}/K_{IIImax}$ or K_{Imean}, K_{IImean}, $K_{IIImean}$. There exists R-curves for the threshold for a given ratio $\Delta K_I/\Delta K_{II}$ and $\Delta K_I/\Delta K_{III}$ for many different mean stress parameters. In addition for a given ratio $\Delta K_I/\Delta K_{II}$ and $\Delta K_I/\Delta K_{III}$ one has to distinguish between in and out of phase loading. For a pure mode II or mode III it may be that one can not reach a constant threshold value (i.e. long crack threshold value) since friction, wear and interlocking of fracture surfaces can unlimited increase with increasing crack extension (Tschegg, 1983).

In order to reduce the number of parameters it is important to find lower limits for the R-curves (or a lower limit for the contribution of contact shielding). For static mode II and mode III and cyclic mode I loading the long crack threshold was investigated by Tschegg et al. (1992). They show that the long crack threshold increases with increasing K_{II} and K_{III} as schematically depicted in Fig. 7. The R-curve should be changed in a similar way, hence the contribution of contact shielding as well as the length where the long crack behavior is reached may be changed.

<u>Variable amplitude loading</u>

Many studies have been made to model and predict crack propagation under variable amplitude loading. These investigations are usually performed at crack growth rates well above the threshold region. Few studies have been made in the threshold region (Stanzl et al., 1986). Most variable amplitude investigations study the amount of retardation or the length where such retardation occurs. At the threshold it is some-what easier because we have take into account only the condition for non-propagation. The physical conditions for non-propagation are simple: a crack does not grown if all effective stress intensity ranges are smaller than the effective threshold. In other words, the maximum values of the stress intensity of a block of variable amplitudes, K^b_{max}, should be equal or smaller than the closure load plus $\Delta K_{eff\ th}$. If we assume that the closure stress intensity is determined by K^b_{max} and the minimum stress intensity in a block K^b_{min}, (since non-propagation is

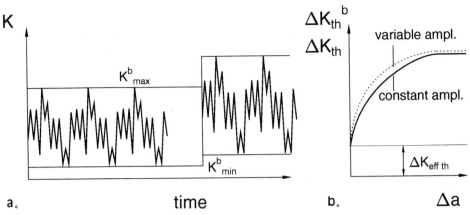

Fig. 8:   Schematic illustration of a step-wise increasing variable load amplitude threshold test
             (a), expected R-curve behavior (b).

considered, the growth rate should play no role), K^b_{max} - K^b_{min} can be used in the same way as in
constant load amplitude test. This is supported by the comparison of the block-loading and constant
load amplitude experiments of Stanzl *et al.* (1986) which shows that $\Delta K^b_{th} = K^b_{max\ th}$ - $K^b_{min\ th}$ is
about equal to the constant amplitude threshold.

One may argue that the variations of loads between K^b_{max} and K^b_{min} may change the contribution of
contact shielding. Nevertheless the main contribution of shielding, the roughness induced closure,
should not be changed and the plasticity and the corrosion debris-induced crack closure may be
somewhat larger. Hence the constant load amplitude R-curve should be a lower limit curve for the
variable load amplitudes.

But one can perform such R-curve tests with rising variable "block" loading or periodic underloads
in the same way as in the constant amplitude case. This is schematically shown in Fig. 8, where the
resulting R-curve should be equal or somewhat larger than in the constant load amplitude test, since
only the contribution of contact shielding may be somewhat larger in the case of variable amplitude
loading.

CONCLUSION

The R-curve for threshold can improve the application of threshold values for engineering design.
A very simple technique to measure such R-curves is the step-wise increase of the load amplitude
on specimens which are pre-cracked in cyclic compression. The resulting R-curve gives the effective
threshold for crack propagation, the minimum length where one can expect a long crack behavior
and the increase of contact shielding as a function of the crack extension. This R-curve
measurement technique is not limited to constant amplitude, mode I loading. It is also applicable to
mixed mode and variable amplitude loading.

Acknowledgement

This work was supported by the "Fonds zur Förderung der wissenschaftlichen Forschung, Projekt 10116" and the "Forschungsförderungsfonds für die gewerbliche Wirtschaft". Thanks are due AMAG-Ranshofen and ARMCO-GmbH, Köln for supplying the material investigated.

REFERENCES

Brook, R. (1984). In: Fatigue Crack Growth Threshold Concepts (D. Davidson and S. Suresh, ed.), pp. 417-429. *The Metallurgical Society of AIME*, Warrendale, Pennsylvania.

Cadman, A.J., Brook, R. and Nicholson (1981). In: *Fatigue Thresholds* (J.Bäcklund, A.F. Blom and C.J. Beevers, ed.) Vol. 1., pp. 59-75, EMAS, Warley, UK.

Marci G. (1992). A fatigue crack growth threshold. *Engineering Fracture Mechanics*, 41, pp. 367-385.

McEvily, A.J. and Z. Yang (1991). The growth of short fatigue cracks under compression and/or tensile cyclic loading. *Metallurgical Transactions*, 22, pp. 1079-1082.

Murakami, Y. and M. Endo (1994). Effects of defects, inclusions and inhomogeneities on fatigue strength. *Fatigue*, 16, pp. 163-182

Nowack, H. and R. Marissen (1987). In: *Fatigue 87* (R.O. Ritchie and E.A. Stark, ed.) Vol. 1., pp. 207-230, EMAS, Warley, UK.

Pineau, A. (1986). In: *Small fatigue cracks* (R.O. Ritchie and J. Lankford, ed.), pp. 191-212, AIME.

Pippan, R., L. Plöchl, F. Klanner and H.P. Stüwe (1994). The use of fatigue specimens precracked in compression for measuring threshold values and crack growth. *Journal of Testing and Evaluation*, 22, pp. 98-103

Pippan, R. and H.P. Stüwe (1986). In: *ECF6* (H.C. Elst and A. Bakker, ed.), pp. 1269-1277, EMAS, Warley, UK.

Pippan, R. (1987). The growth of short cracks under cyclic compression. *Fatigue Fracture Engineering Materials Structures*, 9, pp. 319-328.

Pippan, R. (1991). Threshold and effective threshold of fatigue crack propagation in ARMCO iron, Part. I: The influence of Grain Size and cold working. *Materials Science and Engineering*, AI38, pp. 1-22.

Pook, L.P. (1983). The role of crack growth in metal fatigue. *The Metals Society*, London.

Ritchie, R.O. (1988). Mechanisms of fatigue crack propagation in metals, ceramics and composites: role of crack tip shielding. *Materials science and Engineering*, 103, pp. 15-28.

Romaniv, O., V.N. Siminkovich and A.N. Tkach (1981). In: *Fatigue Thresholds* (J. Bäcklund, A.F. Blom and C.J. Beevers, erd.) pp. 799-808, EMAS, Warley, UK.

Stanzl, S.E., E.K. Tschegg and H.R. Mayer (1986). Slow Fatigue Crack Growth Under Step And Random Loading. *Z. Metallkunde*, 77, pp. 588-594.

Suresh, S. (1985). Crack initiation in cyclic compression and its application. *Engineering, Fracture, Mechanics*, 21, pp. 453-463.

Suresh, S. (1991). *Fatigue of Materials*. Cambridge University Press.

Suresh. S., G.F. Zamiski, and R.O. Ritchie (1981). Oxide - induced crack closure:an explanation for near-threshold corrosion fatigue crack growth behaviour. *Metallurgical Transactions*, 12A, 1435-1443

Suresh. S. and R.O. Ritchie (1984). Propagation of short fatigue cracks. *International Metals Reviews*, 29, pp. 445-476.

Tanaka, K. and Y. Nakai (1983). Propagation and non-propagation of short fatigue cracks at a sharp notch. *Fatigue of Engineering Materials and Structures*, 6, pp. 315-327.

Tanaka, K. and Y. Akinawa (1988). Resistance-curve method for predicting propagation threshold of short fatigue cracks at notches. *Engineering Fracture Mechanics*, 30, pp. 863-876

Tschegg, E.K. (1983). Sliding mode crack closure and mode III fatigue crack growth in mild steel. *Acta Metallurgica*, 31, 1323-1330.

Tschegg, E.K., S.E. Stanzl, H.R. Mayer and M. Czegley (1992). Crack face interactions and near-threshold fatigue crack growth. *Fatigue Fract. Engng Mater.* 16, 71-83.

(NON-) APPLICABILITY OF SOME CONCEPTS IN THE NEAR-THRESHOLD FATIGUE CRACK PROPAGATION REGIME

W. V. VAIDYA* and A. F. BLOM**

*Institute for Materials Research, GKSS Research Center, P.O. Box 1160
D-21494 Geesthacht, Germany
**Structures Department, FFA, The Aeronautical Research Institute, P.O. Box 11021
S-16111 Bromma, Sweden

ABSTRACT

In contrast to the major literature data, the existence of post-threshold non-propagation (hysteresis in fatigue crack propagation) is proved on commercial alloys. Compared to various concepts, microstructural impedance provided a better and a plausible explanation for our results. It is postulated that microstructural impedance prevails in the near-threshold regime and operates (among other effects) through modulus variations.

KEYWORDS

Al-alloys; compliance; compression pre-cracking; crack closure; crack length; load ratio; microstructural effects; post-threshold non-propagation; steels.

INTRODUCTION

Near-threshold fatigue crack propagation (FCP) behaviour is now well-understood (Suresh, 1991) and, among others, three major concepts have emerged for metals tested in air. Firstly, FCP is a unique function of stress intensity range, ΔK, over the entire regime. Secondly, microstructural effects dominate the near-threshold regime and are reflected as crack closure. Although established methods of crack closure determination (Chen et al., 1993) and existing interpretations (Vasudevan et al., 1993) have occasionally been questioned, a majority of opinions in the literature support the view that closure is induced either solely by plasticity (Elber, 1971), oxide and roughness (Ritchie et al., 1980, 1981), Mode II displacement (Minakawa and McEvily, 1981) or a combination of these mechanisms. FCP in such a case can be rationalised by the effective (ΔK_{eff}) rather than by the applied stress intensity range (ΔK). Thirdly, when normalised by Young's modulus, ($\Delta K/E$, the strain intensity), FCP can also be rationalised irrespective of flow properties of materials tested (Pearson, 1966). This concept has not been extensively employed in the literature, but, as shown recently by Ohta et al. (1992), is certainly a convenient tool. Experimental observations that do not fit in such a conceptional framework are rather rare.

Apparently, the observation that a crack exhibits post-threshold non-propagation and leads to so-called hysteresis in FCP (Vaidya, 1986) is such an exception. We confirmed during earlier work that hysteresis occurs even under closure-free testing conditions and

431

irrespective whether the threshold was obtained by load shedding (Vaidya, 1992) or by compression pre-cracking (Vaidya and Blom, 1994). In the present paper additional observations on hysteresis are reported and it is shown that when hysteresis occurs, certain concepts are not strictly applicable. So as to exclude problems related to load shedding rate and, hence, to data validity, threshold was obtained mainly by utilising compression pre-cracking (Suresh, 1985) which is expected to yield the absolute threshold value. Thus, it is assured that the results are not influenced by load interactions as may be the case when utilising load shedding method.

EXPERIMENTAL PROCEDURE

Major results were obtained on Al-2024T3 and a quenched and tempered low alloy ferritic steel (LA-steel), and additionally on Al-7475T761 and a medium plain carbon steel (MC-steel) with ferritic-pearlitic microstructure. As regards FCP behaviour, LA-steel has a lower load-shedded threshold than MC-steel; the same is true for Al-7475T761 and Al-2024T3. C(T)-specimens having the dimensions a_n = 15 mm, W = 60 mm and B = 4±0.1 mm were tested in laboratory air on servo-hydraulic machines operated under peak-load controlled sine wave mode. Threshold was obtained mainly by compression pre-cracking, and also by load shedding. For compression pre-cracking, specimen ligament was loaded between parallel plates and cycled at R = 20 and at 30 or 50 Hz. Subsequently, such specimens were tested under tension-tension. In other case, load ratio was kept constant during load shedding and post-threshold testing. Test frequency was 125±5 Hz; MC-steel was tested at 20 Hz. Data were acquired manually; crack length was measured optically on both faces at a resolution better than 0.02 mm in the threshold regime. Load vs crack opening curves were obtained by a gauge in the load-line, and plotted at 0.1 Hz. The threshold condition was assumed to be reached when no detectable crack growth occurred on both faces after a minimum of 5 million cycles. Thereafter load level was increased in small steps. If FCP was not detected after a minimum of 0.5 million cycles, crack was assumed to exhibit post-threshold non-propagation and, hence, hysteresis in FCP. This procedure was repeated until non-propagation was no more detected. The specimen was then tested without further load increase. In the following, specimens initially compression pre-cracked are denoted by subscript "cpc".

RESULTS AND DISCUSSION

<u>Nature of Hysteresis in FCP</u>

Usually FCP is believed to be a unique function of ∆K. That this is not necessarily the case is shown in Fig. 1. Both specimens were compression pre-cracked at the same load levels; the mean pre-cracked length was 16.18±0.16 and 15.965±0.015 mm (shown by symbols □ and O respectively). Although the pre-cracked length differed, the threshold value was nearly the same, ΔK_{th-cpc} = 2.45 ≤±0.01 MPa√m; therefore, the specimens were expected to exhibit nearly the same FCP behaviour. It turned out that under comparable loading conditions the crack remained dormant in one specimen (∆a ≤ 0.01 mm) whereas a notable crack extension was observed in the other (∆a ≈ 0.5 mm). The ∆K value, (ΔK_{hys}), up to which non-propagation was observed, was slightly

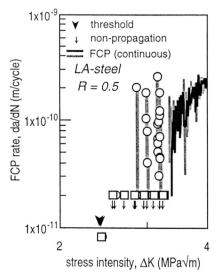

Fig. 1. Nature of hysteresis in FCP.

higher in the former than in the latter case (3.25 against 3.21 MPa√m). Above this range non-propagation was no more observed. Thus, this example clearly shows that hysteresis in FCP is a phenomenon that cannot be explained alone by loading parameters such as ΔK or K_{max}. On the other hand, microstructural impedance which operates through obstructions by interfaces and local work hardening at the crack tip, offers a more plausible explanation. A crack front arrested over its whole length at grain boundaries shall be the most ideal condition for non-propagation. An increase in ΔK shall then work harden the crack tip area and, hence, propagation can be obstructed further.

So as to prove our argument about work hardening, additional experiments were carried out on LA-steel at R = 0.1. The threshold $ΔK_{th-cpc}$ was 2.45 ≤±0.05 MPa√m for three specimens. The specimen tested with stepwise load increase yielded the maximum $ΔK_{hys-cpc}$ value (6.65 MPa√m), although temporary non-propagation (N ≤ 0.3 million cycles) existed up to 7 MPa√m. Indeed, increasing the starting ΔK value from threshold to 5 MPa√m in one step yielded a decreased value ($ΔK_{hys-cpc}$ = 6.4 MPa√m) and on further increase to 7 MPa√m hysteresis was absent. These experiments also indicate that the near-threshold FCP is dependent on the dynamic equilibrium between the microstructural impedance and the loading conditions. Once the microstructural impedance has been overcome, post-threshold non-propagation ceases and continuous FCP commences (Fig. 1). Thus, similar to $ΔK_{th}$, $ΔK_{hys}$ is a material property. The data in Fig. 2 support this conclusion further.

Crack Length Effect on Hysteresis Behaviour

Depending on whether a crack is physically short or long, crack length may affect FCP (Suresh, 1991). Whereas a short crack may exhibit non-propagation over a range of crack extension or ΔK (very similar to that shown in Fig. 1), a long crack either remains non-propagating (threshold condition) or propagates. Note that by hysteresis in FCP we are describing "short crack behaviour" on a long crack (Fig. 1). With a_n = 15 mm, the pre-cracks in Fig. 2 can be classified as physically "short" ($Δa_{cpc}$ ≤ 1 mm) and "long" ($Δa_{cpc}$ ≥ 1 mm), and according to the literature, they should exhibit different types of FCP behaviour. In other words, hysteresis should not occur on long cracks.

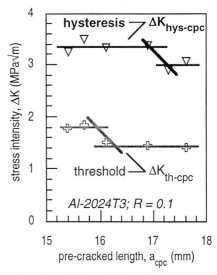

Fig. 2. (In-) Sensitivity of hysteresis in FCP to compression pre-cracked length.

The threshold value varied in the range of 1.6±0.2 MPa√m, (note the enlarged scale), which gives an accuracy of ±12 % in threshold determination. The trend that the transition in threshold occurs at $Δa_{cpc}$ ≈ 1 mm should support the assumption about pre-crack classification. Although pre-cracked lengths differed by a large extent, the post-threshold behaviour was very much comparable (and not different as to be expected from "short" and "long" cracks). All specimens exhibited hysteresis in FCP and the difference in the $ΔK_{hys}$ range was comparable to that at threshold (±0.2 MPa√m). In fact, the trend curve was very much comparable to that for threshold, with the difference that the transition value was shifted diagonally to much higher pre-cracked length. This effect may be resulting from difference in prior loads; loads while compression pre-cracking were higher for longer pre-cracks and some residual damage may have been retained and affected the FCP behaviour during tension-tension test. This had, however, no significant effect on threshold or post-threshold non-propagation.

Crack Closure

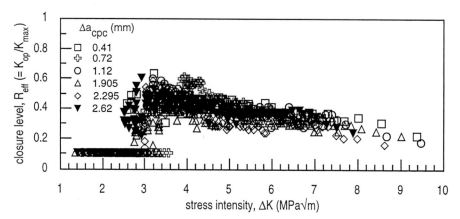

Fig. 3. Crack closure at R = 0.1 in Al-2024T3 initially compression pre-cracked to different lengths.

Crack closure observed in Al-2024T3 is shown in Fig. 3. There was a large variation in the closure level which differed from specimen to specimen, and which could also not be correlated with pre-cracked length or the ΔK level. These data were repeatedly evaluated to find out the subjective error which was found to be much smaller ($R_{eff} \leq \pm 0.05$) than the variation observed. Interestingly, data on hysteresis in FCP could be correlated better with ΔK (see Fig. 2) than with ΔK_{eff}. It was confirmed that the variations in closure were certainly not an artefact resulting from measurement technique.

So as to understand the nature of closure, cracks in Al-2024T3 and LA-steel were carefully observed in-situ during compression pre-cracking. Being under full compression it was expected that a crack should be closed at P_{min} (max. compressive load). However, except at the tip, the crack was found to be open (at a magnification of 120), although crack morphology was zig-zag, rubbing should have occurred and the crack was embedded in the deformed zone. Moreover, scratches sectioned by the crack remained continuous. Therefore, roughness and oxide (Ritchie et al., 1981, 1980), plasticity (Elber, 1971) or Mode II displacement (Minakawa and McEvily, 1981) were certainly not the major mechanisms of crack closure. That the crack should have been closed at the tip is supported by compliance measurement. The data in Fig. 4 suggest that (as to be expected) compliance increases initially. However, as the crack becomes non-

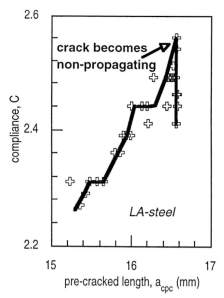

Fig. 4. Compliance variations during compression pre-cracking.

propagating, compliance begins to decrease. This effect cannot be explained by residual stresses alone which, being mainly tensile (Suresh, 1985), should contribute to crack opening and, hence, compliance should in fact increase. Compliance (by definition) is a characteristic of an opened crack (for tension-tension loading). Assuming compliance to be valid also under compression (the present values for a given crack length are lower than those in tension-tension), it is concluded that the process at the crack tip (such as local work hardening and modulus variation, which in turn be dependent on the microstructural impedance) should be the major cause for the decrease in compliance.

Returning to hysteresis behaviour, Fig. 5 shows that being observed at R = 0.7 ($\Delta K = \Delta K_{eff}$), hysteresis is neither dependent on crack closure nor on the test method. The threshold values in compression pre-cracked specimens are nearly constant and are much lower than those obtained by load shedding and the typical load ratio trend is

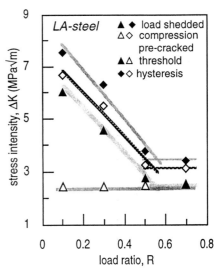

Fig. 5. Occurrence of hysteresis in FCP at various load ratios.

present only in the latter case. Apart from such differences, the trend in ΔK values up to which hysteresis behaviour occurred, is, however, comparable within 0.5 MPa√m. This confirms our postulation that hysteresis in FCP is a material property (see Figs. 1-2).

Microstructural Effect

The above results reveal the occurrence of hysteresis under a number of parameters, irrespective by which method the data were acquired. Whenever a crack exhibits non-propagation, the crack closure concept offers the most probable explanation; the crack, being closed, is not subjected fully to the applied stress intensity. Hysteresis does occur even in the absence of crack closure (at R = 0.7 in Fig. 5) and, hence, crack closure concept is not applicable. According to the nature of near-threshold FCP in Fig. 1, neither ΔK nor K_{max} can rationalise such FCP. On the other hand, microstructural impedance provides certainly an alternative explanation. Its validity was proved by metallography on the surface (Vaidya, 1986, 1992). Its existence over through-thickness is supported by data in Fig. 6.

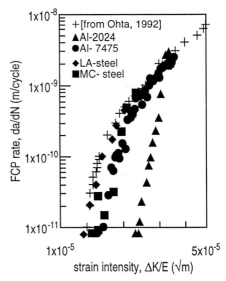

Fig. 6. Remanence in rationalisation

Contents of Fig. 4 lead to the postulation that as long as microstructural impedance prevails, elastic modulus should vary. If this is true, FCP cannot be rationalised by the strain intensity, since the concept assumes the constancy of elastic modulus. The data in Fig. 6 were obtained at R = 0.7 during load shedding to threshold; closure was not observed in any material ($\Delta K = \Delta K_{eff}$). The load shedding rate in steels was faster (C ≈ 0.15 mm^{-1}) than in Al-alloys (C ≈ 0.06 mm^{-1}). Despite this difference, FCP curves in the near-threshold regime of MC-steel and Al-2024T3 (higher threshold; damage tolerant) were clearly shifted to higher $\Delta K/E$ values than those of LA-steel and Al-7475T61 (lower threshold; damage sensitive). The latter curves were nearer to the data by Ohta *et al.* (1992). On going from our results (Figs. 4 and 6) we conclude that microstructural impedance operates through obstructions of interfaces and work hardening at crack tip which in turn increases the elastic modulus. Therefore, some concepts are not strictly applicable in the near-threshold regime.

CONCLUSIONS

1) The phenomenon of post-threshold non-propagation (hysteresis in FCP) exists in the near-threshold regime, and leads to non-applicability of unique da/dN versus ΔK-type of data.

2) Hysteresis occurs even in the absence of crack closure. Therefore, FCP cannot be rationalised solely by crack closure concept, or also by K_{max}-concept.

3) It is postulated that microstructural impedance, local work hardening and variations in elastic modulus control the extent of hysteresis effect.

Acknowledgement: We thank FFA (Bromma, Sweden) and GKSS (Geesthacht, Germany) for making the present investigation possible by providing internal funding.

REFERENCES

Chen, D. L., B. Weiss and R. Stickler (1993). A new concept of closure evaluation for fatigue crack propagation. In: *FATIGUE '93* (J.-P. Bailon and J. I. Dickson, ed.), pp. 555-563. EMAS, Warley, U. K.

Elber, W. (1971). The significance of fatigue crack closure. In: *Damage Tolerance in Aircraft Structures, ASTM STP 486*, pp. 230-242. ASTM, Philadelphia, U. S. A.

Minakawa, K. and A . J. McEvily (1981). On crack closure in the near-threshold region. *Scr. Met.*, 15, pp. 633-666.

Ohta, A., N. Suzuki and T. Mawari (1992). Effect of Young's modulus on basic crack propagation properties near the fatigue threshold. *Int. J. Fatigue*, 14, pp. 224-226.

Pearson, S. (1966). Fatigue crack propagation in metals. *Nature*, 211, pp. 1077-1078.

Ritchie, R. O. and S. Suresh (1981) Some considerations on fatigue crack closure at near-threshold stress intensities due to fracture morphology. *Met. Trans.*, 13A, pp. 937-940.

Ritchie, R. O., S. Suresh and C. M. Moss (1980). Near-threshold fatigue crack growth in 2¼ Cr-1 Mo pressure vessel steel in air and hydrogen. *J. Engng. Mater. Techn. (Trans. ASME)*, 102, pp. 293-299.

Suresh, S. (1985). Crack initiation in cyclic compression and its applications. *Engng. Fract. Mech.*, 21, pp. 453-463.

Suresh, S. (1991). *Fatigue of Materials*. Cambridge University Press, Cambridge, U. K.

Vaidya, W. V. (1986). Near threshold fatigue crack propagation behaviour of a heterogeneous microstructure. *Fatigue Fract. Engng. Mater. Struct.*, 9, pp. 305-317.

Vaidya, W. V. (1992). Fatigue threshold regime of a low alloy ferritic steel under closure-free testing conditions. *J. Test. Eval., (JTEVA)*, 20, pp. 157-179.

Vaidya W. V. and A. F. Blom (1994). A critical verification of post-threshold non-propagation (hysteresis in fatigue crack propagation) utilizing compression pre-cracking on Al-2024T3 alloy. *Scr. Met. et Mater.*, 30, pp. 821-826.

Vasudevan, A. K., K. Sadananda and N. Louat (1993). Critical evaluation of crack closure and related phenomena. In: *FATIGUE '93* (J.-P. Bailon and J. I. Dickson, ed.), pp. 565-576. EMAS, Warley, U. K.

CRACK OPENING LOAD MEASUREMENT DURING VARIABLE AMPLITUDE LOADING

M. Lang , H. Döker

German Aerospace Research Establishment, Institute of Materials Research
D-51140 Köln, Germany

ABSTRACT

The development of K_{op} during variable amplitude loading was studied. Therefore, K_{op}-measurements were conducted during FALSTAFF-loading and during different single peak overload tests, using the near threshold crack growth technique (NTCG). Crack propagation was observed by optical microscopy, potential drop method and SEM. In this way, the retardation phenomenon on the surface and in the interior of the specimens could be investigated separately. The experimental results are presented with respect to the ΔK_{eff}-concept.

KEYWORDS

Fatigue crack propagation, crack closure, crack opening load, variable amplitudes, overload effects, delayed retardation

INTRODUCTION

Since Elber found the crack closure effect, it is known that during constant loading only part of the amplitude is responsible for crack growth. The crack opening load K_{op} and the maximum load K_{max} determine the effective load range ΔK_{eff}. For constant amplitude loading K_{op} depends on K_{max} and the loading ratio R (Elber, 1971). Sudden loading changes cause retardation or acceleration effects. The ΔK_{eff}-concept correctly describes constant amplitude loading. The load interaction effects should also be described by the same concept. For the development of crack growth life prediction models, load interaction effects have to be well understood. The crucial point in the ΔK_{eff}-concept is the determination of the crack opening load K_{op}. Many different methods have been developed, but there is still need of further work, since no specific information exists on K_{op} under variable amplitude loading. It is the objective of this investigation to study the development of K_{op} under variable amplitude loading. K_{op} was measured first at distinct points in a FALSTAFF (Fighter Aircraft Loading STAndard For Fatigue-loading)-sequence, and then single overload tests were conducted to study the retardation phenomenon more carefully.

437

EXPERIMENTAL PROCEDURE

CCT-specimens ($2{\times}W{=}160$mm, B=8mm, LT) and CT-specimens (W=50mm, B=10mm, LT) were cut out of a 7475-T7351 aluminum alloy sheet ($\sigma_y{=}459$MPa). K_{max}-constant tests (Döker et al., 1982; Hertzberg et al., 1987) were conducted to generate da/dN-ΔK data for this material. Furthermore, da/dN-ΔK_{eff}-curves were calculated by eq.1, which was developed for the same alloy, using the NTCG-method (Marci, 1990).

$$K_{op} \,/\, K_{max} = 0.455 + 0.321R + 0.208R^2. \qquad (1)$$

The CCT-specimens for FALSTAFF-loading experiments and the CT-specimens for single overload tests were loaded by a 400kN and 25kN servohydraulic computer-controlled testing machine. To determine the crack opening load, the near threshold crack growth method (NTCG) was used (Döker et al., 1988; Marci et al., 1990). The method is shown in Fig.1. At the position in the loading sequence where K_{op} is to be measured, a cyclic load ΔK_j is applied far below the expected K_{op} (but not below K_{min}). ΔK_j is slightly larger than the intrinsic threshold value ΔK_T. If no detectable crack growth occurres, the mean load is increased by a small amount. This procedure is repeated until crack growth is detected by the potential drop method. The crack opening load is given by $K_{op} = K_{max,j} - \Delta K_T$. $K_{max,j}$ is the average value of the maximum load of the block in which the crack starts to grow, and the maximum load of the previous block.

Kop- measurement during FALSTAFF-loading - The FALSTAFF loading sequence was applied with a loading frequency of 10Hz. The crack length was monitored by DC potential drop method. K_{op} was measured at three points of the spectrum - before and after the highest overload and after a compressive load, which follows a few cycles later (Fig.2). After every FALSTAFF-sequence, the loading factor was reduced to keep the K- range "constant". The K_{op}-measurements were repeated several times and the reproducibility was very good.

Single peak overload experiments - During constant amplitude loading (two different amplitudes: $K_{max}{=}10$MPa$\sqrt{m}$, R=0.1 and $K_{max}{=}8$MPa$\sqrt{m}$, R=0.1), different single overloads were applied (19, 23, 25.5, 28MPa$\sqrt{m}$ for 10MPa$\sqrt{m}$ and 15.2, 17.5, 22MPa$\sqrt{m}$ for 8MPa$\sqrt{m}$, see Tab.1). First, K_{op}-measurements were conduc-ted right after the application of the overloads (except 10-25.5-10-test). To study the retardation phenomenon after the overloads in more detail, two additional crack length measurement techniques besides the potential drop method, were applied. An optical microscope ($125{\times}$) was used to observe the crack length on the surface of the specimen. To study the crack propagation behaviour in the interior of the specimen, the crack surfaces were investigated by SEM. Therefore, markerloads were introduced at distinct load cycles after the overload. Tab.1 shows the marking points for each experiment. To rule out the possibility of load interaction effects due to marker loading, separate tests were conducted for each marking point after the overload. Additionally, a clip gage was mounted at the mouth of the CT-specimens during the overload tests to record the compliance curve.

RESULTS and DISCUSSION

Kop- measurement during FALSTAFF-loading
Results are given in Fig.2. After the highest overload K_{max}, the crack opening load rose from 9.8MPa$\sqrt{m}$ to 10.9MPa$\sqrt{m}$. This behaviour is expected from the ΔK_{eff}-concept, due to the retardation effect after a big overload. After the subsequent compressive load, K_{op} dropped down to 9.9MPa$\sqrt{m}$. This increase in ΔK_{eff} is in agreement with the literature. Fig.2 clearly shows that the experiments

coincide qualitatively with the expectations from the ΔK_{eff}-concept. Because these experiments do not allow a quantitative description of load interaction effects, simpler experiments have to be conducted. Therefore, the retardation after a single overload was investigated in more detail.

Single peak overload experiments

The results of K_{op}-measurements are presented in the right part of Tab.1. The crack opening load after the overload is called $K_{op,afterOL}$. In all cases the crack opening load rises significantly after application of the overload. This behaviour has also been observed during FALSTAFF-loading. In the following, a strict application of the ΔK_{eff}-concept is shown. Since we have the equation for K_{op} under constant amplitude loading (eq.1), we know the position of K_{op} before the overload, $K_{op,beforeOL}$. $K_{op,beforeOL}$ has been checked experimentally and very good agreement with eq.1 was found. Values are given in Tab.1. If we continued to cycle with the overload amplitude, we would get a block loading sequence from low to high. Then, K_{op} had to increase to the equilibrium value for the higher amplitude. This K_{op}, called $K_{op,const.OL}$, can also be evaluated by eq.1. $K_{op,const.OL}$-values are given in tab.1. Eq.2 was used to calculate the increase in K_{op} in relation to the difference of the constant amplitude K_{op}-values. $F_S \times 100$ gives the step in % of the difference in K_{op} of the

$$f_S = (K_{op,after\ OL} - K_{op,beforeOL}) / (K_{op,const.OL} - K_{op,beforeOL}). \qquad (2)$$

two equilibrium conditions. F_S is called step factor and is given in Tab.1. It turns out that in all cases f_S takes very similar values. The deviation is 1.8% and the average is 48.2%. Already after the first overload, nearly half the difference between K_{op} of the equilibrium levels is reached. The effect of an overload is therefore excellently described in terms of K. In contrast to the NTCG-method, the clip gage indicated a drop in K_{op} after the application of the overload, as it is often observed (Elber, 1971; Yisheng, *et al.*, 1995). This seems to be due to blunting of the crack tip. Thus, the point where the compliance curve deviates from the linear course, can not represent the real K_{op}, because if K_{op} would fall short after the overload, at least a second big striation just after the overload has to be found on the fracture surface due to the high ΔK_{eff}. No indications to that were found here or in the literature. Of cause the compliance curve also indicates that the crack "opens", meaning the partitioning of the crack flanks, but not in respect to the ΔK_{eff}-concept, which says that above K_{op} crack propagation occurs (compare Marci, 1990). Going back to the highest overload in FALSTAFF, we try to apply eq.2. We use eq.1 to cal-culate the equilibrium-K_{op} for the overload, and get 12.7Mpa√m (R=0.06). When applying eq.2, a K_{op} of 11.2Mpa√m after the over-load should be found. This is in good correspondence with the actually observed K_{op} of 10.9Mpa√m.

Fig.3 shows the results for the 10-23-10-test using an a-N and da/dN versus normalized cycles plot. The results of crack length measurements with SEM, optical microscope and potential drop signal, are indicated. After the overload the crack continues to grow at the surface, while in the centre of the specimen sudden drop in crack speed is observed (see also Matsuoka *et al.*, 1980; Park *et al.*, 1992). The crack starts to grow again in the middle of the specimen and after some time, the crack on the surface follows with higher crack growth rate. The potential drop signal gives an averaged crack growth rate. The same behaviour was found in all the other tests. Crack propagation behaviour was ruled to a great part by the interior of the specimen and therefore higher crack speed of the surface reached only 0.1 to 0.3mm into the specimen. To show the overload effect more clearly, the crack length was normalized by the monotonic plastic zone of the overload $[r_{p,OL}=1/\pi(K_{max}/\sigma_y)^2$ for plane stress (opt. meas.) and $1/3r_{p,OL}$ for plane strain (SEM-meas.)] and plotted versus da/dN. The region short after the overload for four representative tests is shown in fig.4. The crack velocity due to the overload itself was omitted. The da/dN-value which corresponds to $K_{op,afterOL}$, was taken out of the ΔK_{eff}-curve. At the surface da/dN becomes a minimum between 10-20% of r_p. The potential drop signal shows a later decrease in crack growth rate because it indicates an average crack length. Since the NTCG-method mainly covers the plane strain behaviour, the points from SEM-measurements were connected to the points from $K_{op,afterOL}$. It can be observed that the da/dN-data from SEM and K_{op}-measurements are in good agreement. Even in relation to the plastic zone, the crack growth rate in the middle of the specimen drops down abruptly. If the crack is stopped by the overload (10-28-10, 8-22-8), it also continues to propagate at the surface for some time, until it also stops there. In the 10-23-10-test a higher crack growth rate was found just after the overload (Fig.4). At this moment the crack growth was only some micrometers. The higher crack speed might be due to the damaged zone shortly after blunting. In this zone continuum-mechanics for plane strain conditions is not yet working because of the vicinity to the blunted crack tip. Vargas et al. (1973) also observed a little higher crack growth rate, but also only over some microns after the overload. This is neglectable in comparison to the extent of retardation at the surface and in the interior. Taking this into account, delayed retardation is confined to the plane stress region close to the specimen surface.

SUMMARY

(1)   The variation of K_{op} during variable amplitude loading was found to be in accordance with the ΔK_{eff}-concept, since ΔK_{eff} and the observed crack growth rates correspond well . Tensile and compressive loads cause an immediate rise and drop in K_{op}, respectively.

(2) K_{op}-measurements during FALSTAFF and after single overloads showed, that K_{op} changes selectively from cycle to cycle. Therefore, the mechanism that rules K_{op} during variable amplitude loading, mainly has to be a crack tip effect. (compare Marci, 1971; Hertzberg et al.,1988)

(3)   After single peak overloads K_{op} increases by a certain percentage (f_S) of the the difference between the equilibrium crack opening loads (const. ampl.) of the participating amplitudes (f_S=48.2%). F_S has to be checked with other R-values.

(4)   In the interior of the specimen no indication for delayed retardation was found. Therefore the authors think that the often observed delayed retardation effect is a surface phenomenon.

References:

Döker, H., Bachmann, V., (1988). Determination of Crack Opening Load by Use of Threshold Behaviour. *ASTM STP 982,* 247-259.

Döker, H., Bachmann, V. and Marci, G., (1982). A Comparison of Different Meth. of Determination of Threshold for Fat. Crack Prop. In: *Fat. Thresh., Fund.and Eng. Appl.,* EMAS, 4547.

Elber, W., (1971). The Significance of Crack Closure. *ASTM STP 486,*230-242.

Hertzberg, R. W., Herman, W. A., Ritchie, R. O., (1987). Use of a Constant K_{max} Test Procedure to Predict Small Crack Gr. Behaviour in 2090-T8E41Al-Li Alloy. *Scr. Met.* Vol.21, 15411546.

Hertzberg, R. W., Newton, C. H. and Jaccard R., (1988). Crack Closure: Correlation and Confusion. *ASTM STP 982,* 139-148.

Marci, G., Hartmann, K. and Bachmann, V., (1990). Experimentelle Bestimmung des ΔK_{eff} für Ermüdungsfortschritt. *Mat.-wiss. u. Werkstofftechn. 21,* 174-184.

Marci, G., (1979). Effect o. the Act. Pl. Zone on Fat. Crack Gr. Rates. *ASTM STP 677,* 168-186

Matsuoka, S. and Tanaka, K., (1980). The Influence of Sheet Thickness on Delayed Retardation Phenomena in HT 80 Steel and A5083 Al-Alloy. *Engng. Fract. Mechan.* Vol.13, 293-306.

Park, Y., De Vadder, D. and Francoir, D., (1992). In-Situ Mid-With and Near-Surface Measurements of the Retardation of the Propagation of a Crack After an Overload. *9th European Conf. on Fracture.* EMAS

Vargas, L. G. and Stevens R. I., (1973). Subcritical Crack Growth under Intermittent Overloading in Cold.Rolled Steel. *3rd Int. Conf. on Fracture,* Munich, 1973.

Yisheng A., Schijve J., (1995). Fatigue Crack Closure Measurement on 2024-T3 Sheet Specimens. *Fat. Fracture Engng. Mat. Struct.,* Vol. 18, No. 9, 917-921.

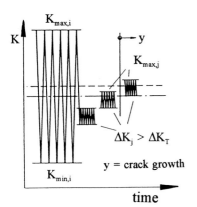

Fig.1: NTCG-method

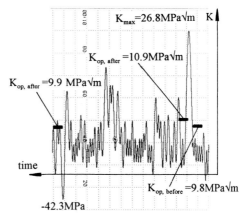

Fig.2: K_{op}-meas. during FALSTAFF-loading. Before and after the highest tensile load and after a subsequent compressive load

Constant Ampl.- Overload - Tests R=0.1,[Mpa√m]	markerload position after overload [cycles]	K_{op} afterOL [MPa√m]	K_{op} beforeOL [MPa√m]	K_{op} const.OL [MPa√m]	f_S [%]
10-19-10	10 19 3×16 2500, 5000, 7500;	6.85	4.89	8.97	47.8
10-23-10	275, 1250, 10150, 25350	7.71	4.89	10.8	47.8
10-25.5-10	100 000	not meas.	4.89	11.93	--
10-28-10	crack stopped	9.08	4.89	13.67	47.7
8-15.2-8	5250, 11500	5.44	3.91	7.18	46.7
8-17.5-8	4000, 10000, 22500	6.05	3.91	8.23	49.5
8-22-8	crack stopped	7.06	3.91	10.27	49.5

Table 1- *Single peak overload tests*

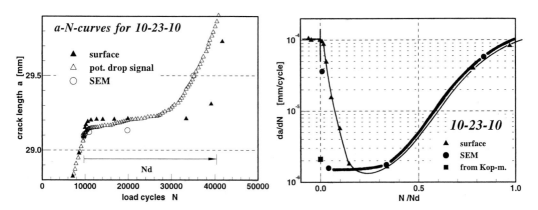

Fig.3: *a-N plot and da/dN versus normalized load cycles N/Nd for 10-23-10.*

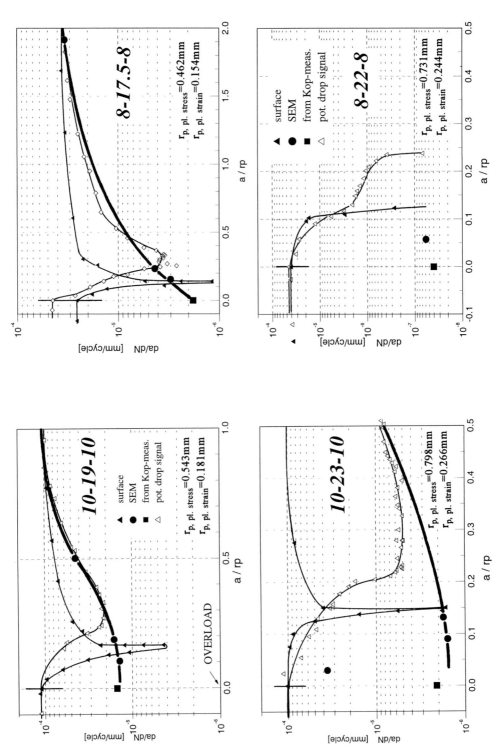

Fig.4: *da/dN versus normalized crack length. (the overload was applied at a/rp=0)*

FUNDAMENTAL THRESHOLD FATIGUE CRITERIA.
NUMERICAL ESTIMATION AND APPLICATION.

Yu.I. RAGOZIN

Machines' Dynamics & Strength Department', Technical University,
Komsomolsky Avenue, 29a, Perm 614600, Russia

ABSTRACT

Threshold fatigue criteria have been proposed, all distinguished by being physically substantiated and defined through fundamental materials' constants due to which these criteria appear to be invariant to a whole set of factors.

KEYWORDS

Fatigue; threshold criteria; estimation; application.

INTRODUCTION

Among the estimation criteria of carrying capacity of materials (and articles of them) there are those ones that are calculated by fundamental constant values and, hence, invariant to quite a number of factors. These criteria are of special interest. Owing to the force of the last circumstance such criteria themselves become fundamental constant values of materials. In conformity with cyclic loading the author developed a number of similar criteria. Their physical nature, numerical estimation and application are set out in the present paper.

THEORETICAL BASIS AND RESULTS

Threshold Microcrack Length at the Fatigue Limit

It is known that there exists a definite microcrack length l_0 that does not affect fatigue limit. Some authors (El Haddad et al., 1979, Lukas et al., 1986) assume that the value of l_0 is independent of the form of a solid and the mode of its loading, and is determined only by the material nature. Proceeding from experimental data the estimation of this value was made for some materials. However, it is extremely difficult to determine the crack size l_0 experimentally, so the quest for analytical methods of determining this constant is of current concern. A method suggested by the author is considered below.

In 1970-th a phenomenon of electromagnetic radiation brought about by plastic strain and fracture of solids was dicovered (Misra, 1975). Thorough study of this phenomenon revealed a series of new facts. Specifically, it was shown (Khatiashvili and Perelman, 1982) that supersonic wave passing through a crystal stimulates (as a result of interaction with the dislocation structure and crack) electromagnetic radiation with the frequency of this wave and with frequency twice lower. The experimental results presented form the basis of the following physical model developed on the base of phonon conception of fracture and allowing not only to explain the nature of a new constant for a structural material, but to suggest a method of calculating the numerical values of l_0 as well.

In accordance with the fracture phonon conception (Ragozin, 1989, 1990) quite a definite descrete spectrum of natural frequencies of atom modes in the lattice corresponds to each particular crystal material. Atom mode frequencies v_i in this spectrum are approximately by an order of magnitude smaller than the Debayev's frequency. They are determined by the type of dislocations specific for a given solid structure and, in principle, can be calculated for any crystal. While passing of originating hypersonic waves through the dislocation structure and cracks, electromagnetic radiation with the frequency of these waves is stimulated in full conformity with the observations set forth and with frequency twice lower ($v_i^0 = v_i / 2$). The last is corresponding to the frequency of zero modes. The signal amplitude will be maximum at the moment of the crack nucleation. In their turn, the originating electromagnetic modes will interact with the crack being formed. However, the like interaction may occur only in case when the length of the originating radiation wave will be less than the crack size, otherwise the waves will not "feel" the crack. Moreover, the crack may sense the wave energy in quants only.

Thus, in comformity to the accepted model a formed crack, when reaching a critical length being quite definite for each material, begins to absorb energy from the surrounding electromagnetic field which results in a sharp increase of the crack growth rate. The analysis of the data available showed that the crack growth mechanizm is proved correct by experiment (Miller, 1992).

The accepted physical model allows to determine the numerical values of the new constant as well, since for many metals the frequency spectra v_i are known (Ragozin, 1990). Assuming that minimum possible threshold length of a microcrack l_i is equal to $\lambda_i / 2$ the spectra of possible values of this criterion were calculated a number of metals. Results of calculation are given in Table 1. It may be assumed that the similar spectra of l_i values will be characteristic of the alloys of these metals as well, and alloying and thermal treatment, practically not influencing the numerical values of l_i in spectrum, will determine the level of l_0 (from spectrum) perculiar to a given specific alloy.

Table 1. Calculated levels l_i for several metals at room temperature

Metal	l_i, mm										
	1	2	3	4	5	6	7	8	9	10	11
Mg	0,232	0,188	0,161	0,157	0,122	0,116	0,086	0,061			
Al	0,412	0,206	0,191	0,139	0,095						
Ti	0,203	0,165	0,159	0,109	0,101	0,077	0,055				
Fe	0,177	0,118	0,107	0,081	0,076	0,047	0,037				
Ni	0,350	0,175	0,169	0,113	0,084	0,063[*]	0,056[*]	0,051[*]	0,04[*]	0,025[*]	
Cu	0,517	0,258	0,223	0,174	0,110	0,096[*]	0,073[*]	0,072[*]	0,052[*]	0,051[*]	0,031[*]

[*] with package defects

An extensive analysis of calculated values of compared to experimental ones presented in literature was performed and showed a good agreement of the theory and the experiment. It has also been demonstrated that the constant of materials l_i is of sensible use first of all in analyzing threshold situations.

Crack-Driving Forces G_0 and G_p

From the above analysis it follows that when the growing crack reaches a certain threshold size ($l_i = \lambda_i / 2$) there appear additional crack-driving forces resulting from its interaction with the surrounding electromagnetic field and sharply increasing the crack growth:

$$G_0 = \frac{h\nu_i}{2} \cdot \frac{\lambda_i}{2} = \frac{hc}{4} \tag{1}$$

and

$$G_p = h\nu_i \cdot \frac{\lambda_i}{2} = \frac{hc}{2}, \tag{2}$$

where G_0 and G_p - crack driving forces from zero and basic modes respectively (dimensions of G_p and G_0: $\frac{J \cdot m}{atom}$ or $\frac{J}{m^2}$ or $\frac{N}{m}$); h - Plank's constant; c - velocity of light.

Conforming to these forces stress intensity factor, e.g. for the conditions of flat deformation, will be equal to:

$$K_I^0 = \sqrt{\frac{G_0 E}{1 - \mu^2}} \tag{3}$$

and

$$K_I^p = \sqrt{\frac{G_p E}{1 - \mu^2}}, \tag{4}$$

where E - modulus of elasticity; μ - the Poisson ratio. Numerical values for the new constants for several metals are given in Table 2.

Table 2. Several constant values of metals.

Metal	G_0, kN/m	G_p, kN/m	E, 10^4 MPa	μ	K_I^0, MPa$\sqrt{m}$	K_I^p, MPa$\sqrt{m}$
Mg	2,14	4,28	4,22	0,33	10,0	14,1
Al	2,99	5,98	7,06	0,32	15,3	21,6
Ti	2,81	5,62	11,0	0,31	18,5	26,2
Fe	4,21	8,42	20,6	0,28	30,6	43,3
Ni	4,56	9,12	21,0	0,31	32,5	46,0

The role of new estimation criteria of the materials carrying capacity and articles of them is the task for future analysis, although their fundamental importance is exemplified in quite a number of cases:
1. Under certain conditions the fatigue crack propagation process assumes discontinuous intermittent character (Forsyth, 1976; Yasny, 1981). New constant fractures make possible to estimate the loading modes resulting in such anomaly.

2. If we represent the dependence of the fatigue fracture crack growth rate v from the maximum stress intensity factor K_{Imax} of cycle in the second stage of the fatigue fracture diagramme as $v = B\left(K_{Im\,ax} / K^{*}_{Im\,ax}\right)^{n}$, where B and n - material's constants, then value K_{Imax} for the single - base alloys appears to be exactly equal $K^{0}_{I}(K_{Im\,ax} = K^{0}_{I})$.

3. Cyclic loading fracture in quaisi-brittle materials (e.g. steels after hardening and low tempering) occurs close to K^{0}_{I} or K^{p}_{I} ($K_{fc} \approx K^{0}_{I}$ or $K_{fc} \approx K^{p}_{I}$).

<u>Minimum Fatigue Striation Spacing, s_{e}</u>

The hypothesis that for metallic materials the striation spacing is multiple wave halflength of high frequency electromagnetic modes occurring with the main crack propagation has been put forward. On the basis of the model assumed minimum striation spacing s_{e} for the metals with known values for plasma frequency ω_{p} (Ashok, 1980) was obtained. In particular calculated values s_{e} for Al, Ti and Fe are correspondingly equal to $4.05 \cdot 10^{-8}$ m, $3.52 \cdot 10^{-4}$ m and $3.92 \cdot 10^{-8}$ m. The analysis of experimental data available showed that minimum striation spacing s_{e} for alloys comes to be close to the value s_{e} for the alloy base.

Striation spacing increases with crack propagation. Therefore, it can be assumed that increasing striation spacing s remains multiple minimum spacing s_{e}: $s = Ns_{e}$, where N=1, 2, 3, ... As is evident from the data cited in Fig.1, this relation really takes place. Where horizontal lines - estimated fatigue striation spacing, (——— - for Al, ------- - for Fe), figures - experimental values of the striation spacing according to different authors.

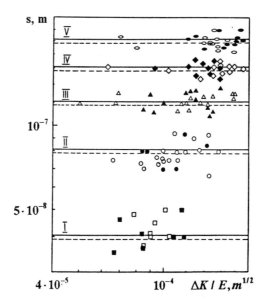

Fig.1. Estimated and experimental values of the fatigue striation
spacing. ■,●,▲,♦,● - aluminium alloys, □,○,▲,◇,○ - steels.

Effective Fatigue Threshold, $\Delta K_{eff,th}$

According to break mechanizm, with fracture by plain strain being independent of loading conditons, the relation between the stresses σ_y perpendicular to crack propagation and the distance r from the crack edge is expressed by the following equation in accordance with the theory of linear elastic fracture mechanics:

$$\sigma_y = \frac{K}{\sqrt{2\pi r}} \tag{5}$$

It was accepted (Ragozin, 1992) that the crack starts its propogation close to the effective threshold stress intensity factor $K_{eff,th}(\Delta K_{eff,th})$ at the moment when in the zone of plastic strain of r_e size, at the crack apex, the energy equal to one of the threshold energy levels W_i calculated on the basis of phonon theory of fracture is absorbed (Ragozin, 1990). For the samples of final dimensions we have (Ragozin, 1991):

$$K_{Ic} = Y\sigma_p\sqrt{\pi l_p} = \sqrt{\frac{0,8LW_iE}{1-\mu^2}} , \tag{6}$$

where σ_p - critical stress with which the crack of l_p length starts its propagation; L - constant having a length dimension and equal to 10^{-3} m; Y - calibration coefficient. Assuming the equation (6) is valid in the scope of microvolumes (Esterling, 1981) we transform it on the basis of some relations derived earlier. Setting up $Y\sigma_p = \sigma_D$ and $l_p = l_i$, we obtain:

$$\sigma_D(\Delta\sigma_D) = \sqrt{\frac{0,8LW_iE}{(1-\mu^2)\pi l_i}} \tag{7}$$

Then the condition of the propagation start can be written as

$$\Delta K_{eff,th} = \Delta\sigma_D\sqrt{2\pi r_e} = \sqrt{\frac{1,6LW_iEr_e}{(1-\mu^2)l_i}} \tag{8}$$

Identifying the value r_e of equation (8) with the minimum spacing of fatigue striations s_e and assuming in accordance with the hypothesis accepted above $s_e = \lambda_p/2$ (λ_p - wave length of plasmon), we obtain the equation for calculating the criterion $\Delta K_{eff,th}$:

$$\Delta K_{eff,th} = 0,894\sqrt{\frac{LW_iE\lambda_p}{(1-\mu^2)l_i}} \tag{9}$$

or in the following form:

$$\Delta K_{eff,th} = K_{Ic}\sqrt{\frac{\lambda_p}{l_i}} \tag{10}$$

Since alloying and thermal treating change the constants W_i, E and, apparently, λ_p only to a small extent, the relations (9) and (10) obtained for pure metals are as well valid for alloys based on them. Their own descrete spectrum of threshold energy levels W_i is typical for each metal and alloys based on them (Ragozin, 1990). Hence, in accordance with equation (9) the descrete spectrum of values $\Delta K_{eff,th}$ is typical for them as well. For a number of metals (and alloys based on them) important for industry the levels $\Delta K_{eff,th}$ calculated by equation (9) are presented in Table 3. Analysis has proved good correlation of the estimated (Table 3) and experimental values $\Delta K_{eff,th}$ obtained by different authors. From the equation (8) and (9) it seems to be easy to derive the dependence $s = f(\Delta K_{eff})$ (Ragozin, 1992) that can be used for developing the fractographic method of expert estimation of pieces intensity level during their fracture in the operation process.

Table 3. Calculated levels of effective threshold stress intensity ranges
in *Mg, Al, Ti, Fe* and *Ni* at room temperature.

Metal	E,	μ	$\Delta K_{eff,th}$, MPa $\sqrt{m}$										
	10^4 MPa		1	2	3	4	5	6	7	8	9	10	11
Mg	4,22	0,33	0,59	0,73	0,85	0,88	1,13	1,19	1,61	2,27			
Al	7,06	0,32	0,42	0,85	0,91	1,26	1,83						
Ti	11,0	0,31	0,97	1,19	1,24	1,81	1,93	2,56	3,60				
Fe	20,6	0,28	1,94	2,91	3,21	4,25	4,52	7,28	9,25				
Ni	21,0	0,31	0,94	1,88	1,94	2,90	3,89	5,20[*]	5,82[*]	6,37[*]	8,20[*]	8,22[*]	13,18[*]

[*] with package defects

REFERENCES

Ashok, K.V. (1980) A chemical approach to the estimation of surface plasmon energies of some metals. *Surface Technol.*, No.10, 277-282.

El Haddad, M.H., T.H. Topper and K.N.Smith (1979). Prediction of non propagating crack. *Eng. Fract. Mech.*, **11**, No.2, 573-584.

Esterling, D.M. (1981). Equivalence of macroscopic and microscopic Griffith conditions for subcritical crack growth. *Int. J. Fract.*, **17**, No.3, 321-325.

Forsyth, P.J.E. (1976). Some observations and measurements on mixed fatigue/tensile crack growth in aluminium alloys. *Scripta met.*, **10**, 383-386.

Khatiashvili, N.G. and M.E. Perelman (1982). Generation of the electromagnetic radiation during the passage of accoustic waves through crystalline dielectrics and some rocks. *Reports of AS USSR*, **263**, No.4, 839-842.

Lukas, P., L.Kunz, B.Weiss and R. Stickler (1986). Non-damaging notches in fatigue. *Fatigue Engng. Mater. Struct.*, **9**, No.3, 195-204.

Miller, K.J. (1992). The effects of environment on short fatigue crack growth. In: *Mechanical Fatigue of Metals: Proc. 11th Int. Colloq.*, (V.T. Troshchenko, Ed.), Kiev, Vol.2, 207-217.

Misra, A. (1975). Electromagnetic effects at metallic fracture. *Nature*, **254**, No.5496, 133-134.

Ragozin, Yu.I. (1989). The quantization effect of mechanical energy absorbed by metals under deformation and fracture. In: *Proc. 7th Int. Conf. Fract. (ICE7)*, (K. Salama, K. Ravi-Chandar, D.M.R. Taplin and P. Rama Rao, Eds.), Houston, Vol.6, 3731-3738.

Ragozin, Yu.I. (1990). Phonon conception of metal and alloy fracture. In: *Proc. 8th Eur. Conf. Fract. (ECF8)*, (D.Firrao, Ed.), Vol. 2, 1150-1156.

Ragozin, Yu.I. and Yu.Ya.Antonov (1991). Predicted values of static fracture toughness (K_{Ic}) for metallic alloys.In: *Mechanical Properties / Materials Desing: Proc. C-MRS International '90*, (B.Wu, Ed.), Beijing, Vol.5, 141-144. Elsevier, Amsterdam, Netherlands.

Ragozin, Yu.I. (1992). Nature of fatigue striation and effective threshold. In: *Prepr. 9th Eur. Conf. Fract. (ECF9)*, (S. Sedmak, A. Sedmak and D. Ruzic, Eds.), Varna, Vol.1, 427-432.

Yasny, P.V. (1981). Study of unstable propogation and inhibition of cracks under cyclic loading. *Problems of strength*, No.11, 31-36.

Prediction of Fatigue Thresholds of Notched Components based on Resistance-Curve Method

Yoshiaki AKINIWA, Keisuke TANAKA and Luoming ZHANG

Department of Mechanical Engineering, Nagoya University,
Furo-cho, Chikusa-ku, Nagoya 464-01, Japan

ABSTRACT

Single-edge-notched specimens of a structural carbon steel (JIS S45C) and a low-alloy steel (JIS SNCM439) were fatigued under axial loading. The propagation behavior of short fatigue cracks emanating from a sharp notch was investigated from a viewpoint of crack closure. The fatigue thresholds of notched components were predicted by a resistance-curve method. The resistance-curve was constructed in terms of the threshold value of the maximum stress intensity factor which was the sum of the threshold effective stress intensity range and the opening stress intensity factor. The predicted values of the fatigue limit of crack initiation, the fatigue limit of fracture, and the non-propagating crack length agreed well with the experimental values. Finally a simplified method to determine the resistance-curve was proposed for design applications.

KEYWORDS

Fatigue; Short Crack; Notch; Fatigue Limit; Resistance Curve; Crack Closure

INTRODUCTION

Fatigue fracture of engineering components usually originates at notches or other stress concentrations. The propagation rate of a short fatigue crack nucleated at the root of a sharp notch first decreases as the crack propagates. Under low stresses, the crack finally becomes non-propagating. On the other hand, the crack accelerates after taking a minimum propagation rate and causes a final fracture under high stresses. This anomalous propagation behavior can not be explained on the bases of ΔK-based conventional fracture mechanics. An increasing number of studies show that the development of crack closure with crack growth plays a key role in the propagation and non-propagation of short fatigue cracks at notches (Tanaka and Nakai, 1983; Nishikawa, et al., 1986; Akiniwa and Tanaka, 1987). In most cases of notch fatigue, the propagation behavior of short fatigue cracks is predictable from the relation between the crack propagation rate and the effective stress intensity factor (SIF)

449

range.   In our previous study (Tanaka and Akiniwa, 1988),   a resistance-curve (R-curve) method was proposed to predict the fatigue limit of crack initiation,   the fatigue limit of fracture and the non-propagating crack length in notched components. The method was also extended to explain the effect of mean stresses (Akiniwa et al., 1990a) and the microstructure(Akiniwa et al., 1989; Akiniwa and Tanaka, 1990b).   However, the validity of the method for high strength steels was not clear.

In the present study,   single-edge-notched specimens of a structural carbon steel and a low-alloy steel were fatigued under axial loading.   The propagation and non-propagation behavior of short fatigue cracks at the notch root was observed.   The closure behavior of short cracks was measured as a function of crack length.   The experimental results of fatigue limits of notched specimen were compared with the model prediction.   Finally a simplified method to determine the resistance-curve was proposed.

EXPERIMENTAL PROCEDURE

Material and Specimen

The materials used in this study were a structural carbon steel (JIS S45C) and a low-alloy steel (JIS SNCM439). Chemical compositions and mechanical properties are shown in Table 1 and 2, respectively.   Specimens of S45C were annealed at 850°C for 1hr in vacuum and cooled in the furnace and those of SNCM439 were tempered at 600°C after oil quenching.   An average ferrite grain size of S45C is 17 μ m and a prior austenite grain size of SNCM439 is 12 μ m. The specimen is a single-edge-notched plate as shown in Fig. 1.   The notch was introduced by an electro-discharge machine.   The surface of all specimens was polished by emery paper, and was finally finished by electropolishing.

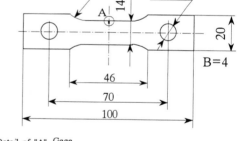

Fatigue Tests

Single-edge-notched specimens were subjected to cyclic axial tension-compression at constant stress amplitude in a computer-controlled servo-hydraulic testing machine.   The specimen was gripped by a ram with wood's metal chuck to obtain good alignment.   The stress ratio was R=−1. The frequency of stress cycling was 15Hz. The crack length was measured by traveling microscope.   The

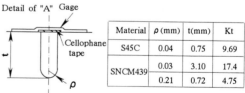

Material	ρ (mm)	t(mm)	Kt
S45C	0.04	0.75	9.69
SNCM439	0.03	3.10	17.4
	0.21	0.72	4.75

Fig. 1. Specimen.

Table 1. Chemical compositions (mass%).

	C	Si	Mn	P	S	Cu	Ni	Cr	Mo
S45C	0.43	0.19	0.81	0.022	0.02	0.01	0.02	0.14	<0.01
SNCM439	0.38	0.27	0.76	0.022	0.008	0.05	1.93	0.95	0.25

Table 2. Mechanical properties.

Material	Yield stress σY(MPa)	Tensile strength σB(MPa)	Vickers hardness Hv	Fatigue limit σwo(MPa)
S45C	316	570	124	223
SNCM439 (T.T.600°C)	970	1047	312	490
SNCM439 (T.T.400°C)	1350	1482	444	700
SNCM439 (T.T.200°C)	1510	1894	496	860

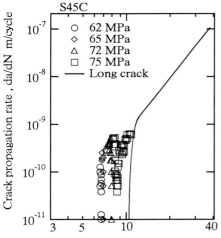

Fig. 2. Relation between crack propagation
rate and stress intensity range.

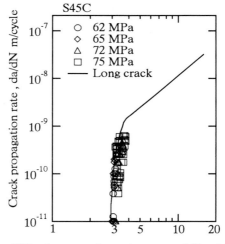

Fig. 4. Relation between crack propagation
rate and effective stress intensity range.

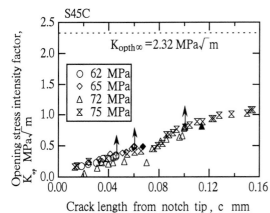

Fig. 3. Change of crack opening stress intensity
factor with crack length.

specimens were broken at liquid nitrogen
temperature after fatigue tests. The length
of a non-propagating crack was measured
with a scanning electron microscope. The
crack opening point of a short fatigue crack
was measured by the compliance method.
The displacement parameter adopted was
the output of the strain gage glued on the
front or back face of the notch.

EXPERIMENTAL RESULTS

Crack propagation behavior

Figure 2 shows the relation between crack
propagation rate, dc/dN, and SIF rage, Δ
K, obtained for S45C. The solid line in the

figure indicates the relation obtained for long cracks. When the stress amplitude was less than 72MPa, the crack propagation rate decreases as the crack propagates. The crack then becomes non-propagating. For the case of 75MPa, the crack accelerate after taking a minimum propagation rate.

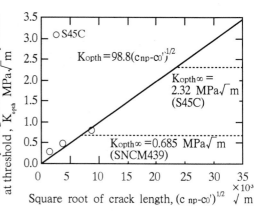

Fig. 5. Realation between crack opening SIF and crack length.

Crack closure behavior

Figure 3 shows the change of the crack opening SIF, K_{op}, with the crack length. The broken line in the figure indicates the threshold value of long cracks. The value of K_{op} obtained for short cracks increases with increasing the crack length. The points with arrows were obtained from non-propagating cracks. Figure 4 shows the relation between dc/dN and effective SIF range ΔK_{eff}. A unique relation can be seen, and it agrees well with the relation for long cracks.

DISCUSSION

Resistance-curve

A resistance-curve method was proposed in our previous study (Tanaka and Akiniwa, 1988). The resistance-curve was constructed in terms of the threshold value of the maximum SIF which was the sum of the threshold effective SIF range and the crack opening SIF. The threshold condition is given by

$$K_{maxth} = K_{opth} + \Delta K_{effth} \tag{1}$$

where K_{opth} is the crack opening SIF at threshold and ΔK_{effth} is the effective component of the SIF range at threshold. ΔK_{effth} is constant irrespective of stress amplitude and crack length, and equal to the value for long cracks, $\Delta K_{effth\infty}$. In our previous study (1990b), K_{opth} (MPa$\sqrt{m}$) was given by

$$K_{opth} = 98.8 \, [c_{np} - c_0']^{1/2} \qquad c_0' \leqq c_{np} \leqq c_2 \tag{2a}$$
$$K_{opth} = K_{opth\infty} \qquad c_2 \leqq c_{np} \tag{2b}$$

where cnp is the non-propagating crack length and c_0' (m) is the characteristic crack length calculated by

$$c_0' = (\Delta K_{effth\infty}/1.122 \, \sigma_{w0})^2/\pi \tag{3}$$

where σ_{w0} is the fatigue limit of smooth specimen. The material parameters used in this method are ΔK_{effth}, $K_{opth\infty}$ and σ_{w0}.

Evaluation of model

Figure 5 shows the relation between K_{opth} and $(c_{np} - c_0')^{1/2}$ obtained for S45C. The solid and broken lines in the figure indicate eq.(2a) and eq.(2b), respectively. The results obtained for S45C can be approximated by eq.(2a). In this study, eqs.(2a) and (2b) are also adopted for

Table 3. Fatigue limits.

Material	Threshold stress intensity range		Fatigue limit	Characteristic crack lengths		Fatigue limit for crack initiation σ_{w1} (MPa)		Fatigue limit for fracture σ_{w2} (MPa)	
	$K_{maxth\infty}$ (MPa$\sqrt{m}$)	$\Delta K_{effth\infty}$ (MPa$\sqrt{m}$)	σ_{w0} (MPa)	c'_0 (mm)	c_2 (mm)	Prediction	Experiment	Prediction	Experiment
S45C $\rho=0.04$ $t=0.75$	5.26	2.94	223	0.044	0.594	55 (56)	60~62	75 (75)	72~75
SNCM439 $\rho=0.03$ $t=3.10$	3.69	3.00	490	0.010	0.058	41 (41)	40~45	41 (41)	40~45
SNCM439 $\rho=0.21$ $t=0.72$	3.69	3.00	490	0.010	0.058	113 (113)	115~120	113 (113)	115~120

( ) : Calculated by Hv ($\Delta K_{effth\infty}=3$MPa$\sqrt{m}$).

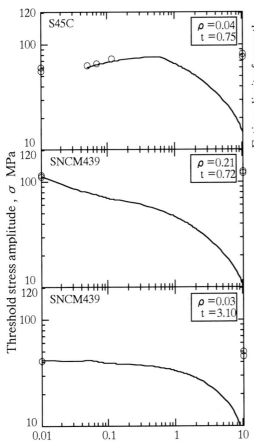

Fig. 6. Limitting curve for non-propagation of fatigue cracks.

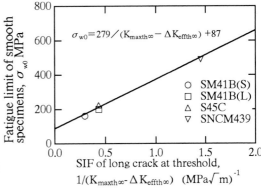

$$\sigma_{w0}=279/(K_{maxth\infty}-\Delta K_{effth\infty})+87$$

○ SM41B(S)
□ SM41B(L)
△ S45C
▽ SNCM439

SIF of long crack at threshold,
$1/(K_{maxth\infty}-\Delta K_{effth\infty})$ (MPa$\sqrt{m}$)$^{-1}$

Fig. 7. Relation between fatigue limit of smooth specimen and threshold values for long cracks.

SNCM439. The relation between the threshold stress amplitude and the non-propagating crack length is shown in Fig. 6. The solid line in the figure indicates the relation predicted by the resistance curve method. The marks on the ordinate on the left and right sides mean no crack initiation and final fracture, respectively. The predicted value of non-propagating crack length and fatigue limits agree very well with the experimental results.

Simplified method

The material parameters used in this method are ΔK_{effth} (MPa$\sqrt{m}$), $K_{opth\infty}$ (MPa$\sqrt{m}$) and σ_{w0} (MPa). By the way, there are relations between σ_{w0} and the hardness, Hv, and between σ_{w0} and the tensile strength, σ_B as follows.

$$\sigma_{w0} = 1.6Hv \tag{4}$$
$$\sigma_{w0} = 0.5\sigma_B \tag{5}$$

If σ_{w0} is unknown, the resistance-curve can be estimated by eqs.(4) and (5) through eq.(3). Yokomaku et al. (1991) reported the following equation for the relation between $K_{maxth\infty}$ and grain size, d.

$$K_{maxth\infty} = \Delta K_{effth\infty} + \alpha\sqrt{d} \tag{6}$$

σ_{w0} has Petch-type relation.

$$\sigma_{w0} = \sigma_0 + \beta/\sqrt{d} \tag{7}$$

From eqs.(6) and (7), we obtain the following relation.

$$\sigma_{w0} = A/(K_{maxth\infty} - \Delta K_{effth\infty}) + B \tag{8}$$

Figure 7 shows the relation between σ_{w0} and $(K_{maxth\infty} - \Delta K_{effth\infty})^{-1}$. $A=279(MPa2\sqrt{m})$ and $B=87(MPa)$ was determined through a least square regression method. By using eq.(8), the resistance-curve is determined from two material parameters. $\Delta K_{effth\infty}$ is generally 2 to 3MPa $\sqrt{m}$ for steel. If $\Delta K_{effth\infty}$ is assumed 3MPa$\sqrt{m}$, the resistance-curve can be roughly estimated from only one material parameter e.g. Hv. The fatigue limits obtained by this simplified method are parenthesized in Table 3. The predicted fatigue limits also agree well with experimental results.

CONCLUSIONS

The resistance-curve method is applied to high strength steels. The fatigue limits and non-propagating crack length predicted by the method are agree very well with experimental results. A simplified method to determine the resistance-curve using hardness or the tensile strength was proposed.

REFERENCES

Akiniwa, Y. and K. Tanaka (1987). Notch-root-radius effect on propagation of short fatigue cracks in notched specimens of low-carbon steel. Trans. Japan Soc. Mech. Engrs, 53, 393-400

Akiniwa, Y., K. Tanaka and M. Kinefuchi (1989). Effect of microstructure on propagation and non-propagation of short fatigue cracks at notches. J. Soc. Mater. Sci. Japan, 38, 1275-1281

Akiniwa, Y., K. Tanaka and N. Taniguchi (1990a). Prediction of propagation and nonpropagation of short fatigue cracks at notches under mean stresses. JSME Int. J., Ser. I, 33, 288-296

Akiniwa, Y.and K. Tanaka (1990b). Microstructural effect on propagation of short fatigue cracks in notched components. Proc. Int. Conf. Fatigue and Fatigue thresholds, II,1121-1126

Nishikawa, I., M. Konishi, Y. Miyoshi and K. Ogura (1986). Small fatigue crack growth at notch root in elastic-plastic range. J. Soc. Mater. Sci. Japan, 35, 904-910

Tanaka, K. and Y. Nakai (1983). Propagation and non-propagation of short fatigue cracks at a sharp notch. Fatigue Fract. Engng Mater. Struct., 6, 315-327

Tanaka, K. and Y. Akiniwa (1988). Resistance-curve method for predicting propagation threshold of short fatigue cracks at notches. Eng. Fract. Mech., 30, 863-876

Yokomaku, T., M. Kinefuchi, and Y. Minokata (1991). Effect of microstructure on fatigue properties in low and ultra low carbon steels. J. Soc. Mater. Sci. Japan, 40, 1415-1420

Influence of Loading Parameters on the ΔK-Threshold Value for Old Steels in Riveted Bridges and for Modern Steels

S. Han[*], W. Dahl[*], P. Langenberg[*], G. Sedlacek[**] and G. Stötzel[**]

[*] Institute of Ferrous Metallurgy, RWTH Aachen, Germany
[**] Institute of Steel Construction, RWTH Aachen, Germany

ABSTRACT

Residual safety and life time prediction are important questions in the course of planning and rehabilitation of steel structures. If, for any reason, cracks have to be assumed due to fabrication failures or fatigue loading in service, the life prediction can be carried out with means of cyclic fracture mechanics using the well known Paris-Erdogan equation. The crack propagation under fatigue loading is strongly connected to the load level. If the load remains below a certain level, no crack propagation will happen. The stress intensity factor calculated for this level is called ΔK-threshold and can be determined experimentally. Results from three different experimental test procedures to measure fatigue crack growth (constant-load-amplitude, R-constant and K_{max}-constant method) for old steels used in riveted bridges (puddled and wrought iron) and for modern steels up to a yield strength 355MPa will be presented. The ΔK-threshold from the R-constant method decreased with increasing R-ratio in region of crack closure. The $K_{max,th}$ corresponding to the ΔK-threshold in this region did not show a constant lower limit but increased linear with increasing R-ratio. The K_{max}-constant method represented a simple and reliable experimental determination of ΔK-threshold with a high K_{max} value. From considering relationship of reversed plastic zone size and grain size the ΔK-threshold in region of crack closure was calculated theoretically. It has shown good agreement with the experimentally determined ΔK-threshold.

KEYWORDS

Residual life time prediction; fracture mechanics; crack propagation rate; ΔK-threshold; region of crack closure; reversed plastic zone size; grain size

INTRODUCTION

A lot of existing steel bridges have been built in the last century. Residual safety of such riveted structures is an important question for the bridge management, because
- the planned service life is often exceeded by far,
- loading conditions have changed due to more traffic with higher loads,
- fatigue cracks might have occurred that are hidden under the heads of rivets or other structural details as connecting angles and therefore cannot be detected during bridge inspection,
- material properties, especially toughness values, of old steel might have changed to lower quality due to ageing processes.

To answer these questions, a method has been developed and applied successfully for several bridges, which combines the fracture mechanic based concept for the calculation of the load bearing capacity under static

455

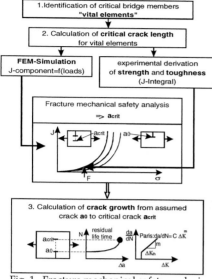

Fig. 1. Fracture mechanical safety analysis

loading and the calculation of the residual life time under cyclic loading (Fig. 1) (Sedlacek *et al.*, 1992). Life time prediction is based on the well known Paris-Erdogan equation

$$da / dN = C \cdot \Delta K^m. \qquad (1)$$

So far the crack growth rate has been calculated in practice on the basis of different constant loading amplitudes $\Delta\sigma$ which were derived from equivalent fatigue loads representing the various loading amplitudes under service conditions. The constants C and m have been estimated conservatively as 5% fractiles. As small loading amplitudes do not contribute to crack growth, the knowledge of threshold values of the stress intensity factor ΔK would help to exclude small loading amplitudes and hence improve the life time prediction.

In this paper both the derivation of the constants C and m and the ΔK-threshold value for old steels (puddled and wrought iron) were presented. In addition structural steels for modern bridges up to 355MPa yield strength were examined in the same way. A mathematical description of the ΔK-threshold value depending on the R-ratio and the grain size has been developed.

MATERIALS AND EXPERIMENTAL PROCEDURE

The steels included in this study were chosen on the basis that they are widely used in old riveted bridge constructions and for modern steel bridges:
i.   A 11.5mm thick angle of puddled iron. This steel was manufactured from the Puddle process so that slag as impurity remains in a inhomogeneous micro structure.
ii. A 18mm thick plate of wrought iron from a main girder of the Hammer bridge. It was manufactured by the Thomas process.
iii. A 13mm thick plate of normalised St 37 (900°C/10min).
iv. A 15mm thick plate of St 52-3 as rolled roughed slab.
v.   A 15mm thick plate of StE 355. This plate material was manufactured by thermo mechanical rolling. A banded structure in rolling direction was observed.

	C	Mn	Si	P	S	Cr	Al	Ni	Mo	Cu
Puddled iron	0.012	0.118	0.061	0.290	0.017	0.004	0.002	-	-	-
Wrought iron	0.051	0.431	0.002	0.040	0.036	0.008	0.002	-	-	-
St 37	0.084	0.658	0.169	0.020	0.009	0.009	0.042	0.035	0.002	0.068
St 52-3	0.222	1.580	0.251	0.022	0.028	0.047	0.039	0.040	0.013	0.015
StE 355	0.065	1.430	0.310	0.013	0.003	0.028	0.024	0.127	0.002	0.116

Table 1. Chemical composition (wt%) of tested steels

	d [μm]	σ_y [MPa]	σ_u [MPa]	A_5 [%]	Z [%]	T_{27J} [°C]
Puddled iron	44	254	354	24	34	20
Wrought iron	27	255	370	37	67	20
St 37	19	277	415	43	74	-42
St 52-3	12	362	566	34	70	-41
StE 355	4.8	435	530	34	74	-119

Table 2. Grain sizes and mechanical properties of tested steels

Table 1 shows the chemical composition of tested steels. The tensile and charpy tests were performed with specimens taken in the longitudinal direction. Grain sizes and characteristic values for tensile and charpy tests are listed in Table 2.

For measuring the fatigue crack propagation, the materials were machined into 12.5mm thick compact tension (CT) specimens with W=50mm following ASTM E647 (1989). For the reason of too thin angles of the puddled iron, the thickness of these CT-specimens was chosen as 8.9mm. The CT specimens were taken from the angles and the steel plates in the L-T orientation.

Tests were carried out on a servo hydraulic PC-monitored test machine. The crack growth was measured with direct current potential drop technique with a sensitivity to changes in crack length of less than 0.01mm. The actual crack length during the experiments was automatically monitored by means of a experimentally derived unique calibration function which is independent of chemical composition and micro structure of each material. The test frequency was 60 Hz.

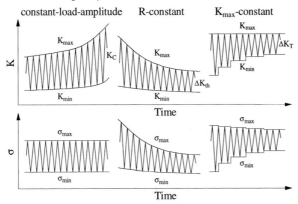

Fig. 2. Test methods to measure fatigue crack growth

To derive the constants C and m the constant-load-amplitude test procedure from ASTM E647 was carried out. Each specimen was tested at a constant load range and a fixed set of loading variables. However, application became increasingly difficult as crack growth rates decrease below 10^{-5}mm / cycle.

In order to reach the ΔK-threshold two test methods with decreasing ΔK have been carried out in laboratory atmosphere: The first one is the R $(R = K_{min} / K_{max})$-constant method in accordance with ASTM E647. Loads were decreased continuously as the crack grows. The rate of load shedding $(C_R = 1 / K \cdot dK / da)$ was chosen as -0.08mm^{-1}. The R-ratio and the rate of load shedding C_R was maintained constant. During K-decreasing the relationships between ΔK, a and C_R are given as $\Delta K = \Delta K_o \cdot exp[C_R \cdot (a - a_o)]$, where ΔK_o is the initial ΔK and a_o is the corresponding crack length. The second one, the K_{max}-constant method, maintains a constant K_{max} while K_{min} is gradually increased. The K_{max}-constant method with decreasing ΔK could be used to measure ΔK-threshold values with a high K_{max} value which helps to avoid crack closure effects in the near threshold region. For this method the choice of increasing rate of K_{min} is not critical according to sequence effects, because the monotone plastic zone size maintains constant (Castro, 1989). In the present paper the rate of load shedding C_K $(\Delta K = C_K \cdot \Delta K_o)$ was chosen as 0.98 per each crack growth of 0.1mm.

In Fig. 2 the three test procedures to measure fatigue crack growth are shown schematically.

RESULTS AND DISCUSSION

Constant-load-amplitude test

With the constant-load-amplitude test the crack growth rate da/dN in the region (da/dN > 10^{-5}mm / cycle) where the Paris-Erdogan equation applies is shown as function of the ΔK in Fig. 3 for all steels.

In a comparison of both old steel types with the results of Wittemann (1993) given as a scatter band of ΔK for puddled iron (R=0.1), the current results show a steeper slope, whereas a good agreement of the measured results of wrought iron with the fatigue crack growth curve with R=0.1 (Wittemann, 1993) was observed. It might be expected that the fatigue crack growth behaviour at a low crack growth rate (da/dN<10^{-5}mm/cycle) is dependent on the micro structure. Inhomogeneous micro structures and coarse

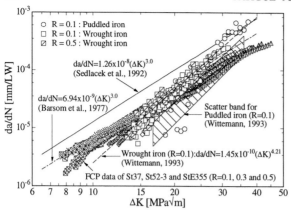

Fig. 3. Fatigue crack growth rates for tested steels

grains of puddled iron cause in this region a slower fatigue crack growth than that for wrought iron. However in the region of da/dN above $5×10^{-4}$mm/cycle the crack growth rates of puddled iron show higher values since the K_{max} of fatigue loading should approach the fracture toughness K_{IC} earlier than that of wrought iron. As might be expected, the micro structure causes the different behaviour in fatigue crack growth for both old steels even though they have the same strength level.

The comparison of the old steels with the three modern steels revealed a different crack growth behaviour with a steeper slope for the old steels. The overlapping of the results for the tested modern steels at R-ratios of 0.1, 0.3 and 0.5 in this diagram confirms independence of R-ratio. In a region of da/dN below $5×10^{-6}$mm / cycle the measured da/dN scattered strongly, because they approach the near threshold region. In addition, this results are compared with fatigue crack growth curves for the safe side assumption to predict fatigue life time from Sedlacek *et al.* (1992) and for general ferritic-pearlitic steels from Barsom *et al.* (1977). The results for the modern steels are estimated conservatively in view of fatigue life evaluation.

R-constant method

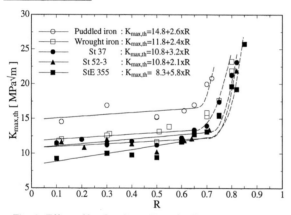

Fig. 4. Effect of load ratio on $K_{max,th}$ by R-constant method

The limiting condition for taking the measured ΔK that has been decreased continuously with constant R-ratio as ΔK-threshold level was chosen as $da / dN ≤ 10^{-7}$ mm / cycle. It was observed that the ΔK-threshold decreases with R-ratio in the region of crack closure until to ca. 0.7 and maintains independent for higher R-ratios.

Fig. 4 shows $K_{max,th}$ corresponding to K_{max} of each ΔK-threshold over the whole R-ratio range. Results from Schmidt and Paris (1973) showed that the $K_{max,th}$ maintains a constant value in the region of crack closure. However, for the examined five steels $K_{max,th}$ has not shown the such lower limit. It increased linear in the region of crack closure (R ≤ 0.7). The linear relationship between $K_{max,th}$ and R-ratio can be described with

$$K_{max,th} = K_{max(R=0)} + λ·R, \qquad (2)$$

where $K_{max(R=0)}$ and λ were derived experimentally. According to Fig. 4 the $K_{max(R=0)}$ varies from 14.8 to 8.3MPa√m, and the slope λ for puddled iron, wrought iron, St 37 and St 52-3 is given as a mean value of 2.6 with small scatter. In case of StE 355 a higher λ of 5.8 is obtained, that might be explained with the typical banded micro structure from the thermo mechanical rolling process.

From considering the linear increasing $K_{max,th}$ with increasing R-ratio in the region of crack closure the ΔK-threshold over R-ratio can be derived as follows:

$$ΔK_{th} = (K_{max(R=0)} + λ·R)·(1-R). \qquad (3)$$

A similar relationship of the ΔK-threshold over R-ratio Döker (1993) has carried out mathematically to represent the measured ΔK-threshold for austenite steel 1.4909.

Kmax-constant method

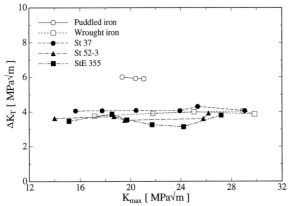

Fig. 5. ΔK-threshold by K_{max}-constant method

Calculation of ΔK-threshold considering the grain size

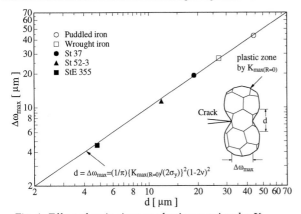

Fig. 6. Effect of grain sizes on plastic zone sizes by $K_{max(R=0)}$

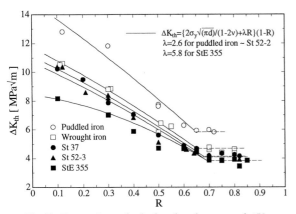

Fig. 7. Comparison of calculated and measured ΔK_{th} in region of crack closure

Again da/dN≤10⁻⁷ mm/cycle was taken as limiting condition for ΔK to be taken as ΔK_T without effects of crack closure. The experimental results of the ΔK_T for tested steels in the K_{max}-range from 14 to 30 MPa√m are shown in Fig. 5. An influence of K_{max} on ΔK_T is not detectable. In case of puddled iron, the mean value of $\Delta K_T = 5.9$ MPa√m is higher than for the other steels. ΔK_T for the other steels are in a same level with the following mean values: 3.9 for wrought iron, 4.1 for St 37, 3.7 for St 52-3 and 3.5 MPa√m for StE 355.

The influence of R-ratio on ΔK-threshold is associated with crack closure phenomena arising from plasticity-induced closure and closure due to crack face oxide debris. At very low crack growth rate (da/dN<10⁻⁶ mm/cycle), approaching the ΔK-threshold, significant closure effects occur in plane strain (Suresh *et al.*, 1983). Then in near threshold region the reversed plastic zone size $\Delta \omega_{max}$ at R=0 ($K_{max(R=0)} = \Delta K_{th}$) is given by

$$\Delta \omega_{max} = \frac{1}{\pi} \cdot \left(\frac{K_{max(R=0)}}{2 \cdot \sigma_y} \right)^2 \cdot (1 - 2 \cdot v)^2, \qquad (4)$$

where v is Poisson's ratio. In Fig. 6 the calculated reversed plastic zone sizes $\Delta \omega_{max}$ by means of $K_{max(R=0)}$ from Eq. (2) are compared with grain sizes d of tested steels. The agreement is excellent so that the plastic zone size is equal to the grain size ($\Delta \omega_{max} = d$) and by Eq. (2) and (4) $K_{max(R=0)}$ and $K_{max,th}$ can be calculated theoretically from this relationship.

In the present work using Eq. (3) and (4) the theoretical ΔK-threshold for the tested steels submitted to a fatigue stress characterised by R-ratio can be derived with the following general equation:

$$\Delta K_{th} = \left(\frac{2 \cdot \sigma_y \cdot \sqrt{\pi \cdot d}}{1 - 2 \cdot v} + \lambda \cdot R \right) \cdot (1 - R). \qquad (5)$$

In Fig. 7 the calculated ΔK_{th}-R curves from Eq. (5) are compared with the measured ΔK_{th} in the region of crack closure. For the slope λ of the ΔK_{th}-R curves of puddled iron, wrought iron, St 37 and St 52-3 was obtained a mean value of 2.6. In the case of StE 355 as a higher value of 5.8 has been derived.

The results obtained with Eq. (5) show good agreement with the measured ΔK_{th}. Thus, when the material properties σ_y, d and the material constant λ of ΔK_{th}-R curve are given, the ΔK-threshold for various R-ratios in the region of crack closure can be calculated theoretically.

CONCLUSIONS

In the safety assessment of old steel bridges or in the robustness assessment of new steel structures according to Sedlacek *et al.* (1992) the fatigue crack growth plays an important role for the life time prediction. So far a safe sided assumption for the constants C and m of the Paris-Erdogan equation has been used for calculating the fatigue crack growth, e.g. fatigue equivalent loads with constant loading amplitudes have been used and ΔK-threshold values have been neglected consequently. This paper gives more accurate material data concerning the theoretically calculated ΔK-threshold for the old steels (puddled and wrought iron) and the modern steels (St 37, St 52-3 and StE 355) thus allowing to perform more accurate calculation of the fatigue crack growth when using measured or simulated stress-time histories.

In detail the results obtained from experiments and theoretical analyses are as follows:

(1)   The comparison of the both old steels with the three modern steels from the constant-load-amplitude test revealed a different crack growth behaviour with a steeper slope for the old steels, since at a low crack growth rate coarse grains of the old steels cause a slow fatigue crack growth and on the other hand in a high da/dN-range the old steels show by means of low fracture toughness higher fatigue crack growth rates than that for the modern steels.

(2)   The $K_{max,th}$ by R-constant method with decreasing ΔK has not shown a lower limit but increased linear with increasing R-ratio in region of crack closure.

(3)   An influence of K_{max} on the measured ΔK_T in accordance with K_{max}-constant method was not detectable.

(4)   Calculated ΔK_{th} considering the reversed plastic zone size $\Delta \omega_{max}$ and grain size d have shown good agreement with the measured ΔK_{th} in region of crack closure.

ACKNOWLEDGEMENTS

Parts of the presented work have been supported by the Deutsche Forschungs Gemeinschaft. The support is gratefully acknowledged.

REFERENCES

ASTM E647-88a (1989). Standard Test Method for Measurement of Fatigue Crack Growth Rate. 646-666.

Barsom, J.M. and S.T. Rolfe (1977). In: *Fracture and Fatigue Control in Structures*. Prentice-Hall, pp. 232-267.

Castro, D.E. (1989). In: *Ermüdungsrißausbreitung bei betriebsähnlichen Belastungen*. PhD Thesis, University of Karlsruhe, pp. 23-27.

Döker, H. (1993). Neue Verfahren zur Messung und Charakterisierung des Schwellwertverhaltens für Rißausbreitung. *Werkstoff-Kolloquium '93 in DLR*, 5-7.

Schmidt, R.A. and P.C. Paris (1973). Threshold for Fatigue Crack Propagation and Effects of Load Ratio and Frequency. *Progress in Flaw Growth and Fracture Toughness Testing, ASTM STP* **536**, 79-94.

Sedlacek, G., W. Hensen, J. Bild and W. Dahl (1992). Ermittlung der Sicherheit von alten Stahlbrücken unter Verwendung neuester Erkentnisse der Werkstofftechnik. *Bauingenieur* **67**, 129-136.

Suresh, S. and R.O. Ritchie (1983). On the Influence of Environment on the Load Ratio dependence of Fatigue Thresholds in Pressure Vessel Steel. *Engineering Fracture Mechanics*, Vol.**18**, No.**4**, 785-800.

Wittemann, K. (1993). Wie sicher sind alte Stahlbauwerke?, *Materialprüfung* **35**, No.3, 53-57.

Fatigue Crack Propagation in Vanadium-Based Microalloyed Steel Including the Threshold Behaviour and R-Ratio Effects

A. Fatemi*, G. Tandon**, and L. Yang**

*Associate Professor, **Research Assistant
Department of Mechanical, Industrial, and Manufacturing Engineering , The University of Toledo
Toledo, Ohio 43606, USA

ABSTRACT

Microalloyed (MA) steels are carbon or low alloy steels to which small amounts of microalloying elements have been added. MA forging steels are considered to be economical alternatives to the traditional quenched and tempered (Q&T) steels, since the need for the post-forging quenching and tempering is eliminated. This paper studies fatigue crack growth and threshold behaviour of a vanadium-based MA steel. Both the as-forged and the Q&T conditions are used, such that a comparison could be made between materials with and without post-forging heat treatment, but at the same hardness. Fatigue crack growth tests were conducted using compact tension specimen configuration and crack length measurements were made using an optical technique. Specimens were tested with load ratios of 0 and 0.5, to investigate the mean stress or R-ratio effects on the FCG behaviour. The threshold values were obtained employing the load shedding technique with both K-decreasing and K-increasing test procedures. Crack closure was measured to facilitate calculation of effective stress intensity factor range in crack growth rate correlations. Several empirical and semi-empirical models proposed in the literature were used to predict or correlate crack growth rates in the entire da/dN - ΔK regime. Macroscopic and microscopic examinations of fracture surfaces were carried out using SEM to evaluate and compare fracture surface characteristics and crack growth mechanisms.

KEYWORDS

Fatigue crack growth; microalloyed steel; R-ratio effect; fatigue crack growth threshold

INTRODUCTION

Microalloyed (MA) steels also known as High Strength Low Alloy (HSLA) steels differ from the conventional low alloy steels due to the small additions of microalloying elements which result in great improvement in their mechanical properties. Over the past three decades, a large variety of MA steels have been developed and been used in applications such as oil and gas offshore structures and pipelines, marine structures, construction material reinforcements, and automobile engine and suspension components. These steels derive their superiority from the microstructural modifications achieved by the addition of small amounts (rarely exceeding 0.1%) of the microalloying elements such as vanadium (V), niobium (Nb), titanium (Ti), and aluminum (Al). The principal strengthening mechanisms operative in these steels are grain refinement, precipitation strengthening, solid solution strengthening, and transformation strengthening. MA steels however, possess inferior impact properties compared to the conventional quenched and tempered steels. Toughness improvement is the key to a successful application of as-forged MA steels for automotive components.

Significant amount of research has been conducted on optimizing the monotonic properties and on improving impact toughness of these steels. Fatigue crack propagation and threshold behaviour of MA steels however, have been accorded little attention over the years. An overview paper by Yang and Fatemi [1993] on fatigue and fracture of MA steels reviewed the results from the few available FCG studies of MA steels in the threshold, intermediate, and high stress intensity factor regions. In view of the increasing applicability of damage tolerant design as a design parameter, additional studies on the fatigue crack growth behaviours of MA steels are needed.

The steel considered in this study was a modified AISI 1141 carbon steel microalloyed with vanadium. The V-based MA steel in the as-forged condition has already been compared with the quenched and tempered steel with the same composition and hardness, in terms of their monotonic and cyclic deformation properties, low and high cycle fatigue, charpy V-Notch impact toughness, and plane-stress and plane-strain fracture toughness [Yang and Fatemi, 1995a; Yang and Fatemi, 1995b]. In this paper, threshold and midregime crack growth Behaviour of this steel in the two conditions are studied.

EXPERIMENTAL PROGRAM

Material

The chemical composition of the vanadium-based MA steel consisted of 0.4% C, 1.5% Mn, 0.25% Si, traces of P, S, Cu, Cr, Ni, Mo, and 0.055% vanadium. The ingots were hot rolled to starting bar stocks followed by hot forging into 25 mm thick pancakes. Subsequently, these pancakes were air cooled to about 425°C. At this stage of processing, pancakes were split into two groups. One group was not subject to any post-forge heat treatment (as-forged, AF), while the other group was subjected to quenching and tempering heat treatment (quenched and tempered, Q&T). The AF steel microstructure consisted of uniformly distributed ferrite and pearlite with the pearlite being fine and acicular. The ferrite phase was much smaller in size and distributed primarily around the pearlitic grain boundaries. The grains were equiaxed. The Q&T steel had a martensitic microstructure with fine precipitates of tempered or secondary martensite. Inclusions were mainly manganese sulphide (MnS) stringers approximately 10 to 15 microns long, running along the rolling direction. The inclusions are unaffected by the heat treatment, and were similarly distributed in both the AF and Q&T steels. The AF and Q&T steels have yield strengths of 524 MPa and 670 MPa, respectively; ultimate strengths of 875 MPa and 777 MPa, respectively; and percent elongations of 26% and 40%, respectively.

Specimen and Equipment

The specimens used were the compact tension (CT) specimens in the LT orientation, machined in accordance with the ASTM standard E647 [1994]. The ratio of initial crack length to the specimen width, a_0/W, was between 0.30 and 0.38 for all specimens. Specimens were fatigue precracked prior to testing using a load-shedding technique. Loads were reduced to the test loading ranges in two to four steps, with a reduction in the maximum load for each step being less than 20%. A sufficient crack length increment in the final step was ensured to avert transient effects. The starting crack length was about 18 mm. The equipment used was a 100 kN closed loop servo-controlled hydraulic test system. The specimens were attached to the loading frame using a set of monoball grips. A clip gage was used to monitor the crack mouth opening displacement. Since the specimen geometry used had a thickness to width ratio, B/W, of less than 0.15, crack length measurements only on one side of the specimen was required. Optical measurements were made by employing a 30X traveling microscope with a least reading of 0.01 mm.

Loading Conditions and Data Reductions

Each material condition was tested under stress ratios of 0 and 0.5, to evaluate the R-ratio effect. In the midregime FCG tests, only the Q&T MA steel specimen tested at R = 0 showed crack closure in the low ΔK region, with the crack opening load decreasing linearly with elapsed number of cycles. An effective stress intensity factor range, ΔK_{eff}, was used to account for the closure effect. Crack length versus cycles data were reduced using the seven point incremental polynomial technique. Validity of the test data were verified based on both local and gross yielding requirements. The coefficient A and the exponent n in the Paris equation were obtained by fitting the equation to the closure corrected data.

In the threshold tests, the load shedding was done manually, with the rate of load shed gradual enough to preclude anomalous data. The K-decreasing test was considered complete when crack growth rates less than or equal to 10^{-10} m/cycle were reached. The K-decreasing test was followed by the K-increasing test. The a vs. N data obtained from the threshold tests were reduced to da/dN values using the secant method. The crack lengths and stress intensity factors, if required, were corrected for curvature effects. The fracture surfaces were examined at the precrack and the terminal fatigue crack lengths, to determine the through thickness crack curvature. If a contour was visible, a three point average crack length was calculated in accordance with the standard practice in ASTM E399 [1994]. The K-increasing data and the K-decreasing data showed a good match in all tests. The threshold stress intensity factor range value for each test was calculated using a best fit straight line from a linear regression through five approximately equally spaced da/dN values between 10^{-9} and 10^{-10} m/cycle, as suggested by the ASTM standard. The ΔK_{th} value was defined as the stress intensity factor range corresponding to a crack growth rate of 10^{-10} m/cycle.

EXPERIMENTAL RESULTS AND DISCUSSION

Crack Growth Rate Comparisons

The results of all FCG tests for the two material conditions and the two stress ratios are summarized in Table 1 and compared in Fig. 1. As can be seen from Fig. 1, the midregime FCG behaviours for the two grades are nearly the same. The slopes (n values) for R = 0 condition are close to that found by Yan et al. [1989] for a low carbon V-based MA plate steel with tempered martensite structure in the same strength category (n = 2.65 in their study, compared to 2.70 for the AF MA steel and 2.96 for the Q&T MA steel, in this study). However, several V-based MA forging steels studied by Farsetti and Blarasin [1988] show n values of about 3.5.

Table 1. Fatigue crack growth test results

	AF steel		Q&T steel	
	R = 0	R = 0.5	R = 0	R = 0.5
A $(\times 10^{-12})$	15.4	4.8	6.0	4.2
n	2.70	3.17	2.96	3.17
ΔK_{th} (MPa $\sqrt{m}$)	3.79	3.71	8.11	2.56

da/dN = A $(\Delta K)^n$, where da/dN is in m/cycle, and ΔK is in MPa $\sqrt{m}$

Mean load or R-ratio effects on FCG behaviour for each grade are compared in Figs. 1c and 1d for the AF and Q&T grades, respectively. Little mean stress sensitivity is observed in the midregime region for either material conditions. This is in accordance with general observations for a variety of other materials. For the AF steel, the ΔK_{th} value was the same for the two R-ratios. For the Q&T steel, the ΔK_{th} value decreased about 68% with increasing R-ratio. Therefore, the Q&T steel showed a strong influence of R-ratio on the fatigue threshold. The AF material condition had a threshold value which was 53% lower than the Q&T material condition at R = 0, but 45% higher than the Q&T steel at R = 0.5.

Crack Growth Rate Correlations

Several empirical and semi-empirical crack growth rate equations have been proposed to account for the load ratio effect. One such equation often used is the Forman equation [Forman et al., 1967] based on the observation that a marked increase in growth occurs at high ΔK values as K_{max} approaches K_C.

This equation is expressed as:

$$da/dN = C \, \Delta K^n / [(1 - R) \, K_C - \Delta K]$$ (1)

Midregime crack growth rate data plotted according to this equation did not correlate the two R-ratio data sets of either material. In fact deviation of data between the two different R-ratios broadend, rather than narrowing. A modification is, therefore, suggested by replacing the denominator term $[(1-R) \, K_C - \Delta K]$ in the Forman equation, with $[(1 - \alpha R) \, K_C - \Delta K]$, in which α can be considered to be a material and/or thickness dependent constant. For the MA steel in this investigation, $\alpha = 0.4$ resulted in satisfactory correlations of the midregime crack growth data with the two load ratios for both material conditions. By using the modified equation, variations between midregime crack growth rate tests from different R-ratios are significantly reduced for both MA forging steel conditions. The R-ratio variations in the midregime crack growth rate region however, do not indicate significant effect on crack growth rates for either material conditions. A similar conclusion was made by Hertzberg and Goodenow [1977] for a V-based MA steel with similar tensile properties.

Over the years, researchers have proposed empirical relations to predict the entire fatigue crack propagation behaviour. Klesnil and Lukas [1972b] modified the original Paris equation in the following manner, in an attempt to incorporate the low ΔK regime by proposing that:

$$da/dN = C \, (\Delta K^n - \Delta K_{th}^n)$$ (2)

Donahue et al. [1972] suggested a slightly different version:

$$da/dN = C \, (\Delta K - \Delta K_{th})^n$$ (3)

Relations combining the departures from the power law behaviour at high and low ΔK values have also been proposed. These relations are generally either empirical or semi-empirical. In an effort to develop an analytical relation, McEvily and Groeger [1977] proposed the following relation:

$$da/dN = [A / (E \, \sigma_y)] \, (\Delta K - \Delta K_{th})^2 \, [1 + \Delta K / (K_{IC} - K_{max})]$$ (4)

where A is an environmentally sensitive material property and σ_y is the yield strength. Further, ΔK_{th} is dependent on the mean stress and environment, therefore McEvily and Groeger suggested that:

$$\Delta K_{th} = [(1 - R) / (1 + R)]^{0.5} \, (\Delta K_o)$$ (5)

where ΔK_o is the threshold value at R = 0. The aforementioned analytical relations were used to formulate the da/dN curves for the AF and Q&T steels at R = 0 as shown in Figs. 2a and 2b. The empirical relation proposed by Klesnil and Lukas best describes the fatigue crack growth curve of both the AF and Q&T steels. The equation proposed by Donahue et al. was also found to be equally effective in predicting the fatigue crack growth curve for both steels. The approach proposed by McEvily and Groeger however, provided poor predictability, as can be seen from the same figures. Similar trends were found for both material conditions at the load ratio of 0.5.

Fractography

Fractography was performed on the fracture surfaces of the specimens using SEM. Microscopic examination of the fracture surfaces indicate that differences exist between material conditions, rather than between R-ratios. Secondary micro-cracking, voids, and inclusions were common features found on all fracture surfaces. The role of inclusions in microcracking is more critical for the Q&T steel, because of the MVC fracture mechanism observed in this material condition in the threshold tests. Microcracking occurs by the coalescence of voids formed around second phase particle carbides and inclusions. Therefore, microcracking in the Q&T steel is strongly dependent on the inclusion size and morphology. On the other hand, cleavage fracture mechanism is operative in the AF steel. Therefore, in this material condition, there are multiple crack paths available to the progressing crack. The crack propagates by the lowest energy path and is not strongly dependent on the inclusions size and shape.

CONCLUSIONS

Fatigue crack growth was studied and the threshold values were obtained at two stress ratios of 0 and 0.5 for the AF and Q&T MA steel. In the midregime fatigue crack growth region, the two materials exhibit the same behaviour in the linear region. The Q&T specimen tested at R = 0 exhibits crack closure, while no crack closure was observed for the AF grade. Mean stress (R-ratio) effects on midregime FCG behaviour for both materials are also very similar. A modified Forman equation resulted in better correlation of the FCG rate data with different R-ratios. The fatigue threshold values decreased with increasing R-ratio for the Q&T steel, but not for the AF steel. The Klesnil and Lukas approach, and the relation proposed by Donahue et al., best described the fatigue crack growth curves for the AF and Q&T material conditions. These empirical relations can be used to predict the fatigue crack growth rate curves for R-ratios other than those experimentally determined in this study.

REFERENCES

ASTM standard E399 (1994). Standard test method for plane strain fracture toughness of metallic materials. *Annual Book of ASTM Standards*, Vol. 3.01, Philadelphia, pp. 488-512.

ASTM standard E647 (1994). Standard test method for measurement of fatigue crack growth rates. *Annual Book of ASTM Standards*, Vol. 3.01, Philadelphia, pp. 648-668.

Donahue, R. J., H. M. Clark, P. Atanmo, R. Kumble and A. J. McEvily (1972). Crack opening displacement and the rate of fatigue crack growth. *Int. J. Fract. Mech.*, Vol. 8, pp. 209-219.

Farsetti, P. and A. Blarasin (1988). Fatigue behaviour of microalloyed steels for hot-forged mechanical components. *Int. J. of Fatigue*, Vol. 10, No. 3, pp. 153-161.

Forman, R. G., V. E. Kearney and R. M. Engle (1967). Numerical analysis of crack propagation in cyclic loaded structures. *ASME J. of Basic Eng.*, Vol. 89, No. 3, pp. 459-464.

Hertzberg, R. W. and R. H. Goodenow (1977). Fracture toughness and fatigue-crack propogation in hot-rolled microalloyed steel. In: *Microalloying 75* (M. Korchysky, ed.), New York, pp. 503-516.

Klesnil, M. and P. Lukas (1972b). Dependence of fatigue threshold stress intensity factor on the R-ratio. *J. of Mat. Sci. and Eng.*, Vol. 9, pp. 231-235.

McEvily, A. J. and J. Groeger (1977). On the threshold for fatigue crack growth. In: *Proc. 4th Int. Conf. on Fracture*, University of Waterloo Press, Waterloo, Canada, Vol. 2, pp. 1293-1298.

Yan, W. B., M. J. Tu, Z. Z. Hu, J. X. Liu and X. H. Tu (1989). Fatigue crack growth behaviour of HSLA steel at low temperature. In: *Proc. ICF 7, 7th Int. Conf. on Frac.*, Houston, pp. 2645-2650.

Yang, L. and A. Fatemi (1995a). Impact resistance and fracture toughness of vanadium-based MA forging steel in the as-forged and Q&T conditions. Appears in: *ASME J. Eng. Mat. Tech.*

Yang, L. and A. Fatemi (1995b). Deformation and fatigue of vanadium-based microalloyed forging steel in the as-forged and Q&T conditions. *ASTM J. Testing and Eval.*, Vol. 23, No. 2, pp. 80-86.

Yang, L. and A. Fatemi (1993). Fatigue and fracture of microalloyed steels: an overview. In: *Proc. ICF8, 8th Int.Conf. on Fracture*, Kiev, USSR.

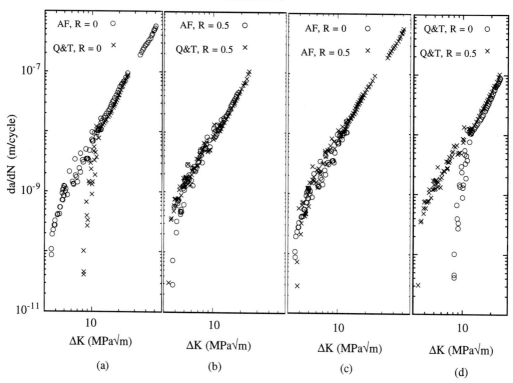

Fig. 1. Crack growth rate comparisons between a) AF and Q&T steels at R = 0, b) AF and Q&T steels
at R = 0.5, c) R = 0 and R = 0.5 for the AF steel, and d) R = 0 and R = 0.5 for the Q&T steel.

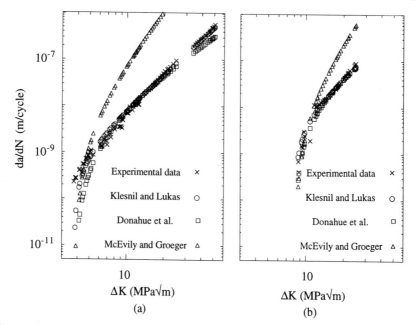

Fig. 2. Comparison of experimental crack growth rates with Klesnil and Lukas, Donahue et al., and
McEvily and Groeger models for a) the AF steel at R = 0, and b) the Q&T steel at R = 0.

OBSERVATIONS AND LIMITATIONS ON THE USE OF
THE GOODMAN DIAGRAM FOR COMBINED HCF/LCF LOADING

J. R. Zuiker and T. Nicholas

Wright Laboratory Materials Directorate,
Wright-Patterson AFB, OH 45433, U. S. A.

ABSTRACT

Combined effects of LCF and HCF are considered in an analytical study to predict the stress states for safe operating conditions. To account for the interactions, initiation criteria as well as fatigue crack growth thresholds for both the LCF and HCF loading are considered. Numerical results are contrasted with the HCF thresholds represented in a Goodman diagram. Experimental data for a Ti-6Al-4V alloy are used to demonstrate the applicability of the analysis and the limitations of the use of the Goodman diagram in service. The predicted safe design space is found to be a subset of the safe design space for pure HCF and decreases with increasing number of LCF cycles.

KEYWORDS

Goodman diagram; high cycle fatigue; low cycle fatigue; life prediction; combined cyclic fatigue.

INTRODUCTION

Design of components for high cycle fatigue (HCF) must generally account for the detrimental effects of a superimposed mean stress. This accounting is often in the form of an alternating versus mean stress (AMS) plot which shows allowable vibratory stress amplitude as a function of applied mean stress for a specified life. In many cases little or no data are available for conditions other than fully reversed loading ($R = -1$) and tensile overload ($R = 1$), and assumptions must be made as to how one interpolates between these limiting cases. The Goodman assumption is often used and postulates a straight line connecting the fully reversed alternating stress amplitude, σ_a, on the y-axis with the ultimate stress, σ_{ult}, on the x-axis, and the resulting plot is referred to as the Goodman diagram. (This assumption has also been referred to as the modified Goodman assumption in deference to Goodman's original proposal for a straight line connecting $\sigma_{ult}/3$ on the y-axis with σ_{ult} on the x-axis. Today the modified Goodman assumption refers to other modifications of the linear relationship between σ_a and σ_{ult}.) A more general AMS plot can be generated using data at various values of mean stress and a specified number of cycles to failure, e.g. 10^7, as obtained from S-N curves and plotting the locus of points. Such a plot differs from the Goodman diagram by allowing the representation of data to be nonlinear. For cases where high frequency data are available only under fully reversed loading conditions (R= -1), extrapolation to non-zero mean stresses is accomplished with some type of functional form or equation, using static yield stress or ultimate stress to anchor the curve at zero alternating stress on the AMS plot. For any of these plots, the number of cycles is typically taken to be those corresponding to a "runout" condition, perhaps 10^8 or even 10^9, but there are few data available to demonstrate that a true runout condition ever exists for a material, particularly for titanium alloys. For convenience and practicality, the number of cycles chosen is taken to correspond to the region where the S-N curve becomes nearly flat with increasing number of cycles, or is selected such that the number of cycles exceeds that which might be encountered in service. In some cases, neither condition may be satisfied. For design purposes, because of the statistical variability of fatigue data, particularly in the long life regime where S-N curves tend to be close to horizontal, AMS plots commonly represent a statistical minimum such as a minus 2σ or minus 3σ limit. For the purposes of the present discussion, only average material property data will be discussed.

Material quality is a key parameter which has to be considered when using the Goodman diagram in the design process. While data are obtained in the laboratory for a given material on smooth specimens, the specifics of the heat treatment, processing, microstructure, surface finish, and specimen size must all be considered when applying the data to structural components. Perhaps the single most critical issue in the use (or misuse) of a Goodman diagram in design is the degree of initial or service induced damage which the material in the component may contain when such damage is not present in the material used for the data base. Examples of this are fretting, galling, foreign object damage, or damage induced by superimposed low cycle fatigue (LCF). If any of these conditions are present, the Goodman diagram is not valid for the material because it represents "good" or undamaged material. A design methodology which considers the development of damage from sources other than the constant amplitude HCF loading must be used to account for the different state of the material. Turbine engine components, for example, which are subjected to HCF, are typically subjected to LCF in addition because the non-zero mean stress is achieved due to the centrifugal loading typical of operation. Each startup and shutdown, therefore, constitutes a LCF cycle. Thus, the component sees combined HCF/LCF and, for design purposes, the effect of LCF loading on the HCF life should be considered. In this paper, some observations on the influence of LCF superimposed on HCF are presented.

ANALYSIS

In order to illustrate HCF/LCF interactions, analytical predictions of the fatigue life in a representative titanium alloy, Ti-6Al-4V, are made for a material experiencing 10^7 HCF cycles divided equally over N LCF loading blocks. Each block consists of a low frequency cycle over which the material is loaded from zero stress to a given mean stress and held while $n = 10^7/N$ high frequency cycles are superimposed about the mean stress. This is shown schematically in Fig. 1. It is assumed that total life can be divided into two distinct phases: a crack initiation phase, and a crack propagation phase.

During initiation the material is initially uncracked. Initiation damage, characterized by the parameter D_i, is accumulated over each HCF and LCF cycle until $D_i = 1$ at which point a crack of depth a_i initiates. The number of LCF cycles required to reach $D_i = 1$ is defined as $N_{i,LCF}$. For LCF-only applied at $R = 0$, a power law function of the applied stress range, similar to the Basquin equation, is used:

$$N_{i,LCF} = \sigma * (2\sigma_a)^r, \tag{1}$$

where σ_a is the alternating stress amplitude and $\sigma*$ and r are constants. Different constants are used over specific ranges of σ_a such that equation (1) forms a piece-wise linear approximation to the actual material response when plotted on a log-log scale. Equation (1), which is written for LCF-only loading where R = 0, can also be used for HCF cycles at $R \neq 0$ by substituting an equivalent alternating stress amplitude, $\sigma_{a,eq}$, which is calculated for any relation representing constant life under various combinations of σ_a and mean stress, σ_m. Here we assume the Goodman relation which postulates that constant life is obtained for any combination of alternating and mean stress that lies on a straight line connecting the fully reversed stress for a given life (R = -1) to the ultimate stress, σ_{ult}, (R = 1.) It is easily shown that, for a given value of σ_{ult} and stress point (σ_m, σ_a), $\sigma_{a,eq}$ is found as

$$\sigma_{a,eq} = 1/(1/\sigma_{ult} + 1/\sigma_a - \sigma_m/\sigma_a\sigma_{ult}). \tag{2}$$

Thus, the initiation life due to HCF cycles, $N_{i,HCF}$, is obtained via (1) by replacing σ_a with $\sigma_{a,eq}$. Here it is assumed that the material does not exhibit an endurance limit. That is, any finite cyclic stress range will induce a finite amount of damage in the material. Thus, 10^7 HCF cycles does not represent a "runout" condition, but rather a typical number of HCF cycles that might be encountered over the life of the part. Depending upon the application, 10^7 cycles may be non-conservative.

To determine the initiation life under combined HCF/LCF loading, the linear damage summation model (Miner, 1945) is used such that the initiation life, in HCF/LCF blocks, is

$$N_{i,CCF} = 1/(1/N_{i,LCF} + n/N_{i,HCF}), \tag{3}$$

where $N_{i,CCF}$ is the initiation life under combined cyclic fatigue (CCF).

During crack propagation, the crack is assumed to grow during LCF and/or HCF cycles following the Paris Law as modified by Walker (1970) to account for different values of R as

$$da/dN* = C(\Delta K)^m/(1-R)^d. \tag{4}$$

Here, C and m are material constants describing the crack growth rate at R = 0, and d is a material constant accounting for the higher crack growth rate at higher R. N* is either N in the case of an LCF cycle, or n in the case of an HCF cycle.

Equation (4) holds for $\Delta K > \Delta K_{TH}$ for individual LCF cycles (R = 0) as well as individual HCF cycles (R > 0, generally.) In the case of tension-compression cycling (R < 0), the crack tip is assumed to be open and the crack growing only when the applied stress is positive. Thus, the minimum effective stress is always positive or zero. ΔK_{TH} is taken as a monotonically decreasing function of R.

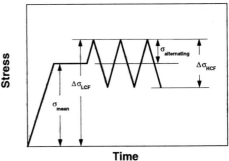

Fig. 1. HCF/LCF loading history.

Crack initiation data on surface-cracked, notched ($K_T = 2$), round bars are reported by Guedou and Rongvaux (1988), and piece-wise linear (in log-log space) power-law fits to the remote stress vs. life data have been determined. The stress intensity factor solution of Raju and Newman (1986) is utilized in predicting crack growth rates. Because the solution assumes a smooth bar ($K_T = 1$), the remote stress applied to the bar is multiplied by two to determine crack growth rates and the onset of crack growth in relation to ΔK_{TH}. While valid initially, this assumption will overestimate the crack growth rate as the crack grows out of the notch. The specimen is assumed to have failed when K_{max} surpasses K_{IC}, or when the crack depth exceeds the diameter of the bar, whichever occurs first. Crack growth is estimated for each HCF and LCF cycle in the order in which they occur for combined HCF/LCF loading. Thus, growth increments are computed for one LCF cycle, n HCF cycles, one LCF cycle, and so on.

RESULTS

A limited number of HCF/LCF tests on notched bars were conducted by Guedou and Rongvaux (1988). Notched Ti-6Al-4V bars were cycled at room temperature with $n = 1800$ HCF cycles per LCF block and $R_{HCF} = 0.85$. The results are shown along with those for LCF-only tests in Fig. 2. All results are plotted as a function of maximum stress ($\sigma_m + \sigma_a$). Predictions of the LCF only tests are good. This is expected as the initiation prediction is simply a correlation with experimental data and initiation life is an order of magnitude larger than propagation life. That the predicted life is, on average, somewhat lower than the experimental life is indicative of the fact that crack propagation life is underestimated, as noted above. Predictions for HCF/LCF loading are accurate at lower values of maximum stress. At high values of maximum stress life is underestimated by nearly an order of magnitude.

Fig. 3 shows the computed values of allowable σ_a as a function of σ_m for failure in 10^7 HCF cycles (and one LCF cycle.) The differences between the predicted solution and the Goodman diagram are due to assumptions in the analysis and are discussed below. Line 1 indicates the alternating and mean stress combinations which will cause initiation in 10^7 HCF cycles with no LCF cycles ($N = 1$). If the number of cycles to initiation is increased from 10^7 to 10^8, the line moves down as shown in the figure. Another curve (line 2) can be drawn representing stress states above which a crack will propagate under HCF based on the initiation crack size (50 µm here) and the value of ΔK_{TH} as a function of R. At low values of mean stress, this line has a slope which increases significantly and follows a line along which $\sigma_a + \sigma_m$ = constant. This value of maximum stress corresponds to $\Delta K = \Delta K_{TH}$ at the initiating crack length. Note that in this region, corresponding to negative R, the crack can initiate in less than 10^7 HCF cycles at stress states above line 1 but will never propagate if the stress state is below line 2. Conversely, there is a region between approximately 200 and 600 MPa mean stress where the crack will not initiate (below line 1), yet a 50 µm crack could propagate (above line 2) based on the assumptions in this analysis. Following the same logic, curves can be drawn for the boundaries below which LCF cycles will not cause initiation or below which a 50 µm crack will not propagate. The initiation curves for $N = 1$ and $N = 10^4$ LCF cycles (R = 0) are shown in Fig. 3. Line 3, corresponding to $N = 1$, represents $\sigma_a + \sigma_m = \sigma_{ult}$. Any stress state above or to the right of line 3 will cause tensile failure on the first cycle. Finally, above and to the right of line 4, crack growth will occur under LCF loading. Note that line 4 coincides with the line above which HCF crack growth will occur (for a 50 µm crack) at low values of σ_m. These lines are coincident because LCF crack growth (R = 0) is merely a subset of HCF crack growth (any value of R). Under the assumptions of this analysis, the region where failure can occur due to either HCF or LCF is above the heavy line in Fig. 3.

Fig. 4 shows solutions for various values of N and n where Nn = 10^7. As N increases, the allowable alternating stress is decreased at higher values of mean stress. As expected, a finite number of LCF cycles reduces the maximum mean stress that may safely be applied to the structure.

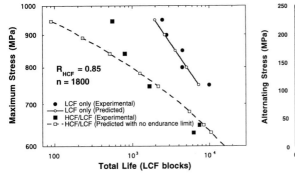

Fig. 2. Life predictions for LCF only
and HCF/LCF loading

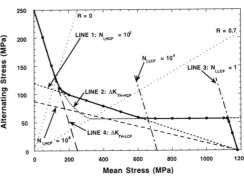

Fig. 3. Predicted AMS plot

CLOSURE

The numerical results presented here are based on simple models of crack initiation and propagation. Many potentially important phenomena are neglected such as: the possible non-linear accumulation of initiation damage from HCF and LCF; the effect of previous cycling on the instantaneous endurance limit; acceleration in the HCF crack growth rate due to periodic underloads (LCF cycles); reduction in ΔK_{TH} as a function of the number of LCF cycles; small crack effects; hold time effects on LCF cycles; and many others. Therefore these results must be viewed as preliminary, giving only a qualitative indication of how LCF and HCF cycling interact to reduce overall life. This method satisfactorily predicts the effects of combined HCF/LCF loading despite the limited amount of experimental data available for calibration. Additional experimental results should allow for better calibration and will pave the way for incorporation and assessment of many of the above phenomena.

Although simple, this analysis provides insight into designing for HCF/LCF loading. It has been common practice to use the Goodman diagram to design for allowable vibratory stress in the presence of a mean stress in metal components. This assumption has been observed to give good predictions for brittle metals and to be conservative for ductile metals. For high frequency, low-amplitude fatigue the crack propagation life is generally observed to be a small fraction of the total life. (In the numerical simulations presented above, crack propagation life was generally less than 1% of total life.) Thus the N = 1, n = 10^7 initiation line is a good approximation to the Goodman diagram for the Ti-6Al-4V material under investigation. This is shown in Fig. 5 along with the numerical predictions from Fig. 4 for N = 10^4. The allowable mean stress at σ_a = 0 is, by definition, the stress range for a life of N LCF cycles (10^4 in this case) with no superimposed vibratory loading. The line for N = 10^4 in Fig. maintains a constant value of $\sigma_a + \sigma_m$ as σ_a increases. That is, for a given value of N, $\sigma_a + \sigma_m = \Delta\sigma_{LCF}$, where $\Delta\sigma_{LCF}$ is the stress range causing failure in N cycles at R = 0, implying that superimposed HCF cycles have absolutely no detrimental effect life. This relationship was found to hold in both Ti-6Al-4V and Inconel 718 smooth bar specimens at low values of alternating stress (Guedou and Rongvaux, 1988). The fact that the numerical solution lies to the left of the $N_{LCF} = 10^4$ line indicates that the HCF cycles do adversely affect life. Thus, it is hypothesized that the safe design space for combined HCF/LCF loading is below the Goodman line, and to the left of the line of constant maximum stress corresponding to the appropriate number of LCF cycles. For a significant number of LCF cycles, this removes a sizable region at high values of R from the safe design space. It has been the USAF experience that many HCF failures occur in the high R regime. Note that the limits of the safe design space proposed are *non-conservative*. The discrepancy is greatest at the intersection of the two lines and its magnitude is dependent upon the details of the numerical analysis. Prediction in this regime will require further refinement of the numerical model.

In summary, predictions have been made for the safe design space under combined HCF/LCF loading in terms of allowable values of mean and alternating stress using data from the literature on Ti-6Al-4V. The numerical algorithm provides good correlation with limited experimental data on HCF/LCF loading. The predicted safe design space is a subset of the safe design space for pure HCF and decreases with increasing number of LCF cycles.

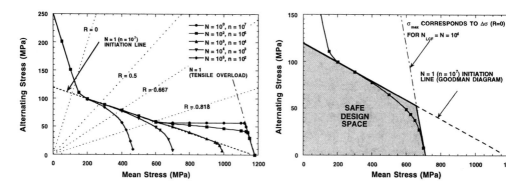

Fig. 4. Effect of LCF cycles on HCF   Fig. 5. Safe design space in combined
capability.   HCF/LCF loading.

REFERENCES

Guedou, J.-Y. and Rongvaux, J.-M. (1988). Effect of Superimposed Stresses at High Frequency on Low Cycle Fatigue. In *Low Cycle Fatigue, ASTM STP 942* (H. D. Solomon, G. R. Halford, L. R. Kaisand, and B. N. Leis, eds.), pp. 938-960. American Society for Testing and Materials, Philadelphia.

Miner, M. A., (1945). Cumulative Damage in Fatigue. *J. App. Mech.*, 6, 359-365.

Raju, I.S. and Newman, J. C. (1986). Stress Intensity Factors for Circumferential Surface Cracks in Pipes and Rods under Tension and Bending Loads. In *Fracture Mechanics: Seventeenth Volume, ASTM STP 905* (J. H. Underwood, R. Chait, C. W. Smith, D. P. Wilhem, W. A. Andrews, and J. C. Newman, eds.), pp. 789-805. American Society for Testing and Materials, Philadelphia.

Walker, K., (1970). The Effect of Stress Ratio During Crack Propagation and Fatigue of 2024-T3 and 7075-T6 Aluminum. In *Effects of Environment and Complex Load History for Fatigue Life, ASTM STP 462*, pp. 1-14. American Society for Testing and Materials, Philadelphia

FATIGUE CRACK GROWTH IN ADVANCED MATERIALS

A. K. Vasudevan and K. Sadananda[#]

Office of Naval Research, Code-332, Arlington, VA 22217
[#]Naval Research Lab., Code-6323, Washington, D. C. 20375

ABSTRACT

Fatigue crack growth in advanced materials has been examined using the new concepts developed by the authors. These concepts have been extended to generalize the overall behavior in terms of five main classes. It is shown that such a classification of fatigue crack growth behavior can be readily applied to variety of materials under various conditions. Applications of these general concepts and their implications are briefly discussed using the available literature data on the advanced materials, such as polymers, intermetallics, metal matrix composites and ceramics.

KEYWORDS

fatigue crack growth, advanced materials, mechanisms

INTRODUCTION

In the recent years we have been critically examining the current understanding of the fatigue damage processes and have come to a conclusion that the *fatigue damage has to be described in terms of two intrinsic loading parameters.* These two parameters are the basic unambiguous driving forces that govern the progress of a crack to initiate and grow, independent of the conventional extrinsic crack closure approaches. This provided a basis for new interpretation to describe the experimental results using a ΔK vs K_{max} curve (called the fundamental fatigue curve), instead of the conventional ΔK vs R. At the near threshold fatigue crack growth region, the form would be in terms of ΔK_{th} vs K_{max}. Such a plot gives two asymptotic values for ΔK_{th} and K_{max} defining two critical values : ΔK^*_{th} and K^*_{max}. At higher growth rates, different ΔK vs K_{max} curves can be obtained, with their respective critical asymptotic values. Thus, these concepts provide the basis for describing the fatigue behavior over the entire range of crack growths (Vasudevan, Sadananda & Louat 1994; Vasudevan & Sadananda 1995). The universality of these concepts can be applied to a broad spectrum of materials encompassing brittle,

473

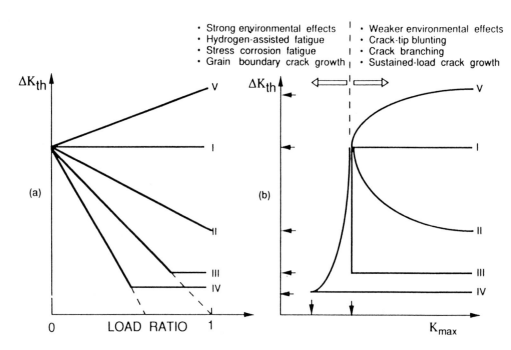

Fig. 1. Schematic representation of the overall fatigue mechanisms to indicate the *five* main classification in terms of (a) ΔK_{th} vs R , and (b) ΔK_{th} vs K_{max}; along with the regions affected by the crack-tip environment.

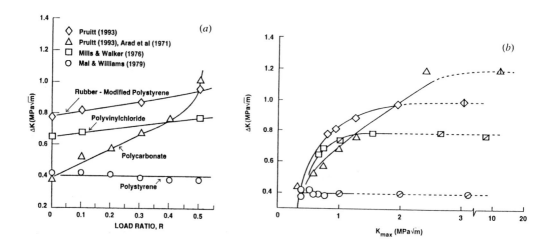

Fig. 2. Experimental results from a selected number of polymeric materials: (a) (a) ΔK_{th} vs R , and (b) ΔK_{th} vs K_{max}; illustrating the Class-V fatigue behavior.

semi-brittle and ductile materials: ceramics, intermetallics, metals, alloys, composites under inert and corrosive environments. These two parametric description has been extended to account for the behavior of short cracks, retardation effects under overloads, underloads, etc. (Sadananda & Vasudevan, 1996 this conference).

Analysis of the extensive fatigue data (in terms of ΔK_{th} vs R) from the literature led us to classify all fatigue crack growth behavior into *five* groups. This gives an identification of the relative roles of the two stress intensity parameters from near threshold region to all crack growth rates. Such a classification method helps in recognizing the individual fatigue behavior of different family of materials and understand the relative roles of the environment on fatigue. Fig. 1a shows the basic classification scheme on the basis of ΔK_{th} vs R, for R>0 region. In the Class-V behavior, ΔK_{th} increases with R, a trend observed in polymeric materials, but seldom in metals. Class-I behavior, where ΔK_{th} is independent of R, is observed in some commercial alloys of aluminum, nickel, titanium and steels. This trend pertains to materials which are insensitive to environmental effects; commonly observed under vacuum. Class-II (ΔK_{th} vs R extrapolates to an intercept of ΔK_{th} >1 at R=1) is observed in low strength/high ductility materials, showing lesser (than Class-I) environmental effects where crack-blunting type of process could be the operating mechanism. Class-III is a *normal* fatigue behavior, commonly observed in aluminum and steel alloys. Here ΔK_{th} vs R curve extrapolates to R=1 at ΔK_{th} =0. Finally, Class-IV shows the behavior when ΔK_{th}<1 at R=1. Experimentally, the ΔK_{th} vs R curve levels off to a plateau prior to ΔK_{th} approaching to zero., as in Class-IV and -III. Thus, the classification shows the role of environmental effects on fatigue giving rise to different slopes in ΔK_{th} vs R; moving them from Class-I (as in inert environment) to Class-IV that corresponds to relatively more aggressive environment (Vasudevan & Sadananda 1995). As the fatigue mechanisms change with growth rates, the classification hierarchy moves toward lower classes. Fig. 1b illustrates the corresponding ΔK_{th} vs K_{max} version of Fig. 1a. For the sake of brevity, all curves in Fig. 1a begin at the same ΔK_{th} level. The figure also shows the overall fatigue mechanisms that can operate in different regions of classification. Following examples give validity to the application of these concepts to advanced materials. Alloy data used for analysis are for only those cases where systematic data were available in the literature.

EXAMPLES

Fig. 2a & 2b shows the ΔK_{th} increasing with R and K_{max}, for a limited number of polymeric materials. Even in such cases, the data still exhibit the asymptotic behavior at the low ΔK_{th} and at the high K_{max} ends of the curve. The high load ratio (R>0.5) data is lacking, probably due to experimental difficulties in these soft materials. As the mechanistic understanding is not clear, one can infer from the results that with K_{max} increasing, the polymer chains can rearrange to become stiffer requiring higher ΔK_{th} for crack growth (Pruitt 1993, Mills & Walker 1976, Mai &Williams 1979, Arad et al 1971).

Fig. 3a shows a systematic ΔK data on a planar slip ($\alpha_2+\beta$) Ti-aluminide alloy at four R-ratios, at threshold region and at higher crack growth rates. The linear decrease in ΔK between 0<R<0.8 is

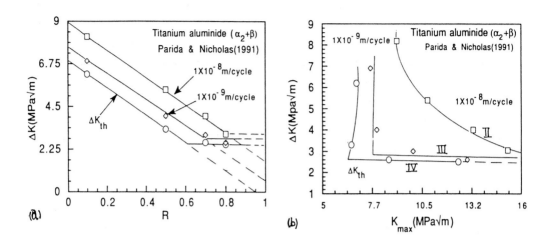

(a)   R   (b) K_{max}(MPa√m)

Fig. 3. Fatigue crack growth characteristics of $(\alpha_2+\beta)$Titanium aluminide: (a) ΔK_{th} vs R , and (b) ΔK_{th} vs K_{max}.

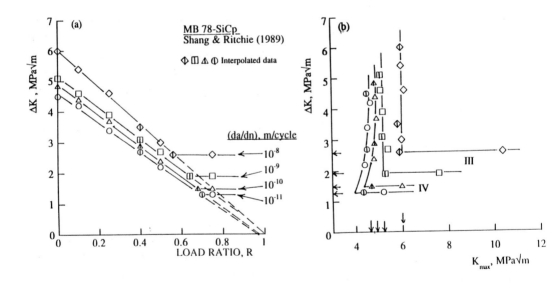

Fig. 4. Threshold fatigue results are extended to higher crack growth rates showing the transition from Class-IV to Class-III in an Aluminum matrix composite: (a) ΔK_{th} vs R , and (b) ΔK_{th} vs K_{max}.

representative for many other materials. Based on the classification, Fig. 3b shows the transition from Class-IV to II, as the crack growth rates become higher, in terms of ΔK vs K_{max} curves. In each classes, there are two distinct parameters of ΔK vs K_{max} controlling the crack advance. The negative curvature at the threshold region (Class-IV) is attributed to greater environmental effects where K_{max} role is reduced by the superimposed stress corrosion mechanism. At higher growth rates above threshold region, some crack blunting process can become dominant making the curvature more positive. One can also observe that with increase in growth rates resulted in smaller changes in the cyclic component ΔK than the static component K_{max} (Parida & Nicholas, 1991). Similarly, TiAl matrix alloy reinforced with ductile 20% TiNb phase also exhibit the plasticity-induced Class-II ductile crack growth behavior at room temperature (Rao & Ritchie 1994). The overall behavior in such planar slip TiAl-type alloys depends on the spacing of the ductile reinforcements (Sadananda & Vasudevan 1995).

The case of ductile aluminum matrix composites, such as MB-78-SiC$_p$ composite, results are shown in Fig. 4a & 4b. The alloy starts with Class-IV environmental type of behavior at threshold region and transitions toward Class-II at higher crack growth rates. As the crack growth increases from 10^{-11} to 10^{-8} m/cycle, K_{max} component tends to become dominant. This is due to the role of plasticity increasing (at higher growth rates) with an attendant decrease in the role environment (at lower growth rates) on fatigue. The observation is consistent with relative role of each mechanism listed in Fig. 1 (Shang & Ritchie 1989).

Finally, an example of threshold fatigue result in a monolithic sintered Si$_3$N$_4$ doped with yittria and alumina is shown in Fig. 5. The brittle ceramic was tested in a 60% humid air environment under various frequencies and R-ratios. From the Fig. 5a & 5b it is clear that the fatigue process in this ceramic is of the type Class-IV (see FIG 1). It is interesting to see that the material ΔK_{th} begins at R=0 at a stress intensity of about 5MPa$\sqrt{m}$ (which is close to K_{Iscc}) and deviates from it to a value of 3.8MPa$\sqrt{m}$ and then remains constant. This indicates that $\Delta K^*_{th}=0$(with no cyclic component), but with a static component $K^*_{max} =3.8$MPa$\sqrt{m}$. Hence, crack extension in these type of materials is primarily controlled by the environment (Ueno et al 1991). This trend is similar to the Class-IV type of behavior in aluminum composite at near threshold region in Fig. 4.

SUMMARY

Thus fatigue crack growth phenomena is intrinsically a two parametric problem, that depends on the material deformation characteristics (such as ductile or brittle) and on the crack-tip environment. The entire fatigue behavior can be classified into *five* main groups, each governed by a given fatigue mechanism. The basic concepts are applicable to the advanced materials, showing varied fatigue mechanisms, each material falling into one class or the other in its behavior. It is important to emphasis that it is necessary to obtain systematic sets of data at various load rates and growth rates to render complete analysis of the phenomena.

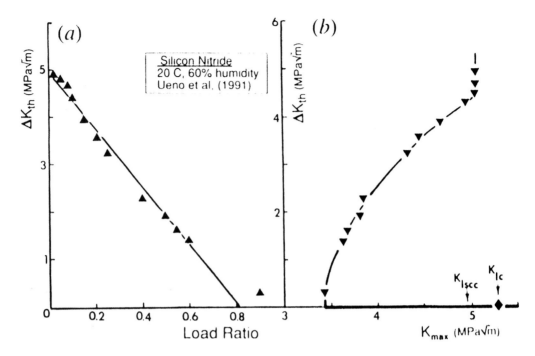

Fig. 5. Experimental threshold data from a monolithic silicon nitride ceramic illustrating the Class-IV
behavior: (a) ΔK_{th} vs R , and (b) ΔK_{th} vs K_{max}.

REFERENCES

Vasudevan, A. K., Sadananda, K and N. Louat (1994).*Mat. Sci. Eng.* A188, 1-22.

Vasudevan, A. K. and Sadananda, K(1995).*Met. Trans.,* 26A, 1221-1234.

Sadananda, K. and Vasudevan, A. K. (1995) *ASTM STP,* submitted for publication.

Pruitt, L. (1993) Ph. D. Thesis, Brown University, Providence, RI.

Mills, N. J. and Walker, N. (1976) *J.Mat. Sci.,* 17, 335.

Mai, W. and Williams, J. G. (1979) 14, 1933

Arad, S., Radon, J. C., and Culver, L. E. (1971) 13, 75.

Parida,B. K. and Nicholas, T. (1991) *Int. J. Frac.,* 52, R51.

Rao, K. T. V. and Ritchie, R. O. (1994) *Acta Metall.,* 42, 893.

Sadananda, K. and Vasudevan, A. K. (1995) *Mat. Sci. Eng.* A192, 490-501.

Shang, J. K. and Ritchie, R. O. (1989) *Met. Trans.* , 20A, 897.

MACROCRACKS AND DAMAGE TOLERANCE DESIGN

DEVELOPMENT OF TWO-PARAMETRIC CRITERION FOR FATIGUE CRACK GROWTH IN DUCTILE MATERIALS

H.M. NYKYFORCHYN, O.Z. STUDENT, I.D. SKRYPNYK

Department of Corrosion and Metal Protection, Karpenko Physico-Mechanical Institute of NAS of Ukraine, 290601, Lviv, Ukraine

ABSTRACT

A two-parametric deformation criterion of fatigue crack growth in high-plastic steels is proposed. Its application allows to describe short fatigue crack growth as well as to take account of the stress ratio and load cycles frequency influence in high-temperature testing of materials.

KEYWORDS

Ductile steels; fatigue crack growth; high temperature; short cracks; two-parametric criterion.

INTRODUCTION

Investigation of fatigue crack growth in high-plastic materials requires application of non-linear fracture mechanics approaches: J-integral, CTOD. However, for assessment of high-plastic material fracture, a two-criteria approach is used, as a rule, in fracture toughness assessment (Newman, 1973, Morozov, 1975).

This paper is aimed at developing the two-parametric criteria for fatigue crack growth in plastic materials, specifically in high-temperature testing.

EXPERIMENTAL PROCEDURE

Cantilever bending testing of aluminium beam specimens (5x23x180mm, 99,5% Al, σ_{ys} = 52MPa, δ= 29%) have been performed under loading with frequency 0.1 Hz and stress ratio R=-1. A model of non-linear-elastic beam bending was used for evaluation of nominal stresses near the crack tip.

The fatigue crack growth was investigated in heat-resistant chromium-nickel HK-40 steel at 870 °C too. The rings-specimens cut out from the pipes of the reforming furnace were tested. The local heating by electric current was performed. The testing procedure of short cracks and calculation of nominal stresses for these specimens was described earlier (Romaniv et al., 1990). A crack closure was registered during sequence of the experiments.

RESULTS OF MODEL EXPERIMENT

Crack growth rate da/dN in aluminium, depending on ΔK_{eff} or $\Delta\sigma$ (a range of effective SIF and nominal stresses in a net section) for different crack lengths can not be described by a common dependence (Fig. 1, 2). However, let us consider the variation of these values at constant crack growth rate in a wide range of crack lengths (Fig. 2). They show a reverse response to variation of stress state at the crack tip caused by change of the crack length a. Evidently, together with ΔK_{eff} and $\Delta\sigma$ they can check the fatigue fracture kinetics.

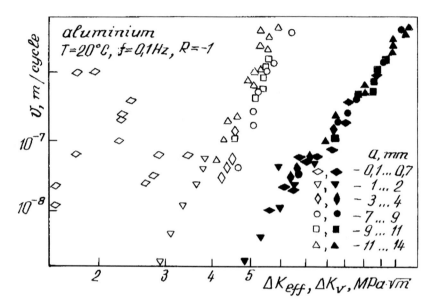

Fig. 1. The curves da/dN-ΔK_{eff} (light symbols) and da/dN-ΔK (dark symbols) for various crack lengths.

This supposition agrees with the conception of two-parametric approach according to which the plastic deformation in the developed plastic conditions is caused both by the influence of singular and regular compounds of the stress tensor. In case of ductile materials cyclic loading it can be presented as follows: instead of the curve $da/dN=f(\Delta K)$ we seek a dependence $da/dN=f(\Delta K, \Delta\sigma)$.

To obtain the above dependence we construct the curves of constant crack growth rate da/dN, which are the fracture curves analogues and are the surface level lines of $f(x,y)$. Equation of the constant crack growth rate was assumed in implicit form like (Vasiutin, 1988):

$$\left(\frac{\Delta\sigma^{+}}{\Delta\sigma_{v}^{+}}\right)^{\frac{1}{n}}+\left(\frac{\Delta\sigma^{+}}{\Delta\sigma_{ys}^{+}}\right)^{\frac{1-n^{*}}{n^{*}(1+n^{*})}}*\left(\frac{\Delta K^{+}}{\Delta K_{v}^{+}}\right)^{\frac{2}{1+n}}=1,\tag{1}$$

$$n^{*}=1\text{ if }|\sigma|<|\sigma_{ys}|;\quad n^{*}=n\text{ if }|\sigma|\geq|\sigma_{ys}|$$

In this case $\Delta\sigma_{v}^{+}$, ΔK_{v}^{+} is the range of nominal stresses and SIF values in a tensile semicycle, that cause crack growth rate $da/dN = v$ in the following conditions:

$$v=f(\Delta K_{v},0),\quad v=f(0,\Delta\sigma_{v})\tag{2}$$

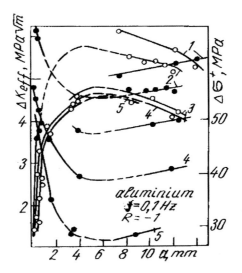

Fig. 2. The plots of ΔK_{eff} - a ($\circ$) and $\Delta\sigma$ - a ($\bullet$) for different rates of fatigue crack growth: 1 - 4.5x10^{-6} m/cycle, 2 - 1x10^{-6} m/cycle, 3 - 1.2x10^{-7} m/cycle, 4 - 6x10^{-8} m/cycle, 5 - 3x10^{-8} m/cycle.

Thus, these curves describe combinations of loading parameters $\Delta\sigma_v^{+}$, ΔK_v^{+} which give rise to a certain crack growth rate v in a wide range of crack lengths and loading conditions (up to fulfilment of LEFM conditions). For restoration of function f a hypothesis has been adopted that the constant crack growth rates curves are similar. They can be built basing on one constant rate curve ($da/dN = v$), considered to be a basic one, and using fracture kinetics data in case of other crack growth rates, obtained in a narrow range of crack lengths when LEFM requirements were not satisfied. For this purpose we use equation (1) and an equation:

$$\Delta K_v^{+} = \left\{ \left(\frac{\Delta\sigma^{+}}{\Delta\sigma_{\bar{v}}^{+}} \right)^{\frac{1}{n}} + \left(\frac{\Delta\sigma^{+}}{\Delta\sigma_{ys}^{+}} \right)^{\frac{1-n^*}{n^*(1+n^*)}} * \left(\frac{\Delta K^{+}}{\Delta K_{\bar{v}}^{+}} \right)^{\frac{2}{1+n}} \right\}^{\frac{n+1}{2}} * \Delta K_{\bar{v}}^{+}. \tag{3}$$

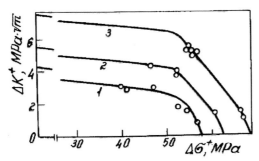

Fig. 3. The curves of constant crack growth rate: 1 - 6x10^{-8} m/cycle, 2 - 1.8x10^{-7} m/cycle, 3 - 1x10^{-6} m/cycle.

BOUNDARY CASES ANALYSIS

Construction of dependence of fracture kinetics using the two parametric approach, allows prediction of the kinetic fatigue fracture curves shape in the near boundary cases: when the LEFM condition is satisfied or when the crack length is small.

In the first case, experimentally obtained curves da/dN - ΔK_v were transformed into calculational ones da/dN - ΔK_{eff}, which are invariant with respect to the crack length, including short cracks (Fig. 1). Therefore, a specific behaviour of short crack in plastic materials is determined not only by absence of a closure effect but also by non-effective usage of LEFM approaches.

It is also interesting to analyse in terms of two-criteria approach a second case, when a crack in the ideal case does not form stress razors and fracture is defined only by nominal stresses range in a net-section. In this case it can be neglected and the loaded specimen is considered to be smooth, a section of which corresponds to a net section.

As in the first case, intersection of a constant rate curve with axis $\Delta\sigma$ caused a wide range of $\Delta\sigma_v^+$ values, that describe the fatigue fracture process with a crack growth rate $da/dN = v$ from the smooth surface. In other words, it monitors the crack nucleation. It was conditionally assumed that crack initiation stage goes on for N loading cycles, during which the crack of length $a = 0.5$ mm is formed from the smooth surface. Proceeding from the above said the kinetic diagram da/dN - ΔK_{eff} was replotted into a durability curve (the Veller curve type) in coordinates $\Delta\sigma_v^+$ - N.

This approach was experimentally proved by the fatigue smooth specimens (specimen cross-section 10x5 mm, $f = 0.1$ Hz, $R = 1$) testing. The agreement of the calculated and experimental data (Fig. 4) makes it possible to use the two-criteria approach for prediction of the crack initiation stage in the low-cycle fatigue crack using the data of evaluation of fatigue crack growth rate in the high-amplitude loading area.

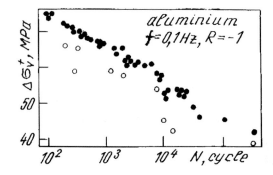

Fig. 4. Estimated (●) and experimental (○) dependences of the cycle number before 0.5 mm crack initiation on nominal stress range.

HIGH-TEMPERATURE TESTING OF HK-40 STEEL.

Influence of Loading Frequency.

The obtained effective kinetic diagrams, which took into account crack closure were replotted, using the deformation two-parametric fracture criterion, into da/dN - ΔK_v curves which correspond to the conditions of legitimate application of LEFM criteria (Fig. 5).

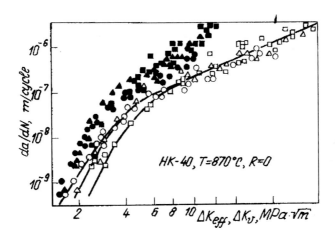

Fig. 5. Dependencies da/dN - ΔK_v (○, △, □), da/dN-ΔK_{eff} (●, ▲, ■) at $f = 0.08$ Hz (■, □), 1 (▲, △), 10 (○, ●)

The difference between experimental and calculated data is observed in the overage tested range of velocities. Even at near threshold hold velocities application of LEFM approaches for evaluation of high-temperature fatigue crack growth is incorrect. Irrespective of v value, the calculated crack propagation rate data at high loading levels form one band in the da/dN -ΔK_v coordinates. This shows that in case of high temperature fatigue crack growth rate at high ΔK values could be frequency insensitive. However, the deformation two-parametric fracture criterion application does not allow to describe the fatigue crack growth rate on the first region of kinetic diagram.

Influence of Stress Ratio.

Taking into account the crack closure, da/dN-ΔK_{eff} curves (Fig. 6) plotted for various R coincide only in the nearthreshold region. With ΔK increase the crack closure effect diminishes and at the beginning of the second region it disappears. The R effect on crack closure is substantial in spite of the crack closure absence. Alongside the obtained due to two-parametric fracture criterion da/dN-ΔK_v curve appears to be independent on R in the overall range of fatigue crack growth rates and can be considered material characteristics.

Thus, only in case of legitimate application of LEFM approaches we can obtain in effective coordinates the invariant (in reference to stress ratio) kinetic diagram of fatigue fracture.

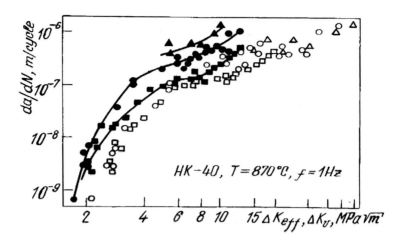

Fig. 6. Depedences da/dN - ΔK_{eff} (●, ▲, ■) and da/dN - ΔK_v (○ , △, □)
at R = -1 (□, ■), 0 (○, ●), 0.4 (△, ▲).

CONCLUSIONS.

1. A two-parametric fatigue fracture criterion is proposed based on constant rate curves of crack growths.

2. Such approach allows to describe the growth rate of fatigue cracks in a large range of crack lengths (including short cracks), construct calculated kinetic diagrams of fatigue fracture for LEFM conditions and predict life time of initiation stage of a fatigue crack nucleating from the smooth surface.

3. The use of the proposed approach when studying high-temperature fatigue crack growth in heat-resistant steel HK-40 makes it possible to take into account the effect of frequency and amplitude of loading cycles at high SIF level when a variation of crack growth rate is induced by a variation of plastic deformation intensity.

REFERENCES.

Morozov, Ye.M. (1975). Strength calculation in the presence of crack. In: *Strength of Materials and Structures* (G.S. Pysarenko, ed.), pp. 323-330. Naukova Dumka, Kyiv.

Newman, J.C. (1973). Fracture analysis of surface - and throughcracked sheets and plates. *Eng. Fract. Mech.* - 5, 667-689.

Romaniv, O.M., Nykyforchyn, H.M., Student, O.Z. and Skrypnyk, I.D. (1990). Analysis of high-temperature fatigue crack growth in corrosion resistant steel using the two-parametric fracture criterion. *Physicochemical Mechanics of Materials*, 26, No 5, 9-19.

Vasjutin, A.N. (1988). On the strength criterion of material with short cracks. *Physicochemical Mechanics of Materials*, 24, No 3, 68-74.

EFFECTS OF OVERLOADS ON THE FATIGUE STRENGTH OF STEEL COMPONENTS

L. JUNG [1], G. BREMER [1] and H. ZENNER [2]

[1]Volkswagen AG, Betriebsfestigkeit, 1712, 38436 Wolfsburg, Germany
[2]TU Clausthal, Institut für Maschinelle Anlagentechnik und Betriebsfestigkeit
Leibnizstraße 32, 38678 Clausthal-Zellerfeld, Germany

ABSTRACT

Overloads can be measured on every cyclically loaded construction. The effects of overloads range from static failure and additional damage to changes in the local stress-strain state. Random fatigue tests with and without overloads under torsion, bending and axial loading show lifetime reductions through overloads, all except one test series. Relatively these reductions are much more severe for the failure criterium crack initiation than for fracture. The nominal stress concept as well as the local stress concept can not catch all observed effects of overloads.

KEYWORDS

Overloads; random fatigue tests; lifetime prediction; nominal stress concept; local stress concept; damage accumulation

INTRODUCTION

The analysis of load histories measured on vehicles, machines or structures shows "special events" which lead to much higher loads than the "normal" service loads (Westermann and Zenner, 1988), Fig 1. They are very rare and only longtime measuring can determine a probability of their occurrence. In (Gehlken, 1992) for example, the probability of occurrance is numbered at 1: $1.3 \cdot 10^{-6}$ for the designed lifetime. The reasons for these special events or overloads respectively, are accidental situations, operating trouble, misuse or resonance. The fact that overloads can not be avoided raises up the question as to how these overloads influence the lifetime of components and how they can be taken into account for a lifetime calculation. In the literature a lot of investigations deal with overloads during crack propagation. They were considered predomenantly in the design process of aircraft or reactor constructions, especially to guarantee the damage tolerant design and show lifetime reduction as well as increasing lifetime. Only a few investigations deal with the design for mechanical engineering parts under service loads with and without overloads, but show lifetime reduction due to overloads.

GENERAL EFFECTS OF OVERLOADS

• Overloads can be seen as static loads which can cause plastic deformation, instability or static failure. This has to be considered during the design process.

- According to Miner's Rule, overloads cause additional damage at cyclic loading which depends on the slope of the S-N-curve, Fig 2. Only a few cycles cause more damage for a flat S-N-curve of k = 20 (torsion) than for a steeper S-N-curve with k = 5 (axial loading, bending). During cyclic loading the endurance limit decreases with the result that the S-N-curve becomes increasingly steeper.

- Overloads can change the local stress-strain situation in a notch. Tension overloads will lead to compressive residual stresses and compression overloads will lead to tensile residual stresses. Depending on the material, these residual stresses will be relaxed more or less through the cyclic loading that follows and lead to more or less transient effects.

- During crack propagation overloads increase or decrease the crack opening stress S_{op} so that for further crack growth an effective stress S_{eff} is relevant. The crack opening stress S_{op} changes after the overload cycles during the cyclic loading.

EXPERIMENTAL INVESTIGATIONS

To study the effects of overloads on lifetime random fatigue tests without (= normal loads) and with interspersed blocks of overloads were carried out for torsional and axial loading as well as for bending, (Jung, 1993; Bremer and Zenner, 1996). The different parameters are summerized in Table 1. The experiments show that overloads influence crack initiation lifetime and crack propagation lifetime very differently. With the above named effects it is possible to explain the observed differences in the different lifetime phases (crack initiation, crack propagation).

Under torsional loading the lifetime for tests with overloads decrease significantly compared to those without overloads. The factor of lifetime reduction depends on the number of cycles of the overload spectrum (H_s), the load direction for the overload cycles of the ratio $R_s = -0,5$ and the load level. Because of the flat S-N-curve the damage caused by the overloads is much greater than the positive transient effect through changes in the residual stress state.

The results of the investigated welded cruciform joint under axial loading show no significant change in lifetime through tension overloads ($R_s = 0$), but do show a lifetime reduction for compressive overloads ($R_s = -\infty$), (Jung, 1993). In this case two effects superpose. For both test series the additional overloads are damaging so that lifetime reduces, but for the tension overloads there is a positive effect through the induced compressive residual stresses. For the tests with compressive overloads the negative effect will be intensified through induced tensile residual stresses.

The observed small scatter band of the results under bending emphasises that the changes in lifetime resulting from different overload parameters are significant, (Bremer and Zenner, 1996). Overloads with the stress ratio $R_s = -1$ reduce lifetime while overloads with the stress ratio $R_s = 0$ increase lifetime. The change in lifetime is moderate so that safety factors which will be presumed for a lifetime estimation will take them into account.

In Fig 3 all results for the failure criterium "fracture" are gathered by the way of presenting relative experimental data (N_s/N_o) as a function of the calculated damage sum of the overload spectrum D_s divided by the damage sum of the normal load D_o according to Miner's Rule (elementary). The ratio D_s/D_o is influenced basicly by:

- the modifications of Miner's Rule, which consider the loads below the endurance limit S_D differently (Miner original, Miner elementary, Miner modified by Haibach, Miner modified by Liu and Zenner)

- the slope k of the S-N-curve, Fig 2

- the ratio of the maxima of the spectra $\hat{S}_s/\hat{S}_o$

- the ratio of the maxima of the spectra $\hat{S}_s/\hat{S}_o$ towards the endurance limit S_D
- the ratio of the number of cycles of the spectra $\hat{H}_s/\hat{H}_o$

The bold line represents the theoretical relationship between relative damage D_s/D_o and relative lifetime N_s/N_o for the nominal stress concept. Points above the line indicate that the decrease in lifetime through overloads is not as great as one would expect according the calculated relative damage and vice versa, (Bremer and Zenner, 1996). Within a scatterband of ± 2 of this theoretical line the nominal stress concept gives good approximations of the observed experimental lifetime with overloads. The test series with a flat S-N-curve (torsional loading) as well as those with axial compressive overloads $R_s = -\infty$ can not be caught by the nominal stress concept sufficiently. For further informations on the effects of compressive overloads see (Buschermöhle et al., 1996).

For the failure criterium "first technical crack" Fig 4 gives an easy approximation how to consider overloads for a lifetime calculation for crack initiation.

CONCEPT TO TAKE OVERLOADS INTO ACCOUNT

Lifetime calculations for all test series show that neither the nominal stress concept nor the local stress concept, although the calculation were done cycle by cycle, can catch all observed results because a) the nominal stress concept only calculates additional damage through overloads, b) the local stress concept has no standardized way of handling the numerous input-parameters. The range of calculated damage sums ($N_{experiment}/N_{calculation}$) from 0.1 until 25 documents this. To improve the scatter of the damage sums in (Jung, 1993) a concept to consider overloads in lifetime calculation for the criterium "fracture" was suggested and modified in (Bremer and Zenner, 1996) for lifetime calculation until "first technical crack".

This concept quantifies the effect of overloads to the damage for the normal load spectrum globaly by calculating the change of the local mean stress $\Delta\sigma_m$ through the maximum overload cycle for a cycle which most often occurs in a spectrum. In connection with the stress concentration factor K_t and the residual stress sensivity m^* of the material, (Macherauch and Wohlfahrt, 1985), a nominal stress ΔS_o is calculated which has to be added to the nominal stress of the normal load spectrum, Fig 5. For this new effective normal load spectrum a damage calculation according to the nominal stress concept can be carried out. Depending on what sign $\Delta\sigma_m$ gets the damage sum will be greater or less than the damage sum of the original normal load spectrum and a negative or positive effect can be quantified.

The aim of the future work is to extend the proposed concept to the phase of crack propagation.

ACKNOWLEDGEMENT

Financial support for this investigation by the "Bundesministerium für Wirtschaft (BMWi)" and "Arbeitsgemeinschaft industrieller Forschungsvereinigungen e.V. (AIF)" is gratefully acknowledged. The authors wish to express special thanks to the members of the working group "Lastkollektive" of the "Forschungsvereinigung Antriebstechnik (FVA)" under guidance of Dr. W. Fischer and of the working group "FA 9" of the "Deutscher Verband für Schweißtechnik (DVS)" under guidance of Prof. Dr. D. Radaj for valuable contribution to discussions.

REFERENCES

Westermann-Friedrich, A. and H. Zenner (1988). Sonderereigniskollektive - Kennzeichnende Zeitfunktionen und Kollektive für Anlagen und Arbeitsprozesse. Forschungsheft FVA, Heft 274

Gehlken, Ch. (1992). Analyse von Betriebsbeanspruchungen zur lebensdauerorientierten Auslegung verfahrenstechnischer Maschinen. Dissertation TU Clausthal

Jung, L. (1993). Einfluß von Überlasten auf die Bauteillebensdauer. Dissertation TU Clausthal

Bremer, G. and H. Zenner (1996). Einfluß von Überlasten auf die Bauteillebensdauer. Forschungs-heft FVA

Macherauch, E. and H. Wohlfahrt (1985). Eigenspannung und Ermüdung in "Ermüdungsverhalten metallischer Werkstoffe". Hrsg. D. Munz, DGM-Informationsges. Verlag

Buschermöhle, H., D. Memhard, D and M. Vormwald (1996). Fatigue crack growth acceleration or retardation due to compressive overload excursions. Fatigue '96, Berlin 1996

Kloth, W. and H. Stroppel (1936). Kräfte, Beanspruchungen und Sicherheiten in den Landmaschinen. VDI Zeitschrift, Bd. 80 Nr. 4, page 85 -92

	investigation 1	investigation 2	investigation 3
loading	torsion	axial tension	bending
specimen and stress concentration factor K_t	notched shaft $K_t \approx 1.3$	welded cruciform joint $K_t \approx 2.0$	notched round bar $K_t \approx 1.8$ and 2.9
material	42 CrMo 4 V	S355J2G3 (St 52-3)	42 CrMo 4 V
load history	a) woehler test b) random load with gaussian standard	a) woehler test b) random load with gaussian standard	a) woehler test b) random load with gaussian standard
stress ratio normal loads	$R_o = 0$	$R_o = -1$	$R_o = -1$
stress ratio overloads	$R_s = 0$ and $R_s = -0.5$	$R_s = 0$ and $R_s = -\infty$	$R_s = -1$ and $R_s = -0.5$
cumulative cycles of the normal load spectrum H_o	10^5	$5 \cdot 10^5$	$5 \cdot 10^5$
cumulative cycles of the overload spectrum H_s	$10^0, 10^1, 10^3$	$5 \cdot 10^1, 5 \cdot 10^3$	$5 \cdot 10^0, 5 \cdot 10^1, 5 \cdot 10^3$
ratio of spectrum maxima $\hat{S}_s / \hat{S}_o$	1.23	1.5	1.5 and 1.25
measured lifetime	crack initiation => N_A and fracture => N_B		

Table 1: Test parameters of three investigations to analyse the effects of overloads

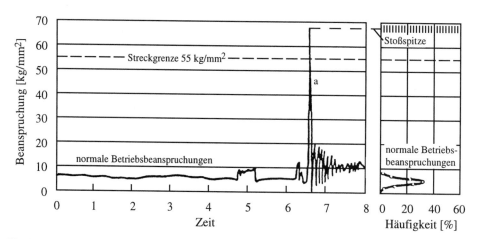

Fig 1: Measured overload at a frame of a plough as a result of a collision with a stone, (Kloth 1936)

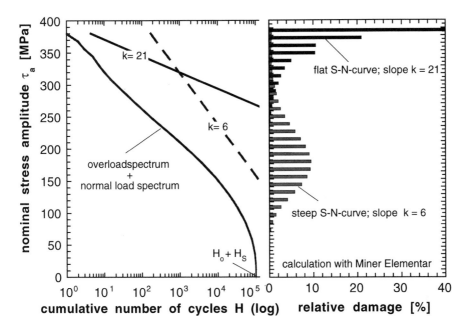

Fig 2: Influence of the slope of the S-N-curve to the relative damage of a load spectrum

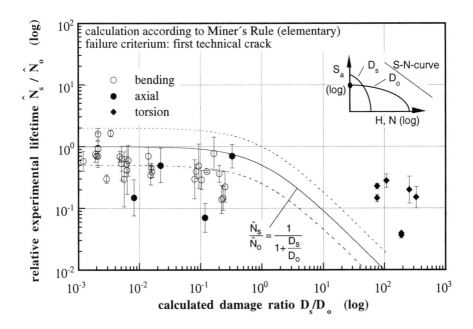

Fig 4: Effects of overloads to lifetime until first technical crack

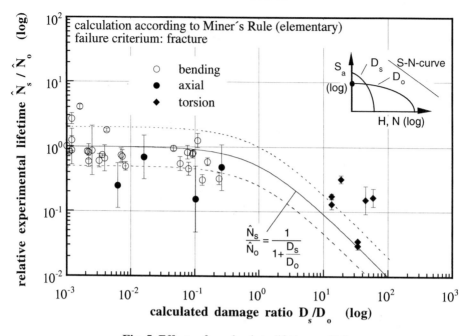

Fig. 5: Effects of overloads to lifetime until fracture

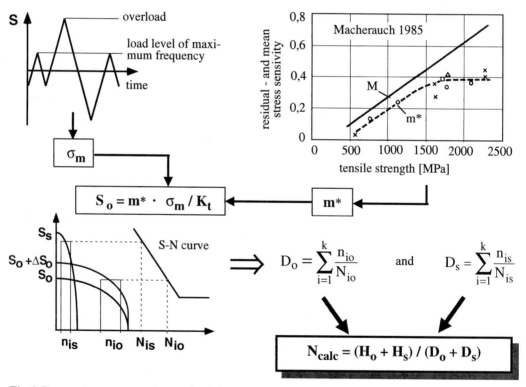

Fig 6: Proposed concept to take overloads into account for lifetime calculation until crack initiation

FAILURE MODE BELOW 390 K WITH IMI 834

G. Marci

DLR - German Aerospace Research Establishment
Institute for Materials Research
D-51140 Cologne

Abstract

An abnormal crack growth behavior is observed for the IMI 834 turbine disk material if K_{max} is higher than $^oK_{max}$ => 26 MPa$\sqrt{m}$ and the cyclic plastic zone is smaller than that coresspponding to the material property ΔK_T.

Introduction

It is a very well known fact that crack parts made of Ti-alloys can faile substantially below the plane strain fracture toughness K_{IC} (ASTM E 399). If Ti-structures are loaded under sustained load for a longer time, failure can occur at loads which are much below the load previously carried by the structure for some time (for example, for an hour). But to the author's knowledge there is no mention in the scientific literature that a low temperature (below approximately 390 K) a different failure mechanism might exist for Ti-alloys. The present publication reports on preliminary findings which, at least, warrant more research efforts to clarify this particular failure mode.

Test Procedures and Materials

C (T)-specimens were used in this investigation; they were machined out of a IMI 834 disk forging with B = 10 mm and W = 50 mm and RC- and CR orientation. FCP rates da/dN versus ΔK for IMI 834 were determined in accordance with ASTM E 647 (R-constant) and with the K_{max}-method (see Fig. 1). The test frequency for determination of FCP-rates and threshold values was 50 Hz. Automated test procedures for FCP rates and threshold testing were used [1] with on-line data acquisition, data processing and machine control. A DC-Potential Drop method was used to measure the crack length. The R-constant test method was used to determine da/dN-ΔK data for R = 0.1, 0.3, 0.5, 0.6, 0.7 and 0.75. In an initial series, da/dN-ΔK data were deter-mined with the K_{max}-constant method for K_{max} = 15, 20, 25 and 30 MPa√m. The lowest da/dN value was determined over 0.1 mm FCP in 10^6 or more cycles, i.e., da/dN < = 10^{-7} mm/cycle. Subsequently, a second series of K_{max}-constant tests were executed with K_{max} = 24, 26, 28, 30 and 32 MPa√m. For this test series, the lowest da/dN to be determined was da/dN < = $5 \cdot 10^{-8}$ mm/cycle, i.e., 0.1 mm FCP in 5 • 10^6 or more cycle. The plane strain fracture toughness K_{IC} for IMI 834 ranged from 38 to 42 MPa√m, the higher K_{IC} values were found in the midsection of the forging .

Results

Figure 2a shows the da/dN-ΔK data from R-constant test method while Fig. 2b shows these obtained with the K_{max}-constant test method. In Figure 2b, some of the da/dN-ΔK data for K_{max} = 30 MPa√m i.e., for low ΔK's, were omitted because they deviated considerably from their expected value based on the results obtained for the da/dN-ΔK curves with lower K_{max}. The total da/dN-ΔK curve for K_{max}= 30 MPa√m is given in Fig. 3, where it has to be observed that the abscissa (Δ K) starts with log 0.1.

Because of the unusual curve in Fig. 3, tests with K_{max}-constant method were repeated to obtain da/dN-ΔK data for K_{max} = 24, 26, 28 and 30 MPa√m. The da/dN-ΔK-curves are shown in Fig. 4. In order to separ-ate the individual curves, the da/dN-ΔK-curves with K_{max} = 30 MPa√m (same curve as in Fig. 4) and the one with K_{max} = 32 MPa√m are shown in Fig. 5.

Discussion of Results

There is still an ongoing discussion about the meaning of da/dN-ΔK data from the K_{max}-constant method compared to that obtained with the

R-constant method (ASTM E 647). Comparing Fig. 2a with 2b, one finds a visible difference between the two data sets in terms of da/dN versus ΔK. But is generally accepted by the researchers in the field of FCP that the ΔK_{eff}-concept furnishes a more precise description of the FCP process. The data in Fig. 2 had been converted into the ΔK_{eff}-concept in Ref. 2, namely into da/dN versus $(K_{max}-K_{op})$ curves, and were shown to be identic. As far as the topic of the present investigation is concerned, it is important to note that the R-constant and the K_{max}-constant test procedures are equivalent. The details of the FCP curves as shown in Fig. 2 are only to be considered as a basis for comparison for the following discussion.

Since all tests were conducted with the cycle frequency of 50 Hz, every cycle corresponds to a fixed time. If sustained loading was responsible for crack propagation shown in Figs. 3, 4 and 5 all tests with K_{max} >
$^oK_{max}$ = 26 MPa$\sqrt{m}$ should result in a constant da/dN for ΔK < FCP threshold. That this is not the case is clearly visible in Fig. 3, 4 and 5. On the other hand, all da/dN-ΔK curves obtained with K_{max}-constant test method for K_{max} < = 26 MPa$\sqrt{m}$ approach da/dN values < 10^{-7} mm/cycle (Fig. 2b) or approach < $5 \cdot 10^{-8}$ mm/cycle (Fig. 4) with no indication that sustained load crack propagation is contributing to the total crack propagation.

Looking at the curves in Figs. 3, 4 and 5, one sees the for ΔK > 4 MPa$\sqrt{m}$, the da/dN data from K_{max}-constant tests with K_{max} = 28, 30 and 32 MPa$\sqrt{m}$ are in line with data for the same ΔK from other test. It must be concluded that the da/dN-ΔK curves for K_{max}-constant > 26 MPa$\sqrt{m}$ with ΔK < 4 MPa$\sqrt{m}$ are not containing any part of the "conventional" sustained load induced cracking. It should be documented here that the da/dN-ΔK curves shown in Figs. 3 and 5 contradict the fracture mechanics concept of FCP. With the fracture mechanics concept of FCP it is the basic assumption that FCP is caused by and proportional to the cycle by cycle load excursions (for FCP rates roughly larger than 10^{-7} mm/cycle, i.e., a Burger's vector per cycle). As can be seen in Figs. 3 and 5, the crack propagation rates are larger than 10^{-7} mm/cycle, but there is no general correspondence to the load excursions.

In the author's opinion, it is not so much troubling that $^{o}K_{max}$ = 26 MPa$\sqrt{m}$ for IMI 834 is very low, but more concern is caused by the fact that the metallurgical and mechanical conditions connected with this abnormal crack growth behaviour are completely unknown.

Conclusion

Abnormal crack growth behaviour is observed in IMI 834 under high mean load and very small cyclic load excursion. For monotonic plastic zones corresponding to load levels above $^{o}K_{max}$ and cyclic plastic zone sizes smaller than a certain, limit size, an abnormal crack growth mechanism can start, where the crack growth rates are not any more proportional to the amplitudes of cyclic loading. Similarly, for small cracks in a uniformly strained body which, because the small crack size, have a small ΔK, i.e., a small cyclic plastic zone, the abnormal crack growth behaviour is observed, too.

References

1.   Bachmann, V., Marci, G. and Sengebusch, P. (1994). Procedure for Automated Tests of Fatigue Crack Propagation. STP 1231, p. 146, ASTM, Philadelphia (1994)

2.   Marci, G. Indien-German Workshop, Bangalore, India (1994) Fatigue Crack Propagation: Testing, Parameters and Concepts. to be published in Int. Journ. for Engineering Analysis and Design.

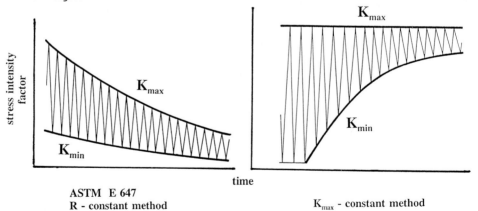

Fig. 1 Test methods for determination of FCP .

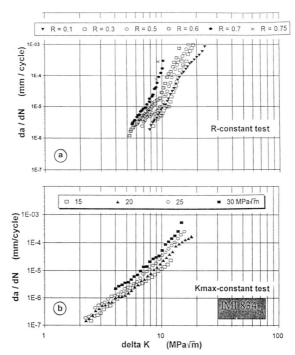

Fig. 2 FCP-rates as a function of ΔK for IMI 834,

a - obtained with the R-constant method,

b- obtained with the K_{max}-constant method.

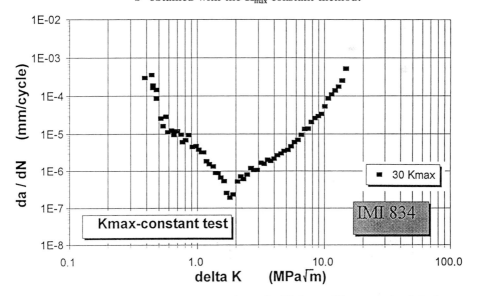

Fig. 3 FCP-rates as a function of ΔK for a K_{max}-constant test

with K_{max} = 30 MPa√m.

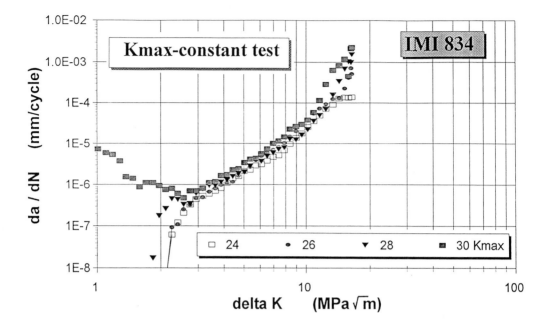

Fig. 4 FCP-rates as a function of ΔK for IMI 834 of K_{max}-constant tests with K_{max} = 24, 26, 28 and 30 MPa√m.

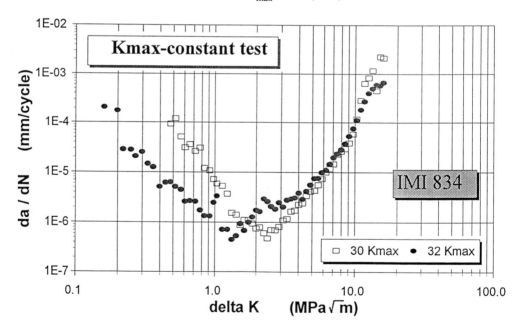

Fig. 5 FCP-rates as a function of ΔK for IMI 834 of K_{max}-constant tests with K_{max} = 30 and 32 MPa√m.

COMPLEX STRESS-TIME CYCLES INFLUENCE ON AIRCRAFT ENGINE PARTS FATIGUE STRENGTH

V.I.Astafiev, D.G.Fedorchenko and I.N.Tzypkaikin

Samara State University, Pavlov St. 1, Samara 443011, Russia
SSTC "NK Engines", Lazo St. 1, Samara 443026,Russia

ABSTRACT

An investigation was conducted to determine the influence of complex stress-time cycles on the fatigue strength of aircraft engine parts. The design method and the results of experimental investigation are presented. Comparison are made of the test results with the results of calculations.

KEYWORDS

Fatigue; biharmonic stress-time cycle; stress-time cycle with endurance.

INTRODUCTION

Heavy-duty machine parts operating life, in particular aircraft engine parts, depends on their low-cycle fatigue (LCF) strength. Machine parts are subject to forces which induce complex stress-time cycles. Fig. 1 shows an example of typical stress-time cycle of engine parts during a flight.

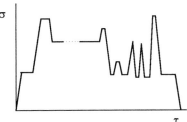

Fig.1.   Typical stress-time cycle of engine parts during a flight

Real stress-time cycle can be represented as some harmonic components sum, as biharmonic cycle for example:

$$\sigma(t)= \sigma_1 Sin(\omega_1 t)+\sigma_2 Sin(\omega_2 t-\rho), \tag{1}$$

where σ_1, ω_1 - amplitude and frequency of low-frequency ($\omega_1<\omega_2$) component; σ_2, ω_2 - amplitude and frequency of high-frequency component; ρ - phase angle.

FATIGUE LIFE VARIATION
UNDER COMPLEX STRESS-
TIME CYCLE

Many authors investigated fatigue strength of specimens under biharmonic stress-time cycle (1). Their investigations analysis shows that fatigue life variation under biharmonic cycle as compared to fatigue life under harmonic cycle with the same amplitude practically does not depend on material properties, stress concentration, test temperature, residual stress and loading type. Most important factors are the parameters of stress-time cycle (1)– stress ratio $k=\sigma_2/\sigma_1$, frequency ratio $f=\omega_2/\omega_1$ and phase angle ρ. Therefore cycle (1) can be represented as follows:

$$\sigma(t)=\sigma_1[Sin(t)+kSin(ft-\rho)]. \tag{2}$$

s_1–s_2 Diagrams

Experimental data can be represented on s_1-s_2 diagram, where
$s_1=\sigma_1/\sigma_{f1}$;
σ_{f1}– amplitude of a harmonic cycle (with frequency ω_1) with the same fatigue life as biharmonic cycle has;
$s_2=\sigma_2/\sigma_{f2}$;
σ_{f2}– amplitude of a harmonic cycle (with frequency ω_2) with the same fatigue life as biharmonic cycle has.

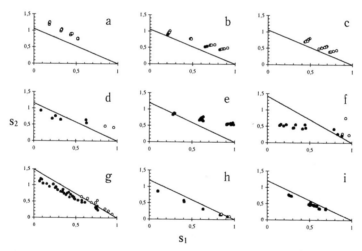

Fig.2. s_1-s_2 diagrams. a- Steel 20, f=2, $\rho=0°$; torsion (Garf, 1960); b-
SAE4340, f=2, $\rho=0°$; c- SAE4340, f=2, $\rho=90°$; tension (Starkey et al.,
1957)); d- Steel 00H12N3D, f=2.7, $\rho=0°$; bending (Zaitzev et al.,
1975); e- Steel 45, f=3; bending (Filatov, 1966); f- S45C, f=11.6;
bending (Tanaka et al., 1969); g- SS41, f=12.5; bending (Tanaka,
1968); h- Steel TS, f=16.5; tension (Buglov, 1975); i- Steel 0.23%C,
f=40; bending (Yamada et al., 1966)

Fig.2 shows s_1-s_2 diagrams based on various experimental data. Straight lines are plotted in Fig.2 on the basis of the following equation:

$$s_1=1-f^{-1/m}s_2, \tag{3}$$

where m- power rate in S–N curve equation $\sigma^m N = C$. The straight lines are borders of zones with fatigue life lowering and growth. The point on diagram above the line (3) means that fatigue life under cycle (1) is greater than the fatigue life under harmonic cycle:

$$\sigma(t) = (\sigma_1 + \sigma_2)Sin(\omega_1 t) \qquad (4)$$

and vice versa. The size of fatigue life growth zone decreases with increasing of the frequency ratio f. It is necessary to note that stress range S_{max} of biharmonic cycle (1) depends on phase angle ρ and S_{max} usually less than double sum of component amplitudes $2(\sigma_1 + \sigma_2)$. Therefore fatigue life growth under cycle (1) as compared to cycle (4) does not mean that fatigue life is greater than fatigue life under cycle:

$$\sigma(t) = (S_{max}/2)Sin(\omega_1 t). \qquad (5)$$

The light points in Fig.2 correspond to fatigue life growth under cycle (1) as compared to cycle (5), the dark points correspond to life lowering. Fatigue life growth zone exists as shown in Fig.2, especially for small frequency ratio (f<20).

Authors' Experimental Data

LCF tests were performed for investigation of the influence of the complex stress-time cycle on the fatigue strength. Stress-time cycle used in the tests is shown in Fig.3. This cycle has endurance $\tau_c = 30$ sec. for simulating of a starting and a warming of the aircraft engine. Round (Ø5 mm) nickel base alloy specimens with V-type notch (stress-concentration factor $\alpha_\sigma = 3.08$) were tested. Test temperature– 700°C.

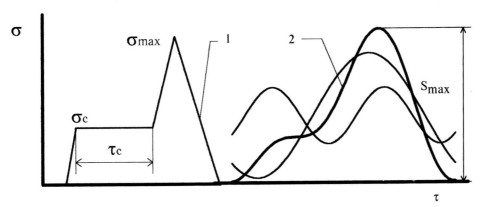

Fig.3.   Test's stress-time cycle (1) and its biharmonic approximation (2)

Experimental S–N curves are shown in Fig.4. The experimental data are given in Table 1 (non-failure specimens are signed *). The experimental data for other nickel alloy specimens with temperature 650°C ([1]) and titanium alloy specimens with temperature 450°C ([2]) are also given in Table 1.

Statistical Analysis of Experimental Data

All experimental fatigue lives were reduced to one stress level with the help from follow equation:

$$N^* = N(\sigma^*/\sigma)^m, \qquad (6)$$

where σ– stress level; N– fatigue life; σ^*– most representative stress level ($\sigma^* = 700$ MPa); N^*– life that reduced to stress level σ^*.

Table 1.   Experimental data

$\sigma_c/\sigma_{max}=0$		$\sigma_c/\sigma_{max}=0.3$		$\sigma_c/\sigma_{max}=0.6$		$\sigma_c/\sigma_{max}=0.8$	
σ_{max}, MPa	N, number of cycles	σ_{max}, MPa	N, number of cycles	σ_{max}, MPa	N, number of cycles	σ_{max}, MPa	N, number of cycles
1000	252	1200	180	1200	80	1200	120
1000	225	1000	440	1100	177	1100	183
900	663	900	504	1000	598	1000	244
900	573	800	1253	900	543	900	415
800	857	700	1942	850	950	800	464
750	1366	700	1614	800	844	700	1314
700	1507	680*	2005	780	1060	700	1190
670*	2002	680*	2000	750	1710	680	1674
650*	2005			750	1095	650*	2000
600*	2070			720	990	650	2470
				720	980	600*	2042
				700	1445		
				700*	2050		
				680*	2107		
750[1]	474	750	902				
650[2]	1097	650	1222				

It can be assumed that logarithm of life has Gauss distribution. Therefore Student t–criteria can be used for comparison of the mean values of reduced lives N^*. The results of statistical analysis are given in Table 2. As result of statistical analysis it can be concluded that fatigue life growth under "complex" (with endurance) stress-time cycle with $\sigma_c/\sigma_{max}=0.3$ and $\sigma_c/\sigma_{max}=0.6$ as compared to "simple" (without endurance) cycle takes place with significant level $\alpha=5\%$ and fatigue life lowering under cycle with $\sigma_c/\sigma_{max}=0.8$ takes place with significant level $\alpha=10\%$.

Table 2.   Results of statistical analysis

σ_c/σ_{max}	Mean value of $\ln N^*$	Variation of $\ln N^*$	Value of Student t-criteria	Degrees of free-dom number
0	7.331	0.0520		
0.3	7.507	0.0164	-2.067	16
0.6	7.534	0.1156	-1.750	24
0.8	7.179	0.0400	1.617	20

Table 3.   Parameters of biharmonic cycle that used for the approximation

σ_c/σ_{max}	k	ρ	S_{max}
0.3	0.45	140°	2.74
0.6	0.50	200°	2.78
0.8	0.35	260°	2.45

Biharmonic approximation of "complex" stress-time cycle

"Complex" stress-time cycle with endurance can be approximated by the biharmonic cycle as shown in Fig.3. The parameters of biharmonic cycle (2) that used for the approximation are given in Table 3, where S_{max}- stress range of the biharmonic cycle with $\sigma_1=1$, $\sigma_2=k$. This approximation describes only a shape of the "complex" cycle. Using the biharmonic approximation s_1-s_2 diagram can be plotted for the cycles with the endurance. Those cycles s_1-s_2 diagram is shown in Fig.5. The experimental data (Garf, 1960 and Starkey et al., 1957) are also shown in Fig.5.

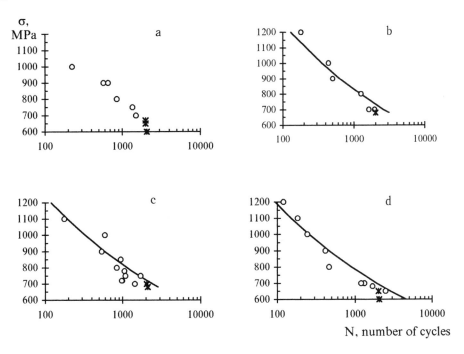

Fig.4.   Experimental data (as points) and results of calculation (as curves) with the help from equation (8). a- "simple" cycle (without endurance); b- $\sigma_c/\sigma_{max}=0.3$; c - $\sigma_c/\sigma_{max}=0.6$; d - $\sigma_c/\sigma_{max}=0.8$

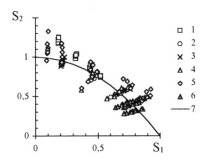

Fig.5.   s1-s2 diagram for frequency ratio f=2. 1- Steel 20, $\rho=0°$ (Garf, 1960); 2- SAE4340, $\rho=0°$; 3- SAE4340, $\rho=90°$; 4- 76S-T61, $\rho=0°$; 5- 76S-T61, $\rho=90°$ (Starkey et al., 1957); 6- authors' data; 7- curve corresponding to equation (7).

Tzeitlin and Fedorchenko (1980) proposed following equation for estimation of the influence of superimposed high-frequency stress (vibration) on the low-cycle fatigue strength:

$$s_2 = 1 - s_1^\beta, \tag{7}$$

where β– empirical coefficient. The experimental data for frequency ratio f=2 can be described using equation (7) with β=2. Therefore S–N relationship can be represented as follows:

$$(\sigma/\varphi)^m N = C, \tag{8}$$

where m and C– coefficients of S–N curve for "simple" stress-time cycle (σ_c/σ_{max}=0); φ– coefficient that can be found from a solving of the equation (7) with β=2:

$$\varphi = S_{max}\left(\sqrt{f^{2/m} k^2 + 4} - f^{1/m} k\right) / 4.$$

S–N curves are shown in Fig.4 calculated using the equation (8). Results of calculation and experimental data has satisfactory agreement as seen in Fig.4.

CONCLUSION

On the basis of above mentioned results it can be concluded:

–The equations (7), (8) can be used for estimation of the influence of the "complex" stress-time cycle on aircraft engine parts fatigue strength.

–"Complex" stress-time cycle can have a greater fatigue strength than "simple" cycle with the same stress range. Therefore engine parts operating life can be increased not only by stress level lowering, but also by the shape of stress-time cycle optimisation.

REFERENCES

Buglov, E.G. (1975). Low-Cycle Fatigue Strength and Some Properties of Structural Materials Histeresis under Biharmonic Loading. In: *Strength of Materials and Structures* (G.S. Pisarenko, Ed.), pp.148-159. Naukova Dumka, Kiev.

Filatov, M.Y. (1966). The Influence of Stress-Time Cycle Shape on Fatigue Damage Cumulation. *Prikladnaya Mehanika*, **11**, 83-89.

Garf, M.E. (1960). Fatigue Strength under Complex Stress-Time Cycle. *Zavodskaya Laboratoriya*, **26**, 94-98.

Starkey, W. and S. Marco (1957). Effects of Complex Stress-Time Cycles on the Fatigue Properties of Metals. *Transactions of ASME*, **9**, 1329-1336.

Tanaka, T. (1968). Effect of the Superimposed Stress of High Frequency on Fatigue Strength. *Bulletin of JSME*, **11**, 77-83.

Tanaka, T. and S. Denoh (1969). Effect of Superimposed Stress of High Frequency on Fatigue Strength of Annealed Carbon Steel. *Bulletin of JSME*, **12**, 1309-1315.

Tzeitlin, V.I. and D.G. Fedorchenko (1980). The Estimation of Operating Life under Simultaneously Acting of Low-Frequency and Vibrational Loadings. *Problemy Prochnosty*, **1**, 14-17.

Yamada, T. and S. Kitagawa (1967). Investigation of Fatigue Strength of Metals under Actual Service Loads (With Two Superimposed Cyclic Loadings). *Bulletin of JSME*, **10**, 245-252.

Zaitzev, G.Z. and A.Y. Aronson (1975). *Fatigue Strength of Hydro-Turbine Parts.* Mashinostroenie, Moscow.

FATIGUE CURVILINEAR CRACK IN TURBOJET BLADE BEHAVIOR

M.Nikhamkin

Perm State Technical University,
Komsomol str.29A, GSP-45, Perm, 614600,Russia

ABSTRACT

The problem of fatigue crack propagation in turbojet blades is an application of crack mechanics to the important and complex problem of engine reliability. Mathematical model of curvilinear crack propagation process in blades based on fracture mechanics methods is developed. Special experimental technique and equipments are described. The main particular feature of the technique is the ability to study crack behavior in natural blades under loading and conditions being near to real ones. Crack propagation history, life time and crack nonpropagation conditions are discussed. It is found out that curvilinear front of the fatigue crack in blade tends to stable form in its propagation.

KEYWORDS

Turbojet blades; fatigue crack; crack propagation.

INTRODUCTION

Fatigue fracture is the most frequent cause of turbojet blades break. Fatigue crack is usually the result of service induced or technological defect (outer object impact notch, stamping crack). It is impossible to exclude such defects completely. For engine serviceability provision it is important to know what material, technology, heat treatment is to be chosen in order to make a crack grow slower, what loading a crack is not dangerous at. Besides it is necessary to estimate life crack propagation to choose diagnostic check method and frequency.

The solution of these problems must be based on cyclic fracture toughness characteristics and crack propagation process analysis. The given work tasks are: to develop experimental technique for these characteristics determination; to work out material and technology choice recommendations; to develop a crack growth mathematical model for life time prediction and description of fracture kinetics.

EXPERIMENTAL TECHNIQUES

Characteristic features of turbojet blades are specific surface layer structure and residual stress. These factors are the result of blades manufacturing method and heat treatment so it is important to reproduce them in an experiment. Therefore the characteristic feature of experimental technique is use of natural blades inside the specimens.

Other characteristic feature of the experimental technique is reconstruction of loading near to real. High frequency vibration loading is combined with quasi-static one (gas-dynamic and centrifugal). Before this dynamic stress cycle is asymmetric. The blade exploitation loading consists of determined by a typical flight cycle blocks. Parameters of the block are variable because flight conditions are variable.

The experimental plant is designed to study the crack propagation processes (Nikhamkin *et al.*, 1993). The main characteristic feature of the plant is the ability to study the crack behavior of natural details under loading and conditions near to real. The plant basis is the electromagnetic vibrator. It creates high frequency loading of a tested blade under resonance vibration regime. Static loading and asymmetric loading cycle is produced by a special device attached to the blade. The plant control, information record and result calculation is produced by IBM PC type computer with special interface. This system is adaptable and allows modification of experimental procedure easily. Modification of computer program is to be done only. The software for computer experiment control is designed.

The plant provides the following test conditions: arbitrary time-dependent regime of vibration amplitude loading and heating; random loading; asymmetric loading cycle. The crack propagation rate and blades life time can be obtained under these conditions.

Several groups of turbojet high pressure compressor blades are studied. Groups differed in alloy mark, heat treatment regimes and manufacturing technology; blades form and dimensions were identical.

Crack propagation rate - stress intensity factor swing ΔK diagrams for different blades groups are obtained in the result of the experiments. Characteristic feature of all diagrams is the crack propagation threshold existence. It is found out that there is considerable difference of cyclic fracture toughness indicators (particularly - threshold stress intensity factor swing ΔK_{th}) for different blades groups. This difference is considerable even in the case of identical alloy mark. This difference is the result of different technology and heat treatment.

MATHEMATICAL MODEL OF FATIGUE CRACK PROPAGATION

Blades life time from the moment of crack initiation to breaking is the clearest characteristic for comparison of materials, manufacturing methods, heat treatment variants, blades form. It reflects the interaction of the marked technological, in-service (static and dynamic stress, duration of regimes) and structural factors. Life time was studied on the base of experimental data with the using of created mathematical model and computer program.

Crack propagation mathematical model is based on the following

simplifying assumptions. The crack grows in the plane of the most
stressed blade cross section. The crack ridge is experimentally
determined and its growth can be described by its length L only.
The crack propagation rate is determined by ΔK stress intensity
factor and is described by kinetic equation

$$dL/dN = V_o(\Delta K - \Delta K_{th})^n/(\Delta K^* - \Delta K_{th})^n \tag{1}$$

with initial condition

$$L(0) = L_o \tag{2}$$

where L - crack length, L_o - its initial value, N - high frequency
loading cycle number, ΔK - stress intensity factor swing, ΔK_{th} -
threshold stress intensity factor swing, ΔK^*, V_o, n - blade
toughness parameters.

Equations parameters are obtained in the described experiments with
natural blades. Effects of cycle asymmetry and overloading are
taken into account by means of these parameters variation. The
ability of static and dynamic stress time variation in accordance
with typical flight cycle and random regime variations are realized
in this model too.

Stress intensity factor swing is to be calculated as the result of
tree-dimension elasticity stress analyses boundary problem.
Turbojet blade has a complex geometry form, its boundary surfaces
can not be represented analytically; developed crack has a
curvilinear ridge (see figure 1). The finite elements method is
used for calculation. For a general case stress intensity factor
has three components, bat if the crack length L is smaller than
0.2-0.3 of the blade chord-line B, the normal break stress
intensity factor component is determining one for crack
propagation.

Dialogical program system for developed mathematical model
realization and blades life prediction was worked out. This
software has special data bases (loading regimes, blades
parameters, materials, results) and convenient interface. It dives

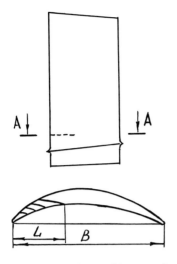

Fig. 1. Blade with crack

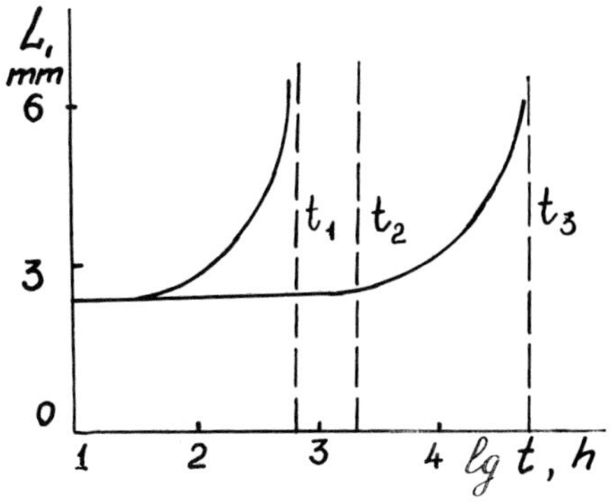

Fig. 2. Crack propagation
process   in blade

an ability of quick life prediction in the condition of engine
exploitation if the crack was initiated.

Life time analysis for different blade groups demonstrates the
nonlinear effect of structural, technological and in-service
factors interaction. This effect is the result of crack propagation
rate and threshold difference for different blade groups. Definite
operation stress and initial crack size are dangerous for some
blades and are not dangerous for others. The example of such
results are presented in figure 2. The crack grows from the initial
length step by step in every loading block to break in t_1
moment (curve 1). This calculation result turned out to be near to
the in-service blade break point t_2 observed for a real engine.
Other materials or manufacturing methods or heat treatment choice
can make this initial crack not dangerous (curve 2).

CONCLUSION

Experimental data demonstrate the considerable influence of alloy
kind, manufacturing method, heat treatment on the crack propagation
rate and threshold stress intensity factor. Threshold stress for

different variants of marked factors are distinguished in 2-3 times
even for the identical blades form and dimensions. This difference
is very important for blades life time extension: the initial crack
length may be dangerous for one type of blades and non-dangerous
for other types.

Proposed in this manuscript experimental technique and computer
programs can be useful for the choice of blades material,
manufacturing method, heat treatment regimes with the aim of blades
life time extension.

REFERENCES

Nikhamkin M., Voronov L., Konev I. (1993). Computer driven Plant
    for Research of Crack Propagation under High Frequency Loading of
    Natural Details and Specimens. In: *Proceedings of the ICF-8,*
    Vol.2, pp. 607-608. Kiev.

SMALL CRACK GROWTH AND FATIGUE LIFE PREDICTIONS FOR LY12cz CLAD Al - ALLOY UNDER SPECTRUM LOADING

H. Yu, X. R. Wu and C. F. Ding

Institute of Aeronautical Materials
P. O. Box 81 - 23, Beijing, 100095, P. R. China

ABSTRACT

The objectives of the present work were to investigate naturally - occurring small crack and large crack growth behavior of LY12cz clad alloy and to evaluate the ability of a closure - based crack - growth model, FASTRANII by Newman, to predict fatigue crack growth rates and total fatigue lives under Mini - TWIST spectrum loading. The small crack grew well below the large crack threshold and grew faster than large cracks at the same stress - intensity factor ranges under Mini - TWIST spectrum loading. The predicted small crack, large crack growth behavior and total fatigue lives were in reasonable agreement with experimental data.

KEYWORDS

Small crack; spectrum loading; crack closure; fatigue life prediction

TEST PROGRAM

The test program conducted on LY12cz alloy consisted of small - crack tests on single - edge notched tension (SENT) specimens under constant amplitude loading at $R=0.5$ and under Mini - TWIST spectrum loading, large - crack tests on center - crack tension (CCT) specimen under both constant amplitude ($R=0.5$, 0 , -1) and Mini - TWIST spectrum loading, and standard fatigue tests on SENT specimens under Mini - TWIST spectrum loading.

Material and Specimen

LY12cz clad Al alloy is similar in chemical composition and tensile properties to 2024 - T3 alloy. Its chemical composition and tensile test results are given in Table 1 and Table 2, respectively. The LY12cz sheet has a cladding layer 60 to 80 μm thick on each surface of the sheet. The cladding layer thickness was reduced to 40 - 60 μm after chemical polishing. A single - edge - notched tensile specimen (Wu and Newman, $et\ al.$ 1994) was used. The notch was semi - circular with a radius of 3.2mm and the width (w) was 50 mm. The stress - concentration factor is 3.15, based on gross - section stress. All large - crack tests were conducted on 100mm wide center - crack tension specimens. The LY12cz sheets had about 70 μm thick cladding on each surface of the specimens.

Test Procedures

Small - Crack Test. Small - crack tests were conducted at $R=0.5$ for maximum stress levels of 220MPa and 260MPa, and under Mini - TWIST spectrum loading. Teflon lined anti - buckling guide plates were loosely fastened against the specimen under Mini - TWIST spectrum loading. Crack length measurements along the bore of the notch were performed using a plastic - replica technique. Replicas were taken at 80% of the maximum load (to insure that any crack present would be open) periodically during the test at $R=0.5$ constant amplitude loading. To take replicas under the Mini - TWIST load sequence, the test machine was programmed to stop and hold at a specified peak level, number VIII in the Mini - TWIST spectrum. After the desired number of cycles had been completed, the replica was taken along the bore of the notch. The number VIII level corresponded to about 60% of the maximum applied stress. After the replica was removed from the notch, the test machine was restarted and continued from the specified peak level.

Large - Crack Test. Large - crack tests (crack length greater than 4mm) were conducted under each of the constant amplitude conditions ($R=$ -1, 0, 0.5) and under the Mini - TWIST spectrum loading. In all tests, cracks initiated and grew about 2mm at each end of the central slot before growth rate data were recorded. Crack length measurements were made visually using a micrometer slide. Two types of tests were conducted: constant amplitude and threshold. In the threshold tests, crack - growth rate data were obtained using a load - shedding (decreasing ΔK) procedure (load was decreased by 10% every 0.5mm of crack growth for rates greater than 2E-09m/cycle and 5% for rates lower than 2E-09m/cycle). After reaching threshold, some specimens were tested under load - increasing conditions to obtain more crack - growth rate data.

ANALYSIS PROGRAM

Calculation of Stress - Intensity Factors

The stress - intensity factor for a quarter - elliptical corner crack located at the edge of a semi - circular notch and the stress - intensity factor for a through crack emanating from the semi - circular notch subjected to remote uniform stress or uniform displacement developed previously(Wu and Newman et al. 1994) were used in the analysis.

Effect of Cladding

In the LY12cz clad alloy, cracks initiated in the cladding layer and grew as corner cracks into the core material. To account for the effect of the cladding, the stress - intensity factor solution for corner cracks was modified to approximately model cracks in a multiple layer medium (Wu and Newman et al. 1994). The clad - correction factor Gc was applied to all locations along the corner - crack configuration. Thus, the boundary - correction factor for the corner crack (boundary and clad) became $Fcn'=Fcn*Gc$. This clad - correction was added to FASTRANII in the present calculation.

Crack - Growth Model

A crack - growth model that accounts for crack - closure effects was used to predict small - and large - crack growth rates and total fatigue lives for LY12cz clad alloy. The crack growth model was first developed for a central through crack in a finite - width specimen subjected to remote uniform stress (Newman, 1981), and was later extended to a through crack emanating from a circular hole in a finite - width specimen also subjected to remote uniform stress (Newman, 1982). It was applied to analyze small cracks in a SENT specimen(Newman and Edwards, 1988; Wu and Newman, *et al.* 1994). The growth model is based on the Dugdale plastic - zone model, but modified to leave plastically deformed material in the wake of the crack tip.

TEST RESULTS AND PREDICTIONS

Large - Crack Tests and Analysis Under Constant Amplitude Loading

Crack growth rates against ΔK for the three constant amplitude ($R= -1, 0, 0.5$) test conditions are presented in Fig. 1. Application of the closure - based crack growth model required a ΔK_{eff} - rate relationship as input. According to Wu and Newman (1994), the large crack data generated in the present work under constant - amplitude loading conditions on CCT specimens were used to establish this relationship (ΔK_{eff} - dc/dN)as shown in Fig. 2. The solid line segments were generated using a visual fit and the end points of these segments are listed in Table 3. The corner - crack data generated on the SENT specimen under constant - amplitude ($R = 0.5$) loading gave da/dN in the thickness direction. Because the crack - opening behavior of small fatigue cracks stabilized after an extremely small amount of crack growth at high R - values, these data were used to help establish crack - growth behavior in the depth (a - direction). A ΔK_{eff} - da/dN baseline curve was thus developed for cracks growing in the a - direction. The end points of its segments are also listed in Table 3.

Large - Crack Tests and Predictions Under Mini - TWIST Spectrum Loading

One test was conducted on a center - crack tension specimen to monitor large - crack growth under the Mini - TWIST load spectrum at three different mean flight stress (Smf) levels. The tested and predicted results are shown in Fig. 3. They agreed well. The constraint factors in predicting large crack growth behavior under Mini - TWIST spectrum loading were the same as that used in analyzing large crack growth data under constant amplitude loading. The constraint factor (α) was 1.73 for crack - growth rates less than 9E-08m/cycle and 1.1 for rates greater than 7.5E-07m/cycle. For intermediate rates, α varied linearly with the logarithm of crack - growth rate.

Small - Crack Tests and Predictions Under Mini - TWIST Spectrum Loading

The small - crack data generated from the LY12cz clad alloy SENT specimens under Mini - TWIST spectrum loading are shown in Fig. 4. These data were analyzed using the clad correction. The small - crack growth rate data were compared to large - crack data (dc/dN) generated from the CCT specimens under the same loading conditions. Predictions using the crack - closure model, along with the experimentally derived ΔK_{eff} - rate relations and the assumed initial material defect size ($a_i=64\mu m$, $c_i=64\mu m$, $b=0.5\mu m$) were compared with the tested data. Note that ΔK was calculated using the highest and lowest stresses in the spectrum sequence. The crack - growth rates were the average rates over about 30,000 cycles. The option of α in predicting small crack growth rates was the same as that used in the analysis of large - crack growth data. Both experimental and predicted results displayed the " classical small crack effect ". This was attributed to the fact that small corner cracks lack crack closure in their early stage of crack initiation and growth as shown in Fig. 5 and 6. The predicted small crack growth rates were higher than the tested small and large crack growth rates.

The predicted and tested a-N curves of small corner cracks under Mini - TWIST spectrum loading were shown in Fig. 7 (here $Nt = 1,000,000$ cycles was obtained from standard fatigue tests conducted on SENT specimens); the agreement was reasonable . They also showed that most of the fatigue life was consumed by crack growth starting from the initial material defect ($a_i = 64 \mu m$, $c_i = 64 \mu m$, $b = 0.5 \mu m$). Thus, for engineering applications, it is reasonable to make total fatigue life prediction solely based on crack growth starting from the initial material defect.

Total Fatigue Life Predictions

The crack - closure model and the experimentally derived ΔK_{eff} - rate relations were used to predicted total fatigue lives based solely on crack growth starting from the initial material defect to failure. In this approach, a crack was assumed to initiate and grow at the notch root on the first cycle. The measured and predicted fatigue lives for LY12cz clad alloy under Mini - TWIST spectrum loading are shown in Fig. 8. The predicted fatigue lives using elastic stress - intensity factors denoted by " Δ " are compared with those using elastic - plastic stress - intensity factors denoted by " solid curve ", it is seen that plasticity effects on the predicted fatigue lives are very small.

CONCLUSIONS

1 The calculated crack opening stresses of small and large cracks confirmed that small corner cracks did lack crack closure in their early stage of crack initiation and growth.

2 Both experimental and calculated small crack growth rates displayed the small crack effects in LY12cz clad Al - alloy under Mini - TWIST spectrum loading.

3 With the baseline of ΔK_{eff} - da/dN (dc/dN) relationship obtained from constant - amplitude tests of large (for dc/dN) and small (for da/dN)cracks as input, the closure model can predict small, large crack growth rates and total fatigue lives in reasonable agreement with experimental data under Mini - TWIST spectrum loading.

REFERENCES

Newman, J. C. Jr (1981), A crack - closure model for predicting fatigue crack growth under aircraft spectrum loading, *Methods and models for predicting fatigue crack growth under random loading. ASTM STP*748. J. B. Chang and C. M. Hudson eds. American Society for Testing and Materials. Philadelphia, pp53-84.

Newman, J. C. Jr. (1982) A nonlinear fracture mechanics approach to the growth of small cracks, *Behavior of short cracks in airframe components*, AGARD conference proceedings No.328 pp6.1- -6.28.

Newman, J.C. Jr and P. R. Edwards (1988) *Short - crack growth behavior in an aluminum alloy*. An AGARD cooperative test program, AGARD Report No.732.

Wu, X. R. and Newman, J. C. Jr. *et al.* (1994) *Small crack effects in high strength aluminum alloys*. A CAE/NASA Cooperative Program. Selected papers in scientific and technical international cooperative program (5), Chinese Aeronautics Establishment (in chinese); see also Newman, J. C. Jr and Wu, X. R. *et al.* (1994) NASA RP1309.

Table 1.   Nominal chemical composition (wt%) of LY12cz clad Al - alloy

Element	Aluminum	Manganese	Copper	Iron	Magnesium	Silicon	Zinc
wt%	Balance	0.3-0.9	3.8-4.9	0.5	1.2-1.8	0.5	0.3

Table 2.   Average tensile properties of LY12cz clad Al - alloy

Thickness B, mm	Ultimate tensile strength, MPa	Yield stress, MPa (0.2 - percent offset)	Modules of elasticity, MPa	Elongation percent
2.4	443.5	332.5	65400	21.7

Table 3.   The end points of the segments of ΔK_{eff} - da/dN and ΔK_{eff} - dc/dN baselines

ΔK_{eff} MPa*m$^{1/2}$	dc/dN m/cycle	da/dN m/cycle
0.9	1E-11	1E-11
1.26	1E-9	1E-9
3.1	5.5E-9	5.5E-9
4.21	2.6E-8	
9	2.5E-7	
11.22		2.5E-7
13.04	1.14E-6	
16.52		1.14E-6
17.83	2E-5	
22.18		2E-5

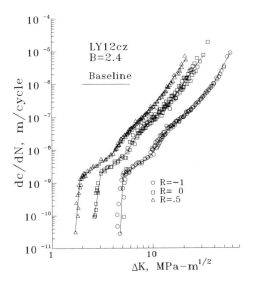

Fig.1 Crack - growth rate against ΔK under constant amplitude loading

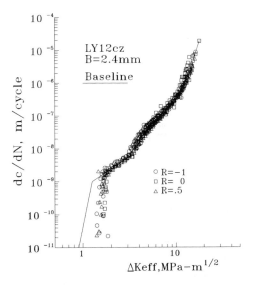

Fig.2 ΔK_{eff} - dc/dN of large cracks under constant - amplitude loading

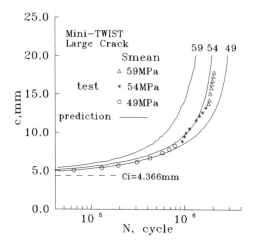

Fig.3 Experimental and predicted large
       crack growth behavior

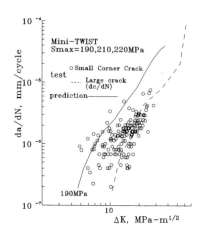

Fig.4 Experimental and predicted small
       corner crack growth behavior

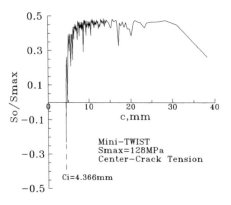

Fig.5 Calculated crack-opening stresses for
       large cracks under Mini-TWIST loading

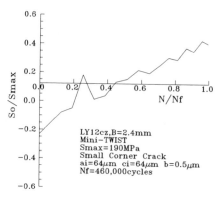

Fig.6 Calculated crack-opening stresses for
       small corner crack under Mini-TWIST loading

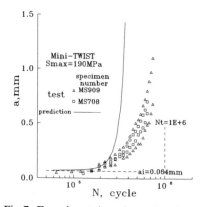

Fig.7 Experimental and predicted a - N
       curves for small corner cracks

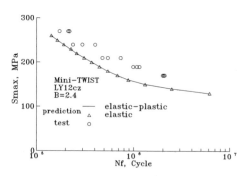

Fig.8 Experimental and predicted total fatigue
      lives under Mini-TWIST loading

A NEW FORMULA FOR PREDICTING FATIGUE
CRACK PROPAGATION RATE

LING CHAO   LI FENG and SHA JIANG BO
Department of Material Science and Engineering, Guangdong
University of Technology, GuangZhou, 510090, P. R. China.
ZHENG XIU LIN
Department of Material Science and Engineering, Northwestern
Polytechnic University, Xian, 710072. P. R. China

ABSTRACT

In the present study a new formula for predicting fatigue crack propagation (FCP) rate was put forward, the given formula reveals that the FCP rates of metals may be predicted from material properties constants and the loading conditions. The 30CrMnSiNi2A steel was used to substantiate the validity of the formula, it is shown that the FCP rates predicted by the formula are in good agreement with the test measuring results.

KEYWORDS

Fatigue; crack propagation; life prediction; crack; ultra − high strength steels.

INTRODUCTION

The FCP rate is one of very important parameters in the studying of reliability of structural fatigue and fracture, and is also essential in estimate of residual life of structures. At the moment, the FCP rates are still measured by experiments, which are expensive and time consuming. Therefore, a new attempt has been made to predict the FCP rates from material properties constants and loading conditions based on the developed FCP expression (Zheng and Hirt, 1983) and ΔK_{th} formula (Ling, 1990) in this study. The developed FCP rate formula will provide a simple and practical basis for the selecting of materials in designing and for the estimating of residual life of structures.

THE CORRELATION BETWEEN THE FCP RATE AND
MATERIAL PROPERTIES CONSTANTS

Based on the static fracture model, Zheng and Hirt(1983) developed a FCP formula where ΔK_{th} is

$$\frac{da}{dN} = B(\Delta K - \Delta K_{th})^2 \tag{1}$$

the FCP threshold, ΔK is stress − intensity factor range and the coefficient B is a material constant relating to tensile properties and the FCP mechanism in the intermediate region. Equation (1) has been proved to be applicable in the range of $da/dN \leq 10^{-6}$ m/cycle, as indicated (Zheng 1983, 1987, and Ling et al., 1991). For the mechanism of bands (Zheng and Hirt, 1983):

$$B = 1/2\pi(0.1E)^2 \tag{2}$$

and for the other FCP mechanisms:

$$B = 1/2\pi E \sigma_f \varepsilon_f \tag{3}$$

where E is elastic modulus, σ_f is fractural stress and ε_f is fractural ductility. It can be seen from eq. (1) to eq. (3) that, when the value of ΔK_{th} and tensile properties are obtained, the FCP rate will be calculated. So that many formulas and methods for estimating ΔK_{th} have been put forward based on empiral results and theoretical analysis(Ling and Zheng, 1990; Zheng, 1987). Based on the fracture mechanics and strain fatigue analysis, Ling and Zheng(1990) developed a expression for predicting the value of ΔK_{th} as follows:

$$\Delta K_{th} = \frac{K_{IC}\sqrt{\alpha \cdot \pi}}{\sqrt{E\sigma_Y}}(2\sigma_{-1} - \frac{E\varepsilon_f}{10^{3.5}})\frac{1}{(1-2\nu)(1+\nu)}\sqrt{\frac{1-R}{1+R}} \tag{4}$$

where K_{IC} is fracture toughness, σ_Y is yield stress, σ_{-1} is fatigue endurance limit and ν is poisson ratio, for metals $\nu = 0.20 \sim 0.33$, α is a material constants and range from $0.05 \sim 0.10$.

Put eq. (4) into eq. (1), it can be drawed the formula for predicting da/dN as follows:

$$da/dN = B\left[\Delta K - \frac{K_{IC}\sqrt{\alpha \cdot \pi}}{\sqrt{E\sigma_Y}}(2\sigma_{-1} - \frac{2E\varepsilon_f}{10^{3.5}})\frac{1}{(1-2\nu)(1+\nu)}\sqrt{\frac{1-R}{1+R}}\right]^2 \tag{5}$$

So it can be seen from above formula, the FCP rate may be directly predicted from material properties constants and the loading conditions.

TEST VERIFYING

The 30CrMnSiNi2A ultra − high strength steel was used as test material, . Its heat − treatment technique and corresponding material properties constants are listed in Table I.

Table I. Heat − treatment technique and material properties
constants of 30CrMnSiNi2A steel

Heat − treatment technique \ Properties constants	σ_b MPa	σ_y MPa	σ_f MPa	ε_f	E MPa	σ_{-1} MPa	K_{IC} MPa $\sqrt{m}$
900℃ Heating 250℃ Isotherming 250℃ Tempering	1584	1324	2147	0.71	206000	645	78.42

The microstructures of 30CrMnSiNi2A steel are mainly Martensite by above heat − treatment, and the fatigue crack grows along with the Martensite lathes, the value of B may be calculated from eq. (3)(Zheng and Hirt, 1983; Zheng, 1994), substituting the values in Table I. into eq. (5), it can obtained $B = 5.1 \times 10^{-10}$ MPa^{-2}. Therefore, under loading conditions, using the material properties constants in Table I. and eq. (5), the da/dN can be developed for 30CrMnSiNi2A steel as follows:

$$\frac{da}{dN} = 5.1 \times 10^{-10} \left[\Delta K - 5.51 \sqrt{\frac{1-R}{1+R}} \right]^2 \tag{6}$$

The da/dN were measured on Mayes Fatigue Tester, the testing frequency is 720 P.M., the minimum FCP rate was measured in 10^{-6} mm/cycle quantitative class, during the testing procedures, the up and down loading methods are both used (Zheng et al., 1987). The FCP rates measured are shown in Fig. 1. (a) and Fig. 1. (b), and the curves predicted from eq. (6) are also shown in Fig. 1. (a) and Fig. 1. (b), so it can be seen from Fig. 1. (a) and Fig. 1. (b) that the FCP rates curves obtained based on material properties constants of 30CrMnSiNi2A steel and the loading conditions as well as the developed FCP rate formula eq. (6), are in good agreement with the test measuring results. However, the values of FCP rate predicted from eq. (6) are little large than those of test data, so that when the developed formula is used to estimate the FCP life of 30CrMnSiNi2A steel, the conservative and safety results will be obtained, which are good for engineering application.

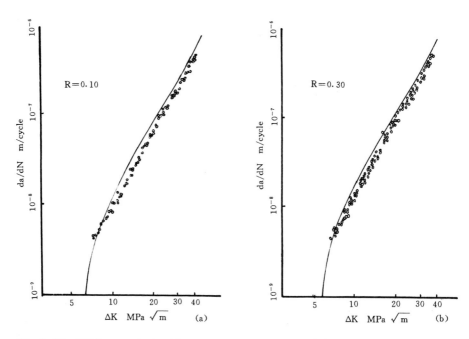

Fig. 1. The FCP rates obtained by test measuring and eq. (6) for 30CrMnSiNi2A steel

CONCLUSIONS

The correlation between da/dN and material properties constants can be expressed as follows:

$$da/dN = B\left[\Delta K - \frac{K_{IC}\sqrt{\alpha \cdot \pi}}{\sqrt{E\sigma_Y}}(2\sigma_{-1} - \frac{E\varepsilon_f}{10^{3.5}})\frac{1}{(1-2\nu)(1+\nu)}\sqrt{\frac{1-R}{1+R}}\right]^2$$

above equation reveals that, the FCP rate may be predicted from material properties constants and loading conditions, and no need to test measuring, so that it can save most expensive and time, meanwhile, above − mentioned method for predicting FCP rates also provides a basis for resonable selecting materials in engineering and estimating the residual life of structures.

for 30CrMnSiNi2A steel and above − mentioned heat − treatment technique, its da/dN can be written as follows:

$$\frac{da}{dN} = 5.1 \times 10^{-10} [\Delta K - 5.51 \sqrt{\frac{1-R}{1+R}}]^2 \quad m/cycle$$

REFERENCES

Zheng, X. L., and Hirt, M. A. (1983). Fatigue crack propagation in steel. Eng. Fract. Mech., $\underline{18,}$ 965 – 979.

Ling, C., and Zheng, X. L. (1990). A new method for predicting fatigue crack propagation threshold. Mechanic strength, $\underline{11,}$ 42 – 46. (in chinese)

Zheng, X. L. (1987). A simple formula for fatigue crack propagation and a new method for determination of ΔK_{th}. Eng. Fract. Mech., $\underline{27,}$ 465 – 475.

Ling, C., and Zheng, X. L., (1990). Prediction of fatigue propagation rates of aluminium. NPU Journal, $\underline{8,}$ 115 – 120.

Zheng, X. L. (1994). Metal Fatigue Quantitative Theory. Northwestern polytechnic University press. Xian. P. R. China.

TRANSIENT CRACK GROWTH BEHAVIOR UNDER TWO-STEP VARYING LOADS

Y. Katoh[*], H. N. Ko[*], T. Tanaka[**] and H. Nakayama[***]

[*]Nakanihon Automotive College, 1301 Fukagaya, Sakahogi-cho, Kamo-gun, Gifu, 505, Japan
[**]Dept. Mech. Enging., Fac. Sci. & Enging., Ritsumeikan Univ., Kusatsu-shi, Shiga, 525, Japan
[***]Executive Office, Osaka Sangyo University, 3-1-1, Nakagaito, Daito-shi, Osaka, 574, Japan

ABSTRACT

Transient crack growth behavior was investigated on aluminum alloy A2017-T3 under two-step varying amplitude loads. It was found that the prediction of the crack growth rate by the linear cumulative law was valid only for limited cases. Transient crack growth behavior consisted of the deceleration and the subsequent acceleration of the growth rate was observed at the lower level of the two-step load patterns. The equations governing these processes were formulated as functions of the stress intensity factors at high and low load levels. A reasonable crack growth model was proposed based on the experimental results.

KEYWORDS

Fatigue crack growth; aluminum alloy; varying amplitude load; linear cumulative law; crack growth retardation; crack growth model.

INTRODUCTION

Fatigue crack growth behavior under varying amplitude loads is quite different from that observed under steady loads of constant amplitude, because the crack growth rate is strongly affected by the load history, especially by the preceding overload, and various kinds of transient phenomena are observed. Two types of load conditions have been usually used for studying the fatigue crack growth behavior under varying load; one is an application of overload during constant amplitude loading, and another one is a two-step varying amplitude load. The crack growth behavior under the former load condition has been observed macroscopically on specimen surfaces, and several crack growth models were proposed by using the parameters such as the size of plastic zone at crack tip and the variation of crack closure level during retardation period after the overload (Wheeler, 1972, Matsuoka et al., 1976, Broek, 1978). Under the latter load condition, since the amount of crack growth at each stress level in respective blocks are too small to observe by an optical microscope, linear cumulative law has been proposed to estimate the crack growth on the assumption that the crack growth rate is steady at each level (Kikukawa et al., 1981). Thus, the crack growth behaviors under the above two load conditions have been treated as different phenomena, and mutual relationship between them is not clear at present.

For this reason, the authors carried out a series of crack growth tests under several types of two-step varying loads using compact tension specimens of aluminum alloy A2017-T3, and formulated analytical expressions governing the transient crack growth behaviors caused by the variation of load. These results are described in this paper together with a crack growth model proposed to understand the transient phenomena, which is consistent with the results reported by many investigators cited above.

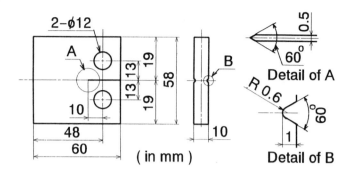

Fig.1. Shape and sizes of specimen.

MATERIAL, SPECIMEN AND EXPERIMENTAL PROCEDURE

Aluminum alloy A2017-T3 was used in the experiments. Ultimate tensile strength and 0.2% proof stress are 443 and 340 MPa, respectively. The shape and sizes of specimen is shown in Fig.1. The tests were carried out by using an electro-hydraulic servo type fatigue testing machine with a loading capacity of 10kN under stress ratio R=0 and a loading frequency within the range from 2 to 20 Hz. Crack length was measured by an optical microscope, but in some cases, microfractographs were used to determine the crack length.

TRANSIENT CRACK GROWTH BEHAVIORS UNDER TWO-STEP VARYING LOADS

Previously, the authors showed that the simple methods based on the linear cumulative law provided no reasonable estimations of the crack growth rates under two-step varying load excepting limited cases (Tanaka *et al.*, 1994). This fact suggests that the crack growth behavior at low intensity level is much complicated, and detailed observations are necessary to understand the total phenomenon.

Load patterns A, B, C and D shown in Fig.2 were used in the experiments, among which the load pattern A is the simplest one with only one step from K_H to K_L. The pattern B is a regular two-step varying load pattern with constant N_H and N_L. In the pattern C, load P_L is kept constant, and therefore, the low intensity level K_L is increased gradually with the increase of the crack length, but, since the increment of crack length at K_L level in each block is small, the value of K_L is assumed to be practically constant in each block. The pattern D is different from the pattern B in the point that the N_L is gradually prolonged. In the previous paper, the authors pointed out that the crack growth amounts at K_L under load pattern B could not be neglected even though the crack length increment was too small to observe on specimen surface by an optical microscope (Tanaka *et al.*, 1994). Consequently, fatigue crack growth tests under load pattern C and D were carried out to observed the small crack growth at K_L in each block by means of microfractography.

It was assumed that, by repetition of N_H cycles of high level load with stress intensity K_H, the effect of the preceding load history before the application of K_H was completely erased, and as was pointed out in the previous paper (Tanaka *et al.*, 1994), the crack growth rate at high stress intensity level in each load pattern was assumed to be the same as the value under constant ΔK conditions. Accordingly, the observation was focused on the crack growth at low intensity level and especially on its transient behaviors just after the variation of the load from K_H to K_L.

The crack growth at K_L level was observed under the load patterns A, C and D. In Fig.3, the crack growth a_L at K_L was plotted as a function of the number of cycles N_L after the variation of the load from K_H to K_L in the case of $K_H=24MPa\sqrt{m}$, where a_L is put to zero at $N_L=0$. Square symbols represent the results

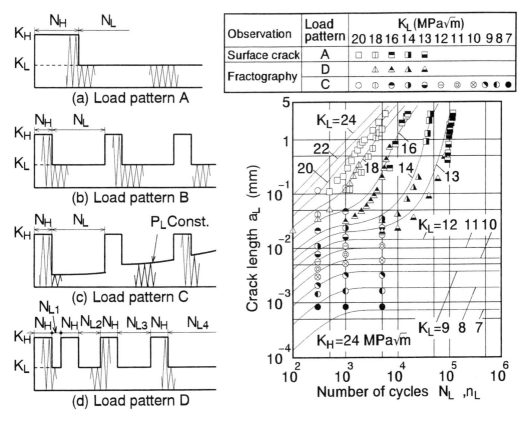

Fig.2. Load patterns.   Fig.3. Crack growths at K_L level.

of macroscopic measurements of the crack length in the case of the load pattern A, and triangles and circles indicate the crack lengths measured on microfractographs. It is observed in Fig.3 that the crack initially grows with the number of cycles N_L, but its growth rate gradually decreases, and when K_L is low, it ceases to grow after the crack length a_L reaches a certain value. In the case of $K_H=24 MPa\sqrt{m}$ shown in Fig.3, the upper bound of K_L, below which the crack growth terminates, is about $12 MPa\sqrt{m}$, The incremental of the crack length was not observed though K_L applied more than 10^6 cycles under load pattern A ($K_H/K_L=24/12$). When K_L is higher than this value, the crack growth gradually gains its speed after the initial deceleration process, and the increase of the growth rate continues until it reaches the growth rate obtained under the corresponding constant ΔK condition.

ESTIMATION OF THE AVERAGE CRACK GROWTH RATE AT K_L

Figure 4(a) illustrates the trend of crack growth shown in Fig.3. Point (n_t, a_t) is a transient point at which the crack growth changes its trend from deceleration to acceleration, and r_t is the crack growth rate at this point. r_0 is the crack growth rate just after the load variation from K_H to K_L, and r_c is the final growth rate under constant K_L conditions. When K_L is low, the acceleration process disappears, as shown in Fig.3. This trend of crack growth at K_L level was formulated as follows (Tanaka et al., 1994). First, the relationships between carck length and number of cycles are given by:

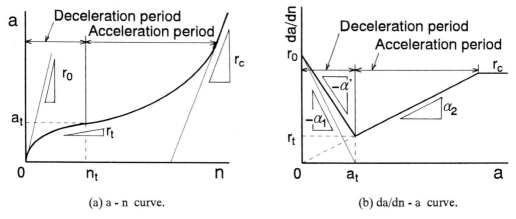

(a) a - n curve. (b) da/dn - a curve.

Fig.4. Schematic crack growth behavior after load variation from K_H to K_L.

Deceleration process: $a(n) = (r_o/\alpha')(1 - e^{-\alpha'n})$, $\alpha' = \alpha_1 - \alpha_2$, $(n \leq n_t)$, (1)

Acceleration process: $a(n) = c_1 e^{\alpha_2(n - n_t)}$, $(n > n_t)$, (2)

where subscript L is omitted from a_L (the crack length) and n_L (the number of cycles) for simplicity. These equations satisfy the following differential equations of crack growth in each process.

Deceleration process: $da/dn + \alpha' a = r_o$, $\alpha' = \alpha_1 - \alpha_2$, $(a \leq a_t)$. (3)

Acceleration process: $da/dn - \alpha_2 a = 0$, $(a > a_t)$. (4)

In the equations, parameters c_1, α_1, α_2 are independent of n but are, in general, functions of K_H and K_L, whose explicit forms were determined from the experimental results as follows.

$$\left. \begin{array}{l} c_1 = C_1 \{(\Delta K_L)_{eff}\}^{m_1}, \ C_1 = 1.01 \times 10^{-4}, m_1 = 2.51. \\ r_o = C_0 \{(\Delta K_L)_{eff}\}^{m_0}, \ C_0 = 1.59 \times 10^{-7}, m_0 = 2.92. \\ \alpha_2 = C_2 \{(\Delta K_L)_{eff}/(\Delta K_H)_{eff}\}^{m_2}, C_2 = 1.47 \times 10^{-2}, m_2 = 6.58. \end{array} \right\}$$ (5)

In Eq.(5), if it is assumed that the crack opening level remains constant throughout the loadings at K_H and K_L levels, then $(K_L)_{op} = (K_H)_{op}$, and the value of $(\Delta K_L)_{eff}$ is replaced by $(K_L)_{max} - (K_H)_{op}$. In the deceleration process α' is nearly eqeal to α_1, and α_1 is given by $r_0 = c_1 \alpha_1$. When the acceleration process does not occur, $\alpha_2 = 0$. Here the parameters to be determined are a_t, n_t and r_t, among which a_t is approximated by c_1 and r_t is given by $r_t = c_1 \alpha_2$. Finally, n_t is calculated by Eq(1) as the value of n when $a = c_1$. Interesting findings by the above formulations are that the parameters c_1, α_1 in Eq.(3) for the deceleration process are the functions of $(\Delta K_L)_{eff}$ alone, and that the parameter α_2 in Eq.(4) for the acceleration process is a function of not only $(\Delta K_L)_{eff}$ but also of the maximum values at K_H and K_L levels. In Fig.4(b) the variation of the crack growth rate da/dn during retardation period expressed by Eqs.(3) and (4) are drawn schematically by the thick line against crack length a. In the figure the fine line shows the trend of the crack growth behavior when the acceleration process does not occur. Respective curves in Fig.3 provide the relations between a_L and n_L during the retardation period, and it is seen that these curves represent the trends of the experimental crack growth, excepting the case of extremely prolonged retardation period.

It is now possible to calculate the average crack growth rate during N_L cycles at K_L level, and, in Fig.5, thus calculated crack growth rates for various values of N_H/N_L are drawn as functions of $(\Delta K_L)_{eff}$

$(da/dN)_L$ from the experimental data by the linear cumulative law, assuming that the crack growth rate at K_H is eqaul to the value obtained under the constant ΔK condition, and solid symbols represent the growth rates $(da/dN)_m$ measured on the microfractographs. In the figure, the fundamental crack growth rate curve is drawn by the thick line,while a dashed line is the extension from the upper part (II_B stage) under constant amplitude loading. It is seen that the estimations by the linear cumulative law provide reasonable values comparing with the direct measurements, and there are good agreements between the trends of the experimental points and the curves drawn by the analysis. Furthermore, in Fig.5, while the experimental points for large values of N_L are located far below the fundamental growth rate curve, those points for small values of N_L are located above the fundamental curve, and in the limit of N_L tending to zero, it can be assumed that the crack growth rates are approximately located around the thick broken line. The relationship between the initial growth rate r_0 at $N_L=0$ and $(\Delta K_L)_{eff}$ given in Eq.(5) was formulated by this assumption, which suggests that the linear cumulative law based on ΔK_{eff} may be valid when N_L is small and the crack growth is not saturated (Kikukawa et al., 1981).

CRACK GROWTH MODEL UNDER TWO-STEP VARYING LOADS

In order to clarify the transient crack growth behavior during the retardation period after the load variation from K_H to K_L, three types of plastic deformed zones are considered in front of crack tip as shown in Fig.6. In this model, the occurrence of the monotonic and cyclic plastic zones was widely confirmed in early studies of the fatigue crack growth, and the existence of the highly deformed region was also confirmed by various techniques (Hahn et al., 1972). The fact that the crack growth rate of strain hardened materials is comparable to that of the virgin material also indicates the existence of this region (Kobayashi et al., 1977).

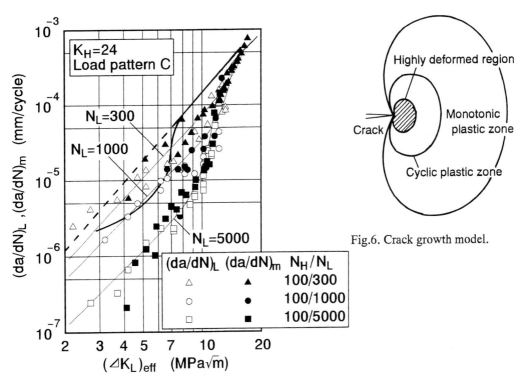

Fig.5. Estimated crack growth rate at K_L.

Fig.6. Crack growth model.

It is assumed that the following two events take place by an application of one load cycle:
[I] Fracture at the crack tip inside the highly deformed region.
[II] Damage accumulation in the cyclic plastic zone to form the highly deformed region.
Under the constant ΔK or constant load conditions, it is obvious that these events continuously take place, and a kind of steady state is realized for each load condition. But this is not the case when the applied load is suddenly changed, because, just after the load variation from K_H to K_L, the stress state in front of the crack tip is affected by the residual compressive stress induced by K_H. This phenomenon obviously delays the damage accumulation in the cyclic plastic zone, and therefore, the formation of the highly deformed region. This is thought to be the reason why the retardation takes place after the load variation. On the other hand, just after the load variation from K_H to K_L, since the highly deformed region was already formed by K_H, the crack growth rate is fully determined by the fracture mechanical parameters such as $(\Delta K_L)_{eff}$, as was indicated in the previous sections. Furthermore, if it is assumed that the event [II] does not occur, the event [I] gradually decays with the crack growth, and the crack ceases to grow before it reaches the end of the highly deformed region. However, in practice, this region gradually develops even at K_L level, and the crack growth behavior such as that shown in Fig.3 is observed

Finally, as was mentioned in Introduction, there are two categories in the methods for predicting the crack growth under varying load conditions :
(1) Application of the linear cumulative law by the use of the crack growth rates under constant load conditions.
(2) Prediction by models or empirical formulas derived by taking into account the interaction of the crack growths at different load levels.
Linear cumulative law (Kikukawa et al., 1981) belongs to the former category, and retardation models (Wheeler, 1972, Matsuoka et al., 1976, Broek, 1978) belong to the latter one. The model presented here also belongs to the latter, but in the case when the variation of load is so rapid that the crack growth at K_L level occurs only in an early stage of the deceleration process, the growth rate at K_L level depends on $(\Delta K_L)_{eff}$ alone, and therefore, the prediction by the linear cumulative law is possible. It is only noted that, in the models by Wheeler and Willenborg the deceleration process as shown in Fig.3 was not dealt with (Wheeler, 1972, Broek, 1978), and that the variation of K_{op} during the retardation period suggested by Matsuoka et al. was not observed in the present study (Matsuoka et al., 1976) .

CONCLUSIONS

Main conclusions obtained are as follows.
(1) In the retardation period due to the load variation, deceleration and acceleration processes were observed as shown in this text.
(2) Experimental formulas for the two processes were derived. There are good agreements between experimental data and analytical results.
(3) Crack growth model during retardation period was proposed. This model can be widly applied to the studies by other investigators concerning the crack growths under varying loads.

REFERENCES

Broek, D. (1978). *Elementary Engineering Fracture Mechanics*, Martinus Nijhoff Publishers, 250-287.
Hahn, G. T., Hoagland, R. G. and Rosenfield, A. R. (1972). Local yielding fatigue crack growth. *Met. Trans.* **3**, 1189-1202.
Kikukawa, M., Jono, M. and Kondo, Y.(1981). An estimation method of fatigue crack propagation rate under varying loading conditions of low stress intensity level. *Trans. JSME* **A-47**, 468-482.
Kobayashi, H., Sugiura, M., Murakami, R., Nakazawa, T., Katagiri H. and Iwasa, H. (1977). The influence of tensile pre-strain on fatigue crack propagation behaviour in low carbon steel plates. *Trans.Trans.JSME*, **A-43**, 416-425.
Matsuoka, S. and Tanaka, K. (1976). The retardation phenomenon of fatigue crack growth in HT80 Steel. *Eng. Frac. Mech.* **8**, 507-523.
Tanaka, T., Katoh, Y. and Nakayama, H. (1994). Transient crack growth behaviors under two-step varying loads. In: *Handbook of Fatigue Crack*, Elsevier, **2**, 999-1026.
Wheeler, O. E.(1972). Spectrum loading and crack growth. *Trans. Trans. ASME*, **Ser. D 94**, 181-186.

EFFECTS OF LOAD SPECTRUM AND HISTORY LENGTH ON FATIGUE CRACK CLOSURE AND GROWTH BEHAVIOR UNDER RANDOM LOADING

Chung-Youb Kim* and Ji-Ho Song**

* R&D Center, Mando Machinery Co., P.O.Box 2, Deokso, Kyongki-Do, Korea
** Dept. of Automation and Design Engng, Korea Advanced Institute of Science and Technology, 207-43, Cheongryangri-Dong, Dongdaemoon-Gu, Seoul, Korea

ABSTRACT

The effect of load spectrum and history length on crack closure and growth behavior is investigated. Special attention is given to the effects of the relative magnitude of crack growth increment during a random loading block to the monotonic plastic zone size due to the largest load cycle in a random loading block. Consistent effects of random load spectrum and history length on crack closure and growth behavior are not observed. However, The relative magnitude of crack growth increment during a random loading block to the monotonic plastic zone size due to the largest load cycle in a random loading block seems to have a slight influence on the crack closure and growth behavior.

KEYWORDS

Fatigue crack growth; crack closure; random loading; history length; load spectrum.

INTRODUCTION

Many models to predict fatigue crack growth under random loading have been developed. Of these prediction models proposed until now, the model based on fatigue crack closure concept can physically well account for load interaction effects, and may lead to the simpler prediction procedure if crack closure behavior under random loading can be represented simply. Elber(1976) reported that the crack opening stresses are nearly constant irrespectively of crack length through the random loading test. On the other hand, FE analysis by Newman Jr.(1981) showed that crack opening points under flight simulation loading vary in an irregular pattern, oscillating about a mean value. Kikukawa, Jono and their coworkers(1984, 1985) have obtained the results that crack opening point is controlled by the maximum range pair load cycle in a random load history and is identical with constant amplitude opening data. Recently, Sunder(1992) has proposed a CCZT(constant closure zero threshold) model which assumes that crack opening is constant and there is no threshold stress intensity factor range under random loading.

Quite recently, the authors(Kim and Song, 1994) have investigated fatigue crack closure and growth behavior under random loading by performing various random spectra and simple variable amplitude loading tests and obtained the following conclusions: The crack opening point is governed by the largest load cycle in a random load history, but is different from that under constant amplitude loading. It indicates that the crack closure behavior under random loading cannot be estimated from constant amplitude tests.

In this study, the effects of load spectrum and history length on crack closure and growth behavior is investigated by performing narrow and wide band random loading tests of various history lengths up to 16000 cycles.

EXPERIMENTAL PROCEDURE

The material used is 2024-T351 aluminum alloy. In this study, centre-cracked tension(CCT) specimen were used. The CCT specimen is 184 mm long, 70 mm wide, 10 mm thick, and an initial notch of $2a_0$ = 16 mm is introduced by electrical discharge machining. The semi-circular side grooves with radius of 1 mm were machined to obtain plane strain crack growth data, and a circular hole of radius 8 mm was machined to attach a clip-on gage for measuring the crack opening displacement at the center of the specimen. Two types of random spectra, namely, narrow and wide band random spectra as shown in Fig. 1, were generated by computer simulation for random loading test. Simulations were performed by superimposing sinusoidal functions of random phase with uniform distribution(Kondo, 1980). For each type of random spectrum, five different random loading blocks having history length of N_h = 500, 1000, 2000, 8000, and 16000 cycles, respectively, were constructed. Random loading tests were performed by repeatedly applying each random loading block. The stress ratio corresponding to the largest load cycle in a block was kept at 0. All tests were conducted using a closed servo-hydraulic fatigue testing machine at a frequency of 7 Hz.

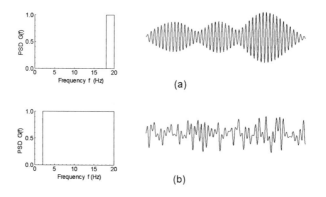

Fig.1 Random loading histories (a) narrow band, and (b) wide band.

EXPERIMENTAL RESULTS

Figure 2 shows the variation of the ratio, $(\Delta a)_{\text{unit block}}/\omega_p^{\text{max}}$, as a function of $K^{\text{rp}}_{\text{max}}$ for N_h = 8000 and 16000, where $(\Delta a)_{\text{unit block}}$ denotes the crack growth increment during a random loading block, ω_p^{max} is the plastic zone size due to the largest load cycle in a random loading block and $K^{\text{rp}}_{\text{max}}$ is the stress intensity factor of the largest load cycle in a random loading block. ω_p^{max} is defined as $(K^{\text{rp}}_{\text{max}}/\sigma_y)^2/(3\pi)$, where σ_y is yield strength of the material. The maximum value of the ratio

$(\Delta a)_{\text{unit block}}/\omega_p^{\text{max}}$ is 2.1 and 1.2 for narrow and wide band random loading respectively within the range of this study. The values of $(\Delta a)_{\text{unit block}}/\omega_p^{\text{max}}$ for $N_h = 500 \sim 2000$ were found to range from 0.05 to 0.4.

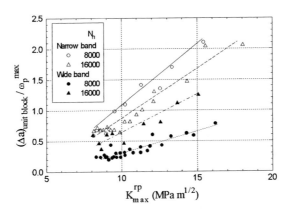

Fig. 2   Crack growth increment during a random loading block.

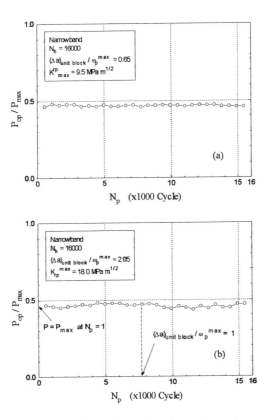

Fig. 3   The variations of crack opening points
under narrow band random loading.

Figures 3 (a) and (b) show the variations of crack opening point under a narrow band random loading of history length of 16000 cycles for $(\Delta a)_{\text{unit block}}/\omega_p^{\text{max}} < 1$ and $(\Delta a)_{\text{unit block}}/\omega_p^{\text{max}} > 1$, respectively. In this figure, P_{op} and P_{max} are the crack opening load and the maximum load in a random block, respectively, and N_p represents the sequential numbers of cycles in a random loading block. The crack opening points for $(\Delta a)_{\text{unit block}}/\omega_p^{\text{max}} < 1$ are nearly constant during a random loading block as has been already reported in the previous study(Kim and Song, 1994). For $(\Delta a)_{\text{unit block}}/\omega_p^{\text{max}} > 1$, the variation of crack opening behavior during a random loading block is discernibly enhanced. It is found that the crack opening point is slightly reduced when the crack increment is greater than ω_p^{max}.

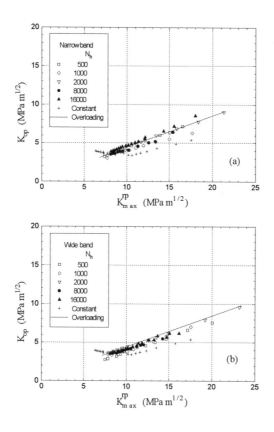

Fig. 4 Crack opening stress intensity factor K_{op} as a function of K^{rp}_{max}.

In Fig. 4, the crack opening stress intensity factor K_{op} are plotted against K^{rp}_{max}, and the crack opening behavior under various overloadings obtained in the previous work(Kim and Song, 1994) is also plotted as a straight line. In this figure, K_{op} is the averaged one over a random block. A consistent effect of loading block length on crack closure is not found and the crack opening values of random loading is nearly identical with that of overloading, irrespectively of type of random loading. This result implies that the crack opening point is mainly governed by the largest load cycle in a random loading block.

Fatigue crack growth under random loading was predicted using the following equation,

$$da/dN = C(\Delta K_{eff})^n, \tag{1}$$

where C and n are material constants dependent on the growth rate regime and were obtained from the constant amplitude loading test. The effective stress intensity factor range, ΔK_{eff}, was determined based on the measured crack opening data. The experimental and analytical lives for each test are compared in Fig. 5. The analytical / experimental life ratios, N_{pred}/N_{test}, are ranging from 0.75 to 1.7 and from 0.58 to 1.3 for narrow and wide band random loadings, respectively. This result indicates that the crack closure concept can account well for fatigue crack growth under random loading. However, the predictions for random loadings of history length of 8000 and 16000 cycles appear to be slightly non-conservative.

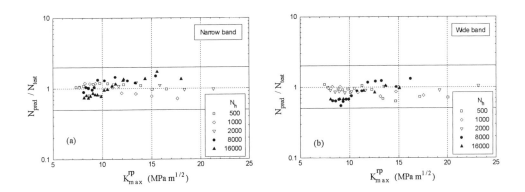

Fig. 5 Prediction results of fatigue crack growth under
random loadings based on the measured opening data.

Utilizing the results obtained in this study, fatigue crack growth under random loading can be predicted in a simple method. As can be seen in Fig. 4, the crack opening behavior under random loading is proportional to K^{rp}_{max}, and the value is nearly equal to that under various overloadings. Therefore, the crack opening value under random loadings can be estimated from the overloading tests under appropriate test conditions, as follows,

$$K_{op} = \alpha K^{rp}_{max}, \tag{2}$$

where α is the slope of the straight line in Fig. 4 and $\alpha = 0.43$ is used in this study. Using the value of K_{op}, crack growth under random loading can be estimated from eq. (1). Figure 6 shows the prediction results by using the method described above. The prediction ratios, N_{pred}/N_{test}, are ranging from 0.54 to 1.84 and 0.58 to 2.12 for narrow and wide band random loadings, respectively. The results agree relatively well with the prediction results based on the measured crack opening data. However, a relatively large scatter is found in this case. A close inspection reveals that the predicted data for $N_h = 8000$ tends to be relatively non-conservative. This tendency may be partly attributed to the lack of consideration of the variation of crack opening point when the crack increment is greater than the plastic zone size.

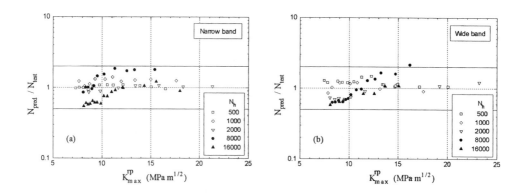

Fig. 6 Prediction results of fatigue crack growth under random loadings
by using the crack opening data obtained from overloading tests.

CONCLUDING REMARKS

The effects of load spectrum and history length on crack closure and growth behavior is investigated by performing narrow and wide band random loading tests of various history lengths up to 16,000 cycles. Consistent effects of random load spectrum and history length on crack closure and growth behavior are not observed. However, when the history length is so long that the growth increment exceeds the plastic zone during a random block, the variation of crack opening points is discernibly enhanced and the predictions based on measured opening data appear to be slightly non-conservative.

REFERENCES

Elber, W. (1976). Equivalent constant amplitude concept for fatigue crack growth under random loading. *ASTM STP* **595**, 236-250.

Jono, M., Song, J. H. and M. Kikukawa (1984). Fatigue crack growth and closure of structural materials under random loading. *Proc. ICM.*, **6**, 1735-1742.

Jono, M., Song, J. H. and A. Sugeta (1985). Prediction of fatigue crack growth and closure under non-stationary random loading. *Proc. ICOSSAR*, **4**, 465-474.

Kim, C.Y. and J. H. Song (1994). Fatigue crack closure and growth behavior under random loading. *Engng Fracture Mech.*, **49**, 105-120.

Kondo, Y. (1980). Study on the behavior of fatigue crack growth and closure under service loading. Ph.D Thesis, Osaka University, Osaka, (In Japanese).

Newman, J. C. (1981). A crack closure model for predicting fatigue crack growth under aircraft spectrum loading. *ASTM STP* **748**, 53-84.

Sunder, R. (1992). Near threshold fatigue crack growth prediction under spectrum loading. *ASTM STP* **1122**, 161-175.

FATIGUE CRACK GROWTH UNDER VARIABLE AMPLITUDE LOADING WITH A LOAD-TIME HISTORY FOR URBAN BUS STRUCTURES

C. DOMINGOS* and M. de FREITAS#

* Escola Superior de Tecnologia do Instituto Politécnico de Setúbal,
Estefanilha, 2900 Setúbal, Portugal
#Departamento de Engenharia Mecânica, Instituto Superior Técnico,
Av. Rovisco Pais, 1096 Lisboa Codex, Portugal

ABSTRACT

Fatigue crack growth on CCT specimens of DIN 17 100 St 37 steel were carried out, on computer controlled servo-hydraulic machines under variable amplitude loading representative of the service loading of urban buses operating in the city of Lisbon. The load-time history was acquired in real service by bonding strain gages to an urban bus frame, and the data was simplified according to the rainflow cycle counting method. From this simplified spectrum, three new random spectrum were generated, rainflow equivalent, and used for the program testing. Curves of crack growth against the number of spectrum were determined and showed significant repetitive results. Experimental results are then compared with computer crack growth analysis using crack growth Forman law. Calculated crack growth rate predicts non conservative results for this spectrum, enhancing the importance of testing.

KEYWORDS

Fatigue (materials), fracture mechanics, variable amplitude, crack growth, rainflow method, load spectra, bus structures, test automation, data analysis

INTRODUCTION

A major problem in the structural design of buses lies in the complexity of their geometry and of the dynamic loading conditions. Design rules used in the bus industry, (Beerman, 1987) are usually based on stress levels induced by static loads (bending and torsion), multiplied by safety and dynamic factors, that are then compared with allowable stress levels based on constant amplitude fatigue strength of the materials. An urban bus structure is composed mainly of a frame of welded hollow steel beams with rectangular cross sections and the constant amplitude fatigue behavior of welded joints for bus frames for different joint configurations and materials has been reported in the literature (Ferreira *et al*, 1983).

However, buses are subjected to very complex load histories due to different types of loading, for example passengers, road conditions, speed, traffic, etc. Hence, the corresponding continuous but variable load spectrum must be known in order to design vehicles with improved fatigue performance.

In a previous work (Freitas *et al*, 1994), the load-time history representative of the average real service bus working conditions in the city of Lisbon, bus line no. 42, was determined. Dynamic strain gage bridges and a computer-aided data acquisition system were used to measure, control, and store the information collected, during real traffic service. For each strain-gage, a large amount of stored data was processed in order to generate a typical in-service spectrum. The characteristic stress loading was obtained by using the rainflow counting method.

In this paper, the procedures for reducing the measured characteristic loading, and the generation of new rainflow equivalent spectrum are presented. Variable amplitude tests on CCT specimens of the steel used on hollow beams of the bus frame, using the new spectra generated from the previous one, were carried out on a computer-aided servo-hydraulic machine. Crack growth predictions were performed using Forman's law of crack propagation determined from constant amplitude crack growth tests.

EXPERIMENTAL PROCEDURE

The material used in the structural elements of the urban buses is DIN 17 100 St 37-2 steel and was adopted for the tests with a yield strength of 200 MPa and an ultimate tensile strength of 300 MPa.. Crack growth tests were performed on 60-mm-wide center-cracked tension (CCT) specimens of 2 mm thickness, according to ASTM Test Method for Measurement of Fatigue Crack Growth Rates (E-647-93). Crack length measurements were made using the compliance method measured by a clip gage attached to one side of the specimens with a calibration procedure using a 20 to 50X optical microscope reading a scale attached to the other side of the specimen just below the growing crack. As the tests involved compressive loading, freely sliding anti buckling plates were placed around the thin specimens. Tests were performed on a computer aided servo-hydraulic testing machine INSTRON of ±250 kN capacity, using the software FLAPS, Fatigue Laboratory Applications Software. This software allows the programming in a micro-computer of a sequence of tasks to be performed by the servo-hydraulic machine, then programming a variable amplitude test using a pre-defined spectrum. It was adapted in order to acquire every previously determined number of cycles, the load-displacement data, necessary for crack length measurements using the compliance method (Domingos, 1995).

The measured load-time history of the most loaded point in the urban bus frame, is very long and has a complex shape, then can't be used as input in the servo-hydraulic machine. This load-time history was simplified and analyzed through the rainflow cycle counting method (Freitas *et al*, 1994) but it still contains about 40 000 cycles per block, considered as one typical travel around Lisbon. A further spectrum simplification was then introduced, omitting the lowest stress amplitudes, calculated as those that in any case could cause on the specimen a crack growth because the threshold of the stress intensity factor ΔK_{th}, was not reached. This new spectrum is presented as a sequence of blocks of variable stress amplitude and mean value with about 2 000 cycles per block corresponding to the most loaded point on the urban bus frame and is represented on figure 1, which shows the cumulative frequency of peaks and valleys for the spectrum considered in the tests.

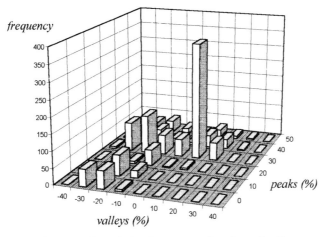

Figure 1 - Cumulative frequency of peaks and valleys

From this stress distribution a new load-time history was designed. For this purpose three new random distribution of peaks and valleys were obtained with the constraint that they should be rainflow equivalent, i.e., a rainflow analysis of each new load-time sequence should present the same cumulative frequency of peaks and valleys as shown in figure 1.

RESULTS AND DISCUSSION

The crack length during testing was obtained against the number of blocks of loading, and the crack growth rates per block were calculated and plotted as a function of the maximum stress intensity factor obtained in each block of loading. These crack growth results can then be presented as the classical Paris law plotting and are represented in figure 2. It shows that the crack growth rates obtained for each load-time history are very similar and can be considered perfectly equivalent. A fourth load-time history was generated from further rainflow analysis of one of the previous load-time history, following the same random sequences. The crack growth obtained, also represented in figure 2 is equivalent. For these tests a repeatability of the results was obtained for the same load-time history (at least two specimens, results not shown in figure 2 for clarity) and for the four (3+1) load-time histories.

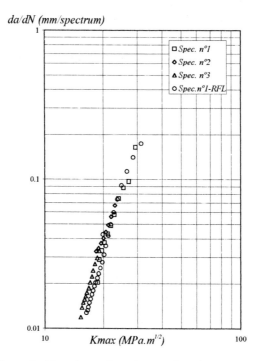

Figure 2 - Experimental crack growth under spectrum loading

In order to analyze the experimental results and further application on more complex structures such as the urban bus frame, a crack growth prediction was performed. For this purpose, a computer program was performed, based on Forman's law of crack growth:

$$\frac{da}{dN} = \frac{C_F(\Delta K)^{n_F}}{(1-R)K_c - \Delta K}$$

where a is the crack length, N the number of cycles, ΔK the stress intensity factor range, R the stress ratio and K_c=33 MPa√m the critical stress intensity factor. C_F=3.6E-10 and n_F=2.7 where determined from constant amplitude tests for different stress ratios. For CCT specimens the stress intensity factor is given by:

$$K = \sqrt{\sec\frac{\pi a}{W}} \sigma\sqrt{\pi a}$$

where W is the specimen width..

The numerical integration of Forman's law was made cycle by cycle because the sequence is organized in peaks and valleys and it is possible to aid to the computation any interaction effects, as the Wheeler or any other model. The results obtained are a sequence of crack lengths against the number of cycles or the number of block loading. Figure 3 represents the comparison between the experimental results

obtained when testing with the random sequences considered and the crack growth predictions
obtained with the computer simulation, for the same initial crack length 2a=10 mm and identical
maximum load in each block $P_{máx}$= 15 000 N. When no interaction effects between load sequences are
considered, crack growth predictions are identical for all the spectrum considered.

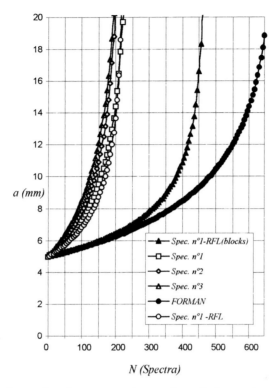

Figure 3 - Comparison between crack growth predictions and experimental results

Figure 3 shows another test not described before, where crack growth tests were carried out on the
CCT specimens, with a random sequence of cycles of the same amplitude and mean stress in order to
show the evidence of the difference in crack growth. With no retardation or acceleration effect due to
variable amplitude loading, the crack growth prediction is the same, but experimentally different results
are obtained. Crack growth predictions for this load sequences are unconservative as they predict lower
crack growth rates than the experimentally obtained, which means that no retardation effects were
obtained but an acceleration effect. This is due to the fact that the random sequences do not present
higher overloads, but they present several underloads with a global high stress ratio, as shown in figure
4, which represents a small segment of one of the load-time histories considered. This unconservative
result, enhances the importance of experimental testing with the in time measured spectrum, in the
crack growth prediction of structures, for design or maintenance purposes.

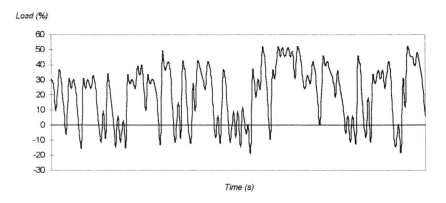

Figure 4 - Segment of the random loading sequence

CONCLUSIONS

From crack growth tests on CCT specimens under variable amplitude loading, representative of the spectrum load in an urban bus frame, we can conclude:
- crack growth tests under spectrum loading may be described by exponential laws, where the number of cycles is replaced by the number of blocks.
- identical constants of the crack growth equations are obtained for rainflow equivalent loading.
- for this particular spectrum, crack growth calculations using Forman's law based on constant amplitude loading, predict lower crack growth rates than those obtained with experimental tests.

REFERENCES

Beerman, H. J. (1987). *The Analysis of Vehicle Structures.* (English translation) Ed. Mechanical Engineering Publications, London

Domingos, C. (1995). *Fatigue Crack Growth in Mild Steel under Variable Amplitude Loading.*(in Portuguese), MSc Thesis, Instituto Superior Técnico, Lisboa

Ferreira, J. M., Branco, C. M. and Radon, J. C. (1983). Fatige life assessment in welded rectangular hollow sections using fracture mechanics. *Proceedings of the International Conference on Application of Fracture Mechanics to Materials and Structures,* Freiburg, Germany.

Freitas, M., Maia, N., Montalvão, J. and Silva, J. D. (1995). Applying contemporary life assessment techniques to the evaluation of urban bus structures. *Automation in Fatigue and Fracture Testing and Analysis, ASTM STP 1231.* American Society for Testing and Materials, Philadelphia, 428-442.

VARIABLE AMPLITUDES AND PREDICTION

FATIGUE LIFE PREDICTION - ACCELERATION OF FATIGUE CRACK GROWTH UNDER VARIABLE AMPLITUDE LOADINGS

Masahiro JONO

Dept. of Mechanical Engineering, Osaka University,
2-1, Yamada-oka, Suita, 565, Japan

ABSTRACT

Fatigue crack growth rates under variable amplitude loadings for life prediction were discussed on metallic and ceramic materials associated with crack closure behavior. Acceleration of crack growth rates were found in the cases of periodic superposition of compressive peak load on pulsating random loading, load reduction at very small fatigue crack length(both on metallic materials) and overload and/or high-low two-step loading on a ceramic material. Crack opening points were measured by using a refined unloading elastic compliance method and above mentioned acceleration of crack growth rates was evaluated quantitatively by the effective stress intensity range. It was concluded that the acceleration of growth rates on metallic materials could be well explained by crack closure behavior, while that on ceramic material could not be expressed quantitatively only by the variation of crack opening point measured by the macroscopic method.

KEYWORDS

Fatigue Crack Growth; Variable Amplitude Loading; Life Prediction; Acceleration of Crack Growth Rate; Crack Closure.

INTRODUCTION

Fatigue crack growth behavior is strongly affected by the load history and growth rates under complex loadings can not be expressed only by the instantaneous macroscopic fracture mechanics parameter such as the stress intensity range, ΔK(For example; Wei and Stephens,1976; Jono,1990). In many cases, however, variable amplitude loads reduce the fatigue crack growth rates to give a conservative effect to life prediction and may not be seriously considered in design works except in light-weighting designs. On the other hand, it is recognized that a low-to-high level(Lo-Hi) load sequence accelerates the fatigue crack growth rate, although a small effect as a whole, and also the threshold condition to the constant amplitude loadings disappears under variable amplitude loadings and fatigue cracks can propagate at the ΔK values even below the threshold condition(Kikukawa *et al.*,1981), to suggest the necessity of sophisticated method to predict fatigue crack growth rates under variable amplitude loadings.

Some models have been proposed to estimate the crack growth retardation on the basis of the compressive residual stresses developed ahead of the crack tip(Wheeler,1972), whereas the recent understanding is that the main reasons to cause crack growth retardation should be crack closure(Elber,1971). Author and others have conducted a large number of experimental works on crack closure under varying loading conditions(For example Jono *et al.*,1981; Jono,1994) and found, as shown in Fig.1(Jono,1993), that the load history affected crack closure behavior and crack growth increments including those below the threshold condition could be well estimated by linear integration of the *da/dn-ΔKeff* relationship of constant amplitude data by extrapolating the curve to the region below the threshold condition for stationary variable amplitude loadings(Kikukawa *et al.*,1981). In this paper some examples of acceleration of fatigue crack growth due to variable amplitude loadings will be demonstrated and discussed associated with crack closure behavior.

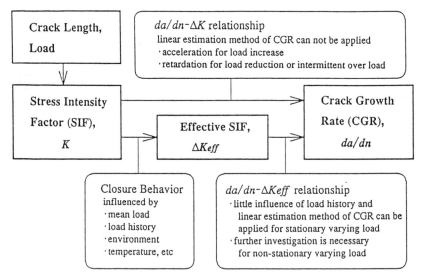

Fig.1. Approach to evaluate fatigue crack growth rate under variable amplitude loadings.

FATIGUE CRACK GROWTH ACCELERATION DUE TO COMPRESSIVE PEAK LOAD

The effect of compressive peak load on fatigue crack growth rate was investigated on a welding structural steel, SM50, by using the load history shown in Fig.2 where Nc cycles of compressive peak load (in this figure $Nc = 1$) are applied every repetition of N units of pulsating pseudo-random loading (a unit of random loading is composed of 2000 peaks with a broad-band Gaussian distribution). The peak value of compressive peak load, $(K)min$, was chosen equal to the maximum stress intensity involved in the random loading, $(K)max$, and the number of units of random loading to be continuously repeated, N, which represents the interval of compressive peak load application was changed within the range of 1 to 5.

Fatigue crack growth rates are evaluated in terms of an acceleration ratio, α, which is a ratio of the observed crack growth increment to the expected growth increment under the stationary pulsating random loading of the identical magnitude without compressive peak load, and shown in Fig.3 by solid circles as a function of peak load interval. α is found to show the higher values than unity, indicating an acceleration of growth rate due to compressive peak load, when the peak load interval, N, is small, and decreases to approach unity as N increases.

Crack opening points under random loadings were monitored by using the unloading elastic compliance method(Kikukawa *et al.*,1976) and examples of load-differential displacement hysteresis loops during random loadings of $N = 1$ and 5 are illustrated in Fig.4. For both cases, the crack opening points indicated by the short horizontal bars on the hysteresis loops are found constant during one block of the random loading. However, the crack opening point observed for the case of $N = 1$ is found to be lower than that observed for $N = 5$. Crack opening stress intensities, K_{op}, are plotted against the maximum stress intensity of random loading, $(K)_{max}$, in Fig.5 with those for constant amplitude tests of $R = -1$ and 0 and for stationary pulsating random loading tests. In the case

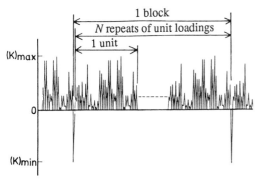

Fig.2. Pulsating random load history with periodic compressive peak load.

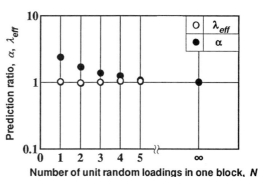

Fig.3. Effect of interval of compressive peak loads on prediction ratios.

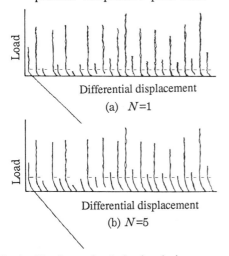

Fig.4. Crack opening behavior during one block of random loadings.

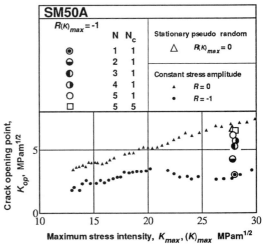

Fig.5. Variation of crack opening point under random loading with interval of compressive peak loads.

of $N = 1$ where the compressive peak loads are frequently applied, crack opening stress intensity under random loading is found to coincide with the constant amplitude test of $R = -1$ having the identical maximum stress intensity range. This means that the crack opening stress intensity is controlled by the maximum stress intensity range-pair (equal to the maximum range obtained by the rain-flow counting) and its stress ratio as observed for the stationary random loadings(Kikukawa et al.,1981). However, it is found that K_{op} increases with increase of the interval of compressive peak loads, and K_{op} under the random load of $N = 5$ coincides with that of constant amplitude test of $R = 0$. The effect of the number of continuously applied cycles of compressive peak load, N_c, was also investigated and crack opening point for $N_c = 5$ is plotted by a open square symbol in Fig.5. The increase in N_c appears to have little effect to lower the crack opening point, compared with the case of $N_c = 1$.

Fatigue crack growth rates are again evaluated by a prediction ratio, λ_{eff}, which is a ratio of the observed crack growth increment to the predicted growth increment on the basis of the modified-Miner rule type, linear accumulation of crack growth in terms of the effective stress intensity range-pair incorporating closure measurements, i.e. calculated by using the extrapolated da/dn-ΔK_{eff} relationship of constant amplitude loading to the region below the threshold condition, and indicated in Fig.3 by open symbols. λ_{eff} accounting for the variation of K_{op} with N is found almost unity for all N values investigated, indicating the acceleration due to the compressive peak load comes from the reduction of crack opening point and can be well accounted for with the crack closure concept as have been proved for the stationary random loadings.

BEHAVIOR OF SMALL FATIGUE CRACK UNDER VARIABLE AMPLITUDE LOADING

In order to investigate the effect of load variation on the growth and closure behavior of small fatigue cracks Hi-Lo(high-to-low level) two-step loading tests were conducted at different crack lengths of as small as possible by using the specially designed testing apparatus at a frequency of 25 Hz(Jono and Sugeta,1995) and crack opening points were monitored in detail by the computer-aided unloading elastic compliance method(Jono and Song,1987) during the load cycling of variable amplitude.

Load amplitude of $R = -1$ was reduced to about 70 % at different crack lengths on a high tensile strength steel, HT80. Fatigue crack growth rates under Hi-Lo two-step loading were plotted against the maximum stress intensity factor in Fig.6 with the scatter band of small cracks under constant amplitude loading. Large symbols denote the crack growth rates to high level loadings which located near the upper limit of scatter band of constant amplitude test and small symbols indicate those to subsequent low level loadings. It is worthy to note that when load was reduced at a very short crack length, e.g. less than 0.1 mm, acceleration of crack growth was found compared to the growth rate of constant amplitude test with identical stress intensity, whereas retardation was observed under Hi-Lo two-step loading tests at longer crack lengths, although crack growth rates decreased with crack advance during low level load cycling, showing well known delayed retardation, in all cases irrespective of crack length.

Figure 7 represents load-differential displacement hysteresis loops which were obtained during low level load cycling preceded by high level loading. Hysteresis loop immediately after reduction of load to low level (Fig.7(a)) shows width probably due to the plastic strain cycling of high level loading and gives a relatively low crack opening stress intensity which was defined by the deflection point of hysteresis curve or by the tangential point in loading cycle of hysteresis loop to the vertical line parallel to the unloading elastic compliance line, indicated by short horizontal bars in the figure. Thus obtained crack opening points were found increase with crack advance during low level load cycling as shown in Figs.7(b) to (d).   Crack opening stress intensities in Hi-Lo two-step tests at different crack lengths were plotted in Fig.8 where arrows indicate the load reduction points. Similar tendency as observed in Fig. 7 was found at all test conditions, but crack opening stress intensity, as a whole, increased with increase of crack length, rapidly in the region of shorter crack length than 0.1 mm and gradually in longer crack regime than 0.2 mm as observed under constant amplitude tests(Jono and Song,1987).

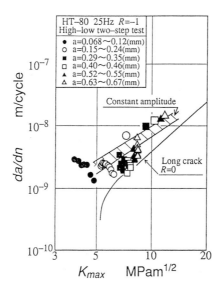

Fig.6. Growth rates of small fatigue cracks under Hi-Lo two step
loadings as a function of maximum stress intensity factor.

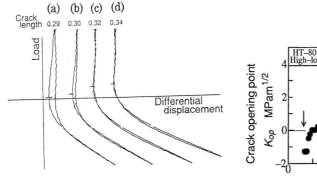

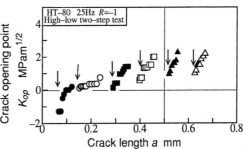

Fig.7. Example of load-differential displacement
hysteresis loops during low-level load
cycling under Hi-Lo two step test.

Fig.8. Variation of crack opening stress
intensity as a function of crack length
under Hi-Lo two step test.

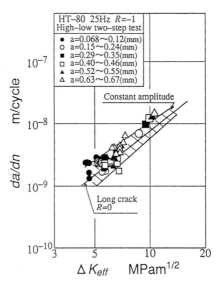

Fig.9. Growth rates of small fatigue cracks under Hi-Lo two step
loadings as a function of effective stress intensity factor.

Figure 9 shows the relationship between crack growth rate under Hi-Lo two-step loading and the effective stress intensity range. Growth rates under both high and low level loadings in Hi-Lo two-step loading tests were found to lie within a scatter band of constant amplitude loading test indicating that the acceleration and retardation behavior could be well expressed by the crack closure behavior. From these experimental findings it may be concluded that the crack growth rates, irrespective of small or long crack and of under constant or variable amplitude loading, are principally governed by the effective stress intensity range, and the effect of load variation might be explained as follows; fatigue cracks of longer than 0.2 mm usually have a residual tensile deformed region in the wake of crack tip, and behave like a long sized crack to show the retardation by load reduction due to the difference of crack opening levels. On the other hand, for the smaller crack than 0.1 mm closure is insufficient to control the crack growth behavior, and damage zone introduced by the high level loading at the crack tip may give an acceleration effect of crack growth behavior.

CRACK GROWTH RATE UNDER VARIABLE AMPLITUDE LOADING ON CERAMICS

Crack growth rates were investigated on a sintered silicon nitride, Si_3N_4, under cyclic loadings at 1.6 Hz, using CT (Compact Type) specimen. Figure 10 shows the load-differential displacement hysteresis loop measured by a back face strain gage. Crack closure is clearly observed in the ceramic as well as in metals. This crack closure is thought to be induced by fracture surface roughness and debris on a fracture surface. The effect of stress ratio on crack growth rate, however, could not be explained solely by the difference of crack opening point and crack growth rates were found to be governed by both of the maximum and the effective range of the stress intensity so that the equivalent stress intensity range such as $\Delta K_{eff,eq} = K_{max}^{0.85} \cdot \Delta K_{eff}^{0.15}$ might be necessary to evaluate the crack growth rate even under constant amplitude loadings of different conditions(Sugeta *et al.*,1996).

A crack growth acceleration was observed under variable amplitude loadings such as multiple overloads and repeated two-step loadings. Figure 11 shows the fatigue crack growth behavior for the constant amplitude loading with multiple overloads(arrow indicates the location of putting overloads) where the ratio of overload to a low-level load was set at 1.13 with $R = 0.1$ and the number of the applied overload was 10. The vertical axis represents the acceleration ratio to the constant amplitude loading of identical value to the low level loading. The acceleration ratio was found higher than unity immediately after the overload and decreased with crack growth and recovered within the crack growth increment of about 50 μm. It is thought that the overload excursion crashed more severely the rough fracture surface and the debris produced under the low-level loading before overload, decreasing the crack opening point to result in the acceleration of crack growth.

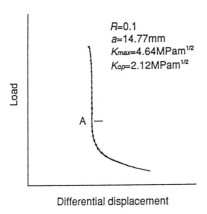

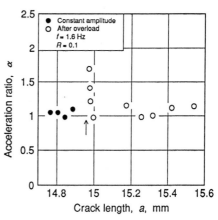

Fig.10. Example of load-differential displacement hysteresis loop under constant amplitude loading on sintered silicon nitride.

Fig.11. Variation of crack growth rate expressed by prediction ratio during low-level load cycling after multiple over load.

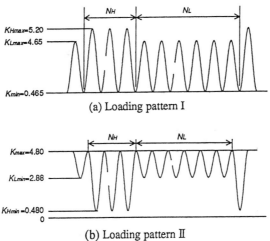

Fig.12. Repeated two-step loading patterns.

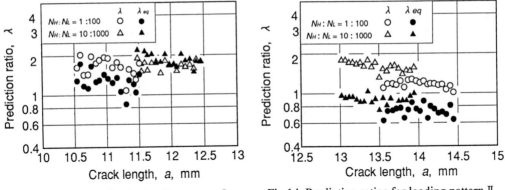

Fig.13. Prediction ratios for loading pattern I.   Fig.14. Prediction ratios for loading pattern II.

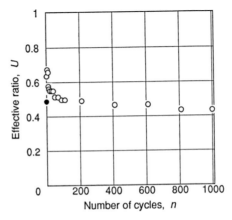

Fig.15. Variation of effective ratio during low-level load cycling under
two-step loading pattern I(N_L =1000).

A quantitative evaluation was made for repeated two-step loadings where the crack opening behavior was more easily monitored. In Fig.12 are shown the two kinds of loading pattern, where the stress ratio of high-level loading nearly equals to 0.1 and the ratio of the number of high-level loading, N_H, to that of low-level loading, N_L , is maintained 1/100 and fatigue tests were performed under the K_{max} controlled condition. Fatigue crack growth rates were evaluated by the growth increment prediction ratio,λ, defined by $\Delta a/\Delta a^*$, where Δa is an observed crack increment and Δa^* is the crack increment predicted by a linear summation rule from constant amplitude growth rate data in terms of the stress intensity range with the identical stress ratios, and are shown by open symbols in Figs.13 and 14 for load patterns of I and II, respectively. λ shows the values higher than unity irrespective of the loading patterns and N_L values, showing that the acceleration of the fatigue crack growth in terms of ΔK occurred under the repeated two-step loading as well as the multiple overload. The variation of the effective ratio, U, under low-level loading for the loading pattern I (N_L = 1000) is shown in Fig.15. After U increased immediately after high-level load, U decreased with crack growth and returned to the stationary value under constant amplitude loading about at 100 cycles. It seems that this increase of U induced by the high-level load excursion resulted in the acceleration of crack growth. Then the crack growth rates were evaluated based on the aforementioned equivalent stress intensity range considering the closure effect and are shown in Figs.13 and 14 by solid symbols. In loading pattern II, where the maximum load of low-level load is same as that of high-level load, the prediction ratio considering crack closure behavior, λ_{eq}, took the values from 0.7 to 1.0, indicating that the estimation method of crack growth increment based on λ_{eq} gave a good or conservative prediction. However, in loading pattern I, where the minimum load of low-level load is same as that of high-level load, λ_{eq} took the higher values than unity showing an unconservative estimation of crack growth. It may be concluded that the crack closure behavior measured by the macroscopic method can explain the acceleration of fatigue crack growth qualitatively but can not always account for that quantitatively on ceramic material depending on loading patterns. Detailed test results and discussion are described in another paper of this Conference(Sugeta *et al.*,1996).

CONCLUSIONS

Fatigue crack growth rates under variable amplitude loadings, especially acceleration phenomena were discussed associated with crack closure behavior. The conclusions obtained are summarized as follows;

1) The periodic compressive peak load excursion superposed on the pulsating random loading accelerated the fatigue crack growth rates, although the degree of acceleration was reduced when the interval of compressive peak load application was increased.

2) The effect of load variation was found to depend on the crack length, and acceleration of crack growth by reduction of reversed cyclic load amplitude was observed for the smaller crack than 0.1 mm, while retardation was found for longer crack than 0.2 mm as in usual case of long fatigue cracks.

3) Above mentioned acceleration of crack growth rates could be well explained by the difference of crack closure stress intensities between variable amplitude loadings and constant amplitude loading, and it was concluded that the main reason to cause crack growth acceleration is also the crack closure as well recognized for crack growth retardation in metallic materials.

4) Crack closure behavior was clearly observed in the sintered silicon nitride as well as in metals, and the over load and/or high level load excursions decreased crack closure point under low-level loadings to result in the crack growth acceleration. However, the effect of load variation on crack growth in this material was concluded to be generally so complicated that the acceleration could not be explained quantitatively only by crack closure behavior measured by the macroscopic method.

REFFERENCES

Elber,W.(1971). The Significance of Fatigue Crack Closure. ASTM STP 486. 230-242.

Jono,M.,Song,J. and Kikukawa,M.(1984). Fatigue Crack Growth and Crack Closure of Structural Materials under Random Loadings. Proc. ICF-6, 1735-1741.

Jono,M. and Song,J.(1987). Growth and Closure of Short Fatigue Cracks. Current Japanese Mat. Res.(Tanaka,T. *et al* ed.), Elsevier Appl. Sci., 1, 41-65.

Jono,M.(1990). Fatigue Crack Growth under Variable Amplitude Loadings. Proc. Fatigue 90. 1485-

1498.

Jono,M.(1993). Direct SEM Observation and Growth Rate of Fatigue Crack under Variable Amplitude Loadings. Proc. Fatigue 93, 545-554.

Jono,M.(1994). Fatigue Crack Closure and Its Related Problems. Computational and Experimental Fracture Mechanics(Nisitani,H. ed.), 317-345.

Jono,M. and Sugeta, A.(1995). Crack Closure and Effect of Load Variation on Growth Behaviour of Small Fatigue Cracks. Int. J. Fatigue and Fract. of Eng. Mat. and Str., to be appeared.

Kikukawa,M.,Jono,M. and Tanaka,K.(1976). Fatigue Crack Closure Behavior at Low Stress Intensity Level. Proc. 2nd Int. Conf. Mech. Beh. Mat.(ICM-2). Special Vol.,254-277.

Kikukawa,M.,Jono,M. and Kondo,Y.(1981). An Estimation Method of Fatigue Crack Propagation Rate under Varying Loading Conditions of Low Stress Intensity Level. Proc. ICF-5, 1799-1805.

Sugeta,A.,Jono,M. and Koyama,A.(1996). Fatigue Crack Growth Behavior under Variable Amplitude Loadings in Sintered Silicon Nitride. Proc. Fatigue 96, to be appeared.

Wei,R.P. and Stephens,R.I.(ed.)(1976). Fatigue Crack Growth under Spectrum Loads. ASTM STP 595, American Society for Testing and Materials.

Wheeler,O.E.(1972). Spectrum Loading and Crack Growth. ASME Transaction Journal of Basic Engineering. 94, 182-186.

A COMPUTER CONTROL SYSTEM
FOR SPECTRUM LOADING FATIGUE TEST

X.Y. Huang[1], X.R. Wu[1], B.X. Ge[2], H. Ouyang[1] and S.L. Liu[1]

[1] Institute of Aeronautical Materials, Beijing, 100095, P.R. China
[2] Beijing University of Aeronautics and Astronautics, Beijing, 100083, P.R. China

ABSTRACT

This paper presents a microcomputer control system MTA92 based on IBM PC386 (or its compatible). This system is aimed at improving control accuracy and implementing the fatigue and crack propagation test under spectrum loading in servohydraulic fatigue test machines with which no computer control system is equipped before. The hardware has achieved the required functional ability and durability as proven by systematic testing.

KEYWORDS

Fatigue test; computer; test automation; spectrum loading; servohydraulic machine.

INTRODUCTION

Laboratory material tests are integral parts of structural design in aircraft industry. In fact, the "design process" could be viewed as the way of transforming materials properties (from simple, short duration laboratory test) into a specified structural response. With the development and application of damage tolerance concept, aircraft structure designs need the fatigue crack propagation data under spectrum loading.

Many test machines, however, only have regular waveform signal generators and it is difficult to do the spectrum loading fatigue tests with these machines. The demands of programmable function generator and data acquisition system are increasing. Nowadays, some new digital control systems are available. These systems are developed by some corporations and have strong functions and so, are costly devices. The efforts to develop a real-time, low cost and easy-to-use automatic control system are keeping going on in many research organizations (ASTM STP 613,1092, 1231).

A spectrum loading fatigue test means long time (comparing with the static tests, namely elongation test) and a huge amount of data. A common method is to build up an independent system in which data acquisition board, signal generator board, event detector and communication adapter are installed. The computer only works as a host to send commands out to the independent system by serial communication cable. The problem is this system not working in a real-time control mode.

553

There is always a delay when a new command comes out because it needs time to transmit information. Another method is used also in some laboratories. With this method, two or more computers are used together. One is used as a control signal generator, and the other as a data acquisition device. In this paper, a compact computer control system MTA92 is described which has several advantages: real-time control, closed loop, easy-to-use and reduced cost.

CONFIGURATION

The system consists of a PC386, an analogue/digital (A/D) converter and a programmable function generator (FG), as shown in Fig. 1, which are designed based on the requirements of variable amplitude fatigue tests. The programmable function generator is a special Digital/Analogue converter. The central component is an 8031 micro controller that can work independently. Both function generator and A/D converter are installed in PC slots and data communication between them is in PC BUS. So the data transmit speed is the same as inner-computer chips. One signal output channel is used for test control and four A/D channels for data acquisition. The highest test frequency can be 50Hz. Hardware parameters of this system are listed in Table 1.

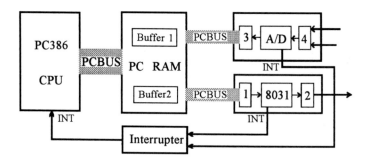

1. Command registers   3. Data registers
2. D/A converter   4. Sample holding circuits

Fig.1 The hardware architecture of MTA92 system

Table 1. The parameters of MTA92 computer test assistant system

FG board	A/D converter
Data Port: 12bits	Data Port: 12bits
D/A Convert Speed: 1μs	Convert Speed: 15KHz
Waveform: havensin, ramp	Channels: 4
Frequency: 0~50Hz	
Mode: interrupted	Mode: interrupted

SOFTWARE DESIGN

Spectrum loading fatigue test is very time-critical. Because the testing task requires parallelism to allow the controlling, data acquiring and on-line conditional processing run together, the time schedule is very important. In order to improve the single-task DOS system of PC computer and to implement multitask procedures, the TSR (Terminate and Stay Resident program) technique is used.

Based on the analysis to PC ROM-BIOS and the DOS operating system, the interrupted service routine can work as a parallel procedure. PC DOS reserves many interrupted vectors for users. But it is tricky because some mistake in the user's interrupted routines can cause deadlock of the whole computer system. It is found that the mistake is a kind of wrong data transition between interrupted routines and test subprograms but the mechanism is not clear. Here the TSR technique is used and the interrupted service routines stay in RAM of PC386. For the sake of correct data transition, two buffers, Buffer1 and Buffer2, are built up. Control subprograms are developed to convert the load (or strain, displacement etc.) data into FG commands and A/D signals into feedback data. The main program is a procedure that can work with TSR routines almost simultaneously so the spectrum signal can be sent out continuously and at the same time, the data of load, strain and displacement can be taken in and stored in harddisk. TSR routines and control subprograms are written by assembler language and the main programs by QUICK BASIC. Fig. 2 is the schematic drawing of the MTA92 software.

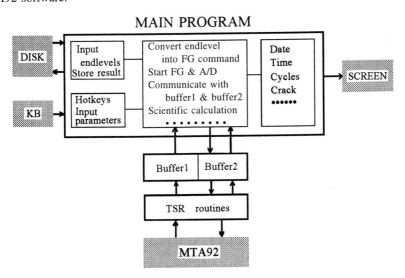

Fig.2 The structure of test program and the function of software for MTA92

With the TSR technique, the MTA92 test assistant system works well. Besides control and data acquisition, many procedures - such as date and time display, hot key response and various scientific calculations that are time-consuming for variable amplitude fatigue tests, can work together. Of course, it doesn't mean that all these procedures can run at the exactly same time, but they can work in their assignment time sectors. The trouble of getting into operational deadlock of the system has been solved completely.

A SPECTRUM LOADING TEST

Mat 92 system has been used for several types of fatigue test machines, such as INSTRON 1343, SCHENCK and MAYES, which have external signal input ports. This paper gives the results of flight spectrum fatigue tests conducted in MAYES machine.

Specimen type is CCT by ASTM standard E647. The specimen is 75mm wide and 6.35mm thick. The initial crack length is about 3.0mm. Some mechanical properties of the 7475-T7351 aluminum alloy plate are listed in Table 2. A random flight spectrum is used with two stress ratios, R=0 and R<0. The crack length was measured by optical microscopes at two sides.

Table 2.   Mechanical properties of Al 7475-T7351

Yield strength	Ultimate strength	Fracture toughness
460MPa	520MPa	50MPa√m

The random spectrum amplitudes change so quickly in the test that it is impossible to check the load peaks especially for lower levels. The signals of input and output were checked in static loading process before the tests began. In spectrum loading tests, some higher load peaks were checked one by one. The difference is within ±3%.

The MAYES test machine has an external analogue signal input port for complex waveform. In this experiment, MAT92 was used as a signal generator. The FG output port of MTA92 was linked to the external signal port of MAYES test machine, shown in Fig.3. The spectrum data were stored in disk with a series of peaks and valleys. The test program read these endlevels partly and sent them to Buffer2 in PC RAM by calling related subprograms. TSR routines transmitted each endlevel to FG board from Buffer2 and FG made a half waveform between two endlevels then sent the waveform out in steps. A half waveform consists of 512 steps so the waveform is smooth. The frequency of each half waveform can be changed according to the machine frequency response. For higher endlevels a lower frequency is used.

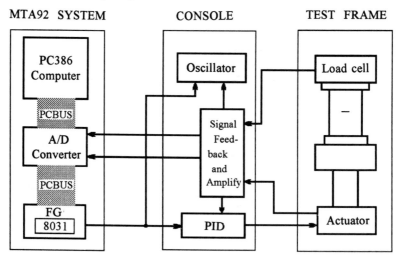

Fig. 3 Schematic diagram of connections between MTA92 and MAYES test machine

Three tests were conducted in each stress ratio but only one of them was used to get a whole a-N curve. For the others, the tests were stopped when crack growth rate was as high as 1mm per block because the interest was on the low crack growth rate regime. Fig.4 shows the a-N curves of test results, a) for R=0 and b) for R<0. They exhibit good coincidence.

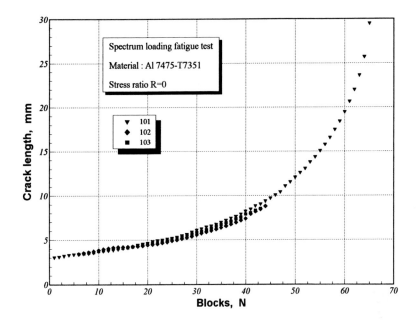

a)

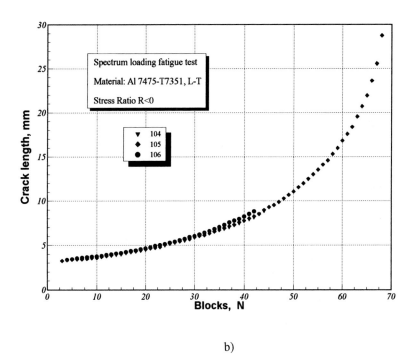

b)

Fig.4   a-N curves of fatigue test under spectrum loading

CONCLUSION

A computer assistant system MTA92 is developed for spectrum loading fatigue test. With this PC based control and data acquisition system, it is possible to generate a random spectrum loading signal and get good load accuracy. This system employs TSR technique and interrupted service of PC computer and the problem of system deadlock is solved. MTA92 has not only achieved most of the designed features, but also featured reliability and easy-to-use.

REFERENCE

Application of Automation Technology to Fatigue and Fracture Testing (1990) , ASTM STP 1092, A. A. Braun, N.E. Ashbaugh and F.M. Smith, Eds., American Society for Testing and Materials, Philadelphia.

Automation in Fatigue and fracture: Testing and Analysis (1994), ASTM STP 1231, C. Amzallag, Ed., American Society for Testing and Materials, Philadelphia .

Use of Computers in the Fatigue Laboratory (1976), ASTM STP 613, H. Mindlin and R. W. Landgraf, Eds., American Society for Testing and Materials, Philadelphia.

EXPERIENCES WITH LIFETIME PREDICTION UNDER SERVICE LOADS

J. LIU, K. PÖTTER and H. ZENNER

TU Clausthal, Institut für Maschinelle Anlagentechnik und Betriebsfestigkeit

Leibnizstraße 32, 38678 Clausthal Zellerfeld

ABSTRACT

A new modification of the Miner rule leads to an improvement in lifetime calculation. An evaluation of a comprehensive data set indicates a damage sum near to unity with a decreased scatter range.

KEYWORDS

Lifetime prediction; modification of the Miner rule; random loads; blocked progamme loads; fictious fatigue limit; crack initiaton

INTRODUCTION

The damage sum at failure, that is, the ratio of the experimentally determined lifetime to the calculated lifetime, is not equal to unity; as a rule, it is scattered over a large range (scatter range) (Schütz and Zenner, 1973). The scatter range is decisive for appraising the confidence level of the approach of lifetime prediction. The total scatter of the lifetime calculation results from the superposition of the uncertainty factors in the computational chain (from the component S-N curve, the load spectrum up to the damage analysis). In Fig. 1, the superposition of the scatter due to various effects is represented diagrammatically. If the effects are uncorrelated, the total scatter can be determined from the quadratic sum of the individual scatter.

As indicated by W. Schütz (Schütz and Heuler, 1993), a distinction can be made between scientific and industrial lifetime prediction. In the case of scientific lifetime prediction, the S-N curve for the components and the load spectrum, or load-time function, are usually well known, Fig. 1a. As a rule, the variable amplitude test is performed on specimens with the same shape as those for the S-N curve test. The accuracy is determined by the damage analysis as well as the load characteristic. Moreover, scatter from the test procedure is included in the final results. In contrast, S-N curves and load spectra are usually subject to larger errors in the case of industrial lifetime prediction, Fig. 1b. The confidence level depends especially on the accuracy of the load assumption and the assumption of load bearing capability.

In the present article, the scientific lifetime prediction is considered. If an improvement is achieved in the damage analysis, the overall lifetime calculation can be improved, but the confidence level of the industrial lifetime prediction is not considerably affected.

MODIFICATIONS OF THE MINER RULE

At present, the lifetime is still predicted from the linear damage accumulation hypothesis, the Palmgren-Miner rule (Miner, 1945). Nonlinear damage accumulation hypotheses have not yet resulted in any appreciable improvement in the accuracy of the lifetime prediction, as far as any known comparisons with test results are concerned. In accordance with the Palmgren-Miner rule, the damage during each individual load is calculated from the ratio of the class frequency and the endurable number of load cycles. The total damage results from the sum of the partial damages for the individual classes.

$$D = \sum_{i=1}^{k} \frac{n_i}{N_i} \tag{1}$$

As indicated by Miner, the application of equation (1) is subject to the limitation that all loads must exceed the fatigue limit. In practical applications, however, this limitation is not considered. In accordance with the so-called original Miner rule (MO), the loads below the fatigue limit are neglected.

For considering the loads below the fatigue limit, a number of hypotheses have been proposed in the past. In the Federal Republic of Germany, especially the elementary Miner rule (ME: the fictitious S-N curve is extended with the original slope k below the fatigue limit) and the modified Miner rule as given by Haibach (Haibach, 1970) (MM: the S-N curve is extended with the fictitious slope 2k-1 below the fatigue limit) are applied, Fig. 2. Furthermore, the so-called consistent Miner rule (MK) is known; this rule takes into account a continuous decrease in the fatigue limit with progressive damage (nonlinear damage accumulation) (Haibach, 1989).

In 1992, Liu and Zenner (Zenner and Liu, 1992) developed and proposed a new modification of the Miner rule. This modification (LZ) includes a new definition of the reference S-N curve for the damage analysis. The reference S-N curve is specified as follows:

- The reference S-N curve is rotated with respect to the component S-N curve. The centre of rotation for the reference S-N curve is located on the component S-N curve at the height of the maximum for the load spectrum.
- The slope of the reference S-N curve is calculated from the average value of the slopes for the component and crack propagation S-N curves, $k^* = (k+m)/2$, with $k > m$ and $m \approx 3.6$.
- The fictitious fatigue limit is decreased to $S_D^* = S_D/2$.

This new Miner modification is similar to that given by Corten and Dolan (Corten and Dolan, 1956) as far as the procedure is concerned. The decisive difference, however, is the specification of the fictitious slope for the reference S-N curve. The damage analysis is thus corrected to a larger extent in the case of weakly notched specimens with a flat S-N curve than for sharply notched specimens. This approach takes into consideration the experience that the damage sum is close to unity in the case of sharply notched specimens (slope of the S-N curve near $m \approx 3.6$), whereas the damage sum is lower for weakly notched specimens (flat S-N curve slope).

The fictitious fatigue limit is initially set equal to one-half the fatigue limit. This specification is derived especially from the fact that the loads less than one-half the fatigue limit do not cause any damage, as confirmed by omission tests (Heuler and Seeger, 1986). However, the effects of the loads below the fatigue limit have not yet been conclusively investigated. Two circumstances must be taken into account: Loads below the fatigue limit contribute appreciably to crack propagation especially in the zone of the short cracks (Miller, 1991); the fatigue limit decreases as damage progresses; see the following section.

LIFETIME CALCULATION FOR RANDOM LOADS

Test results, mainly from a comprehensive study of the LBF supported by BMFT and industry (LBF, 1979), have been compiled and evaluated in (Zenner and Liu, 1992). The evaluated test results involve random load sequences with the stress ratio, $\bar{R} = -1$, and the irregularity factor, $I = 0.5 \sim 1$. The S-N curve for the mean stress for the process has always been determined experimentally for the purpose. The mean stress fluctuations are relatively small for an irregularity factor I greater than 0.5. The distribution of the damage sum for fracture is plotted in Fig. 3. On the average, the damage sums indicated by Miner original, Miner modified, and Miner elementary are situated on the unsafe side with relatively severe scatter, whereas the modification given by Liu and Zenner indicates a damage sum close to unity. In particular, the scatter range is decreased considerably.

Kotte and Eulitz et al. (Eulitz et al., 1994) have compiled and evaluated a more comprehensive set of data. In this case, results for which the S-N curve for the mean stress of the process has not been determined experimentally have also been evaluated. In the evaluation, results associated with pronounced variations in mean stress (for instance, I < 0.5) have also been recorded. By way of departure from Fig. 3, the scatter is more severe for all modifications. The modification given by Liu and Zenner indicates a decrease in scatter range, but not so clearly as in Fig. 3, since other uncertainties predominate in the computational chain; see Fig. 1.

FICTITIOUS FATIGUE LIMIT

Calculation of Lifetime in the Range near the Fatigue Limit

As a rule, variable amplitude tests are performed within a lifetime range up to 10^7 load cycles, in order to limit the experimental effort and expense. However, many machine and vehicle components are designed for a lifetime exceeding 10^8 load cycles. The maximum for the load spectrum is thereby located in the range near the fatigue limit. In this range, essential load cycles of the load spectrum are situated at the level of the fatigue limit or just below this limit. The lifetimes calculated by means of various Miner modifications differ considerably from one another within this range. In Fig. 4, the lifetime curves calculated from various modifications are plotted with the use of corresponding test results from (Eulitz et al.,1993) for an initial S-N curve with a slope of 5. The area of a circle is proportional to the respective number of the specimens. In the range near the fatigue limit, the available test results are not sufficient as a whole for permitting a conclusive appraisal of the modifications. On the average, the modification due to Liu and Zenner tends to yield good results.

For a possible correction of the lifetime calculation in the range near the fatigue limit, the fictitious fatigue limit can be adapted. The lifetime curves resulting from the Liu and Zenner modification with fictitious fatigue limits of 0.5, 0.7, 0.8, 0.9, and 1.0 times the actual fatigue limit are plotted in Fig. 5. For specifying the fictitious fatigue limit, the different decrease in fatigue limit as damage progresses, for instance, must be taken into account for crack initiation and for fracture.

Crack Initiation as Failure Criterion

For technical crack initiation as failure criterion, the first difficulty involves the definition of a standard crack length. For a reliable measurement, a crack length of about 1 mm is necessary. During the phase up to technical crack initiation, therefore, a portion of the crack propagation phase is already included. In accordance with Liu and Zenner, the slope of the reference S-N curve can be selected as average value for the slopes of the component S-N curve and of the S-N curve for crack propagation. In the case of failure associated with technical crack initiation, however, the component still has a fracture-mechanical fatigue limit. Logically, therefore, the fictitious fatigue limit for technical crack initiation must be set higher than that for fracture.

In Fig. 5, also test results for one material and two specimens shapes from (Bremer et al., 1995) for the failure criterion of crack initiation are plotted. With a fictitious fatigue limit equal to 90% of the original fatigue limit, the experimental lifetime curve is well defined. However, the extent to which the fictitious fatigue limit is to be specified as a function of the component geometry and material must still be checked.

CONCLUSIONS

With the use of the new modification of the Miner rule given by Liu and Zenner, the lifetime calculation for random loading and for blocked programme loading can be regarded as accurate if the experimentally determined S-N curve is available for the mean stress of the process and the load sequence is regular. In the range near the fatigue limit and for technical crack initiation as failure criterion, the lifetime calculation with the new Miner modification depends decisively on the value of the fictitious fatigue limit. Further investigations are necessary for specifying the fictitious fatigue limit as a function of the component geometry, the loading type, and the material.

REFERENCES

Schütz, W. and H. Zenner (1973). Schadensakkumulationshypothesen zur Lebensdauerberechnung bei schwingender Beanspruchung - ein kritischer Überblick. Z. f. Werkstofftech. 4, S. 25-33 u. S. 97-102

Schütz, W. and P. Heuler (1993). Miner´s Rule Revisited. AGARD Report R-797

Miner, M. A. (1945). Cumulative damage in fatigue. Trans. ASME, J. Appl. Mech. 12, A159-A169

Haibach, E. (1970). Modifizierte lineare Schadensakkumulations-Hypothese zur Berücksichtigung des Dauerfestigkeitsabfalls mit fortschreitender Schädigung. LBF-Ber. TM 50/70, Darmstadt

Haibach, E. (1989). Betriebsfestigkeit - Verfahren und Daten zur Bauteilberechnung. Düsseldorf: VDI-Verlag

Zenner, H. and J. Liu (1992). Vorschlag zur Verbesserung der Lebensdauerabschätzung nach dem Nennspannungskonzept. Konstruktion 44, S. 9-17

Corten, H.T. and T.J. Dolan (1956). Cumulative fatigue damage. Proc. Int. Conf. Fatigue of Metals, Inst. Mech. London, New York, S.235-246

Heuler, P. and T. Seeger (1986). A criterion for omission of variable amplitude loading histories. Int. J. Fatigue 8, pp. 225-230

Miller, K. J. (1991). Metal Fatigue - past, current and future. Proc. Instn. Mech. Engrs. 205 S. 1-14

LBF (1979). Ringbuch Bd. 1 zur Gemeinschaftsarbeit BMFT-Industrie-LBF. Darmstadt

Eulitz, K.-G., H. Döcke, K. L. Kotte, A. Esderts and H. Zenner (1994). Lebensdauervorhersage I, Abschlußbericht FKM Heft 189

Bremer, G., L. Jung and H. Zenner (1995). Sonderereigniskollektive - Auswirkungen von Sonderereignissen auf die Lebensdauer. Forschungsheft FVA, Vorhaben Nr. 131/III

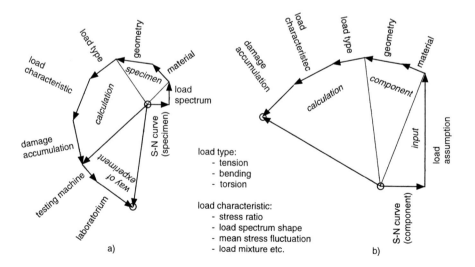

Fig. 1: Scatter arrangement in Scientific and Industrial Liftime Prediction

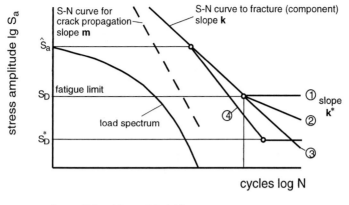

①   **MO:**   Miner original, $k^* = \infty$
②   **MM:**   Miner modified by Haibach, $k^* = 2k-1$
③   **ME:**   Miner elementary, $k^* = k$
④   **LZ:**   Miner modified by Liu and Zenner, $k^* = \dfrac{k+m}{2}$; $S_D^* = \dfrac{1}{2} S_D$

Fig. 2:   Modifications of the Palmgren-Miner Rule

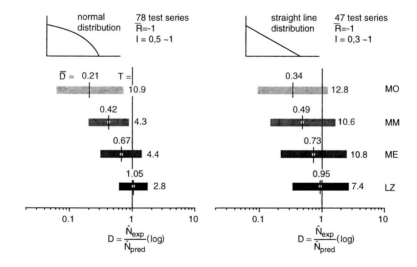

Fig. 3:   Distribution of Scatter Range with Random Load
(Zenner and Liu, 1992)
D: mean damage sum, T: scatter range
($T = D_{90\%}/D_{10\%}$ log. normal distribution)

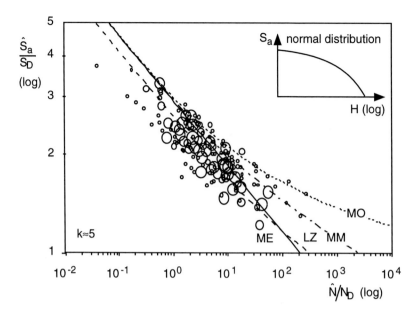

Fig. 4: Referred Test Results and Calculated Lifetime Curves
for Fracture

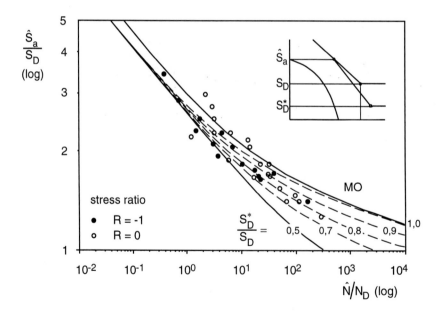

Fig. 5: Referred Test Results for Crack Initiation and Calculated Lifetime Curves according to Liu and Zenner with
varying Fictitious Fatigue Limits

MODELING FATIGUE DAMAGE IN TERMS OF CRACK LENGTH IN THE WHOLE FATIGUE AND FRACTURE DOMAIN, BY USE OF THE GENERAL FATIGUE DIAGRAM

Menachem P. Weiss

Technion, Faculty of Mechanical Engineering
Haifa 32000, Israel

ABSTRACT

It has been known for some time that fatigue damage can be expressed in terms of crack length throughout the whole fatigue domain. In practice the switch from the classical dimensionless damage concept, to the comprehensible crack length concept, has been slow. The whole fatigue domain has been displayed by the author on one general fatigue diagram, with crack length and stress amplitude as axes. The domain was divided into six zones and regimes, and an appropriate crack propagation relation was assigned for each zone. This sort of presentation makes it easy to comprehend the damage as crack length concept, and will be shortly reviewed. In this present study, the influence of variable stress amplitude and of mean stresses on crack propagation in the short and long cracks is discussed. Additional fatigue life influencing factors are considered in context of the new diagram.

KEYWORDS

Fatigue modeling; Fatigue & Fracture; Fatigue Diagram; Fatigue Cracks; Damage as Crack Length.

INTRODUCTION

Fatigue failures and research have been around for more than 150 years now, and numerous very valuable studies have advanced the theoretical and experimental knowledge in the field. It would be improper to mention only a few of the excellent contributions, and it is impossible to mention them all here. But it is still a fact that even now, with the thorough calculation and prediction procedures, fatigue failures do sometimes happen, some with fatal results, and that the old Wohler`s S-N curve is still an important part of our prediction procedures. All that follows in this study deals with the simple, mode I fatigue behavior in metals, with different loading functions and material parameters. It does not indicate anything about the more complex fatigue phenomena like mixed mode, multiaxial or other materials fatigue behavior, nor temperature, corrosion and other influences.

The S-N curve includes the alternating stress amplitude - σ_a, which was and will always be, a major parameter in fatigue studies, and the number of cycles till total failure - N, which is a controversial number. ΣN is an arithmetic sum of the number of stress cycles, many with totally different damaging characteristics, in distinct fatigue regimes, and even modifying the sum by using a log N cannot properly compensate for this type of scatter in fatigue life results.

To complicate things even more, the cumulative linear damage concept, proposed by Miner in 1945, made the importance of N invincible. Studies that showed that the linear summation of the Miner's rule differ from unity by more than an order of magnitude in each direction, were not likely to discredit it, simply because there was nothing better available. In spite of the enormous scatter of estimated and experimental fatigue life in most cases, the strength and influence of the *damage concept* was so overwhelming, that even people working in theoretical mechanics borrowed it to form the new field of continuum damage mechanics - CDM, as reviewed (Krajcinovic 1989), that has flourished since. In retrospect, it did not make the use of N as a fatigue parameter, nor the use of the cumulative damage concept, more justified. It hardly helped to predict the expected fatigue life, with cosiderable scatter, and therefore statistical fatigue values and methods came into use.

With the calculation of stresses at the vicinity of cracks (Mushkelishwili 1933), and later the introduction of the, then new, Fracture Mechanics, the fatigue and fracture community started to treat fatigue as crack initiation and crack propagation. With the introduction of the Paris' law (Paris et. al. 1963) and the improved life predictions it made possible, in certain cases, the future research direction seemed clear. The vast amount of theoretical work needed to calculate stress intensity factors - SIF and other fracture mechanics relations kept the research community busy for many years. Only at a later stage it became evident that Linear Elastic Fracture Mechanics - LEFM, can reasonably supply good predictions, only in the low stress amplitude regime below the endurance limit and above the SIF threshold range. Tests out of these limitations end with greater scatter than expected, though smaller than the S-N results scatter. *LEFM came out as a reasonably good method for life predictions* in welded structures, and cracked parts with low stress amplitudes, *but not for the classical smooth specimen, loaded cyclically above the fatigue limit.*

FATIGUE DAMAGE IN TERMS OF CRACK LENGTH

It is fully accepted that at the late stage of the fatigue process, one major crack extends till separation, in many cases when reaching the fracture toughness value of the material, and in some cases, when the stresses are high, in gross yielding. Tracking *the process reversed*, the crack was smaller and smaller, at some stage it crossed the stress intensity threshold range - ΔK_{th} and became very small indeed. At earlier stage, several microstructurally small cracks developed in parallel, until one of them became the major one, by void coalescence or some other crack formation mechanism, and from that stage and on, only this one is part of the process, as described above. But even this certain very small crack, can theoretically be tracked back till its origin in some defect, dislocation or local stress raiser. In reality one cannot locate or distinguish this special micro-crack in real time, nor measure its propagation until it grows to a certain size, but nevertheless the process takes place, and the instantaneous size of the crack can be calculated or simulated through the whole formation and propagation. For engineering purposes, only the instantaneous crack length and not its exact identification is important.

As can be seen the whole fatigue process can now be estimated in terms of crack length, as was shown earlier (Miller 1991). The switch from Damage to Crack Length, as the main parameter of fatigue life seems simple and clear, but has not yet been fully apprehended by the scientific community, and only a few studies on the short cracks range prediction procedures have been published. The whole fatigue domain was classified (Weiss 1992) on a new fatigue diagram. Only this classification enabled to clearly specify distinct crack propagation relations to each zone and fatigue regime.

THE FATIGUE DIAGRAM

The whole fatigue and fracture domain has been classified and divided into six zones and fatigue regimes, on one comprehensive diagram. Fatigue damage from the onset of cracking, is expressed in terms of crack length (or respectively micro crack length), in the whole range of very short, short and long cracks ranges. The diagram is shown in Fig. 1, and a short description is called for.

The fatigue domain is divided into zones by three constant stress amplitude lines: the endurance limit S_e, the yield strength line S_y and the ultimate tensile strength S_u, and by two constant stress intensity factor (CIF) lines: the plane strain fracture toughness K_{1c} and the effective threshold CIF range ΔK_{theff}. The zones differ by the fatigue regimes that they contain, as follows:

1. The safe zone. It can contain non-propagating cracks. It lays below both the endurance limit and the CIF threshold range, under the Kitagawa (Kitagawa et al. 1976) line .
2. Linear Elastic Fracture Mechanics (LEFM) regime. Here mostly striation fatigue mechanisms prevails, and the Paris' law shows good predictions.
3. High Cycle Fatigue (HCF) regime, very short and short propagating cracks. This is the old HCF crack initiation zone.
4. Both HCF and LEFM regimes are active here. Crack propagation can be predicted by a linear combination of two relations and is formed by a combination of mechanisms.
5. Low Cycle Fatigue (LCF) regime, very short and short propagating cracks in the plastic regime.
6. Both LCF and Elasto-Plastic Fracture Mechanics (EPFM) regimes are active in this, highly plastic regime. Crack propagation can be predicted by a combination of relations, and formed by a combination of mechanisms.

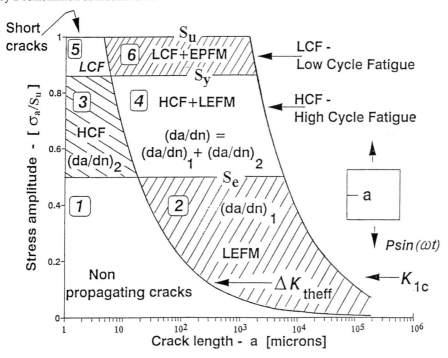

Fig. 1. The Fatigue Diagram, divided into different Fatigue and Fracture regimes and six zones.

It has been shown that a convenient way of crack propagation prediction, in zone 4, is given by :

$$\frac{da}{dN} = (\frac{da}{dN})_1 + (\frac{da}{dN})_2 \tag{1}$$

$$\frac{da}{dN} = C\Delta K^m_{eff} \cdot F_1\{K_{1c}; \Delta K_{theff}\} + \frac{a^\alpha}{N_i} F_2\{\sigma_a; S_u\} \tag{2}$$

Here **a** is the crack length, N_i is the number of stress cycles to failure, as measured from the S-N curve for a given stress amplitude - σ_a, **da/dN** is the crack propagation in one cycle, $(da/dN)_1$ is the calculated value for zone 3, and $(da/dN)_2$ is the calculated value in zone 2.   **C** and **m** are material parameters, and F_1 and F_2 are parametric functions, that take care of the boundary effects near S_u, K_{1c}, and S_e, namely they stop the propagation process below S_e , and hasten it till the critical FCP rate, above S_u and K_{1c} is reached . A more detailed description of the diagram, the zones, the regimes, the fatigue life prediction procedures and the mechanisms was shown (Weiss 1992, 1993).

VARIABLE AMPLITUDE LOADING

Fatigue behavior under variable amplitude loading was one of the main weaknesses of the linear damage concept, which was not affected by it at all . It is well known that sequence effect does exist, and a High-Low sequence loading causes, in most cases, a shorter fatigue life than a Low-High sequence. This phenomenon has been treated in detail by (Weiss et al. 1996), and demonstrated with the help of the fatigue diagram. A description of a High-Low sequence fatigue loading is depicted in fig. 2. It can be seen on the left diagram on scale of crack length, and on the right part as crack propagation rate.

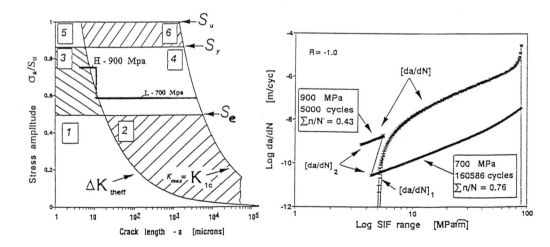

Fig. 2. Crack length and fatigue crack propagation rate, shown for a High-Low sequence of loading, as simulated on a certain AISI 4340 steel specimen (Weiss et al. 1996).

INFLUENCE OF MEAN STRESSES

The influence of mean stresses on fatigue life has been studied by many scholars and the state of the art has not changed much since reviewed, among others, by Buch (Buch 1988). Several relations have been proposed, most of them to calculate the new endurance limit and safety margins. The main proposals were those by Soderberg, and Goodman, who suggested to reduce the estimated safe endurance limit linearly from its basic value S_e (with no mean stress), gradually to zero (when the mean stress amplitude is increased up to the yield or the ultimate tensile strength respectively), and by Gerber, who proposed a failure parabola line that is expressed by eq. 3. This relation describes a line that falls, more or less in the middle of the scatter of the failure points, for various mean and alternating stresses. Here σ_a is the stress amplitude and σ_m is the mean stress of the cycle. The value of i equals 1 for the regular case, close to the endurance number of cycles - $5*10^6$. Larger values of i will be needed for smaller N values.

$$\frac{\sigma_a}{S_e} + \frac{\sigma_m}{S_u^2} = i \tag{3}$$

Diagrams for mean stresses and constant values of N, have been constructed experimentally and semi-analytically in many studies. Buch also shows a method to find loading stress amplitudes with zero mean, that are seemingly equivalent (in damage?) to other loading functions that include mean stresses, based on the Goodman or Gerber lines. A similar method has been used (Weiss 1996) to predict crack propagation in zones No. 3 and 4 in the Fatigue Diagram. One can use the extended Gerber lines to estimate equi-damage stress amplitudes, or different stress amplitudes that cause the same crack extension in one cycle. The idea is to estimate values of stress amplitudes with zero mean stress in a certain specimen, that are seemingly equivalent in the crack extension rates (or "damage" per cycle) that they cause in the specimen, to the applied stress amplitudes that include mean stresses, based on the Gerber lines. Utilizing this idea one can now use the Fatigue Diagram and the fatigue crack propagation rate models that are used, and calculate the crack extensions using the equivalent stress amplitudes, instead of the real ones which include mean stresses.

ADDITIONAL INFLUENCING FACTORS

Very many loading and material parameters have been found to influence fatigue life. The prediction of practical fatigue life expectation of a component has always been a problem due to the variety of these parameters. The proposed classification to fatigue zones and regimes has somewhat clarified the specimen behavior in these regimes, and helped to predict crack propagation. There are additional parameters that have not yet been studied in context of the Fatigue Diagram, and that can help to explain and validate quantitatively a few of the previously reported experimental studies.

One such factor is the *surface finish of the specimen* material, that has been extensively studied in the past, and was reviewed (Buch 1988). The surface finish influence on the short and very short crack propagation, can be calculated by the proposed procedures, and supply a simulated basis to the experimental results. For very rough surfaces, local stress concentrations will have to be utilized. *The size effect* can also be explained by the proposed methods, fatigue *behavior under random loading* may be an interesting application as well. It would be effective to reconsider and express the influence of all parameters that affect fatigue life, with terms of crack length rather than damage, by the proposed methods.

DISCUSSION AND CONCLUSIONS

1. The use of *crack length instead of damage* for all fatigue problems has been proposed before, as mentioned earlier, but has not caught up, and many "fatigue damage" studies are still being reported. It is the aim of this study to encourage the use of *crack length instead of damage* and the zones classification in a broader extent. Due to lack of space, the influence of the additional factors could have been mentioned only, but not displayed.

2. Due to the many parameters involved in fatigue problems, it is possible to show seemly correlation between prediction and tests, in almost any case. It would be appropriate now to look again on the numerous previous experimental studies, and fit them in, to the proper fatigue zone. This will *reduce considerably the unrelated scatter*, by comparing results that have been measured or estimated, only for the same fatigue zone. It is claimed that many of the seemingly statistical scatters, are in reality test results in different fatigue regimes, that should not be compared nor evaluated on the same basis at all. By refining our experimental results group classification, we will be able to smooth our prediction procedures better, by using only valid results. One very good example is comparing LEFM crack propagation rates for zones 2 (low stresses below the endurance limit, the true LEFM zone) and zone 4 (stresses above the endurance limit), and getting considerable scatter at the predicted values. *Narrow scatter can be expected systematically only for results estimated for one zone.*

3. Through the whole simulation, *effective SIF range* has been used, by neglecting the compression part of the loading cycle. It is the authors opinion, that this way of estimate of the effective value is appropriate and with relatively small flaws.

4. Study of the influence of additional parameters on fatigue life, as well as new audit of previous results, by use of the proposed new fatigue diagram method, can contribute to the state of the art of fatigue and fracture, and may advance the trend of expressing fatigue damage in terms of crack length.

REFERENCES

Buch A. (1988). Fatigue Strength Calculation. Trans. Tech. SA, Switzerland.
Kitagawa H. and Takahashi S. (1976). Proc. 2nd. Int. Conf. Mech. Behavior of Mater. ASM 627-631.
Krajcinovic D. (1989). Damage Mechanics. Mechanics of Materials 8, Elsevier 117-197.
Miller K.J. (1991), Metal Fatigue, past current and Future , Proc. Inst. Mech. Engng. 205, 291-304.
Mushkelishvili N.I., (1953). Some basic problems of the Mathematical Theory of Elasticity,
    P. Noordhoff Ltd. Groningen, The Netherlands (published originally in Russian in 1933).
Paris P.C. and Erdogan F. (1963). A critical analysis of crack propagation laws.
    Trans. ASME J. Basic Engng. 85, 528-534.
Weiss M.P. (1992). Estimating Fatigue Cracks, from the onset of loading, in smooth AISI 4340
    specimens, under cyclic stresses. Int. J. of Fatigue 14 (2), 91-96.
Weiss M.P. (1993). The use of classification zones for fatigue behavior in steels - 2. ASME
    J. Engng. Mater. and Structures 14 (2/3), 329-336.
Weiss M. P. and Hirshberg Z. (1996). Crack extension under variable loading, in the short and long
    cracks regime, using the general fatigue diagram. accepted for publication by the Fat. & Fract.
    Engng. Mater. Struct.
Weiss M.P. (1996) Modeling Fatigue Damage in terms of Crack Length, for stress cycles including
    mean stresses, in the whole fatigue & fracture domain. In preparation.

Acknowledgement:
This study has been supported by the fund for the promotion of Research at the Technion.

FATIGUE CRACK PROPAGATION BEHAVIOR OF AN ALUMINUM ALLOY UNDER VARIABLE AMPLITUDE LOADING CONDITIONS

I. TROCKELS, G. LÜTJERING, and A. GYSLER

Technical University Hamburg-Harburg
21071 Hamburg, Germany

ABSTRACT

The fatigue crack propagation behavior of a high-purity X-7075 alloy was studied in vacuum by periodically applied tensile overloads after various numbers n (10 to 10^5) of constant amplitude cycles with R = 0.1. The results showed significant crack growth retardation with increasing numbers of n, together with a pronounced increase in through-thickness crack front tortuosity. Load sequence variation from n = 1000 (straight profile) to n = 20000 (rough profile) resulted in delayed retardation within a short crack extension interval to adjust the high growth rate of the previous to the very low equilibrium growth rate of the following loading procedure. For the reversed sequence variation the propagation rate increased immediately after the loading change, however the much higher equilibrium growth rate for n = 1000 was only established after a rather large crack extension distance.

KEYWORDS

Fatigue crack propagation; variable amplitude loading; Al-alloy; crack front geometry; load sequence effects.

INTRODUCTION

The effect of periodically applied tensile overloads superimposed on constant amplitude fatigue loading has been studied in the past for different alloy systems (1-3). The studies have shown that considerable crack growth retardation occurred if the numbers n of intermittent baseline cycles between consecutive tensile overloads were increased. However, the reasons for this pronounced retardation phenomenon are discussed in the literature in controversial ways. While in (1) a combination of variations in crack tip blunting and mean stress with increasing numbers of n was assumed to be rate controlling, in (2) variations in crack closure levels were considered as the dominating parameter. Recently published results (3) revealed significant variations in the through-thickness crack front geometry configuration by varying the number n of intermittent baseline cycles between tensile overloads, which were thought to contribute to the observed crack propagation behavior, possibly in addition to the other mechanisms mentioned in the literature (1,2). It was found (3) that the through-thickness crack front tortuosity increased considerably with increasing numbers of intermittent baseline cycles between overloads, resulting in pronounced deviations of the crack front from a straight line, usually assumed in fracture mechanics concepts, due to local crack extension along slip bands which were activated by the preceding overload.

In the present investigation it was mainly tried to evaluate how these crack front geometry variations are influencing the crack propagation behavior under load sequence variations, for example by changing from a periodically applied overload sequence, having a straight crack front geometry to one which exhibited a very tortuous geometry and vice versa. These tests were performed on a high purity Al-Zn-Mg-Cu alloy with an equiaxed grain size in an underaged condition.

EXPERIMENTAL PROCEDURE

The tests were performed on a high-purity X-7075 alloy without Cr-dispersoids (Al-5.8Zn-2.6Mg-1.5Cu, wt.%), supplied by ALCAN, Banbury, UK, as a 30 mm thick hot rolled plate. Blanks of this plate were homogenized, water quenched and unidirectionally cold rolled to 9 mm thickness. The subsequent recrystallization treatment at 440°C resulted in an equiaxed grain size of 50 µm. The final age-hardening (24h 100°C) produced an underaged condition. All mechanical tests were performed at room temperature with the loading axis being parallel to the rolling direction. Tensile properties were determined in air on cylindrical specimens (20 mm gage length, 4 mm gage diameter) using an initial strain rate of 8×10^{-4} s^{-1}.

Fatigue crack growth tests were performed in vacuum on CT-specimens (8 mm thick, 32 mm wide) under load control at 30 Hz (sinusoidal wave form) using a computer controlled servohydraulic testing machine. The crack length was measured at the specimen surface with a travelling light microscope. The variable amplitude loading conditions consisted of constant amplitude baseline cycles ($R = 0.1$) with periodically applied tensile overloads (overload ratio 1.5). The number n of intermittent baseline cycles between consecutive overloads was varied between 10 and 100 000. Load sequence effects were studied by changing the number n at a defined ΔK-value within a CT-specimen from 1000 to 20 000 or vice versa. Crack closure values were determined for some selected specimens applying conventional back face strain technique. Crack front profiles in the through-thickness direction, taken from sections perpendicular to the crack growth direction, were analysed by light micrographs. Fracture surface studies were performed by SEM. In addition the fracture surface roughness along the through-thickness direction was measured at defined ΔK-values using a profilometer.

EXPERIMENTAL RESULTS

The microstructure of X-7075 consisted of fully recrystallized equiaxed grains with a size of about 50 µm, hardened by coherent precipitates (4). The tensile properties are summarized in Table 1.

Table 1. Tensile properties

Alloy	$\sigma_{0.2}$ (MPa)	UTS (MPa)	σ_F (MPa)	TE (%)	ε_F
X-7075	415	535	820	20	0.60

The fatigue crack propagation rates as a function of ΔK baseline are shown in Fig. 1 for different numbers n of intermittent baseline cycles ($R = 0.1$) between consecutive tensile overloads. These curves exhibited significant crack growth retardation with increasing numbers of n, especially in the low and intermediate ΔK regime. For example at $\Delta K = 15$ MPa·m$^{1/2}$ the propagation rates decreased almost three orders of magnitude by increasing n from 10 to 10^5 cycles. At high ΔK-values the curves were approaching each other (Fig. 1). The crack extension Δa between consecutive overloads along the main crack growth direction, calculated from the curves in Fig. 1, is shown in Fig. 2 as a function of ΔK baseline, together with the maximum overload plastic zone size under plane strain. These curves revealed that the tip of the propagating fatigue crack for all variable amplitude loading conditions was always confined within the maximum plastic zone size up to high ΔK-values. It should be noted that the crack extension between two overloads was similar for periodically applied overloads with $n = 1000$ and $n = 20000$ (Fig. 2), which will be discussed later with regard to the crack front profiles.

Examples of through-thickness crack front profiles at ΔK of about 13 MPa·m$^{1/2}$ are shown in Fig. 3. It can be seen that the profiles are fairly straight for periodically applied overloads with up to 1000

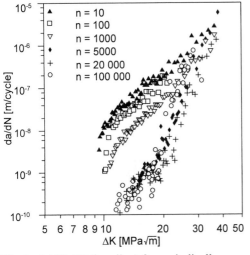

Fig. 1. da/dN-ΔK (baseline) for periodically applied overloads after n intermittent baseline cycles, (5).

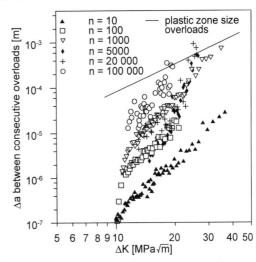

Fig. 2. Crack extension Δa between consecutive overloads and plane strain maximum overload plastic zone size vs. ΔK (baseline).

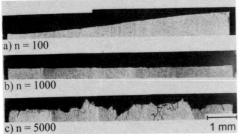

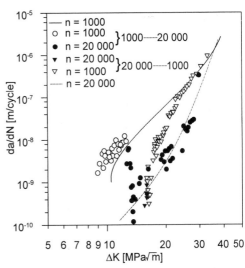

Fig.3. Through-thickness crack front profiles at ΔK≈ 13 MPa√m (LM), (5).

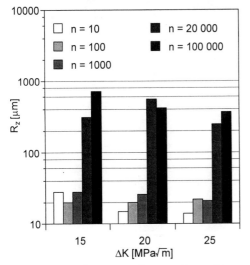

Fig. 4. Average fracture surface roughness R_z along through-thickness direction.

Fig. 5. Load sequence variations and equilibrium crack growth curves.

Fig. 6. Fracture surface: load sequence
variation n = 1000 → 20 000.

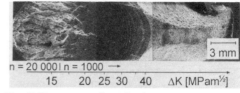

Fig. 7. Fracture surface: load sequence
variation n = 20 000 → 1000.

Fig. 8. Fracture surface, transition region n = 1000→20 000

a) ΔK = 15.9 MPa·m$^{1/2}$ b) ΔK = 21.5 MPa·m$^{1/2}$ n = 1000
Fig. 9. Fracture surface, transition region n = 20 000 → 1000.

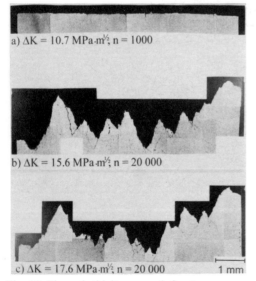

Fig. 10. Through-thickness crack front
profiles, n = 1000 → 20 000.

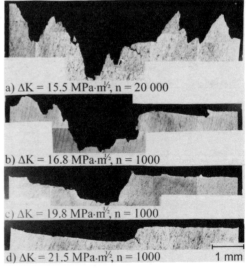

Fig. 11. Through-thickness crack front
profiles, n = 20 000 → 1000.

intermittent baseline cycles (Figs. 3a und b). For higher numbers of n the profile roughness increased drastically with increasing n (Figs. 3c to 3e). A more quantitative evaluation of the surface roughness, obtained by a line scan with a profilometer along the through-thickness direction, is shown in Fig. 4 at three different ΔK-values. These roughness values R_z are average distances between neighboring major peaks and valleys of the profiles shown in Fig. 3. These results revealed a tendency for a decreasing roughness with increasing ΔK, the absolute values at a constant ΔK being below about 30 µm for fracture surfaces of specimens tested with n-values up to 1000 cycles, while much higher R_z-values (above 400 µm) were found for fracture surfaces from tests with higher numbers of n (Fig. 4).

The effects of load sequence variations by changing from an equilibrium loading condition with n = 1000 to one with n = 20000 intermittent baseline cycles and vice versa are shown in Fig. 5, together with the corresponding equilibrium curves without load sequence variations. It can be seen that the crack propagation rates dropped by almost two orders of magnitude with a change from n = 1000 to n = 20000 intermittent baseline cycles between consecutive overloads. The necessary crack extension along the main growth direction to establish the new equilibrium propagation rate for n = 20000 was about 0.8 mm. On the other hand, for the reversed load sequence variation (n = 20000 to n = 1000) the crack growth rates immediately increased after the transition point (Fig. 5). However the necessary crack extension to reach the new equilibrium crack propagation rates for the loading condition with n = 1000 was comparatively long (about 4 mm). Overviews of the fracture surfaces for the two load sequence variations are shown in Fig. 6 (n = 1000 $\rightarrow$ n = 20000) and in Fig. 7 (n = 20000 $\rightarrow$ n = 1000), where it can be seen in Fig. 6 that the fracture topography changed after a very short crack extension from very flat (n = 1000) to very rough (n = 20000), while the transition region in Fig. 7 from the rough surface (n = 20000) to the equilibrium smooth surface (n = 1000) extended over a much longer distance. Details of the transition region for the sequence 1000/20000 are shown by the SEM micrograph in Fig. 8, where in the specimen interior (plane strain regime) the local crack growth rates, calculated from the distances of the overload markers before and immediately behind the load sequence change until the next overload appeared, were approximately the same. However, the fracture surface topography already changed significantly within this short distance of about 300 µm (Fig. 8) from a very flat to a much rougher appearance. It should be noted that the local growth rate dropped by a factor of two after the occurrence of the first overload in the n = 20000 sequence. The transition region for the sequence 20000/1000 can be seen in Fig. 9a. In this case the local crack growth rate increased by a factor of about three immediately after the load sequence was changed. It should be mentioned that the observation of overload markers in the highly facetted transition region (compare Fig. 7) was rather difficult. The isolated area in Fig. 7 which exhibited still a facetted appearance is shown in Fig. 9b at higher magnification. It can be seen that the overload markers are heavily curved in the vicinity of this area, indicating a retardation of the overall propagation rate.

Corresponding through-thickness crack front profiles for the two load sequence variations are shown in Figs. 10 and 11. For the sequence change from n = 1000 to n = 20000 the transition from the very smooth profile for n = 1000 (Fig. 10a) to the very rough profile for n = 20000 (Fig. 10b) occurred within a rather short crack extension region along the main growth direction of about 0.8 mm (Fig. 6). In contrast, for the reversed sequence variation the transition from the very rough profile for n = 20000 (Fig. 11a) to the flat profile for n = 1000 a much longer distance was observed (Fig. 7). The corresponding profile exhibited already shortly after the transition point individual flat parts (Fig. 11b), which however were displaced against each other parallel to the loading axis by large steps. These displacements prevailed (Figs. 11c and d) until the crack front extended about 4 mm beyond the transition point (Fig. 7).

The closure levels K_{cl}/K_{max} (baseline) of about 0.16 were found to be rather low, exhibiting no significant changes before and after the transition points for both load sequence variation procedures.

DISCUSSION

It has been found in previous studies (1-3) that the fatigue crack propagation rates decreased considerably if the number n of intermittent constant amplitude baseline cycles between periodically applied tensile overloads was increased (e.g. Fig. 1). This retardation behavior is thought to result mainly from the observed concomitant increase in through-thickness crack front tortuosity with increasing n (Fig. 3). The reason for the high crack growth rates and the very flat crack front geometries for frequently applied overloads (n ≤ 1000) seems to result from the high number of slip systems repeatedly activated ahead of the crack tip, which allow the crack front to maintain a very straight through-thickness configuration (Figs. 3a and b). By increasing the number n of intermittent baseline cycles the local crack front segments are able to propagate for longer distances along previously activated slip systems out of the main crack plane into less strain hardened regions, thus producing very rough crack front profiles (Figs. 3c to e). The underlying retardation mechanisms are thought to result from the local reduction of the crack driving force at each out-of-plane crack segment and from the observed unfractured ligaments between neighboring crack segments (3). This explanation seems to be supported by the coinciding curves of calculated crack extension Δa between consecutive overloads along the main crack growth direction for n = 1000 and n = 20000 (Fig. 2), while the corresponding crack front profiles were completely different (Figs. 3b and d) and the concomitant propagation rates revealed a difference of at least one order of magnitude (Fig. 1). Crack closure effects at R = 0.1 were found to be too small to explain the observed retardation phenomena, especially since crack propagation measurements at R = 0.5, where no closure was observed, resulted in similar crack growth curves and crack front geometries as a function of n as at the low R-ratio (3). Crack closure variations as being the sole mechanism to explain the propagation behavior under periodically applied overloads has been questioned before in the literature (6).

The load sequence variation results obtained in the present study are thought to depend significantly on the different equilibrium crack front geometries for the two concomitant loading procedures, while the very low and approximately unchanged closure levels were unable to explain the observed behavior. The variation from the equilibrium loading condition with n = 1000, having a straight crack front geometry (Fig. 10a), to that with n = 20000, exhibiting a very tortuous configuration (Fig. 10b), occurred within a rather short crack extension interval along the main growth direction (~ 0.8 mm, Fig. 6). The adjustment from the high propagation rate at n = 1000 towards the more than one order of magnitude lower equilibrium rate for n = 20000 followed a so-called delayed retardation curve (Fig. 5). The nearly identical local propagation rates determined for the last blocks of n = 1000 and the first 20000 constant amplitude cycles following the sequence transition, suggest that the straight crack front geometry of the preceding sequence was dominating within the following crack extension regime, although the fracture surface appearance already changed (Fig. 8). However, the occurrence of the first overload in the new sequence already reduced the growth rate by a factor of two, which can be explained with a further adjustment of the crack profile towards the rough equilibrium configuration (Fig. 10b). While the total crack extension to reach the low equilibrium growth rate for n = 20000 was rather short, the necessary number of cycles was very high due to the continuously decreasing propagation rate as a consequence of the increasing profile roughness.

In contrast, for the reversed load sequence variation from n = 20000 with the rough profile (Fig. 11a) to n = 1000 with the straight equilibrium configuration (Fig. 3b) a much longer crack extension interval was necessary (~ 4 mm, Fig. 7) to reach the new equilibrium growth rate (Fig. 5). The concomitant crack profiles indicated that individual areas exhibited a fairly straight configuration shortly after the transition point (Fig. 11b), however with significant displacements against each other parallel to the loading axis (Figs. 11c and d) as a result of the large deviations of individual local crack segments produced during the preceding load sequence with n = 20000 (Fig. 11a). The unfractured ligaments between the displaced flat parts (Figs. 11b - d) seemed to be effective barriers against propagation of the overall through-thickness crack front (Fig. 9b), thus reducing the growth rate as compared to the equilibrium rate typically observed for the fully straight profile configuration (Fig. 5).

REFERENCES

1. N.A. Fleck, Acta Metall. 33 (1985) 1339-1354.
2. S. Zhang, R. Marissen, K. Schulte, K.K. Trautmann, H. Nowack, and J. Schijve, Fatigue Frac. Engng. Mater. Struct. 10 (1987) 315-332.
3. I. Trockels, A. Gysler, and G. Lütjering, in Aluminum Alloys, Georgia Institute of Technology, Atlanta (1994) 717-724.
4. J. Lindigkeit, A. Gysler, and G. Lütjering, Metall. Trans. 12A (1981) 1613-1619.
5. J. Kiese, A. Gysler, and G. Lütjering, in FATIGUE '93, EMAS, Warley (1993) 1569-1574
6. J. Schijve, in FATIGUE '87, EMAS, Warley (1987) 1685-1721.

STRESS-STRAIN ESTIMATION AND A FATIGUE DAMAGE RULE

YASUSHI IKAI

Department of Mechanical Engineering, Kobe University
Rokko, Nada, 657 KOBE, JAPAN

ABSTRACT

A cumulative fatigue damage rule is proposed for a wide range of random load on the basis of plastic strain of the material. In connection to obtain strain response against random stress sequence, an interpolating technique NAIS is applied, which describes strain response as the function of stress and its history. The rule is compared to a variety of fatigue tests.

KEYWORD

Life estimation; random load; plastic strain; hysteresis; interpolation.

INTRODUCTION

Fatigue life evaluation is one of the most important problems in a fatigue study. Under varying stress conditions at high cycle fatigue region, however, the fatigue life evaluation is often insufficient, especially when stress levels vary in a wide range. For an example, fatigue failure occurred at stress levels below the fatigue limit when then are frequently changed(Iwasaki, 1986). Many attempts have been made to revise Miner's rule by modifying S-N curve such as Corten and Dolan(1956), they have not been successful under varying stress conditions. The main reason of these difficulties may be inferred to arise due to the fact that their treatments do not pay attention to the mechanical behavior of the material against stress level shifts. In that case the stress-strain behavior is widely different from the behavior after a constant loading condition.

The same state of affairs is observed for varying strain conditions in a low cycle fatigue condition. Manson-Coffin's rule(1954) is not well developed to deal with range pair $\Delta \varepsilon_{pr}$ with mean strain and also to deal with a random strain sequence.

The purpose of this report is to obtain a fatigue life evaluation method for a wide range of random load on the basis of plastic strain of the material. This is because the fatigue damage is caused by plastic strain of the material. In order to obtain strain response against random stress sequence, an interpolating technique NAIS(Ikai,1991) is applied, which describes strain response as the function of the stress and its history. After the variety of loading tests, fatigue lives are evaluated against a given stress sequence.

MATERIAL AND METHOD

The material used for tests is S35C steel with yield stress 343MPa and elongation 33.6%, whose chemical composition   is 0.35mass%C-0.20Si-0.79Mn-0.027P-0.021S-0.02Cu-0.02Ni-0.16Cr. Hourglass type specimens are used for higher strain conditions and smooth specimens are for lower strain conditions. Push-pull fatigue tests are carried   out with a hydraulic testing machine installed with an A/D-D/A converter to control machine and stress-strain measurements.

Higher dimensional interpolation technique NAIS makes possible to obtain the smoothest curved surface which passes through all of the data points composed of N variables.   Referring to Fig.1, the stress-strain relation of i+1th half cycle depends on $\Delta\sigma$, (σ_i, ε_i) and the foregoing (σ_{i-1}, ε_{i-1}), and also σ_{i+1} which represents stress velocity for sinusoidal stress wave, and

$$\Delta\varepsilon = f(\sigma_{i-1}, \varepsilon_{i-1}, \sigma_i, \varepsilon_i, \sigma_{i+1}, \Delta\sigma). \tag{1}$$

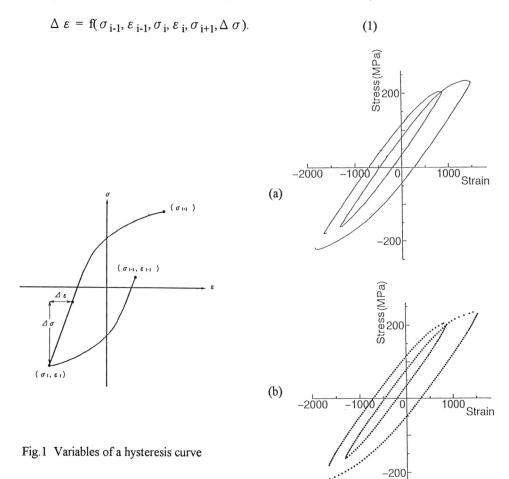

Fig.1 Variables of a hysteresis curve

(a)

(b)

Fig.2 Comparison of the observed (a) and estimated hysteresis curve (b).

Table I Fatigue lives for constant strain amplitude

No	$\Delta \varepsilon p(\%)$	Number of cycles	ΣD
1	0.02	>10000000	
2	0.03	>10000000	
3	0.04	>10000000	0.592760
4	0.05	1832485	0.592561
5	0.06	2214404	1.315666
6	0.08	1184277	1.951152
7	0.1	456242	1.512523
8	0.2	37000	0.785711
9	0.3	22000	1.205406
10	0.5	2377	0.413309
11	1.3	1162	1.259282
12	2.2	660	1.606202
13	4.5	226	1.470692

Table II Fatigue lives for increasing mean strain

No	plastic stain(%) $\Delta \varepsilon p+$	$\Delta \varepsilon p-$	number of cycles	ΣD
14	3.08	1.54	18	1.130593
15	2.66	1.33	19	0.935122
16	0.63	0.31	165	0.750383
17	0.05	0.04	28108	1.123893
18	0.04	0.035	237037	0.913980
19	0.18	0.72	17	0.413567
20	0.10	0.20	103	0.379205
21	0.05	0.06	5100000	1.825851
22	0.04	0.035	6110000	0.645654

Table III Fatigue lives for program stress sequences

σ (MPa)	No. 23	24	25	26	27	28
σ_1	229.6	235.4	190.3	180.5	186.4	300
σ_2	-229.6	-206.0	-197.2	-191.3	-186.4	-300
σ_3	182.5	191.3	176.6	191.3	225.6	250
σ_4	-212.3	-231.5	-206.0	-197.2	-225.6	-200
σ_5	212.3	191.3	220.7	206.0	196.2	250
σ_6	-182.5	-180.5	-180.5	-206.0	-196.2	-250
σ_7	182.5	206.0	220.7	221.7	215.8	
σ_8	-229.6	-223.7	-231.5	-223.7	-215.8	
σ_9	199.7	221.7	206.0	231.5	206.0	
σ_{10}	-199.7	-197.2	-188.4	-235.4	-206.0	
ΔD	1.95377	1.04671	1.728	0.54807	1.9112	0.9723

The function f is determined by interpolation technique with experimental values of six points on each halfcycle of ($\sigma - \varepsilon$) curve for various stress cycles. Then, $\Delta \varepsilon$ is estimated for a given six variables (σ_{i-1}, ε_{i-1}, σ_i, ε_i, σ_{i+1}, $\Delta \sigma$) to draw a half a hysteresis curve. In this way stress-strain hysteresis curves are obtained for a given stress sequence. For a steel with yield phenomenum, the estimation is insufficient at the early stage of stress cycles untill cumulative strain reaches two per cents. A comparison of observed hysteresis curve and estimated one is shown in Fig.2.

RESULTS AND DISCUSSION

Fatigue tests for constant strain amplitude are shown in Table I. In Table II are shown the results for increasing mean strain tests, which means gradual migration of a hysteresis curve to right or left on the $\sigma - \varepsilon$ plane, are shown in Table II. Fatigue tests for the program stress sequences are also shown in Table III versus their stress levels. Since there is no fatigue damage criterion to describe these results, the following criterion is proposed and examined:

> The fatigue damage given in one half cycle is smaller and independent
> on mean strain, if the strain retraces within that of the foregoing half
> cycle. If the strain exceeds the foregoing one, the excess quantity
> ($\Delta \varepsilon_{pm,i}$) gives larger damage depending on mean strain. There
> exist the lower limit of strain which gives no damage to the material.

Fatigue damage rule is expressed in the equation given below. The coefficients were determined through an optimization technique to satisfy the experimental results.

D_i = (damage depending on $\Delta \varepsilon_p$) + (damage depending on mean strain)
    = $A(\log(\Delta \varepsilon_p - f(i)+1.0))^B + G\ X^H(\Delta \varepsilon_{pm,i}/33.6)^I$,
    where $f(i) = C\exp(\log(i)+D)^E$,
    and A=0.018119 B=2.507876 C=0.000091
       D=-1.572801 E=0.010542,
    for positive mean strain
       G=10.26674 H=0.000248 I=1.663092,
    for negative mean strain
       G=0.146667 H=0.132471 I=1.955505,
    and $\Delta \varepsilon_{pm,i}$ means the excess strain, and X is its cetral position.

The cummulated damage evaluations are shown in Tables I and II. The damage evaluations are also shown in Table III. The evaluations correspond to the experimental results within the factor of two.

Since fatigue damage is caused of plastic strain, the fatigue damage should be evaluated on plastic strain. While a material reconstructs its substructure against applied stress shift, the behavior has not been fully formulated yet in dynamics. The interpolation technique makes possible to describe the strain response against a given stress sequence. The damage rule introduced here showed good applicability for high- and low-cycle stress levels and/or under random stress conditions.

CONCLUSIONS

1) The stress-strain relation of S35C steel is described for a given stress sequence by means of the interpolating technique NAIS.

2) A fatigue damage rule based on plastic strain concept is proposed and its applicability for a given stress sequence with wide stress range under random loading conditions is demonstrated.

REFERENCES

Coffin, Jr. L.F.(1954). A Study of the Effects of Cyclic Thermal Stresses on a Ductile Metal. *Trans. ASME,* 76,931

Corten, H.T. and T.J. Dolan, (1956). Cumulative Fatigue Damage. *Proc. Int. conf. Fatigue of Metals,*235

Iwasaki, C. and Y. Ikai. Fatigue Failure under Stresses below Fatigue Limit.*Fatigue Fract. Eng. Mater. Struct.* 9,117

Ikai, Y. *et al.*(1991). In: *Computer Aided Innovation of New Materials* (M. Doyama ed.), pp127-130. Elsevier Science Publishers B.V. North-Holland.

FATIGUE CRACK GROWTH ACCELERATION OR RETARDATION DUE TO COMPRESSIVE OVERLOAD EXCURSIONS

H. Buschermöhle[1], D. Memhard[2], M. Vormwald[3]

[1] Inst. f. Maschinelle Anlagentechnik u. Betriebsfestigkeit. Technical University Clausthal TUC
[2] Fraunhofer Institut für Werkstoffmechanik, FhG/IWM, Freiburg
[3] Fachhochschule Jena, formerly Industrieanlagen - Betriebsgesellschaft, IABG, Ottobrunn

ABSTRACT

Experimental investigations of the sensitivity of fatigue crack growth rates on sporadically applied compressive overloads have been performed using thick-walled specimens of 40 CrMoV 4 7 as a member of the group of quenched and tempered steels. While in most cases compressive overloads have only a vanishing effect on crack growth rates, some experiments with single edge crack tension specimens reveal a marked growth retardation. Predictions with the Newman strip-yield-model give an acceleration. Although contrary to these predictions, which take the crack closure mechanism into account, the same experimental result has been obtained in a few other investigations on the subject. It is concluded that different crack growth mechanisms must be active additionally and that they are dominating in certain cases. Unfortunately, they have not been quantifiable in prediction algorithms up to now.

KEYWORDS

Fatigue crack growth, overload, underload, acceleration, retardation, damage accumulation.

INTRODUCTION

Since components in machines, power systems, as well as in aircraft structures are subject to complex load histories, the stress interaction effects on fatigue crack growth are of great interest. The beneficial effect of tensile overloads on fatigue life has been widely established (Mills and Hertzberg, 1975). In contrast with tensile overloads it has been found that compressive overloads (underloads) tend to accelerate crack growth (Hertzberg, 1989). Whereas quite a number of investigations (see references in Table 1) have been performed on materials and structures typically used in aircraft applications, little is known for materials and geometries typical in power systems and engines. Especially in this field compressive overload histories are rather common. They are composed of a large number of small amplitudes due to normal operation which are occasionally interrupted by a few large amplitudes due to short circuit or missynchronization loads at turbine parts. The present investigation was aimed to supply information on fatigue crack growth due to compressive overloads in thick-walled steel specimens as they are representative for power system application.

STATE OF THE ART SURVEY

Accelerated crack growth due to compressive overloads is dominating and should therefore be considered in the design process. However, some papers note retarded crack growth. Attempting to explain this somewhat strange behaviour, leads us to note the underlying mechanisms. A more detailed survey of relevant mechanisms is given by Fleck (1987).

- Crack closure is regarded as an important mechanism in the explanation of load interaction effects during crack growth and is discussed below.
- Compressive overloads are thought to resharpen the crack tip which was blunted during tension loading. This should in general lead to accelerated crack growth.
- Opinions are split on whether or not crack tip branching due to overloads will accelerate or retard crack growth.
- Strain hardened material in front of the crack tip is thought to be predamaged. Thus, overloads of any sign should produce a high amount of predamage resulting in an acceleration.
- Mean stress relaxation occurs at the reversed plastic zone near the crack tip. During overload tests the mean stress of the major instead of the minor cycle is likely to relax to zero. Therefore, minor cycles have a tensile mean stress which will lead to faster growth.
- The crack growth mechanism at different amplitudes or R-ratios may be different. High amplitudes tend to produce "irregular" crack paths in zig-zag-mode or shear lips. The effect of load interaction effects due to this crack front mismatch mechanism is not completely understood, although some researchers see it as reason for retardation (Zuidema et al., 1991).

The following table gives a survey on some of the investigations on the subject.

Table 1: Literature survey

Material	Spec. type	Loading			Result	Ref.
	thickness	R_{minor}	R_{major}	n/N	factor f	
Ti-6Al-4V (TR)	CT 13mm	0.75, 0.82, 0.90	$\approx 0.$ $\leq$	$10^3/1$ to $10^5/1$	LDA	Powell and Duggan, 1987
HT 80 and HT 80 weld	CCT 20mm	R = 0.4; 0.9 R = 0; $\Delta K_{major} / \Delta K_{minor} = 1.3$		$10^4/10^4$	ACC $f\approx 2$; LDA RET $f\approx 1.5$	Ohta et al., 1987
BS1501 32A BS 4360 50B Al 2014-T4	CT 3mm, 24mm 6mm	0.75	0.5	1/1 to 1000/1	ACC f between 1.0 and 1.8;	Fleck, 1985
Al 2091-T8 Al 2024-T3	CCT 12mm	0.7	0.175	10/1	ACC $f\approx 2$	Ohrloff et al., 1987
Al 7150 sev. heat treatmts.	CT 6.4mm	threshold test at R=0.1	-1 to -3 -5	1 major cycle	no effect; growth after crack arrest	Zaiken and Ritchie, 1985
Al2024-T351 SAE 1045 CSA G40.21	CCT 2.54mm	0	-3	200000/1 to 1/1; $\infty/1$	ACC and threshold reduced	Topper and Yu, 1985, Yu et al.,1984
AISI 542 Class 3	CCT 2.54mm	$R_{minor} \approx 10$ (fully compr.); OL factor 2 and 3 in compr.		100/1 and only one initial OL	ACC; growth after crack arrest	Aswath et al., 1988
Al2024-T351	cylindrical LCF-spec.	-1	< -1 S_{min}=con	50000/1 to 10/1	ACC	Pompetzki et al., 1990

Al2024-T351	CCT 6mm	0.8; 0.7; 0.575; 0.32	0.44; 0.18; 0.32; 0;	overload blocks of various lengths	RET	Zuidema et al., 1991
WASPALOY M50 NiL	Corner crack specimen	0.1	-2	blocks, lenghts not given	RET as transient effect	Carlson and Kardomateas, 1994
SAE 1026	SENT 2.54mm	0	-1	$10^2/1$ to $10^4/1$	ACC at large strain	McClung and Sehitoglu, 1988
AISI 4335	K_{Ic}-spec. ASTM 399 C 25.4-38.1mm	0.1	-2	1 major cycle	ACC; $f\approx1.25$ to 2;	Underwood and Kapp, 1981
Al2219-T851	CCT	0.7	-0.3	2500/1	ACC $f\approx1.3$	Chang et al., 1981

LDA ≡ linear damage accumulation; no interaction	CT ≡ Compact tension specimen
ACC ≡ acceleration compared to LDA	CCT ≡ Centre cracked tension specimen
RET ≡ retardation compared to LDA	SENT ≡ Single edge notched tension spec.
OL ≡ Overload	$f = \begin{cases} N_{EXP}/N_{LDA} & \text{for RET} \\ N_{LDA}/N_{EXP} & \text{for ACC} \end{cases}$

EXPERIMENTAL INVESTIGATION

Tests have been carried out with SENT-, CCT- and notched (K_t=2.3) cylindrical specimens. The material tested was the alloyed steel 40CrMoV 4 7. The ratio of the number of periodic minor cycles n to periodic compressive overload cycles N (n/N = 10, 1000, and 10000) has been varied. The ratio of the compressive overload cross section stress to the tensile 0.2-yield strength ($\sigma_{min,major}/R_{p0.2}$ = - 0.25 and -0.5) has been varied as well. The stress ratio was R_{minor}= 0 or 0.5 and R_{major}= -1 or -2. Additional strain measurements were also performed to investigate the influence of compressive overloads on the development of crack opening stresses. For more details see Buschermöhle et al. (1995). The test results can be summarised as follows:

- Compressive overloads with the ratio $\sigma_{min,major}/R_{p0.2}$ > -0.5 have no influence on the threshold value ΔK_{th} for all specimen thicknesses tested (10mm, 25mm).
- Compressive overload effects are closely related to specimen type and geometry. Overloads with the ratio $\sigma_{min,major}/R_{p0.2}$ = -0.25 caused a significant retardation on the fatigue crack growth rate depending on the overload frequency n/N with the 10mm thick SENT-specimens, (Figure 1). Increasing the compressive overload ($\sigma_{min,major}/R_{p0.2}$ = -0.5) resulted in a somewhat smaller retardation due to a compressed plastic wake profile and the thereby reduced crack opening stress. No overload effect was found for the 25mm thick SENT-specimens.
- Contrary to the results of the SENT specimens, tests results of the 10mm and 25mm thick CCT-specimens as well as notched specimens showed no significant effect due to compressive overloads except for a small reduction in fatigue crack growth life for the overload frequency n/N=10.

A significant acceleration due to compressive overloads could not be found in any of the investigated cases (3 specimen types, several load parameters). Some tests performed for comparison purposes showed that 40 CrMo V 4 7 is less sensitive to compressive compared with tensile overloads. Figure 2 gives the crack opening loads measured as stiffness changes by a series of strain gauges mounted in the line of crack growth as close as possible to the crack tip. Measurements have been performed before, at, and after overloading. The smaller values are not necessarily due to measurements after overloading. So, considering scatter due to the measurement technique, a significant transient effect could not be resolved. The general trend is simply a reduction in opening stress with increasing crack length. This gives reason neither for acceleration nor retardation, because the same behaviour can be found in constant amplitude tests. Summarising, the following Table 2 can be added to Table 1 as the result of the present investigation.

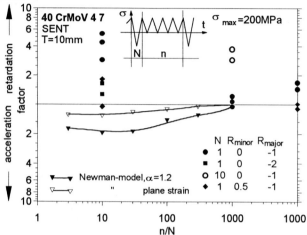

Fig. 1. Compression overload test results for 40 CrMoV 4 7 SENT-specimens

Table 2. Present Results

Material	Spec. type	Loading			Result
	thickness	R_{minor}	R_{major}	n/N	
40CrMoV 4 7	SENT	0; 0.5	-1; -2	10/1 to	RET
	10mm			10^4/1	
40CrMoV 4 7	SENT; 25mm	0; 0.5	-1; -2	10/1 to	LDA
	CCT; 10mm; 25mm			10^4/1	
	notched shaft 50mm diam.				

NUMERICAL INVESTIGATION

Figure 2 also contains the predicted crack opening stresses applying the Newman crack growth model (Newman, 1981) which is available as commercial software (Newman, 1984). Formulated as a strip-yield-model, it takes into account the elastic-plastic deformation in front of the crack tip and in the wake of the crack. Thus it is possible to predict load history dependent opening stresses. Together with Elber's (1971) concept of effective stress ranges being responsible for crack growth, the overload effect on crack growth should be predictable as far as it is related to the crack closure mechanisms. Figure 1 contains two lines predicted with this model. The first one uses a constraint factor α of 1.2 which indicates a near plane stress situation at the crack tip. Using this value it is possible to predict the constant amplitude crack growth curves for different R-ratios in the plates with 10mm thickness from the da/dn-ΔK_{eff} master curve. At least for shorter crack lengths a more plane strain dominated situation would have been expected. A second calculation gives the prediction for the pure plane strain situation. In both cases acceleration is predicted which is more pronounced for the plane stress situation. The maximum acceleration factors on the order of $f=2$ are found for cycle ratios n/N between 10 and 30. The acceleration effect has vanished for n/N approaching 1000 indicating that the crack opening stresses of approximately a couple of hundred cycles after the major cycle are considerably decreased by the compressive overload.

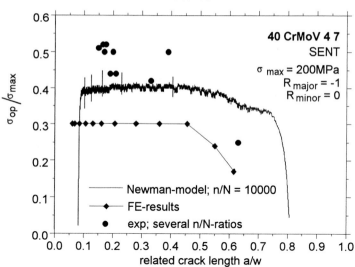

Fig. 2. Measured and calculated crack opening stresses

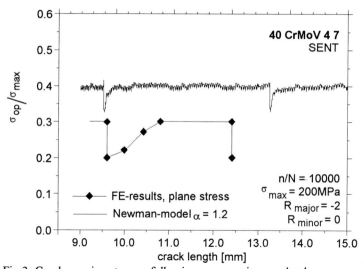

Fig 3. Crack opening stresses following compressive overload

Furthermore, finite-element calculations have been performed in order to compare crack opening loads. The calculations were done with the ADINA-code (Bathe, 1980) subjoined by a module for the determination of fracture mechanics parameters and for the simulation of fatigue crack growth (IWM, 1988). For more details of the calculation again see Buschermöhle *et al.* (1995). While Figure 2 gives the general trend of crack opening stresses as the crack grows through the ligament, Figure 3 shows the development of the crack opening load after a compressive overload. The prediction using the Newman model is also included for comparison. There is a good agreement between the results of the different models and experiment as far as the overall trend is concerned. However, actual values differ quite a lot. One reason for the differences between calculated and measured crack opening loads is the two-dimensional modelling of the three-dimensional problem, this especially due to crack surface asperities.

DISCUSSION AND CONCLUSION

Nevertheless, the numeric investigation on opening stresses and acceleration factors clearly shows that compressive overloads give rise to acceleration via the crack closure mechanism. The general trend as outlined by Table 1 is predictable. The corresponding acceleration factors seem reasonable. The present experimental investigation revealed retardation which is in accordance only with the minority of previous papers. This is only to be explained by the argument that other mechanisms than crack closure are operative. The crack front mismatch mechanism is thought to be the relevant one, here. According to Figure 1 this mechanism operates on about 10 to 100 cycles after a single compressive overload but on several thousand cycles when compressive overload blocks are applied. Unfortunately the mechanism has not been quantifiable up to now. For practical purposes the retardation should therefore be neglected in the design process and linear damage accumulation should be assumed when dealing with thick-walled steel components.

ACKNOWLEDGEMENT

Financial support of this investigation by the "Bundesminister für Wirtschaft (BMWi)" and the "Arbeitsgemeinschaft industrieller Forschungsvereinigungen (AIF)" under contract no. 8636 is gratefully acknowledged. The authors wish to express special thanks to the members of the working group "Bauteilfestigkeit" of the "Forschungskuratorium Maschinenbau (FKM)" under the guidance of Dr. Ch. Berger for valuable discussions. The material was supplied by Siemens AG, UB KWU.

REFERENCES

Aswath, B.P. , Suresh, S., Holm, D.K., Blom, A.F (1988). *J. Engng. Mater. Tech.*, 110, , pp.278-285

Bathe, K.L. (1980). ADINA, report 82 448-1, MIT, Cambridge, Mass USA

Buschermöhle, H., Memhard, D., and Vormwald, M. (1995). Rißfortschritt an Bauteilen bei Drucküberlasten. FKM-Report, Vol 195, in German

Carlson, R.L. and Kardomateas, G.A. (1994). *J. Fatigue*, 16, , pp. 141-146

Chang, J.B, Engle, R.M., and Stolpestadt. J. (1981). In: *ASTM STP 743*, pp.3-27

Elber, W. (1971). In: *ASTM STP 486*, pp. 230-242

Fleck, N.A. (1985). *Acta Metall.* 33, No 7, pp. 1339-1354,

Hertzberg, R. W. (1989). Deformation and Fracture Mechanics of Engineering Materials, 3rd edition, . Wiley, New York.

IWM-Crack (1988). A Software package for crack problems, Fraunhofer Inst. für Werkstoffmechanik, Freiburg FRG,

McClung, R.C. and Sehitoglu H. (1988). *ASTM STP 982*, pp. 279-299

Mills, W. J. and Hertzberg, R. W. (1975). *Engng. Frac. Mech.*, 11, , pp. 705-711.

Newman, J.C. (1981). In: *ASTM STP 748*, pp. 53-84

Newman, J.C. (1984). Instructions for Use of FASTRAN, COSMIC, Athens, GA 30602 USA,

Ohta, A., Konno T., and Nishijima, S. (1985). *Engng. Frac. Mech.*, 21, No 3, 1985, pp. 521-528,

Ohrloff, N., Gysler, A., and Lütjering, G. (1987). *J. de Physique, Colloque C3*, pp. 801-807,

Pompetzki, M.A., Topper, T.H., DuQuesnay, D.L., and Yu, M.T. (1990). *JTEVA*, 18(1), , pp. 53- 61

Powell, B.E. and Duggan, T.V. (1987). *Int. J. Fatigue* 9, No 4, , pp. 195-202

Topper, T.H. and Yu, M.T. (1985). *Int. J. Fatigue* 7, No 3, pp. 159-164,

Underwood, J.H. and Kapp, J.A. (1981). *ASTM STP 743*, , pp. 48-62

Yu, M.T., Topper, T.H., Au, P. (1984). *Proc. Int. Conf. Fatigue 84*, pp. 179.190,

Zaiken, E., Ritchie, R.O. (1985). *Eng. Fract. Mech.* Vol. 22, No l, pp. 35-48,

Zuidema, J., Wu Yi Shen, Janssen, M. (1991). *Fat. Fract. Eng. Mater. Struct.*, 14(10), pp. 991-1005

Crack Opening Stress Reductions Due to Underloads and Overloads in 2024-T351 Aluminum and SAE 1045 Steel

A.A. Dabayeh, C. MacDougall and T.H. Topper

Department of Civil Engineering, University of Waterloo
Waterloo, Ont., N2L 3G1, Canada

ABSTRACT

Changes in crack opening stress after underloads and compression-tension overloads were examined in a 2024-T351 aluminum alloy and an SAE 1045 steel. A variable amplitude block loading history consisting of high, near yield stress, underloads or compression-tension overloads followed by constant amplitude small cycles was used. The crack opening stress reductions due to underloads and compression-tension overloads and the subsequent build-up to a steady state level were measured for seven different R-ratios of the small cycles using a 900 power short focal length optical microscope. Results indicated a more rapid crack opening stress build-up for 2024-T351 aluminum than for SAE 1045 steel. The crack opening stress build-up is described by an empirical formula in terms of the ratio of the difference between the instantaneous crack opening stress of the small cycles (S_{op}) and the post underload or post compression-tension overload crack opening stress level (S_{opol}) and the difference between the steady state crack opening stress of the small cycles (S_{opss}) and the post underload or post compression-tension overload crack opening stress levels, $(S_{op}\text{-}S_{opol})/(S_{opss}\text{-}S_{opol})$. For simplicity both underloads and compression-tension overloads will be referred to as overloads in the remainder of this paper.

KEYWORDS

Crack opening stress, steady-state crack opening stress, variable amplitude loading, underloads, compression-tension overloads

INTRODUCTION

Fatigue underloads are known to have a detrimental effect on crack initiation and growth. The large compressive plastic zone produced in the wake of the crack causes flattening of the crack surface asperities which in turn reduces the crack closure level.

Pompetzki et al [1990a, 1990b] studied the load interaction effect of large stress cycles on the fatigue behavior of subsequent smaller cycles. They calculated the interactive damage per cycle for block histories containing either compressive underloads or tensile overloads. The underloads used were higher than the yield stress and overloads were equal to the yield stress. The interactive damage decayed as a power law function of the number of cycles following the overload or underload. Results indicated that small cycles including those below the constant amplitude fatigue limit, contributed significantly to damage accumulation. It was concluded that the presence of large compressive underloads or high tensile stress overloads decreased the average crack opening level and resulted in increased damage for cycles below the fatigue limit. Duquesnay [1991] measured crack opening stresses in smooth specimens at cyclic stress levels up to and slightly beyond the cyclic yield stress. He used strain gauges across the crack to measure the crack closing and opening stresses and verified the results using replica measurements. He found a linear decrease of opening stress (Sop) with decreasing minimum stress (Smin) and a non-linear relationship between the crack opening stress (Sop) and the maximum stress (Smax). As the maximum stress approached the yield stress the crack opening stress level decreased to negative opening stress levels.

McClung and Sehitoglu [1988] observed a decrease in the normalized crack opening stress level with increasing strain amplitude for constant amplitude histories. Their results showed that crack closing stress levels are significantly lower than crack opening stress levels at high strains. Large strain excursions in their simple block histories were observed to have a major impact on crack opening behavior during subsequent smaller strain cycles. They noticed that the crack tip remained entirely open during subcycles immediately following the major cycle causing a significant acceleration of crack growth.

Topper and Yu [1985] investigated the effect of tensile and compressive overloads on the threshold stress intensity level and the crack closure behavior of an aluminum alloy and three steels. They found that after compressive overloads the crack opening stress of an annealed SAE 1010 steel was decreased and it took more than 10,000 cycles for it to return to its stable level. The squeezing of the material at the crack tip due to the compressive overload, resulting in a low or negative crack opening stress and strain was the reason given for the decrease in the crack closure level. During the period after an overload, in which the crack opening stress returned to its stable level, an initially high crack propagation rate gradually decreased to a stable value. The threshold stress intensity decreased as the frequency of overload applications increased.

Gamache and McEvily [1993] who studied crack closure development in three ferrous steels due to the wake of a newly formed crack, showed that the crack opening stress could be described by the equation of Minakawa et al [1984]:

$$K_{op} = K_{opmax} (1 - e^{-kl}) \qquad (1)$$

where

K_{opmax}	is the level of steady state crack opening stress intensity factor at the crack tip for a long crack.
k	is a material constant of dimensions mm^{-1} which reflects the rate of closure development.
l	is the crack length.

They also found that the rate of increase of the crack opening stress increased with increasing material strength level.

The objective of this paper is to present test data on crack opening stress build-up after overloads and to describe it by an empirical formula which can be used to calculate effective stresses and predict fatigue life for variable amplitude load histories.

EXPERIMENTAL PROCEDURE

Material, equipment and test technique

Materials used in this study were a 2024-T351 aluminum alloy and an SAE 1045 steel with the chemical composition given in table 1. Flat specimens with dimensions of 19 mm in width and 2.54 mm in depth, for the 2024-T351 aluminum, were machined from 19 mm diameter drawn rods of the material in the as received condition and flat specimens with dimensions of 44 mm in width and 2.54 mm in depth were machined for the SAE 1045 steel specimens. The mechanical properties of the materials are listed in table 2.

Table 1. Chemical composition (% by weight)

2024-T351	Aluminum						
Si	Fe	Cu	Mn	Mg	Cr	Zn	Ti
0.5	0.5	4.35	0.6	1.5	0.10	0.25	0.15
SAE 1045	Steel						
C	Si	Mn	P	S	Fe		
0.46	0.17	0.81	0.027	0.023	remainder		

Table 2. Mechanical properties

Material	2024-T351 Aluminum	SAE 1045 Steel
Elastic Modulus	72400 MPa	206000 MPa
Cyclic yield stress (0.2 % offset)	450 MPa	400 MPa
Ultimate tensile strength	498 MPa	745 MPa

The specimens were prepared in accordance with ASTM standard E606 for constant amplitude low-cycle fatigue tests. The preparation included hand polishing of the gauge section in the loading axis direction with emery paper of grades 400 and 600. Final polishing was done using a metal polish which left a highly reflective surface which aided crack observations. All tests were carried out in a laboratory environment at room temperature (23°c) using a uniaxial , closed-loop, servo-controlled electrohydraulic testing machine. A small notch of about 0.01 inch was introduced on one side of the specimen, at mid length, to produce a stress raiser which served to localize the crack initiation site. This size was small enough that, once initiated, the crack rapidly grew out of the zone of influence of the notch. A traveling microscope with a high magnification (900X) was mounted on the machine facing the specimen. A vernier with an accuracy of 0.0005 inch was attached to the microscope to measure changes in crack length. A digital process control computer was used to output both constant-amplitude small cycles and periodic overloads in the form of a sinusoidal loading wave.

The program used in studying the changes in crack opening stress involved running at a constant load amplitude until a steady state crack opening stress is reached, then applying an overload cycle, followed by cycling at the initial test condition until the crack opening stress level returned to the value it had prior to the overload cycle.

Crack opening stress measurements

Crack opening stresses were measured optically during a load history consisting of repeated blocks of a large overload followed by constant amplitude small cycles. The small cycles had four stress ratios of R=-1.0, R=0.0, R=0.5 and R=0.8 for 2024-T351 aluminum, and had four stress ratios of R=-0.2, R=0.3, R=0.7 and R=0.8 for SAE 1045 steel. The maximum stress level for the constant amplitude cycles was always the same as the maximum stress of the overload cycle for SAE 1045 steel testing, while it was varied for the 2024-T351 aluminum.

The crack tip region was examined throughout a stress cycle using a 900x microscope at cycle numbers 1, 5, 10, 20, 50, 100, 200, 500, 1000, 2000, 5000,10000, 20000 and 50000 after the overload and the crack opening and crack closure stress levels were recorded. Crack opening stress levels which fell below the minimum stress of the constant amplitude small cycles were determined by decreasing the load level until the crack closed. A set of readings was repeated five times for each small cycle stress level and the average of the five readings at a given cycle number after the overload was taken as the crack opening or crack closure stress level. The crack opening and crack closure stress levels are defined as the levels at which the crack surfaces just open or close respectively at a distance of 0.25 mm behind the crack tip.

EXPERIMENTAL RESULTS

Figure 1 shows the 2024-T351 aluminum alloy crack opening stress as a function of the number of cycles after an overload for R=0.5. The overload stress cycle had a maximum stress (S_{max}) of 100 MPa and a minimum stress (Smin) of -350 MPa. The maximum and minimum stresses of the constant amplitude small cycles were 59.0 MPa and 29.5 MPa respectively. The crack length when the measurements were made was 2.74 mm. The crack opening stress level decreased to a minimum immediately after the overload. The crack opening stress increased from a post overload value of 14.6 MPa to a steady state stress level of 37.4 MPa. The increase is roughly exponential in form and at this stress ratio the steady state crack opening stress was higher than the minimum stress of the small cycles and consequently the steady state small cycles were partly closed. The opening stress reached the level of the minimum stress at about 100 cycles after the overload. Thus during the first 100 cycles after the overload the crack is fully open throughout the stress cycle.

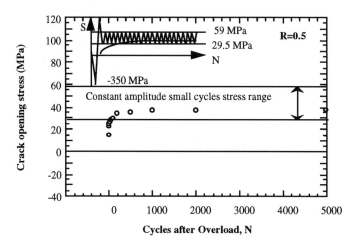

Fig. 1. 2024-T351 aluminum crack opening stress build up measurments
for R=0.5

Similar crack opening stress build-up measurements were taken for different overload levels and different constant amplitude cycle stress ratios. The results obtained are tabulated in table 3.

Table 3. Number of cycles to reach the steady state crack opening stress

Material	Overload levels (MPa)	constant amplitude R-ratio (stress levels (MPa))	No. of cycles to reach the steady state crack opening stress
2024-T351 Aluminum	-350 to +100	0.8 (+100 to +80)	2,000
	-350 to +100	0.5 (+59 to +29.5)	5,000
	-350 to +100	0.0 (+41.4 to 0)	2,000
	-350 to +100	-1.0 (+ 35.2 to -35.2)	2,000
	-400 to +250	0.8 (+250 to +200)	30,000
	-400 to +100	0.0 (+100 to 0)	3,000
SAE 1045 Steel	-300 to +300	0.8 (+300 to +240)	15,000
	-350 to +350	0.7 (+350 to +245)	15,000
	-350 to +350	0.3 (+350 to +105)	7,500
	-370 to +370	0.3 (+370 to +111)	1,000
	-370 to +370	-0.2 (+370 to -74)	30

CRACK OPENING STRESS BUILD-UP EQUATION

All crack opening stress measurements for 2024-T351 aluminum and SAE 1045 steel are plotted on a semi-log. scale in figures 2 and 3 respectively in terms of $(S_{op} - S_{opol})$ divided by $(S_{opss} - S_{opol})$ versus the number of cycles following the overload, N, normalized by the number of cycles at which this ratio reaches 80% of the steady state level, $N_{0.8}$. The following equation gives a reasonably good fit to the data:

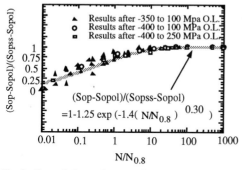

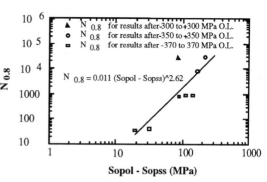

Fig. 2. Curve fitting to the normalized crack opening stress build-up data for 2024-T351 aluminum

Fig. 3. Curve fitting to the normalized crack opening stress build-up data for SAE 1045 steel

$$\frac{(S_{op} - S_{opol})}{(S_{opss} - S_{opol})} = 1 - \psi \, Exp \left(- b \, (N / N_{0.8})^a \right) \qquad (2)$$

where

ψ, b and a are material constants.

N is the number of cycles following the overload.

$N_{0.8}$ is the number of cycles following the overload at which the normalized recovered stress $(S_{op} - S_{opol})/(S_{opss} - S_{opol})$ reaches 80% of its steady state level and,

ψ, b and a constants are derived from curve fitting and have the values of 1.25, 1.4 and 0.30 respectively for 2024-T351 aluminum and have the values of 1.0, 1.0 and 1.7 respectively for SAE 1045 steel..

Values of $N_{0.8}$ are plotted versus $(S_{opol}-S_{opss})$ on a log-log scale in fig's 4 and 5 for 2024-T351 aluminum and SAE 1045 steel respectively. A simple linear curve fit to the data is shown giving $N_{0.8}$ as:

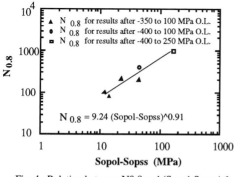

Fig. 4. Relation between N0.8 and (Sopol-Sopss) for 2024-T351 Aluminum

Fig. 5. Relation between N0.8 and (Sopol-Sopss) for SAE 1045 Steel

$$N_{0.8} = 9.24\,(S_{opol} - S_{opss})^{0.91} \qquad \text{For 2024-T351 aluminum} \qquad (3)$$

and

$$N_{0.8} = 0.011(S_{opol} - S_{opss})^{2.62} \qquad \text{For SAE 1045 steel} \qquad (4)$$

DISCUSSION

There is a common trend for the crack opening stress build-up after overloads for 2024-T351 aluminum and SAE 1045 steel. The crack opening stress in both materials increased in an exponential manner until it reached the steady state stress level. However, in case of the aluminum alloy the crack opening stress starts to build up immediately following the overload while in case of steel there is a delay before the crack closure stress increases.

It is of interest to note that the empirical equation 2 which was first used for the 2024-T351 aluminum alloy [Dabayeh and Topper, 1995] (Fig. 2) also gives a good fit to the quite differently shaped curve for the SAE 1045 steel (Fig. 3).

CONCLUSIONS

1- The crack opening stress level immediately dropped after underloads and compression-tension overloads having near yield level compressive stresses and then gradually increased with subsequent constant amplitude cycling.

2- The crack opening stress build-up for 2024-T351 aluminum and SAE 1045 steel can be described by an empirical formula in terms of ratio of the difference between the instantaneous crack opening stress of the small cycles (S_{op}) and the post overload crack opening stress level (S_{opol}) and the difference between the steady state crack opening stress of the small cycles (S_{opss}) and the post overload crack opening stress levels, $(S_{op}-S_{opol})/(S_{opss}-S_{opol})$.

3- The crack opening stress build-up starts immediately after an overload in 2024-T351 aluminum but is delayed for SAE 1045 steel.

REFERENCES

Duquesnay, D.L. (1991). Fatigue damage accumulation in Metals Subjected to High Mean Stress and Overload Cycles. Ph.D. Thesis, University of waterloo, Waterloo, Ontario, Canada.

Gamache, B. and A.J. McEvily (1993). On the Development of Fatigue Crack Closure", Proceedings of the fifth international conference on fatigue and fatigue thresholds, 577-582.

McClung, R.C. and H. Sehitoglu (1988). Closure behavior of small cracks under high strain fatigue histories. ASTM STP, 982, 279-299.

Minakawa, K., H. Nakamura and A. J. McEvily (1984). On the Development of Crack Closure with Crack Advance in a Ferretic Steel. Scripta metallurgica, 18,1371-1374.

Pompetzki, M.A., T.H. Topper and D.L. Duquesnay (1990a). The Effect of Compressive Underloads and Tensile Overloads on Fatigue Damage Accumulation in SAE 1045 Steel. Int. J. Fatigue, 12, 207-213.

Pompetzki, M.A., T.H. Topper, D.L. Duquesnay and M.T. Yu (1990b). Effect of compressive underloads and tensile overloads on fatigue damage accumulation in 2024-T351 aluminum. J Test Eval, JTEVA 18.

Topper, T.H. and M.T. Yu (1985). The Effect of Overloads on Threshold and Crack Closure. Int J. Fatigue , 7, 159-164.

Dabayeh, A.A. and T.H. Topper (1995). Changes in crack-opening stress after underloads and overloads in 2024-T351 aluminum alloy. Int. J. Fatigue, 17, 261-269.

COMPARISON OF ANALYTICAL AND ALGORITHMICAL METHODS FOR LIFE TIME ESTIMATION IN 10HNAP STEEL UNDER RANDOM LOADINGS

Cyprian T. Lachowicz, Tadeusz Łagoda, Ewald Macha

Technical University of Opole
ul. Mikołajczyka 5; 45 - 233 Opole, Poland

SUMMARY

The results of fatigue tests of 10HNAP steel under cyclic and non-Gaussian random loadings with zero mean values and wide-band frequency spectra have been presented. The experimental data were applied for verification of the rain-flow algorithm and analytical methods using Rajcher, Miles and Kowalewski equations. The authors have found that under non-Gaussian random loadings the rain-flow algorithm is efficient but the analytical methods do not give the satisfactory results. It is possible, however, to correct the equations proposed by Rajcher and Miles using the excess coefficient. It allows to obtain efficient estimations of fatigue life under the considered random loadings.

KEYWORDS

Fatigue life prediction; non-Gaussian loadings.

INTRODUCTION

The well-known methods of estimation of fatigue life of machine elements under random loadings can be divided into two groups. An algorithmic approach, based on numerical methods of cycle counting, is typical for the first group. The other group uses an analytical approach based on the spectral analysis of stochastic processes. In this paper the authors have compared efficiency of the chosen methods for estimations of long-life time under stationary and non-Gaussian random loadings. The specimens made of 10HNAP steel were tested [1].

THE METHODS OF FATIGUE LIFE ESTIMATION

The rain-flow algorithm was chosen for schematization of the stress history. Damages were summed cycle by cycle according to Palmgren - Miner hypothesis and the amplitudes σ_{ai} less than the fatigue limit σ_{af} (a = 0.5) were taken into account. The damage degree $S(T_o)$ at time of observation T_O of stress history $\sigma(t)$ was calculated from

$$S(T_o) = \sum_{i=1}^{k} \frac{n_i}{N_G \left(\dfrac{\sigma_{af}}{\sigma_{ai}} \right)^m} \qquad \text{for } \sigma_{ai} \geq a\sigma_{af} \tag{1}$$

$$S(T_o)=0 \qquad \text{for } \sigma_{ai} \leq a\sigma_{af}$$

Fatigue life T_{ALG} was calculated according to the following relationship

$$T_{ALG} = \frac{T_0}{S(T_0)} \tag{2}$$

From the group of analytical methods the authors chose [2]
- Rajcher method of spectral summation of damages

$$T_R = \frac{A}{\sigma_{std}^m (\sqrt{2})^m \Gamma\left(\dfrac{m+2}{2} \right) \left[\displaystyle\int_0^{\infty} G(f) f^{2/m} df \right]^{m/2}} \tag{3}$$

- Miles method

$$T_M = \frac{A}{\sigma_{std}^m M^+ (\sqrt{2})^m \Gamma\left(\dfrac{m+2}{2} \right)} \tag{4}$$

- Kowalewski method

$$T_K = \frac{A}{\sigma_{std}^m M^+ \left(\dfrac{N_0^+}{M^+} \right)^m (\sqrt{2})^m \Gamma\left(\dfrac{m+2}{2} \right)} \tag{5}$$

where:
$A = \sigma_a^m N$ - fatigue characteristic of the material $\sigma_a - N$
σ_{std} - standard deviation $\sigma(t)$
$\Gamma()$ - gamma function
$G_0(f) = G(f)/\sigma_{std}^2$ - normalized one-sided power spectral density function of $\sigma(t)$,
M^+ - expected rate of peaks,
N_0^+ - expected rate of zero crossing with (+) slope
$I = N_0^+ / M^+$ - irregularity factor

THE RESULTS OF TESTS, CALCULATIONS AND ANALYSES

The fatigue tests were done on a hydraulic machine SHM250 with the controlled loading. The plane specimens were made of 10HNAP steel. It is a higher corrosion - resisting steel of ferritic-pearlitic structure with predominance of ferrite. It contains C - 0.115%, Mn - 0.71%, Si - 0.41%, P - 0.082%, S - 0.028%, Cr - 0.81%, Ni - 0.50%, Cu - 0.30% and Fe - the rest. The steel has the following mechanical properties: yield point $R_e = 389$ MPa, strength limit $R_m = 566$ MPa, elongation $A_{10} = 31\%$, contraction $Z = 29.1\%$, Young modulus $E = 215$ GPa, Poisson coefficient $v = 0.29$. As a result of the fatigue tests under cyclic loadings with frequency $f = 20$Hz we obtain the following fatigue characteristic σ_a-N

$$\text{LgN} = c - m \cdot \lg \sigma_a = 29.689 - 9.8 \lg \sigma_a \qquad (6)$$

where $c \in (22.393, 36.984)$, $m \in (6.9, 12.7)$ for the assumed confidence interval 0.95. The fatigue limit was determined with the stair method. We obtained $\sigma_{af} = 252.33 \pm 18.75$ MPa corresponding to $N_G = 1.28 \times 10^6$ cycles. The fatigue tests were done under random loadings with the zero expected values and at five levels of σ_{std} (four specimens at each level). The random loadings were characterized by the probability density function (PDF) $p(\sigma)$ (Fig.1), the power spectral density function (PSD) $G(f)$ (Fig.2) and the chosen parameters given in Table 1.

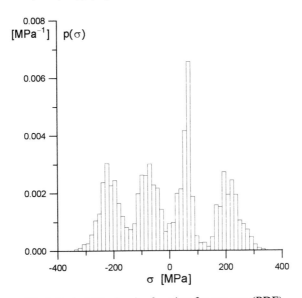

Fig.1 Probability density function for stresses (PDF)

It can be observed that PDF of stresses is the four-modal function and it distinctly differs from the normal distribution. It has a small asymmetry, $q_2 < 0.1$, and a distinct flatness, $q_1 < -1$ (Table 1). From Figure 2 it appears that random stress has the wide-band frequency spectrum and a low irregularity factor, $I \leq 0.27$ (Table 1).

The results of fatigue tests of 20 specimens under the above random loadings were compared with the calculation results obtained according to one algorithmical method and three analytical methods. Figure 3 shows the points corresponding to the calculation lives, T_{ALG} - see equations (1) - (2) and the experimental lives, T_{exp}. Time of observation $T_O = 649$s and sampling time of the stress history $\Delta t = 2.641 \times 10^{-3}$ s were assumed.

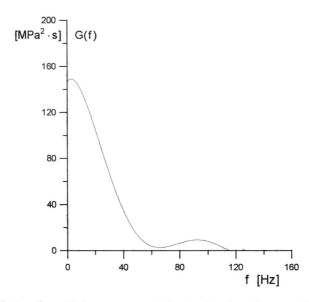

Fig.2   One-sided power spectral density function of stresses (PSD)

Table 1. Parameters of random stresses

	σ_{std} [MPa]	σ_{min} [MPa]	σ_{max} [MPa]	q_1	q_2	I	
1	159	-349	351	-1.06	0.071	0.26	q_1 - coefficient of excess
2	165	-360	362	-1.059	0.094	0.27	
3	172	-384	370	-1.087	0.061	0.25	q_2 - coefficient of skewness
4	177	-386	386	-1.063	0.081	0.26	
5	186	-398	399	-1.157	0.001	0.23	

From Figure 3 it appears that the lives T_{ALG} calculated with the rain-flow algorithm can be accepted because they are included in the scatter band with the factor of 3 in relation to the experimental lives T_{exp}. The calculations with the range pair and hysteresis loop and both Palmgren - Miner and Haibach hypotheses are not discussed here but they also give acceptable results. The lives T_R, T_M, T_K according to three analytical methods are not realistic and they must be neglected. (Table 2) This result is not unexpected because all the considered analytical methods are based on the assumption that random stress has normal probability distribution and Miles equation (4) is valid for stresses with narrow-band frequency spectra [2].

Similarly as Liu and Hu [1], the authors tried to modify equations (3)-(5) with the well-known statistic parameters in order to obtain fatigue life estimations close to the experimental ones. As a consequence of searches and analyses it has been found that there is the mutual correction coefficient, K for Rajcher (3) and Miles (4) equations. The coefficient K depends on the absolute value of the sum of the coefficient of excess q_1 and the unity.

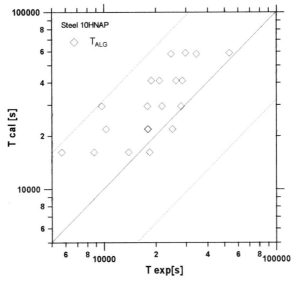

Fig.3   Comparison of calculated lives according to the algorithmical method, T_{ALG},

$$K = \frac{1}{|1 + q_1|} \; ; \; T_R^{'} = KT_R \quad \text{and} \quad T_M^{'} = KT_M \tag{7}$$

The fatigue lives $T_R^{'}, T_M^{'}$ corrected with the coefficient K are compared with the experimental lives T_{exp} in Figure 4.

Table 2.   The experimental and calculation lives for 10HNAP steel

	σ_{std}	T_{exp}	T_{ALG}	T_R	T_M	T_K	$T_R^{'}$	$T_M^{'}$
	[MPa]	[s]	[s]	[s]	[s]	[s]	[s]	[s]
1	159	24563, 34586, 53382, 29875	58094	2480	1227	1905505	41168	20368
2	165	18846, 21013, 28584, 26323	41171	1968	1016	1414564	33337	17211
3	172	21764, 9713, 28116, 17865	29500	1298	643	1096360	14914	7388
4	177	17926, 18035, 24993, 10279	21875	875	441	715459	13886	6998
5	186	13905, 18379, 5695, 8770	16172	533	266	414404	3394	1691

From Figure 4 and Table 2 it appears that the corrected Rajcher and Miles equations allow to estimate the fatigue life of the steel tested efficiently under random loadings with the non-Gaussian probability distribution used in this paper. There is only one exception, i.e. $(\sigma_{std} = 186 MPa)$.

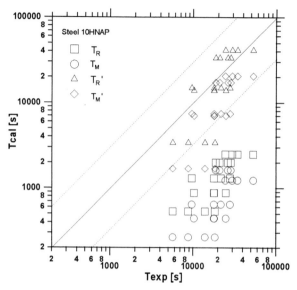

Fig.4   Comparison of the calculation lives according to the analytical methods, T_{cal} and the experimental lives, T_{exp}

CONCLUSIONS

1. Under random loadings with non-Gaussian probability distribution and the wide-band frequency spectrum the best fatigue life estimations are obtained with the algorithmic method in which the stress history is schematized with the rain-flow algorithm and damages are cumulated with Palmgren-Miner hypothesis taking into account amplitudes greater than a half of the fatigue limit. The known analytical methods based on Rajcher, Miles and Kowalewski relationships are incorrect for the considered loading conditions.

2. It is possible to correct Rajcher and Miles equations with the coefficient of excess. It leads to efficient estimations of fatigue life under random loading with non-Gaussian probability distribution.

REFERENCES

[1] Liu H.J., Hu S. R., (1987) Fatigue under nonnormal random stresses using Monte - Carlo method, FATIGUE' 87,Eds.R.O.Ritche and E.A.Starke Jr. EMAS (U.K) Vol III . ,pp.1439-1448.
[2] Rajcher W.L., (1969) Gipoteza spektralnogo summirovaniya i jego primenenia dla opredeleniya ustalostnoj dolgovechnosti pri dejstvii sluchaynykh nagruzok, Trudy Centralnogo Aero-Gidrodinamicheskogo Instituta, Vypusk 1134, Moskwa, p.38.

LOW-CYCLE FATIGUE CUMULATIVE DAMAGE
IN DUPLEX STAINLESS STEELS

S. DEGALLAIX *, H. CHTARA *, J.C. GAGNEPAIN **

* Laboratoire de Mécanique de Lille, URA CNRS 1441
Ecole Centrale de Lille, BP 48, Cité Scientifique
F-59651 Villeneuve d'Ascq Cedex
** CRMC, Creusot-Loire Industries, 71202 Le Creusot Cedex

ABSTRACT

Tensile, low-cycle fatigue and low-cycle fatigue cumulative damage tests were performed at room temperature on two stainless steels alloyed with nitrogen. The tensile test results were analysed in terms of nature of the materials ; the LCF results were interpreted using previous scanning and transmission electron microscopic observations in terms of cyclic plastic strain localisation, more pronounced in the ferrite than in the austenite ; the LCF cumulative damage results were analysed in terms of Miner cumulative damage. Low influence of the loading application order was observed.

INTRODUCTION

Duplex stainless steels combine high mechanical properties (high strength and toughness) with an excellent corrosion resistance, especially against pitting corrosion (Desestret and Charles, 1990). They are preferred to austenitic and ferritic stainless steels because of their better mechanical and chemical properties, in particular for parts of systems working in a strong corrosive environment. Unfortunately, their relatively recent development means that their mechanical properties are not still very well known, especially in cyclic loading conditions.

The aim of this work is to analyse the influence of loading history on the LCF behaviour and strength of duplex stainless steels type 2205 alloyed with nitrogen.

In the first stage, the influence of nitrogen content on tensile and LCF strength was studied. The influence of loading history was then studied in one of the two DSS.

MATERIALS AND EXPERIMENTAL PROCEDURES

Materials

Two DSS, referred to industrially UR35N and UR45N respectively, were studied. Compositions and heat-treatments are given in Table 1. They were supplied in sheets by hot rolling. They differ essentially by their N and Mo contents, higher in UR45N, and by their grain sizes, higher in UR45N. The microstructures are shown in Fig. 1. They consist of composite structures with 50 % α and 50 % γ phase volume fractions, and look like polycristal austenitic islands distributed in a polycristal ferritic matrix, with a morphological texture which has not been analysed. Because of the low solubility limit of N in the ferrite, the nitrogen atoms are essentially concentrated in the austenite.

Table 1. Chemical compositions (in wt %) and heat-treatments

Material	Cr	Ni	N	Mo	C	Mn	Si	Thick.(mm)	H.T.
UR35N	22.78	4.58	0.094	0.17	0.016	1.45	0.38	(190 →) 16	950°C
UR45N	21.99	5.25	0.158	2.8	0.02	1.33	0.44	(160 →) 28	1050°C

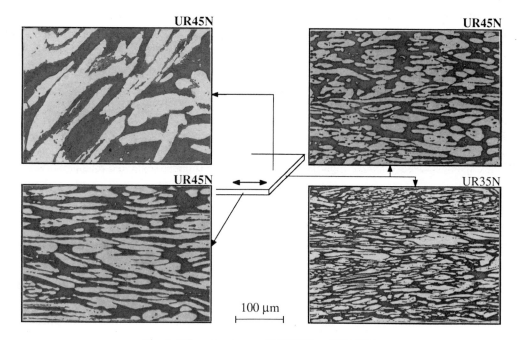

Fig. 1. Microstructures of UR35N and UR45N.

Tensile and low-cycle fatigue tests

Tensile and LCF tests were carried out at room temperature, in both longitudinal and transverse sheet directions. The specimens were cylindrical and button-headed with a 10 mm diameter, and a 25 mm (tensile) and 10 mm (LCF) useful lengths ; they were solution treated, machined, then mechanically polished up to 1 μm, and chemically etched in the useful part (in order to distinguish the α and γ phases). In LCF, only the results in the transverse direction are reported here.

The tests were conducted on a servo-hydraulic machine, with strain control at the total strain rate $\dot{\varepsilon}_t = 4.10^{-3} s^{-1}$, in a fully reverse push-pull mode with a constant total strain amplitude for LCF tests. In LCF, the failure and then the number of cycles to failure were defined at a 10 % decrease in the tensile stress amplitude, compared to the stabilised stress amplitude.

Low-cycle fatigue cumulative damage tests

Two types of LCF cumulative damage tests were carried out on UR45N steel in the transverse direction, with the same conditions as the LCF tests, at two successive strain levels :

> - high, then low strain levels,
> - low, then high strain levels.

Low and high strain levels were respectively 0.6, and 2 or 3 %. The choice of these low and high strain levels comes from previous results (Degallaix *et al*, 1995, Kruml *et al*, 1995), as explained further. N_1 is the number of cycles applied at the first level, calculated from the LCF strength law and chosen close to 20 %, 50 % and 80 % of the fatigue life at that first strain level. N_2 is the number of cycles at the second level, up to failure.

EXPERIMENTAL RESULTS AND DISCUSSION

Tensile properties

The tensile properties obtained in the longitudinal and transverse directions are given in Table 2. A small difference can be observed between the properties in both directions, noticeable only in yield stress (slightly higher in transverse direction) and elongation (slightly higher in longitudinal direction), and a higher resistance of UR45N compared to UR35N, which results from the strengthening effect due to the higher N and Mo contents in UR45N, in spite of the strengthening effect by grain refinement in UR35N compared to UR45N.

Table 2. Tensile properties at 20°C and $\dot{\varepsilon}_t = 4.10^{-3} s^{-1}$

Material	$Re_{0.2\%}$ (MPa)	R_m (MPa)	$A_{2.5d}$ (%)
UR45N(L)	545	730	44
UR45N(T)	555	730	41
UR35N(L)	475	660	46
UR35N(T)	500	670	40

Low-cycle fatigue strength

The LCF strength curves obtained are given in Fig. 2. In terms of strain, a slightly higher fatigue resistance of UR45N can be observed, clearly more pronounced in terms of stress. The LCF strength laws are defined by a mean square method, and are given in Fig. 2. The scatter of the experimental results around the calculated lives is lower than defined by a 1.67 factor.

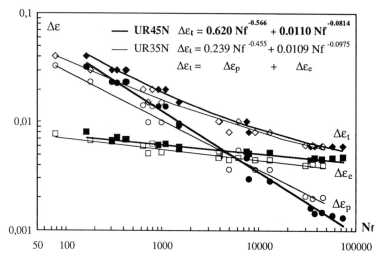

Fig. 2.   LCF curves for UR35N and UR45N in transverse direction.

A previous study on two forged DSS (Degallaix *et al*, 1995, Kruml *et al*, 1995) enabled the role of the plastic strain localisation in LCF damage of these steels to be emphasised. In particular, the two following results were shown : (i) although a 0.18 wt % N content in a 22-05 DSS is not sufficient to make the austenite hardener than the ferrite (Nyström and Karlsson 1995), the nitrogen atoms in the austenite play a beneficial role in its cyclic plastic deformation. Indeed, the nitrogen in solid solution in the austenite favours the planar dislocation slip and thus the homogeneous distribution of the cyclic plastic strain in DSS as in ASS (Degallaix *et al,* 1988). On the other hand, the plastic strain localisation in the ferrite is intense, and the fatigue microcracks are initiated in the slip bands in the ferrite, before propagation in the next austenite grains. Therefore, the beneficial effect of the nitrogen in solid solution on the LCF strength observed in ASS (Degallaix *et al,,* 1988) is hardly perceptible in DSS ; ii) although the austenite remains softer than the ferrite, there is a transition strain below which the plastic deformation, observed in both austenitic and ferritic phases, is much more localised in intense intrusions/extrusions in the ferrite ; above it, the plastic deformation is almost as localised in both phases, leading to microcrack initiation in both phases, but always with some preference in the ferrite.

Approximately the same results were observed in the DSS presently studied, with a tendency of microcracks to be initiated rather in the ferrite (also at high strain level) and at phase boundaries.

Low-cycle fatigue cumulative damage

Low and high strain levels in fatigue cumulative damage tests were chosen such as : i) during cycling at low strain level, the plastic strain is localised in the ferrite, while it is homogeneously distributed in the austenite ; ii) during cycling at high strain level, the plastic strain is localised in both ferrite and austenite.

The table in Fig.3 defines all the tests performed up to the present : five low-high strain level tests and five high-low strain level tests. For each test, the life fraction applied at the first level, the life fraction at the second level and the Miner cumulative damage (i.e. the sum of the life fractions) were defined. In the case where the strain level application order has no influence, the cumulative damage is equal to 1, nevertheless with some scatter around 1, because of the scatter in the one strain level LCF lives. On the contrary, in the case where the first strain level is detrimental for the second one, the cumulative damage is lower than 1, and, if it is beneficial, the cumulative damage is higher than 1.

In the present case, it can be observed in Fig. 3 that the cumulative damage for H-L tests is always lower than 1, while for L-H tests, it is close to 1, or lower than 1. Nevertheless, all two level test results almost lay within the scatter band defined by the one level tests, close to the lower limit for H-L tests, and more usually close to the Miner curve for L-H tests.

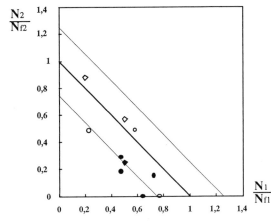

Test	$\Delta\varepsilon_t(H)$	$\dfrac{N_1}{N_{f1}}$	$\dfrac{N_2}{N_{f2}}$	$\sum\dfrac{N_i}{N_{fi}}$
H-L	• 2 %	0.47	0.19	**0.66**
		0.47	0.30	**0.77**
		0.72	0.16	**0.88**
		0.64	0.00	**0.64**
	◆ 3 %	0.50	0.26	**0.76**
L-H	○ 2 %	0.23	0.49	**0.72**
		0.58	0.50	**1.08**
		0.76	0.00	**0.76**
	◇ 3 %	0.20	0.88	**1.08**
		0.50	0.57	**1.07**

Fig. 3. Cumulative damage H-L and L-H tests results.

Figure 4 gives two examples of stress evolution during two level tests (H-L and L-H) with the same strain levels. It can be observed, after a transition evolution at the beginning of each strain level, that the stabilised stress level at a given strain level is almost the same, whatever the first strain level applied. This result is a remarkable property of the DSS. Indeed, in DSS, the CSS curve is almost independent of the strain or stress history, unlike the CSS curve of ASS, in which the history effect is very important.

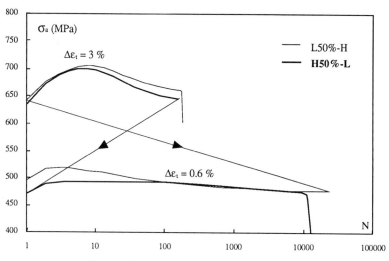

Fig. 4.   Stress evolutions for two-level tests.

CONCLUSION

Tensile and LCF tests were performed on two DSS alloyed with nitrogen. The beneficial effect of nitrogen on the plastic behaviour of these steels is relatively weak ; in particular, the beneficial effect of nitrogen on the plastic fatigue strength of the austenite is counterbalanced by the detrimental effect of the plastic strain localisation in the ferrite. The stress evolution shows a very weak influence of the strain history on the CSS behaviour of the DSS. Finally, the Miner cumulative damage tests show that, unlike many other materials, there is no beneficial effect of a first low strain level on the LCF strength of DSS.

REFERENCES

Degallaix, S., G. Degallaix, J. Foct (1988). Influence of nitrogen solutes and precipitates on low-cycle fatigue of 316L stainless steels. In: _Low-Cycle Fatigue_ ASTM STP 942, (H.D. Solomon, G.R. Halford, L.R. Kaisand and B.N. Leis Eds.), 798-811.

Degallaix, S., A. Seddouki, G. Degallaix, T. Kruml, J. Polak (1995). Fatigue damage in austenitic-ferritic duplex stainless steels. _Fatigue Fract. Engng Mater. Struct.,_ 18, 65-77.

Desestret, A. and J. Charles (1990). Les Aciers inoxydables austéno-ferritiques. In: _Les Aciers Inoxydables_ (P. Lacombe, B. Baroux, G. Béranger, Ed.), 633-678.

Kruml, T., J. Polak and S. Degallaix (1995). Dislocation microstructures in fatigued duplex steels. in "EUROMAT '95", Symposium D and E (Associazione Italiana di Metallurgica, Ed.), 2, 23-28.

Nyström, M. and B. Karlsson (1995). Plastic deformation of duplex stainless steels with different amounts of ferrite. In: _Duplex Stainless Steels' 94_ (T. G. Gooch Ed.), Abington Publishing, 1, Paper 111, 1-12.

FATIGUE LIFE PREDICTION BASED ON LEVEL CROSSINGS AND CRACK CLOSURE COMPUTATION

B L JOSEFSON*, T SVENSSON** and R OGEMAN***

*Division of Solid Mechanics, Chalmers University of Technology, S-412 96 Göteborg, Sweden
**Swedish National Testing and Research Institute, Division of Mechanics, Box 857, S-501 15, Borås, Sweden
***Division of Marine Structural Engineering, Chalmers University of Technology, S-412 96 Göteborg, Sweden

ABSTRACT

The fatigue life of structures subject to variable amplitude loading is predicted. A level crossing approach is proposed based on the effective load process, as defined by the Elber plasticity induced crack closure concept. The crack opening stress level is calculated using an analytical model proposed by Newman, which is based on the Dugdale model. Our approach is applied to some variable amplitude load histories from the literature with promising results. Conditions for the stabilization of the crack opening stress is assessed in particular.

KEYWORDS

Crack opening stress, level crossing, fatigue life

INTRODUCTION

The solution of the problem of fatigue life prediction at variable amplitude (spectrum fatigue) depends on a proper modelling of damage accumulation. The established models for fatigue damage accumulation are based on the linear Palmgren-Miner rule. According to this rule the damage is computed as the sum of the contributions of individual loads. The damage caused by each load is determined by the experimental Wöhler curve or by an experimentally obtained crack growth law. A commonly used approach is to count the loads as cycles (Rain Flow Count) and assume that the contribution from each cycle is independent of the order of cycle application. In another approach, Holm *et al.* (1995), investigated by the authors, the load are specified by its level crossing distribution and load intervals are assumed to be independent of the order of application.

Experiments have shown that these accumulation laws are inaccurate in many cases and it seems like the assumption of independence is the most crucial point, i.e. the order of applied loads is indeed important Holm *et al.* (1995). Based on this observation a modified level crossing model was developed in Holm *et al.* (1995) that take the sequential effects into account through a state variable.

In the long crack regime the Elber crack closure concept has shown to be successful in modelling sequential effects (Elber, 1971). Therefore, by identifying the crack opening level with the state variable in the modified level crossing model we achieve a very useful model. Using this model the load process is converted to a suitable form and the Palmgren-Miner rule is applied on the effective load process, thus avoiding the crucial independence assumption (Holm et al., 1995; Svensson, 1995). Actually, if the damage can be described by a crack growth law of the type, (a = crack length, and S = remote stress and the effective stress $\tilde{S}$ is defined below)

$$\frac{da}{dN} = Cf(a)(\Delta\tilde{S})^n \tag{1}$$

then the level crossing method can be used with parameters determined by this crack growth law. This is demonstrated in Svensson (1995), where prediction results from the simple level crossing approach is compared to predictions based on cycle by cycle calculations.

Also in the short crack regime recent investigations have shown that crack closure has an important influence (Vormwald et al., 1994; Jono, 1994). This implies that the crack closure phenomenon is important for the whole fatigue process and the modelling of the crack closure behaviour at variable amplitude is of vital importance for the development of fatigue life prediction methods.

Crack closure behaviour at constant amplitude loads is well investigated and a great number of empirically based formulas have been proposed. At variable amplitude the closure behaviour has primarily been studied for simple load processes like single or multiple overloads and block loadings. For stationary random load processes it is often assumed that the crack closure level stabilizes at a certain level determined by the distribution of maxima and minima in the process (the max-min approximation).

Present investigation. In this paper we address the crack closure problem in relation to the modified level crossing model presented in Holm et al. (1995). The crack opening levels are computed with the analytical crack closure model Newman (1981). The relevance of the modified level crossing method is demonstrated on some load sequences from the literature in the long crack growth regime. Further, the relevance of the assumption of a stabilized opening level for stationary loadings is investigated.

THE LEVEL CROSSING MODEL

The level crossing approach as proposed in (Holm et al., 1995; Svensson, 1995) will here be briefly described. We have a load process with the remote stress, S, and the crack opening stress at the crack surfaces, S_{op}. We introduce the effective load process $\tilde{S} = (S - S_{op})_+$, where the + sign indicates the positive part. The accumulated damage D at a certain time T is then

$$D(T) = \int_0^\infty N_+(u;\tilde{S},T)\tilde{g}(u)du \tag{2}$$

where $N_+(u;\tilde{S},T)$ is the number of up-crossings of the effective load process at the stress level u and $\tilde{g}(u)$ is a damage exhaustion function determined from constant amplitude tests expressed by the Wöhler equation (3). One finds, see (Svensson, 1995; Svensson and Holmgren, 1993) for details,

$$N = \tilde{\alpha}(\Delta \tilde{S})^{-\tilde{\beta}} \qquad \tilde{g}(u) = \frac{\tilde{\beta}}{\tilde{\alpha}} u^{\tilde{\beta}-1} \qquad\qquad (3,4)$$

CRACK OPENING STRESS CALCULATION

The crack opening stress, S_{op}, quantifies the crack closure, a phenomenon believed to explain experimentally observed load sequence effects in variable amplitude loadings. One important closure mechanism is plasticity induced crack closure (Elber, 1971), that is the crack surface remains closed even for tensile remote loadings, S, until a certain stress level, S_{op}, is reached. To model effects of plasticity induced crack closure we have here adopted the analytical crack closure model by Newman (1981) and the crack growth program FASTRAN-II (Newman, 1992). This model is based on the Dugdale concept where the plastic zone and contact stresses behind the crack tip is obtained by superposing two elastic problems. The plastic region is modelled by parallel uni-axial rigid-plastic bars. At a given remote stress, the normal stresses in the bars are calculated iteratively under the constraint that bars subject to high tensile stresses will break and carry no load. Details of the application of the Dugdale model, the definition of the crack opening stress, S_{op}, and the crack growth algorithm can be found in Newman (1981, 1992).

APPLICATIONS

In the first example we apply our level crossing approach to experimental results from Zhang *et al.* (1987) where crack closure and growth of a through-the-thickness crack in a CCT-specimen subject to eight different block and single over (and under) loadings are presented. The specimen material is Aluminium 7475-T7351. One important parameter in FASTRAN-II is the constraint factor α introduced to correct for multi-axial stress state effects into the uni-axial strip model employed for the crack surfaces (Newman, 1981). It was found in Newman and Dawicke (1989) that $\alpha = 1.7$ correlated experimental results from Zhang *et al.* (1987) well for positive load ratios, $R > 0$. This value for α is used also in the present calculations. Figure 1 shows a calculated part of the crack opening stress history for two load types defined in Zhang *et al.* (1987): Type 05 = single overload-underload cycle repeated every 39 cycles and Type 07 = 50 high R-ratio and 50 low R-ratio cycles repeated. One finds results in agreement with Zhang *et al.* (1987) and Newman and Dawicke (1989).

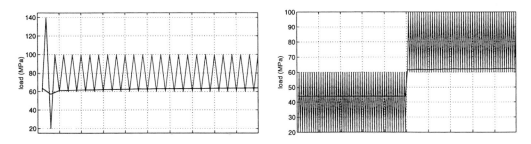

Fig. 1   Calculated crack opening stress history for the load types 05 (left) and 07 (right) as defined in
        Zhang *et al.* (1987). CCT-specimen made of aluminium

Fatigue life predictions are calculated using the level crossing approach (2). The material parameters $\tilde{\alpha}, \tilde{\beta}$ are determined from constant amplitude crack growth data reported in Zhang *et al.* (1987), see Svensson (1995) giving

$$\tilde{\alpha} = 7.61 \cdot 10^9 \quad \tilde{\beta} = 2.972$$

The crack opening levels are calculated employing FASTRAN-II. The predictions are made in two ways, first by using the whole crack opening solution for each block of loading, second by using the mean value of the crack opening level over each block. Our calculations are also compared with results (Newman and Dawicke, 1989) obtained from a crack growth analysis using the same tool for the crack opening stress (Newman, 1981). The predicted results are shown in Table 1, where also comparisons can be made between the calculated mean crack opening level and the max-min approximation. It is seen that the stabilized level is higher for the max-min-approximation for load types 1 - 4, where the minima are constant, and lower for load types 5 - 8, with varying minima.

Table 1. Comparison of predicted fatigue life using present level crossing approach, fracture mechanics based crack growth (Newman and Dawicke; 1989) and experiments (Zhang *et al.*, 1987).

Load type	Test life (cycles)	Sop/Smax max-min- approx.	Sop/Smax calc.mean	Npred/Ntest Newman	Npred/Ntest HdM (Sop)	Npred/Ntest HdM, calc. mean Sop
1	474,240	0.42	0.38	1.40	1.12	1.13
2	637,730	0.42	0.38	1.19	1.00	0.99
3	251,210	0.42	0.39	1.36	1.15	1.15
4	409,620	0.42	0.39	1.28	1.07	1.09
5	179,320	0.39	0.45	1.03	0.75	0.77
6	251,050	0.39	0.46	0.78	0.62	0.57
7	253,840	0.42	0.53	1.11	1.14	0.67
8	149,890	0.39	0.46	0.92	0.71	0.77
		mean		1.13	0.95	0.89
		standard deviation		0.22	0.22	0.23

In the next example centre cracked aluminium specimens studied in Chang (1981) are considered. The crack opening stress levels are calculated for the thirteen spectrum load sequences presented in Chang (1981). These sequences represent five different types with each type scaled with two or three different scale factors. All sequences are block sequences with sizes ranging from 300 to 2187 cycles. One finds, as expected, that the crack opening stress stabilizes at a certain level, except for M93 and M94

Within the same sequence type the stabilized level is lower at higher scale factors. Hence increasing maximum stress levels (for the same load history) will lead to a lower crack opening stress level. In Figure 2 fatigue life predictions using three prediction methods are compared for the load sequences defined in Chang (1981). These three methods are: (a) The level crossing method with the crack opening stress assumed to stabilize on a level (max-min approximation) determined for an equivalent constant amplitude case having the same maximum and minimum stresses as the load sequences (see Zhang *et al.*, 1987), denoted HdM old, (b) The present level crossing approach with a stabilized crack opening stress determined from a cycle-by-cycle calculation using FASTRAN-II, denoted HdM new, (c), Newman's numerical predictions (Newman, 1981) and Chang (1981) using a detailed cycle-by-cycle calculation of crack opening stress and corresponding crack growth.

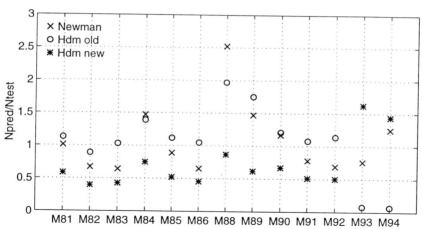

Fig 2. Comparison between predicted and experimentally determined fatigue life for spectrum load
sequences defined in Chang (1981).

One finds that all three methods predict the fatigue life rather well. Note however, that for the
sequences M93 and M9 extremely conservative predictions are obtained with approach (a), HdM old.
In Svensson (1995) this is assumed to be due to the difficulty of estimating a constant crack opening
stress level for these cases containing long blocks of roughly constant stress levels interrupted by strong
single underloads. This assumption is confirmed by the results using method (b), HdM new, where the
calculated stabilized crack opening level is higher giving more realistic predictions.

CONCLUSIONS AND FURTHER WORK

From the results in this investigation we can draw two main conclusions,
1) *Fatigue life predictions made with the modified level crossing model are as accurate as cycle by
cycle crack growth calculations under the presumption that the crack opening levels are known.* The
determination of the crack opening level demands at present a cycle by cycle calculation and not much
simplicity is achieved compared to the cycle by cycle crack growth calculations. However, using the
mean crack opening levels for the blocks give similar results, which shows that it should be possible to
use a global approach with an approximately stabilized crack opening level. In such a situation the level
crossing approach would provide a considerable simplification, without essential loss of accuracy.
Consequently, we would like to have a method to determine this approximate stabilized level. One such
method is the max-min approximation used in Zhang et al., (1987). The results presented in Table 1
show that this method is not valid in this case. The results in Figure 2 also point out that the max-min-
approximation is not satisfactory, especially for the load types M93 and M94. These results lead to our
second main conclusion,
2) *The crack opening level can be approximated by a stabilized level, but new methods need to be developed in order to estimate it.* This may be achieved by a modified max-min-approximation or by using
an auto-regressive model as proposed in Holm et al. (1995) and we intend to make further
investigations regarding this subject.

The conclusions made here are based on a few results in the long crack regime, but should be valid for
other cases if sequential effects originate from crack closure.

ACKNOWLEDGEMENTS

The work presented here is funded by the Swedish Research Council for Engineering Sciences (TFR) through their contract 94-526. Discussions with Dr J C Newman Jr at NASA Langley, Hampton, VA, USA regarding details in the FASTRAN-II code are highly appreciated.

REFERENCES

Chang, J.B., (1991). Round-robin crack growth predictions on center-cracked tension specimens under random spectrum loading. In: *Methods and Models for Predicting Fatigue Crack Growth Under Random Loading, ASTM STP* 748, (J.B. Chang, and C.M. Hudson, eds.), pp. 3-40. American Society for Testing and Materials, Philadelphia, PA, USA

Elber, W., (1971). The significance of fatigue crack closure. In *Damage tolerance in Aircraft Structures, ASTM STP 486*, pp. 230-242. American Society for Testing and Materials, Philadelphia, PA, USA.

Holm, S., Josefson, B.L., de Maré, J., and Svensson, T., (1995). Prediction of fatigue life based on level crossings and a state variable. *Fatigue and Fracture of Engineering Materials and Structures*, **18**, 1089-1100.

Jono, M., (1994). Fatigue crack closure and its related problems. In: *Computational and Experimental Fracture Mechanics,* (H. Nisitani, ed.), pp. 317-345. Computational Mechanics Publications, Southampton, UK.

Newman Jr, J.C., (1981). A crack-closure model for predicting fatigue crack growth under aircraft spectrum loading. In *Methods and Models for Predicting Fatigue Crack Growth Under Random Loading, ASTM STP 748*, (J.B., Chang, and C.M. Hudson, eds.), pp. 53-84. American Society for Testing and Materials, Philadelphia, PA, USA.

Newman, Jr, J.C., (1992). FASTRAN-II - A fatigue crack growth structural analysis program, NASA Technical Memorandum 104159, NASA Langley Research Center, Hampton, VA, USA.

Newman, Jr, J.C., and Dawicke, D.S., (1989). Prediction of fatigue-crack growth in a high-strength aluminum alloy under variable-amplitude loading, In: *Advances in Fracture Research*, (K. Salama, K. Ravi-Chandar, D.M.R., Taplin, and P. Rama Rao, eds.), pp. 945-952. Pergamon Press, Oxford, UK.

Svensson, T., and Holmgren, M., (1993). Numerical and experimental verification of a new model for fatigue life. *Fatigue and Fracture of Engineering Materials and Structures*, **16**, 481-493

Svensson, T., (1995). Fatigue damage calculations on blocked load sequences, Studies in Statistical Quality, Control and Reliability, Division of Mathematical Statistics, Chalmers University of Technology and The University of Göteborg, Göteborg, Sweden, Report 1995:3.

Vormwald, M., Heuler, P., and Krae, C., (1994). Spectrum fatigue life assessment of notched specimens using a fracture mechanics based approach. In: *Automation in Fatigue and Fracture: testing and Analysis, ASTM STP 1231*, (C. Amzallag, ed.), pp. 221-240. American Society for Testing and Materials, Philadelphia, PA, USA.

Zhang, S., Marissen, R., Schulte, K., Trautmann, K.K., Nowack, H., and Schijve, J., (1987). Crack propagation studies on Al 7475 on the basis of constant amplitude and selective variable amplitude loading histories. *Fatigue and Fracture of Engineering Materials and Structures*, **10**, pp. 315-332.

ON THE EQUIVALENCE OF FATIGUE DAMAGES DEVELOPED AT DIFFERENT STRESS LEVELS

L. Y. XIE and W. G. LU

Dept. of Mechanical Design, Northeastern University, Shenyang, 110006, P. R. China

ABSTRACT

The equivalence of fatigue damages developed at different cyclic stresses is studied based on rotating-bending fatigue tests of three metallic materials. At the so called "equivalent damage state", residual fatigue life tests are carried out and different stress-residual life curves are obtained. The test results indicate that no "equivalent damage state" can be developed by different stress levels, despite the fact that same residual life can be yielded in specified condition. Besides, fatigue test results under high-low and low-high two level stress show that different equations are needed in order to reasonably describe cumulative fatigue damage laws under different loading sequences. These test results also indicate that no "equivalent damage state" exists.

KEYWORDS

Fatigue damage accumulation, equivalent damage state, residual life, loading sequence, fatigue test

INTRODUCTION

Almost all of the cumulative fatigue damage models are based on some kinds of damage equivalent assumption (Sarkani, 1990, Leis, 1988, Jen et al., 1994, Fang et al., 1994, Manson et al., 1986, Miller, 1977). Generally speaking, such assumption believes that for two different cyclic stress σ_1 and σ_2, if n_1 cycles of σ_1 produces a damage D, an equivalent cycles n_2 can certainly be found by which σ_2 also produces the same damage D. That is to say, the two cycle numbers are equivalent.

In the quest for a physical justification of the term "state of fatigue of materials subjected to cyclic loading", J. Burbach (1973) indicated that if states of fatigue exist which can be characterized by a single number, they can be determined experimentally, and the mathematical relationships of cumulative damage with regard to irregular loading are then fixed. If it is not possible for states of fatigue to be characterized by a single number, not even for a limited selection of types of loading, then it is likewise impossible to give any simple or closely accurate mathematical relationship for the actual cumulative damage. J. Burbach also indicated the fact that "all the known theories of cumulative fatigue damage assume explicitly or implicitly that states of fatigue can be characterized by a single number, although this assumption is neither trivially nor thermodynamically justified".

It is obvious that if fatigue damage state can be characterized by a single number, the "equivalent damage state" does exist. If fatigue damage state can not be characterized by a single number, the existence of "equivalent damage state" is also dubious.

EXPERIMENTAL STUDY ON RESIDUAL S-N CURVE

In order to study the problem whether "equivalent damage state" exists or not, residual life tests are carried out at rotating-bending test machine. Smooth specimens of normalized 0.45% carbon steel are used. The test program includes fatigue life tests at two given stress levels, cumulative damage tests of two level stress and residual life test under traditional "equivalent damage state". First of all, two cyclic stress levels are selected. The amplitudes are σ_1=366MPa and σ_2=309MPa respectively. In order to determine fatigue lives, more than a dozen specimens are tested at each of the stress level. The test results are listed in Table 1.

Table 1. Fatigue lives at constant cyclic stress (100 times)

stress amplitude	fatigue lives								expectation
366 MPa	395	403	499	344	431	462	552	304	
	306	260	377	492	518	370	545	366	411
309 MPa	5893	5519	7074	9625	9618	5633	7342		
	5058	6752	7089	5532	5089	7852	10578		7047

According to the traditional damage equivalence concept, cumulative fatigue damage tests of two level stress can tell the equivalent cycle numbers of the two stresses by which the same damage state would be produced. Proceeding from this point of veiw, two groups of fatigue tests are done. The first group of test is programmed that let the high stress σ_1 cycle 10900 times (the cycle ratio n_1/N_1 =0.267) first, then the low stress σ_2 continues cycling until failure occurs, by which the expectation (mean value) of the residual life at the low stress σ_2, presented by N_{2P}^{1-2}, is obtained equating to 288400 times. Consequently, the 10900 cycles of σ_1 is equivalent to 416300 ($N_2 - N_{2P}^{1-2} =$ 704700-288400) cycles of σ_2. According to the traditional damage equivalence viewpoint, the 416300 cycles of σ_2 is certainly equivalent to the 10900 cycles of σ_1 if the lower stress σ_2 is applied first. While the test shows that the first 416300 cycles of σ_2 is in fact equivalent to 200 ($N_1 - N_{1P}^{1-2} =$ 41100-40900, N_{1P}^{1-2} is the residual fatigue life under σ_1 after 416300 cycles of σ_2) cycles of σ_1. The second group of test is programmed that let the low stress σ_2 cycle 250000 times (the cycle ratio n_2/N_2 = 0.354) first, then the high stress σ_1 continues cycling until failure occurs, by which the expectation of the residual life at the high stress σ_1, presented by N_{1P}^{2-1}, is obtained equating to 36700 times. Consequently, the 250000 cycles of σ_2 is equivalent to 4400 ($N_1 - N_{1P}^{2-1} =$ 41100-36700) cycles of σ_1. According to the traditional damage equivalence viewpoint, the 4400 cycles of σ_1 is certainly equivalent to the 250000 cycles of σ_2 if the high stress σ_1 is applied first. While the test shows that the first 4400 cycles of σ_1 is in fact equivalent to 350600 ($N_2 - N_{2P}^{2-1} =$ 704700-354100, N_{2P}^{2-1} is the residual fatigue life under σ_2 after 4400 cycles of σ_1) cycles of σ_2. Obviously, these test data have negated the traditional assumption about existing equivalent damage state at different stress levels. Table 2 and Table 3 list these test results, in which loading sequence means which stress is applied first and which one is followed. In Table 2, in the sequence of $\sigma_1 - \sigma_2$, σ_1 is applied 10900 cycles first and the data under the title of "residual life at the second stress" are the residual lives at stress σ_2 after the 10900 cycles of σ_1. In sequence $\sigma_2 - \sigma_1$, the data are residual fatigue lives at σ_1 after 416300 cycles of σ_2. In Table 3, in the sequence of $\sigma_2 - \sigma_1$, σ_2 is applied 250000 cycles first and the data under the title of "residual life at the second stress" are the residual life at stress σ_1 after the 250000 cycles of σ_2. In sequence $\sigma_1 - \sigma_2$, the data are residual life at σ_2 after 4400 cycles of σ_1.

Table 2. Residual fatigue lives of the carbon steel (100 times)

loading sequences	Residual lives at second stress						expectation
$\sigma_1 - \sigma_2$	3967	3262	3509	2305	2912	2587	
	2209	2736	3661	2709	3222	1527	2884
$\sigma_2 - \sigma_1$	401	546	521	270	307		409

Table.3 Residual fatigue lives of the carbon steel (100 times)

loading sequences	Residual lives at second stress					expectation
$\sigma_2 - \sigma_1$	317	329	375	352	462	367
$\sigma_1 - \sigma_2$	3489	4694	2795	2588	4159	3541

Further than that, if equivalent damage state does exist, there will be an identical stress-residual life curve for the material after 10900 cycles of σ_1 or 416300 cycles of σ_2, or after 250000 cycles of σ_2 or 4400 cycles of σ_1. While the test results show no such an identical curve existing. Fig.1 shows these test data as well as the fitted stress -residual life curves. In Fig.1(a) the signs "×" stand for the residual fatigue lives after 10900 cycles of σ_1, and the signs "O" stand for the residual lives after 416300 cycles of σ_2. In Fig.1(b) the signs "×" stand for the residual lives after 4400 cycles of σ_1, and the signs "O" stand for the residual life after 250000 cycles of σ_2. Curve 1 in Fig.1(a) is the fitted curve of stress - residual fatigue life after 10900 cycles of σ_1; curve 2 is the fitted curve of stress - residual life after 416300 cycles of σ_2. Curve 1 in Fig.1(b) is the fitted curve of stress - residual fatigue life after 4400 cycles of σ_1; curve 2 is the fitted curve of stress - residual life after 250000 cycles of σ_2. Obviously, the two stress-residual life curves in both of the figures are far from identical, therefore it supports the conclusion that no traditional "equivalent damage state" exists.

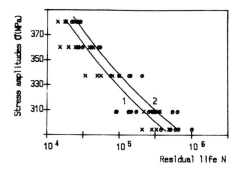 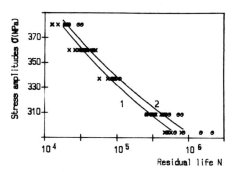

(a) signs "×" and curve 1 -residual
 lives after 10900 cycles of σ_1,
 signs "O" and curve 2 - residual
 lives after 416300 cycles of σ_2.

(b) signs "×" and curve 1 - residual
 lives after 4400 cycles of σ_1,
 signs "O" and curve 2 - residual
 lives after 250000 cycles of σ_2.

Fig. 1. Residual life test data under traditional "quivalent damage state" and
the fitted stress-residual life curves of the 0.45% carbon steel

LOADING SEQUENCE AND CUMULATIVE FATIGUE DAMAGE LAW

In addition to the tests mentioned above, high-low and low-high two level stress tests are carried out to study the existence of equivalent damage state as well as the compatibility of cumulative damage model to different loading sequences. Test condition is the same as described above, the materials are alloy steels 16Mn and 60Si2Mn. The test program is that after a certain cycles of the first level stress, the second level stress continues until failure occurs. For each stress group (a high stress and a low stress), the test data under different loading sequences are drawn in one figure, in which the signs "O" stand for the test data in high-low loading sequence and "×" stand for those in low-high sequence. Special coordinates are used to verify if the "equivalent damage state" exists and analyze the difference between the cumulative fatigue damage laws in high-low and low -high loading sequences. In such figures , the coordinate system $((n_H/N_H)_{H-L}-O-(n_L/N_L)_{H-L})$ is used for describing the test data and fitted cumulative fatigue damage curve in high-low loading sequence, while the coordinate system $((n_L/N_L)_{L-H}-P-(n_H/N_H)_{L-H})$ is used for describing those in low-high loading sequence. $(n_H/N_H)_{H-L}$ and $(n_H/N_H)_{L-H}$ stand for the cycle ratioes of high level stress in high-low and low-high loading sequences respectively, $(n_L/N_L)_{H-L}$ and $(n_L/N_L)_{L-H}$ stand for the cycle ratioes of low level stress in high-low and low-high loading sequences respectively. If the traditional "quivalent damage state" exists, all of the test data, no matter of high-low or low-high loading sequences, should locate on an identical curve in such a figure (almost all of the traditional cumulative fatigue damage models yield an identical curve for any given two stress levels, no matter of high-low or low-high loading sequences). Take a high stress σ_H and a low stress σ_L for example. If σ_H cycles a cycle ratio n_H/N_H , and n_L/N_L is the residual cycle ratio of σ_L , then $(n_H/N_H$, $n_L/N_L)$ is a point on the cumulative damage curve in the loading sequence of high-low. According to the traditional "equivalent damage" concept, if the low stress σ_L cycle a cycle ratio $(1-n_L/N_L)$ first, then the residual cycle ratio of the high stress σ_H will be equal to $(1-n_H/N_H)$. i.e. $(1-n_L/N_L, 1-n_H/N_H$) will be a point on the cumulative damage curve in the loading sequence of low-high. Using the special coordinate system as shown in Fig.2 and Fig.3, the two points will coincide each other. Test results do not show such a tendency. Therefore, these test results also indicate that no traditional "equivalent damage state" exists. At the same time, these test results show the difference between the cumulative fatigue damage laws under different loading sequences and the incompatibility of traditional cumulative damage models. Fig.2 shows the test data and the fitted curves of alloy steel 16Mn , and Fig.3 shows those of alloy steel 60Si2Mn. It can be seen that there are obvious distance between the cumulative fatigue damage curves in high -low loading sequence and in low-high sequence. Only when the high stress is near to the low stress, the two curves coincide together approximately (see Fig.3a).

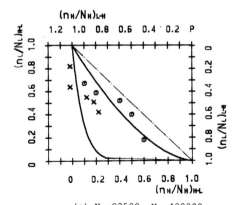
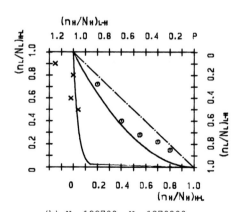

(a) $N_H=93500$, $N_L=402200$ (b) $N_H=199700$, $N_L=1370200$

Fig. 2. Cumulative fatigue damage test data of alloy steel 16Mn

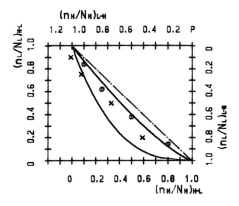
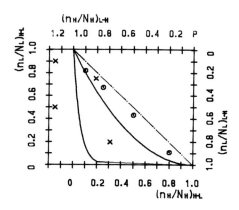

(a) N_H=99025, N_L=187800   (b) N_H=99025, N_L=564000

Fig. 3. Cumulative fatigue damage test data of alloy steel 60Si2Mn

DISCUSSION ON THE EXISTENCE OF "EQUIVALENT DAMAGE STATE"

Micro-aspect of Fatigue Damage

Klesnil (1980) describeed the evolution of the dislocation structure of a pure polycrystalline iron with increasing number of cycles. The observation shows that the saturation structure is developed during the first few per cent of the total number of cycles to fracture. In the case of low-amplitude loading, the saturation structure consists of dislocation bands and loops; while in the case of high-amplitude loading, the saturation structure is cell structure. The evolution of the dislocation structure and the type of the saturation structure depends on the loading amplitude. As a consequence, it seems reasonable to say that there is not equivalent damage state under different stress levels because different dislocation structures, which can characterize fatigue damage state in a sense, will be produced.

Relationship Between Damage Equivalence and Failure Probability Equivalence

Consider a high-low two level cyclic loading history. The pdfs (probability density functions) of fatigue lives are f_1 and f_2 for the high stress σ_1 and low stress σ_2 respectively. After n_1 cycles of σ_1, failure probability equals to P_1. According to the principle of equivalent failure probability, when stress level changes from σ_1 to σ_2, the effect of n_1 cycles of σ_1 is assumed to equivalent to n_{2p} cycles of σ_2, by which the failure probability produced by n_{2p} cycles of σ_2 equals to p_1, too.

Fatigue is a damage accumulation process. According to the viewpoint of damage equivalence, the equivalent loading cycles n_{2d} of stress σ_2 to n_1 cycles of σ_1 should be calculated by Miner's rule. Generally, $n_{2p} \neq n_{2d}$, so the damage equivalence does not coincide with the failure probability equivalence. This also indicates that it is questionable to believe that there is equivalent damage state under different cyclic stress levels.

Complexity of Cumulative Fatigue Damage Law

General opinions about cumulative fatigue damage law believe that the critical damage $\Sigma n_i/N_i$ is greater than unity for low-high loading sequence and smaller than unity for high-low sequence. In fact, cumulative fatigue damage law is quite complex. Even in the conditions that the general law is available, there are also some cases in which no equivalent cycle numbers between different stress levels. Taking a low-high loading sequence for example, the low stress applied firstly may strengthen the material (coax effect), or in other word, caused a negative damage to the fatigue process under the high stress. In such situation, it is obviously that no equivalent damage state exists under the two stress levels. The reason is that the early applied low stress produced a negative damage at the fatigue process under the higher stress, this damage state can never appear if the material is loaded by the higher stress.

CONCLUSION

There is no "equivalent damage state" exists under different loading histories. According to the traditional "quivalent damage" concept, there should be an identical stress-residual fatigue life curve for the materials in the "equivalent damage state", but test results show that such concept is not correct. For example, n_1 cycles of stress σ_1 may be equivalent to n_2 cycles of stress σ_2 ($\sigma_1 \neq \sigma_2$) in the meaning of residual life, but this does not mean that the n_2 cycles of σ_2 is equivalent to the n_1 of cycles of σ_1. That is to say, if σ_1 acts n_1 cycles first and then σ_2 follows to acts, the residual life is equal to n_{2p}. While if the stress σ_2 acts (N_2-n_{2p}) cycles first, the residual life under the stress σ_1 is not equals to (N_1-n_1). Therefore, the "quivalent damage" concept can not be generalized in a common sense. Besides, in probability sense, even though n_1 cycles of σ_1 or n_2 cycles of σ_2 can lead to a same expectation of the residual lives under a third stress σ_3, the deviations of the residual lives after the different loading histories are different. This also means that "quivalence damage" concept can only be used in a specified situation. Besides, there are different cumulative fatigue damage laws for different loading sequences. Therefore, it is not reasonable to use only one equation to describe cumulative fatigue damage law under different loading sequences, even though the equation contains stress or life dependent parameter.

ACKNOWLEDGEMENT

Financial supports by the Natural Science Foundation of Liaoning Province is gratefully acknowledged.

REFERENCES

Sarkani, S. (1990). A sequence dependent damage model for stochastic fatigue damage calculations. In: *Fatigue and Fracture Mechanics* (M. Aliabadi *et al* Eds), Portsmouth, UK, pp. 15-26.

Leis, B. N. (1988). A nonlinear history-dependent damage model for low cycle fatigue. In: *Low cycle fatigue, ASTM STP 942* (H. D. Solomon *et al* Eds), pp. 143-159.

Jen, M.-H.R., Y. S. Kau and I. C. Wu (1994). Fatigue damage in a centrally notched composite laminate due to two-step spectrum loading. *Int J fatigue*, 16, 193-201.

Fang, D. and A. Berkovist (1994). Evaluation of fatigue damage accumulation by acoustic emission. *Fatigue Fract. Engng Mater. Struct.* 17, 1057-1067.

Manson, S. S. and G. R. Halford (1986). Re-examination of cumulative fatigue damage ananlsis, an engineering perspective. *Engng Frac. Mech.*, 25, 539-571.

Miller, K. J. (1977). Cumulative damage laws for fatigue crack initiation and stage I propagation. *J. of Strain Analysis*, 12, 263-270.

Burbach, J. (1973). On the physical justification of the term "state of fatigue of materials under cyclic loading". In: *Cyclic stress-strain behavior-analysis, experimentation, and failure prediction, ASTM STP 519,* pp. 185-212.

Klesnil, M. and P. Lukas (1980). *Fatigue of metallic materials*. Elsevier Scientific Publishing Company, Amsterdam.

A METHOD FOR DERIVING AN ALGORITHM TO EVALUATE THE LOADING LEVEL AND FATIGUE DAMAGE OF AIRCRAFT

V. M. Adrov

The State Science Research Institute of Aircraft
Maintenance and Repair, Lubertsy-3, Moscow.

ABSTRACT

A metod for deriving algorithms to evaluate the airframe
load level and fatigue damage is presented. This
algorithms may be useful for in-service aircraft fatigue
damage control in units of "equivalent working".
Applications of this method to obtain the main
relationships for maneuvering and non-maneuvering
aircraft are given.

KEYWORDS

Airframe fatigue damage, in-service lifetime control.

INTRODUCTION

One of the main directions of the aircraft science
maintenance accompaning is to control of the structure
fatigue life potential. The investigations of aircraft
operation loading during long period of airplanes maintenance show wide variance plane-to-plane load spectrum for
the same airplane kind and, hence, cumulative rate of
airframe fatigue damage. The last one has differencies not
only between different flight missions but for the same
flight mission. So actual cumulative rate of airframe fatigue damage even for the same airplane kind may be differ at 15 - 20 times.

There are many approaches to decide this problem both by
complex mathematic treatment of flight parameters records after flight and by equiping airplane special devices. However its using reqires some involve period and
significant finansical efforts. Present method may be
rather effective for aicraft fatigue damage control as
during above involve period temporary as to obtain express evaluations - continuously.

THE MAIN FEATURES OF THE METHOD

The approach to obtain fatigue life assessments of
"critical" structure piecies under maintenance conditions
is based on the use of stress load spectrum. Proposed
method is based on probabilistic approach which includes
main ideas that inherent for nominal stress life predictions under deterministic loading, using: S-N fatigue
curve, any damage event determinating technique for variable loading, any form to obtain equivalent constant amplitude stress cycles for every damage event and linear
rule for damage summation.

In most practical situations the expression for a load
spectrum in i-th part of a flight (number of occurrences of loads more than σ) may be approximated as:

$$H_i(\sigma) = N_o^i \exp(-\alpha_i \sigma). \tag{1}$$

The mathematic basis and main specialities used to obtain
analitical relationships for fatigue damage are recently
developed (Adrov, 1994a). The final formulaes for fatigue
damage assessments and its RMS have been obtained by
substitution correspondent expresiions in equation

$$\bar{\xi} = \iint_{\Omega} \frac{f(x,y)}{N(x,y)} \, dx \, dy, \tag{2}$$

$$S_{\xi}^2 = \iint_{\Omega} \frac{f(x,y)}{N^2(x,y)} \, dx \, dy - (\bar{\xi})^2, \tag{3}$$

where x and y are independent increments of max and min
load, $f(x,y)$ is mutual density distribution function,
$N(x,y)$ is expression for S-N curve with account mean load
value.

Example. Let for S-N curve we have

$$N(x,y) = C_o(x+\sigma_m)^{-m/2}(x+y)^{-m/2} \tag{4}$$

For airplanes whose structures are loaded in general by
air turbulence, and so values x and y have the same
probabilstic distribution, the averange value of fatigue
damage when m is even we get

$$C_o\bar{\xi} = \frac{1}{2m} P(m, \alpha\sigma_m), \tag{5}$$

$$P(m,z) = \begin{cases} 3 + 2z, & \text{when } m=2; \\ 40 + 24z + 6z^2, & \text{when } m=4; \\ 1260 + 720z + 180z^2 + 24z^3, & \text{when } m=6. \end{cases} \tag{6}$$

To discribe this relationship for any value of m next function may be used (Adrov, 1994a):

$$P(m,z) \approx 10^{a(m,z)}; \quad a(m,z) = 0.24\left[z + \frac{m}{8}\left(\frac{m}{2}-1\right)\right]^{1/3} + 0.55\frac{m-2}{2} + 0.3. \quad (7)$$

Comparison the predictions using Eqs.5,7 with ones which obtained by other known methods for random Gaussian load processes show that accuracy of presented method are better and it is close to accuracy for deterministic approaches (cycle-by-cycle counting) (Adrov, 1994b). Also resonable good results were reached by comparising predictions with experimental lives of notched samples under a quasi-random sequenses SAE.

DERIVATION OF THE RELATIONSHIPS FOR IN-SERVISE AIRFRAME DAMAGE ASSESSMENTS

Above presented method may be used for in-service assessments of airframe fatigue damage. Consider applications of this method to deriving the relationships for some kind aircraft.

Non-maneuvering airplanes.

Firstly note that presented approach is built on the probability properties of random loading ranges that do not exclude the max GAG (ground-air-ground) cycle. It is nesessary to account separately, i.e.

$$C_0 \gamma_{GAG}^{GAG} = \left[\left(\sigma_{max}^{GAG} - \sigma_{min}^{GAG}\right)\sigma_{max}^{GAG}\right]^{m/8} \quad (8)$$

Here σ_{max}^{GAG} is the max stress value in the air, while σ_{min}^{GAG} is the min stress on the ground. The sum of fatigue damage per τ- hour flight is

$$\bar{\gamma}_{fl} = \sum_{i=1}^{n} N_0^i \bar{\gamma}_i \tau + \gamma_{GAG}, \quad (9)$$

where $\bar{\gamma}$ is obtained from Eq.2 for every i-th part of the flight.

The long expirience of airframe fatigue damage control shown that its more useful form is control in units of damage, cumulatived during the "tipical flight". Thus, fatigue damage per every flight are expressed by number (or fraction) of tipical flights which sum airframe fatigue damage is equivalent to one per given flight. This value is named "equivalent working". To determine the value of equivalent working it is nesessary to calculate the damage per tipical flight and then to determ the contribution of one flight hour and one GAG cycle to the sum flight damage.

An example of the derivating the formulae for approximate assessment of the fatigue in-service damage is
just presented for two aircraft wing zones. Calculating
the fatigue damage for two-hour four-parth tipical flight
in related values we get

$$C_0\,\xi_{fe}^{(1)} = \sum_{i=1}^{4} C_0 N_0^i \bar{\xi}_i \tau + C_0 \xi_{(1)}^{GAG} = 2\cdot12,200 + 15,700 = 40,100; \quad (10)$$

$$C_0\,\xi_{fe}^{(2)} = \sum_{i=1}^{4} C_0 N_0^i \bar{\xi}_i \tau + C_0 \xi_{(2)}^{GAG} = 2\cdot20,300 + 24,100 = 64,700. \quad (11)$$

Thus, the contribution of the one flight hour to sum
damage is about 30% for first zone and 31.5% for second
zone, while for GAG cycle it is about 40% and 37% respectively. Consequently we obtain formulae for recalculating
the number of flight hours and the number of landing to
equivalent working

$$N_1^{eq} = 0.3\,\tau + 0.4\,N_{fe}, \quad (12)$$

$$N_2^{eq} = 0.315\,\tau + 0.37\,N_{fe}. \quad (13)$$

Significance of the equivalent working for airplane
fatigue life control may be understood from analisys of
the results the data treatment of fatigue crack detecting
in zone 1 on considered airplane fleet. Table 1 present
statistical data about aircraft lives with detected
cracks expressed in different units. The biggest values
of ratio "signal/noise" criterium for airplane lifetime
to crack detection event corresponds to equivalent working units that proves the rationality of its use for
in-service airframe fatigue life control.

TABLE 1. Effectiveness of different units for in-service
aircraft fatigue damage control. Ratio "sig-
nal/noise" criterium.

Crack detecting place, number of cracks	Landing numb.	Flight	Equivalent working
Hole 1.All fleet. 14 cracks	5.95	6.81	7.23
Hole 1.Airplene group No.1. 8 cracks	8.17	9.80	10.09
Hole 2.All fleet. 93 cracks.	4.72	5.08	5.34
Hole 2.Airplane group No.1. 33 cracks	7.19	7.99	7.98

Maneuvering aircraft.

For this tip airplane the values of x and y in Eq. 1,2 have different distribution parameters because the main load source for this airplanes is maneuvering but not turbulence. The basis for deriving an lifetime control algorithm is the "individual load spectrum" during given flight. This spectrum in coordinates "overload increment – log number of occurrences" is proposed break line with two parts. First part is commonly for all flights, second part is individual and it determine by max value of overload increment per flight. Then this spectrum must be translated to terms of nominal stresses by relationships

$$\sigma_{nom} = r\, M_{bend}, \qquad\qquad\qquad (14)$$

$$M_{bend} = K_1 G n_y + K_2, \qquad\qquad (15)$$

where M_{bend} is value of bending moment, G – airplane weight, r, K_1, K_2 – known coefficients.

Calculating the value of fatigue life by Eq. 1, 2, 14, 15, we get the nomogram (fig. 1) for training airplane. The analisys of the mutual situating of the nomogran scales and field shows that equivalent working is practically independent from flight time for little values of n and, expesially, for little airplane weight. This result reflects the fact that fatigue lifetime control in flight hours units is uncorrect and so uneffective.

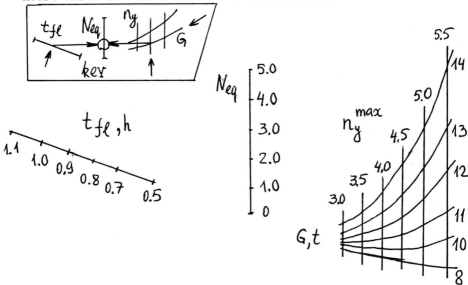

Fig. 1. Nomogram to evaluate the airframe fatigue damage in terms of "equivalent working" N_{eq} using flight time t_{fl}, max overload n_y^{max} and weight G for training airplane.

In conclusion note that all final relationships for aircraft fatigue lifetime in-service control are rather

effective and maximally simplified. So its using in ser-
vise practice do not lead to complex and long learning by
maintenance operators.

REFERENCES

Adrov, V.M.(1994a).The   use   of   a load spectrum in
    predicting airframe fatigue damage. Fat. Fract. Eng. Mat.
    Struct., 17, 1397-1403.
Adrov, V.M.,(1994b). The comparising analysis of the
    metods for fatigue   life assessment of a structure
    elements under random in-service loads. J.   of
    Rus. Aeronautics, No.4, 48-51.

A LOCAL EQUIVALENCE CRITERIUM FOR FATIGUE LIFE PREDICTING UNDER VA-LOADS

V.M. Adrov

The State Science Research Institut of Aircraft
Maintenance and Repair, Lubertsy-3, Moscow.

ABSTRACT

The criterium of local fatigue damage equivalence of different cycles with VA-loads is proposed. The relationships between cycle parameters such as range and mean value, which contribute the same damage, in terms of nominal stresses are obtained accounting the local stress –
strain material behaviour. The parameters in equations of
criterium conditions determine by material deform properties.

KEYWORDS

Fatigue life predicting, VA-loads, local strain analysis.

INTRODUCTION

In order to predict the fatigue life of airframe element
under VA-loads the generalized Oding's formulae is often
used for comparising the load cycles with damage equivalent pulse cycles

$$
S_{eq} = \begin{cases} S_{max}^{\varkappa} \, \Delta S^{1-\varkappa}, & \text{when } S_m > 0; \\ 2^{1-\varkappa}\left(\dfrac{\Delta S}{2} + 6\,S_m\right), & \text{when } S_m < 0,\, S_{max} \geqslant 0; \\ 0, & \text{when } S_{max} < 0; \end{cases} \quad (1)
$$

where Seq - max value of equivalent pulse cycle, Smax, Sm,ΔS are max, mean values and cycle range,$\varkappa$, b are parameters.

Eq.1 wide used at predicting practice in aircraft both design stage and maintenance stage for assessment of airframe load conditions and determining of its remaining lifetime. However, it is need to note some troubles of approaches which use Eq.1.

1. Parameters$\varkappa$and b are empirical and it depend from material properties, design expesialities of elements, etc.

2. Eq.1 used as a rule for integral loads - tension, bending moment, nominal stress, and it do not accounts elasto-plastic behaviour of material near zone with stress concentrator.

In this paper the relationship for calculating damage equivalent nominal stresses in form, which is analogue to Eq.1, is obtained based on local stress- strain analisys and proposed local equivalent criterium.

DERIVING THE MAIN RELATIONSHIP

It is known that the elasto-plastic matereial behaviour under VA-loads may approximated by expressions

$$\frac{\Delta \varepsilon}{2} = \frac{\Delta \sigma}{2E} + \left(\frac{\Delta \sigma}{2K'} \right)^{1/n'} \tag{2}$$

$$\Delta \varepsilon \Delta \sigma = 4 \frac{K_f^2 \Delta S^2}{E} \tag{3}$$

where $\Delta \varepsilon$, $\Delta \sigma$ are ranges of local strain and stress, ΔS is range of nominal stress, K_f, E, n', K' are known parameters. As more wide and valid damage parameter is as proposed Smith's parameter, let it will be criterium of load equivalence. Thus, cycle 1 and cycle 2 are equal (in terms of damage) if corresponding Smith's parameters are equal:

$$\sigma_{max}^{(1)} \Delta \varepsilon^{(1)} = \sigma_{max}^{(2)} \Delta \varepsilon^{(2)} \tag{4}$$

Uniting Eq.2,3,4 we get for σ_{max} :

$$\beta \sigma_{max}^{1+1/n'} + \sigma_{max}^2 - a^2 = 0, \tag{5}$$

$$\beta = \frac{2E}{(2K')^{1/n'}} ; \quad a = K_f S_{max}.$$

To simplify the solving of Eq.5 consider deform slide on $\sigma - \varepsilon$ diagram and determine five tipical sets of and valu-

es:
  a) load cycle is loop with $\Delta\varepsilon_p > 0$, σ_{max} is in plastic deforming area;
  b) loading is elastic, σ is in plastic deforming area;
  c) load cycle is loop with $\Delta\varepsilon_p > 0$, σ_{max} is in elastic area;
  d) loading is elastic, σ_{max} is in elastic area;
  e) $\sigma_{max} \leq 0$.

Solving Eq.5 with accounting Eq.2 and criterium Eq.4 for above mented sets we get the equivalent conditions of load cycles in terms of nominal stresses

$$S_{eq} = \begin{cases} S_{max}^{\frac{n'}{n'+1}} \, \Delta S^{\frac{1}{n'+1}}, & \text{when } \Delta\varepsilon_p > 0, \ S_{max} > \dfrac{\sigma_y}{K_f}; \quad (6.1) \\[2mm] c^{-0.5} \, S_{max}^{\frac{n'}{n'+1}} \, \Delta S^{0.5}, & \text{when } \Delta\varepsilon_p \sim 0, \ S_{max} > \dfrac{\sigma_y}{K_f}; \quad (6.2) \\[2mm] c^{0.5} \, S_{max}^{0.5} \, \Delta S^{\frac{1}{n'+1}}, & \text{when } \Delta\varepsilon_p > 0, \ S_{max} \leq \sigma_y/K_f; \quad (6.3) \\[2mm] S_{max}^{0.5} \, \Delta S^{0.5}, & \text{when } \Delta\varepsilon_p \sim 0, \ S_{max} \leq \sigma_y/K_f; \quad (6.4) \\[2mm] 0, & \text{when } S_{max} \leq 0, \quad\quad\quad\quad (6.5) \end{cases}$$

where $\quad c = \left(K_f/2 \right)^{\frac{1-n'}{1+n'}} \left(E^{n'}/K' \right)^{\frac{1}{n'+1}}$

Fig.1 shows predictions by Eq.6 for alloy 2024 T351 with E=69.0 GPa, n'=0.254, K'= 1517 MPa. There are considred two values of max nominal stresses $S_{max,1}$= 100 MPa, $S_{max,2}$ =200 MPa.

CONCLUSIONS

The main advantages of new criterium are follow.
  1. The form of load cycle equivalence conditions is the same that one for known relationships.
  2. Dividing the sets of cycle parameters at kinds of sets of local strain cycle parameters have more phisical mind and so rather correspond to the phisical material behaviour during deforming.
  3. The parameters in equations of criterium conditions have clear phisical base and may be determine by material deform properties parameters.
  4. The criterium conditions reflect the influence of kind and form of notch that accounted by Kf.
  5. The known equivalence relationships more closed with conditions of new criterium for samples with Kf=4.

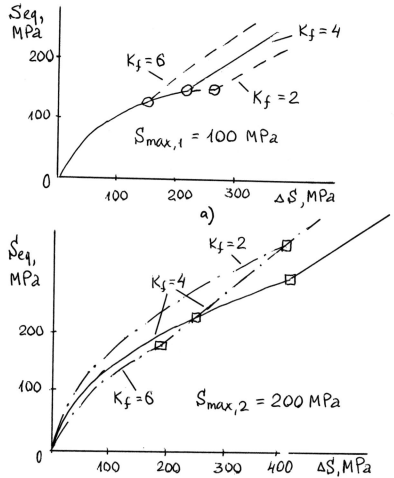

Fig. 1. Graf showing the relationship the altitude of equi-
         ivalent pulse load cycle for different values of
         max stress and range of given load cycle:
         ────────   Eq. 1;
         ──·──   Eq. 6.1, 6.2;
         ── ── ──   Eq. 6.3, 6.4;
         □ -   knee points from Eq. 6.2 to Eq. 6.1;
         O -   knee points from Eq. 6.4 to Eq. 6.3.

EVALUATION FORMULA OF NOTCH LOCAL STRAIN RANGE AFTER AN OVERLOADING

LING CHAO   LI FENG and SHA JIANG BO

Department of Material Science and Engineering, Guangdong
University of Technology, Guangzhou, 510090, P. R. China.

ABSTRACT

An approximate formula for the computation of the notch local strain range after overoading is introduced in this paper. Theoretical analysis show that, although a tensile overloading lead to the residual strain (ε') at the notch root, the notch local strain range ($\Delta\varepsilon$) after the overloading is independent of the ε' and stress ratio (R), and is only dependent on the submitted norminal stress range(Δs) and notch geometry dimension(K_t). The curves of the $\Delta\varepsilon$ of 16Mn and LY12CZ aluminium alloy calculated by using the given formula are in good agreement with test results measured after overloading.

KEYWORDS

Notch; local strain; overloading; cyclic loading; 16Mn steel; aluminium alloys

INTRODUCTION

Fatigue crack often initiates at notch root of the elements. The fatigue crack initiation life of notched elements is predicted by using the local strain approach(Buch., 1989). Therefore, the computation of the local strain range($\Delta\varepsilon$)at notch root is the critical step in the prediction of the fatigue crack initiation life and has a great effect on the prediction accuracy by using the local strain approach (Jones et al., 1989). Meanwhile, overloadings are unavoidable in practical application of engineering stractures and have also great effect on the fatigue life (zheng et al., 1988, Ling, 1990). As there exist many kinds of overloadings (Ling, 1990), the computation of the $\Delta\varepsilon$ was investigated in this paper under being subjected to a simple overloading.

APPROXIMATE FORMULA FOR CALCULATING THE
$\Delta\varepsilon$ AFTER AN OVERLOADING

Under a single overloading, the norminal overloading stress ($S_{O \cdot L}$) is below the yield stress and the element as a whole elastic, the plastic deformation, however, may occure at the notch root because of stress concentration. According to the basic assumption of "Local stress – strain approach" as shown in Fig. 1 (Ling, 1990), The stress – strain relationship of the material element at the notch root can be assumed corresponding to point B on the material tensile curve in plastic zone, as shown in Fig. 2. On unloading, the elastic part will recover and the plastic part will return to point E along the OA parallel line, so that there exists a definit residual strain ε', which is induced by the over-loading. On following subsequent cyclic stresses, and without considering the material cyclic hardening and softening behavior, the stress – strain relationship of the material element at the notch root may be approximately changed within the EB line. The corresponding local stress (σ) and norminal stress (S) may be expressed as follow:

$$\sigma = K_t \cdot S \tag{1}$$

the $\Delta\varepsilon$ can be calculated as follows:

$$\Delta\varepsilon^n = \varepsilon^{n+1} - \varepsilon^n = (\varepsilon' + \sigma^{n+1}/E) - (\varepsilon' + \sigma^n/E)$$
$$= (\sigma^{n+1} - \sigma^n)/E \qquad (n = 2, 3, 4 \cdots\cdots) \tag{2}$$

It can be seen from eq. (2) that the $\Delta\varepsilon^n$ is in direct propertion to the norminal stress range (ΔS), and is independent on the residual strain ε' and stress ratio R.

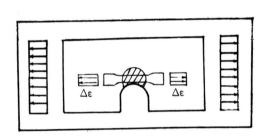

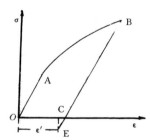

Fig. 1. Schematic diagram of material   Fig2. Schematic diagram of $\sigma - \varepsilon$ relation –
element at notch root   ship of the material elements

EXPERIMENTAL WORK

The materials used for measuring the local strain range were LY12CZ aluminium alloy sheets and 16Mn steel plates in the as – received condition. The former belongs to a continuous strain harden-ing material and the later belongs to a noncontinuous strain harding material (Ling, 1990), the e-lastic modulus of the two testing material are 69375 MPa and 206000 MPa respectively.

The specimens with a central hole (16Mn) and with double notches (LY12CZ) were machined re-spectively from 16Mn steel plates and LY12CZ alloy sheets. The stress concentratioon factor, K_t, were estimated from (Peterson, 1962) and are 2.72 and 3.0 respectively. Cyclic loading pattern were used in measuring the local strain range is shown in Fig. 3. The strain gauge was used to mea-sure smaller strain and the Moire method to measure larger strain, Ling(1990) indicated the details of experiments.

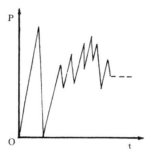

Fig. 3. Schematic diagram of the cyclic loading pattern

TEST RESULTS

The local strain range measured after overloading can be expressed as a linear relationship with the norminal stress range as shown in Fig. 4. The regression analysis gives following expression: For LY12CZ alloy notch elements,

$$\Delta s = 23611 \Delta \varepsilon \ (\mathrm{MPa}) \tag{3}$$

and for 16Mn steel notched elements,

$$\Delta s = 73500 \Delta \varepsilon \ (\mathrm{MPa}) \tag{4}$$

Substituting the values of K_t and E into eq(2). we have

$$\Delta s = E\Delta\varepsilon/K_t = 23125\Delta\varepsilon \ (MPa) \qquad for \ LY12CZ \qquad\qquad (5)$$

$$\Delta s = E\Delta\varepsilon/K_t = 75735\Delta\varepsilon \ (MPa) \qquad for \ 16Mn \qquad\qquad (6)$$

Compare eq. (3) ~ (4) with eq. (5) ~ eq(6), one can obtaine that the relative error of the corresponding coefficients are within 10%, which means that the metal in notch root plastic zone induced by an overlaoding is subjected to the elastic cyclic stress at the lower subsequent cyclic norminal stress and the given eq. (2) can be used to approximately calculate the values of the $\Delta\varepsilon$ after overloading.

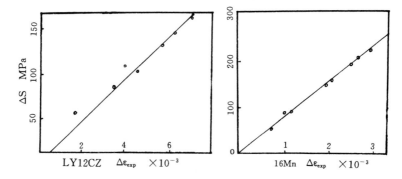

Fig. 4. linear relation between the $\Delta\varepsilon$
and Δs after overloading

DISCUSSIONS

Crews (1971) indicated that the experimental values of the $\Delta\varepsilon$ have no relation with cyclic loading numbers when the subjected norminal stress is lower than yield strength, which means the $\Delta\varepsilon$ is independent of cyclic hardening phenomenon. The test results measured in this study show that, after an overloading, the values of the $\Delta\varepsilon$ can be approximately calculated by using eq. (2) and without considering the material cyclic hardening and softening behavior.

For noncontinuous strain hardening materials, the stress − strain relationship at the notch root elements can be assumed as shown in Fig. 5. One can see from Fig. 5. that , the residual strain ε' induced by an overloading is larger, but it has no influence on the subsequent stress − strain relationship, which are good agreement with the experimental results.

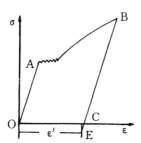

Fig. 5. Schematic diagram of $\sigma - \epsilon$ relation for a
noncontinuous strain − hardening material.

CONCLUSIONS

After an overloading, the local strain range can be approximately calculated from eq. (2). The value
of the $\Delta\epsilon$ is dependent on the norminal stress range and stress concentration factor as well as elasticity modulus, and is independent on the residual strain induced by the overloading and stress ratio
and of the material cyclic hardening and softening behaviors.

REFERENCES

Buch, A. (1989). Prediction of fatigue life under aircraft loading with and without use of material
    memory rules. Int. J. of Fatigue, 11, 97 − 105.
Jones, R. L, Phoplonker, M. A, and Byrne, (1989). Local strain approach to fatigue crack initiation
    life at notches. Int. J. of. Fatigue, 12, 255 − 259.
Zheng, X. L. and Ling, C. (1988). On the expression of fatigue crack initiation life considering of
    overloading effect. Eng. Fract. Mech. , 31, 959 − 966.
Ling, C. (1990). On the models for predicting FCIL under variable amplitude loading and the effects of surface strain hardening on fatigue crack initiation life. PH. D. Dissertation , Northwesten Polytechnic University.
Crews, J. H. (1971). Effects of loading sequence for notched specimens under high − low two −
    steps fatigue loading. NASA Tech. No. D − 6558.

ORIGINS FOR FATIGUE CRACK-TIP FIELD PERTURBATION AS MANIFESTED BY GROWTH RATE TRANSIENTS

Y. KATZ, H. ALUSH and A. BUSSIBA

NRCN, P.O.Box 9001, Beer-Sheva 84190, ISRAEL

ABSTRACT

A single overload may retard or even arrest cyclic subcritical crack growth depending on the specific overload strength. Load interactions achieved by prior overloads (pulses or continuous variable amplitude blocks) provide an extrinsic crack-tip shielding which develops by the production of an affected zone ahead of the original crack-tip.

The present investigation adopted a material approach with attempts to identify various origins for crack extension transients. While recognizing several mechanisms which might alter crack stability, the current study mainly support the excess residual compressive stress field to be the dominant factor. Related problems, demonstrated by Warm Prestressing (WPS) or Warm Precracking (WPC) were included. This in order to extend the views on crack-tip shielding effects under various residual stress field situations.

KEYWORDS

Fatigue crack propagation, overloads, residual stresses, warm prestressing/precracking.

INTRODUCTION

For the generalized case in which the life of structural elements are not characterized solely by crack initiation controlled processes, both crack initiation and propagation stages have to be established. Thus, load interaction studies in fatigue are motivated by at least two objectives. First, on the fundamental level, by taking advantage of crack-tip perturbations such as changes in localized dislocation structures, a more physical view of the local micromechanisms of crack stability may be obtained. Second, better understanding of how irregularities in cyclic spectra affects growth would assist improvements in life prediction methods for real service conditions. Consequently, extensive research activity has been invested in post-overload effects (Matsuoka and Tanaka, 1976, Chanani and Mays, 1977, Gan and Weertman, 1981, Fleck, 1985, Chen et al., 1990). In the framework of an extensive research program, Fe-3%Si single crystals were also included. Prior activities in iron based crystals (Chen et al., 1990) have revealed theoretically and experimentally, the importance of the non-primary stresses, as related to the high sensitivity of cyclic microcracking to the local stress field variations. A modified superdislocation model has been developed (Lii et al., 1989), simulating the sequence of overload, unload and reload processes and how it reflects on the crack-tip field. The crack tip shielding effects as analyzed, were consistent with the experimental findings. It was shown that the fracture toughness at 273K could be increased by nearly a factor of three with a prior room temperature overload. The monotonic loading behavior has been extended to cyclic conditions in the case of subcritical crack extension on the {100} plane (Chen et al., 1990). Under such conditions where other probable mechanisms involved in crack tip shielding were minimized, the residual stress analysis actually predicted the transient behavior following a single overload.

The present study was constructed by centering on the case of transients in metastable austenitic stainless steel while emphasizing phase stability effects. In addition other related aspects in crack stability under residual stress field situations were developed. To assist here, the influence of WPS and WPC were selected in high and a low symmetry crystal structure materials in order to provide some generalization at least in polycrystalline systems.

EXPERIMENTAL PROCEDURES

The current investigation is centered only on a partial section of a broader program regarding crack tip field perturbation effects on cyclic crack growth and fracture resistance. The search for transient origins in the framework of the global activities are summarized in Table 1. This kind of material approach beside phenomenological findings provided a sound background and refinements regarding load interactions effects.

Specifically, for the phase stability aspects, AISI 304 metastable austentic stainless steel was selected. The chemical composition, mechanical properties, fracture toughness parameters and the steady state FCPR have been described elsewhere (Y. Katz et al., 1981). Phase stability aspects in terms of austenite (γ) decomposition and martensitic phases formation have been addressed previously (Y. Katz et al., 1982). The post overload behavior was studied by performing single overload with intensification factor of q (K_{OL}/K_{MAX}) = 2.5 with frequency of 10 Hz and load ratio R≅0.

For comparison studies, phase stability effects were examined by performing the overload at 296K and 198K, keeping in both cases the overload plastic zone size constant. This enabled to achieve two variants of near crack-tip process zones of deformed γ and transformed martensitic zone. The subsequent fatigue runs were carried out at 296K in constant ΔK in order to eliminate the effects of increasing ΔK due to crack extension. Crack length was measured using both, COD gauge and electropotential method. In addition, another variant was selected of overloads above M_d with subsequential cycling at 296K and 198K

For the WPS and WPC, lean uranium alloy (0.75wt.% Ti), and mild steel-ULC (Mn-0.02wt.%, C-0.007wt.%) were selected. For the first, WPS and WPC have been performed at various levels of β [(K^{WPS}, K_f^{WPC})/K_{Ic}] = 0.5, 0.75 and 0.85 at 296K. For the mild steel, pre-fatigue values at 296K, varied in the range of 11 - 50 MPa·m$^{1/2}$, while the toughness tests were conducted at 4K. For the finite element (FE) analysis, PATRAN pre-processor and ABAQUS software have been utilized to evaluate numerically the stress field distribution after loading-unloading at various temperatures.

Table 1. Material approach and experimental programs.

Materials	Tests/Variables	Specimen geometry	Suspected transient origins and mechanisms
Single crystals			
Fe-3%Si	Cyclic tension-tension with overloads. Overload intensification factor.	Mini compact- disc specimens.	Residual-stress in a subcritical growth confined to the cleavage plane.
Polycrystal systems			
Al, Iron base alloys, U-Ti alloys	Cyclic tension-tension or compression-compression with overloads. Various crystal structures, Overload intensification factor. Under subsequential fatigue, constant or variable amplitude range.	CT (Compact Tension) and three point bending specimens.	Blunting, closure, residual-stresses.
Metastable austenitic stainless steels	Cyclic tension-tension with overloads. Thermal effects. Overload intensification factor.	CT, SEN (Single edge notched) and three point bending specimens.	Dilational stresses due to martensitic transformations.
U-Ti alloy AISI 304 metastable stainless steel	Transients activated by overloads, with environmental interactions.	SEN, three point bending, and tapered specimens.	Residual stresses. Deformation/ corrosion effects.
Superplastic model alloy Zn-22Al	Transients by overloads in visco elastic-plastic model material. Thermal effects on fracture modes.	CT specimens.	Residual stresses. Time dependent crack-tip shielding.
U-Ti alloy orthorhombic low symmetry crystal structure Mild steel AISI 4340 steel Al-Li planar slip model material	Monotonic WPS (Warm Pre-stressing) or CPS (Cold Prestressing as also WPC (Warm precracking).	CT and three point bending specimens.	To establish consistency regarding the role of crack tip field perturbations on the toughness values.

Optical and SEM metallography were added for deformation feature observations along the overload affected zone. Finally, AE technique including energy distribution modules with gain of 90 dB was supplemented to explore cumulative damage mechanisms associated with WPS and WPC processes.

EXPERIMENTAL RESULTS

The metastable 304 austenitic stainless steel case

Although similar behavior of the transient plastic zone (dynamic) was observed for the γ deformed case, the more accentuated features of the transformed plastic zone are illustrated in Figs. 1 and 2. Clearly this behavior reveals the role of the localized ΔK_{eff} that was measured and modified solely by the instantaneous variations of the reversed plastic zone. Thus, the evaluation of the local K_{eff} (depending on the exact position along the overload affected zone) provided a simplified measure for the residual stress distribution. Fine features along the wake of the crack are illustrated in Fig. 2 (a, b).

The retardation curve for the transformed martensitic zone is illustrated in Fig. 3. In addition, we found that the best semi-empirical function while recognizing also the Wheeler and Willenborg models could be formulated by; (see Fig. 3)

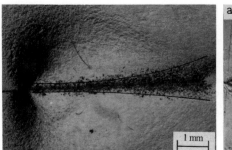
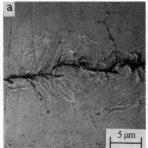
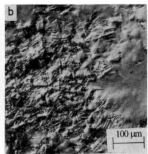

Fig. 1. Overload in γ transformed above M_d Fig. 2. Deformation features along the wake of the crack.
with subsequential cyclic at 198K.   (a) Cycling at 296K. (b) Cycling at 198K.

$$\left(\frac{da}{dN}\right)_i = \left(\frac{da}{dN}\right)_{min} + \left(\frac{da}{dN}\right)_{CA}\left[1 - e^{-(a_i - a_0)/\lambda}\right] \tag{1}$$

where CA corresponds to constant amplitude and the parameter λ indicates the degree of recovery. Along this , some physical rationale could be proposed solely based on the excess of residual compressive stresses provided by the martensitic transformation at low temperatures. Notice the similarity of equation (1) to the expression addressed by (Gamache and McEvily, 1993), for the development of fatigue crack closure. As such differences between the retardation behavior of both cases was indicated by λ (λ=6 for transformed and 4 for deformed) that might be related to dilational stresses at low temperature.

Selected examples affected by residual stress fields

The mild   steel and the lean uranium alloy cases. The WPC case was analyzed by FE analysis as shown in Fig. 4. At this stage only loading and unloading cycle was attempted which allowed a partial evaluation of the residual effects. This is summarized in Fig. 5 indicating the remote fracture toughness values at 4K vs. the prefatigue stress intensity factor conditions K_{max}^f. The load-displacement curves at 4K are demonstrated in Fig. 6.

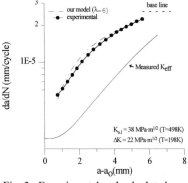
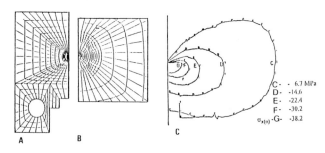

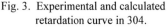

Fig. 3. Experimental and calculated   Fig. 4. FE mesh for (a) CT specimen (b) refinement of the crack tip
retardation curve in 304.   (c) Stress field distribution for K=49.4 MPa·m$^{1/2}$.

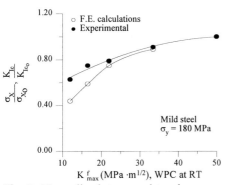

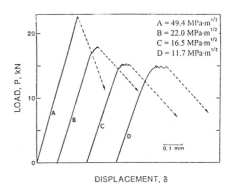

Fig. 5. Normalized stress and toughness vs.
K$_{max}^f$ obtained by FE and experimental.

Fig. 6. Load - displacement curves obtained at 4K
for increasing K$_{max}^f$ in mild steel (ULC)

From the U-Ti alloy findings a comparison between WPS and WPC is shown in Fig. 7. The tendency of increasing toughness due to WPS is well observed. However WPC resulted in some higher values of toughness. The cycling residual are more pronounced compared to the loading-unloading situation as observed also in the AISI 304 stainless steel and the mild steel analysis (Figs. 3 and 5).

Finally, some remarks concerning the AE tracking are in order. It appears that the characterization of damage accumulation mechanisms might shed light on the role of residual stresses. Just for illustration, regarding the specific AE set up which was utilized, an example of its resolution or sensitivity is demonstrated in Fig. 8. The environmental interaction in U-Ti alloy is depicted which provided physical insight into damage evolution while evaluating the relative energy emission distribution and its time dependency. For the lean U-Ti alloy tested at 228K, AE activity for WPS and WPC was compared. As shown in Fig. 9 at least two points emerged. First, for the WPC, AE onset occurred at the early stage of loading in contrast to delayed activity in WPS. Second, more events were distributed along the energy range for the WPC process. These findings served assessment tool for AE in post over-load fatigue crack extension rate transients.

Two more examples are described for damage accumulation in post WPS of U-Ti. Specimens with sharp cracks were tracked at 296K and 228K. Here two relative energy cells were monitored for group emitted waves [see low cell (2) and high (101) in Fig. 10]. While comparing both temperatures the development of cleavage like microcracking at 228K became evident while suppressed at 296K.

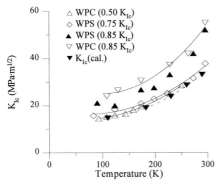

Fig. 7. Toughness vs. Temperature in U-Ti under
various WPC/WPS conditions.

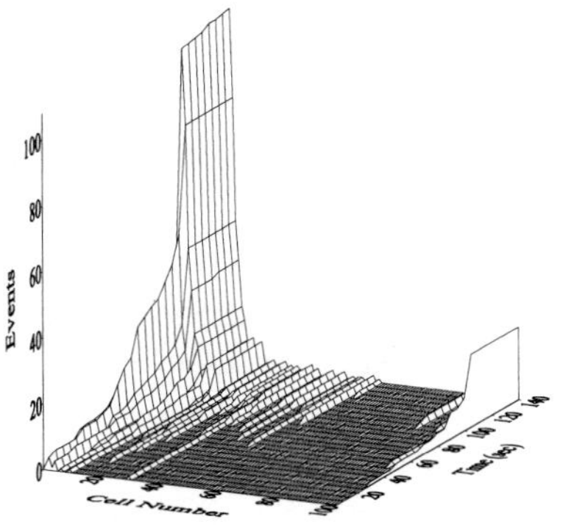

Fig. 8. Damage accumulation by AE in uranium
alloy affected by environment.

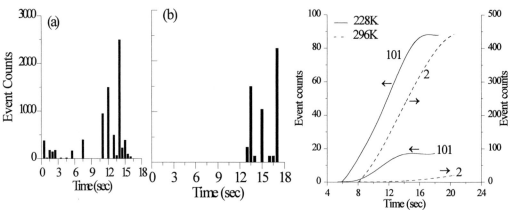

Fig. 9. Damage accumulation during loading in U-Ti alloy affected by: (a) WPC, (b) WPS.

Fig. 10. Damage accumulation in U-Ti while tracking high and low energy cells.

DISCUSSION AND CONCLUSIONS

Experimentally based, it become more agreeable to associate the overload effects on cyclic crack extension rates to the non-primary, compressive stresses. Studies on related problems of WPC and WPS support this underlying major origin while recognizing other influences too. As such, the nature of residual stresses implies that, beside their initial magnitude, additional understanding of their redistribution or time-dependency is required. Variation here clearly affect the crack tip shielding capacity. The importance of the residuals remain consistent on different levels, even though the actual cause may be partially from closure stresses. Thus the following specific conclusions were found.

(1) Post overload effects on cyclic transients are mainly dominated by non-primary stresses.
(2) This underlying origin is also supported by related problem of WPC and WPS activating residual field situations.
(3) Crack tip compressive residual stresses provides excesses energy release as confirmed by refined tracking of damage accumulation.

REFERENCES

Chanani, G.R. and B.J. Mays (1977). *Engng. Frac. Mech.,* 9, 65.
Chen, X.F., Y. Katz and W.W. Gerberich (1990). *Scripta Metall.,* 24, 2351.
Fleck, N.A. (1985). *Engng. Frac. Mech.,* 33, 1139.
Gamache, B. and A.J. McEvily (1993). In: *Fatigue 93* (J.P. Bailon and J.J. Dickson, Eds.), p. 577 EMAS, Warley, UK.
Gan, D. and J. Weertman (1981), *Engng. Frac. Mech.,* 15, 87.
Katz, Y., A. Bussiba and H. Mathias (1981). In: *Materials, Experimentation and Design in Fatigue* (F. Sherratt et al., Eds.), p. 147, Westbury House, Surrey, UK.
Katz, Y., A. Bussiba and H. Mathias (1982). In: *ECF4* (K.L. Maurer and F.E Matzer, Eds.), Vol. 2, p. 503. EMAS, Warley, UK.
Lii, M, T. Foecke, X. Chen, W. Zielinski and W.W. Gerberich (1989). *Mater. Sci. And Engng.,* A113, 327.
Matsuoka, S. and T. Tanaka (1976). *Engng. Frac. Mech.,* 8, 507.
Tobler, R. L. and A. Bussiba, private communication (Fig. 6).

CORROSION FATIGUE, DATA AND DESIGN

CORROSION FATIGUE, DATA AND DESIGN
FROM AN AIRCRAFT MANUFACTURER'S POINT OF VIEW

H.–J. SCHMIDT and B. BRANDECKER

Fatigue and Fracture Mechanics Department,
Daimler–Benz Aerospace Airbus, Hamburg, Germany

ABSTRACT

Corrosion of metallic materials with its effects on the economic and airworthiness aspects is an issue throughout the life of the aircraft structure. Furthermore the environmental conditions influence the durability and the damage tolerance behavior of the structure. Manufacturers' and operators' major activities to control and improve both, damage tolerance and corrosion behavior, are described in this report, i.e. material choice, initial surface protection, maintenance programs and restoration of surface protection.

KEYWORDS

Corrosion; fatigue; damage tolerance; crack growth, surface protection, environmental effect.

INTRODUCTION

The present aircraft generation was designed for an economic operational life of at least 20 years and up to 90 000 flight cycles. These design goals will be exceeded by many operators of jet–powered airplanes and turboprops. According to the latest statistics from 1995 approximately 5700 jets and 2100 turboprops are now 15 years old or more. The fleet leaders of different jet airplane types, except the recently certified aircraft, are between 15 and 35 years old and have accumulated up to 100 000 flight cycles.

Therefore corrosion of metallic materials is one of the most important issues affecting the life of the airplane structure with consequences to economic and airworthiness aspects. In general the occurrence of corrosion damage depends on several parameters, such as:
- The degree of corrosion protection during the manufacturing and assembly process.
- The degree of maintenance performed during the operational life.
- The environmental conditions in which the airplane is operated.
- The type of cargo.

In addition the environmental conditions (temperature and humidity) have an effect on the fatigue life (crack initiation life) and the crack growth in aluminium structure and consequently repercussions on the structural inspection program necessary to maintain the airworthiness of the airplanes. The different aspects of the corrosion issue are described in the following from the point of a structures engineer involved in design and product support activities for all Airbus airplanes at Daimler–Benz Aerospace Airbus (DA). All examples presented here are related to Airbus aircraft.

CORROSION PROBLEMS

In general corrosion is the deterioration of metals due to the reaction with the environment. It is a phenomenon of metal breakdown into oxides, chlorides or carbonates due to chemical or electrochemical reactions with the environment.

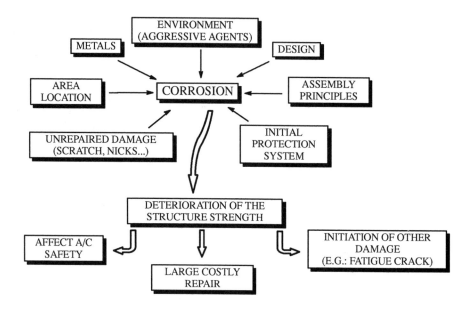

Fig. 1:   Corrosion problems

Figure 1 shows that corrosion is the result of seven major parameters, i.e. material, design, assembly, protection system, area location, accidental damage and environment. Finally the corrosion damage leads to a deterioration of the strength of the structure which may affect the airworthiness of the aircraft, cause costly repairs and initiate other damages, such as fatigue cracks.

The corrosion issue is addressed in the current airworthiness regulations FAR and JAR 25.571 requiring an evaluation of the strength, detail design and fabrication to verify that catastrophic failure due to fatigue, **corrosion** or accidental damage will be avoided throughout the operational life of the airplane. This evaluation must include the consideration of temperature and humidity expected in service.

Types of corrosion

Seven different types of corrosion have to be considered during design and operation of the aircraft, i.e. pitting, intergranular, exfoliation, galvanic, fretting, filiform, biological. The first two of them are briefly described in this report.

Pitting Corrosion. Pitting corrosion occurs when local damage effects the surface protective layer in presence of an electrolyte and when the protective layer is made of a material nobler than the material protected. The metal becomes the anode and the protective layer becomes the cathode. In that case the metal will corrode. The process is shown in Fig. 2.

Pitting corrosion is a localized form of corrosion which extends vertically into the material. It can be detected by the presence of white or grey powder deposits on the material surface. When the surface is cleaned the corrosion appears via small pits and holes on the surface.

This type of corrosion can be the starting point for intergranular corrosion. It can be dangerous due to its vertical extension which reduces the material strength.

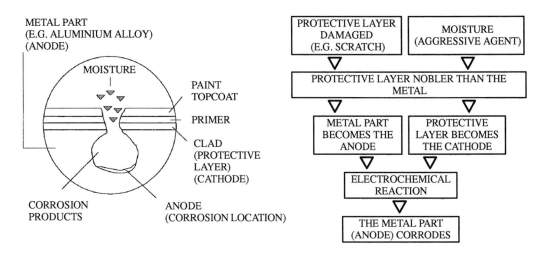

Fig.2:   Pitting corrosion

Intergranular Corrosion. The internal structure of metals consists of numerous individual grains having clearly defined boundaries. The properties of the boundary differ from those for the grain center. These differences are due to changes that occur during material processing: heating, cooling and rolling. In the corrosion process, the grain is the cathode and grain boundaries are the anode and corrode, see Fig. 3.

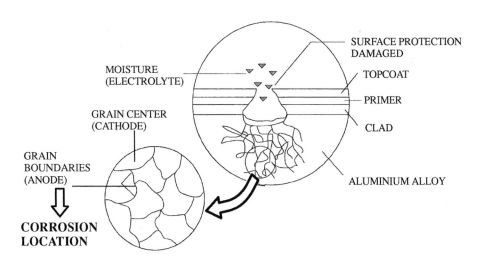

Fig.3:   Intergranular corrosion

Intergranular corrosion can be initiated by pitting corrosion. It may exist with no or small surface effect, and is very hard to detect. Some high tensile strength aluminium alloys are especially sensitive to intergranular corrosion under stress (e.g.: 7079 and 7178 aluminium alloys). Intergranular corrosion, when severe, can cause exfoliation corrosion.

MANUFACTURERS' ACTIVITIES FOR CORROSION CONTROL

Corrosion control affects all stages of design, production and maintenance of the aircraft as presented in Fig. 4. The Airbus Industrie policy, in terms of corrosion control, complies with the IATA (International Air Transport Association) general guidelines to aircraft manufacturers on the design and maintenance of the structure to prevent corrosion.

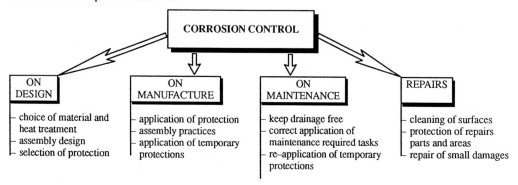

Fig. 4   Basis of corrosion control

The manufacturer provides a major contribution to an adequate corrosion control by correct material and heat treatment choice, assembly design and selection of protection during the design process. Furthermore during the manufacturing process the adequate application of the protection and assembly practices as well as application of temporary protections have a significant influence on the corrosion control.

Material Choice

As an example the material choice for the skin of the fuselage of future large transport aircraft is discussed. From the designer's point of view the static strength, the durability, the damage tolerance behavior and the corrosion resistance are the major technical aspects for the material choice. The skin of the pressurized fuselage of the next Airbus generation may be made from new material, e.g. 6013 or advanced 2024 (C188) instead of conventional 2024T3. The reasons for the possible material change are to reduce manufacturing costs by changing the assembly process or by simplifying the design and to comply with the forthcoming regulations.

Damage Tolerance Behavior. One of the key issues is the crack growth behavior under real environment. Any new material for skin in areas where damage tolerance is the dimensioning case, can only be accepted, if the crack growth behavior under real environment and real frequency is not worse compared with the present 2024T3 material. If the new material would not fulfill this requirement, the skin thickness would have to be increased which is unacceptable due to weight reasons.
Table 1 presents the crack growth behavior of different materials obtained by tests with CCT specimens performed at 20 Hz and 3.5 percent sodium chloride (NaCl) with inhibitors at room temperature. The comparison for the two lower ΔK values reveals that e.g. for 6013 and laboratory air the da/dn values are similar to the 2024T3 values. The same comparison for an environment of 3.5 percent sodium chloride shows significant differences.

Table 1:   Crack growth data for a frequency of 20Hz

Material (L–T direction)	Environment	da/dn (µm/cycle·10^{-3})		
		$\Delta K = 13$ MPa$\sqrt{m}$	$\Delta K = 20$ MPa$\sqrt{m}$	$\Delta K = 27$ MPa$\sqrt{m}$
2024T3	laboratory air	0.20	0.65	2.40
6013T6	laboratory air	0.28	0.55	1.00
7475T76	laboratory air	0.45	1.00	1.90
2024T3	3.5 percent NaCl	0.27	0.65	2.30
6013T6	3.5 percent NaCl	0.65	0.80	1.50
7475T76	3.5 percent NaCl	0.85	1.50	2.20

Considering the above mentioned comparison only it would not be recommended to use these new materials. However, the test conditions are quite different from the reality, therefore new test procedures are to be developed and applied before a final decision is made.

DA has accomplished crack growth tests using new test procedures. In general the effect of the frequency and the environment on the crack growth behavior has to be evaluated for longitudinal and circumferential cracks in a pressurized fuselage. The load spectra for these crack types are quite different, i.e. a constant amplitude spectrum due to internal pressure is driving the longitudinal cracks and a complex flight–by–flight spectrum due to internal pressure and external loading has to be considered for the circumferential cracks.

Up to now DA has analyzed the frequency and environmental effect for longitudinal cracks. Constant amplitude tests (R=0.1) have been carried out with CCT specimens (L–T direction) with a width of 100mm and a length of 200mm and the stress–time–histories shown in Fig. 5.

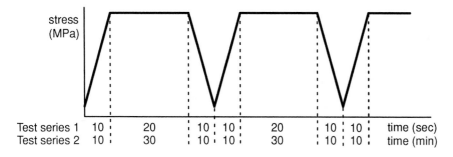

Fig. 5:   Stress–time histories for crack growth tests

Figure 6 contains the results of the DA test series 1 compared with the frequency of 20 Hz. The left hand figures show the da/dn data obtained under laboratory air. Independent from the test frequency the material 6013 shows a better crack growth behavior than 2024. The right hand figures valid for the environment of sodium chloride show different results:

In case of a frequency of 0.025 Hz the crack growth behavior of 6013 is better. For a frequency of 20 Hz and lower ΔK values ($\Delta K \leq 22$ MPa$\sqrt{m}$) the crack growth behavior of 2024 is significantly better. For higher ΔK values ($\Delta K > 22$ MPa $\sqrt{m}$), which are less important, 6013 is superior to 2024.

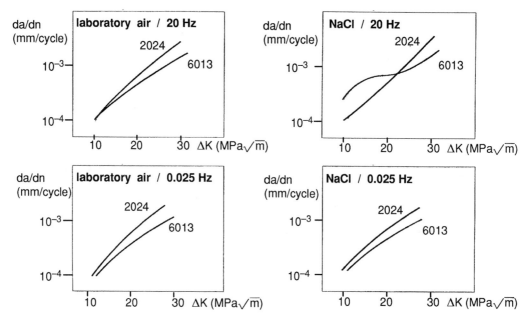

Fig. 6:   Crack growth test results for 2024 and 6013 and frequencies of
20 Hz and 0.025 Hz.

Test series 2 contains similar tests with a frequency of $0.33 \cdot 10^{-3}$ Hz, see Fig. 7. As for test series 1 the material 6013 has a better crack growth behavior than 2024 for the frequency of $0.33 \cdot 10^{-3}$ Hz and 3.5 percent sodium chloride. The same behavior, i.e. 6013 better than 2024, is observed for laboratory air and the low frequency.

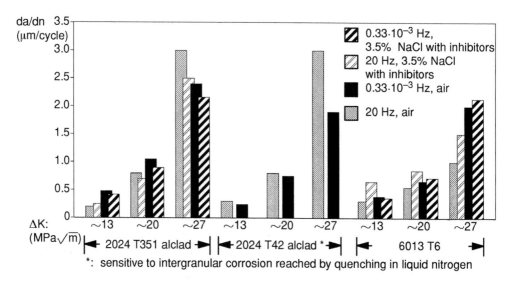

Fig. 7:   Crack growth test results for 2024 and 6013 and frequencies of
20 Hz and $0.33 \cdot 10^{-3}$ Hz.

Figure 8 presents a comparison of crack growth test results for specimens tested at low frequency of $0.33 \cdot 10^{-3}$ Hz in laboratory air and two temperatures (room temperature (RT) and $-55°$C). For the lower temperature of $-55°$C the crack growth is less than for RT for the two lower ΔK values of ~ 13 and ~ 20 MPa$\sqrt{m}$. For the highest ΔK value no decrease is observed for 2024 or even an increase for 6013.

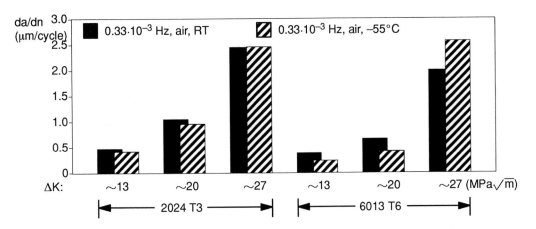

Fig. 8:   Crack growth test results for 2024 and 6013 for low frequency
           and RT/$-55°$C.

The Technical University of Hamburg–Harburg, Germany, has carried out similar tests and investigated the crack growth versus the frequency for the materials 2024, 6013 and 7475 at two ΔK values. In principle the test results correlate with the DA investigations, but a maximum crack growth has been found at 1 Hz for all three materials and 3.5 percent sodium chloride environment, see Fig. 9. It is assumed at present that a re–passivation of the surface may be the reason.

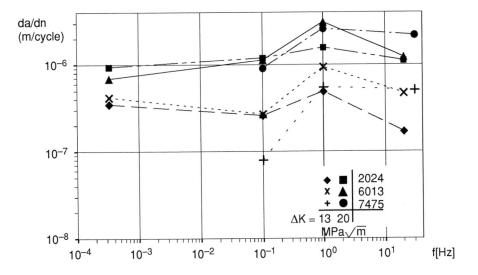

Fig. 9:   Influence of frequency for 2024, 6013 and 7475.

Both investigations revealed the complexity of the problem and that a material selection using the standard crack growth test procedure, i.e. frequency approx. 20 Hz, RT, air and 3.5 percent sodium chloride, resp., may be misleading. The differences of the crack growth rates comparing 2024, 6013 and 7475 depend on the test frequency.

Therefore it seems necessary to develop new standardized test procedures which allow a sufficient prediction of the crack growth behavior for real conditions. A key issue is the standardization of the spectra to be applied for these tests, which should contain information about the stress–time–history, the corresponding frequencies, and the temperature distribution. The standardized spectra should be developed by an international co–operative approach. As an example Fig. 10 contains the DA proposal for the spectra to assess the skin of the pressurized fuselage of a widebody transport aircraft.

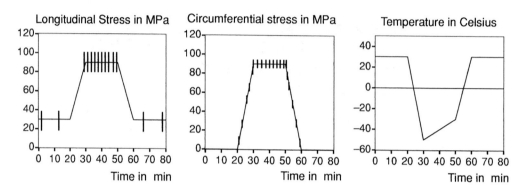

Fig. 10: Proposal of standardized spectra for fuselage skins.

Fatigue Life. Many investigations have been carried out to determine the effect of corrosive environment on the fatigue life (crack initiation). For example DA tested several coupons representing the longitudinal lap joint. The test specimens were manufactured according to the production standard, i.e. with surface treatment (CAA, primer, top coat), wet assembly and wet riveting. The tests performed for two materials, the current skin material 2024T3 and the Al–Li alloy 8090T8, revealed no influence of the different environments, i.e. laboratory air and sodium chloride, see Fig. 11.

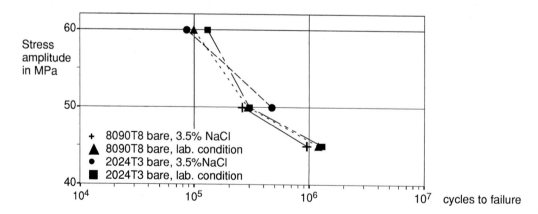

Fig. 11: Influence of the environment on the fatigue life of longitudinal
lap joints.

These results are in line with those presented by Thomas and Perras (1995), where the results from comparative coupon tests of lap joints with and without a pre–aging of 1500 h salt spray are described. For the test specimens, where the rivets (2024, 2117 or 6013) were installed with sealant, the fatigue life was not adversely affected by corrosion exposure. An increase in fatigue resistance was observed for the corroded specimens which was explained by the authors that this increase is probably due to some sort of compression state caused by the development of aluminium oxide during corrosion.

In contrast to the above mentioned results corrosion has a significant effect on the fatigue life of lap joints investigated by Müller (1995). It has to be recognized that these specimens were manufactured without surface treatment and dry assembled. The described results are independent from the test frequency as shown in Fig. 12.

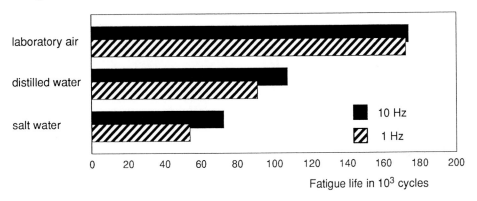

Fig. 12: Influence of the environment on the fatigue life of lap joints
without surface treatment.

These test series lead to the conclusion that no reduction of the fatigue life has to be expected at in–service aircraft as long as the surface protection system is intact.

Corrosion Protection during Manufacturing

The Airbus aircraft are designed and manufactured with the maximum possible resistance to corrosion. The resistance to corrosion is provided by the interaction of different types of protections, i.e. pre–treatments, paint coatings, special coatings, sealants and also adequate aircraft drainage. To select for all areas the adequate corrosion protection Airbus aircraft structures are divided into specific categories according to their sensitivity to corrosion. The type of protection applied depends on the category concerned.

Figure 13 presents the Airbus protection system for the skin of the pressurized A320 fuselage. The pre–treatment consists of chromic acid anodizing (CAA). The CAA treatment is a electrolytical treatment, i.e. the surface is coated with oxide leading to an anodic protection.
In addition to the surface protection described above special coatings are applied in areas were it is required (in principle these are the bilge area and the door surrounding areas). The special coating may be a water repellent coating or a corrosion preventive compound.
Another important contribution to the corrosion protection is made by the sealant, i.e. interfay sealing, sealant beads and protective layers. The sealant is installed for three reasons:
  • sealing for corrosion prevention – non pressurized areas
  • sealing to prevent air leaks and corrosion – pressurized areas
  • sealing to prevent fuel leaks and corrosion – fuel tank areas

Figure 13 presents a typical example of the corrosion protection system used for the longitudinal lap joint areas in the pressurized A320 fuselage showing the internal and external protection, the application of sealant and an additional protection of the internal sealant beads in hydraulic fluid areas, e.g. in the bilge area. All rivets are wet installed and the driven rivet heads are protected by a paint touch–up.

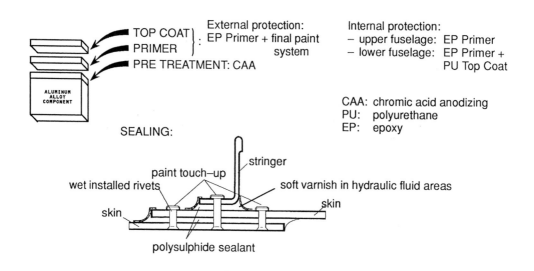

Fig. 13: Corrosion protection system for longitudinal lap joint area in A320 pressurized fuselage.

OPERATORS' ACTIVITIES FOR CORROSION CONTROL

Corrosion Control

In a joint effort the aircraft manufacturer and the operators develop a structure maintenance program prior to certification to avoid unacceptable structural degradation of in–service aircraft. This structure maintenance program considers for metallic structure three kinds of degradation, i.e.:
- accidental damage
- environmental deterioration
- fatigue damage.

For each structural significant item (SSI) the necessary inspections are defined separately for fatigue, corrosion and accidental damage. For Airbus aircraft the environmental deterioration analysis procedure follows the philosophy and the principles of the MSG3 (Maintenance Steering Group No. 3).
Figure 14 contains details of the environmental deterioration analysis for metallic structure. This analysis comprises the following steps for each SSI:
- Determination of the material sensitivity to the types of corrosion (filiform, uniform, etc.)
- Determination of the material sensitivity to stress corrosion
- Determination of the surface protection rating
- Determination of the environmental rating (probability of exposure to corrosive products)
- Consideration of possible deterioration of surface protection
- Combination of the ratings leading to determination of the inspection interval.

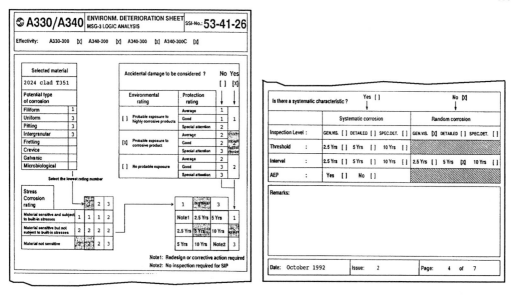

Fig.14: environmental deterioration analysis procedure

For corrosion having systematic characteristics the in–service experience of Airbus or other aircraft could allow the introduction of a threshold for initial inspection. If the corrosion has a random characteristic, e.g. in case of stress corrosion sensitive material or in areas prone to accidental damage, no inspection threshold is applied.

As a result of this analysis inspection intervals of either 2 to 2.5 years or 4 to 5 years are defined. The required inspection level, i.e. general visual, detailed visual or special detailed (NDI), is dependent on the type of possible corrosion relevant for the inspection intervals.

In the aging aircraft activities framework, the Federal Aviation Administration (FAA) requires that all operators must have an effective Corrosion Prevention and Control Program (CPCP) incorporated into their maintenance programs. For the Airbus A300 aircraft Airbus Industrie has issued a separate CPCP in 1990 and for other aircraft types, e.g. A310, A320, A330/A340, the incorporation of the CPCP tasks in the maintenance program revisions is in progress. The CPCP contains information such as location to be inspected including access, inspection thresholds and intervals, inspection level and the possible application of water repelling coating or corrosion preventive compound.

The FAA requires all operators to have a reporting system in their corrosion prevention and control programs. When the corrosion inspection is completed operators must report any level 2 and level 3 findings to the manufacturer of the aircraft. These results are used by the manufacturers and authorities to reconsider inspection intervals and other requirements in order to keep corrosion findings to level 1 or better. For classification of the corrosion findings three corrosion levels were defined as summarized below:

Level 1 corrosion:
- local corrosion within allowable limits
- local corrosion outside of allowable limits, but not typical for other aircraft in the same fleet
- light corrosion exceeding allowable limits due to cumulative blend–out

Level 2 corrosion:
- corrosion requiring repair or (partial) replacement of the primary structural component
- widespread corrosion requiring blend–out approaching the allowable limits

Level 3 corrosion:
- corrosion that is a potential urgent airworthiness concern requiring expeditious action.

Level 2 corrosion can be either local or widespread. The limits for local corrosion are defined in the relevant Structural Repair Manual (SRM) for the primary structure, e.g. for the skin of the pressurized fuselage 5 percent of the original skin thickness are allowed to be reworked for riveted areas and 10 percent are allowed for unriveted areas.

Corrosion is defined as widespread where an interaction between adjacent damages can not be excluded.

Protection of Repair Parts and Areas

The application of an adequate protection system is one of the most important task to prevent corrosion occurrence on metallic structures. This applies to repair parts, such as doublers, and to repair areas on aircraft before assembly. Unfortunately several original treatments cannot be used 'in situ' due to different reasons, e.g. if bath immersion is required.

The chromic acid anodizing (CAA) and sulphuric acid anodizing (SAA) are galvanic procedures which increase the corrosion resistance of the material significantly. This treatment is only applicable for detachable repair parts, not assemblies. For the reworked aluminium structure of the aircraft either the chemical conversion coating (known as ALODINE) or wash primer may be used, since it can be applied easily, e.g. by brush.

The application of adequate sealant is required similarly as for the aircraft assembly.

CONCLUSION

The corrosion resistance of the aircraft is the result of adequate design, manufacturing, maintenance and repair accomplishment. Therefore the manufacturer and the operators are mainly responsible, but so also are the airworthiness authorities with respect to the regulations, the maintenance programs and the conclusions from in–service experience.

The Airbus concepts presented above lead to a sufficient corrosion resistance verified by evaluation of in–service results.

The effect of the environment on the fatigue life of the major joints is insignificant as long as the protection system is intact. The crack growth rate of current and future aluminium alloys are significantly influenced by the environment, the test frequency and the temperature as shown by the latest Daimler–Benz Aerospace Airbus coupon tests. Therefore further research and a standardization of test procedures (stress–time histories, frequencies, temperatures) is necessary to allow an adequate material selection with respect to damage tolerance behavior. The current standard crack growth tests with high frequency and at room temperature may lead to decisions which are not in line with the conclusions from more realistic tests.

REFERENCES

Müller, R.P.G. (1995). An experimental and analytical investigation on the fatigue behaviour of fuselage riveted lap joints. *Ph.D. thesis*.

Thomas, B. and M. Perras (1995). Comparison between automatic and hand riveted 6013 AeroLock rivets. *Society of Automotive Engineers, Inc.*, 952168, 11–18.

INFLUENCE OF FREQUENCY ON FATIGUE CRACK PROPAGATION BEHAVIOR OF ALUMINUM ALLOYS IN AGGRESSIVE ENVIRONMENT

I. TROCKELS, A. GYSLER, and G. LÜTJERING

Technical University Hamburg-Harburg
21071 Hamburg, Germany

ABSTRACT

The fatigue crack propagation behavior of two medium-strength Al-alloys (6013 T6, 2024 T351) and a high-strength alloy (7475) was studied under constant amplitude loading conditions at three different testing frequencies (20 Hz, 1 Hz, 0.1 Hz) in aqueous 3.5 % NaCl solution and in vacuum (6013). The results showed a general trend for all three alloys tested in salt-water, the fatigue crack propagation rates first increased with decreasing frequency (20 Hz till 1 Hz), passed over a maximum and then decreased considerably at the lowest frequency. It is suggested that the initial increase in propagation rates with decreasing frequency resulted from the prolonged time interval in each rising load cycle for which the aggressive environment could react at slip steps, probably causing hydrogen embrittlement, while the significantly decreasing propagation rates at the lowest frequency are thought to be due to the formation of a protective oxide layer by repassivation as a consequence of the long load rising time. Concomitant fracture surface observations from specimens tested in salt-water and in vacuum are supporting this argumentation.

KEYWORDS

Fatigue crack propagation; Al-alloys; frequency effects; NaCl solution; repassivation

INTRODUCTION

The influence of loading frequency on the fatigue crack propagation behavior in aggressive environments (e.g. salt water) was studied extensively in the past, including different alloy systems such as for example Al-alloys (1,2), Ti-alloys (3,4), and steels (5). The results exhibited a general tendency for increasing crack propagation rates with decreasing frequency for tests in salt-water, the amount of degradation in crack resistance being however dependent on the frequency interval and on the ΔK-regime for different alloys. For example, only a small increase in growth rate was found for the Al-alloy 2219 by decreasing the loading frequency from 4 Hz to 0.017 Hz in salt-water, while within the same ΔK-regime for the alloy 7079 an increase in propagation rates of at least two orders of magnitude was observed with regard to the same frequency variation (2).

The present investigation was stimulated by preliminary results (6), which exhibited unexpected differences in the fatigue crack propagation behavior of two medium strength Al-alloys (2024 and 6013) and a high strength alloy (7475) for constant amplitude tests in salt-water at high and very low testing frequencies.

EXPERIMENTAL PROCEDURE

The tests were performed on three commercial Al-alloys (in wt.%): 6013 T6 (Al-0.95Mg-0.75Si-0.9Cu-0.35Mn), 2024 T351 (Al-4.4Cu-1.5Mg-0.6Mn), and 7475 (Al-5.8Zn-2.3Mg-1.68Cu-0.23Cr) in an underaged condition (UA: 24h 100°C). The alloys 6013 and 2024 were supplied as 1.6 mm thick sheet material, while 7475 was a 30 mm thick rolled plate. All mechanical tests were carried out at room temperature, the loading axis being parallel to the long transverse direction (T) for 6013 and 2024, and parallel to the rolling direction (L) for 7475.Tensile properties were determined in air on rectangular test specimens (8 mm width, 40 mm length) for 6013 and 2024, and on cylindrical specimens (20 mm gage length, 4 mm diameter) for 7475, using an initial strain rate of 8×10^{-4} s^{-1}.

Constant amplitude fatigue crack propagation tests were carried out on CCT specimens (30 mm width, 60 mm length) for 6013 and 2024, and on CT specimens (8 mm thick, 32 mm wide) in L-T orientation for 7475 under load control at $R = 0.1$ (sinusoidal wave form) using a servohydraulic testing machine. The loading frequency was varied from 20 Hz to 1 Hz and to 0.1 Hz. The aggressive environment consisted of an aqueous 3.5 % NaCl solution with an inhibitor of 0.3 % $Na_2Cr_2O_7$ + 0.2 % Na_2CrO_4 in order to prevent the fracture surfaces from local pitting. These tests were performed under open circuit conditions at the free corrosion potential. A few additional fatigue crack growth measurements were carried out in vacuum. The crack extension was monitored by a travelling light microscope, where for the CCT specimens an average length was calculated from the two cracks. Crack closure was measured by strain gages mounted in front of the propagating fatigue cracks for CCT specimens or by conventional back face strain technique (CT).

EXPERIMENTAL RESULTS

The three alloys exhibited a pancake grain structure with average dimensions along the L-, T- and S-directions of 80x60x20 µm for 6013, 25x20x10 µm for 2024, and 1500x250x30 µm for 7475. Alloy 6013 in the peak-aged T6 condition is known to be hardened by coherent and semi-coherent precipitates, while for the underaged conditions of 2024 T3 and 7475 only coherent precipitates are present. The tensile properties of these three alloys are summarized in Table 1.

Table 1. Tensile properties

Alloy	Test-Direction	$\sigma_{0.2}$ [MPa]	UTS [MPa]	σ_F [MPa]	TE [%]	ε_F
6013-T6	T	345	395	490	14	0.34
2024-T351	T	330	490	645	20	0.32
7475-UA*	L	460	580	700	17	0.20

*from ref. (7)

The fatigue crack propagation curves for 6013 in salt-water are shown in Fig. 1 for the three different frequencies (20 Hz, 1 Hz, and 0.1 Hz), together with corresponding reference curves obtained in vacuum at two frequencies (20 Hz, 0.1 Hz). A reduction in loading frequency in salt-water from 20 Hz to 1 Hz resulted in a significant increase in crack propagation rates in the intermediate ΔK regime between 7 and 20 MPa·m^{½}. However, by further decreasing the frequency to 0.1 Hz the propagation rates dropped drastically even below the curve obtained for 20 Hz (Fig. 1). At low and high ΔK values all three curves approached each other. On the other hand, the crack propagation curves measured in vacuum at the high and the low frequency coincided, the propagation rates being below those in salt-water. At high ΔK values (above 20 MPa·m^{½}) the curves from tests in vacuum and in salt-water approached each other.

The crack propagation curves for 2024 in salt-water are shown in Fig. 2. Reducing the loading frequency from 20 Hz to 1 Hz this alloy also revealed an increase in propagation rates, however less

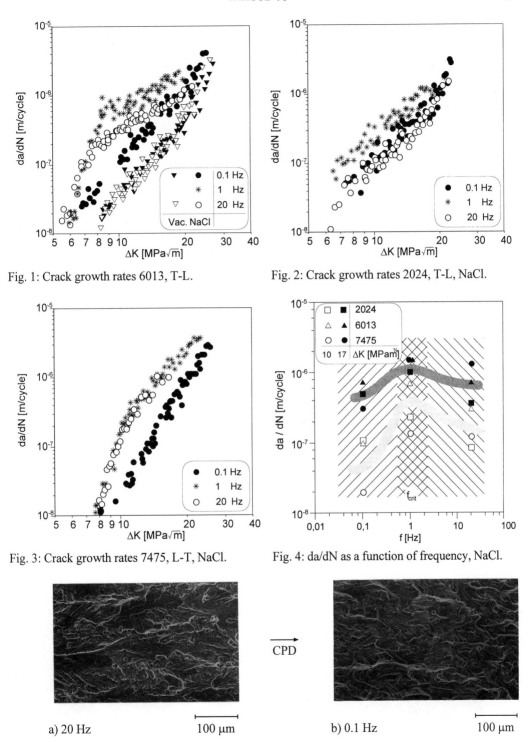

Fig. 1: Crack growth rates 6013, T-L.

Fig. 2: Crack growth rates 2024, T-L, NaCl.

Fig. 3: Crack growth rates 7475, L-T, NaCl.

Fig. 4: da/dN as a function of frequency, NaCl.

a) 20 Hz   100 μm

CPD

b) 0.1 Hz   100 μm

Fig. 5: 6013, NaCl, $\Delta K = 11$ MPa·m$^{1/2}$, Fracture surfaces (SEM).

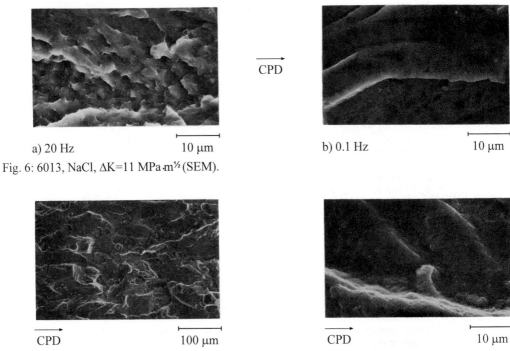

a) 20 Hz   10 μm   b) 0.1 Hz   10 μm

Fig. 6: 6013, NaCl, ΔK=11 MPa·m½ (SEM).

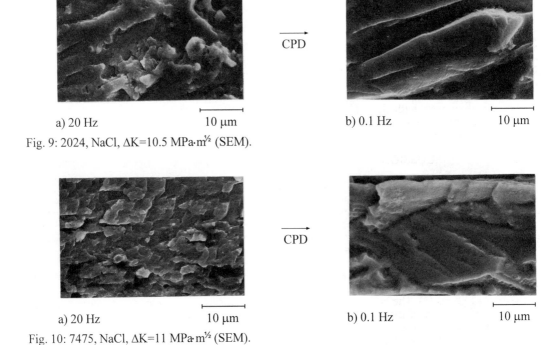

CPD   100 μm   CPD   10 μm

Fig. 7: 6013, NaCl, ΔK=18 MPa·m½, 0.1 Hz (SEM).   Fig. 8: 6013, Vac., ΔK=10 MPa·m½, 20 Hz (SEM).

a) 20 Hz   10 μm   b) 0.1 Hz   10 μm

Fig. 9: 2024, NaCl, ΔK=10.5 MPa·m½ (SEM).

a) 20 Hz   10 μm   b) 0.1 Hz   10 μm

Fig. 10: 7475, NaCl, ΔK=11 MPa·m½ (SEM).

pronounced as compared to alloy 6013. A further reduction in frequency to 0.1 Hz again resulted in a drop of the crack propagation rates below those for 1 Hz, but not below the crack growth curve obtained for 20 Hz (Fig. 2).

The corresponding fatigue crack propagation results in salt-water for 7475 are summarized in Fig. 3. The frequency reduction from 20 Hz to 1 Hz did not significantly affect the crack propagation behavior, both curves lying on top of each other. However, with further decreasing the frequency to 0.1 Hz again a pronounced decrease in propagation rates was observed.

A summary of the effect of frequency on the fatigue crack propagation rates at two ΔK levels (10 and 17 MPa·m$^{1/2}$) is given in Fig. 4 for all three alloys, indicating that the resistance against crack propagation in salt-water showed a minimum at the intermediate frequency of 1 Hz.

It should be noted that the crack closure values obtained from specimens tested in salt-water were very low ($K_{cl}/K_{max} \approx 0.13$) and almost identical for all three alloys, regardless of testing frequency.

Fracture surface studies revealed exclusively transgranular fracture modes for all three alloy systems, independently of loading frequency and of environment. Examples of fracture surfaces for 6013, tested in salt-water with the highest and lowest frequency, are shown in Figs. 5 (overview) and 6 (higher magnification) for the ΔK regime where the most significant differences in growth rates were observed (Fig. 1). The fracture surfaces from tests at the higher frequencies (20 Hz, 1 Hz) exhibited a fairly brittle appearance (Fig. 5a) with rather large flat areas elongated along the growth direction, exceeding the grain dimension in L-direction. In contrast, the corresponding fracture surface from tests at the low frequency of 0.1 Hz showed a slightly rougher, however more ductile appearing topography exhibiting smooth facets (Fig. 5b). At higher magnification the elongated flat areas in Fig. 5a revealed much smaller flat regions of a few μm in size (Fig. 6a), whereas the fracture surface of the specimen tested at the low frequency (0.1 Hz) exhibited a rather smooth appearance (Fig. 6b). At high ΔK-values, where the da/dN-ΔK curves for all three loading frequencies were approaching each other (Fig. 1), the corresponding fracture surface topographies were found to be nearly identical, an example is shown in Fig. 7.

The fracture surface appearance of specimens tested in vacuum at the highest and lowest frequency was similar, as shown in Fig. 8, resembling approximately the topography obtained for specimens tested at an equivalent ΔK in salt-water with the lowest frequency (compare Figs. 6b and 8).

The fracture surfaces of the two other alloys, 2024 and 7475, exhibited qualitatively a similar characteristic variation with regard to tests in salt water at the three different loading frequencies as found for 6013. The elongated flat areas, typically observed on fracture surfaces from tests at high and intermediate frequencies (20 Hz, 1 Hz), also exhibited small flat regions as found for 6013, somewhat less pronounced for 2024 (Fig. 9a) but very similar for 7475 (Fig. 10a). The fracture surface appearance from tests with the low frequency of 0.1 Hz was nearly identical for all three alloy systems, as can be seen by comparing Fig. 6b (6013) with Fig. 9b (2024) and Fig. 10b (7475).

DISCUSSION

The results of the present study will be discussed mainly for the intermediate ΔK regime where a tendency for all three alloys was found that with a decrease in testing frequency the fatigue crack propagation rates in salt-water first increased, passed over a maximum and then decreased significantly (Figs. 1 to 4). The increasing growth rates with a frequency reduction from high (20 Hz) to intermediate values (1 Hz) are generally explained by the increasing time within the rising part of each load cycle, permitting the aggressive aqueous NaCl solution to react for increasing time periods with the bare metal produced at the crack tip by slip bands shearing off the protective oxide layer. The underlying mechanism for the observed concomitant crack growth increase is thought to be hydrogen embrittlement, a process which is generally accepted to occur not only in steels (5) and Ti-alloys (3), but also in Al-alloys (2). The hydrogen atoms formed at the unprotected slip steps are transported into the plastic zone ahead of the crack tip by moving dislocations (8), thus lowering the local fracture stress within slip or cleavage planes. A previous study (7) on 7475 has shown, by applying an etch pit technique, that the fatigue fracture surface obtained from tests at 30 Hz in salt-water consisted partly of {111} slip planes and partly of {100} cleavage planes, having an identical appearance as shown in Fig. 10a of the present study. Crack propagation is thought to occur along these locally embrittled planes, changing frequently from one plane to the other (Figs. 6a, 9a, 10a), which allows the overall crack front to maintain a rather straight profile resulting in the flat fracture surface appearance shown in Fig. 5a.

The pronounced decrease in propagation rates with a further reduction in testing frequency from 1 Hz to 0.1 Hz (Figs. 1 to 4) is thought to result from an increasing contribution of repassivation forming a protective film on the emerging slip steps as a consequence of the longer rise time in each load cycle (5 sec). A similar frequency dependent crack propagation behavior, as found in the present study for Al-alloys, was also observed for steels (5), where those authors discussed the decreasing growth rates with decreasing testing frequency also on the basis of repassivation, but in addition with regard to variations in crack tip blunting and crack closure. However, the very low and frequency independent closure levels found in the present study for Al-alloys are not in favor of a closure dominated process.

The observed significant changes in fracture surface topography from a rather brittle mode for tests at 20 Hz and 1 Hz (Figs. 5a, 6a, 9a, 10a) to a fairly ductile appearance at 0.1 Hz (Figs. 5b, 6b, 9b, 10b), similar to that one obtained in vacuum (Fig. 8), are supporting repassivation as playing an important role.

The tendency of the crack propagation curves for the three testing frequencies to approach each other in the high ΔK regime is thought to result from the large crack extension per load cycle, which obviously reduced the detrimental environmental attack significantly.

Although all three alloys exhibited qualitatively a rather similar crack propagation behavior in salt-water at the three different frequencies, there exist still some differences. For example, the increase in crack propagation rates with a reduction in frequency from 20 Hz to 1 Hz varied for the three different alloys, the highest resistance against environmental assisted crack propagation within this frequency range was observed for 7475 (Fig. 3), followed by alloy 2024 (Fig. 2), while 6013 exhibited an inferior behavior (Fig. 1). Furthermore, the crack propagation rates for tests at the low frequency of 0.1 Hz as compared to those at 20 Hz were considerably lower for 7475 (Fig. 3) and 6013 (Fig. 2), while almost no difference was observed for alloy 2024 (Fig. 2). However these differences are not readily explained mainly because no attempt was made to determine the critical frequency for each alloy (Fig. 4) where a maximum degradation occurred in crack growth resistance, and because the repassivation times are not known for all Al-alloys investigated in the present study.

ACKNOWLEDGEMENT

The authors are indebted to Mr. N. Ohrloff of Daimler Benz Aerospace Airbus (DA), Hamburg, for provision of the sheet material and assistance in specimen preparation.

REFERENCES

1. R.J.H. Wanhill, Int. J. Fatigue, 16 (1994) 99-110.
2. M.O. Speidel, in Problems with Fatigue in Aircraft, ICAF 8, Eds. J. Branger, F. Berger, Lausanne, Switzerland (1975) Paper 2.2, 1-17..
3. D.B. Dawson and R.M. Pelloux, Met. Trans. 5 (1974) 723-731.
4. H. Döker and D. Munz, Inst. Mech. Eng. Conf. Publ. 4 (1977) 123-130.
5. J.D. Atkinson and T.C. Lindley, Inst. Mech. Eng. Conf. Publ. 4 (1977) 65-74.
6. N. Ohrloff, H.-J. Schmidt, Daimler Benz Aerospace Airbus (DA), private communication.
7. F.-J. Grau, A. Gysler, and G. Lütjering, in Aluminum Alloys, Eds. E.A. Starke, T.H. Sanders, Georgia Institute of Technology, Atlanta (1994) 709-716.
8. J.K. Tien, A.W. Thompson, I.M. Bernstein, and R.J. Richards, Met. Trans. 7A (1976) 821-829.

ENVIRONMENTALLY ENHANCED INITIATION AND RE-INITIATION OF FATIGUE CRACKS UNDER FULLY COMPRESSIVE CYCLIC LOAD

Y. N. LENETS

Institute for Materials Research, GKSS Research Center
Max-Planck-Street, D-21502 GEESTHACHT, Germany*

ABSTRACT

The influence of 3.5% NaCl aqueous solution on fatigue performance of 7075 Al-alloy under compressive cycling was evaluated for situations where damage would never occur in laboratory air. The situations included natural crack initiation from a smooth surface and re-initiation of a crack that remained dormant under compressive cycling in air. In both cases crack extension was detected. Such behavior conforms to the earlier proposed mechanism whereby, in a corrosive environment, tensile stresses ahead of a compression fatigue crack arise due to the crack tip blunting caused by anodic dissolution process. Moreover, this mechanism also appears to be applicable to surface flaws, inasmuch as it provides reasonable explanation for experimentally observed initiation of cracks from corrosive pits.

From an engineering point of view, it seems plausible that environmentally enhanced compression fatigue damage may occur in structures even if they do not originally contain local stress-concentrators.

KEYWORDS

7075 Al; corrosive pitting; crack tip blunting; cyclic compressive load; fatigue; residual stress.

BACKGROUND

Detrimental consequences of fully compressive cyclic load were reported long ago (Katz, 1965, Gerber and Fuchs, 1968). Subsequent studies (Hubbard, 1969, Suresh, 1985, Suresh et al., 1986, Pippan, 1987, Romaniv et al., 1987) have attributed this phenomenon to the dislocation plasticity that develops tensile residual stress associated with local stress-concentrators. In the above experiments, "notch-dependent" compression fatigue cracks grew at progressively decreasing velocity until being completely arrested at a certain saturation size. Therefore, from an engineering standpoint the phenomenon seemed to be non-catastrophic and thus has hitherto evoked rather little practical interest.

Recent studies (Lenets, 1995a, 1995b) have revealed an environmental sensitivity of the compression fatigue crack growth in metallic alloys. In the previous tests only notched specimens were utilized, so that a crack initiation stage was primarily governed by the residual stress field associated with a machined notch. Results obtained that way do not allow one to judge convincingly the role of environment in fatigue crack initiation under cyclic compression. Therefore, one of the objects of the present study was to investigate the influence of corrosive environment on fatigue crack initiation from a smooth surface.

---

*Present address: WL/MLLN, Wright-Patterson AFB, OH 45433, USA.

In contrast to the generally known saturation behavior in air, continual propagation of the compression fatigue cracks was observed in an aluminum alloy tested at low load frequency in 3.5% NaCl solution (Lenets, 1995a, 1995b). Such behavior was rationalized in terms of anodic dissolution of material immediately adjacent to the crack front. According to the proposed mechanism (Lenets, 1995a), corrosion results in fatigue crack blunting and concomitant generation of tensile residual stresses ahead of the blunted crack, provided that the latter is subjected to compressive cycling. The mechanism implies that merely the presence of a fatigue crack may be sufficient for further damage to occur under the conjoint action of fully compressive load and corrosive environment. To examine such a possibility, compression fatigue crack re-initiation tests were also conducted on specimens containing dormant cracks of different sizes.

EXPERIMENTAL

All specimens used in the present study were taken from a rolled plate of 7075 Al-alloy and stress relieved for 2 hours at 190°C after machining. Resulting tensile properties were $\sigma_y = 310$ MPa and σ_{UTS} = 380 MPa. The microstructure consisted of elongated grains of 200 by 50 μm. Hour-glass shaped cylindrical specimens of a nominal diameter, $d = 6$ mm, were subjected to grinding at final stage of the machining process. Some of them were additionally circumferentially notched (notch depth = 0.5 mm; angle = 60°; tip radius = 0.1 mm). More details on specimen geometry and test set-up are available elsewhere (Lenets, 1995a, c).

A resonance testing machine was used for specimen pre-cracking in air at 110 Hz. A servo-hydraulic Schenck system of 10 kN capacity was used for fatigue crack initiation and re-initiation tests in air and in 3.5% NaCl. The solution was dripped upon the gage section of specimen with a frequency of about 1 Hz. The load ratio, $R = 20$, and the minimum nominal stress, $\sigma_{min} = -232.5$ MPa $= -0.75 \cdot \sigma_y$, were maintained constant for each specimen.

For crack initiation, unnotched specimens were fatigued to different number of load cycles in a certain environment. Several specimens were subjected to corrosive environment without any loading. The duration of the exposure was equal to the duration of the fatigue crack initiation tests which were conducted at 2 Hz. After each crack initiation test, the specimen's surface was examined in SEM. For crack re-initiation tests, notched specimens were pre-cracked in air for $N \geq N_s$ cycles, where N_s = 5.0×10^5, is a number of compressive load cycles required for the crack to reach saturation depth under given testing conditions (Lenets, 1995a). Pre-cracked specimens were subjected to additional compressive cycling in air or 3.5% NaCl solution at 2 or 20 Hz and unchanged value of σ_{min}. Several crack re-initiation tests were duplicated under identical parameters to check for reproducibility. After mechanical testing and (if appropriate) surface examination, the specimens were opened in tension (in case of unnotched specimens, liquid nitrogen was used) and fracture surfaces were examined using Philips SEM 505. The crack depths reported below represent average values of at least 12 measurements.

RESULTS

Crack initiation

Although no damage was expected to occur on a smooth surface due to compression fatigue in air, two unnotched specimens were fatigued under such conditions for 2.0×10^5 and 2.0×10^6 cycles, respectively. Subsequent scrupulous scrutiny in SEM failed to reveal any discernible changes in surface morphology which could be treated as a consequences of fatigue damage.

Under stress-free influence of 3.5% NaCl solution, surface degradation of microscopic scale was already noticeable after 28 hours (this duration corresponds to an application of 2.0×10^5 cycles at 2 Hz). The surface grew dim and separated corrosive pits were formed locally. In general, the pits had a uniform shape with aspect ratio of about unity. Maximum size of these pits on the surface ranged from 5

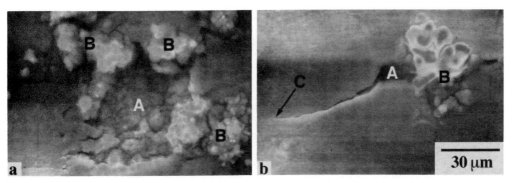

Fig. 1. Surface degradation in 3.5% NaCl solution after (a) 280 hours under
stress-free condition and (b) 28 hours under cyclic compression.
A - corrosive pit; B - corrosion deposits; C - fatigue crack tip.

to 20 μm. After 280 hours (corresponding to $2.0×10^6$ cycles at 2 Hz), darker spots on the surface could
be observed with the naked eye. They had different intensity and almost circular shape with an average
size ranging from 0.5 to 4.0 mm. Closer examination revealed two types of a change in surface
morphology associated with such spots (Fig. 1a). The first type was associated with corrosive pits (A)
similar to those described above, but of increased size varying between 50 and 100 μm. The second one
was represented by corrosion product deposits (B) on the surface which otherwise seemed unchanged.

A different type of damage has been induced by the corrosion process in the presence of alternating
stresses. Evidently due to preferential crevice corrosion, most of the pits detected after $2.0×10^5$ cycles
had an elongated shape with the longer dimension oriented perpendicular to the load direction. Several
such pits had served as fatigue crack initiation sites (Fig. 1b). It is seen that originally the crack was
inclined at about 45° to the load direction. However, as a result of several successive kinks, it eventually
assumed another orientation, with final crack growth direction on the specimen surface being
perpendicular to the applied load. The length of the cracks on the surface ran up to 70 μm. Their depth,
however, could not be estimated since none of them had been found in the plane of final fracture when
the specimen was eventually opened in tension. Much longer surface cracks (up to 500 μm) were found

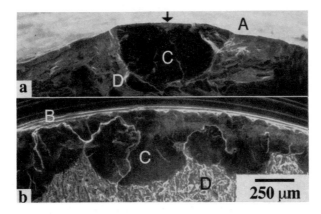

Fig. 2. Fracture surfaces obtained in (a) smooth and (b) circumferentially
notched cylindrical specimens after 10^6 compressive cycles in
3.5% NaCl solution at 2 Hz and $\sigma_{min} = -0.75·\sigma_y$. A - surface;
B - notch; C - fatigue crack; D - final fracture.
An arrow indicates location of the initial corrosive pit.

after 2.0×10^6 cycles. One such crack was completely revealed after final fracture of the specimen and can be seen in Fig. 2. Also shown for comparison is a fracture surface obtained under identical load and environmental conditions, but with a circumferentially notched specimen (Lenets, 1995a).

Crack re-initiation

Final crack depths obtained after the crack re-initiation tests in air are summarized in Table 1. Specimens 1 and 2 were broken open immediately after pre-cracking to provide a basis for comparison. An average value of the pre-crack depth (91 μm) obtained for them was in a good agreement with the saturation crack depth reported earlier for the same specimen geometry and load conditions (Lenets, 1995a). In view of these findings, it was assumed that the remaining specimens also contained pre-cracks of a similar size. It can be seen from the results of Table 1 that no increase in total crack depth was detected after crack re-initiation tests in air, irrespective of the number of additional compressive cycles and their frequency (specimens 3 through 6).

Table 1. Results of compression fatigue crack re-initiation tests in laboratory air.

Specimen's number	Number of additional cycles	Load frequency, Hz	Total crack depth, μm
1	0	-	90
2	0	-	92
3	5.0×10^5	2	90
4	5.0×10^5	20	89
5	1.0×10^6	2	91
6	1.0×10^6	20	92

In contrast, additional cycling in corrosive environment resulted in noticeable crack extension (Table 2). A crack depth increment has been calculated by subtracting the average (assumed) pre-crack depth from the total crack depth. The increment increased with the number of additional compressive cycles. For identical number of additional compressive cycles, the increment also increased as load frequency decreased.

Table 2. Results of compression fatigue crack re-initiation tests in 3.5% NaCl solution.

Specimen's number	Number of additional cycles	Load frequency, Hz	Total crack depth, μm	Crack depth increment, μm
7	5.0×10^5	2	139	48
8	5.0×10^5	2	135	44
9	5.0×10^5	20	98	7
10	5.0×10^5	20	96	5
11	1.0×10^6	2	186	95
12	1.0×10^6	20	108	17
13	1.0×10^6	20	105	14

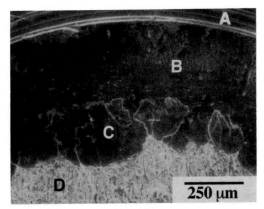

Fig. 3. Fracture surface of a specimen pre-cracked with periodic
overloads obtained after 10^6 compressive cycles in
3.5% NaCl solution at 2 Hz and $\sigma_{min} = -0.75 \cdot \sigma_y$.
A - notch; B - pre-crack; C - re-initiated fatigue crack;
D - final fracture.

To obtain a pre-crack of a greater depth, periodic tensile-compressive overloads with absolute extreme values not exceeding $0.9 \cdot \sigma_y$ were applied to the specimen during otherwise fully compressive loading in air. This technique (Lenets, 1995c) provided a pre-crack of an average depth of 300 μm. Subsequent application of 10^6 constant amplitude compressive load cycles in corrosive environment resulted in crack re-activation and propagation, so that the final crack depth was about 420 μm (Fig. 3). The fracture surface formed during pre-cracking (area B) and that formed by propagation of the re-activated crack (area C) can be easily distinguished by their appearance.

DISCUSSION

As expected, no detrimental influence has been detected during the tests conducted in laboratory air. In case of corrosive environment, however, fully compressive cycling resumed propagation of dormant pre-cracks. Two different sizes of the pre-cracks have been studied, one of them being definitely greater than the residual stress field associated with the notch. Under such conditions, resumption of fatigue crack growth cannot be explained by residual tensile stresses caused by a local stress-concentrator (Hubbard, 1969). Therefore, the results obtained support the earlier proposed mechanism (Lenets, 1995a, b) whereby, in corrosive environment, tensile stresses ahead of a compression fatigue crack arise due to the crack tip blunting via anodic dissolution process. As a result, the presence of any geometrical discontinuity (*e.g.*, a fatigue crack, even though theoretically completely closed and practically initially dormant) is sufficient for further damage under the conjoint action of fully compressive load and corrosive environment.

The mechanism (Lenets, 1995a, b) also provides a reasonable explanation for experimentally observed initiation of cracks from smooth surface. In this case, damage starts as corrosion pitting. A minor influence of mechanical stresses manifests itself in the shape of corrosive pits (compare Figs. 1a and b). As soon as these surface defects become large enough to concentrate mechanical stresses, microcracks initiate and grow to the boundary of the tensile residual stress field associated with corrosive pits. At this stage, in any non-aggressive atmosphere (*e.g.*, laboratory air) the existing microcracks would become non-propagating. In corrosive environment, however, the damage process is maintained by contributions of chemical and mechanical origin which complement each other and keep the crack growing. The overall view of the surface crack (Fig. 2a) is quite similar to that of individual convex-shaped segments constituting the crack initiated from notch (Fig. 2b). Moreover, the extension of both cracks in the main propagation direction is almost the same, indicating that subsequent crack growth was insensitive to the nature of initial defect (corrosive pit or fatigue crack) as well as to its shape and size.

In general, the results obtained support the assumption (Lenets, 1995a, b) that in corrosive environment, compression fatigue damage may attain a continuous character. From a practical standpoint this implies that final (catastrophic) fracture of engineering components may arise from fatigue damage induced by fully compressive cyclic load. It is important that under the conjoint action of cyclic compression and corrosion, such damage can initiate from any geometrical discontinuity or even from a smooth surface.

CONCLUSIONS

Compression fatigue crack resistance of 7075 Al-alloy during crack initiation and propagation stages was studied in laboratory air and 3.5% NaCl aqueous solution. The conclusions derived are as follows:

1. In corrosive environment, fully compressive cycling far below general yielding may initiate a crack from a smooth specimen surface as well as resume propagation of a dormant crack-like defect.

2. The results support the earlier proposed mechanism whereby, in a corrosive environment, tensile stresses ahead of a compression fatigue crack arise due to the crack tip blunting caused by anodic dissolution process.

3. This mechanism appears to be also applicable to the behavior of surface flaws, inasmuch as it provides an explanation for experimentally observed initiation of compression fatigue cracks from a smooth surface.

4. Due to the possible catastrophic character of environmentally enhanced compression fatigue, the phenomenon may represent a problem for metallic structures even if they do not originally contain local stress-concentrators.

ACKNOWLEDGMENT - The author is most grateful to Prof. K.-H. Schwalbe of GKSS Research Center, Geesthacht and Prof. J.K. Gregory of Martin Luther University, Merseburg for providing the opportunity to conduct this work and continual encouragement.

REFERENCES

Gerber, T.L. and H.O.Fuchs (1968). Analysis of nonpropagating fatigue cracks in notched parts with compressive mean stresses. *Journal of Materials, JMLSA,* **2,** 359-374.

Hubbard, R.P. (1969). Crack growth under cyclic compression. *J. Basic Engng, Trans. ASME.* **91,** 625-631.

Katz, J. (1965). Exploratory experiments in the effect of mean stress on the fatigue life of notched parts. *M.S. thesis.* University of California - Los Angeles, .

Lenets, Y.N. (1995a). Compression fatigue crack growth behaviour of metals: effect of environment. Submitted to *Egng Fract. Mech.*

Lenets, Y.N. (1995b). Environmentally assisted fatigue damage of metallic materials under fully compressive cyclic loading. *Book of Abstracts, 7th International Conference on Mechanical Behaviour of Materials* (A. Bakker, Ed.), Delft: Delft University Press, pp.715-716.

Lenets,Y.N. (1995c). Compression fatigue crack growth behaviour of metals: effect of overloading. Submitted to *Fatigue and Fract. of Egng Mater. Struct.*

Pippan, R. (1987). The growth of short crack under cyclic compression. *Fatigue and Fract. Engng Mater. Struct.,* **9,** 319-328.

Romaniv, O.N., Tkach, A.N. and Y.N. Lenets (1987). Fatigue crack propagation under compression and its application for fatigue crack growth resistance evaluation. *Physicochemical mechanics of materials,* **6,** 57-63. (In Russian).

Suresh, S. (1985). Crack initiation in cyclic compression and its application. *Engng Fract. Mech.,* **21,** 453-463.

Suresh, S., Christman, T. and C. Bull (1986). Crack initiation and growth under far-field cyclic compression: theory, experiments and application. *Small Fatigue Cracks* (R.O.Ritchie & J.Lankford, Eds.), Warrendale, pp.513-540.

SIMULTANEOUS EVOLUTION OF SURFACE DAMAGE AND OF CURRENT AND POTENTIAL TRANSIENTS DURING CORROSION FATIGUE OF ZIRCALOY-4

Jacques STOLARZ

Ecole Nationale Supérieure des Mines, Centre SMS, URA CNRS 1884
158, cours Fauriel
42023 SAINT-ETIENNE CEDEX 2
FRANCE

ABSTRACT

The corrosion fatigue damage of the annealed h.c. Zircaloy-4 in methanolic solutions is analysed using the simultaneous observation of the surface microcracking process and of the cyclic evolution of current and/or potential transients. The method is successfully applied to detect different modes of crack initiation and evolution and to study the influence of surface treatment of Zircaloy-4 on corrosion fatigue damage. The method prooves the electrochemical character of reactions operating at the metal surface during cyclic straining in iodated methanol.

KEYWORDS

Zircaloy-4; low cycle fatigue; corrosion fatigue; microcracking,;cyclic current and potential transients.

INTRODUCTION

Zircaloy-4 cladding of the fuel rods in pressurized water reactors (PWR) is subjected to cyclic strains in a iodine containing environment leading to a strong damage through stress corrosion cracking and/or corrosion fatigue. Previous researches have shown that the behaviour of Zircaloy-4 subjected to free iodine at 350°C is effectively reproduced during SCC or corrosion fatigue tests carried out in iodated methanol at ambient temperature (Cox, 1972). However, controversy persists about the nature of the metal-solution interaction (chemical or electrochemical) leading to SCC or corrosion fatigue of Zircaloy-4 in iodated methanol (Cox and Wood, 1974).

The method of detection of the surface corrosion fatigue damage from the potential or current transients was developed by Magnin (Magnin, 1987) and it was mainly applied to the studies of the aqueous corrosion fatigue of passive metals and alloys (Magnin, 1995). This very sensitive method applied to a metal-solution system in which the nature of surface reactions during cyclic straining is not known can be supposed to give supplementary informations about corrosion fatigue damage processes at the metal surface.

The aim of this study is to analyze the evolution of electrochemical parameters (potential and current) during low cycle fatigue tests on Zircaloy-4 in methanol with or without iodine additions and to correlate the results with the observations of the metal surface at different stages of the fatigue process.

EXPERIMENTAL METHOD

All tests were carried out on the annealed (505°C/1h) industrial Zircaloy-4 (Zr - A) supplied by FRAMATOME. Some annealed specimens were preoxydized (Zr - Ox) before fatigue testing.
Symmetrical tension-compression tests were performed on a servo-hydraulic machine on smooth specimens (ø 6 mm) under plastic strain control ($\Delta\varepsilon_p/2 = \pm 2 \cdot 10^{-3}$) and at constant strain rate ($d\varepsilon/dt = 10^{-3}$ s^{-1}) under vacuum ($< 10^{-3}$ Pa), in methanol and in iodated methanol (iodine concentration: $10^{-3}g/g$). One part of corrosion fatigue tests was carried out at free electrochemical potential, for other specimens the potential was fixed at its equilibrium value.
A sensitive testing equipment is used to record simultaneously the cyclic evolution of the mechanical parmeters and of the potential or current transients. The transients are recorded as a function of the applied strain and of the number of cycles. Figure 1 shows a typical evolution of current density recorded during one strain cycle at an imposed electrochemical potential for a passivated material.

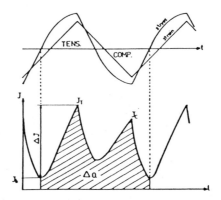

Fig. 1. Current transient characteristics at imposed potential of a passivated material

Following parameters are used to describe the pertrurbations of the metal-solution interface :
-      J_T (or E_T) and J_C (or E_C) - peak values of the current density (or potential) respectively in tension and in compression,
-      J_b - the base current density which reflects the overall activity of the surface between strain reversals,
-      ΔE_T - the difference between E_T and E_b which characterizes the maximal depolarisation within one strain cycle and the evolution of the depolarisation with the number of cycles N.

RESULTS

Fatigue life and crack propagation mode

Results of low cycle fatigue tests are presented in Table 1. In spite of differences between fatigue life observed under vacuum, in ambient air and in methanol, the microcrack nucleation and propagation modes remain the same in all these cases, e.g. intragranular nucleation on slip lines and mainly transgranular microcrack evolution through coalescence and growth of existing microcracks with simultaneous creation of new cracks (Fig. 2a). In all these cases, the nucleation of first surface microcracks with the length which does not exceed the main grain diameter (<25μm for Zircaloy-4), occurs at about 25% N_R. In view of this result, it can be assumed that potential or current transients recorded during fatigue tests in methanol without iodine additions represent the state of surface damage of a sample under experimental conditions where no corrosion fatigue phenomena occur. On the other hand, the results of fatigue tests in iodated methanol show not only a very important decrease of fatigue life, measured by the ratio N_{R-V}/N_{R-IM}, but also a changement of crack nucleation and propagation mode into exclusively intergranular (Fig. 2b, 2c).

Table 1.   Results of low cycle fatigue tests: number of cycles to failure and crack nucleation
and propagation modes

	σ_{max}	Vacuum	Air	Methanol	Iodated methanol (IM)	N_{R-V} N_{R-IM}
Zr - A	520 MPa	4000 (Tr)	2900 (Tr)	2700 (Tr)	220 (IG)	0,056
Zr - Ox	520 MPa	4000 (Tr)	-	-	>500 (IG)	>0,125

Crack nucleation and propagation modes:
(Tr)   - intragranular crack nucleation and mainly transgranular propagation,
(IG)   - intergranular nucleation and propagation.

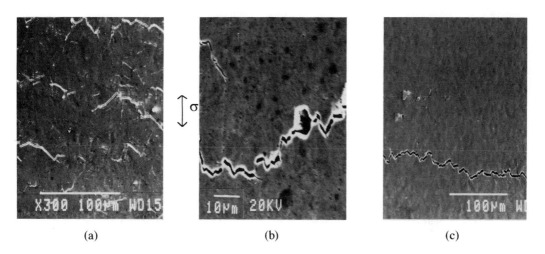

(a)   (b)   (c)

Fig. 2.   (a) Transgranular crack propagation in Zircaloy-4 under vacuum at 86% N_R
(b) Intergranular crack propagation in Zircaloy-4 in iodated methanol at 86% N_R
(c) Intergranular crack propagation in the preoxydized Zircaloy in iodated methanol
at 90% N_R

In all cases, the results of low cycle fatigue tests (N_R, crack nucleation and propagation modes) are the same at free potential and at imposed equilibrium potential. Thus, it is possible to compare directly the potential and the current transients recorded for the same couple metal-environment.

Current and potential transients in methanol without iodine additions (no corrosion fatigue)

Figure 3 illustrates the current and potential transients recorded in methanol without iodine addition. The current density peaks J_T and J_C are present during the whole test. Both peak values increase during first 5 cycles, then they decrease and stabilize near to N=50 (2% N_R). Between N=50 and N=500 (25% N_R), no evolution of peak values of current is observed. In this region, the amplitude of the compression peak becomes very small compared to the tension peak. At the same time, the base current density J_b does not change which reflects a stable overall activity of the surface between strain reversals. Beyond N=500, all values increase continuously until final failure.

The evolution of the electrochemical potential during a free potential test is characterized by ΔE_T, the difference between E_T (peak potential in tension) and base potential E_b (Fig.3). The depolarisation of the metal surface is the lowest in the region of the stability of peak and base current densities (between 50 and 500 cycles). An increase of ΔE_T beyond N=500 reflects a growing electrochemical activity of the metal surface.

Observation of metal surface at different stages of the fatigue tests allows to correlate the evolution of electrochemical parameters with the surface damage of the sample. The initial growth of peak densities is related to the creation of sites at which electrochemical reactions occur. These sites are mainly intersections of slip lines with metal surface which can be observed since the first cycles at imposed plastic strain. The plastic deformation is then rapidy localized in slip bands and the overall area on which depassivation occurs becomes smaller than during first cycles. The density of slip bands remains almost constant within a large range of N which reflected by a stability of all electrochemical parameters recorded. The nucleation of first surface microcracks takes place in Zircaloy-4 at about 25%N_R (near to N=500) when no corrosion fatigue phenomena occur (Stolarz and Beloucif, 1995). This corresponds to the end of the stability region of all electrochemical parameters recorded near to 25%N_R. The evolution of surface damage progresses then rapidly through microcrack growth with simultaneaous nucleation of new microcracks. The growing amount of surface damage is reflected by the increase of electrochemical parameters. In particular, a constant increase of the base current density prooves that the repassivation process becomes more and more difficult due to a growing microcrack density and length.

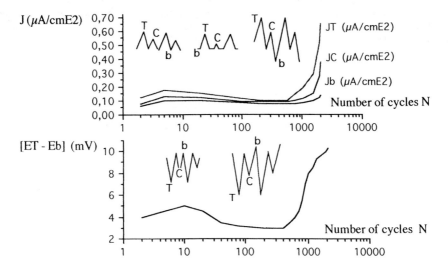

Fig.3. Evolution of current and potential transients of Zircaloy-4 in methanol

Potential and current transients in iodated methanol (corrosion fatigue)

Figure 4 represents the evolution of current density transients for annealed Zircaloy-4 tested in iodated methanol. The evolution of peak current densities is very different from the one observed in pure methanol with no corrosion fatigue effect (cf. Fig. 3). During first 15 cycles, current transients exhibit well defined tension and compression peaks. Beyond N=15, the amplitude of the tension peak starts to increase rapidly and the compression peak disappears completely for N>50. In the remaining part of the fatigue test only a tension peak is recorded.

Microscopic observations of metal surface indicate that the nucleation of first intergranular surface microcracks occurs in iodated methanol near to N=20. From this moment until the end of the test, surface damage occurs through creation of new intergranular microcracks and the growth of already existent ones (cf. Fig. 2). Thus, it seems that the modification of the shape of current transients near to N=20 can be related to the variation of the nature of surface damage from intragranular during first cycles to purely intergranular.

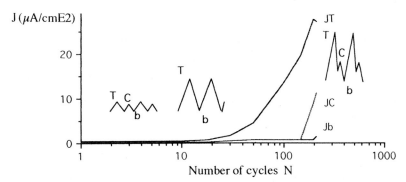

Fig. 4. Cyclic evolution of current transients in annealed Zircaloy-4 (Zr - A) in iodated methanol

Free potential corrosion fatigue test confirms the above observations (Fig. 5a). The evolution of ΔE_T, the difference between E_T (peak potential in tension) and base potential E_b reflects the progressing localisation of fatigue damage during first cycles. ΔE_T reaches then a minimal value which does not change in a significant way until first intergranular microcracks are created. For N>20, a continuous growth of ΔE_T is recorded in a similar way like for current densities.

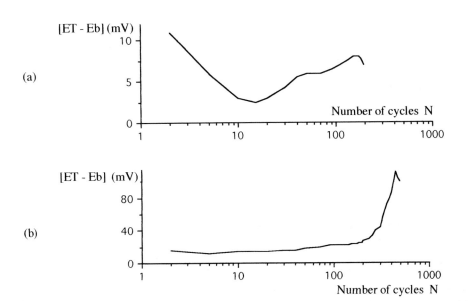

Fig. 5. Cyclic evolution of potential transients on Zircaloy-4 in iodated methanol
    (a) - annealed (Zr - A)
    (b) - annealed and oxydized (Zr - Ox)

Figure 5b illustrates the evolution of potential transients in preoxydized Zircaloy-4 (Zr - Ox). Since the corrosion fatigue resistance of this alloy is greatly improved compared to the alloy without any surface treatment (cf. Table 1), it is interesting to analyse what kind of information about fatigue damage can be obtained from the recording of electrochemical parameters.

Two distinct regions on the $\Delta E_T=f(N)$ curve can be observed: one corresponding to first 200 cycles which is qualitatively similar to the one observed on the only-annealed alloy (Zr - A) in pure methanol and another one with a strongly increasing potential amplitude ΔE_T. In this second region, the compression peak tends to disappear as it was found for Zr - A in iodated methanol (Fig. 5a).

The presence of two distinct parts on the $\Delta E_T=f(N)$ curve can be related to the state of surface damage revealed by microscopic observations. Until 200 cycles, only local damage at intersections of slip lines and of slip bands with the metal surface takes place. This kind of damage is reflected by the presence of tension anf compression peaks on the $E=f(N)$ curve as it was observed at the first stage of fatigue test in pure methanol. Crack nucleation starts after about 200 cycles on the preoxydized alloy in iodated methanol. Crack nucleation and propagation is purely intergranular and it is not surprising that potential transients recorded in this region are very similar to the ones observed on the only-annealed alloy.

CONCLUSION

The presence of iodine in methanol leads to a very important decrease of fatigue life of the annealed Zircaloy-4 through the intergranular corrosion fatigue damage. Iodine induces intergranular microcracking from the very beginning of the fatigue test with a rapid development of the intergranular damage.

Surface oxydation prior to corrosion fatigue tests improoves the resistance of the alloy to crack nucleation. Nevertheless, surface crack propagation occurs at the same rate as in only-annealed alloy. This result is fully confirmed through the analysis of current and potential transients.

The electrochemical method allows to detect different corrosion fatigue processes like localization of plastic deformation or nucleation of first surface microcracks. Moreover, it has been shown that the crack propagation mode can be determined through analysis of current and potential transients.

Finally, the presence of current transients during corrosion fatigue tests in methanolic solutions prooves the existence of anodic dissolution at the metal surface. This observation confirms the hypothesis of electrochemical character of corrosion fatigue damage of Zircaloy-4 in methanolic solutions even if the experimental method used does not allow to determine the exact nature of reactions which occur at the metal surface.

Acknowledgements   -   Thanks are owed to Mr. J. Joseph,   Société FRAGEMA, Lyon, France, for the provision of material and helpful discussions

REFERENCES

Cox, B. (1972). Environmentally induced cracking of zirconium alloys - a review.
    Journal of Nuclear Materials, **170,** 1-23.
Cox, B. and Wood J.C. (1974). Iodine induced cracking of zircaloy fuel cladding.
    In: *Corrosion Problems in Energy Conversion* (Craig. S. Tedruon Editor).
Magnin T. (1987). Simultaneous evolution of microcracks and electrochemical reactions during
    corrosion fatigue. *ICMS, Beijing, 1987.*
Magnin T. (1995). Recent advances for corrosion fatigue mechanisms.
    ISIJ International, **35,** 223-233.
Stolarz J. and Beloucif A. (1996). Low cycle fatigue of Zircaloy-4. *Proceedings "Fatigue '96"*
    Berlin 6 - 10 May 1996.

CORROSION FATIGUE CRACK GROWTH RESISTANCE OF STEELS: GENERAL CONSIDERATION OF THE PROBLEM

O. ROMANIV, A. VOLDEMAROV

Karpenko Physico-Mechanical Institute of the National Academy
of Sciences of Ukraine, 5, Naukova Str., Lviv, 290601, UKRAINE

ABSTRACT

The paper presents general view on the regularities of fatigue crack growth resistance of low alloyed steels with various level of strengthening under the influence of convienient water environment. Special attention to the near threshold behaviour is given. Influence of environment on fatigue crack growth resistance strongly depends on the level of steel strength as a function of carbon content and thermal treatment. In high strength steels corrosive environment sufficiently lowers this resistance and vice veresa in low strength steels environment may improve crack growth resistance. Such effects can be intensified in the range of low frequences of loading.

Near threshold behaviour depends on the electrochemical mechanism of the interaction of material with environment and on the mechanical situation near the crack tip (morphology of the crack, level and kind of crack closure). The paper presents experimental method of evaluation of effective stress intensity factors of corrosion fatigue cracks which takes into account crack branching, crack tip blunting and crack closure.

Hydrogen embrittlement and anodic dissolution as leading mechanism of environment-metal interaction during crack growth are discussed. Using special modelled aproton environment it was shown that adsorption factor which was ignored by previous investigators, can be a driving force of crack acceleration. To evaluate drastic crack growth acceleration connected with intergranular rupture of high strength steels acoustic emission control was used.

KEYWORDS

Corrosion fatigue crack growth resistance; low alloyed steels; near threshold behaviour; effective stress intensity factor; electrochemical mechanisms.

MATERIALS AND EXPERIMENTAL PROCEDURE

This investigation is devoted to the comparative analysis of fatigue crack growth behaviour in high and low strength alloyed steels under the influence of corrosive water and usual air environment. The generalization are made on investigations of autors which were made on the analysis of various classes of alloyed structural steels (Romaniv and Nikiforchyn,

1986; Romaniv et al., 1990). But in this investigation as typical examples are used only two steels with chemical content and yield strength which are presented in a Table 1. The first of them which is conventionaly marked as HS steel belongs to the high strength steels. The second one conditionally marked as LS steel belongs to the low strength steels of the same type of alloying as mentioned HS steel. Both steels were used in quenched and tempered state. They differ substantially in the level of strength (Table 1).

Table 1. Chemical content and yield strength of tested steels

Steel	Chemical content, wt.%									$\sigma_{0,2}$,
	C	Cr	Mn	Ni	S	Mo	V	Si	P	MPa
HS	0.45	1.02	0.61	2.20	0.03	0.45	0.30	0.30	0.020	1800
LS	0.16	0.50	0.43	2.75	0.01	0.20	-	0.29	0.008	550

Fatigue crack growth curves were determined in the wide range of stress intensity factors (SIF) K and ΔK. Laboratory air, distilled water, and 3% water solution of NaCl were used as working medium. The investigation were performed at testing frequency f=10, 1.0 and 0.1 Hz using two regimes of stress ratio R=0.05 and R=0.5.

The changes in the crack growth kinetics under the influence of corrosion fatigue are connected not only with the influence of the electrochemical driving forces at the crack tip but also with the changes of the mechanical situation near a crack tip. Such changes include crack branching, crack blunting and crack closure, especially in the range of SIF which concerns to the near threshold behaviour (Romaniv et al., 1985). For evaluation of such mechanical changes in this study the proposed earlier method of determination of effective SIF (Romaniv et al., 1990) was used.

The principles of this method are as follows. The evaluation of the changes of crack morphology may be done using comparable fracture toughness tests of samples with fatigue cracks created in air K_{Ic} and in corrosion conditions K^*_{Ic}. Effective SIF K^α_{eff} can be recieved from equation

$$K^\alpha_{eff}=\alpha K,$$

where $\alpha=K_{Ic}/K^*_{Ic}$ is named as the factor of relaxation α, and K is the nominal value of SIF of straightline fatigue crack. Determination of α is connected with preserving of special condition of fracture tests when the plastic zone size is equal to the size of this zone during fatigue tests. Such conditions are achieved by lowering testing temperature to the level T_L when

$$K_{max}/\sigma_{0,2}=K^*_{Ic}/\sigma^L_{0,2}$$

The pecularities of this testing procedure are explained in (Romaniv et al., 1990).
Taking into account formation on crack sides during fatigue of corrosion products it is also important to evaluate properly changes of crack closure (CC). Such evaluation of CC is connected with determination of factor of the crack opening U

$$U=(K_{max}-K_{op})/\Delta K$$

Final evaluation of ΔK_{eff} takes into account crack morphology and CC factors

$$\Delta K_{eff} = \ae \Delta K = \alpha U \Delta K,$$

where æ is a factor of integral influence of crack branching (blunting) and CC on the stress level at the tip of corrosion fatigue crack.

CRACK GROWTH KINETICS - COMPARISON OF HIGH AND LOW STRENGTH STEELS

According to the results of our previous investigations typical differences in the corrosion fatigue crack growth behaviour of high and low strength steels are shown in Fig.1. In

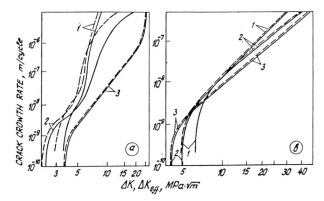

Fig.1. Crack growth rate vs ΔK (solid lines) and vs ΔK_{eff} (dashed lines) of HS (a) and LS (b) steels tested in distilled water (1, 2) and in air (3) at R=0.5 and f=0.1 Hz (1), 1 Hz (2) and 10 Hz (3).

corrosive environment HS steel revealed essential acceleration of crack growth and descreapancy of fatigue thresholds. And contrary, in LS steel influence of water environment is slight, in near threshold region we observe even essential growth of the threshold. Such tendencies are intensified when the testing frequency is lowered from 10 to 1 Hz. For LS steel we observe cumulative effect of two types of blunting connected with increased plasticity and anodic dissolution of the crack tip. At the same time acceleration of crack growth in HS steel is observed despite of distincly expressed macro- and micro-branching revealed on the lateral surfaces of the specimens. Micro-branching is typicaly intergranular (across boundaries of the previous austenite grains).

Let us evaluate the stress intensity relaxation on the basis of determination ΔK_{eff}, taking into account α, U and æ factors. For HS steel after low tempering ($\sigma_{0,2}$=1800 MPa) stress relaxation connected with crack branching (factor α) and CC (factor U) depends very specifically on the loading frequency and ΔK level (Fig.2a). For maximal used frequency (10 Hz) crack branching is very weak, and CC is absent as it is typical of high strength steels. Transition to tests at 1 Hz introduces interesting changes in the phenomena of branching and CC. With increasing ΔK crack brancing essentially increases and achieves its saturation in the middle part of Paris region. On contrary to the tests in air (Fig.2a), increased CC connected with corrosion products in high strength steels and for high level of stress ratio is observed. The supplementary reason of CC increase may be connected with the crack surface roughness arising due to intergranular rupture of alloys.

For LS steel ($\sigma_{0,2}$=550 MPa) we observe essential drop of α which progresses with diminishing of ΔK and frequency f (Fig.2b). In this case arised stress relaxation is not

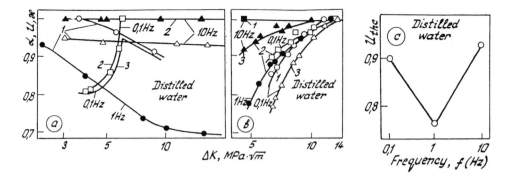

Fig.2. Dependency of α (1), U (2) and æ (3) coefficients on ΔK range (a, b) and U_{thc} dependency on frequency (c) for HS (a), LS (b, c) steels at R=0.5.

connected with branching but with essential blunting of a crack, as a result of anodic dissolution of the crack tip. This phenomenon progresses with diminishing ΔK and frequency f. Corrosive environment also stimulates crack closure in LS steel. A specific dependence of U_{thc} on the frequency of loading of LS steel is observed (Fig.2c). Supposed tendency to decreasing U_{thc} with diminishing loading frequency is disrupted at f=0.1 Hz. It is supposed that in such conditions the process of formation on the surface of corrosion products is combined with a process of their dissolution and it leads to the drop of CC.

Taking into account the level of factors α, U and æ in Fig.1 crack velocity curves related to $ΔK_{eff}$ are presented. In HS steel despite of crack branching and crack closure corrosive environment accelerates crack growth and diminishes $ΔK_{th\,eff}$ values. Indicative tests of corrosive fatigue of investigated HS steels under imposed cathodic polarization show supplementary decreas of crack growth resistance and such influence is interpreted as prevailance of hydrogen embrittlement as leading mechanisms of influence of corrosive environment.

Taking into account our investigations concerning formation of $ΔK_{thc}$ and $ΔK_{th}$ levels in the used Cr-Ni-Mo-V steels with various levels of strengthening after quenching and

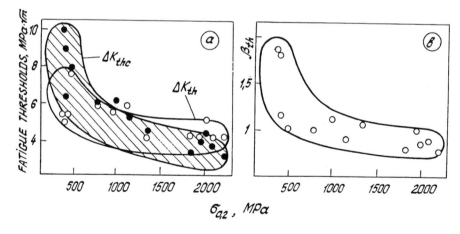

Fig.3. Dependencies $ΔK_{th}-σ_{0,2}$, $ΔK_{thc}-σ_{0,2}$ (a) and $β_{th}-σ_{0,2}$ (b) of various structural steels.

tempering (Romaniv *et al.*, 1985) on the Fig.3 revealed ΔK_{thc}, ΔK_{th}-$\sigma_{0,2}$ dependences are presented. The influence of corrosive environment on fatigue threshold can be also demonstrated with a ratio $\beta_{th}=\Delta K_{thc}/\Delta K_{th}$ (Fig.3b). In the area of $\sigma_{0,2}$ levels near 600-700 MPa we observe transition from the positive to the negative influence of corrosive environment on the fatigue thresholds. Such level of strength characterizes the change of the dominating electrochemical mechanisms from anodic dissolution to hydrogen embrittlement.

ADSORPTION FACTOR IN THE CORROSION FATIGUE CRACK PROPAGATION

Presented analysis dose not take into account adsorption factor and his possible primary role in the corrosion fatigue of metals as it was declared in studies of Karpenko (1963). His idea concerning adsorption decay of fatigue strength in various oil and acid environments was neglected by Nickols and Rostocker (1969) that connected presumable adsorption influence with the presence of oil protodonor admixtures which acts after the hydrogen mechanism.

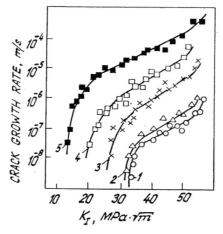

Fig.4. SCG kinetics in HS steel tested in the environments of: 1 - DMSO, 2 - DMSO+1wt.%H_2O, 3 - DMSO+10wt.%H_2O, 4 - DMSO+40wt.%H_2O, 5 - H_2O.

To check G.Karpenko hypothesis the special experimental study was performed (Agladze *et al.*, 1985) which concerns investigation of static crack growth of HS steel in dimethyl-silfoxide - the high purity aprotonic polar solvent. Mentioned steel did not reveal susceptibility to subcritical crack growth (SCG) in vacuum. Such susceptibility to SCG was detected in dimethylsilfoxide environment (curve 1 on Fig.4). This is an evidence of crack resistance decay which is connected with adsorption effect. It is also evident that such influence can be magnified by introducing into the solvent protodonic components (curves 2-5, Fig.4). But some primary influence of adsorption factor must be not ignored in the corrosion fatigue tests.

ACOUSTIC EMISSION CONTROL DURING CORROSION FATIGUE CRACK GROWTH TESTS

On the mentioned HS steel special investigation of acoustic emission (AE) control was carried out during fatigue crack growth tests. Simultanously electron fractographic control of fracture surfaces was performed with special attention to intergranular crack growth.

For AE studies special apparatus APh-20A was used with determination the number of events of AE on the unit of surface $\overline{N}$ and amplitude level $\overline{A}$ (Romaniv *et al.*, 1987).

A directly proportional correlation exists between the number of events $\overline{N}$ and the surface percent B of intergranulare facets in the unit of the fracture surface (Fig.5a,b). This permits to recieve determination of the intergranular rupture volume using AE control of fatigue tests. As it is shown on Fig.5b increasing of stress ratio R increases also fraction of intergranular cleavage. The same influence on the crack propagation mechanisms supplies corrosive water environment (Fig.5c). The phenomena of transgranular cleavage achieve

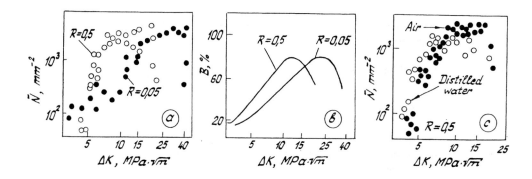

Fig.5. Dependencies $\overline{N}$ (a, c) and percentage of intergranular cleavage B (b) on ΔK range for HS steel.

maximum level in the upper part of Paris region, further increasing of ΔK intensifies the ductile mechanism of fatigue crack propagation. There is an opinion that volume of intergranular cleavage rupture is controlled by K_{max} and properly by $\overline{N}$, as ΔK correlates with amplitude level of AE signals (Romaniv *et al.*, 1987).

REFERENCES

Agladze, T.R., Ya.M. Kolotyrkin, O.N. Romaniv and G.N. Nikiforchyn (1985). The effect of adsorption-chemical interaction for metal-environment system in stress corrosion cracking. In: *Extended Abstracts of the Fourth Japan-USSR Corrsion Seminar,* pp. 256-269, Tokyo.

Karpenko, G.V. *Steel Strength in Corrosive Environment.* Mashgiz Publishers, Kiev.

Nickols, C. and W. Rostocker (1969). Brittle fracture of steel in the presence of organic liquids. In: *Susceptibility of Mechanical Properties to Environment Effect,* pp. 231-254. Mir Publishers, Moscow.

Romaniv, O.N. and G.N. Nikiforchyn (1986). *Corrosion Fracture Mechanics of Engineering Alloys.* Metallurgia Publishers, Moscow.

Romaniv, O.N., G.N. Nikiforchyn and A.V. Voldemarov (1985). Cyclic corrosion crack resistance: generalities of threshold formation and life assessment of various engineering alloys. *Fiz.-Khim. Mekhanika Materialiv,* **3,** 7-20.

Romaniv, O.N., K.I. Kirilov, Yu.V. Zima and G.N. Nikiforchyn (1987). Relation between acoustic emission and kinetics and micromechanism of fatigue fracture of high strength steel of martensitic structure. *Fiz.-Khim. Mekhanika Materialiv,* **2,** 51-55.

Romaniv, O.N., S.Ya. Yarema, G.N. Nikiforchyn, N.A. Makhutov and M.M. Stadnyk (1990). *Fracture Mechanics and Strength of Materials.* Naukova Dumka, Kyiv.

A NEW METHOD FOR PREDICTION OF INHIBITOR PROTECTION AGAINST CORROSION FATIGUE CRACK INITIATION IN STEELS.

H.M. NYKYFORCHYN[†], M. SCHAPER[‡], O.T. TSYRULNYK[†], V.M.ZHOVNIRCHUK[†], A.I. BASARAB[†]

[†] Department of Corrosion and Metal Protection, Karpenko Physico-Mechanical Institute of NAS of Ukraine, 290601 Lviv, Ukraine

[‡]Chair on Material Reliability, TU Dresden, Germany

ABSTRACT

A new approach is presented for prediction of influence of inhibitors at corrosion fatigue that is based on usage of charge of juvenile surface oxidation during its repassivation. The experiments showed good correlation between this parameter and a number of cycles to crack initation.

KEYWORDS

Corrosion fatigue; crack initation; juvenile surface; nonstationary electrochemical processes.

INTRODUCTION

The crack initiation under the corrosion fatigue conditions is provided by the kinetics of damage accumulation under the conditions of cycle loading and aggressive environment effect. There are various damage definitions, but concerning the corrosion fatigue this notion consists, first of all, in failure of the protective surface film resulting in appearance of a new juvenile metal surface. On the other hand, it is known that the corrosion fatigue crack is normally forming by anodic dissolution localised just at juvenile surfaces. Here the intensity of anodic dissolution is determined by the rates of activation and repassivation of juvenile surfaces. Thus, not every inhibitor can effectively arrest the fatigue crack initiation, but only that adsorbing quickly on the juvenile surfaces and improving the surface repassivation. Hence the nonstationary adsorption and electrochemical processes at juvenile surfaces (Beck, 1973, Rosenfeld et al., 1980) are of a current interest.

In the this paper an approach for prediction of inibition effect on corrosion fatigue crack initiation is presented. This approach is based on correlation between parameters of juvenile surface repassivation and number of load cycles critical for crack initiation.

EXPERIMENTAL PROCEDURE

A 0.4C-1Cr steel (0.41% C; 0.91% Cr; 0.60% Mn) in H_2SO_4 solutions (of pH=1.6 and 3.5 respectively) containing 0.02 mole/litre of iodides of potassium, pyridine, propylepyridine, and amilepyridine. The characteristics of relaxation electrochemical processes at juvenile surface were investigated using the reverse torsion specimens of a diameter of 5 mm and a length of 25 mm; the torsion rate was 0.5 rad/s. Such load technique provides high plastic deformation (and thus large juvenile area) without fracture. The registration of electrochemical parameters was started after a certain strain level was reached. To estimate the effect of juvenile surfaces on environment composition, the specimen was exposed to long-cycle torsion.

Before the torsion the specimen was put into a chamber coaxially to an auxiliary reference silver chloride electrode. The electrochemical measurements were performed with potentiostate. The variation of potential and current during the relaxation was fixed with oscilograph with memory. To evaluate pH level in the nearsurface layer a antimony-oxide micro-electrode was used.

The stage of corrosion fatigue crack initiation was investigated by using single-edged-notched beam bend specimens (6x18x160 mm). The bending cycling has a frequency of 10Hz and a load ratio of 0.05. The crack was considered to be nucleated when its length was 0.05 mm.

EXPERIMENTAL RESULTS

The corrosion rate in the investigated systems was evaluated by value of polarisation anode R_p which was determined from impedance measurements. The R_p and corrosion rate exponent γ values (the latter was estimated by weight method) are presented in Table.

Table. Electrochemical properties of the Steel in inhibited H_2SO_4 solutions.

Inhibitor	pH 1.6		pH 3.5		σ^*
	γ, g/sm^2 h	R_p, Om/sm^2	R_p, kOm/sm^2	k_a	
None	6.2×10^{-4}	70	1.65	-0.46	-
Pyridineiodide	1.1×10^{-4}	320	6.60	-0.55	0
Propilepyridineiodide	5.2×10^{-5}	780	11.0	-0.58	-0.115
Amilepyridineiodide	3.4×10^{-5}	990	14.0	-0.63	-0.162
Alilepyridineidide	4.9×10^{-5}	850	12.0	-0.65	0.180

The Table manifests a definite correlation between these parameters, and this confirms correctness of usage of polarisation resistance for the evaluation of the corrosion rate. The curves corresponding to dependencies of γ and R_p on the Taft constants for pyridine iodides are V-like (Fig. 1) and consist of two straight lines crossed in a point of nonsubstituted pyridine.

When the specimen was exposed to elastic torsion, E changed just a little (some mV), while the plastic deformation caused abrupt growth of negative E value (Fig. 2). It increases with increase of plastic strain and also is affected by the environment composition: it increases by increasing pH of the solution and decreases by adding J$^-$ ions and inhibitors.

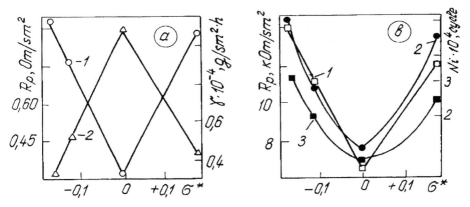

Fig. 1. Dependencies of R_p (1a, 1b), γ (2a), and N_i (2b, 3b) on σ^* in inhibited H_2SO_4 solutions with pH1.6 (1a,2a,2b) and pH3.5 (1b,3b).

Plastic deformation of the steel substantially intensifies both anodic and cathodic reactions. The
polarisation curves for deformed surface contain Tafel linear areas bounded by areas of limited current.
The rate of these reactions depends on the exposure period τ, polarisation potential E_p, pH of
environment, and inhibitor (and J) share in the solution. By increasing solution acidity from pH3.5 to
pH1.6 the rates of the anodic and cathodic reactions increases.

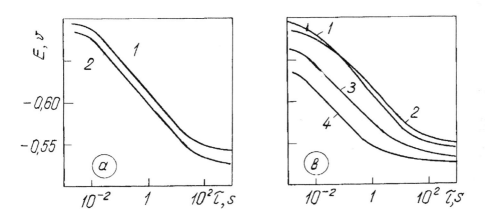

Fig. 2. Time dependence of E in H_2SO_4 solutions with pH 3.5 at various
torsion strain degrees: ε = 8% (1a, 1b), 32% (2a), and
environment compositions: with addition (2b), pyridineiodide
(3b), amilepyridineiodide (4b).

Adding of the inhibitors results in slowing these reactions and changes the slope of Tafel areas.
Relaxation of polarisation current i is described by a straight line in coordinates lg i - lg τ at various
angle coefficient R which is proportional to the electrode reaction overvoltage $\Delta E = E_p - E_0$ (E_0 is a
potential deformed surface at $\tau \sim 10^{-2}$ s).

The decrease of pH of the solution retards the relaxation of the electrode reactions, while adding of the inhibitors accelerates them. The Table shows values of k_a (for the anodic reaction) at $\Delta E = 75$ mV for various environment compositions.

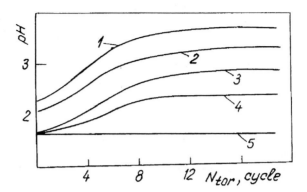

Fig. 3. Dependence of pH on N_{tor} in H_2SO_4 solution (1, 4, 5), and with addition of KJ (2), amilepyridinumiodide (3), near the deformed surface (1, 2, 3), and a distance of 0.5 mm (4), and 1 mm (5).

The measurements of pH values showed that from some distance with electrode approaching the specimen surface, pH increases and has its maximum value in the near-surface solution layer (Fig. 3). The more torsion cycles is exposed to the specimen the more rapidly pH increases. Adding of the inhibitors into the solution retards this process. Similarly to the torsion loading, the cyclic loading accelerates the alkalification of the near-surface solution layer at the notch bottom, and this effect increases by increasing $\Delta\sigma$. In the case of solution inhibition, the number of cycles to fatigue crack initiation N_i at a certain load $\Delta\sigma$ correlate with Taft σ^* constants of the investigated compounds.

DISCUSSION

It is known (Grigoriev and Osipov, 1966), the adsorption of pyridine molecule is caused by electron-donor interaction between excessive electron density of nitrogen atom and uncompleted electron d-levels of the metal. When substituents of different polarity are added to the pyridine molecule the electron density of nitrogen atom changes substantially that affects on reaction and adsorption ability of pyridine salts. Quantitative value of the electron density change of nitrogen atom is characterised by the Taft constants which can be determined basing on the principle of free energy linear variation (FELV).

In the paper of Grigoriev and Osipov, it was suggested to use FELV principle for assessment of effect of a certain reaction series on the corrosion rate, hydrogenation, corrosion-mechanical fracture. Our work confirms good correlation between γ, R_p and σ^* (Fig. 1). Although the correlation between N_i and σ^* is not linear, the FELV principle has good prospects for prediction tendency of change of protective effect of inhibitors of the specific reaction series in the specific corrosion solution. However, it can hardly be predicted in other solutions or in other reaction series.

The displacement of E to the negative side (Fig. 2) during the deformation of metal electrode is caused by failure of the protective surface films. The bigger the strain the wider the juvenile surface, and in its turn, the more negative is E value. The relaxation can be explained by growth of passivising films on the juvenile surface. However it cannot be argued that only anodic process is localised on the juvenile surfaces (Logan, 1966), since at all exposition times the cathodic process is facilitated.

Acceleration or arrestment of the cathodic process is clearly manifested by pH_s change. Evidently, the change of pH_s reflects the process of accumulation of nucleations and depassivation of the juvenile surfaces, or in other words, the process of accumulation of protective film damage during fatigue.

Adding the inhibitors into the solution affects on the electrode reactions on the juvenile surfaces. Our experiments manifested this influence at the exposition times of $\tau \sim 10^{-2}$ s, and in the work (Rosenfeld et al., 1980) this effect was detected at $\tau \sim 10^{-2}$ s. This means that primary inhibitor adsorption acts can be quite fast.

The results obtained do not agree with the approaches of Smith et al. (1970) and Panasyuk et al. (1983), based on the unambiguous relationships between corrosion-mechanical fracture kinetics and E and pH environment parameters. According to such approaches at narrow potential range the decrease of pH increases the aggressive environment influence. For the investigated steel - environment systems an opposite effect takes place (Fig. 4): the higher the pH_s, the smaller the number of cycles to crack initiation. Evidently, the electrode reaction rate is affected not only by E and pH but also by presence of activators, passivators, oxidisers, and inhibitors.

The fullest evaluation of these effects can be made by analysis of nonstationary processes during the interaction of the deformed surface and the environment. Activity of the juvenile surface can be characterised by amount of charge q_0 necessary for oxidation of the surface during its repassivation. The charge q_0 was evaluated by integration of corrosion current relaxation curves obtained in experiments at torsion loading during the time interval of $10^{-2}..10$ s. The obtained values of q_0 are in satisfactory correlation with N_i (Fig. 4).

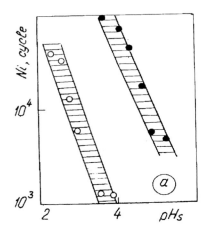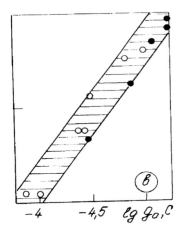

Fig. 4. Dependence of N_i on pH_s (a) and $\lg q_0$ (b) at $\Delta\sigma=100$ MPa in inhibited H_2SO_4 solutions; light symbols - pH1.6, dark ones - pH3.5.

It should be mentioned that the proposed approach has no limitations peculiar to FELV approach and allows to range environments by their corrosion effect, depending on their pH, presence of stimulating and inhibiting additions.

CONCLUSION

The parameters E and pH have ambiguous effect on the crack initiation resistance in inhibited water solutions: unlike generally known relationships, there is a bigger time to crack initiation in the inhibited solution at a lower pH level in the nearsurface layer.

The correlation is found between the Taft constants for pyridine salts and their inhibiting effect on corrosion and corrosion-fatigue fracture.

The new approach is proposed for prediction of the protective effect of inhibitors on the corrosion-fatigue crack initiation that uses the charge value necessary for the oxidation of the juvenile surface during its repassivation.

REFERENCES

Beck, T.R. (1973). Electrochemistry of freshly-generated titanium surfaces. Rapid fracture experiments. *Electrochem. Acta*, 18, 815-827.

Rosenfeld, I.L., Afanasiev, K.I., and Marychev, V.A. (1980). Study of electrochemical properties of freshly-generated metal surfaces in electrolyte solutions. *Phys.-Chem. Mech. of Materials*, 16, 49-54.

Logan, H.L. (1966). *The stress corrosion of metals*. Academic Press, New York.

Grigoriev, V.P. and Osipov, O.A. (1966). Relations between compositions of some organic compounds and their inhibiting properties. In: *Proc. III International Conference on Corrosion of Metals*, Vol. 2, pp. 27-53. Mir, Moscow.

Smith, J.A., Peterson, M.H., and Brown, B.F. (1970). Electrochemical conditions at the tip of an advancing stress corrosion crack in 4340 steel. *Corrosion*, 26, 539-542.

Panasyuk, V.V., Ratych, L.V., and Dmytrakh, I.M. (1983). Dependence of fatigue crack growth rate on electrochemical conditions at a crack tip. *Phys.-Chem. Mech. of Materials*, 19, 33-37.

INFLUENCE OF OXIDE LAYER ON THE FATIGUE DAMAGE BEHAVIOUR IN A 316L TYPE STAINLESS STEEL.

C. VERNAULT, P. VIOLAN and J. MENDEZ

Laboratoire de Mécanique et de Physique des Matériaux
URA 863 / CNRS
ENSMA BP 109
FUTUROSCOPE Cedex (France)

ABSTRACT

In order to determine the intrinsic role of oxide layers on cyclic damage, fatigue tests were performed on a 316L type stainless steel at 300° C under vacuum using pre-oxidized or non pre-oxidized specimens.
Results have established that oxide layers obtained at different temperatures play an important role on the emergence of Intense Slip Bands and the initiation of early microcracks.
These effects, however, remain limited compared to those obtained with continuous oxidation by performing tests in air.
Furthermore, neither static oxide layers nor continuous oxidation in air affect the cyclic stress-strain response on a macroscopic scale.

KEYWORDS

Stainless Steel, Oxide layers effects, Environment, Low Cycle Fatigue, Mechanical behaviour, Cracks initiation.

INTRODUCTION

It is well known that environment can drastically reduce the fatigue resistance of metallic materials. This deleterious effect is frequently discussed in terms of hydrogen (Williams et al., 1992) and oxygen (Achter, 1967) embrittlement at the crack tip or in terms of mechanical properties modifications in oxidized areas.
However, very few studies have been focussed on the interactions between the static oxide layers and the near surface cyclic deformation and cracking processes.
The aim of this paper is to illustrate how oxide layers obtained at high temperature can affect the subsequent fatigue damage behaviour of a 316L type stainless steel. In order to establish the intrinsic role of the oxide layers, the behaviour of pre-oxidized and non pre-oxidized samples, both cycled under vacuum at 300° C, were compared. Moreover, some tests have been conducted in air at the same temperature in order to consider the dynamic oxidation effects.

EXPERIMENTAL CONDITIONS

The material used in this study is a 316L type austenitic stainless steel. Its main mechanical properties at 300° C are a yield strength of 156 MPa and a tensile strength of 460 MPa. The chemical composition has been indicated elsewhere (Driver et al., 1988). Cylindrical fatigue specimens were annealed for 1h 30mn at 1050° C in vacuum and then waterquenched. This heat treatment confers an austenitic microstructure with a mean grain size of 60 µm. Specimens were mechanically polished down to a 1 µm grade diamond paste. Some of them were pre oxidized at 300° C or 600° C during 100h, which respectively induces an homogeneous yellow and a spalled blue superficial oxide layers.

Cyclic deformation tests were performed in a symmetrical uniaxial push-pull mode on a electromechanical machine at 300° C in laboratory air or in a high vacuum (pressure lower than 10^{-5} mbar). Tests were conducted under constant plastic strain amplitude $\Delta\varepsilon p/2 = \pm 2.10^{-3}$ with a constant plastic strain rate of 2.10^{-3} s^{-1}. For these conditions the fatigue life, established in a previous study (Driver et al., 1988), is about 19000 cycles in air and 215000 cycles under vacuum. However, in the present study tests were interrupted early in the life, after 3000, 6000, 13200 cycles in vacuum or 3000 cycles in air, in order to investigate the very early stages of crack initiation.

For each test, the conventional $\Delta\sigma/2$-N curves have been plotted and fatigued specimens have been observed by S.E.M. These observations permit to establish a qualitative analysis (nature of the initiation sites) and a statistical quantification of superficial damage.

Two types of quantification have been performed. The first concerns intense slip bands. In this case the slip band population has been characterized by counting at a magnification of 600 the number of intense slip marks within twenty different areas regularly distributed over the specimen surface. This examined area corresponds to a minimum of about 480 grains in each specimen. We have calculated a mean value of the number of slip bands divided by the total analysed surface area; that will be named slip density.

The second statistical quantification concerns microcracks. It is based on microcrack population founded over the area scanned during three circumferential bands using a magnification of 400. This analysed area corresponds to about 7.5 per cent of the specimen useful surface. Hence, a crack density has been obtained by dividing the number of cracks by the total observed area.

RESULTS AND DISCUSSION

Cyclic Hardening.

Figure 1. shows some examples of the curves $\Delta\sigma/2$-N corresponding to tests conducted on pre-oxidized or non pre-oxidized specimens cycled in air or in vacuum. All curves follow the cyclic behaviour already observed in a previous study (Belamri, 1986). It is characterized by a short period of primary hardening during the first 60th cycles followed by a softening until the 1000th cycle. The secondary hardening period observed then is a characteristic behaviour of the 316L in the temperature range of 300-400° C (Alain, 1993).

It is clear from these results that no significant differences are observed on the different specimens. That means that at the opposite of some results obtained on single cristals (Bowman et al., 1988), oxide layer and environment do not affect the total cyclic behaviour of the polycrystalline 316L type stainless steel. Moreover, considering the relationship between cyclic stress behaviour and dislocation structure (Gerland et al., 1989), it is evident that environment and oxide layer do not affect dislocation structure in the specimen bulk even if we can admit that significant modifications could be induced within the near surface grains due to dislocations-oxide film interactions.

Superficial damage.

In Fig. 2., we have plotted the evolution of the slip band density as a function of the number of cycle N, for both non pre-oxidized or 300° C pre-oxidized specimens cycled in vacuum. Each value reported on these curves corresponds to the average of two different tests performed in identical conditions.

It can be seen that both curves follow the same evolution. The formation of the first slip bands occurs early in the life and then their number only increases slightly. For the number of cycles investigated here, up to 13200 cycles, more of 80 per cent of the intense slip bands were already formed at 6000 cycles. However, whatever the number of cycles was, the presence of the oxide layer built at 300° C (~ 0.1 µm thick) produces the formation of a higher number of intense slip bands compared to the non oxidized specimen.

Figure 3. illustrates the aspect of the intense slip bands on the reference specimen. The slip bands are constituted of fine and continuous intrusions and rather thick extrusions.

In contrast, on the 300° C pre-oxidized specimen, at 3000 cycles a high number of grains exhibit very fine and discontinuous extrusions (Fig. 4.). Moreover, numerous slip bands do not emerge at the surface and give rise to straigth marks. As it can be seen in Fig. 5., in some of these marks extruded material are just growing out.

The effect and the behaviour of the oxide layer appears therefore to be very similar to the one associated with PVD and ion mixed thin coatings on fatigued materials (Villechaise et al., 1993).

Such behaviour can be explained by the preventing action of the oxide layer on the emergence of slip bands but also, by the edge dislocation nucleation (Bowman et al., 1987, Grosskreutz, 1967) and the role of the image force on dislocations near the surface film (Head, 1953).Oxide layer-substratum interface can act as a source of edge dislocations under applied stress and exert a long range attraction and a short range repulsion force on the near surface dislocations. So, combining these effects, dislocation bundles can be formed near the surface which induces local stresses enhancement in the superficial grains. As a consequence, the deformation becomes much more homogeneous than in the non oxidized samples. Now, it is clear that the 0.1 μm thick oxide layer obtained at 300° C is not sufficiently resistent to impeed its shearing by the slip band activity.

In contrast, when the oxide layer becomes thicker, as it is the case with a pre-oxidation at 600° C, the majority of the intense slip bands are then contained.

This could explain why for the 600° C pre-oxidized specimen the number of intense slip bands, observed after 13200 cycles, is very lower than in the non pre-oxidized sample (3 times lower than in the 300° C oxidized one). The oxide film is in this case homogeneously deformed.

Figure 6. gives the evolution of the microcrack density as a function of the number of cycles for pre-oxidized or non pre-oxidized specimens cycled in vacuum or in air. It can be seen that the presence of the oxide film favours crack initiation since it induces an increase of the surface crack density. However, this effect does not appear in terms of the number of cycles leading to the initiation of the first microcracks. Indeed, at 3000 cycles, no microcracks were observed on the 300° C pre-oxidized sample whereas several microcracks were already formed on the reference specimen. This enhancement of crack initiation increases with the nature or the thickness of the oxide layer. For instance, the specimen pre-oxidized at 600° C exhibits a higher crack density than the 300° C pre-oxidized one, twice higher than the reference sample.

It must be noted that the nature of the initiation sites are also weakly modified. For the testing conditions investigated most of the microcracks are intergranular, even on the non pre-oxidized specimen cycled in vacuum; however, in this case, a third part of cracks are transgranular.

On the pre-oxidized samples the increase of the microcrack density, with regard to the reference specimen, is associated to an intergranular cracking mode (Fig. 7.); indeed, the number of transgranular microcracks does not change significantly or even decreases, when the fatigue specimen is pre-oxidized at 600° C.

However, it has not been clearly established if this intergranular damage can be attributed to the deformation modifications in the superficials grains (Kim et al., 1978) or to embrittlement due to preferential grain boundary oxidation (Mc Mahon et al., 1970, Paskiet et al., 1972).

Histograms of Fig. 8. give the distribution of the frequency of counted microcracks as a function of their length for pre-oxidized or non pre-oxidized specimens. These results concern tests performed at 3000 cycles in air and at 13200 cycles under vacuum.

It must be noted that the main cracks, on vacuum tested samples, do not have significant differences in length. Moreover, no difference on the microcracks frequency distribution is observed between the 300° C pre-oxidized and the reference specimens. For instance, the majority of their cracks are distributed in the length classes lower than 30 μm.

In contrast, 600° C pre-oxidized sample exhibits a more homogeneous distribution. In fact, nearly half of the total number of cracks have a length between 30 and 70 μm, whereas in the other cases of tests in vacuum less than 20 % of cracks are longer than 30 μm.

As a consequence, the oxide layer obtained at 600° C, in addition to induce a greater increase of the intergranular microcracks initiation with regard to the 300° C pre-oxidized one, it favours early stages of growth. Indeed, it has not been clearly established if this can be explained by a greater propagation rate or by a larger crack initiation length. However, both cases can be linked with a more important oxide penetration in the grain boundaries .

Results obtained on specimen tested in an aggressive environment (Fig. 6. and Fig. 8.) show an easier crack initiation and a faster propagation rate than tests performed under vacuum.
Therefore, the acceleration of surface damage due to oxide layers remains limited compared to the effect of continuous oxidation.

CONCLUSION

The following conclusions from this study on the influence of oxide layers on low cycle fatigue damage behaviour of a 316L type stainless steel can be drawn.

1). Oxide layers or environment do not modify the whole cyclic mechanical behaviour and therefore the dislocation structure in the specimen bulk. Their influence only takes place in the superficial grains.
2). Oxide layers prevent the emergence of slip bands and the deformation at the specimen surface.
3). Oxide layers promote intergranular microcrack initiation. This effect is increased by the oxide thickness which also favours early stages of growth.
4). Continuous oxidation has a stronger effect on the cracking processes than static oxide layers.

ACKNOWLEDGEMENT

We are very greatefull to Région Poitou-Charentes for the grant N° 95/RPC-R-151 and to EDF and Framatome companies for financial support.

REFERENCES

Achter, M. R. (1967).Effect of environment on fatigue cracks. In: Fatigue Cracks Propagation. *ASTM STP* **415**. Am. Soc. Testing Mats., 181.
Alain, R. (1993). *PhD Thesis*. University of Poitiers (France).
Belamri, B. (1986). *PhD Thesis*. University of Poitiers (France).
Bowman, K.J.,V.K. Sethi, I. Rusakova and R. Gibala. (1988). *ICSMA8*. (P.O. Kettunen, T.K. Lepistö and M.E. Lethonen Ed.). Pergamon. **1**. 199.
Bowman, K.J. and R. Gibala. (1987). *Fatigue 87*. (R.O. Ritchie and E.A. Stakre Ed.). EMAS. **1**. 63.
Driver. J.H., C. Gorlier, C. Belamri, P. Violan and C. Amzallag. (1988). Influence of temperarure and environment on the fatigue mechanisms of single-cristal and polycristal 316L. In: Low Cycle Fatigue. *ASTM STP* **942**. *Am. Soc. Testing Mats.*, 438-455.
Gerland, M., J. Mendez, P. Violan, and B. Ait Saadi. (1989). Influence of temperature and environment on the fatigue mechanisms of single-cristal and polycristal 316L. *Mater. Sci. Engng*. **A118**. 83-95.
Grosskreutz, J.C. (1967). The effect of oxide films on dislocation-surface interactions in aluminium. *Surface Science*. **8**. 173-190.
Head, A.K. (1953). The interaction of dislocations and boundaries. *Phil. Mag.*, **44**. 92.
Kim, W.K. and C. Laird. (1978). Crack nucleation and stage I propagation in high strain fatigue. *Act. Metall*. **26**. 789-799.
Mc Mahon, C.J. and L.F. Coffin. (1970). Mechanisms of damage and fracture in high-temperature, low cycle fatigue of a cast nickel-based superalloy, *Metall. Trans.*, **1**, p. 34, 1970.
Paskiet, G.F., D.H. Boone, and C.P. Sullivan. (1972). Effect of aluminide coating on the high-cycle fatigue behaviour of a nickel-base high-temperature alloy. J. Inst. Metal.. **100**. 58.
Villechaise, P., J. Mendez and P. Violan. (1993). *Fatigue 93*. (J.P. Bailon and J.I. Dickson Ed.). EMAS. **1**. 141-145.
Williams, D.P., III, P.S. Pao, and R.P. Wei., (1992). Environment-sensitive fracture of engineering materials. *U.M.I*. 3.

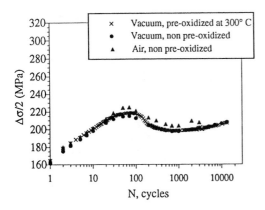

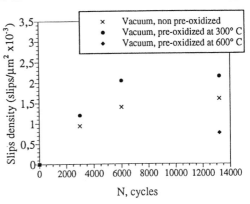

Fig. 1. Evolution of the stress amplitude $\Delta\sigma/2$ as a function of number of cycles N (test conditions: T=300° C and $\Delta\varepsilon p/2 = \pm 2.10^{-3}$).

Fig. 2. Evolution of the slip density as a function of number of cycle N (test conditions: T=300° C and $\Delta\varepsilon p/2 = \pm 2.10^{-3}$).

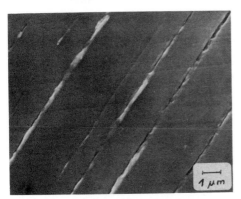

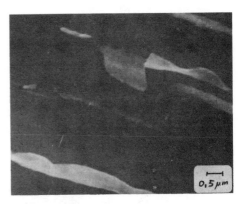

Fig. 3. Aspect of the intense slip bands on the non pre-oxidized specimens after 6000 cycles in vacuum.

Fig. 4. Aspect of the extrusions on the 300° C pre-oxidized specimen after 3000 cycles in vacuum.

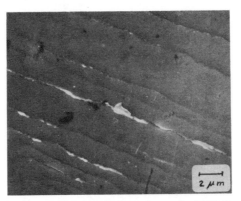

Fig. 5. Aspect of the straigth marks on the 300° C pre-oxidized specimen after 3000 cycles in vacuum.

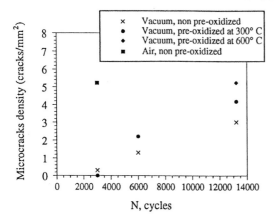

Fig. 6. Evolution of the microcrack density
as a function of the number of cycle
N (test conditions: T=300° C and
$\Delta\varepsilon p/2 = \pm 2.10^{-3}$).

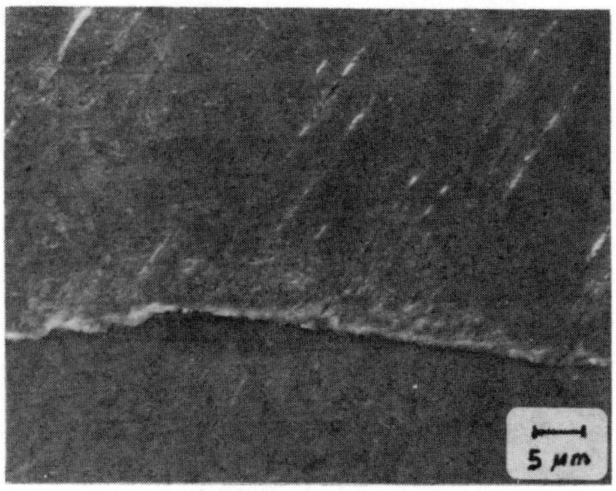

Fig. 7. Intergranular microcracks
already formed after 13200
cycles under vacuum on the
600° C pre-oxidized
specimen.

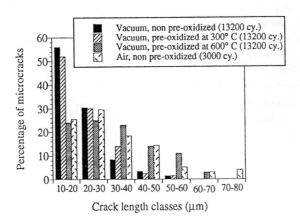

Fig. 8. Microcracks distribution in the different length
classes for pre-oxidized and non pre-oxidized
specimens. (test conditions: T=300° C and
$\Delta\varepsilon p/2 = \pm 2.10^{-3}$).

HIGH FREQUENCY TORSIONAL AND AXIAL FATIGUE LOADING OF AL 2024-T351

H.R. MAYER AND S.E. STANZL-TSCHEGG

Institute of Meteorology and Physics, University of Agriculture,
Türkenschanzstr. 18, 1180 Vienna, Austria

ABSTRACT

A new experimental procedure for high frequency torsional fatigue loading has been used to determine the fatigue behaviour of 2024-T351 aluminium alloy. With this method specimens are excited to a torsional resonance vibration at a frequency of about 20 kHz. The high testing frequency allows an economical investigation of the regime of very high numbers of cycles. Fatigue data for cyclic torsional loading are compared to high frequency axial fatigue loading experiments in two environments: ambient air and distilled water. For both environments torsional loading S-N curves are shifted about 15 - 20 % towards lower stress amplitudes in comparison with axial loading S-N curves. Numbers of cycles to failure are reduced by a factor 10 - 100 for water in comparison with air environment for axial as well as torsional loading. Hydrogen embrittlement is assumed to be the main reason for early crack formation in water.

KEYWORDS

Fatigue, high cycle, high frequency, torsion, 2024-aluminum alloy

INTRODUCTION

Long testing times make the determination of the fatigue properties of structural materials in the regime of very high numbers of cycles rather uneconomic. The high frequency resonance testing method is an experimental technique to obtain numbers of cycles as high as 10^9 within a few days. Loading frequencies in the range of 20 kHz have been used to measure axial (tension-compression) loading S-N curves first (Neppiras, 1960) and to measure fracture mechanical properties later (Mitsche *et al.*, 1973). Meanwhile testing devices for axial high frequency loading have obtained a high technical standard (Stanzl *et al.*, 1989). The use of the high frequency resonance testing method for torsional fatigue loading experiments is relatively new (Stanzl-Tschegg *et al.*, 1993). An application of this technique for cyclic torsional and superimposed static axial loading in order to determine the fatigue properties of a ceramic material is described by Mayer *et al.* (1994). In the present work, the high frequency resonance testing method was used to determine both, the axial and the torsional fatigue properties of 2024-T351. Experimental procedure as well as results are compared.

MATERIAL

Fatigue experiments were performed using the high strength aluminum alloy Al 2024. The chemical composition is (in weight-%): Cu 4.5%, Mg 1.5%, Si 1%, Fe 0.2%, Mn 0.7%, Cr < 0.05%, Zn < 0.05%, Ti < 0.03%. Heat treating was according to T351 (solution annealing, quenching, cold work and age hardening). Static tests of 2024-T351 yield an elastic limit of $R_{p0.2} = 352$ MPa, a static strength of $R_m = 460$ MPa, $A_5 = 18$ %, $K_{IC} = 35$ MPa$\sqrt{m}$ and a Young's modulus E of 72.5 GPa and a Poisson ratio of $\nu = 0.3$.

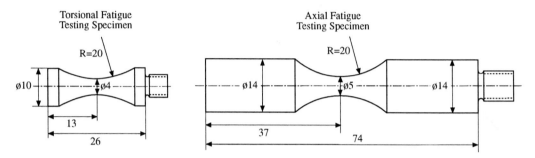

Fig. 1: Specimen shapes used for fatigue tests

The material was supplied as wrought sheets of 20 mm thickness. The average grain size of the material in L-direction is 230 μm, in T-direction (normal to the rolling direction) 90 μm and in S-direction (throughout the sheet) 45 μm. The specimens were cut out from the sheets such that the orientation of the tensile axis was parallel to the rolling direction (L-direction) and machined afterwards. Specimen shapes for axial and torsional fatigue tests are shown in Fig. 1. The specimen surfaces were polished prior to the measurements.

EXPERIMENTAL PROCEDURE

The torsional and axial fatigue experiments were performed with the high frequency resonance testing method. With this method, specimens are excited to a resonance vibration at a frequency of approximately 20 kHz. Piezoelectric transducers transform the high frequency AC-signal into a mechanical (torsional or axial) vibration which then is amplified using an amplification transducer. The mechanical parts of the equipment (amplification transducer, fatigue specimen) must be dimensioned such that a torsional or axial resonance vibration, respectively, can be generated. The (torsional or axial) vibration displacement of a specimen is symmetric around the specimen center in length direction, and the displacement maximum is obtained at the specimens ends. Since the wavelength for torsional vibration is smaller than for axial vibration, the specimen length for torsional fatigue testing is shorter than for axial loading. Details for the design of specimen and mechanical parts are described elsewhere (Stanzl-Tschegg *et al.*, 1993).

The strain amplitude is maximum in the specimen center and becomes zero towards both specimen ends. Strain amplitudes are measured using strain gauges applied in the center of the specimen. Stress amplitudes are calculated using the measured strain amplitudes and the elastic constants according to Hooke's law. The cyclic strain amplitudes were selected prior to the fatigue experiments and no additional static preload was superimposed in the present constant amplitude experiments. Therefore cyclic loading was under fully reversed loading conditions (R=-1) for axial as well as torsional loading.

The load amplitude is measured with an electromagnetic coil. It detects the (axial or torsional) vibration amplitude at the specimens ends, which is exactly proportional to the strain amplitude in the center of the uncracked specimen. An amplitude control-unit guarantees that pre-given and actual displacement amplitude agree within 99 % during the experiment. Since crack initiation increases the compliance of the specimen and reduces the resonant frequency, control of the loading frequency is necessary. Monitoring the resonance frequency serves to interrupt the experiment after specimen failure and makes automatic operation possible.

Axial and torsional fatigue tests were performed in two environments:
1. Ambient air (20 - 22 °C, relative humidity 40 - 60 %)
2. Distilled water (20 - 22 °C, saturated with oxygen by permanent blow-by of ambient air)

Cyclic torsional and axial load is applied in pulses, each consisting of 1000 cycles in the present investigation. Pauses between the pulses serve to lead off the heat produced by internal friction in the specimen during loading. In air environment a fan is used and the pause length is chosen such that heating of the specimen is inhibited. In distilled water environment, the pause length was chosen 100 ms irrespective of the load level.

RESULTS

Results of fatigue tests under cyclic torsional or axial loading in ambient air and distilled water are summarized in Fig. 2. The fatigue number of cycles to failure are presented vs. torsional (open symbols) or axial stress amplitude (closed symbols). Arrows indicate unbroken specimen.

S-N curves for axial and torsional loading in ambient air do not show a pronounced fatigue limit, but the slope of the curves decreases within the whole investigated regime. Nevertheless, a "technical fatigue limit" may be defined on base of 10^9 cycles, since cycles to failure above 10^9 are of minor technical importance. Thus, the technical fatigue limit for axial loading of 2024-T351 in ambient air is about 96 MPa. For torsional fatigue loading, this technical fatigue limit is about 75 MPa. At same axial and torsional stress amplitudes in the range of 120 to 170 MPa, failure for axial loading needs typically 5 - 10 times longer than for torsional loading.

At cyclic torsional load amplitudes above approximately 170 MPa, the slope of the S-N curve in air decreases. This result may be attributed to the fact that this loading amplitude is close to the torsional yield strength. The yield strength for shear loading was not determined, but calculation according to the Mises criterion $\sigma_{eq} = \sqrt{\sigma^2 + 3\tau^2}$ yields a value of approximately 200 MPa.

The results of the axial loading fatigue tests in distilled water are indicated by closed triangles in Fig. 2. Failure occurs approximately 10 to 100 times earlier than in ambient air. The tension-compression fatigue curve measured in water approaches the S-N curve measured in ambient air neither for low nor for high stress amplitudes. Two specimens, which were loaded with a stress amplitude of 72 MPa did not break within more than $2x10^9$ cycles. Torsional loading fatigue loading experiments in distilled water (open triangles in Fig. 2) yield a fatigue limit of 57 MPa. The S-N curve for torsional loading is approximately parallel to the S-N curve for axial loading. The number of cycles to failure are about a factor 3 lower for torsional cycling for same torsional and axial stresses.

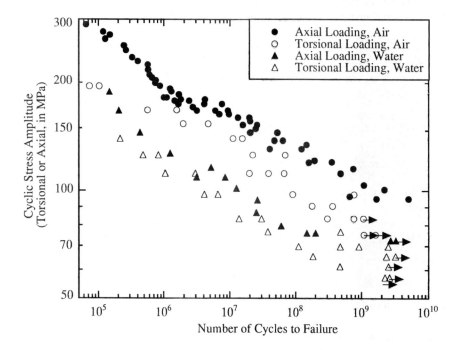

Fig. 2: S-N curves of Al 2024-T351 for axial loading in ambient air (dots) and in distilled water (closed triangles) and for torsional loading in air (circles) and in water (open triangles)

After loading with more than 10^8 cycles in distilled water, the specimen surfaces were covered with a grey aluminium oxide layer. This oxide layer is formed earlier and is more pronounced for axial than for torsional loading, and extensive flow of water around the specimen surface caused by the axial vibration of the specimens ends is probably the reason for this.

DISCUSSION

The S-N curves for torsional loading conditions are shifted about 15 - 20 % towards lower stress levels in comparison to axial loading. According to the Mises criterion a larger difference, namely by approximately 40 % ($1/\sqrt{3}$) had been expected. Several reasons are responsible for the relatively high number of cycles to failure in the torsional experiments in comparison to the axial loading experiments:

1. SEM investigations showed that fatigue crack initiation followed both directions of maximum shear stress i.e. normal and parallel to the specimen axis. The initial cracks were longer and more frequent in axial than in circumferential direction (second direction of maximum shear stress), which may be caused by the elongated shape of the grains. After crack formation, crack propagation likewise takes place in the planes of maximum shear stress (Fig. 3). Axial cracks, however, cannot cause specimen

failure, since the load bearing area is not reduced by this kind of cracks. They rather cause crack branching and secondary crack formation. Crack growth in radial direction is forced into different crack extension planes, which results in a "lamella type" fracture surface (Tschegg, 1985). Interaction of the crack fronts causes sliding mode crack closure, and crack branching as well as formation of multiple crack tips additionally reduce the effective stress intensity of the crack tip.

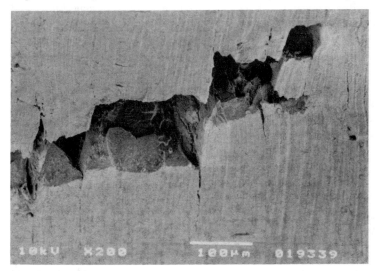

Fig. 3: Specimen surface after torsional loading in air ($\tau = 140$ MPa, 1.10×10^7 cycles)

2. Besides crack closure, the method of load control influences the number of cycles to failure in fatigue tests. In the present investigation, the displacement amplitude of the specimens ends was kept constant at a pregiven value during the whole fatigue experiment. The specimen compliance increases during fatigue crack propagation, and thus the nominal stress amplitude in the center of the specimen is reduced. In torsional loading experiments the crack extension period contributes significantly to the number of cycles to failure and therefore longer lifetimes were measured for the present displacement controlled fatigue tests in comparison to analogous load controlled experiments. In axial loading fatigue tests, however, the crack extension period is negligible small so that the life times for stress or displacement controlled fatigue experiments will be almost identical.

3. Using solid and hollow specimens, Tanaka et al. (1984) showed that the stress gradient towards the interior of a solid specimen causes an increased number of cycles to failure in torsional fatigue experiments. An analogous result is expected for the present torsional fatigue tests with solid specimens.

The influence of water on the fatigue lives is similarly pronounced for torsional as well as axial loading. It is assumed, that hydrogen embrittlement is mainly responsible for the reduction of fatigue lifetimes (Wei et al., 1989). Oxidation of aluminium surfaces produces free hydrogen, which is swept into the grains by diffusion and dislocation movement. Embrittlement of the material then leads to earlier crack initiation as well as a shorter crack propagation period. As a second mechanism, the formation of a protective oxide layer on the specimen surfaces probably plays some role for both loading conditions especially in the high cycle regime. It cannot be decided if the observed fatigue limits for axial as well as torsional S-N measurements in water are "intrinsic fatigue limits" or if the oxide layer prevents the surface from further environmental attack.

CONCLUSION

The high frequency resonance testing method has been used for torsional and axial fatigue testing of Al 2024-T351 in the very high cycle regime. Following results have been obtained:

1. The S-N curves for torsional loading conditions are shifted about 15 - 20 % towards lower stress levels in comparison to axial loading whereas the Mises criterion predicts a difference of about 40 %. Relatively high numbers of cycles to failure in torsional fatigue loading tests are caused by relatively long periods of mode II and mode III crack growth. Crack closure effects, the influence of load control and the stress gradient in the solid specimen cause slow crack growth velocities.

2. Water environment shifts the S-N curves for axial as well as torsional fatigue loading to about 1/3 lower stress amplitudes. Hydrogen embrittlement is assumed to be the reason for earlier crack initiation as well as a shorter crack propagation period.

ACKNOWLEDGEMENTS

The authors gratefully acknowledge financial support by the BMWF, Wien.

REFERENCES

Mayer H.R., E.K. Tschegg and S.E. Stanzl-Tschegg (1994). High Cycle Torsion Fatigue of Ceramic Materials Under Combined Loading Conditions (Cyclic Torsion + Static Compression). Proc. of 4th Int. Conf. on Biaxial/Multiaxial Fatigue, Paris, 2, 357-368

Mitsche R., S.E. Stanzl and D.G. Burkert (1973). Hochfrequenzkinematographie in der Metall-forschung. Wissenschaftlicher Film, 14, 3-10

Neppiras E.A. (1960). Very High Energy Ultrasonics. Brit J. Appl. Physics, 11, 143-150

Stanzl, S.E., M. Czegley, H.R. Mayer, H.R. and E.K. Tschegg (1989). Fatigue Crack Growth Under Combined Mode I and Mode II Loading. ASTM STP 1020 Eds. R.P. Wei and R.P. Gangloff 479-496

Stanzl-Tschegg S.E., H.R. Mayer and E.K Tschegg (1993). High Frequency Method for Torsion Fatigue Testing. Ultrasonics, 31, 4, 275-280

Tanaka K., S. Matsuoka and K. Kimura (1984). Fatigue Strength of 7075-T6 Aluminium Alloy Under Combined Axial Loading and Torsion. Fatigue Engng. Mater. Struct., 7, 3, 195-211

Tschegg E.K. (1985). Fatigue Crack Growth in High and Low Strength Steel Under Torsional Loading. Theor. and Appl. Fract. Mech., 3, 157-178

Wei R.P. and R.P. Gangloff (1989). Environmentally Assisted Crack Growth in Structural Alloys: Perspectives and New Directions. ASTM STP 1020, Eds. R.P. Wei and R.P. Gangloff, 233-264

A MECHANICAL CONDITION OF FATIGUE CRACK INITIATION FROM CORROSION PITS

M. NAKAJIMA* and K. TOKAJI**

* Department of Mechanical Engineering, Toyota College of Technology,
2-1 Eisei-cho, 471 Toyota, Japan
** Department of Mechanical Engineering, Gifu University,
1-1 Yanagido, 501-11 Gifu, Japan

ABSTRACT

Corrosion fatigue tests in 3%NaCl solution were conducted on a low carbon steel, S10C, a stainless steel, SUS304, and an aluminum alloy, 2024-T4, in order to clarify the crack initiation behavior from corrosion pits. The sizes of corrosion pits which concerned with crack initiation were measured, and the stress intensity factor was calculated by regarding the corrosion pit as a crack. The results were plotted on a log-normal probability paper. It was found that the stress intensity factors ranged $0.51 \sim 2.33 MPa\sqrt{m}$ in S10C, $0.37 \sim 1.40 MPa\sqrt{m}$ in SUS304, and $0.21 \sim 1.21 MPa\sqrt{m}$ in 2024-T4.

KEYWORDS

Corrosion fatigue; Corrosion pit; Crack initiation; Statistical treatment; Low carbon steel; Stainless steel; Aluminum alloy.

INTRODUCTION

Active path corrosion fatigue process consists generally of the initiation and growth of corrosion pits, followed by the initiation of fatigue cracks, growth and coalescence [Kitagawa et al., 1978]. Therefore, the corrosion fatigue life is defined as the sum of these processes. Since the initiation life of corrosion pits and the life after crack coalescence can be ignored practically, if the growth characteristics of corrosion pits and fatigue cracks are obtained [Nakajima et al., 1995], then the corrosion fatigue life can be estimated [Nakajima et al., 1993]. However, in this prediction, the transition condition from a corrosion pit to a fatigue crack needs to be known. Accordingly, it is particularly important to clarify the condition of fatigue crack initiation from corrosion pits. None the less, because of the difficulties in observations, the condition has not yet been studied in detail [Komai et al., 1987]. In the present study, fatigue tests in 3%NaCl solution have been conducted on three kinds of materials in order to establish the condition of fatigue crack initiation from corrosion pits.

EXPERIMENTAL PROCEDURES

Three kinds of materials, i.e. low carbon steel, S10C, stainless steel, SUS304, and aluminum alloy, 2024-T4, were used in this study. Table 1 shows the chemical compositions of these materials. After heat treatment, specimens with a minimum diameter of 5.5mm were machined and then electro-polished. Mechanical properties after heat treatment are shown in Table 2. Cantilever type rotating bending fatigue testing machines operating at a frequency of 19Hz were used for fatigue tests. Tests were performed in 3%NaCl solution whose temperature, pH and resolved oxygen content were 30 °C, 6.0 and 6.8ppm, respectively. Two kinds of measurements were employed for

697

Table 1. Chemical compositions (wt.%) of materials.

Material	C	Si	Mn	P	S	Cu	Ni	Cr	Mo	Mg	Zn	Ti	Fe	Al
S10C	0.12	0.17	0.50	0.014	0.025	0.02	0.02	0.03	-	-	-	-	Bal.	-
SUS304	0.05	0.48	1.50	0.027	0.003	-	9.45	18.60	0.13	-	-	-	Bal.	-
2024-T4	-	0.13	0.63	-	-	4.57	-	0.01	-	1.65	0.09	0.02	0.24	Bal.

Table 2. Mechanical properties of materials.

Material	Upper yield point σ_u MPa	Lower yield point σ_l MPa	Tensile strength σ_B MPa	Elongation δ %	Reduction of area ϕ %
S10C	337	223	370	36	72
SUS304	217*		587	78	82
2024-T4	419*		551	11	13

* 0.2% proof stress

the sizes, *i.e.* surface length, $2c$, and depth, a, of corrosion pits. One is the measurement on fracture surfaces after fatigue test using SEM. The other is the direct measurement on specimen surfaces during fatigue test. In the latter case, cyclic loading was applied to specimens for the number of cycles corresponded to the cyclic ratio, N/N_f, at which crack initiation was observed. Subsequently, fatigue tests were interrupted, and $2c$ and a were measured by a laser microscope. It is assumed that fatigue cracks are generated from corrosion pits when the stress intensity factor, K, calculated by regarding a corrosion pit as a sharp crack exceeds the threshold value. The K values for corrosion pits were calculated with the analytical solution developed by Shiratori *et al.* [1987].

RESULTS AND DISCUSSION

S-N curves

The S-N curves of S10C, SUS304 and 2024-T4 in room air and in 3%NaCl solution are shown in Figure 1 (a), (b) and (c), respectively. As shown in Fig.1(a), the fatigue limit for S10C in room air is approximately 180MPa, but is not observed in 3%NaCl solution. Fatigue lives less than 10^5cycles in both environments are almost the same. In SUS304 (Fig.1(b)), the fatigue limits in room air and in 3%NaCl solution are 255MPa and 245MPa, respectively. However, the fatigue lives at the stress level more than $\sigma=260$MPa in 3%NaCl solution are longer than those in room air. This is due to the cooling effect of 3%NaCl solution. As can be seen in Fig.1(c), while the fatigue limits for 2024-T4 are not observed in both environments, the fatigue strengths at 10^7cycles are 130 and 40MPa in room air and in 3%NaCl solution, respectively. Moreover, fatigue lives in 3%NaCl solution are considerably reduced at all stress levels compared with those in room air.

Macroscopic views of fracture surfaces

Macroscopic observations at two stress levels were made on fracture surfaces of S10C, SUS304 and 2024-T4, and the corrosion pits which seemed to lead the crack initiation were observed by SEM, then the K values of those corrosion pits were calculated. The results are shown in Fig.2~4. As shown in Fig.2 (a) and (b), macroscopic observation for S10C reveals that fractures at $\sigma=220$MPa and 80MPa were brought about by the initiation and growth of multiple cracks which seemed to be generated from many corrosion pits. The stress dependence of the K values is found, as can

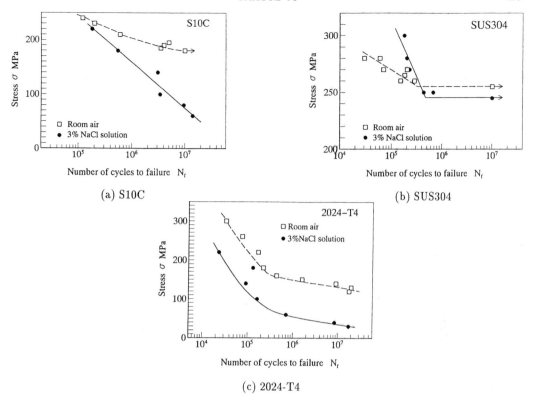

Fig.1. S-N curves.

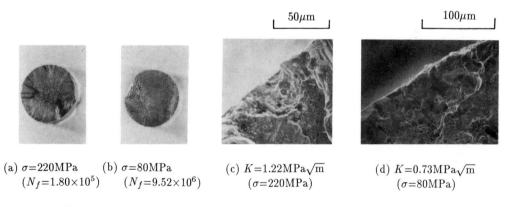

(a) $\sigma=220$MPa   (b) $\sigma=80$MPa   (c) $K=1.22$MPa$\sqrt{m}$ (d) $K=0.73$MPa$\sqrt{m}$
   ($N_f=1.80\times10^5$)   ($N_f=9.52\times10^6$)   ($\sigma=220$MPa)   ($\sigma=80$MPa)

Fig.2. Macroscopic views of fracture surfaces and corrosion pits in S10C.

be seen in Fig.2 (c) and (d). From macroscopic observations of fracture surfaces on SUS304 (Fig.3 (a) and (b)), it is found that fracture at higher stress level (280MPa) resulted from a few cracks, while at lower stress level (250MPa) from the growth of a single crack. However, it is also observed in SUS304 that cracks were generated from the sites other than corrosion pits. Therefore, it is not clear whether the cracks shown in Fig.3 (a) and (b) were generated from corrosion pits or not.

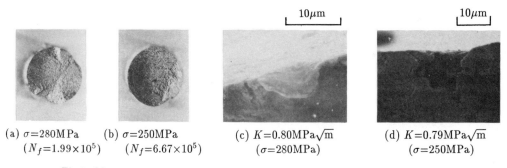

(a) σ=280MPa   (b) σ=250MPa   (c) K=0.80MPa$\sqrt{\mathrm{m}}$ (d) K=0.79MPa$\sqrt{\mathrm{m}}$
  (N_f=1.99×10^5)   (N_f=6.67×10^5)   (σ=280MPa)   (σ=250MPa)

Fig.3. Macroscopic views of fracture surfaces and corrosion pits in SUS304.

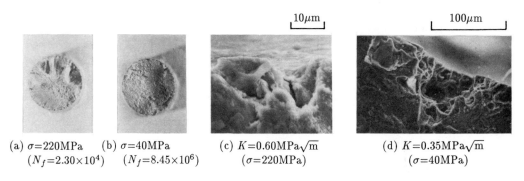

(a) σ=220MPa   (b) σ=40MPa   (c) K=0.60MPa$\sqrt{\mathrm{m}}$ (d) K=0.35MPa$\sqrt{\mathrm{m}}$
  (N_f=2.30×10^4)   (N_f=8.45×10^6)   (σ=220MPa)   (σ=40MPa)

Fig.4. Macroscopic views of fracture surfaces and corrosion pits in 2024-T4.

Corrosion pits observed on fracture surfaces are shown in Fig.3(c) and (d). Figure 4 shows the results for 2024-T4. Macroscopic observations reveals that fractures at both stress levels were caused by multiple cracks. Moreover, the stress dependence of the K values can be found, as observed in S10C.

Configurations of corrosion pits

Corrosion pits which generated fatigue crack were measured on fracture surfaces and on specimen surfaces in three materials. The sizes, $2c$ and a, of corrosion pits measured were plotted on a log-normal probability paper. Consequently, it was found that the distributions of $2c$ and a in three materials followed a log-normal distribution. Since much time is needed at lower stress levels until cracks were generated from the corrosion pits, $2c$ and a became larger at lower stress levels than at higher stress levels. Based on the above measurements, the aspect ratios, a/c, of corrosion pits were calculated and plotted on a log-normal probability paper as shown in Fig.5. The distributions of a/c in three materials follow a log-normal distribution. The stress dependence is not observed in all materials, but the a/c values obtained from fracture surface are somewhat larger than those from specimen surface.

Conditions for crack initiation

The K values in all materials plotted on a log-normal probability paper are shown in Fig.6. As can be seen in the figure, the K values increase with increasing stress level in S10C and 2024-T4, but they do not depend on stress level in SUS304. For instance, paying attention to the average K values obtained from specimen surface observation, the average K values at higher and lower

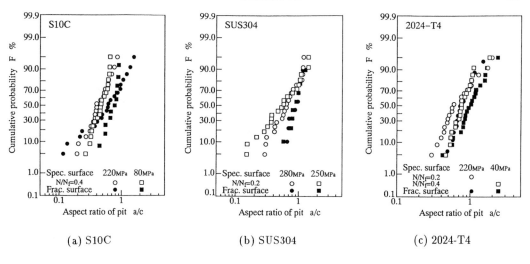

(a) S10C   (b) SUS304   (c) 2024-T4

Fig.5. Distributions of aspect ratios plotted on a log-normal probability paper.

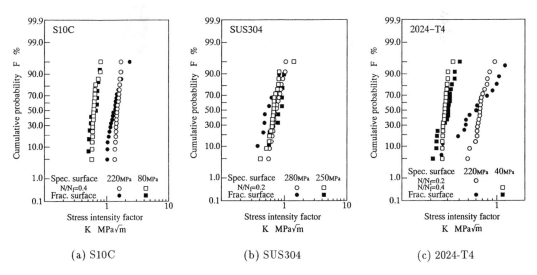

(a) S10C   (b) SUS304   (c) 2024-T4

Fig.6. Distributions of K values at which crack initiated from corrosion pits
plotted on a log-normal probability paper.

stress levels in each material are 1.48MPa$\sqrt{\mathrm{m}}$ and 0.62MPa$\sqrt{\mathrm{m}}$ for S10C, 0.74MPa$\sqrt{\mathrm{m}}$ and 0.67MPa$\sqrt{\mathrm{m}}$ for SUS304 and 0.67MPa$\sqrt{\mathrm{m}}$ and 0.28MPa$\sqrt{\mathrm{m}}$ for 2024-T4, respectively. The results in SUS304 is considered to be due to a small difference between two stress levels examined. The ranges of the K values obtained in three materials are summarized in Table 3, which as a whole are 0.51~2.33MPa$\sqrt{\mathrm{m}}$ for S10C, 0.37~1.40MPa$\sqrt{\mathrm{m}}$ for SUS304 and 0.21~1.21MPa$\sqrt{\mathrm{m}}$ for 2024-T4. Furthermore, it can be seen that the K values are almost the same between those observed on specimen surfaces and on fracture surfaces. Based on these results, it is possible for prediction to use the data obtained from fracture surfaces after fatigue tests. In a previous study [Nakajima *et al.*, 1993], the K values for SNCM439 (4340 steel) ranged 0.60~0.98MPa$\sqrt{\mathrm{m}}$ at a low stress of 100MPa, and 1.23~2.39MPa$\sqrt{\mathrm{m}}$ at a high stress of 400MPa. These values are in a good agree-

Table 3. K values at which crack initiated from corrosion pits.

Material	Stress σ (MPa)	Stress intensity factor K (MPa$\sqrt{m}$)	
		Specimen surface	Fracture surface
S10C	220	1.33 ~ 1.69	1.01 ~ 2.33
	80	0.57 ~ 0.79	0.51 ~ 0.77
SUS304	280	0.55 ~ 1.04	0.37 ~ 0.81
	250	0.41 ~ 1.40	0.56 ~ 0.98
2024-T4	220	0.49 ~ 0.95	0.26 ~ 1.21
	40	0.26 ~ 0.34	0.21 ~ 0.41

-ment with the results in this study. Finally, it will be possible to predict the fatigue crack initiation period using the corrosion pit sizes measured on specimen surface or fracture surface, provided that the growth characteristics of corrosion pits are already known.

CONCLUSIONS
Fatigue tests in 3%NaCl solution were conducted on three kinds of materials, $i.e.$ low carbon steel, S10C, stainless steel, SUS304, and aluminum alloy, 2024-T4, in order to establish the condition of fatigue crack initiation from corrosion pits. The conclusions obtained are as follows;
(1) The sizes of corrosion pits which concerned with crack initiation were larger at lower stress levels than at higher stress levels. The size distributions of corrosion pits in three materials followed a log-normal distribution.
(2) The stress intensity factor, K, was calculated by regarding the corrosion pit as a crack. The K values, at which cracks initiated from corrosion pits, at higher and lower stress levels were 1.01~2.33MPa$\sqrt{m}$ and 0.51~0.79MPa$\sqrt{m}$ for S10C, 0.37~1.04MPa$\sqrt{m}$ and 0.41~1.40MPa$\sqrt{m}$ for SUS 304, and 0.26~1.21MPa $\sqrt{m}$ and 0.21~0.41MPa$\sqrt{m}$ for 2024-T4, respectively.
(3) The K values increased with increasing stress level in S10C and 2024-T4, but they did not depend on stress level in SUS304. The results in SUS304 were due to a small difference between two stress levels examined.
(4) The K values were almost the same between those obtained on specimen surface and on fracture surface. Therefore, it is possible to predict the fatigue crack initiation period using the corrosion pit sizes obtained on specimen surface or fracture surface, provided that the growth characteristics of corrosion pits are already known.

REFERENCES
Kitagawa,H., T.Fujita and K.Miyazawa (1978). Small randomly distributed cracks in corrosion fatigue. In: *ASTM STP 642*, 98-114.
Nakajima,M. and K.Tokaji (1995). Fatigue life distribution and growth of corrosion pits in a medium carbon steel in 3%NaCl solution. *Fatigue Fract. Engng Mater. Struct.*, **18**, 345-351.
Nakajima,M., H.Kunieda and K.Tokaji (1993). A simulation of corrosion fatigue life distribution in low alloy steel. In: *Fatigue Design*, ESIS **16**, 269-281.
Komai,K., K.Minoshima and G.Kim (1987). Corrosion fatigue crack initiation behavior of 80kgf /mm^2 high-tensile strength steel weldment in synthetic sea water. *J.of Soc.Mater.Sci., Jpn.*, **36**, 141-146. (in Japanese)
Shiratori,M., T.Miyoshi, Y.Sakai and G.-R.Zhang (1987). Analysis of stress intensity factors for surface cracks subjected to arbitrarily distributed surface stresses; 3rd report. *Trans.Jpn.Soc. Mech.Eng.*, **53(A)**, 779-785. (in Japanese)

FATIGUE CRACK PROPAGATION BEHAVIOR OF MAGNESIUM-ALUMINUM ALLOY IN AIR AND ARGON

YASUO KOBAYASHI, TOSHINORI SHIBUSAWA and KEISUKE ISHIKAWA

Department of Mechanical Engineering,Toyo University
2100 Kujirai-Nakanodai, Kawagoe, Saitama, Japan 350

ABSTRACT

We have studied the effect of the atmosphere on the fatigue crack propagation as well as of the testing conditions and of the microstructures of the alloys. We have obtained the interesting results of the intrinsic and extrinsic influences on the crack fatigue propagation of the Mg-Al-Zn alloys. The inert atmosphere (argon) retards the fatigue crack propagation rate in comparison with the air. The precipitates produce the acceleration of the fatigue crack propagation, in spite of the improvement in the strength of the alloy. The effect of the loading frequency is remarkable. The lower the frequency, the higher the fatigue crack propagation rate. The fatigue behaviors of the Mg-Al-Zn alloy strongly depend upon the environment and the oxygen in particular.

KEYWORDS

Magnesium alloy; fatigue crack propagation; atmosphere; heat-treatment; microstructures.

INTRODUCTION

Magnesium alloys are very important materials since the specific strength is high. Recently, magnesium alloys are applied to the transportation system for reducing the weight from the viewpoint of energy saving. However, the alloys are much chemically active and so sensitive to the environment, especially on the fresh surface. Fatigue produces repeatedly fresh surfaces at the propagating crack tip. It is practically important to understand the interaction with the atmosphere. Hence, we have studied the effect of the atmosphere on the fatigue crack propagation as well as of the testing conditions and of the microstructures of the alloys.

EXPERIMENTAL

Material end Heat-treatment

The magnesium alloy for this experiment is AZ91D, which is a modified pore free Mg-Al-Zn alloy.

Therefore, this alloy is heat-treatable. The chemical composition is shown in Table 1 and the applied heat-treatments are also given in Table 2. The microstructures contain neither void nor pore but a small number of magnesium oxides are observed in the matrix. The most microstructures are dual phases composing of lamellar nodules and α-matrix except for the solid solution.

Table 1. Chemical composition of AZ91D (mass%)

Al	Zn	Mn	Si	Fe	Be	Ni	Cu	Mg
9.0	0.68	0.17	0.051	0.002	0.0015	0.001 >	0.001 >	Bal.

Table 2. Heat tretment of AZ91D

Simbol	Heat treatment
F	as cast
T4	693K , 7.2×10^{3}s (water quenched)
T6a	693K , 7.2×10^{3}s (water quenched) + 448K , 5.76×10^{4}s (air cooled)
T6b	693K , 7.2×10^{3}s (water quenched) + 448K , 4.6×10^{5}s (air cooled)

Tests

The fatigue test was carried out at a room temperature. The specimen dimension used for the fatigue crack propagation test is shown in Fig. 1. The thickness of the tested specimen is 8 mm. The system for the fatigue test in an argon atmosphere is illustrated in Fig.2. The fatigue crack length (a) was estimated by the Saxena's equation (Saxena and Hukad, 1978) through the crack opening displacement directly measured with a clip gauge. The relationship between the fatigue crack propagation rate (FCPR), da/dN and the stress intensity amplitude, ΔK was obtained with the aid of the microcomputer.

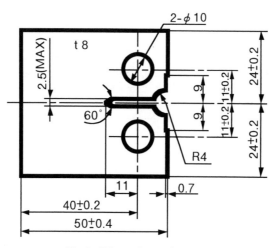

Fig.1. CT specimen (in mm)

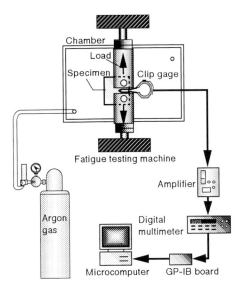

Fig.2. Measurement system of fatigue
crack propagation behavior

EXPERIMENTAL RESULTS

We have investigated the effect of heat-treatment (microstructures), environment, loading frequency and stress ratio on the relationship between FCPR and ΔK for the magnesium alloy.

Effect of heat-treatment

Figure 3 shows the effect of the heat-treatment on the fatigue crack propagation behaviors.The alloy in the solid solution condition (T4) yields the lowest FCPR among them. The as-cast (F) and heat-treated alloys (T6a and T6b) are dual phases with the α–matrix and the nodule. The existence of the nodule in itself accelerates the FCPR. Besides, the precipitates in the matrix further promote the FCPR. Both the nodules and the precipitates reduce the FCPR as well as the S-N behaviors (Ishikawa et al., 1995). In spite of the increase in the yield and the tensile strength, they do not improve the fatigue properties of the alloy. They might be more brittle.

Effect of environment

The effect of the environment on the FCPR is very remarkable as shown in Fig.4. The inert atmosphere reduces the FCPR. Oxygen would be an important factor. Magnesium has an affinity for oxygen. Especially, the fresh surfaces produced through cyclic loading are much sensitive to the oxygen. The rapid formation of the oxide thin films on the fatigue cracked surfaces could bring the irreversibility of the crack front under the unloading. However, the brittleness of the oxide film would not contribute to the effect of the closure compared with steels and aluminum alloys.

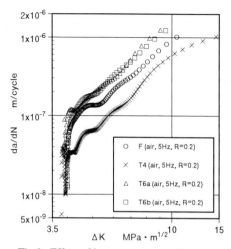

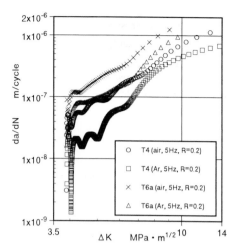

Fig.3. Effect of heat treatment on fatigue
crack propagation rate for AZ91D

Fig.4. Effect of environment on fatigue
crack propagation rate for AZ91D

Effect of loading frequency

The effect of loading frequency on the FCPR are shown in Fig.5. The lower the frequency is, the higher the FCPR is. The behavior was observed for both heart-treated and solution treated materials. But the effect in the latter is larger than in the former. The nodule would appear to be more sensitive to the environment. Although the nodule accelerates the FCPR, the morphology of the fractured surfaces do not change. The fractography reveals a mixture of the striation and the brittle flat pattern.

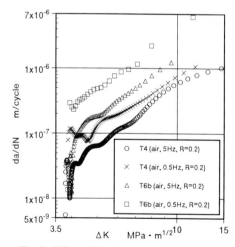

Fig.5. Effect of frequency on fatigue crack
propagation rate for AZ91D

Effect of stress ratio

The effect of applied stress ratio, $R = P_{min}/P_{max}$ on the FCPR is shown in Fig.6. The minimum ΔK, so-called ΔK_{th} reduces with decreasing in R. Furthermore, the FCPR increases a little, too. We can expect the effect of the crack closure. The effect could be associated with the properties of the oxide films produced on the fresh surfaces, which are separated through the cyclic loading. But, the brittleness of the oxide films could not bring the clear effect of the closure.

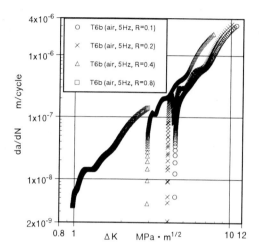

Fig.6. Effect of stress ratio on fatigue crack
propagation rate for AZ91D

DISCUSSION

We summarized the experimental result in Figs. 7 and 8 with the other important engineering materials (Liaw et Al., 1983). The fatigue data of magnesium alloys are very rare and so we can not evaluate our results in comparison with the others. These experimental results reveal that the effect of the environment is outstanding as well as the testing conditions. The main and common cause would be the existence of oxygen. Mainly, the oxidation on the fresh surfaces brought through the fatigue crack propagation plays the important role.

CONCLUSION

We have got the following conclusions in this experiment.
1. The aging, which bring the nodules and the precipitates, accelerates the FCPR of Magnesium-
   Aluminum-Zinc alloy in air and argon.
2. The inert atmosphere (argon) retards the FCPR of the alloy.
3. The lower the loading frequency is, the higher the FCPR is both in air and argon.
4. The higher mean stress (the lower stress ratio) brings the lower ΔK_{th}.
5. The FCPR of the magnesium alloy strongly depends upon the environment, namely the oxidation
   on the fresh surfaces made through the cyclic loading.

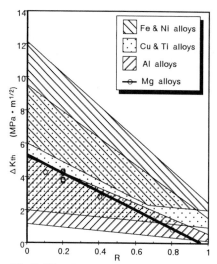

Fig.7. Threshold stress intensity factor range v.s stress ratio for various alloys

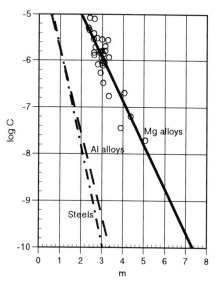

Fig.8. Characteristic parameters, logC and m in Paris law for various alloys

ACKNOWLEDGEMENTS

We are grateful for the support provided by the Iketani Science and Technology Foundation. We also thank Nippon Light Metal Co., Ltd. for the supply of the magnesium alloy.

REFERENCES

Ishikawa, K., Y. Kobayashi and T. Shibusawa (1995). Characteristics of fatigue crack propagation in heat treatable cast magnesium alloy. to be published in Light Metals '95. TMS.

Laiw,P.K., T.R. Leax and W.A. Logsdon (1983). Near-threshold fatigue crack growth behavior in metals. Acta Metallurgica, 31, 1581-1587.

Saxena, A. and S.J. Hukad, Jr. (1987). Review and extension of compliance information for common crack growth specimen. International Journal of Fracture, 14, 453-468.

FATIGUE AND CORROSION FATIGUE MECHANISMS IN HIGH PURITY NICKEL AND Ni-Cr-Fe ALLOYS

T. MAGNIN, N. RENAUDOT and J.STOLARZ

Ecole des Mines, Centre SMS, URA CNRS 1884
158, Cours Fauriel, 42023 Saint-Etienne Cedex 2, France

ABSTRACT

Low cycle fatigue and corrosion fatigue mechanisms of ultra high purity Ni and Ni base alloys in H_2SO_4 solutions are analysed. A particular attention is paid on the effect of chromium on both corrosion and cyclic plasticity and on the effect of cathodic hydrogen on damage processes.

KEYWORDS

Fatigue; corrosion fatigue; ultra high purity Ni base alloys; hydrogen embrittlement; persistant slip bands; corrosion-deformation interactions.

INTRODUCTION

One of the major problem for steam generator tubes in Pressurized Water Reactors (PWR) is the resistance of Ni base alloys to stress corrosion cracking (SCC) and corrosion fatigue (CF). Recent studies of mechanisms (Rios et al., 1995, Magnin, 1995) involve the effect of hydrogen on the damage process.

Thus the aim of the present paper is to analyse the effect of cathodic hydrogen on the corrosion fatigue damage of ultra pure Ni and Ni base alloys (corresponding to the fcc matrix of Inconel 600 and Inconel 690) in a 0.5 N H_2SO_4 solution at room temperature. Localized corrosion-deformation interactions are taken into account (Magnin, 1995) through the relation between cyclic plasticity, strain localization and hydrogen absorption. Moreover the effect of chromium which is known to increase the corrosion resistance of Ni base alloys on such interactions is emphasized.

EXPERIMENTAL PROCEDURE

Ultra high purity (UHP) Ni, Ni-16Cr-9Fe (Inconel 600 matrix) and Ni-30Cr-9Fe (Inconel 690 matrix) specimens at the mill annealed state have been tested in low cycle corrosion fatigue at prescribed plastic strain $\Delta\varepsilon_p/2$ and total strain rate $\dot{\varepsilon}_t$ in symmetrical tension-compression ($\Delta\varepsilon_p/2 = 2 \times 10^{-3}$, $\dot{\varepsilon}_t = 2 \times 10^{-3}$ s^{-1}). Thus the effect of impurities, such as sulfur, on localized corrosion and grain boundary embrittlement is avoided. Ultra pure alloys contain less than 10 ppm of C, O, S, N.

A 0.5 N H_2SO_4 solution at room temperature is used for tests at free potential and at applied cathodic potential to promote hydrogen reduction. A previous study (Jui-Ting Ho and Ge-Ping Yu, 1995) has shown that a galvanic cell may be established, in the power plant, upon contact due to the fact that the Ni base alloy is cathodic relative to the carbon steel support plate. Moreover, a particular attention is paid on Persistant Slip Bands (PSB) formation, using the scanning electron microscope to examine intrusions-extrusions.

RESULTS

Figure 1 shows the evolution of the tensile maximum stress of the three ultra high purity alloys as a function of the number of cycles and the electrochemical conditions.

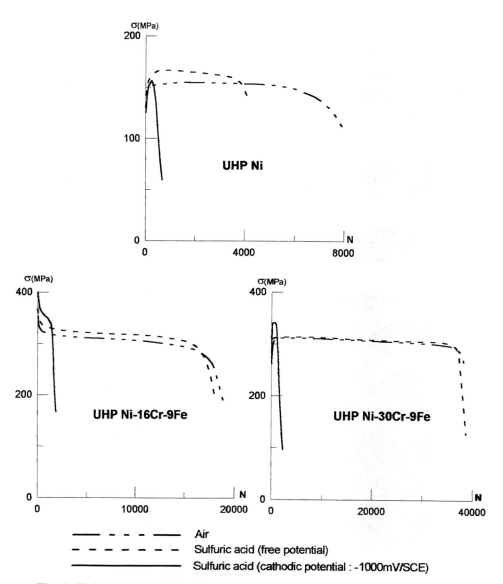

Fig. 1. Fatigue tests on UHP Ni and 2 UHP Ni base alloys in 3 different environmental conditions.

Three important points can be noticed :

(i) The increase of chromium content increases the fatigue life in air.

(ii) The low cycle fatigue life of Ni is reduced at free corrosion potential in comparison to air, in contrast with Ni-16Cr-9Fe and Ni-30Cr-9Fe alloys.

(iii) Tests at applied cathodic potential induce a very marked reduction of the fatigue life whatever the material.

DISCUSSION

Effect of Chromium Content

This effect can be related both to corrosion resistance and to cyclic plastic deformation processes. Cr_2O_3 formation on Ni base alloys induces a more marked protection against corrosion than NiO. The electrochemical rest potential of Ni base alloys increases with Cr content. Thus, it is not surprising to observe an increase of the low cycle corrosion fatigue life of Ni alloys at free potential.

The influence of Cr on cyclic plasticity can be related to the decrease of the cross slip ability with the increase of Cr content. It has been shown (Clément, 1975) that the Stacking Fault Energy (SFE) of Ni alloys decreases when Cr content increases. Thus, cross slip is easier in pure Ni than in Ni base alloys. Moreover the formation of PSB and of related intrusions-extrusions (which are crack nucleation sites) during low cycle fatigue is promoted by the cross slip ability, because dislocation annihilation is then favoured (Magnin, 1991).

Figure 2 shows an example of such intrusions-extrusions at the specimen surface of pure Ni, with crack nuclei at PSB/matrix interfaces. This crack nucleation is accelerated for pure nickel in comparison to Ni-Cr alloys, which explains the increase in fatigue life in air with Cr content.

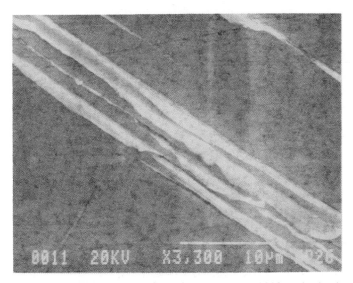

Fig. 2. Extrusions on the surface of UHP Ni after 1000 cycles in air

Another coupled effect on crack initiation must be considered. It is related to the grain size effect and the correlated PSB/grain boundary interactions. In fact, observations of the specimen top surfaces show that crack initiation in air is intergranular in pure Ni (see fig. 3) but transgranular in Ni base alloys.

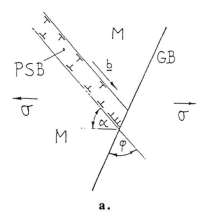

a.

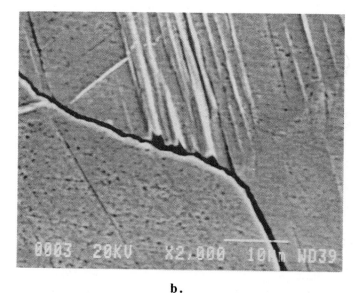

b.

Fig. 3. PSB-grain boundary interaction (it leads to an intergranular crack
initiation) : **a.** Geometry of optimal interaction ($\alpha = 45°$ and $\varphi =$
70.5°) (Mughrabi, 1985); **b.** PSB-grain boundary interaction on
the surface of UHP Ni after a test in sulfuric acid (free potential).

This can be explained by the difference in grain size (more than 250 μm for pure Ni, less than 150 μm
for Ni-Cr alloys) through the interaction between PSB and grain boundaries illustrated by fig. 3.

It has been shown (Esmann *et al.*, 1981, Mughrabi, 1982, 1985) that PSB-grain boundary interaction
is governed by their angle and the height of the protusion formed in PSB. This height increases with
grain size, which increases the incompatibilities of deformation at grain boundaries. Cracks can then
initiate more easily at grain boundaries as clearly shown on fig. 3b. This effect will be of importance for
localized corrosion-deformation and hydrogen embrittlement, as we are now going to discuss.

Effect of Hydrogen

The marked reduction in the fatigue life of the three ultra pure materials tested in H_2SO_4 at cathodic potential is due to hydrogen effects, as it has been previously suggested (Was et al., 1981, Totsuka et al., 1987, Jui-Ting Ho and Ge-Ping Yu, 1995). Complementary tests have been performed to emphasize this hydrogen effect on fatigue life decrease. After about 25 % of the fatigue life at free potential (to be sure that surface microcracks are formed), a cathodic potential is applied (fig. 4). For pure Ni with an intergranular crack initiation process, the effect of hydrogen is quite rapid, which can be related to a rapid diffusion at grain boundaries (short-circuit effect). Nevertheless, in the case of transgranular crack initiation (for Ni-Cr alloys with low grain size), the hydrogen effect is delayed even if it finally induces a marked damage.

It must be also noticed that the potential changes made on Ni base alloys after 90% of the life time during the bulk propagation (see fig. 4), show a faster crack velocity at cathodic potential than at free potential.

CONCLUSION

Chromium has a beneficial effect on the fatigue and corrosion fatigue resistance at free potential of the Ni base alloys in 0.5 H_2SO_4. This effect is related both to a better protection against corrosion through Cr_2O_3 formation and to a delayed extrusion-intrusion formation in PSB during cyclic plasticity.

Nevertheless, UPH Ni and Ni-Cr-Fe alloys are very sensitive to hydrogen effect when cathodic potentials are applied. In this case, one cannot involve the influence of impurities on the embrittlement. Grain boundaries act as short-circuits for hydrogen diffusion.

Moreover, the cyclic plastic properties of alloys must be taken into account particularly through the grain size effects and the correlated PSB-grain boundary interactions. Small grain sizes delayed intergranular embrittlement because transgranular fatigue crack initiation is then favoured.

So, to improve Ni base alloys fatigue resistance in H_2SO_4 solutions, it is necesary to increase the Cr content in the alloys and to favour small grain size.

REFERENCES

Clément N. (1975). Thesis, n°655, France, 86

Esmann U., U. Goesele and H. Mughrabi (1981). Phil. Magazine A, 44, n° 2, 405-426

Jui-Ting Ho and Ge-Ping Yu (1995). Effect of hydrogen on the fatigue behavior of alloy 600 at cathodic potential, Scripta Metallurgica & Materiala, 32, n°11, 1845-1849

Magnin T. (1991). Développements récents en fatigue oligocyclique sous l'angle de la métallurgie physique, Mémoires et Etudes Scientifiques, Revue de Métallurgie, 88, n° 1, 33-48

Magnin T. (1995). Recent advances for corrosion-fatigue mechanisms, Isij International, 35, n° 3, 223-233

Mughrabi H. (1982).A model of high cycle fatigue crack nucleation at grain boundaries by persistant slip bands, Second International Symposium on Defects, Fracture and Fatigue and 7th Canadian Fracture Conference, 1-3

Mughrabi H. (August 1985). Cyclic deformation and fatigue : some current problems, Plenary Paper, ICSMA 7, 3

Rios R. And T. Magnin, D. Noël and O. de Bouvier (1995). Critical analysis of alloy 600 Stress Corrosion Cracking mechanisms in primary water., Metal. and Mat. Trans. A, 26A, 925-939

Totsuka N., E. Lunarska, G. Gragnolino, Z. Szklarska-Smialowska (1987). Corros., 43, 505

Was G.S. and H.H. Tischner, R.M. Latanision, R.M. Pelloux (1981). Metal. Trans. A, 12A, 1409

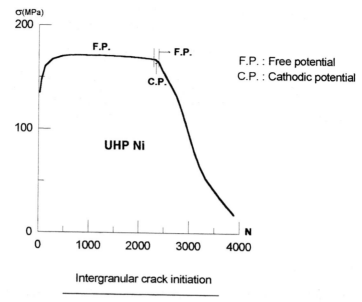

Intergranular crack initiation

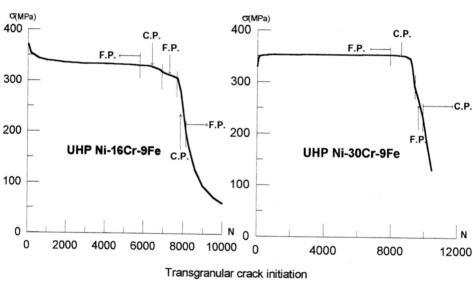

Transgranular crack initiation

Fig. 4. Fatigue tests on UHP Ni and UHP Ni base alloys in sulfuric acid
(0.5N) with modifications of potential during testing.

ENVIRONMENTALLY-ASSISTED FATIGUE CRACK PROPAGATION: SOME CRITICAL ISSUES

Gilbert HÉNAFF and Jean PETIT

Laboratoire de Mécanique et de Physique des Matériaux - URA CNRS 863
ENSMA - Site du Futuroscope - BP 109 - Chasseneuil du Poitou
F-86960 FUTUROSCOPE CEDEX
Tel: (33) 49.49.82.33 - Fax: (33) 49.49.82.38 - Email: henaff@ensma.univ-poitiers.fr

ABSTRACT

The present paper is addressing the understanding of environmentally-assisted fatigue crack propagation in the near-threshold region. The difficulty to separate the respective role of different parameters is first highlighted. Then the role of moisture in promoting fatigue crack growth is emphasised. In particular it is shown that traces of moisture in any gaseous atmosphere are sufficient to significantly enhance crack growth. The micromechanisms are also discussed.

KEYWORDS

environmentally-assisted propagation; threshold; crack closure; water vapour; adsorption; hydrogen.

INTRODUCTION

A literature survey of the studies dedicated to the fatigue resistance of engineering alloys during the last two decades brings to the fore a pronounced interest for the near-threshold fatigue crack propagation behaviour of engineering alloys. The reason for this is twofold. From an engineering point of view there was a need of reliable and reproducible data in order to assess the damage tolerance of components which had to withstand a high number of low-amplitude loading cycles. More fundamentally the basic fatigue crack growth micromechanisms in this region were ill-known in sharp contrast with the so-called Paris regime formerly extensively investigated. As a result, different parameters were identified to influence more or less deeply the near-threshold propagation:
    - some intrinsic parameters like grain size, cyclic behaviour, etc....
    - plus extrinsic parameters like the load ratio (and its connection with crack closure (Elber, 1971)) and the environment.
The main difficulty encountered when tackling the issue of thoroughly understanding the respective role of these parameters resides in their complex interaction. The development of an interdisciplinary approach encompassing physical metallurgy, fracture mechanics or surface chemistry is therefore required. The present paper is addressing this issue and aim to shed a new light on the role of environmentally-assisted fatigue crack propagation in the near-threshold region.

ON THE DIFFICULTY TO SEPARATE THE RESPECTIVE ROLE OF DIFFERENT PARAMETERS

The need to clearly separate the respective influence of different parameters involved in the near-threshold propagation has been highlighted in a previous paper. In particular it was shown that neglecting crack-closure effects or sequence effects in the case of variable amplitude loadings could turn into confusion when trying to understand the specific role of environment. Another example, introducing the next part of the paper, is presented here and is provided by the work completed by Ritchie and Suresh (Ritchie *et al.*, 1980; Suresh *et al.*, 1983; Suresh *et al.*, 1981) in the early 80's. They perform testings on a pressure vessel steel in so called dry environments (helium or hydrogen) and a moist environment, namely ambient air, at different load ratios.

Literature is well documented on the atmospheric moisture effects on the fatigue resistance of metals (Bennet, 1963; Petit *et al.*, 1994). The observed decrease in the fatigue strength of steels and aluminium alloys is generally considered to result from a kind of hydrogen embrittlement. The hydrogen would evolve from a former surface reaction between adsorbed water vapour molecules and the base metal. Therefore moist atmospheres like ambient air are expected to substantially enhance the growth process.

The authors reached the following conclusions:

- at near-threshold levels, for R=0.05 crack growth rates in dry atmospheres are two orders of magnitude higher than in air, resulting in substantial differences in the threshold values;

- conversely at R=0.75 the threshold values are similar whatever the environment.

In order to explain this behaviour and the interaction between the influence of R and the environmental effect, the authors introduced the concept of *oxide-induced closure*. Thus in a moist environment an oxide film forms on the rupture surfaces and for low R ratios is subsequently thickened by fretting oxidation. The crack is wedged and the effective stress intensity amplitude ΔK_{eff} consequently lessened. Conversely, the lower the moisture level, the thinner the oxide layer and the higher the crack growth rates. It is however worth noticing that the role of closure was largely inferred and no measurement assessed these conclusions.

Liaw et al. (Liaw *et al.*, 1982; Liaw *et al.*, 1983) performed the same kind of experiments on higher strength Ni-Mo-V and Ni-Cr-Mo-V steels, but by performing closure measurements. They found that the oxide-induced closure was unable to fully account for the observed behaviours. Actually, after closure correction, the crack growth rates measured in dry argon and dry hydrogen were similar and lower than those obtained in air. They concluded that in these gaseous environments crack growth rates were mostly controlled by the residual moisture content. The same conclusion had been reached previously by Wei and co-workers (Spitzig *et al.*, 1968; Wei *et al.*, 1972). However no sound explanation based on micromechanisms was provided to explain differences between air and argon or hydrogen atmospheres containing residual moisture. Recent progress in the understanding of environmentally-assisted fatigue crack propagation sheds a new light on these results.

THE EXISTENCE OF AN ADSORPTION-ASSISTED REGIME

In the sequential process proposed by Wei (Wei *et al.*, 1981; Weir *et al.*, 1980), *hydrogen embrittlement* is the mere mechanism involved in the deleterious effect of a hydrogenous atmosphere, although some previous step like transport of active species and surface reaction kinetics might be rate-controlling. In particular this means that surface reaction by itself does not play a significant role in promoting crack growth.

Results obtained under very low water vapour partial pressure (10^{-3} Pa) and under closure-free conditions in the case of a high-strength low-alloy steel clearly demonstrated that this residual moisture content was sufficient to induce a significant increase in crack growth rates especially when the load frequency was decreased (fig. 1)(Hénaff *et al.*, 1993). The curve obtained at low frequency (0.2Hz) exhibits a linear relationship on the explored with a Paris law exponent m=4 like the intrinsic stage II regime(Petit *et al.*, 1991). It was deduced that, in this saturated regime, the basic mechanism is not deeply affected and could still be described by the formulation derived from Weertman's model

(Hénaff, *et al.*, 1993; Weertman, 1966): $\dfrac{da}{dN} = \dfrac{A}{D^*}\left(\dfrac{\Delta K_{eff}}{E}\right)^4$ where D* denotes the critical

displacement leading to rupture. Within this framework it was assumed that the *adsorption of water vapour molecules* could by itself lessen the value of the energy required to create a unit area of cracked surface or the value D^* and thus enhance the propagation. D^* is thus closely related to the surface coverage rate θ.

A model has been proposed to provide a comprehensive picture of the processes involved in the behaviour observed on figure 1 (Hénaff *et al.*, 1995). Actually the pressure at the crack tip results from a compete between transport of active species and consumption by adsorption of water vapour molecules on fresh surfaces. The transport of active species to the crack tip is calculated by taking into account the crack impedance effect. The impedance is actually mostly governed by the crack opening displacement and therefore the stress-intensity factor value. This crack impedance effect induces a pressure in a control volume located at the crack tip lower than the nominal pressure. The consumption by adsorption is estimated on the basic assumption that once a molecule impinges the surfaces, it sticks to it provided that the collision site is free. The saturating regime is reached as soon as all available sites are occupied ($\theta=1$). The lower regime exhibited at higher frequency (typically 35Hz) corresponds to the case $\theta=0$. The results provided by this model are in fair agreement with experimental data (solid line on figure 1).

A compendium of data from ancillary testings and from literature proves that the adsorption-assisted regime might be encountered for a wide variety of materials and environmental conditions (figure 2). When the moisture content is increased a hydrogen-assisted failure process is assumed to take place and then induces a much more dramatic enhancement in crack growth rates. The detailed mechanisms are discussed elsewhere. The purpose of the next section is to show that this distinction between two distinct environmental effects offers a new tentative interpretation for the behaviours obtained in gaseous atmospheres containing traces of water vapour.

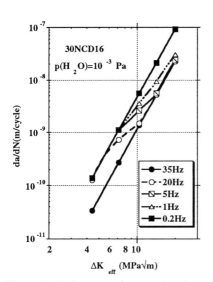

Figure 1: Fatigue crack growth enhancement under very low water vapour pressure.

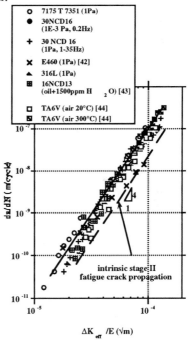

Figure 2: compendium of fatigue crack growth data related to adsorption-assisted propagation.

A RE-ASSESSMENT OF RESULTS IN MOIST ATMOSPHERES

We now reconsider the results obtained on the Ni-Mo-V steel in dry argon (Liaw, *et al.*, 1983). The results obtained in the 448 kPa argon atmosphere at R=0.5 (crack closure effects are here supposed to be negligible) are compared in figure 3 to the two characteristic regimes previously discussed. The behaviour in each case is similar to what is observed on the 30NCD16 under very low water vapour pressure at intermediate frequency. This similarity assesses the presence of residual moisture in the test environment and its active role. At low ΔK values the adsorption process saturates; then, as the crack grows faster as the ΔK increases, the adsorption influence progressively decreases (transient regime) until reaching the intrinsic behaviour. The results provided by the proposed model are in good agreement with these experiments. The same remarks apply to the case of the Ni-Cr-Mo-V steel (Hénaff, et al., 1995; Liaw, et al., 1982).

On the opposite it is worth noticing that the moisture content in these experiments is not sufficient to trigger a hydrogen-assisted regime as observed on propagation in nitrogen containing traces of water vapour (Hénaff, et al., 1993).

Liaw and co-workers conducted the same kind of experiments under 448kPa of hydrogen at 93°C on the Ni-Mo-V steel (Liaw, et al., 1983). Similar testings have also been performed by Stewart at room temperature (Stewart, 1980) using a procedure assumed to produce closure-free data. The results are compared on figure 4 with the saturating adsorption regime described above. Obviously all these curves follow this adsorption-assisted regime but are always below the curve obtained at ambient air where hydrogen effects are assumed to take place. The underlying physical significance is then under question. An explanation could be that the preferential adsorption of water vapour with respect to hydrogen would prevent, at least at slow growth rates, the triggering of the hydrogen-related process (figure 5). However in the case of the Ni-Cr-Mo-V steel (Liaw, et al., 1982) it seems that the propagation at slow growth rates is assisted by the hydrogen-assisted process. Nevertheless this effect takes place at a lower growth rate than in air and is less pronounced than in this environment. As the moisture content is not controlled, an explanation could be that the water vapour content is higher in this latter case than in the former one. The similarity between these results and those obtained under nitrogen containing 15 to 150 ppm. of water vapour is noteworthy.(Bignonnet et al., 1983; Hénaff, et al., 1993).

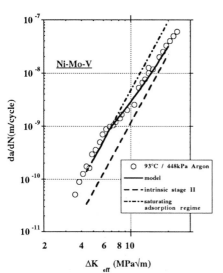

Figure 3: comparison of the adsorption model results with experimental data of Liaw et al. (1983).

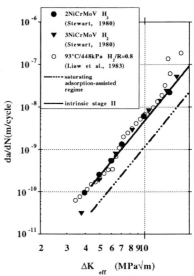

Figure 4: fatigue crack growth in gaseous hydrogen containing traces of moisture.

Conversely, at higher growth rates (da/dN>10^{-9}m/cycle in (Liaw, et al., 1982; Liaw, et al., 1983)), hydrogen would substitute to water vapour when the number of adsorption sites at the crack tip increases to ensure a formation of a complete adsorbed layer. However the sound conclusion is that water vapour has in this case a beneficial effect in preventing or in delaying and lessening hydrogen embrittlement presumably by means of competitive adsorption. This role of competitive adsorption has been considered by other authors in the past (Frandsen et al., 1977; Hofmann et al., 1965).

Finally, for these experiments under gaseous hydrogen, rupture surfaces exhibit intergranular facets identical to those observed in ambient air (Cooke *et al.*, 1975; Hénaff *et al.*, 1992; Stewart, 1980). The proportion of facets in air seems in some cases to be deeply related to the load ratio (Hénaff, *et al.*, 1992; Stewart, 1980) but this issue is not clear and will be discussed in a next paper. Such facets are generally considered as a manifestation of hydrogen embrittlement. Liaw et al. formerly noticed that "there is no one-to-one correspondence between the cracking propensity and the percentage of intergranular fracture", suggesting that there is no relationship between the number of facets and the intensity of environmental effects. The present observations further suggest that there is no relationship between intergranular facets and hydrogen-assisted propagation since there are encountered in conditions where the hydrogen-related process is assumed to be inoperative.

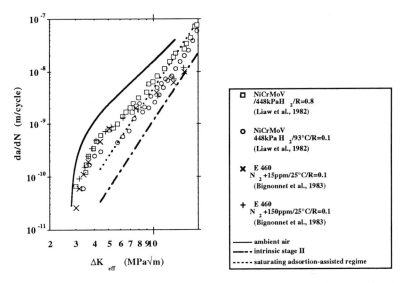

Figure 5: comparison between the results obtained in gaseous hydrogen with residual moisture in the NiCrMoV steel (Liaw, *et al.*, 1982) and results under nitrogen containing traces of water vapour (Bignonnet, *et al.*, 1983).

CONCLUSIONS

Recent progress in understanding environmentally-assisted fatigue crack propagation offers new interpretations on the near-threshold fatigue crack propagation. In particular the oxide-induced closure concept is not sufficient to fully account for differences observed in fatigue crack growth behaviour under various atmospheres. It has been proved that generally the humidity level in the test atmosphere controls the overall behaviour. Therefore a detailed examination of the interaction of the stressed material at the crack tip and water vapour molecules present in the test environment is needed. On the basis of both experimental data and modelling results it is suggested that moisture might act by two distinct processes, namely:
- a propagation regime assisted by the adsorption of water vapour molecules;
- another regime related to the hydrogen evolved from the dissociation of adsorbed molecules.

However it is shown that in presence of traces of water vapour the adsorption step is the mere mechanism acting in the slow growth rate regime. It is also suggested that water vapour would adsorb preferably to hydrogen and would thus prevent or at least delay the hydrogen related process. Finally it is concluded that intergranular facets encountered in the near-threshold might definitively no longer be considered as a proof and/or a measurement of the action of hydrogen in environmentally-assisted fatigue crack propagation.

REFERENCES

Bennet, J. A. (1963). Effect of reactions with the atmosphere during fatigue of metals. *Fatigue: an interdisplinary approach, 10th Sagamore Army materials research conference* Proc. ed.), 209-227.

Bignonnet, A., Loison, D., Namdar Irani, N., Bouchet, B., Kwon, J. H. and Petit, J. (1983). Environmental and frequency effects on near-thresholds fatigue crack propagation in a structural steel. *Fatigue Crack growth Tresholds Concepts* Proc. (D.L. Davidson and S. Suresh, ed.), 99-113.

Cooke, R. J., Irving, P. E., Booth, G. S. and Beevers, C. J. (1975). The slow fatigue crack growth and threshold behaviour of a medium carbon steel in air and vacuum. *Engng. fract. Mech.*, 7, 69-77.

Elber, W. (1971). The Significance of Crack Closure. *Damage Tolerance in Aircraft Structures, ASTM STP 486* Proc. ed.), ASTM STP 486, 230-242.

Frandsen, J. D. and marcus, H. L. (1977). Environmentally Assisted Fatigue Crack Propagation in Steel. *Metall. trans.*, 8A, 265-272.

Hénaff, G., Marchal, K. and Petit, J. (1995). On Fatigue Crack Propagation Enhancement by a Gaseous Atmosphere: Experimental and Theoretical Aspects. *Acta Metall et Mater*, 43, 2931-2942.

Hénaff, G. and Petit, J. (1993). Pure corrosion fatigue crack propagation, "Corrosion-deformation interaction". *Proceedings CDI92* Proc. (T. Magnin and J.M. Gras, ed.), 599-618.

Hénaff, G., Petit, J. and Bouchet, B. (1992). Environmental influence on the near-threshold fatigue crack propagation behaviour of a high-strength steel. *Int. J. Fatigue*, 14, 211-218.

Hofmann, W. and Rauls, W. (1965). Ductility of steel under the influence of external high pressure hydrogen. *Weld. Res. Suppl.*, 225s-230-s.

Liaw, P. K., Hudak Jr, S. J. and Donald, J. K. (1982). Influence of gaseous environments on rates of near-threshold fatigue crack propagation in NiCrMoV steel. *Metall Trans*, 13A, 1633-1645.

Liaw, P. K., Hudak Jr., S. J. and Donald, J. K. (1983). Near-threshold fatigue crack growth investigation of NiMoV Steel in Hydrogen Environment. *Fracture mechanics: fourteenth symposium-volume II: testing and applications, ASTM STP 791* Proc. (J.C. Lewis and G. Sines, ed.), II.370-II.388.

Petit, J., De Fouquet, J. and Hénaff, G. (1994). Influence of ambient atmosphere on fatigue crack growth behaviour of metals. In: *Handbook of fatigue crack propagation in Metallic structures* (A. Carpinteri, ed.), 2, Section VI on Influence of Environmental condition, 1159-1204/ Elsevier.

Petit, J. and Hénaff, G. (1991). Stage II intrinsic fatigue crack propagation. *Scripta Metall.*, 25, 2683-2687.

Ritchie, R. O., Suresh, S. and Moss, C. M. (1980). Near-threshold fatigue crack growth in 21/4Cr-1Mo pressure vessel steel in air and hydrogen. *J. Engng. Mat. Tech.*, 102, 293-299.

Spitzig, W. A., Talda, P. M. and Wei, R. P. (1968). Fatigue crack propagation and fractographic analysis of 18Ni(250) maraging steel tested in argon and hydrogen environments. *Engng. Fract. Mech.*, 1, 155-166.

Stewart, A. T. (1980). The influence of environment and stress ratio on fatigue crack growth at near threshold stress intensities in low-alloy steels. *Engng. Fract. Mech.*, 13, 463-478.

Suresh, S. and Ritchie, R. O. (1983). Near-threshold fatigue crack propagation: a perspective on the role of crack closure. *Fatigue Crack growth Tresholds Concepts* Proc. (D.L. Davidson and S. Suresh, ed.), 227-261.

Suresh, S., Zamiski, Z. A. and Ritchie, R. O. (1981). oxide-induced closure closure: an explanation for near-threshold fatigue crack growth behavior. *Metall. Trans.*, 12A, 1435-1443.

Weertman, J. (1966). rate of growth of fatigue cracks calculated from the theory of infinitesimal dislocations distributed on a plane. *Int J Fract Mech*, 2, 460-467.

Wei, R. P. and Ritter, D. L. (1972). The influence of temperature on fatigue crack growth in a mill annealed Ti-6Al-4V alloy. *JMLSA*, 7, 240-250.

Wei, R. P. and Simmons, G. W. (1981). Recent progress in understanding environment-assisted fatigue crack growth. *Int. J. Fract.*, 17, 235-247.

Weir, T. W., Simmons, G. W., Hart, R. G. and Wei, R. P. (1980). A model for surface reaction and transport controlled fatigue crack growth. *Scrpta Metall.*, 14, 357-364.

INFLUENCE OF ENVIRONMENT ON FATIGUE PROPERTIES OF AL8090

H. HARGARTER[1], G. LÜTJERING[1], J. BECKER[2] and G. FISCHER[2]

[1]Technical University Hamburg-Harburg, 21071 Hamburg, Germany
[2]Otto Fuchs Metallwerke, 58528 Meinerzhagen, Germany

ABSTRACT

Environmental fatigue, (i.e. crack initiation and crack growth of long and short cracks) of Al8090 (T852) was investigated for the ST-direction of a heavy section forging in vacuum, air and salt water with and without chromate containing inhibitors. Different stress ratios were applied. High propagation rates of long fatigue cracks as well as of small surface cracks in the aggressive environments were related to hydrogen embrittlement. Although flat crack fronts developed, pronounced crack closure was observed for long cracks in air and in NaCl-solution with inhibitors, which was explained by the formation of corrosion products on the fracture surface. Independently of crack front geometry and crack closure the environment caused alterations of the fatigue crack propagation behavior with regard to stress ratio variations. This is explained on the basis of the reduced fracture stress as a result of the hydrogen uptake. The fatigue crack initiation resistance was only reduced in the NaCl-solution without inhibitors as compared to the other three environments tested.

KEYWORDS

Al-alloys; S-N curves; fatigue crack propagation; environmental effects; crack closure; small surface cracks

INTRODUCTION

The Aluminum-Lithium alloy 8090 was, besides for other applications, mainly intended as a replacement material for aerospace forgings, and good fatigue properties are therefore essential. Commonly, in heavy section forgings the short transverse (ST) direction is most critical. As for environmental influences, it is well known, that Al8090 is susceptible to hydrogen embrittlement (1). Still, further insight into the mechanisms by which the environment controls fatigue in Al-Li alloys has to be gained. For example, a controversy exists, whether the high susceptibility to oxidation as a result of the high lithium content affects the crack propagation behavior of these alloys in air (2,3). Although fracture toughness, ductility, and resistance against stress corrosion cracking have been found to be the lowest in ST-direction (4), there is, however, little information about the fatigue behavior in this critical direction. This paper presents results of an investigation in which the effects of different environments and stress ratios on S-N curves and on the fatigue crack growth behavior of Al8090 were studied in ST-direction.

EXPERIMENTAL PROCEDURE

The Al8090 material was produced by Alcan, UK, as a conventional ingot casting, with a composition of Al-2.3Li-1.1Cu-0.8Mg-0.12Zr (wt.-%). The fatigue properties at room temperature were investigated on a 80mm thick rectangular forging (T852) in the following environments: vacuum ($< 10^{-6}$ Pa), ambient laboratory air, and two aqueous solutions of NaCl at free corrosion potential: 3.5 wt% NaCl (NaCl-pure), 3.5 wt% NaCl + 0.3% $NaCr_2O_7$ + 0.2% Na_2CrO_4 (NaCl+inhibitor). Stress ratios of R=0.1 and 0.5 were applied at about 30 Hz. CT-specimens with ST-L-orientation were used to measure the fatigue crack propagation of long cracks (macrocracks). Crack propagation was monitored by light microscopy. Crack tip opening displacement (COD) and/or backface strain (BFS) were measured at selected crack velocities, shortly reducing the frequency to 0.1 Hz. S-N curves were evaluated on electrolytically polished, round, hour-glass shaped specimens, loaded in ST-direction. The same specimen geometry was also used to study the nucleation and propagation of small surface cracks (microcracks) in air. Cracks were investigated by light microscopy, interrupting the fatigue test for each measurement. A more detailed description of the material and the experimental procedures is given in (5). The fracture morphology was studied on all specimens by SEM of fracture surfaces and by investigating the crack front profiles, i.e. sections perpendicular to the crack propagation direction, which were taken at selected crack velocities.

RESULTS

The microstructure consisted predominantly of unrecrystallized pancake grains with average dimensions of 50 (ST) x 100 (LT) x 1000 (L) μm. The alloy was age hardened by a homogeneous distribution of the δ' and S'phase. Incoherent particles of the equilibrium phases were found at the grain boundaries, resulting in precipitate free zones. The tensile properties and the fracture toughness in ST-direction are given in Table 1.

Figure 1 summarizes the fatigue crack propagation data of long cracks (CT-specimens) for all investigated environments and stress ratios as da/dN-ΔK curves. In the aggressive environments, most pronounced in NaCl-pure solution, but also in NaCl+Inhibitor and in air, the resistance against fatigue crack propagation was significantly reduced compared to inert environment. It is also obvious, that the environment had a distinct influence on the effect of the applied stress ratio. In NaCl-pure at low propagation rates almost no difference between the curves at R=0.1 and R=0.5 existed. In the other environments, especially in air and in NaCl+Inhibitor, tests at R=0.5 resulted in enhanced crack growth as compared to R=0.1. It should also be noted, that it proved to be impossible to establish slower crack velocities than 1 x 10^{-8} m/cycle in the NaCl+Inhibitor solution, since complete crack arrest occurred at lower ΔK-values. In the NaCl-pure solution no such crack arrest was observed.

The influence of the different environments on crack closure is demonstrated in Fig. 2, which compares typical plots of load vs. COD/BFS from tests with a nominal R-value of 0.1. At this stress ratio some nonlinearity was observed in all the curves, indicating crack closure effects, but in NaCl-pure as well as in vacuum these effects were negligibly small. In air and in NaCl+Inhibitor pronounced crack closure was observed. In the latter solution even at an applied stress ratio of 0.5 measurable crack closure was found, whereas in the other environments the COD/BFS curves were completely linear at R=0.5. Closure corrected ΔK-values (ΔK_{eff}) were calculated by using $P_{min,eff}$ as indicated in Fig. 2. Also, the resulting effective stress ratios (R_{eff}) were calculated and used to mark the curves of da/dN-ΔK_{eff} in Fig. 3. Since in each load cycle only the actual minimum stress intensity

Table 1. Tensile properties and fracture toughness

Test Direction	$\sigma_{0.2}$ [MPa]	UTS [MPa]	σ_F [MPa]	TE [%]	ε_F	K_{Ic} [MPa m$^{1/2}$]
ST	385	515	535	3.9	0.04	11 (ST-L)

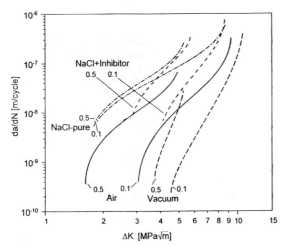

Fig. 1: Nominal da/dN-ΔK curves
for R = 0.1 and R = 0.5

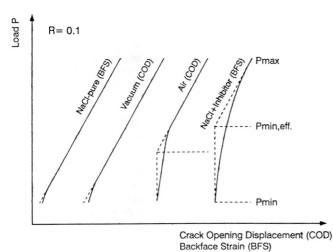

Fig. 2: Typical COD-curves from
tests at R = 0.1

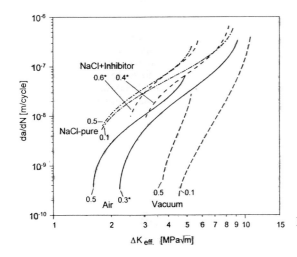

Fig. 3: Closure corrected
da/dN-ΔK$_{eff.}$ curves

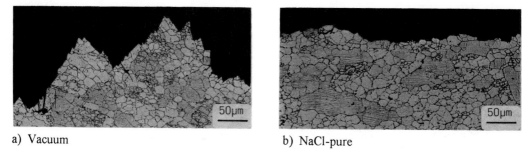

a) Vacuum   b) NaCl-pure

Fig. 4: Crack front profiles at da/dN = 1x10⁻⁸m/cycle, R = 0.1

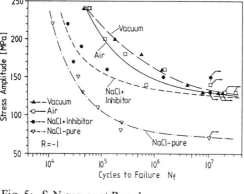

Fig. 5: S-N curves at R = -1

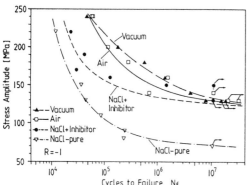

Fig. 6: S-N curves at R = 0.5

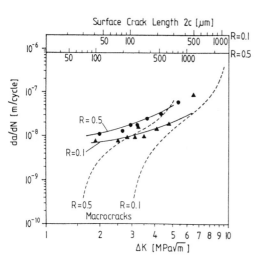

Fig. 7: Microcrack propagation curves in air
         at R = 0.1 and R = 0.5

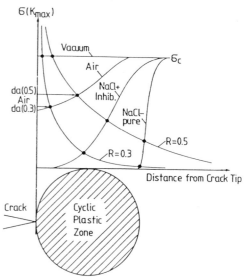

Fig. 8: Influence of environment and stress
         ratio on crack propagation da per cycle
         (schematically)

at the crack tip is influenced by crack closure, but not the maximum stress intensity, it is obvious that not only ΔK changes to ΔK_{eff} but also R to R_{eff}. The given R_{eff} values are only correct for the beginning of the curves, because with rising ΔK crack closure becomes less prominent. Due to crack closure three curves (the two curves for NaCl+Inhibitor at R=0.1 and R=0.5 and the curve for air at R=0.1) are shifted considerably towards lower ΔK, when compared to the nominal curves in Fig. 1.

SEM investigations revealed a facetted fracture surface for the tests in vacuum with predominantly transgranular slip band fracture. In the aggressive environments, including air, fracture occurred mainly along embrittled grain boundaries with some transcrystalline fracture, which, in contrast to the slip band facets in the vacuum tests, was always oriented normal to the loading direction.

A very pronounced difference was found between the crack front profiles in vacuum and the other environments. In vacuum a rough profile was found with local crack deviations of more than 100 μm perpendicular to the plane of propagation (Fig. 4a). In the other environments smoother profiles developed with local deviations below 50 μm (Fig. 4b). In all environments no differences existed between the crack profiles from tests with low and high nominal stress ratios and no significant changes in roughness were observed with increasing ΔK.

Figure 5 represents the S-N curves obtained in the different environments at a stress ratio of R= -1. For the fatigue strength at 10^7 cycles it was found, that testing in laboratory air and in the NaCl solution with inhibitors did not have an influence when compared to testing in vacuum. In these environments a fatigue strength of 130 MPa was established. Testing in the NaCl solution without inhibitors resulted in a significantly lower fatigue strength level of only about 80 MPa. At high stress amplitudes, i.e. in the low cycle fatigue regime, the following order for the number of cycles to failure (N_f) was observed: NaCl-pure<NaCl+Inhibitor<air<vacuum. Tests at a stress ratio of R=0.5 gave similar results (Fig. 6). In all environments crack initiation occurred preferentially at grain boundaries.

Figure 7 presents da/dN-ΔK curves for microcracks in air tested at R = 0.1 with $\Delta \sigma$ = 340 MPa and at R = 0.5 with $\Delta \sigma$ = 220 MPa. For comparison the corresponding nominal curves for long cracks are also given in Fig. 7. A very pronounced difference in the propagation between macro- and microcracks was found at a nominal stress ratio of R=0.1. The small surface cracks propagated already at lower ΔK and much faster than their long counterparts. At R=0.5 microcrack propagation also was significantly enhanced compared to macrocracks. The applied stress ratio not only had an influence on the propagation rates of macrocracks, but microcracks also propagated faster at the high stress ratio than at the low stress ratio.

DISCUSSION

The following discussion of the fatigue crack growth behavior will focus on the interaction between the environmental influences and the factors ductility, crack front geometry and crack closure that control crack growth behavior.

At a nominal R-ratio of 0.5 crack closure had no influence on the results of the fatigue crack propagation tests with the exception of the curve for NaCl+Inhibitor (Fig. 1). Also, identical fracture modes and crack front geometries were found for air and the two different NaCl solutions. Consequently the descending resistance against fatigue crack growth from air to NaCl+Inhibitor to NaCl-pure directly reflects the differences in the amount of embrittlement as a result of hydrogen uptake. In air the hydrogen absorption is a function of the humidity and in saltwater with inhibitors the reaction processes at the crack tip are hindered by the inhibitor (6).

A secondary effect of the hydrogen uptake is the change from slip band fracture in vacuum to a brittle fracture mode in the aggressive environments, including air, which resulted in fairly flat crack front geometries. A contribution of the much larger crack deviations in vacuum to the higher resistance against fatigue crack propagation in vacuum may therefore be assumed.

Keeping in mind that the very rough crack fronts observed in vacuum resulted in only little closure it is obvious that roughness induced closure does not play a dominant role in creating the high closure levels found in air and in the chloride solution with inhibitors. The findings in this work rather hint at a strong influence of oxides or other deposits on the fracture surface which result from the reaction of the freshly exposed metal at the crack tip with the oxidizing environments. The reason why pronounced closure was observed in NaCl solution with inhibitors but not in NaCl-pure seems to be that the protective layer formed at the fracture surface due to the inhibitor addition prevented the dissolution of the deposits, whereas without inhibitors they could be dissolved.

Taking crack closure as well as crack front geometry into account it is obvious that in vacuum, and especially pronounced in air and NaCl+Inhibitor a significant difference remains between the curves from tests with low and high effective stress ratios, whereas in NaCl-pure no such influence is found. To explain this behavior the following model is proposed, which originates from the concept that locally a certain critical rupture stress (σ_c) either in slip planes or at boundaries has to be reached to facilitate crack propagation. In each load cycle the crack will propagate up to the point ahead of the current crack tip where the maximum stress becomes equal to the local σ_c. The influence of the stress ratio is obvious, since at a given ΔK the distance at which σ_c will be reached depends on K_{max}. This is schematically illustrated in Fig. 8. In vacuum σ_c is not influenced by the environment. Because in vacuum the point of intersection of the curve for $\sigma(K_{max})$ and σ_c has a greater distance to the crack tip at R=0.5 crack propagation is faster than at R=0.3. Due to hydrogen which is swept in by moving dislocations σ_c is strongly reduced within the cyclic plastic zone in aggressive environments. Outside the cyclic plastic zone σ_c reaches the same value as in vacuum. Therefore σ_c becomes a function of the distance from the crack tip, depending on how much hydrogen is evolved in a particular environment, and on the size of the cyclic plastic zone. Curves for σ_c as in Fig. 8 appear reasonable to explain the experimental observations. In the pure NaCl-solution enough hydrogen is introduced so that the rupture stress becomes small within the whole cyclic plastic zone. As a result cracks propagate very fast and the growth rates do not depend on the R-value any more. The transport of hydrogen into the plastic zone which is controlled by ΔK becomes the single controlling step of fatigue crack growth. The large influence of K_{max} in air and in NaCl+Inhibitor can be attributed to a partially embrittlement of the plastic zone, because in these environments less hydrogen is generated. As a consequence curves for σ_c as in Fig. 8 may result. No direct proof of the assumed hydrogen distributions or fracture stress distributions ahead of the crack tip can be offered at this point. But further support for the discussed findings is given by the results of the S-N tests and the microcrack propagation tests.

While the fatigue strength at 10^7 cycles is usually indicative of the resistance against fatigue crack initiation, the number of cycles to failure in the low cycle fatigue regime of S-N tests is also determined by crack propagation. For the microcracks in Al8090 the descending lifetime at high stress amplitudes in the sequence vacuum, air, NaCl+Inhibitor and NaCl-pure therefore reflects the same dependence of the propagation rates on the environment as was observed for long cracks.

Whereas for long cracks the crack front geometry and crack closure contribute to a crack retardation it has been found that for very small cracks these factors are not operative, often resulting in much enhanced crack growth of microcracks. This behavior was also found for Al8090. The faster crack growth of the surface cracks in air at R=0.5 than at R=0.1 reflect the same influence of the effective R-value, independent of crack front geometry and crack closure, as was concluded for the long cracks.

REFERENCES

1. N.J.H. Holroyd, A. Gray, G.M. Scamans, and R. Hermann, in Aluminium-Lithium Alloys III, The Institute of Metals, London, (1986) 310-320.
2. J. Petit and N. Ranganathan, in Aluminium Lithium, DGM, Oberursel (1992) 521-532.
3. K.V. Jata and E.A. Starke Jr., Met. Trans. 17A, (1986) 1011-1026.
4. W.E. Quist and G.H. Narayanan, in Aluminum Alloys - Contemporary Research and Applications (Treatise on Materials Science and Technology, Vol.31), Academic Press (1989) 219.
5. H. Hargarter: Ph.D. Thesis (1995), Technical University Hamburg-Harburg, Germany.
6. R. Braun and H. Buhl, in 4th Intern. Aluminium-Lithium Conf., les éditions de physique, Les Ulis (1987) 843-849.

AUTHOR INDEX

SUBJECT INDEX